PROCEEDINGS
FIFTH INTERNATIONAL CONGRESS
INTERNATIONAL ASSOCIATION OF ENGINEERING GEOLOGY

COMPTES–RENDUS
CINQUIEME CONGRES INTERNATIONAL
ASSOCIATION INTERNATIONALE DE GEOLOGIE DE L'INGENIEUR

20–25 OCTOBER 1986 / BUENOS AIRES / 20–25 OCTOBRE 1986

VOLUME 1

PROCEEDINGS
FIFTH INTERNATIONAL CONGRESS
INTERNATIONAL ASSOCIATION
OF ENGINEERING GEOLOGY

COMPTES–RENDUS
CINQUIEME CONGRES INTERNATIONAL
ASSOCIATION INTERNATIONALE
DE GEOLOGIE DE L'INGENIEUR

20–25 OCTOBER 1986 / BUENOS AIRES / 20–25 OCTOBRE 1986

VOLUME 1

1 *Engineering geological investigations of rocks masses for civil engineering projects
and mining operations*
*La géologie de l'ingénieur et l'étude des massifs rocheux pour les projets de génie civil
et les travaux miniers*

2 *Engineering geological problems related to foundations and excavations in weak rocks*
Les études de géologie de l'ingénieur pour les fondations et les fouilles dans des roches tendres

A.A.BALKEMA / ROTTERDAM / BOSTON / 1986

CIP-DATA KONINKLIJKE BIBLIOTHEEK, DEN HAAG

Proceedings

Proceedings of the Fifth International Congress of the International Association of Engineering Geology, Buenos Aires,
20-25 October 1986 = Comptes-Rendus du Cinquième Congrès International de Association Internationale de Géologie
de l'Ingénieur, Buenos Aires, 20-25 Octobre 1986. – Rotterdam [etc.] : Balkema
Text in English and French.
ISBN 90-6191-660-7
Vol. 1. – Ill.
ISBN 90-6191-661-5 bound
SISO 563 UDC 551
Subject heading: geology.

The texts of the various papers in this volume were set individually by typists under the supervision of each of the authors concerned.

Complete set of four volumes: ISBN 90 6191 660 7
Volume 1: ISBN 90 6191 661 5 / Volume 2: ISBN 90 6191 662 3 / Volume 3: ISBN 90 6191 663 1 / Volume 4: ISBN 90 6191 664 X

© 1986 A.A.Balkema, Postbus 1675, 3000 BR Rotterdam, Netherlands
Distributed in USA & Canada by: A.A.Balkema Publishers, P.O.Box 230, Accord, MA 02018
Printed in the Netherlands

Les articles publiés dans ce volume ont été dactylographiés sous la responsabilité de leurs auteurs.

Collection complète de quatre volumes: ISBN 90 6191 660 7
Volume 1: ISBN 90 6191 661 5 / Volume 2: ISBN 90 6191 662 3 / Volume 3: ISBN 90 6191 663 1 / Volume 4: ISBN 90 6191 664 X

© 1986 A.A.Balkema, Postbus 1675, 3000 BR Rotterdam, Pays Bas
Distribué aux USA & Canada par: A.A.Balkema Publishers, P.O.Box 230, Accord, MA 02018
Imprimé aux Pays Bas

Foreword

In 1980, on occasion of the 26th International Congress of Geology held in Paris, the Argentina Association of Engineering Geology (ASAGAI) proposed at the council meeting of the IAEG, that an international congress be held in the city of Buenos Aires. Subsequently during the 4th International Congress of the IAEG held in New Delhi in 1982, Buenos Aires was nominated as venue of the 5th International Congress. Hereby the Organizing Committee has prepared the following subjects for the meeting:
1. Engineering geological investigations of rock masses for civil engineering projects and mining operations
2. Engineering geological problems related to foundations and excavations in weak rocks
3. Engineering geological aspects of foundations in soils
4. Engineering geological problems related to hydraulic and hydroelectric developments
5. Engineering geology in the development of road, railroad, coastal and offshore projects
6. Engineering geological aspects in environment planning and urban areas
Colloquium A. Engineering geology in geothermal engineering projects
Colloquium B. Engineering geology related to nuclear waste disposal projects.
A very favourable reply, that is over 263 papers was received from 32 countries. The papers cover 29 topics of the main subjects foreseen. Six new topics had to be added due to the wide scope of papers received.

The Organizing Committee has pleasure in presenting to you in advance the proceedings of the congress, distributed into three volumes. The papers of Colloquia A and B will be published in Bulletin Nr. 34 of the IAEG in view of their importance and in order to have the best dissemination of same.

After the meeting a post-congress volume will be edited. It will include the papers that arrived after the Congress, a summary of the discussions held during the meeting and special reports. The Organizing Committee of the 5th International Congress of the IAEG wishes to express its gratitude to all the professionals who have made contributions in this field and to A.T. Balkema for making the publication of the congress-proceedings possible.

Horacio V. Rimoldi
Chairman of the 5th International Congress of the IAEG

Avant-propos

En 1980, à l'occasion du 26ième Congrès International de Géologie à Paris, l'Association de Géologie Technique d'Argentine (ASAGAI) proposa à la réunion du conseil de l'IAEG, de tenir un congrès international à Buenos Aires. Consécutivement, au cours du Quatrième Congrès International de l'AIGI à New Delhi en 1982, Buenos Aires fut désigné comme lieu pour le Cinquième Congrès International. Le comité d'organisation a proposé les sujets suivants pour le congrès:

1. La géologie de l'ingénieur et l'étude des massifs rocheux pour les projets de génie civil et les travaux miniers
2. Les études de géologie de l'ingénieur pour les fondations et les fouilles dans les roches tendres
3. Aspects géologiques et géotechniques des fondations dans certains types de sols
4. Problèmes de géologie de l'ingénieur liés aux travaux hydrauliques et hydroélectriques
5. Rôle de la géologie de l'ingénieur dans le choix des tracés routiers et ferroviaires et dans les travaux côtiers et sous-marins
6. La géologie de l'ingénieur, les problèmes d'environnement et les zones urbaines

Colloque A. La géologie de l'ingénieur et les projets géothermiques

Colloque B. La géologie de l'ingénieur et le stockage des déchets nucléaires.

Une très bonne résponse, c'est à dire 263 contributions venant de 32 pays a été obtenue. Les contributions couvrent 29 thèmes sur les sujets principaux prévus. Six nouveaux thèmes ont dû être ajoutés à cause de la diversité des contributions reçues.

Le comité d'organisation se félicite de pouvoir vous présenter à l'avance les comptes-rendus du congrès, repartis en trois volumes. Les contributions des colloques A et B seront publié dans le Bulletin 34 de l'AIGI à cause de leur importance et dans l'intention de leur donner la meilleure propagation possible.

Le congrès terminé, un dernier volume sera publié, contenant les contributions reçues après le congrès, un sommaire des discussions faites pendant le congrès, et des rapports spéciaux. Le comité d'organisation du Cinquième Congrès International de l'AIGI tient à exprimer sa reconnaissance à tous les spécialistes qui ont préparé des contributions dans ce domaine et à A.T. Balkema d'avoir rendu possible la publication des comptes-rendus du congrès.

Horacio R. Rimoldi
Président du Cinquième Congrès de l'AIGI

Scheme of the work
Schéma de l'ouvrage

VOLUME 1

VOLUME 2

VOLUME 3

VOLUME 4

INTERNATIONAL ASSOCIATION OF ENGINEERING GEOLOGY (I.A.E.G.)
ASSOCIATION INTERNATIONALE DE GEOLOGIE DE L'INGENIEUR (A.I.G.I.)
EXECUTIVE COMMITTEE / COMITE EXECUTIF

President / Président	Prof. Dr. M. Langer	F.R. Germany / R.F. Allemagne
Secretary General / Secrétaire général	Dr. L. Primel	France
Treasurer / Trésorier	Dr. M. A. Peter	France
Past Presidents / Anciens présidents	Prof. A. Shadmon	Israel
	Acad. Prof. Qu. Záruba	Czechoslovakia / Tchécoslovaquie
	Prof. Dr. M. Arnould	France
	Acad. Prof. Ye. M. Sergeev	U.S.S.R. / U.R.S.S.

Vice Presidents / Vice-présidents	Western Europe – represented by the President / Europe occidentale – représenté par le Président		
	Eastern Europe / Europe de l'Est	Prof. M. Matula	Czechoslovakia / Tchécoslovaquie
	North America / Amérique du Nord	Dr. D. Varnes	U.S.A. / E.U.
	South America / Amérique du Sud	Dr. H. Rimoldi	Argentina / Argentine
	Asia / Asie	Prof. Wang. Sijing	China / Chine
	Australia / Australie	Dr. D. Bell	New Zealand / Nouvelle Zélande
	Africa / Afrique	Dr. S. Malomo	Nigeria

FIFTH INTERNATIONAL CONGRESS I.A.E.G.
CINQUIEME CONGRES INTERNATIONAL A.I.G.I.
ORGANIZING COMMITTEE / COMITE ORGANISATEUR

President / Président	Dr. Horacio V. Rimoldi
Secretary / Secrétaire général	Lic. Carlos A. Di Salvo
Associate Secretary / Secrétaire associé	Lic. Jorge Casajús
Treasurer / Trésorier	Ing. Nicolás C. A. Di Girolamo
Associate Treasurer / Trésorier associé	Dr. Mario Zapata
Members / Membres	Ing. Eduardo Nuñez
	Dr. Jorge P. Grunbaum
	Ing. Juan G. Meira
	Lic. Gabriel Revuelta López

ADVISING COMMITTEE / COMITE ASSESSEUR
Chairman / Président: Dr. Jorge R. Cuomo

TECHNICAL COMMITTEE / COMITE TECHNIQUE
Chairman / Président: Dr. Bernabé Quartino

FINANCIAL COMMITTEE / COMITE DES FINANCES
Chairman / Président: Ing. José Speziale

PUBLIC RELATIONS COMMITTEE / COMITE DES RELATIONS PUBLIQUES
Chairman / Président: Dr. Jorge Grunbaum

TECHNICAL TOURS COMMITTEE / COMITE DES EXCURSIONS TECHNIQUES
Chairman / Président: Lic. Jorge Casajús

PUBLISHING COMMITTEE / COMITE DES PUBLICATIONS
Chairman / Président: Dr. Juan V. Ploszkiewicz

ARGENTINE ASSOCIATION OF GEOLOGY APPLIED TO ENGINEERING
ASSOCIATION ARGENTINE DE GEOLOGIE APPLIQUEE AU GENIE CIVIL
ARGENTINE COMMITTEE I.A.E.G. / COMITE ARGENTIN DU A.I.G.I.

Chairman / Président	Dr. Víctor E. Mauriño
Vice Chairman / Vice-président	Dr. Horacio Víctor Rimoldi
Secretary / Secrétaire général	Ing. Juan Gerardo Meira
Associate Secretary / Secrétaire associé	Lic. Julio A. de la Vega
Treasurer / Trésorier	Lic. Alfredo N. Del Mónaco
Associate Treasurer / Trésorier associé	Dr. Héctor L. Rosenman
Members / Membres	Ing. José Speziale
	Lic. José M. Cosentino
	Lic. Jorge Casajús

Table of contents
Table des matières

1.2 Geotechnical classifications: Practical applications
Classifications géomécaniques: Applications pratiques

1.3 *Surface and underground excavations of large dimensions: Problems related to stresses*
Fouilles à ciel ouvert ou souterraines de grandes dimensions: Problèmes liés aux contraintes

1.5 Rock mass monitoring: New techniques and results
Surveillance des massifs rocheux: Nouvelles techniques et résultats obtenus

1 Engineering geological investigations of rock masses
for civil engineering projects and mining operations
La géologie de l'ingénieur et l'étude des massifs rocheux
pour les projets de génie civil et les travaux miniers

1.1 Geological and geophysical investigation of rock masses:
Correlation of results
Etudes géologiques et géophysiques de massifs rocheux

Comparaison de diverses méthodes de reconnaissance de la fracturation des massifs rocheux – Approche géostatistique
A comparison of different methods of reconnaissance of fractured rock masses

F.Homand-Etienne, B.Berthout & R.Houpert, *Centre de Recherches en Mécanique & Hydraulique des Sols & des Roches, ENSG-INPL, Nancy, France*
P.Chapot, *Laboratoire des Ponts & Chaussées, Nancy, France*

RESUME : La reconnaissance des massifs rocheux pose le problème général de l'optimisation des supports et de leurs champs d'extension spécifiques. Nous avons choisi une station expérimentale qui est une carrière dont une zone n'est plus exploitée. Ceci nous a permis de disposer de deux fronts de taille perpendiculaires comme référence. Nous avons réalisé une étude de détail de la fracturation de ces fronts et la comparons aux résultats d'essais in situ (sismique par transparence, carottage sonique) et aux mesures d'espacement entre fractures et de R.Q.D. sur des sondages. L'utilisation de méthodes géostatistiques sur des variables régionalisées telles que célérité des ondes et espacement entre fractures, tente de situer les observations effectuées sur le site expérimental par rapport à celles réalisées sur l'ensemble du massif.

ABSTRACT : Investigation of rock masses askes general problem of optimization of the survey mediums and that of their field of extension. We have chosen a field station situated in a quarry of which an area is no longer worked. That mode it possible for us to use two perpendicular rock faces as reference point. We have studied the various fracture groups and compare it to in situ tests (cross-hole method and seismic logging) and to fracture spacings and R.Q.D. measured on core boring. Geostatistical study of regionalized variables such as fracture spacings and compressional velocities tries to place the observations on experimental station in comparaison with those realized on the whole mass.

1 INTRODUCTION

La réalisation de travaux au rocher nécessite la prise en compte de l'existence des discontinuités; la connaissance de la structure des massifs rocheux revêt donc une importance primordiale ; c'est un des objectifs de la campagne de reconnaissance. Celle-ci doit répondre aux principales questions suivantes :
1. Quelle est la nature de la fracturation affectant le massif ?
2. Quelle est la densité des familles de discontinuités rencontrées ?
3. Quelle est la densité totale d'un réseau de surface de discontinuités dans un volume rocheux?
4. Quels sont les espacements moyens des fractures et les lois de distribution correspondantes ?
5. Les propriétés d'un échantillon reflètent-elles celles du massif rocheux ?
Les réponses aux quatre premières questions peuvent être obtenues par une analyse structurale du massif dans la mesure où il existe des affleurements. Dans le cas contraire, des sondages sont nécessaires (détermination des espacements de fracture, du R.Q.D., caractérisation mécanique).Des forages instrumentés peuvent également fournir certains renseignements qui pourront être corrélés avec les résultats obtenus au moyen des sondages. La réponse à la cinquième question implique la réalisation d'essais in situ et de laboratoire.
Ces moyens de reconnaissance sont très coûteux, ce qui amène à se poser les questions complémentaires ci-après :
1. Peut-on,sans risques, extrapoler au massif les valeurs obtenues ponctuellement ?
2. Est-il préférable d'effectuer une reconnaissance fine en un secteur limité au lieu de multiplier les points de prélèvements ?
Pour répondre à ces questions relatives à l'optimisation de la reconnaissance, nous avons choisi une station expérimentale qui est une carrière dont une zone n'est plus exploitée. Ceci nous a permis de disposer de deux fronts de taille perpendiculaires comme référence vis-à-vis des essais réalisés à proximité. L'utilisation de méthodes géostatistiques sur des variables régionalisées telles que célérité des ondes et espacement entre fractures, tente de situer les observations effectuées sur le site expérimental par rapport à celles réalisées sur l'ensemble du massif. Le site choisi est représentatif de la structure des calcaires de la région de Nancy. Il s'agit d'un

calcaire oolithique à stratification subhorizontale, en bancs de 20 à 50 cm, sauf dans les niveaux supérieurs où l'altération aboutit à un débit en plaquettes de quelques centimètres d'épaisseur.

2 ETUDE STRUCTURALE

La structure du massif est abordée de façon à déterminer, d'une part, les orientations des différentes familles de fractures et, d'autre part, leurs espacements. Cette étude est menée à la fois sur le terrain et au moyen de photographies de manière à comparer les résultats.

2.1 Orientation et pendage des discontinuités

Celles-ci ont été mesurées systématiquement à partir de plusieurs lignes implantées de façon aléatoire sur chaque front de taille. Différents types de discontinuités ont été pris en compte :
1. fractures affectant la totalité de l'affleurement (failles, diaclases),
2. fractures qui affectent un ou plusieurs bancs,
3. fractures de dimensions égales ou inférieures au banc.
Les joints subhorizontaux ne sont pas reportés sur le stéréogramme.
   Un répertoire des azimuts et pendages de 800 fractures est établi et un traitement automatique sur micro-ordinateur permet d'obtenir sur un canevas de Schmidt, les pôles des plans des discontinuités, puis les lignes d'équidensité des pôles à partir de comptage (fig. 1).

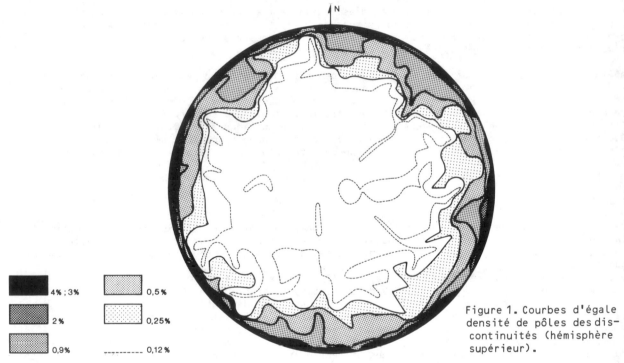

Figure 1. Courbes d'égale densité de pôles des discontinuités (hémisphère supérieur).

4% ; 3%   0,5%
2%   0,25%
0,9%   ----- 0,12%

   Le stéréogramme met en évidence quatre familles principales de fractures. Leur importance relative est précisée au moyen d'histogrammes circulaires. Un traitement séparé des discontinuités majeures (failles et diaclases) et des autres fractures ne montre pas de différences notables dans la répartition spatiale de la fracturation. L'analyse de la cyclographie des principales familles de discontinuités et l'établissement d'un bloc diagramme (fig. 2) nous conduisent à remarquer que la structure du massif cadre bien avec ce que nous connaissons de la géologie aux environs du site. Il s'agit de grandes fractures subverticales à verticales et de joints bien marqués ; les figures géométriques de type dièdre sont absentes.
   Chaque famille individualisée sur le diagramme d'équidensité de la figure 1 se présente sous la forme d'une ellipse dont l'axe majeur se situe sur un grand cercle du canevas de Schmidt et dont l'axe mineur correspond au grand cercle perpendiculaire au premier. Zanbak (1977) a montré que la projection des densités de pôles situés sur ou à proximité de ces axes, en respectant les distances angulaires, permet d'obtenir l'allure générale de la loi de distribution des discontinuités à l'intérieur d'une même famille. Cette méthode, appliquée à nos résultats, montre que la distribution des fractures est normale, ce qui est confirmé par un test de Student-Fisher

4

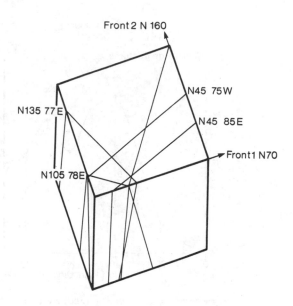

Front 2 N 160

N45 75W

N135 77E

N45 85E

Front1 N70

N105 78E

Figure 2. Bloc diagramme des principales familles
de fractures.

(Berthout 1985). Toutefois, de légères déviations des courbes caractérisent l'influence mutuelle
de familles trop proches.

2.2 Analyse de la photographie de fronts de taille

La photographie des fronts de taille présente l'avantage d'être un support de données dont l'acquisition est facilitée par l'utilisation d'une table à digitaliser. Il est alors possible de calculer
un paramètre plus global que l'espacement entre fracture qui est la densité de fissuration. Celle-
ci est estimée à l'aide d'une méthode mise au point par Quiblier (1978). Il s'agit de superposer
une sorte de canevas de comptage constitué d'une corde animée d'une rotation de 0 à 90° avec un
pas de 10°.

Le système de fracturation du massif est composé de deux grandes familles : il s'agit, d'une
part, de celle des fractures orientées (joints) et, d'autre part, de celle des fractures aléatoires (discontinuités verticales ou subverticales) ; le stéréogramme de la figure 1 ne tient
compte que de la famille des fractures aléatoires.

La méthode de Quiblier (1978) permet, dans ces conditions de distribution de la fracturation, de
calculer les densités respectives de la famille de fractures aléatoires et de la famille de fractures orientées (joints).

La densité de fracturation s'exprime comme la somme des longueurs des traces de fractures contenues dans une aire définie rapportée à cette aire. Le tableau 1 ci-après résume les résultats obtenus.

Tableau 1. Densité de fracturation des fronts 1 et 2
déterminée sur photographie (aire totale observée : $1\,312\,m^2$).

Paramètres mesurés	Front 1	Front 2
Densité totale DT (m/m^2)	1,42	1,31
Ecart-type σ	0,18	0,10
Coefficient de variation $\sigma/\bar{x}$	12,00	7,00
Densité de la famille aléatoire $(L_A)_0$ (m/m^2)	0,25	0,38
Densité de la famille orientée $(L_A)_1$ (m/m^2)	1,17	0,93
Surface de fractures en m^2 par m^3 de volume étudié : S_V	1,49	1,41

Il est possible de calculer la surface totale de fracture par unité de volume à partir des densités des familles aléatoires et orientées. Ce paramètre présente un grand intérêt vis-à-vis de
l'exploitation pétrolière ; dans le domaine du génie civil, il mériterait également d'être utilisé.
A partir de ces données, les histogrammes de la répartition des fractures en fonction de l'orientation

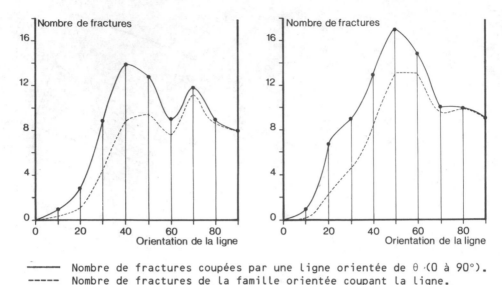

FRONT 1 0 à 10 m.

FRONT 2 10 à 20 m

———— Nombre de fractures coupées par une ligne orientée de θ (0 à 90°).
----- Nombre de fractures de la famille orientée coupant la ligne.

Figure 3. Histogrammes de répartition des fractures selon l'orientation
d'une ligne de mesure.

d'une ligne de mesure sont tracés. Le maximum de fracturation est obtenu pour une orientation de
50° de cette ligne par rapport à l'horizontale ; la figure 3 en donne deux exemples.
   D'autre part, Hudson et Priest (1979) montrent l'intérêt de représenter la variation de la fréquence des discontinuités λ pour des orientations de ligne de mesure variant de 0 à 90°. Généralement, on s'aperçoit qu'il existe une différence entre le graphe des résultats des calculs de
densité par la méthode de Quiblier (1978) et celui obtenu par la méthode de Hudson et Priest
(1979). Cette différence s'explique par le fait que la seconde méthode ne tient pas compte de la
non-régularité des réseaux de discontinuités, ainsi que de l'existence d'une famille prédominante.
   Les espacements entre les discontinuités sont exprimés à partir de leur fréquence (EF = 1/λ) et
suivent une loi exponentielle, c'est-à-dire que le long d'une ligne de mesure, les discontinuités
se répartissent de manière aléatoire. Leurs valeurs moyennes sont les suivantes pour les deux
fronts :

$$F1 = 0,67 \text{ m},$$
$$F2 = 0,70 \text{ m}.$$

   Pour contrôler ces déterminations effectuées à partir de photographies, nous avons tracé deux
lignes de mesure implantées sur chaque front ; celles-ci sont inclinées de 30° à 40° par rapport
à l'horizontale. Les espacements entre fractures ont été systématiquement mesurés et leurs valeurs
moyennes sont celles indiquées ci-dessous :

$$F1 = 0,24 \text{ m},$$
$$F2 = 0,20 \text{ m}.$$

   Cette différence entre les espacements entre fractures mesurés sur photographies et à l'affleurement s'explique par le fait que sur l'affleurement nous prenons en compte un plus grand nombre
de fractures. La digitalisation de photographies implique nécessairement une sélection des discontinuités ; les fractures obliques sont très peu visibles et de plus, il se pose toujours un
problème d'effet d'échelle.

3 MESURES IN SITU

D'une manière générale, les affleurements sont rarement disponibles avant des travaux au rocher
et il faut avoir nécessairement recours à des sondages carottés ou à des essais in situ pour se
faire une idée de la qualité du massif. Nous avons tenu compte des observations effectuées sur les
fronts de taille de façon à implanter au mieux la station expérimentale pour en corréler les résultats.
   Nous avons utilisé des méthodes de reconnaissance in situ de mise en oeuvre rapide et peu onéreuse;

il s'agit d'essais géophysiques et en particulier de la méthode du cross-hole et du carottage sismique.

Les mesures de type cross-hole sont effectuées dans les trous de forage suivant la maille indiquée sur la figure 4. La charge est placée en D5, les capteurs tridirectionnels TR1 et TR2 en D2 et D4 et les capteurs verticaux V3 et V4 en D6 et D3. Pour une position donnée de la source d'énergie (dynamite), une mesure est réalisée sur un ou plusieurs capteurs situés à la même profondeur que la source. L'ensemble du dispositif forme un T, conforme aux directions des fronts d'abattage, de manière à faciliter l'interprétation des temps enregistrés.

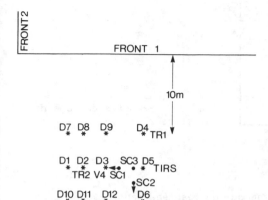

* Forage destructif.
• Sondage carotté (la flèche indique la direction de l'inclinaison).

Figure 4. Plan de position de la maille de reconnaissance.

Les figures 5 et 6 correspondant chacune à une orientation d'un front, présentent une augmentation progressive de la vitesse avec la profondeur. Les fractures verticales et les joints sont facilement repérables ; leurs orientations sont conformes à celles enregistrées lors des mesures structurales. L'utilisation de capteurs tridirectionnels permet d'apprécier la représentativité entre les différents capteurs.

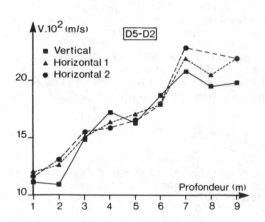

Figure 5. Evolution de la vitesse en fonction de la profondeur (capteur tridirectionnel).

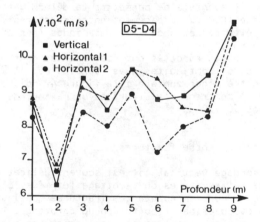

Figure 6. Evolution de la vitesse en fonction de la profondeur (capteur tridirectionnel).

Le système du cross-hole fournit une vitesse moyenne pour une profondeur donnée, mais ne permet pas de préciser par exemple l'extension latérale de zones de vitesse différente.

Nous avons eu recours au carottage sismique pour essayer d'aborder ce point. Cette méthode a été adaptée au problème de la reconnaissance des massifs rocheux ; nous avons employé la sonde utilisée dans les Laboratoires des Ponts et Chaussées (Fourmaintraux 1976). Celle-ci est munie d'un émetteur et de deux récepteurs espacés de 34,5 cm.

Les logs de vitesses qui découlent du traitement de ces données permettent de caractériser la répartition des discontinuités dans le massif sur la distance reconnue (fig. 7). Ils mettent en évidence une grande hétérogénéité de l'état de fissuration de la roche, la gamme des vitesses variant de 800 à 5 000 m/s. Nous distinguons une partie du massif très fracturée et altérée dans laquelle la vitesse des ondes est inférieure à 2 000 m/s et une autre partie plus ou moins fracturée dans laquelle la célérité des ondes est comprise entre 2 000 et 5 000 m/s.

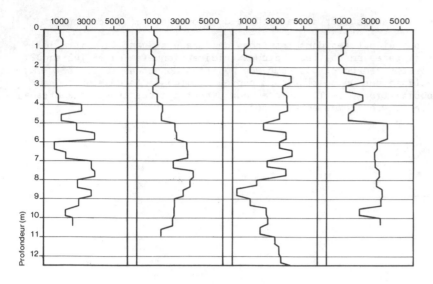

Figure 7. Exemples de logs de carottage sismique.

L'ajustement à une loi exponentielle des valeurs de vitesse du son mesurées dans chaque forage convient bien, ce qui va de pair avec la loi de distribution des espacements entre fractures mesurés sur le front de taille ; ceci confirme l'existence d'une liaison entre ces deux paramètres.

4 ETUDE DES SONDAGES CAROTTES

La seule méthode existant pour visualiser et étalonner les résultats d'essais géophysiques, en l'absence de front de taille, consiste à recourir en l'exécution de sondages carottés. Ceux-ci présentent l'avantage de permettre la détermination d'un certain nombre de paramètres qualitatifs de la roche (R.Q.D., espacement entre fractures), mais également de mesurer des paramètres physiques (célérité des ondes) et mécaniques (résistance à la compression) grâce à la réalisation d'éprouvettes.
Trois sondages carottés ont été réalisés dont deux inclinés et un vertical. Compte tenu de la présence de discontinuités quasi verticales, il a semblé judicieux d'incliner ces sondages à 50° par rapport à l'horizontale ; la figure 4 indique la disposition de ceux-ci. Les histogrammes de la figure 3 montrent d'ailleurs qu'un maximum de fractures sont recoupées pour une inclinaison voisine de 50°.

4.1 Espacement entre fractures

Dans un sondage vertical, il est souvent délicat d'effectuer le filtrage entre les fractures naturelles et celles issues du carottage lui-même. L'intérêt des sondages inclinés est double : il permet de prendre en compte les fractures quasi verticales du massif et d'éliminer les discontinuités provoquées par l'opération de carottage.
Les valeurs de l'espacement entre fractures mesurées pour chaque sondage sont les suivantes :

SC1   0,21 m   (incliné),
SC2   0,26 m   (incliné),
SC3   0,16 m   (vertical).

La comparaison de ces résultats avec ceux obtenus au moyen des lignes de mesures effectuées sur le front montre une corrélation d'autant plus forte que l'angle du support de la mesure est voisin. L'espacement entre fractures est plus faible pour le sondage vertical que pour les sondages inclinés ; la distance entre joints est effectivement plus grande dans ces derniers. Cependant, il devrait y avoir une influence de la fracturation verticale et les espacements devraient être plus petits. En fait, la fréquence de la fracturation verticale est relativement faible et intervient assez peu sur la mesure de l'espacement dans le sondage incliné.

4.2 R.Q.D.

L'indice de carottage R.Q.D. (Deere 1963), calculé à partir de sondages correctement dépouillés, constitue une représentation de l'état de fracturation du massif rocheux. Son estimation sur les trois sondages carottés, pour une longueur de passe de 1 m, conduit aux résultats suivants :

SC1   56 %,
SC2   48 %,
SC3   62 %.

Nous remarquons une évolution inverse entre les valeurs de l'espacement entre fractures et le R.Q.D. Etant donné le principe de la mesure du R.Q.D. qui ne retient que les fragments de dimensions supérieures ou égales à 10 cm, il a semblé intéressant de rechercher s'il n'existait pas une longueur de base optimale de calcul de ce paramètre.

La photographie sans distorsion d'un sondage carotté est digitalisée de façon à permettre un calcul automatique des valeurs du R.Q.D. avec des longueurs de référence LCR variant de 5 cm à 20 cm pour un pas de 2 cm. Les graphes obtenus sont comparés aux logs de vitesse des ondes en carottage sismique. Ces données sont prises comme référence dans la mesure où elles intègrent une partie de massif plus importante que le sondage carotté. L'allure des variations du R.Q.D. et des vitesses soniques est assez semblable dans un domaine de longueur de référence de carotte de 8 à 14 cm. De part et d'autre de ces longueurs, les résultats du R.Q.D. obtenus s'écartent de la réalité du massif (fig. 8). Par conséquent, le calcul du R.Q.D. sur une base standard de 10 cm est tout à fait représentatif pour le massif étudié.

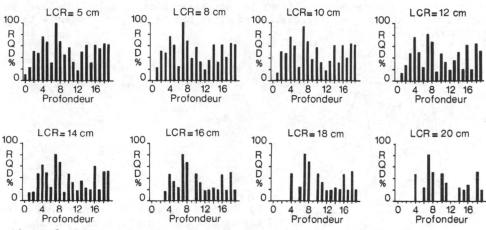

Figure 8. R.Q.D. calculé pour différentes valeurs de longueur de carottes de référence LCR.

Si nous comparons les schémas de variation des valeurs du R.Q.D. (fig. 8), ceux des vitesses obtenues par la méthode du cross-hole (fig. 5 et 6) et par la méthode du carottage sismique (fig. 7), nous constatons une corrélation très acceptable entre ces différentes méthodes, ce qui met en relief l'avantage des méthodes de reconnaissance moins coûteuses basées sur la géophysique.

4.3 Remarque sur le facteur de fissuration C

Il existe une autre méthode d'approche de l'état d'altération et de fracturation d'un massif rocheux; il s'agit du facteur de fissuration proposé par Hansagi (1974), appelé facteur C. Ce procédé de

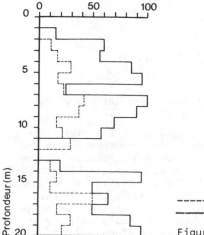

----- Facteur C.
——— R.Q.D.

Figure 9. Comparaison entre le R.Q.D. et le facteur de fissuration C.

quantification du degré de fissuration tient compte de la qualité de la roche grâce à un paramètre lié au nombre d'éprouvettes réalisées dans chaque élément de carotte. La comparaison du facteur de fissuration au R.Q.D. (fig. 9) nous montre que si l'allure des graphes est sensiblement équivalente, en revanche la méthode de Hansagi est trop pessimiste. Le fait de tenir compte du nombre d'éprouvettes réalisables par fragment de carotte, donne une importance peut-être exagérée à la notion de qualité d'un massif.

5 ETUDE GEOSTATISTIQUE

L'analyse au moyen des géostatistiques de variables régionalisées telles que l'espacement entre fractures ou la vitesse du son, doit permettre une approche de l'optimisation des supports de reconnaissance et de leurs champs d'extension. Rappelons que cette méthode a été mise au point dans le cadre de la prospection minière (Matheron 1970 ; Serra 1967).

En ce qui concerne la variable régionalisée espacement de fractures, les lignes de mesures disposées sur les fronts ne sont pas utilisables pour des raisons de disposition. Nous avons choisi d'utiliser les photographies des fronts de taille en leur superposant une maille régulière. Celle-ci permet le calcul des espacements de fractures dans plusieurs directions.

Les demi-variogrammes sont élaborés pour ces différentes directions. Ils présentent tous un effet de trou, caractéristique d'une structure périodique de portée égale à 3 m. La figure 10 en donne un exemple. Ces alternances matérialisent la succession plus ou moins régulière de bancs de bonne qualité avec d'autres très fissurés. Un effet de pépite pur est commun à tous les demi-variogrammes construits.

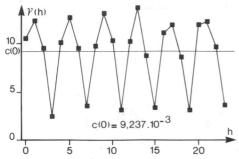

Figure 10. Demi-variogramme des espacements entre fractures (h = 1 correspond à une distance de 1,65 m).

La variable régionalisée vitesse des ondes en carottage sismique est analysée avec comme support géométrique les forages destructifs dont la maille a été présentée sur la figure 4. Les demi-variogrammes obtenus pour chaque forage présentent tous une valeur de la variance semblable ; il nous a donc été possible de construire le demi-variogramme moyen (fig. 11). Celui-ci indique un effet de pépite important dû à une structure de taille inférieure à celle retenue pour la mesure, puisque le pas de mesure en carottage sismique est de 0,50 m. Ce demi-variogramme atteint son palier vers 8 m alors que la profondeur des forages était d'environ 10 m.

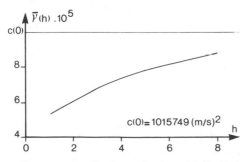

Figure 11. Demi-variogramme moyen vertical des vitesses du son en carottage sismique.

Nous avons également cherché à élaborer des demi-variogrammes horizontaux sur les valeurs de vitesses mesurées chaque 50 cm. La maille n'étant pas régulière, des demi-variogrammes moyens sont construits pour des distances de 3, 4,5, 7,5 et 15 m suivant la direction N70 parallèle au front 1 et pour des distances de 6,5 et 13 m suivant la direction N160 du front 2. La figure 12 représente les valeurs de γ(h) en fonction de la distance. Aucune structure n'est décelable sur un tel schéma et un effet de pépite pur semble être la seule conclusion qui soit envisageable dans ce cas.

Ceci signifierait que la variable vitesse des ondes n'est pas régionalisée dans le plan horizontal. En fait, ceci incombe à l'espacement entre forages qui est beaucoup trop grand par rapport à la zone d'influence de la variable mesurée en carottage sismique. Celle-ci s'apparente à un

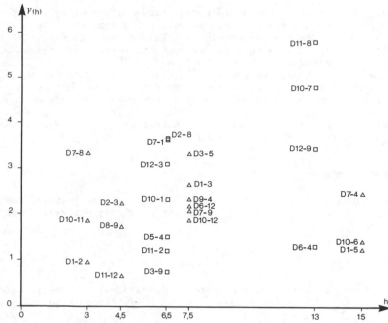

△ Construits dans la direction N70°.
□ Construits dans la direction N160°.

Figure 12. Demi-variogrammes moyens horizontaux des vitesses du son en çarottage sismique.

demi-cylindre dont la hauteur est égale à la base de mesure, soit 34,5 cm et dont le rayon est égal au tiers de cette base, soit 11,5 cm.

Pour affiner le comportement à l'origine du demi-variogramme moyen des vitesses des ondes construit dans le plan vertical, il a été nécessaire de réaliser des mesures de célérité sur éprouvettes. Le support de la mesure est représenté alors pour les éprouvettes de 10 cm de hauteur. Le demi-variogramme moyen des valeurs de la variable régionalisée de la figure 13 montre une croissance régulière et atteint son palier rapidement pour une portée d'environ 1 m. L'utilisation de la mesure sur éprouvette permet de préciser le comportement du demi-variogramme à l'origine.

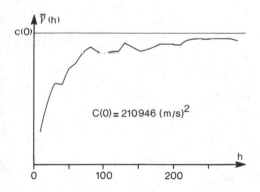

$$C(0) = 210946 \ (m/s)^2$$

Figure 13. Demi-variogramme moyen des valeurs de vitesses des ondes de compression mesurées sur éprouvettes sèches.

La principale difficulté à analyser géostatistiquement ces variables réside dans le fait que les discontinuités varient en orientations, pendages, espacements, ce qui nécessite la prise en compte de supports de grandes dimensions qui peuvent influencer les résultats.

L'emploi de différents supports et, en particulier, la réalisation des mesures sur des éprouvettes ou dans des forages avec un pas relativement faible, peuvent permettre de réduire les effets pépitiques et de déceler ainsi des structures à petite échelle.

6 CONCLUSION

Il apparaît que toute tentative de prévision à partir des demi-variogrammes des variables régionalisées telles que l'espacement entre fractures et la vitesse des ondes telle qu'elle a été mesurée, semble vouée à l'échec. Ceci est dû à la variation en orientations, pendages et espacements des discontinuités. La vitesse des ondes mesurée par carottage sismique n'est pas une variable régionalisée dans le plan horizontal. Il aurait été préférable de pouvoir disposer de

11

mesures de vitesse des ondes entre chaque forage et suivant toutes les directions.

Cette étude a cependant mis en évidence des points très positifs comme une bonne corrélation, d'une part, des espacements entre fractures mesurés sur l'affleurement et dans les sondages et, d'autre part, des valeurs des espacements entre fractures et des valeurs de vitesse des ondes en carottage sismique.

La bonne concordance entre les discontinuités décelées en forage et celles existant sur l'affleurement aurait pu permettre de mettre en oeuvre des méthodes d'estimation de la densité de fracturation par unité de volume, si les directions des discontinuités avaient été repérées en forage. La méthode de Quiblier (1978) présente l'avantage de permettre l'estimation de la densité de fracturation et de la surface de fracture par unité de volume et devrait être testée pour d'autres types de massif rocheux.

REMERCIEMENTS

Les auteurs remercient le Laboratoire Central des Ponts et Chaussées de Paris et le Laboratoire Régional de Nancy pour le financement et l'aide matérielle apportés à la réalisation de cette recherche.

REFERENCES

Berthout, B. 1985. Reconnaissance géomécanique des massifs rocheux. Essai de valorisation des caractéristiques physiques et mécaniques. Approche géostatistique. Thèse Doct. 3ème cycle, I.N.P.L. Nancy, 187 p.
Deere, D.U. 1963. Technical description of rock cores for engineering purposes. Felsmechanik and Ingenieur Geology, vol. 1, p. 16-22.
Fourmaintraux, D. 1976. Quantification des discontinuités des massifs rocheux. In La Mécanique des roches appliquée aux ouvrages de génie civil, éd. M. Panet, doc. formation continue de l'E.N.P.C.
Hansagi, I. 1974. A method of determining the degree of fissuration of rock. Internat. J. of Rock Mech. Min. Sci. and Geomech., vol. 11, p. 379-388.
Hudson, J.A. and Priest, S.D. 1979. Discontinuities and rock mass geometry. Internat. J. of Rock Mech. Min. Sci. and Geomech., vol. 16, p. 339-362.
Matheron, G. 1970. La théorie des variables régionalisées et ses applications. Les Cahiers du Centre de Morphologie Mathématique de Fontainebleau, fasc. 5.
Quiblier, J. 1978. Caractéristiques des réseaux de surface et de plans observables par leurs tracés sur des plans de coupe. Rapport I.F.P.
Serra, J. 1967. Un critère nouveau de découverte de structures : le variogramme. Sci. de la Terre, t. XII, n° 4, p. 277-299.
Zambak, C. 1977. Statistical interpretation of discontinuity contour diagrams. Internat. J. of Rock Min. Sci. and Geomech., vol. 14, p. 111-118.

The regularities of rock massif deformation during considerable water drawdown
Les déformations d'un massif rocheux à l'occasion d'un abaissement important du niveau de la nappe

J.M.Kazikaev, *Technological Institute, Belgorod, USSR*

ABSTRACT: Considerable water drawdown results in deformations and failure of structures in massifs or at land surface. These processes are studied insufficiently. One of the main causes of such situation is the absence of representative observations in field conditions.

RÉSUMÉ: Le rabattement profond est la cause de la deformation et la destruction des ouvrages dans le massif ou a la surface. Ces processus ont ete etudiés assez insuffisant. La cause essentielle de la situation pareille consiste en l'absence des observations representatives in situ.

INTRODUCTION

In spite of numerous studies, being conducted in the field of rock depressive consolidation due to considerable water drawdown, remarkable success has not yet been gained. The main cause of it is the absence of complete knowledge about mechanism of engineering geology phenomena and processes which accompany high hydrostatic pressure relief and lead to observed considerable deformations of rock massifs and land surface.

The most reliable and purposeful way of this problem solving is the accumulation and analysis of actual data obtained in the course of in-situ observations and experimental studies.

The analysis of material of long-term observations concerning rock massif and land surface deformation, hydrogeological condition dynamics and other parameters of rock behavior and mining structures with considerable water drawdown for mining Yuzhno-Belozerskoye iron ore deposit had been accomplished by the author and his collaborators. Simultaneously with it the mechanism and character of depressive consolidation of bedded rock thickness had been studied using series of special physical models.

FIELD OBSERVATIONS

Yuzhno-Belozerskoye iron ore deposit is mined by underground method with hardening fill in complicated hydrogeological conditions. Since 1962 in accordance with a project of dewatering water drawdown measures were being taken which by now had resulted in 160m head reduction in Buchak aquifer and up to 300 m that in the ore crystalline one. Due to developed processes of some rock layer depressive consolidation a zone of deformation of rocks and land surface had been formed at massif vast area. Mining structures, situated in this zone, vertical mine shafts in particular, were also subjected to considerable deformations. Especially it relates to central group of shafts where rock massif developing deformations caused periodical failures of certain mine support elements, reinforcement disturbance, guide deflection from the vertical and horizontal axes etc.

Land surface displacement observations at Yuzhno-Belozerskoye deposit were begun in 1961.

At present an observation station at mine field surface is presented by a network of points located along six profile lines (Fig.1). Total number of points is 165, summary length of profile lines - 17763m. All this enables to cover by observations a surface area nearly 6km along deposit strike and up to 5,8km across it.

In 1964-1965 subsidence trough at land surface began developing due to drainage of the Buchak aquifer and ore crystalline one made since 1962 by means of surface and underground drainage systems. The largest land subsidence up to now has reached value exceeding 2,8m.

Observation data analysis allows to reveal the following peculiarities of a process of rock massif and land surface deformation.

Though surface deformation is caused by rock layer drainage, qualitative and quantitative relationship between the nature of surface point subsidence changes (plots 3, 6, 7 in Fig.2) and head change regime in Buchak aquifer (plots 1 and 2) is not found. Furthermore, considerable head value changes followed later (with amplitude up to 40m, i.e. 25% of initial reduction) didn't affect the regularities of surface point displacement in no way. The evident relation of surface displacement parameters with magnitude of total reduction of underground water when deposit dewatering is not seen from the presented materials.

However, a relationship between surface point subsidence nature (plots 3,6,7 in Fig. 2) and drainage regime (plots 4,5,8), parameter $\sqrt[3]{Q}$ in particular, where Q is the amount of water drained from a layer by an appropriate drainage system, increasing from the dewatering outset, is revealing quite convincing. The plots are sensitive to changes of hydrogeological situation (availability or absence of hydraulic connection between aquifers, drainage intensity etc.).

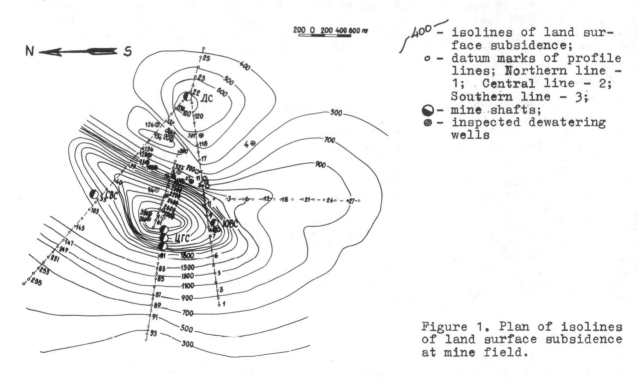

200 0 200 400 600 m

400 — isolines of land surface subsidence;
o — datum marks of profile lines; Northern line - 1; Central line - 2; Southern line - 3;
◑ — mine shafts;
◎ — inspected dewatering wells

Figure 1. Plan of isolines of land surface subsidence at mine field.

The nature and magnitude of land surface deformation reflect clearly geological and structural conditions at deposit depth levels. Presented in Fig.3 plots of subsidence show that a subsidence trough is developing without any restrictions and acquiring "classic" shape at sites of intact (horizontal) contact of loose thickness and bedrock. On the contrary at site of ore deposit bedrock roof raises slightly and ends by upheaving of stiff steeply dipping quartzite layers. Here occurs washed thick layer (up to 50m) of chalkey marlaceous rocks underlying loose thickness. As a result of two said factor influence subsidence trough at this site hadn't reached considerable dimensions (subsidence magnitude is 3 times less, deformation rates are low and so on). It should be noted that it is here that centre of depression cone and the greatest underground water head reduction occur.

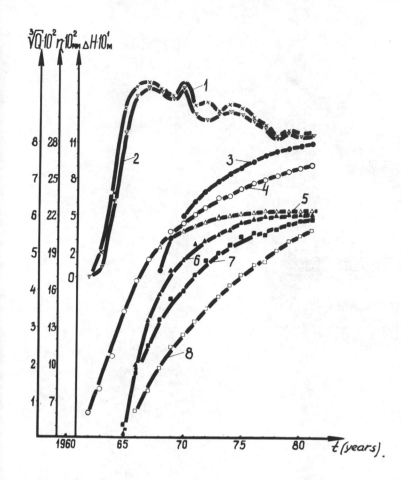

1,2 - plots of hydrostatic head reduction at wells Nos. 47 and 324, respectively; 3,6,7 - plots of subsidence of land surface points at datum mark No.22 of Axial line, datum mark No.8 of Southern line and datum mark No.39 of Northen line; 4 - plot of proportional variation of total discharge of surface and underground drainage systems; 5 - plot of proportional variation of surface drainage system discharge; 8- plot of proportional variation of underground drainage system discharge.

Figure 2. Plots of drainage characteristics and surface displacements

The analysis of plots of surface point subsidence rate changes presents itself an interesting picture (Figs. 4-6).

Initially and during rather long periods points in subsidence trough central part were sinking with the greatest rates (plot 1, Figs. 4-6). Then in 1975-1978 their motion stabilized at various levels (the greatest one is 30-35mm a year). Somewhat earlier the displacement of zone of active subsidence took place towards trough periphery (plot 2). Here subsidence rate proved to be higher than one in the centre of subsidence trough not only relatively but absolutely, having increased for the last 3-4 years up to 65mm a year (the latter is especially typical for the northern and north-western sites of subsidence trough).

The stabilization of surface point subsidence rates over the ore deposit occurred rather early in 1967-1972 (plot 3, Figs. 4-6) and settled on rather low level (5-8mm a year).

In connection with the fact that land surface subsidence occurs at rate 30-35mm a year in the centre of subsidence trough and at rate 60-65mm a year at its wings and analysing also the character of plots in Figs. 4-6 it can be supposed that rock deformation process has been stabilized but it isn't characterized yet by any features of damping. The said rate differences at various subsidence trough sites reflect a contemporary nature of its development, i.e. predominant extension of its area. The comparison of dimensions and shape of subsidence trough middlie part presented as isolines of subsidence by years on surface plan, indicates also this fact.

It should be noted in this connection that judging by the presented plots the central group of shafts is now in the middle part of subsidence trough characterized by the displacement of points mainly in the vertical direction, low values of subsidence trough inclinations and its surface curvature.

In order to estimate rock thickness layer deformation the materials of observations carried out in shafts and dewatering wells, which reflect the processes being studied, were analysed.

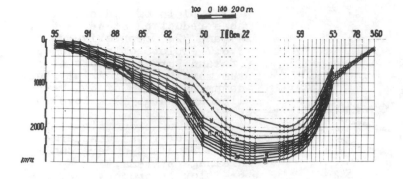

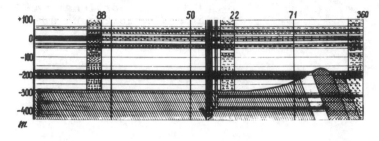

1- quartz-chlorite-sericitic slates;
2- quartzite;
3- serpentine;
4- limestone;
5- ore-bearing thickness; 6- marl;
7- quartz-carbonate-chloritic slates.

Figure 3. Plots of land surface subsidence and geological section along the line "Central".

Observations of mine shaft deformations are being carried out since 1964 and cover the following complex of measurements:
- shaft wall, bunton and guide profiling;
- measurements of values of support yielding units wear;
- measurement of intervals between tubbing support rings;
- observations of underground shaft guide frame settlement;
- measurements of clearances between hoist vessels, reinforcement and shaft lining;
- documentation of places of lining failure and repair.
As a result extensive material covering the nature, magnitudes and dynamics of deformation of shafts and their separate elements has been collected.
The analysis of the said materials indicates the followings:
1. all shafts experience horizontal deformations the magnitude and modes of which depend on their location in subsidence trough;
2. vectors of movement of shaft collar centres reflect satisfactorily the development of land surface deformation in the process of depression cone and subsidence trough formation;
3. unequal (according to depth) distribution of horizontal and vertical sites (sections) displacements according to the magnitudes and directions are registered at all the shafts. The largest deformations (2-3 times greater than the ones in the upper part) are marked for the following rock occurrence interval: Kiev clays - clays of the crust of weathering;

16

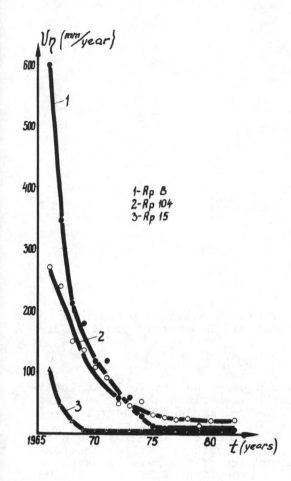

Figure 4. Plots of subsidence rates of "Southern" profile line points

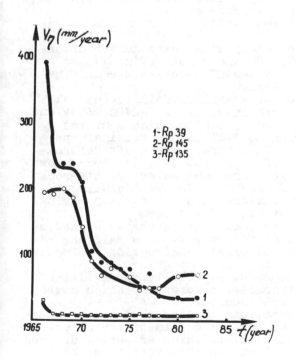

Figure 5. Plots of subsidence rates of "Northern" profile line points

17

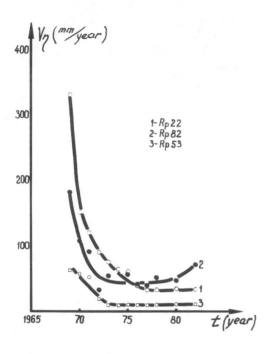

Figure 6. Plots of subsidence rates of points along "Axial-Central" profile line

4. the analysis of shaft deformation dynamics present us the following pattern. The greatest rates and magnitudes of shaft wall horizontal displacements and vertical compression of yielding units are marked in depth interval 250-350m and relate as far as 1964-1970 (up to 70-80% of the registered displacements). Here the evident predominance of displacements in direction west-east in comparison with the ones in direction north-south (2-6 times greater in magnitudes) is marked. Further at the greater part of the said intervals shaft displacements had been stabilized with the exception of sites (300-350m) in the marlaceous thickness lower part and in clays of the crust of weathering where deformation process is going on with the variable rate;

5. units of vertical yielding of tubbing columns instaljed in shafts at various depths work ununiformly.

The results of borehole gauging and technical inspection of a number of deposit dewatering wells Nos. 111, 126, 3p, 402, 200 and others (Fig. 1) are representative for change of characterization of massif stress-state and enclosed in it mine structures during water drawdown.

It is stated that the character and position of stability failure and well lining damage places are characteristic for depth intervals 230-270m (Kiev and Buchak clays) and coincide with the ones observed in shafts which confirms considerable horizontal (shearing) force action as well as the vertical one.

Combined analysis of condition and dynamics of vertical mine shaft deformation and six inspected dewatering wells (Nos. 126, 94, 200,111, 3p, 402) shows that massif deformation is of clearly pronounced zonal character. Mainly rock deformation develops at the extension of depth intervals 70-80m including (if to consider from below to top) a zone of weathered slates, chalkey marlaceous rocks, the Buchak sands and lower part of Kiev clay layer.

As observed in various shafts one and the same rock type is characterized by both absolute and relative to suite of layers different rate of subsidence which is the evidence of differences in hydrogeological, engineering geological and force aspects of deformational process.

The mode and directions of shaft site displacement especially shaft lining failure show that rocks are in the process of complex deformation which combines regimes of compression, tension, bending and shearing.

Data analysis, obtained as a result of observations of land surface and massif displacement, confirms a statement about the largest deformability of watertight and poorly permeable rocks contecting with layers which are being drained under the conditions of considerable head reduction. In this very case these layers are those of the Kiev clay, chalkey marlaceous rocks and clays of the crust of weathering. Plots of rock deformation (for instance, in Fig.2) show that rheological processes play an important part under these circumstances.

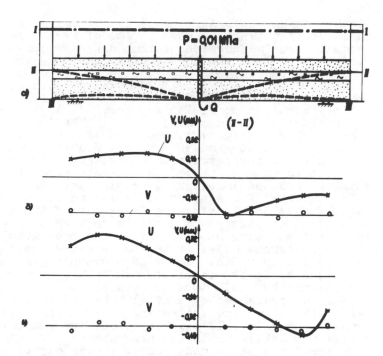

U– horizontal displacements;
V– vertical displacements

Figure 7. Simulation of
rock deformational processes during water drawdown

SIMULATION

The essential complexity of these problem solving consists in the fact that
rock deformational mechanism, manifesting itself during rock drainage, has not
been completely studied and quantitative parameter data of this mechanism revealing are not available. Especially it concerns watertight rocks and clayey
ones in particular. The works, covering said problems, contain contradictory
data and lead to discrepant conclusions. The author doesn't dwell upon this
point in detail but indicates that there are a number of different points of
view concerning one of the most impotant practical problem - that of the relationships between vertical and horizontal displacements of deformable layer
points (it should be noted that the greater part of investigators, working in
this branch of science, doesn't give any solutions about horizontal displacements at all, restricting themselves in considering only the vertical ones).
As a result the obtained horizontal to vertical displacement ratios ranges
within 0 - 1 which naturally can't serve as a scientific basis for predition.
   Practically a degree of problem study, its solution and other characteristics
reflecting rock deformational process during considerable water drawdown
(stress-strain constraint type, time parameter etc.) is at the same level.
   In such situation methods of physical modelling may be fairly of great help.
The author selected models of natural material (clay) out of a big variety of
different types of physical models. Considering all above-said the author pro-
ceeds from the following:
 a) a general problem consists in the study of rock deformational regularities
manifesting themselves in the process of their drainage. Data obtained as a result of field observations and laboratory experiments indicate that the greatest deformations during this process are experienced by clayey rocks (clays,
marls, sandy-clayey and chalkey marlaceous rocks etc.);
 b) more than one half of drainage discharge had been squeezed into drainage
system out of the clayey layers as pore water;
 c)clayey rocks present themselves a unique natural material characterized by
some specific features, physico-mechanical phenomena and processes which had
not been studied yet. The consolidation and rheological processes and phenomena
accompanying them, which are under investigations, relate also to such category.
   Proceeding from these statements and in the absence of sufficient presentation about the said processes a simulation of behavior of a medium itself in
the prescribed conditions should become a necessary stage of investigations.
Here the problem reduces to simulation of physical processes in order to reveal

its mechanism (principle pattern) and register features and parameters of its manifestation. In this case the model isn't demanded much as concerns equivalence towards nature and meeting all requirements relating to similarity criterions, known from the theory of dimensions, is not necessary.

Such methodological technique is justified and admitted at initial stage of studying unknown (or poorly unknown) processes, phenomena and is used when necessary.

The first test model run was carried out in ground chute (duct) made of acrylic plastic with dimensions 150 x 27 x 15cm. Flat model presented by draining rock thickness (Fig. 7[a]) consisted of water-bearing layers 1, including fine pebble and coarse-grained sand and poorly permeable clayey layer 2 disposed between two water-bearing ones. Pottery clays possessing a property which enables it to change its volume together with its moisture content, was taken as material for model deforming layer.

The material for aquifer and lower aquifer thickness (15cm) were selected so as to remove model deforming layer out of the zone of capillary wetting during simulation of a stage of complete underground water level decrease. Clay layer thickness was 4cm. Representative scale of simulation was 1:5000. Adjusted to it the following model pressures were developed: total pressure on clay layer was equal to 0,01MPa, excess pore pressure in clay layer - to 0,005MPa. Bags with lead shot, 5kg each, created total pressure on clay layer which was keeping during the whole test procedure.

Well 3, presented by a metal tube (25cm length, internal diameter o,3cm) perforated along the whole length with holes 0,1cm diameter, occurred in the centre of a model. It allowed to perform simulation under different filtration modes. Simulation included three stages.

At the first one a model after its preparation withstood permanent excess pressure in clay layer equal to 0,005MPa up to the moment of its deformation ceasing.

At the second stage well 3 began to operate and water level in the model decreased up to II-II position (Fig. 7[a]). Depression cone position was determined according to piezometer records. Model was photographed mainly once a day. After deformation stabilization the third stage of simulation began. Complete drainage of model was performed. Depression cone acquired shape III-III due to capillary rise (Fig. 7[a]). Model was also photographed once a day before the moment of deformation stabilization. Altogether seven models had been made and tested.

In Fig. 7(b,c) the most typical results of simulation are presented (model No. 3).

In Fig. 7[b] the second stage final results are presented when piezometric level I-I in the result of well operation had been reduced to position II-II. In Fig. 7[c] the third stage final results are presented when piezometric level had been reduced to position III-III.

It is from these plots that maximum values of vertical and horizontal displacements of model points are close to each other.

Horizontal displacements in well area are close (or equal) to zero. Under the depression cone position II-II maximums of horizontal displacements form near the near the well and the displacements themselves are directed towards it. In the process of further development of depression cone up to III-III position maximums of horizontal displacements move to model boundary increasing in their absolute values. Vertical displacements are practically equal to each other along the whole length of the model and also increase when transferring to the third stage of simulation. Here it should be taken into account that model boundaries are not fixed at edges and this fact results in some deformation process disturbance near them.

The result of simulation allow to make the following principally important conclusions:

1. during clayey rock drainage both vertical and horizontal deformations are characterized by the equal activity and their maximum values are close to each other;

2. a zone of clay maximum horizontal deformation is formed near the water drawdown well and moves towards the boundaries after the depression cone development.

Carrying these result analogy over to nature and considering the above-stated information about the studied processes the author indicates the following:

- an obtained result about the active bulk consolidation of clayey rocks in drainage zone accompanied by comprehensive equal layer point displacement is logically co-ordinated with the clay known properties (high deformational properties, coefficients of lateral thrust is near to 1 etc.);

- translation (transferring) of horizontal displacements from layer to layer upward the rock thickness beginning with the deformed one up to the load surface occurs with damping according to non-linear relationship. It is caused by not only deformational characteristics of rocks in layers but also considerable friction (and tangential stresses) at layer contact (thereby, the postulate of American scientists, concerning the absence of friction between the layers, cannot be recognized a well-founded one);

- unlike horizontal the vertical displacements (settlements) are transferred upward the layers to the surface without any considerable quantitative changes. It can be explained by the fact that a thickness sinks without continuity disturbance and rock failure. Moreover, it can be suggested that rock layers overlying the deformable one sink only as a result of fold bending without internal texture changes;

- it follows from the above-stated that mine structures (for instance, mine shafts) in deforming rock thickness in the condition of considerable water drawdown experience complex ununiform force action revealing in total vertical compression due to vertical consolidation of the whole thickness; in zonal vertical tensions and compressions near the deforming layers; in shearing and bending force developing at contacts with draining and deforming layers; in general unequal shaft displacements and bending as a result of subsidence trough followed by depression cone formation.

Application of engineering geology analogs to support design of hydraulic structures

Utilisation des analogues géologiques pour la justification des projets des ouvrages hydrauliques

A.V.Kolichko & I.A.Parabuchev, *'Hydroproject' Institute, Moscow, USSR*

ABSTRACT: The method of engineering geology analogs helps to exploit actively the wealth of experience in field investigations to support design studies for hydraulic projects. This method operates with similarity criteria which reflect the established notions about dependence of physical-mechanical, seepage and other properties of the rocks on geological factors, which formally express such geological peculiarities of the rock mass as composition and properties of the rocks as sampled, rock mass structure, history of regional structural geology in recent time, etc. Validity of engineering geology prediction to be made using this method is conditioned on the extent of geological investigations conducted at the object in question. The paper describes the adopted procedure including selection of an object-analog and gives a specific example of application of this method.

RESUME: La méthode d'analogies géotecniques permet d'utiliser, d'une manière efficace, l'expérience acquise au cours des études géologiques visant la justification des projets d'ouvrages hydrauliques. Elle est basée sur les critéres de similitude objectifs, qui représentent les idées existants concernant la relation entre les propriétés physiques et mécaniques, la perméabilité et autres caractéristiques des roches et les facteurs géologiques. Ces derniers sont une représentation formelle de telles particularités géologiques du massif que la composition et les propriétés des roches dans les échantillons, la structure du massif, l'histoire du développement géologique de la région au temps récent. La précision des prévisions géotechniques faites à l'aide de cette méthode est déterminée par le degré de la connaissance du terrain étudié. On décrit le processus de la solution du problème, y compris les particularités du choix de l'analogue. Un exemple de l'utilisation de cette méthode est présenté.

Active and task-oriented use of information on the properties of rock masses which has been accumulated to date, would help to optimize the field investigations and to upgrade the safety of engineering projects (Parabuchev, I.A. & I.A. Vanchurov 1981). The method of engineering geology analogs is an effective tool in exploitation of the wealth of such information. Operating with the objective criteria of similarity which are based on the established notion the dependence of physical/mechanical, permeability and other properties of a rock mass on geological factors, this method allows for prediction of the properties of the object in question (geological entity upon which observations have to be made) using the information available on the objects-analogs.

The following characteristics of the rock mass play the part of geological factors such as composition of the constituent rocks, structural tectonic texture, the pattern and intensity of manifestation of physico-geological processes, as well as physical and mechanical properties as sampled and in-situ.

As seen from this listing, the like information about the rock mass in question could be obtained at a certain level of its validity, even during initial phases of the field investigations using results of the geological and geophysical survey, simple laboratory tests on rock specimens or it could be borrowed from the general geological studies of the region.

The method of engineering geology analogs embraces the following procedures - formulation of the task, identification of minimum-required geological factors which would give information on the basic task-specified peculiarities of the object, search, guided by these factors, for objects-analogs to be represented by the geological entities underwent more in-depth studies than the object in question, assessment of difference between the object-analog and that in question using the similarity criteria and prediction of the sought for properties of the object in question by the results of comparison made.

The paper treats the problems related to the determination of the properties of the rock and semi-rock masses which are looked upon as the medium and foundation of the hydraulic structures.

The rock mass properties which are major controls in building and operation of hydraulic structures, are deformability of the rock mass to be characterized by the modulus of deformation E_m, its strength denoted by η and τ, permeability q, K_ϕ. As for other parameters of the rock mass used in the structural design, such as shear modulus (G_m), coefficient of elastic rebound (K_o), design hardness factor used in analysis of the temporary support, i.e. they could be determined with a sufficient level of validity from the following expressions:

$$G_m = \frac{E_m}{2(1+\mu)}; \qquad K_o = \frac{E_m}{100(1+\mu)};$$

$$f_p = \frac{[\sigma_{c\varkappa}]}{100} \times K_1 \times K_2, \text{ etc.}$$

where μ is Poisson's ratio; $[\sigma_{c\varkappa}]$ - ultimate unconfined compression strength of rock as sampled, K_1 and K_2 - coefficients considering the degree of jointing and the depth of excavation. In a general case the solution should contain also information on variability of the rock properties, i.e. it should characterize its heterogeneity. The multitude of the factors characterizing the rock is proposed to be limited to its lithological (petrological) properties, elastic-strength properties, solubility. The lithological (petrological) factor, depending on the availability of an object-analog, may have a different level of validity which is identified with the following line up - rocks of the identical composition belonging to the same genetic type, complex etc. In a general case the application of a factor with a low level of validity as an analog decreases validity of the prediction, but this decrease in its validity would not be felt on such properties of the rock mass as deformability, strength, permeability. By solubility, the rock may fall under three categories. The 1st type embraces easily soluble rocks such as gypsum, halite, etc., or the rocks containing their deposits whose properties could degrade during construction and operation of the hydraulic structure.

The 2nd type embraces the soluble rocks such as carbonate rock in which there may develop solution channels complicating operation of the hydraulic structures.

The 3d type covers practically insoluble rocks.

The type of joint systems developed in the rock mass and the extent of its dissection by the faults of various magnitudes are looked upon as the factors to characterize the peculiarities of structural tectonic texture of the rock mass. There are two types of joint sets - systematic and non-systematic (Kolichko, A.V. 1971). The rock masses with systematic joint sets feature a blocky structure. Its dissection is measured by an average size of the elementary structural block limited by joints and by the number of joint sets (K, K $\geq$ 3). Properties of such masses are essentially governed by the parameters of joint sets. The rock masses with non-systematic joint sets do not form a blocky structures and their properties are conditioned by the properties of the rock as sampled. The stressed-state of the rock mass governs its elastic-strength properties. With rather high stresses, the rock mass features higher elastic characteristics and its strength is conditioned by the strength of the rock as sampled irrespective of the type of the joint set, etc.

The relative strength of rocks as a function of the stressed-state of the rock mass may be characterized by equation $\frac{\sigma_{max}}{[\sigma_{c\varkappa}]} = n$ (Zaslavskiy, Clar, Markov et al.), where σ_{max} - maximum stress existing in the rock mass or caused by the built structure. At n $\geq$ 0.2 the mass composing rocks are considered to be relatively weak - the sampled rock is likely to fail with time. If n > 0.4 the failure may occur during construction. At n < 0.2 the rocks are thought to be relatively strong.

Tectonic faults form the megastructure of the rock mass. In contrast to joints they are accompanied by shatter zones which feature an increased number of joint sets and a smaller block sizes relative to the enclosing rock mass (background). In a general case, the magnitude of the shatter zone is proportional to that of faulting. Degree of the megastructural dissection of the rock mass is described by the parameter A_j = average distance between the faults of j-order.

Among the factors reflecting physical-geological processes occurring in the rock mass attention is primarily given to weathering and relaxation, as well as to karst formation in soluble and hardy-soluble rocks.

Weathering and relaxation cause to degrade the elastic-strength properties and to increase permeability of the rocks. Depending on the nature and extent of such impact, three zones of engineering geology (Kolichko, A.V. & V.N. Fil' 1981; Lykoshin, A.G. et al. 1972) could be delineated - weathering and relaxation (sub-surface) - I, relaxation - II, unaltered rocks - III. Zone I is characterized by decrease in the rock properties as sampled, opening of the existing joints and formation of the new ones.

In zone II the properties of the rock as sampled do not tend to change. In the rock masses with systematic jointing there occurs opening of the joints,formation of destressing joints which inherit the existing joints. In the rock masses with non-systematic jointing, the formation of distressing joints is oriented parallel to the topography features.

The magnitude of the zone of rock mass exogenetic changes (m) depends on lithological peculiarities and strength of the rock and primarily on the depth of erosion incision and the age of the slope (the factor reflecting the history of new relief formation). For the mountain regions of Central Asia where the formation of ground terrain features dates to post-Lower-Quarternary period, the magnitude of the exogenetic changes in the rock mass with systematic jointing may be characterized by the depth of erosion incision (H) over Q_2-Q_4 period. For the rock masses with non-systematic jointing it may be characterized by H in conjunction with the strength indicator (σ).

Relative karstification of the rocks (S), with all other conditions being the same, at each time period of the river valley development is inversely proportional to the rate of erosion incision U_i. The pattern of karst development is characterised by the trends in the hydrogeological process. The procedure of search for an object-analog consists in selection of more thogoughly studied rock masses meeting the following requirements:

- constituent rocks belong to the same type in terms of their solubility as those of the rock mass in question. If the latter is composed of insoluble rocks, rock masses composed of hardly soluble karst-free rocks may serve as an analog;
- same type of joint system and same relative strength (n);
- located in a region having similar geological history and climatic conditions (for example, mountain systems with dry continental climate, etc.).

The last requirement is needed for prediction of the magnitude of zones of the exogenetic changes in rock masses and the intensity of karstification.

In the latter case, if determinations of some properties on an object-analog are not viable enough, or if the object in question contains the rocks with another type of joint sets or relative strength other objects could be employed as complementary analogs.

As said above, the similarity criteria reflect the established notions about the dependency of rock mass properties on geological factors and the validity of the geological prediction is conditioned by the extent of studies made on the object. The last thesis calls for the criteria demanding different levels of accuracy.

Modulus of deformation of rock mass (E_m)

It is known (Zelensky, Ruppeneit, Ukhov et al.) that in the rock masses with systematic sets of joints E_m = f(E_o, b, Δa, ξ) where Δa - joint width and ξ - area of contact along joint surfaces, the value ξ is the most constant one. Assuming ξ = const, we will get the following criteria: 1) $E_o \times \dfrac{b}{\Delta a}$; 2) $E_o b$. It is evident that criterion 1) is more accurate, but it is applicable when there is information on change of Δa with depth. Criterion V_p^2 (Savich et al.) also features high validity, where V_p - velocity of propagation of elastic longitudinal waves in the rock mass.

Resistance of rock mass to shearing force (τ)

In the rock mass with a systematic set of joints the parameter $\tau_i = \sigma_i \, \mathrm{tg}\, \varphi_i + c_i$

is an essentially anisotropic value and it tends to vary within $\sigma_i \, tg \, \varphi_o + C_o \geq \sigma_i \cdot tg \varphi_T + C_T$ (Sander, Muller, Roš, Fisenko et al.), where φ_o, φ_T - angle of friction, C_o, C_T - cohesion respectively as sampled and along joint plane. $\varphi_o \gg \varphi_T$, $C_o \gg C_T$. Assuming that $\sigma_i \, tg \, \varphi_o + C_o \leq [\sigma_p]$, the above expression may be put as $[\sigma_p] \geq \tau_i \gg \sigma_i \, tg + C_T$.

With increase in fracturing of the rock mass (increase in K and decrease in b) value τ_i tends to decrease

$$\varphi_i \rightarrow \varphi_T, \ C_i \rightarrow C_T$$

The criterion reflecting this tendency may be expressed as $\sin^2(\frac{\pi}{2K} - \text{arc } tg\frac{1}{b})$. Values C_i and φ_i could be found from the system of equations $\tau_i = f(\sigma_i)$ for two or more σ_i.

For the rock masses with nonsystematic sets of joints, value σ_p is used as the strength criterion.

Permeability of the rock masses

Permeability of the rock masses with systematic joint sets (q, K_φ) is conditioned by the intensity of its jointing (b) and the width of fissure opening (Δa) (Lomize, Romm, Louis, Zhilenkov et al.). Should information be available on parameter Δa, the expression $\frac{\Delta a^3}{b}$ may serve as the permeability criterion, the criterion $\frac{1}{\sqrt[3]{b}}$ is less viable.

Permeability of the rock masses with non-systematic joint sets may be characterized by the permeability of the rock as sampled $-q_o(K_{\varphi_o})$.

The megastructure of the mass influences tangibly the formation of the mass heterogeneity: the more the faults are in number and the larger they are in size, the more heterogeneous is the rock mass. The criterion characterising the megastructural dissection of the mass (T) is expressed as follows:

$$\sum_{j=1}^{N} \frac{1}{A_j e^j}$$

where A_j - distance between faults of j - order; e - logarithmic base.

The criteria helping to assess the magnitude of the zone of exogenetic changes and degree of karstification are described above. Table 1 gives a list of similarity criteria in the context of the stated problem.

Using the similarity criteria, the parameters are determined from expression $P_u = P_a(\frac{R_u}{R_a})$, where R_a - value of the sought for parameter on the object-analog, R_a and R_u - numerical values.

An example of prediction of rock mass properties made in conformity with the above procedures is given below. Suppose, two projects are located on the same river in western Tien Shan about 30 km apart. One project which is supposed to serve an object-analog has a concrete dam, 100 m high. The project under consideration is proposed to have a dam 70 m high.

Table 2 contains information on the geological factors, which have been obtained at the project in question from the findings of geological survey and single exploratory excavations at the pre-design stage of field investigations.

Table 3 contains numerical values of the criteria and property parameters determined at the object-analog and the project in question. To assess the level of prediction validity, given here are the readings of direct tests of the properties at the project in question, which were conducted at the next stage of field investigations (in three years time).

Analysis of the findings of field investigations conducted at a number of water projects, with the preliminary evaluation of construction conditions made, in a great measure, using the analogs, indicates that the level of validity of the prediction has proved to be adequate for respective stages of design development and provided solutions to the basic problems to be solved by field investigations.

Table 1

Characteristics	Similarity criterion for rock masses with different types of joint sets	
	Systematic	Non-systematic
Relative strength - n Modulus of deformation E_m Resistance to shearing force τ Permeability $q(K_\phi)$ Megastructural dissection of mass - T Magnitude of zone of exegenetic changes (I+II) - m including zone of weathering (I) m_b Degree of karstification, S	1) $E_0 \cdot \frac{b}{\Delta a}$; 2) $E_0 \cdot b \cdot \frac{\sigma_{max}}{[\sigma_{c\varkappa}]}$ 3) V_p^2 $Sin^2(\frac{\pi}{2K} - arc \, tg \, \Psi \frac{1}{b})$ 1) $\frac{\Delta a^3}{b}$; 2) $\frac{1}{\sqrt{b}}$ $\sum_{j=1}^{N} \frac{1}{A_j e_j}$ H $\frac{m}{H}$ $\frac{1}{U_i}$	1) E_0, 2) V_p^2 $[\sigma_p]$ $\frac{H}{\sigma_p}$

Table 2

Geological factors	Project	
	taken as object-analog	under consideration
Lithology, age	Alternation of medium and thin-bedded sandstones and siltstone	Alternation of medium bedded sandstone and siltstone and thick-bedded conglomerate
	C_{1-2}	C_3
Solubility	Insoluble	Insoluble
Strength and deformability as sampled		
$E_m \cdot 10^{-3}$ MPa	25.0	25.0
$[\sigma_{c\varkappa}]$ MPa	90.0	110.0
$[\sigma_p]$ MPa	12.0	16.0
Relative strength, n	0.2	0.2
Type of joint set	systematic	non-systematic
Parameters of joint set K	4/6	4/6
b, cm	8/5	15/5
I	0.8	1.6
Δa_{mm} II	0.5	1.0
III	0.3	0.5
Distance between faults (A, m) of different order of magnitude /j/		
j = III, A_{III}	-	250
j = IV, A_{IV}	100	80
j = V, A_V	50	30
Velocity of elastic longitudinal waves V_p km/s I	1.8/1.7	1.7/1.2
II	2.7/2.2	3.0/2.2
III	3.4/2.5	3.9/2.9
Depth of erosion incision H over Q_2-Q_4 period	320	290

Table 3

Mass properties	engineering geology zones	desing characteristics	value of similarity criterion	value of similarity criterion	predicted characteristics	characteristics determined by detailed investigations
			Project taken as object-analog	Project under consideration		
$E_m \cdot 10^{-3}$ MPa	I	1.3/0.8	1/ 30000/	23430/-	1.0/-	
			2/ 2400/1500	4500/1500	2.4/0.8	1.0/0.6
			3/ 3.24/2.89	2.89/1.44	1.2/0.6	
	II		1/ 48000/-	45000/-	1.7/-	
		1.8/1.2	2/ 2400/1500	4500/1500	3.3/1.2	2.5/1.2
			3/ 7.29/4.84	9.0/4.84	2.2/1.2	
	III		30000/-	90000/-	3.4/-	
		3.0/1.8	2400/1500	4500/1500	5.6/1.8	5.5/1.8
			11.6/6.25	15.2/6.25	4.0/1.8	
	I	0.12/0.17	0.026/-	0.167/-	0.7/-	
			0.5/0.6	0.4/0.6	0.1/0.17	0.7/2.0
q l/min	II	0.08/0.14	0.0062/2	0.04/-	0.5/-	
			0.05/0.06	0.04/0.06	0.07/0.14	0.2/0.8
	III	0.04/0.15	0.0013/-	0.005/-	0.15/-	
			0.5/0.6	0.4/0.6	0.04/0.4	0.1/0.2
	I	0.15/0.07			0.08/0.07	0.10/0.05
		0.8/0.6	0.08/0.005	0.10/0.005	0.9/0.6	0.7/0.5
τ	II	0.15/0.12			0.2/0.12	0.20/0.05
		1.0/0.7			1.0/0.7	0.75/0.5
	III	0.32/0.15			0.38/0.15	0.3/0.1
		1.15/0.8			1.1/0.8	1.0/0.6
T p.u.		1.0	0.0003	0.0006	2.0	-
Magnitude	(I+II)	20-35	320	290	18-31	20-25
m of zone of	(I)	3-5	0.084	0.087	3.2-4.5	5-8
exogenetic changes m_b						

Notes: 1. Ref. to tables 2 & 3: Figures in numerator characterize mass out of zones affected by faults; denominator gives characterization of rock mass in the zone.
2. Ref. to table 3: Index 1/, 2/, 3/ denotes criteria with different level of validity and predicted characteristics obtained using these criteria.

It confirms both viability of use of the analog methods at earlier stages of design-related investigations and the choice of most objective geological criteria to assesss the similarity of the natural conditions (or its separate elements) at the project in question and at projects taken as objects-analogs.

The characteristics depicting composition, state and properties of the foundation and geologic environment of hydraulic structures, proposed as similarity criteria for prediction purpose will form the basis of a computerized system being now developed for search of engineering geology analogs and solving prediction and diagnostic problems.

REFERENCES

Kolichko, A.V. 1971. Determination of shear resistance in jointed rock mass. Field investigations mathematic methods in engineering geology: 92-98 (Russian).
Kolichko, A.V. & V.N. Fil' 1981. Engineering geology at Rogun dam project. Gidrotekhnicheskoye stroitel'stvo, 10: 11-15.(Russian).
Lykoshin, A.G. et al. 1972. Principles of modelling of engineering geology for dam engineering. Gidrotekhnicheskoye stroitel'stvo, 3: 7-11 (Russian).
Parabuchev, I.A. & I.A. Vanchurov 1981. Outlooks for computer application in geological prediction in dam engineering. Proceedings of "Hydroproject" Institute, 75: 134-139 (Russian).

Study on the rock mass structure and the rock slope stability

Etude de la structure d'un massif rocheux et de sa stabilité sur les pentes

Li Yurui & Xu Bing, *Institute of Geology, Academia Sinica, Beijing, China*

ABSTRACT: This paper deals with the relationship between the rock slope deformation and the rock mass structures. The latter determine the scope (limit), type and deepness of the rock mass deformation. According to the basic principles of the engineering geomechanics of rock mass structure and characteristics of Jinchuan slope deformation the slope stability problem has been discussed in this paper.

RESUME: Dans ce papier on a présenté la relation entre la déformation de masses rocheuses et la structure des masses rocheuses, celle-ci determine le domaine (la limite), le type et le fond de déformation de masses rocheuses. D'après le principe fondamental de la géomecanique de l'ingénieur de la structure des masses rocheuse et la caractéristique de déformation de la pente de Jinchuan, le probleme de stabilité d'une pente a été discuté aussi dans ce papier.

The rock slope deformation, controlled strictly by the rock mass structure surface at Jinchuan open pit mine, is considered as a typical example in research of the slope stability.

This paper deals with the theory of rock mass structure and its application to evaluate the possibility, necessity and significance of slope stability.

The slope height at Jinchuan open pit mine is about 310 m. The contour of open pit mine takes an elliptical form with 1000 m in length and 600 m. in width.

The geological conditions are complicated by the various faults, multicycle intrusive bodies, and fractured rock mass. Such a poor and complex engineering geological conditions are rarely encountered in other mines. Therefore, during the excavation of open pit mine, the equilibrium condition of the initial rock stress had been disrupted and the rock slope deformation in the form of toplling-sliding may be induced. The scale of this slope deformation is very large. In order to assure safety in production, the detailed and extensive engineering geological researches have been made at the mining area. On the basis of these researches the mechanism and cause of rock slope deformation can be analysed and discussed. The prediction of the slope deformation and protective measures have been suggested.

1 CHARACTERISTICS OF ROCK SLOPE DEFORMATION

The slope deformation occurred at first on the upper wall of mining area, namely, the I-1 subzone (engineering geological zone) (Fig.1)

The cracks appeared in the primary period of rock deformation and the scale and numbers of these cracks increased with the progress of mining excavation. After a heavy rain the rock mass had topplled and slid along the appeared cracks. The area of rock deformation is very large, covering about 5000 m^2 in the I-1 subzone. The turn upwards cliff of the topplling deformation may rise up to 4 m. and a forest of these cliff were formed. The excavating platform inclined to the pit with the angle $10^{\circ}-20^{\circ}$. The max. velocity of the horizontal displacement was about 44 mm/d. and the velocity of the subsidence was about 29 mm/d. The horizontal displacement amounts to 9 m and the subsidence displacement about 8 m in the 5-6 year period of the slope deformation. (Fig.2). The topplling-sliding is formed in an irregular quadrilateral type on the plan. Due to the many factors,

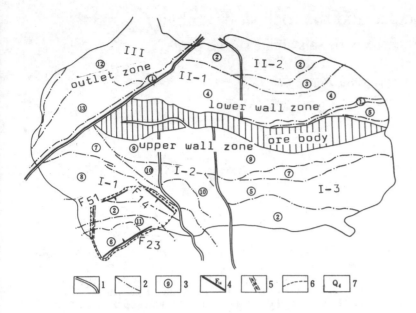

Fig.1 Engineering geological plane sketch map at Jinchuan open pit mine
1. Boundary line of engineering geological zone
2. Boundary line of rock
group
3. Number of rock groups
4. Line of fault and its
number
5. Concentrated joint zone
6. Deformation contour line
7. Quarternary sediments

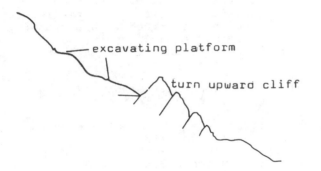

Fig.2 Sketch of the turn upwards
cliff of the toppling deformation.

such as the climate conditions, excavating procedures and the variations of the geological environment, the deformation of rock mass new slows down, new speeds up. Therefore, the slope deformation at Jinchuan mine is characterized by its features and regularity.

The characteristics of the toppling-sliding deformation of the slope rock mass in Jinchuan are:

1. The features of the deformed rock mass are very distinct. The deformation cliff is imbricate in shape. The direction of the imbricate cliff is opposite to the slope. The excavating platform inclines severely, and its angle may reach about 30°. There is a difference of the slope deformation along the slope strike and the dip. The features of the slope deformation are characterized by the irregular sawtooth shape.

2. The displacement of rock mass is of a definite regularity.

The horizontal displacement greater more the vertical one; The direction of rock mass displacement is perpendicular to the slope strike; There is a distinct difference of the rock mass displacement from the periphery of the rock deformation to the toppling centre; The rock mass deformation is characterized by the creep strain.

3. General characteristics of the rock mass deformation.

The feature of rock slope in Jinchuan open pit mine is of a hazardous deformation. However, if the protective measures can be taken the hazard can not apperr in a definite period.

2 STUDY ON THE ROCK MASS STRUCTURE

The important progress of engineering geology in the evaluation of the rock mass stability is that the rock mass is considered as an object with a structure. It is a key problem of the engineering geomechanics of rock mass. Due to the diagenesis and the multiple tectonic movements which the rock mass has been subjected to during its formation and evolution, the rock mass becomes a complex object in the aspects of the material composition and of the structure characteristic. In the evaluation of the rock mass stability the two factors should be considered: the rock mass strength and the rock mass structure. These two factors are closely connected each with other. Due to the difference in material composition the structure characteristics of rock mass are also various. Therefore, a basic study of the rock mass structure is the study of the material components of rock mass, namely, the engineering geological group.

The analysis of the rock mass structure is a core of the engineering geomechanics of rock mass. The rock mass stability is mainly controlled by the rock mass structure. In most cases the rock mass instability may appear mainly by the unfavourable condition of the control structure surface. The components of a rock mass structure are the structure surface and structure body. There is an organic relation between them. The combination of structure surfaces defines the form of structure body. Therefore, the rock mass structure is classified according to form of the structure bodies, it also reflects the combination of structure surfaces.

1. The structure surfaces in Jinchuan mine can be classified into 4 types:
(1) the tectonic structure surfaces (faults, joints, interbedded sliding surface);
(2) the metamorphic structure surfaces (gneissosity, schistosity);
(3) the igneous structure surfaces (contact of fractured zone);
(4) the sedimentary structure surfaces (stratification plane);
2. The development characteristics of the structure surfaces in mine
(1) The striking structure surface is mostly developed. The direction of this structure surface corresponds to the strike of bedding planes. Its scale is large (usually, I. II order) and it can intersect through the whole area of mine. This structure surface is superior in numbers. Its degree of development and characteristics are of great importance in the evalution of rock mass deformation and in the sivision of the rock mass structure types.
(2) The steep structure surface plays an important role. Due to the mine area is subjected by the intensively multiple tectonic movements the tectonic structure surfaces are of the steep angles (70°-90°). These steep structure surfaces are the basic factor determining the "toppling-slide" slope deformation.
(3) The soft structure surface are also widely developed. According to the mechanical characters the structure surfaces can be classified into two types: rigid structure surface and soft structure surface. Because of the multiple tectonic movements the soft structure surfaces are intensively developed. It is of the greatest importance among various structure surfaces. They may essentially influence on the stability of the slope rock mass.
3. Types of rock mass structure in mine.
The rock mass structures can be classified according to the degree of development, orders the arrangement and combination of the structure surfaces. The rock mass structures in Jinchuan mine are classified into 4 types and 7 subtypes, as following:
I Intact blocky structure
II Layered structure
    II1 Layered structure
    II2 Thin layered structure
III Fractured structure
    III1 Mozaic structure
    III2 Layered fractured structure
    III3 Fractured structure
IV Lossened structure
The stability of the various types of the rock mass structure in slope is different depending upon the boundary conditions, the state of stresses and the combination relationship of structure surfaces.

3 THE CONTROLLING EFFECT OF THE ROCK MASS STRUCTURE ON THE SLOPE DEFORMATION

The abovementioned behavior and characteristics of the rock mass structure in Jinchuan mine have laid a foundation of the rock mass deformation and failure.

The rock slope deformation controlled by the structure is very typical and obvious.

1. The boundary of rock slope deformation is controlled by the structure surfaces.
The deformation boundary controlled by the structure surfaces is typical for I-1 subzone. It is an irregular quadrilateral type. The results of the study indicate that they are controlled by following structure surfaces:
1) Fault F_{23} (N70W, NE $\angle 67^o$)
2) Jointed zone of marble (N45E, SE $\angle 34^o$)
3) Shear faults F_{14}, F_{14}^3, F_{14} (N5W, SW $\angle 85^o$)
4) Fault f_{51} (N65-70W, SW $\angle 70^o$)
These structure surfaces can form the tectonic network in "米" type. The striking faults F_{23}, F_{51} are considered to be the direction of the principle tectonic line. The fault F_{14} is the strike-slip fault. The intensively jointed zone is the characteristic structure surface of the marble strata. These structure surfaces are referred to the type of tectonic structure surface. The faults F_{23}, F_{14} are considered to be the structure surfaces of II order, F_{51} — to III order and joints — to IV order.
The observations on the rock mass displacement at ground surface have also proved that the slope deformation is controlled by the abovementioned structure surfaces.
2. The sliding plane of the rock mass deformation is determined by schistosity.
The sliding occurred in the upper part of the slope in I-1 zone. Mainly along the schistosity of the Chlorite biotite-Quartz schist rock group ⑪. In this zone the schist strata are of the overturned occurrence. Therefore, its dip is consistent with the slope direction and the optimum structure surface of the rock mass slide is formed. The schistosity is reffered to the metamorphic structure surface. The strength of schists is very low and is characterized by a soft structure surface.
3. The toppling rock deformation and failure are controlled by the rock mass structure, which is characterized by the steep layered structure with strata inversely and the alternative beds with rigid dipping in relation to slope and soft rock strata.
The rock mass toppling deformation and failure occurred mainly in the migmatitic and marble with magmatic rock groups. These two rock groups are of abovementioned characteristics of the rock mass structure. Therefore, under the action of thrust of the overlying rock mass and the geostress the rock slope has been gradually deformed and topplled from upper to lower (Fig.3).

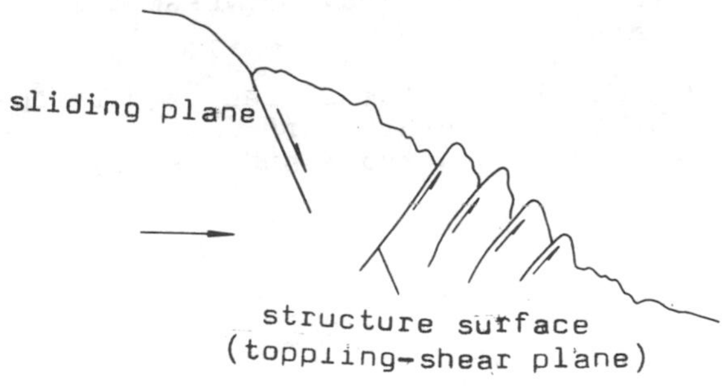

Fig.3 Diagram of the slope "toppling -slide" deformation at Jinchuan open pit mine.

The mechanism of the rock deformation is considered as the action in concert of the upper thrust and the lower loosening force (includes the plastic extrusion of the soft rock bed).

IV CONCLUSION

From the practice in Jinchuan mine the stability of any engineering rock mass may be proved from the two aspects:

1. Study and analtse the rock mass structure behavior. This is a geologic fundamental work. The rock mass structure is the product of the whole geologic process. It has undergone the evolution of the diagenetic process (formation), the tectonic action (transformation) and the secondary process. Therefore, the rock mass characteristics should be studied from the viewpoint of the formation and the evolution of the rock mass. A great attention should be paid to the variation of rock mass in space and time aspects. On this basis the rock mass structure can be classified and recognized.

2. Pay attention to engineering action namely, should consider the effect of force equilibrium condition. The natural rock mass is present in the environment of the human engineering action. It should pay attention not only to the engineering mechanical action on the rock mass, but also to the evolution of the rock mass structure and its engineering geological characteristics. The slope deformation and failure in proper order has close relation to the explosion and excavation at Jinchuan open pit mine.

The latter engineering action has a great influence on the changes of the rock mass behavior. This is one of the researches problem for the environmental engineering geology, namely the environment influence and action of the human activities on the changes of the natural geological mass. The slope engineering and the rock mass structure are determined and influenced each other. The slope engineering action and its deformation will induce the evolution and transition of the rock mass structure. Then, the new equilibrium state can be reached. For the evaluation of the rock mass stability, it should pay attention to this relationship and its transition.

Structural mechanical properties of Shipai shale and their influence on the stability of hydraulic structures of the Qingjiang hydropower project

Schiste de Shipai du complexe hydraulique de Qingjiang – Ses caractéristiques mécaniques de structure et son influence sur la stabilité de l'ouvrage hydraulique

Liu Yu-yi, *Yangtze Water Conservancy & Hydroelectric Power Research Institute, Wuhan, China*

ABSTRACT: The shale of Shipai being fairly developed in structure planes, its stability in shear strength and deformation becomes a matter of major concern to the foundation the structures to be built on. In view of this the author would like to give a relatively systematic analysis of and have a discussion on the following:
1. Types and features of the structure planes and their gradation and influence on the stability of the engineering works;
2. Classification of the types of structures of the rock masses and evaluation of their deformation properties, shear strength, anisotropic characters and engineering stability;
3. Prediction of the stability of the rock slope for the powerhouse and the surrounding rock for the tunnels.
The results of investigation showed that Shipai shale is relatively poor in structural mechanical properties and will probably cause stability problem. It is necessary to reinfoce the rock masses as soon as possible after the excavation during construction.

RESUME: La surface structurale du schiste de Shipai est très développée, sa résistance et sa stabilité de déformation constituent des problèmes principals de la fondation de l'ouvrage. On analyse systématiquement dans cet article ces deux problèmes et présente:
1. Les types de surface structurale ainsi que leurs caractères, la définition de leur ordre de degré et leur influence sur la stabilité de l'ouvrage.
2. La définition des types de la structure du rocher ainsi que leurs caractéristiques de déformation, leur effet de cisaillement, leur anistropic et l'appréciation de la stabilité de l'ouvrage.
3. La prévoyance de la stabilité de la pente de roche de l'usine et de la stabilité du rocher environnant de galerie.
Les résultats d'étude montrent que les caractéristiques de la mécanique structurale du rocher et la stabilité de l'ouvrage hy draulique sont assez faibles. Il faut renforcer à temps le rocher pendant l'exécution des travaux.

1. GENERAL

The Qingjiang project is a proposed large-scale arch-gravity dam. Its power station, located at the toe of the dam, on the right bank, includes an open-type powerhouse and four intake tunnels. Shipai shale the power station is to be founded on (Fig 1.1) is 209 m thick, composed of the low Cambrian interbedded shale and siltstone, occurring with a strike of N70°-80°E and a dip of SE at an angle of 25°-30°. The tectonic pattern of the dam area, finalized during Yenshan movement, belongs to a latitudinal tectonic belt in the middle and lower reaches of the Yangtze River. The results of magnetic survey showed that the magnetic field in the region is normal, its basement is intact, and there is no evidence of either recent tectonics or earthquake-pregnant tectonics. In Shipai shale, however, there exists a considerable amount of bedding structure planes, high-angle joints and minor broken zones, which would be extremely detrimental to the stability of the powerhouse's rock slope and the tunnel surrounding rock. This paper is a special study of the structural mechanical properties of Shipai shale and the stability of the rock slope of the powerhouse and the surrounding rock of the tunnels.

2. STRUCTURES OF THE ROCK MASSES

The rock masses's structure consists of two elements, i.e. the structure planes and the rock mass between them, the " structure body." As the former is the main factor in deformation and failure of the rock masses, so it is especially important to study the behavior of the structure plane and its influence on the stability of the engineering works.

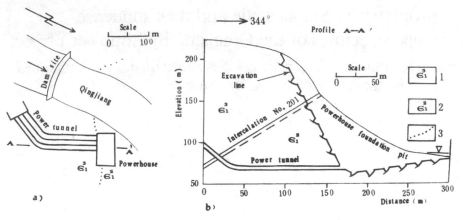

Fig 1.1 A sketch showing the project layout and its relation with Shipai shale.
a) plane;
b) profile.
1. Shilongdong limestone;
2. Shipai shale;
3. dividing line of the rock strata.

2.1 Classification of structure planes

2.1.1 According to their geologic genesis

Those closely related to the engineering works can be grouped into two categories: the sedimentary structure plenes and, mainly, the tectonic structure planes.

1. The sedimentary structure planes: These are mainly stratifications and bedding planes. The stratifications, though well cemented, are liable to crack with desiccation, while the bedding planes, mostly filled with mud, tend to open under tension at the tunnel roof and, when combined with other structure planes, to cause the surrounding rock to collapse. The filling materials between layers are illite, kaolinite, chlorite and other clay minerals, which, under the action of the ground water, would turn the bedding planes into some larger in scale and worse in behavior mudded ones.

2. The tectonic structure planes: Including bedding shear planes, the most developed ones, faults and joints.

Bedding shear planes: According to the degree of their development and behavior the bedding shear planes can be divided into three types: the broken-mudded type, the mudded plane type and the fissured type. The broken-mudded type is a sort of weak intercalations, composed of a series of well-developed interlayer mudded shear planes and broken rock bodies between them. They are large in scale and poor in behavior and occur either on the interfaces between limestone and shale or within the fairly thick-layered shales. For instance, the intercalation No. 201 which occurs on the interface between the top of Shipai shale and the overlying Shilongdong limestone is a major shear zone of up to 0.4 m in thickness. The mudded plane type is referred to as the individual interlayer mudded shear planes, which, dispersed over the bedding surface, frequently occur in thin shale layers within siltstone, on the interfaces between siltstone and shale or inside the relatively thick-layered shale horizons. This type, smaller in scale and smooth in plane, is fairly developed. As a kind of wide-spreaded and well-

developed bedding shear planes, the fissured type is emerging mostly along the siltstone bedding planes, in the form of bedding fissuring, characterized by a slight failure of the rock masses and filled with argillaceous and calcareous materials.

Faults: 34 faults have been found. They can be divided into two principal groups: N5°W50°-80°SW and N65°W60°-85°NE. The first group, mostly sheared, is far bigger in scale than the second, a well developed one. The faulted zone, less than 0.4 m wide in general, is composed for the most part of mylonitic breccia, mylonite and gouge.

Joints: The most developed group is the one with a strike-dip of N45°-65°W45°-80°NE, which, mostly high dipping, is smooth and straight in plane and filled with argillaceous materials.

2.1.2 According to their engineering geologic features

From the engineering geologic point of view importance should be attached not only to the natural features of the structure planes but also to their engineering geologic properties. In that case the structure planes can be further divided, in accordance with their distribution, behavior and direct relations with the safety and stability of the engineering structures, into the following four types: the broken-mudded zone, the continuous mudded plane, the discontinuous mudded plane and the stiff plane. The characteristic features for these types are listed in Table 2.1

2.2 Gradation of structure planes

the mechanical properties and behavior of the structure planes with different sizes are remarkably different. According to their dimensions the structure planes can also be put into four grades: grade I, includes those more than 100 m long with fractured zone over 0.3 m wide; grade II, those with a length from tens to 100 m; grade III, those with a length from several to tens m long; and grade IV, those with a length less than 2m.

Table 2.1 Engineering geologic classification of the structure planes and their characteristics

No.	Name	Tectonic changes	Width of structure plane (cm)	Mudded plane Thickness (cm)	Mudded plane Continuity	Features of tectonite	Range or distribution	Mech. parameters* Deformation modulus D (Gpa)	Friction coefficient μ	Cohesion C (Mpa)	Poisson's ratio ν	Permeability	Influence on the engrg. works	Representative structure planes
1	Broken mudded zone	Acute	>3.0	0.1-1.0	Continuous	Loose	Large-scale tectonic & portions	0.01-0.10	0.19-0.25	0.02-0.03	0.48-0.40	Low	Serious	Intercalation No.201
2	Continuous mudded plane	Strong	0.5-3.0	0.1-1.0	Continuous	Broken or intact	Relatively Wide	0.01-0.05	0.19-0.25	0.02-0.03	0.48-0.45	Lowest	Relatively serious	Interfaces between shale and siltstone
3	Discontinuous mudded plane	Strong-moderate	0.1-2.0	0.1-1.0	Discontinuous	Relatively intact	Wide	0.05-0.10	0.25-0.35	0.03-0.04	0.45-0.40	Relatively high	Relatively slight	Joints and bedding fissures
4	Stiff plane	Slight	<1.0		Discontinuous	Intact	Widest		0.35-0.45	0.04-0.05		High	Slight	Siltstone bedding planes

* The mechanical parameters were determined according to the in situ test results and by way of analogy.

2.3 Types of structures of the rock masses

Due to the fact that the degree of development, the mechanical behavior and the distribution of structure planes are different, the rock masses under discussion should be considered as discontinuous, heterogeneous and anisotropic medium. In this light the structures of the rock masses can again be roughly grouped into four categories: the thin-layered structure, the thin-layered cataclastic structure, the cataclastic structure, and the loosened structure. The characteristic features for the above categories are listed in Table 2.2

3. MECHANICAL EFFECTS OF STRUCTURES OF THE ROCK MASSES

3.1 Correspondence of the types of deformation curves with the types of structures of the rock masses

The rock mass deformation under the hydraulic structure loading consists basically in the compaction and compression of structure planes. According to testing data the pressure-deformation curves can be deduced into four types, i.e. the near-straight line type, the concave-down type, the concave-up type and the near-"S" type (Fig 3.1). All these are occurring respectively in the thin-layered, thin-layered cataclastic, cataclastic and loosened structures. For the parameters concerned see Table 3.1.

It can be seen from Table 3.1 that the various deformation curves and their parameters are more or less corresponding to the respective types of structures of the rock masses. Hence, the deformation properties of various structures of the rock masses can also be characterized by their corresponding deformation curves. This means that the near-straight line type can characterize such a deformation feature that when a load perpendicular to the bedding planes is applied onto a thin-layered structure, which is fairly uniformly distributed both in lithology and structure planes,

Table 2.2 Types of structures of the rock masses and their features

Type		Structure plane		Structure body			Intactness		Mechanical parameters**					
No. Name	Geologic background	Engineering geologic categories	Main grades	Shape	Dimensions (cm³)	Coefficient I*	Degree	Hydrogeology	Sound wave speed V (10³ m/s)	Deformation modulus D (Gpa)	Poisson's ratio ν	Friction coefficient μ	Cohesion C (Mpa)	Evaluation
1 Thin-layered structure	Interbeddings of shale & siltstone and bedding shear planes are apparent.	Continuous mudded planes, stiff planes	Grades II and III	Thin-plate shaped, near-square massive.	<50×30×5	0.50-0.40	Relatively intact	Groundwater, percolating through Grade II structure planes, broken rock masses softened or mudded.	4.0-3.0	2.00-0.80	0.30-0.32	0.85-0.70	0.10-0.08	Good
2 Thin-layered cataclastic structure	Interbeddings of layers are apparent. There occur the broken mudded intercalated layers	Broken mudded zones, continuous mudded planes	Grades I, II, III and IV.	The same as above and rhombic, triangular massive.	<50×30×5	0.45-0.30	Relatively broken	Broken rock masses become water zone, but otherwise the same as No.1.	3.5-2.0	0.50-0.20	0.33-0.35	0.80-0.50	0.09-0.06	Bad
3 Cataclastic structure	interbeddings of layers are not so apparent. Tectonic changes are intense	Discontinuous mudded planes and stiff planes	Grades III and IV	Near-square rhombic, triangular massive	<15×10×2	0.30-0.25	Broken	The effect of groundwater is apparent, with the rock masses softened, mudded or sometimes piped.	2.5-1.5	0.25-0.15	0.35-0.38	0.50-0.35	0.06-0.04	Worse
4 Loose structure	Interbeddings of layers are not apparent. Tectonic changes and weathering are intense	The same as No. 3 but in an irregular order	Grades III and IV.	Lump with mud or mud with lump	<7×5×2	<0.20	Greatly broken	The same as above and easily expanded and disintigrated.	1.5-1.0	0.08-0.05	0.40-0.45	0.40-0.28	0.04-0.03	Worst

*, **: Both the coefficient of intactness and the mechanical parameters were determined according to the in situ test results and by way of analogy.

the structure planes would be compacted and compressed uniformly; that the concave-down type characterizes such a deformation feature that when a load is applied onto a thin-layered cataclastic structure, the structure planes would be deformed at a ratio gradually increased with the transferring of the pressure from the relatively intact rock into the relatively broken one; that the concave-up type characterizes such a deformation feature that when a load is applied onto a cataclastic structure, its structure planes would be deformed at a gradually decreased retio; and

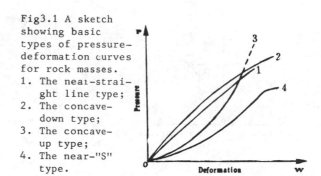

Fig3.1 A sketch showing basic types of pressure-deformation curves for rock masses.
1. The near-straight line type;
2. The concave-down type;
3. The concave-up type;
4. The near-"S" type.

Table 3.1 Corresponding relationship of classification of deformation curves and structures of the rock masses

No.	Types of deformation curves	Types of structures of rock masses	Values of moduli* (Gpa)		Ratio of moduli D/E
			Modulus of deformation D	Modulus of elasticity E	
1	Near-straight line type	Thin-layered structure	2.00-0.80	3.00-1.20	0.7
2	Concave-down type	Thin-layered cataclastic structure	0.50-0.20	1.20-0.50	0.4
3	Concave-up type	Cataclastic structure	0.25-0.15	0.80-0.50	0.3
4	Near-"S" type	Loose structure	0.08-0.05	0.40-0.25	0.2

* Value of moduli are the results of the in situ tests. The load applying direction is nearly perpendicular to the bedding plane and obliquely intersected with the faulted zone.

that the near-"S" type characteizes such a deformation feature that under the action of pressure oblique to the faulted planes, the rock masses of loose structure or the broken faulted zones, which are composed mainly of the loose structures, would have a greater deformation and better plasticity and be more liable to be yielded. The basic patterns of structure-deformation curves for the various rock masses are respectively shown in Fig3.2.

It should be noted that as the rock masses of cataclastic structure are fairly typical in the powerhouse and tunnel area, so the concave-up type curve should be more important from the point of engineering view.

3.2 Shearing effects

The results of the in-situ shear tests made on the rock masses of cataclastic and concurrently thin-layered structure showed that there exists an apparent correspondence between the yielding point for the shear stress-shear displacement curve and that for the shear stress-vertical displacement curve (Fig3.3). The failure surface, composed of mudded bedding planes and high-angle mudded joints, is irregular and shaped like a sawtooth, with a rise-and-down difference of up to 5-20 cm. The strength of the rock masses is nevertheless fairly low, though part of the rock in the specimen was sheared off during the test (Table3.2).

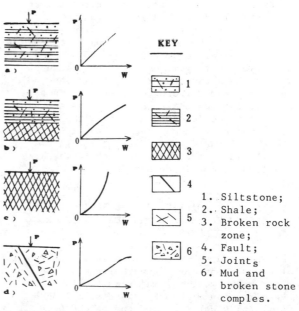

KEY

1. Siltstone;
2. Shale;
3. Broken rock zone;
4. Fault;
5. Joints
6. Mud and broken stone comples.

Fig3.2 A sketch showing the basic types of structure-deformation curve for the rock masses
a) Thin-layered structure-near straight line type;
b) Thin-layered cataclastic structure-concave down type;
c) Cataclastic structure-concave up type;
d) Loosened structure-near "S" type.

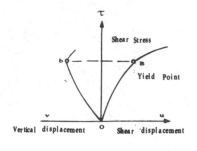
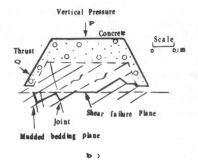

Fig3.3 A sketch showing in-situ
shear tests for the rock masses
a) Stress-displacement curves;
b) A sketch profile of the shear
failure pattern.

Table 3.2 In situ shear test results

Test object	Angle included between shear direction & bedding plane, degree	Failure value		Yield value	
		Friction coef. μ	Cohesion C (Mpa)	Friction coef. μ	Cohesion C (Mpa)
Rock body	25–30	0.73	0.13	0.54	0.07
Mudded bedding plane	0	0.35	0.07	0.26	0.03

Cause for this kind of deformation and failure lies mainly in that the mudded bedding planes and the mudded high-angle joints are respectively parallel and nearly perpendicular to the direction of the shear forces applied.

3.3 Anisotropy

The measured data showed that the moduli of deformation in the direction perpendicular or parallel to the bedding planes are 1.93 Gpa and 4.18 Gpa respectively, with a ratio of 46%, while the moduli of elasticity in the same directions are 1.49–2.80 Gpa and 16.0–7.5 Gpa respectively, with a ratio of 9–37%. The shear test results also showed quite a great difference between the sets tested: one for the rock masses and another for the mudded bedding planes. Either in yield value or in peak value the latter is about ½ of the former (Table 3.2), This utterly indicates that the rock masses are fairly apparent in anisotropy.

4. STABILITY ANALYSES OF THE ROCK SLOPE FOR THE POWERHOUSE AND THE SURROUNDING ROCK FOR THE TUNNELS

The powerhouse is to be located at the toe of the dam, on the ground surface of the right bank, with its major axis being in the direction of N75°E, the design length 94 m, the width 43 m and the height 62.4 m. The axial direction of the power tunnels is 344°.

4.1 Factors influencing the stability of the rock slope and the tunnel surrounding rock

These include mainly in situ stresses, struc-

tures of the rock masses, anisotropic characters, actions of water and explosion effects, etc.

1. In situ stresses: This is the decisive factor that makes the rock slope and the tunnel unstable. In the history of geology the dam area was subjected to the action of tectonic stresses for three times, with their acting direction of 10°–190°, 340°–160° and 290°–110° respectively, for the first two the stresses acting respectively at an angle of about 65° and 86° with the upstream and downstream walls of the powerhouse or the side walls of the tunnels, while for the third one at an angle of about 45° with the side walls of the powerhouse or the axial direction of the tunnels. Practice has proved that an overall support was necessary during the excavation of an exploratory tunnel even of 2 m in width, otherwise the roof falled in; when the span increased to 3 m, the whole tunnel collapsed, causing the work to cease. It is seen that the tunnel surrounding rock was very poor in stability even under its gravity only.

2. Structures of the rock masses: They are the governing factor in causing the slope and the surrounding rock unstable. The approaches of controlling the deformation and failure should be different according as what structures might be met with: the thin-layered structure is governed mainly by the interlayer mudded shear planes; the thin-layered cataclastic structure by the mudded broken intercalation and the broken faulted zones; the cataclastic structure by the dimensions, behavior and quantities of the structure planes and their combinations; and the loosened structure by the composition of materials, the mechanical properties and the state of stress. If they strike in comfority with the tunnel axis or the strike of the slope and dip toward the inside of the tunnel or the foundation pit, the structure planes will have a still greater

influence on the stability of the engineering works.

3. Lithologic characters: Softness and weakness, multilayering, alternating of the soft with the hard, interbedding combination, and clay minerals on bedding planes – these are not only the material basis for the high density of interlayer shear planes, the developed high-angle joints and the minor broken zones to develop, but the essential factor leading the slope and the surrounding rock to lose their stability.

4. Actions of water: Shipai shale, though a water-resisting layer, might be a water passage as well owing to the existence of structure planes. The water, passing through them and entering into the surrounding rock, might cause, first, the structure planes to expand due to its percolation pressure and alteration effect, thereby enhancing the shear stress on the side walls; second, it might reduce the shear resistance of the weakened planes under the lubricating action of the liquid, thus leading the rock masses to slip along the unfavourable structure planes; and thirdly, it might give rise to lixiviation and latent erosion of the fillers contained in the structure planes, resulting in reduced bearing capacity of the rock masses.

5. Explosion effects: It has been revealed by practice that during excavation of the exploratory tunnel the depth of Shipai shale influenced by explosion was about 1 m. However, when large-scale tunnels and slopes are excavated and consequently more powerful explosions are practiced, greater influence depths will be expected. The result will be that this might, especially in case of water filled weakened intercalations or relatively thick-layered mudded structure planes, not only increase the water pressure, but also render the intercalated mudded and softened materials liquified so as to lower the shear strength of the rock masses and endanger the stability of the slopes and the surrounding rock.

4.2 Patterns and mechanisms of the slope deformation and failure

The slopes of the powerhouse are high, the highest being 82 m. In order to study the representative structure planes with stereographic projection, first, we divided the whole slope into 8 parts: the upstream and downstream walls; the interior and exterior side walls; the upper and lower interior angles; and the upper and lower exterior angles. Next, we selected some most detrimental structure planes and listed them in Table 4.1. Using this table, we developed a stereographic projection of structure planes each for the various walls as shown in Fig 4.1. And then, on this very figure we located all the potential sliding bodies (with their sliding direction and angle included) for the various parts and listed them again in Table 4.1.

It is seen from Table 4.1 and Fig 4.1 that there might emerge a total of 7 sets of unstable structure planes on the 8 parts with different strike-dips, the 3 sets of NEE, NWW and NNW of them being a most common sight, and 3 to 4 sets, in general, on a part with the same occurrence.

For patterns and mechanisms of the slope deformation and failure there are four categories as follows:

1. Slide and slump: When cut up by the excavated slope, the mutually combined structure planes would produce various instable clinotetrahedrons and clinoquinhedrons. Their failure patterns and mechanisms were that when an instable clinotetrahedron, for instance, was produced in a sliding bed formed by any two combi-structure planes, it would slide preferentially toward the free surface, in the dipping direction of the intersecting line (Fig4.2, a).

2. Compression failure: At the toe of the slope, a place where the in situ stresses were highly concentrated, the rock masses might be fractured and crushed due to compression (Fig 4.2,b).

3. Dislocation: Both on the slope and at its toe a relative slip of the rock masses along some bedding or joint plane might occur, owing to the lithologic heterogeneity and the horizontal squeezing forces (Fig4.2, c).

4. Upheaval of the floor: Owing to the vertical load releasing and lateral horizontal squeezing force, the rock masses in the floor of the foundation pit might be heaved up (Fig 4.2, d).

4.3 Deduction of the stabilized slope angle and determination of the recommended value for it

The above slope stability anlayses were for the purpose of predicting a safe and rational stable slope angle for use in the design. In this paper the stable slope angle we assumed using the value of plunge angle with the intersecting line formed by every two combi-structure planes for various parts of the slope, which was directly obtained by means of stereographic projection, while the recommeded value for it we determined through integrated analyses, in accordance with the yield value of internal friction of the rock masses given by the in situ shear test results, the natural slope angle, the slope height as well as the various influencing factors related. All these values obtained are summarized in Table 4.2.

Analysing the data listed in Table 4.2 we come to the following views:

1. The stable slope angle of the rock masses is relatively low.

2. The patterns of slope failure may be characterized mainly by slidding, along the structure planes, toward preferentially the free surface as well as upheaving of the floor of the foundation pit, and partly by shear failure and possible compression failure at the toe of the high slope.

3. In case the limestone on the top of the

Table 4.1 Summary of the representative structure planes possible of causing the slope unstable and of the sliding bodies

Sidewall	Name of the set of strike	Set No.	Type of genesis	Strike, degree	Dip, degree	Dip angle, degree	Angle with sidewall, degree	Shape	Combi-structure plane	Slide directions, degree	Plunge angle with the combi-intersecting line, degree	Sub No. of Fig.4.1
Upstream wall	NE	1	faults	40	310	30	35	Clino-tetrahedron	4-3, 1-5.	290-340	32-37-46* / 32	g)
	NE	2	joints	50	320	40	25					
	NEE	3	joints	80	350	40	5					
	NWW	4	faults	95	5	60	20					
	NWW	5	faults	291	21	60	36					
	NWW	6	joints	290	20	45	35					
Downstream wall	NEE	1	bedd.struc.planes	75	165	25	0	Clino-tetrahedron	1-2, 1-3, 1-6, 1-5, 1-4, 3-6.	130-210	26-57 / 26	b)
	NEE	2	faults	60	150	60	15					
	NEE	3	joints	50	140	40	25					
	NWW	4	faults	291	21	60	36					
	NWW	5	joints	290	200	45	35					
	NNW	6	faults	331	241	50	76					
External side wall	NNE	1	faults	20	110	30	35	Clino-tetrahedron	3-5, 3-6, 3-4, 4-6, 8-6, 8-4, 1-4, 5-6, 1-5, 1-6, 2-6.	50-120	10-28-57 / 25	d)
	NNE	2	joints	20	110	40	35					
	NEE	3	bedd.struc.planes	80	170	25	85					
	NWW	4	faults	310	40	60	35					
	NWW	5	joints	290	20	45	55					
		6	faults	330	60	25	15					
	NNW	7	faults	350	80	50	5					
		8	joints	325	55	45	20					
Internal side wall	NNE	1	faults	10	280	30	25		4-2, 4-3, 1-6, 3-6, 2-6, 3-5, 1-5.	210-280	20-37-59 / 30	e)
	NNE	2	joints	0	270	40	15					
		3	joints	20	290	40	35					
	NEE	4	bedd.struc.planes	80	170	25	85					
	NWW	5	faults	291	21	60	54					
	NWW	6	joints	290	200	45	53					
	NNW	7	faults	350	260	50	5					
Upper internal angle	NE	1	faults	41	310	30		Clino-tetrahedron & Clino-quinhedron	1-5, 4-7, 1-4, 1-7, 5-6, 1-3, 3-7, 1-6, 4-6, 5-7, 3-6, 1-2, 2-6, 2-7, 3-6-1 7-4-1	250-330	13-30-46 / 25	h)
	NE	2	joints	50	320	40						
	EW	3	faults	275	5	60						
	NWW	4	faults	291	21	60						
	NWW	5	joints	290	20	45						
	NW	6	joints	315	225	45						
	NNW	7	faults	331	241	50						
Lower internal angle	SN	1	faults	10	280	30		Clino-tetrahedron	5-2, 5-3, 5-1, 5-4, 1-4, 3-4.	190-250	9-30-48 / 30	c)
	SN	2	faults	350	260	50						
		3	joints	0	270	40						
	NEE	4	faults	60	150	60						
		5	bedd.struc.planes	75	165	25						
	NWW	6	faults	300	210	60						
		7	joints	305	215	45						
Upper external angle	SN	1	faults	10	100	30		Clino-tetrahedron	4-1, 4-2, 4-3, 5-1, 5-2, 5-3.	350-60	14-35-55 / 35	f)
	SN	2	joints	0	90	40						
		3	faults	350	80	50						
	NEE	4	joints	50	320	40						
	NEE	5	joints	80	350	40						
	NWW	6	faults	300	30	60						
	NWW	7	joints	305	35	45						
Lower external angle	NEE	1	bedd.struc.planes	75	165	25		Clino-tetrahedron	1-7, 1-5, 1-6, 7-4, 5-4, 6-4, 3-6.	90-140	12-20-47 / 20	a)
	NEE	2	faults	60	150	60						
	NWW	3	faults	291	201	60						
	NWW	4	joints	290	200	45						
		5	joints	325	55	45						
	NNW	6	faults	330	60	25						
	NNW	7	faults	331	61	50						

* $\dfrac{32\text{-}37\text{-}46}{32}$ = $\dfrac{\text{Min.value-medium-max.value}}{\text{Important reference value}}$

42

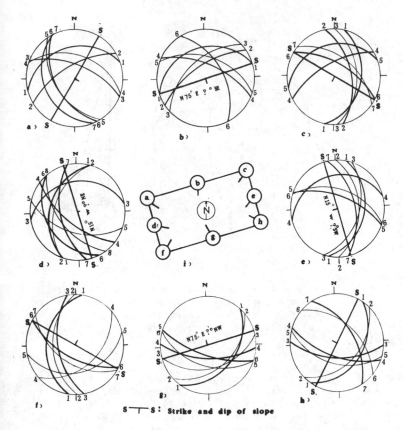

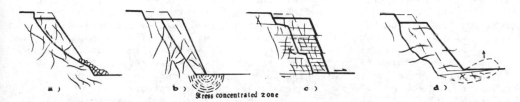

S ⊥ S : **Strike and dip of slope**

Fig.4.1 A stereographic projection of the possible structure planes causing the powerhouse foundation pit slope unstable
a) Lower external angle;
b) Downstream wall;
c) Lower internal angle;
d) External side wall;
e) Internal side wall;
f) Upper external angle;
g) Upstream wall;
h) Upper internal angle;
i) A sketch of the actual position of powerhouse foundation pit and slide direction of its sidewalls.

Fig.4.2 Patterns of the rock slope failure
a) Slide and slump; b) Compression failure; c) Dislocation; d) Upheaving of the floor.

Table 4.2 Stable slope values and their recommended for use values

Sidewall	Stable slope angle by the stereographic projection diagram, degree	Internal friction angle (yield value), degree	Natural slope angle, degree	Stable slope angle recommeded for use, degree
Upstream wall	32–37–46* 32			32
Downstream wall	26–57 26	28	28–35	26
Internal sidewall	20–37–59 30			30
External sidewall	10–28–57 25			25

$* \quad \dfrac{32-37-46}{32} = \dfrac{\text{Min.value-medium value-max.value}}{\text{Important reference value}}$

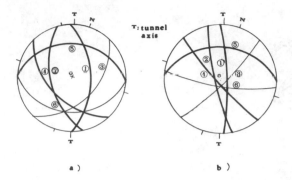

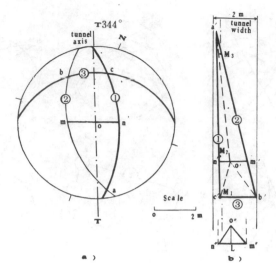

Fig 4.3 Stereographic projection of repersen-
tative structure planes and their relationship
with the tunnel axis for the tunnel surrounding
rock possibly causing unstable
a) Projection of the low dipping structure
   plane in the tunnel roof
b) Projection of the high dipping structure
   plane in the tunnel wall

Fig 4.4 Projection of the labile structure body
in the roof of a horizontal exploration adit
a) Stereographic projection
b) Natural scale projection

slope is of great thickness, the shale would
have a still poorer stability.

4.4 Pattern and mechanism of the tunnel sur-
     rounding rock deformation and failure

The rock formation above the floor of the power
tunnel has a maximun thickness of about 150 m.
A 160 m length of the tunnel passes through
Shipai shale. There are 6 sets of structure
planes, closely related with the stability
of the surrounding rock: (1) $N20^{\circ}W50^{\circ}-80^{\circ}SW$
for faults; (2) $N30^{\circ}W25^{\circ}-83^{\circ}NE$ for faults and
joints; (3) $N25^{\circ}E30^{\circ}-80^{\circ}NW$ or SE for joints;
(4) $N60^{\circ}W45^{\circ}-85^{\circ}NE$ for faults and joints; (5)
$N75^{\circ}E25^{\circ}-30^{\circ}SE$ for bedding structure planes;
(6) $N80^{\circ}E25^{\circ}-75^{\circ}NW$ or SE for joints. The effect
of these drawback upon the stability of the
tunnel are shown in Fig 4.3
   There are four common patterns and mechanisms
of the tunnel surrounding rock deformation and
failure. They are as follows:
   1. Layered collapse or upheaval: The bedding
plane of the tunnel roof would tear open under
its gravity, and the layered collapse would
follow when the rock masses were cut up by
other structure planes. In addtion, the rock
masses of the tunnel floor would upheave the
due to the load releasing induced rebound, the
lateral pressure and the action of ground water
as well.
   2. Upright plate-shaped collapse: The rock
masses in the tunnel roof and walls and their
associated parts would collapse in the shape
of an upright plate, as a result of cutting up
by two or three sets of structure planes. Of
the two or three sets one might be of well
developed high-angle structure planes, while
the other one or two dip toward the tunnel.
   3. Wedge-body shaped collapse: Should the two
sets of structure planes, with their strikes
nearly parallel to or obliquely intersecting
with the tunnel axis, dip oppositely, they

would produce instable rock blocks in the tunnel wall. In that case the low dipping structure planes would mostly fail in the roof and
floor portions of the tunnel, while the high
dipping ones in the tunnel walls. And should
the two sets of structure planes dip toward the
same direction, acute-angled collapsing bodies
would then be formed in the tunnel roof and
floor and their connecting portions.
   Of the above said failure patterns, the second
and the third ones might, for the most part, be
seen in the rock masses of thin-layered cataclastic structure, with the former chiefly emerging in any such horizons as those including
a relatively thick siltstone.
   4. Loose mass shaped collapse: Should the
rock masses of cataclastic and loose structures
be cut up or destroyed during excavation of the
openings, collapse as such would occur. This is
especially true of the surrounding rock composed
wholly of these two structures. In that case
the collapse of the tunnel roof might be all
the more serious, accompanied by inward expansion of its side walls and upheaving of its
floor, thus forming a near circular opening.
   Typical case study: On the portion 158.5-
165m from the inlet of the horizontal exploration adit the roof was cut up by three sets of
structure planes, Through points M_1, M_2 and M_3
producing a wedged body $0'a'b'c'$ with its apex
up (Fig 4.4)
   In this case we studied there existed three
types of structure plane:
1) faults: with a strike-dip of $N18^{\circ}W\ 60^{\circ}SW$;
2) joints: $N30^{\circ}\ W50^{\circ}NE$, with the two of them
   extending discontinuously;
3) bedding shear planes: $N75^{\circ}E25^{\circ}SE$.
   These, fairly even and smooth, were all filled
with argillaceous materials. The projection diagram showed that the wedged body, as a true
tetrahedron, with its apex embeded 0.8 m deep

O"L and its length being 8.4 m long along the
tunnel axis, has a value of Q equal to 46.71
KN, which was estimated using Q=R.V (where
Q=dead load; R=volumetic weight of the rock or
26.07 KN/m^3; and V=volume of the wedged body).
It is seen the body, obviously unstable, is
danger to construction, and needs to be treated.

5. CONCLUSIONS

From the above study we reach a preliminary
knowledge of the following:
   1. Composed of the interbeddings of shale and
siltstone, the rock masses are liable to crack
with desiccation and to disintegrate and soften
with water. The rock blocks contained are rather
small-sized. And the resistance to weathering
is fairly poor.
   2. The bedding structure planes, high-angle
joints and minor broken zones are fairly well
developed, and they, according to their engineering geologic features, can be grouped into
four categories: the broken mudded zone, the
continuous mudded plane, the discontinuous
mudded plane and the stiff plane.
   3. According to the features of development
of the structure planes and bodies, the structures of the rock masses can be divided into
four types: the thin-layered structure, the
thin-layered cataclastic structure, the cataclastic structure and the loose structure.
They are mostly rather poor in engineering
stability.
   4. The deformation properties of the rock
masses can be deducted and described in terms
of four pressure-deformation curves, i.e. the
near straight-line type, the concave-down type,
the concave-up type and the near-"S" type.
These together with their four corresponding
structures of the rock masses, just make four
patterns of structure-mechanics. Hence for
various structures of the rock masses, their
deformation properties can also be characterized
by their corresponding deformation curves.
   5. Being apparently anisotropic both in deformation properties and strength, the rock
masses are rather low in shear resistance.
   6. Being fairly poor in stability, the slope
of the powerhouse has a stable slope angle
ranged from 10° up to 59°. The recommended
value of it for use is of 25°-32°. Should the
stable slope angle be less than the recommended
value, treatment might be necessary for the
relavent structure planes. The failure patterns
and mechanisms of the slope rock masses lie
chiefly in that the sliding body would slump
preferentially, in the plunge direction of the
intersecting line formed by the combi-structure
planes, toward the free surface, and that the
floor of the foundation pit would upheave.
   7. The tunnel surrounding rock being poor in
stability, its failure patterns and mechanisms
lie mainly in slumping of the unstable bodies
produced along the bedding planes, the low-angle
joints and the differently dipping high-angle
joints.
   8. In view of the bedrock being poor in stability prompt strengthening of the rock masses

would be necessary during construction. The same
is especially true of the portions with cataclastic and loose structures, where excavation
should be closely followed by overall support
measures and special supports.

REFERENCE

Dong Xuecheng, 1981, "A discussion on some problems with the powerhouse surrounding rock
   stability analyses." Jouranl of Rock & Soil
   mechanics: Vol.3, No.2, p.51, Nanjing, China.
Gu Dezhen, 1979, "The basis of engineering
   geologic mechanics of the rock masses."
   Beijing, China. The Science Publishing House,
   p.229-244.

Li Di, 1980, "Anlayses of the rock masses'
   pressure-deformation curves." Journal of Rock
   & Soil Mechanics; Nanjing, China; Vol.2, No.2,
   p.49-51.
Research Institute of Geology, ASC, 1976, "An
   Approach of Analysing the Rock Masses's Stru-
   ctures with Stable Slope Angles." Beijing,
   China. The People's Railroad Publishing
   A collection of papers on landslide, p.328-329.
Sun Yuke, Gu Xun, 1983, "The Application of
   Stereographic Projection in the engineering
   Geologic Mechanics of the Rock Masses."
   Beijing, China. The Science publishing House.

Geostatistical investigations of rock masses: The Sierra del Medio case
Etude géostatistique de massifs rocheux: Exemple de la Sierra del Medio

José Augustin Matar, Maria Angélica Matar de Sarquís, Jorge Pablo Girardi & Gustavo Héctor Tabbia, *Institute of Mining Research, University of San Juan, Argentina*

ABSTRACT: The geoestatistical techniques applied for the selection of a minimum-fracturation volume in Sierra del Medio allow to quantify and qualify the variability of mechanic characteristics and density of fractures and also the level of reliability in estimations.

The role of geostatistics is discussed in this work so as to select minimum fracturation blocks as a very important site selection step.

The only variable used is the "jointing density" so as to detect the principal fracture systems affecting the rocky massif. It was used on the semivariograms corresponding to the previously mentioned regionalized variables.

The different results of fracturation are compared with the deep and shallow geological surveys to obtain two and three dimensional models. The range of the geostatistical techniques to detect local geological phenomena such as faults is discussed.

The variability model obtained from the borehole data computations is investigated taking as basis the vertical Columnar Model od Discontinuity (fractures) hypothesis derived from geological studies about spatial behaviour of the joint systems and from geoestatistical interpretation.

RÉSUMÉ: Les techniques géoestatistiques utilisées pour selectioner un volume de minime fracturation dans Sierra del Medio ont permis de caractériser cuantitaivement e cualitativement la variabilité des proprietés méchaniques, de la densité des fractures et aussi du nivel d'exactitude de l'estimation.

La géoestatistique employée pour déterminer des blocs de minime fissurité c'est une trés importante étape dans la sélection d'une place.

On a utilisé seulement une variable pour connaître les systèmes de fissurité principaux afectant la masse rocheuse. Cette variable dènommée "densité de diaclase" fut usée dans les semivariogrames concernants les variables regionalisées q'on a mentionné avant.

Aprés on a comparé les resultats de fissurite avec les études géologiques de surface et de sous-sol pour obtenir des modéles en deux et en trois dimensions et aussi on a analysé le trascendence des techniques géoestatistiques pour conaître des phénomenes géologiques locals par example failles.

Le modéle de variabilité obtenu pour le traîtement en ordinateur, des données des sondages fut investigué avec l'appui du modéle de colonne vertical de l'hypothese de discontinuité (fractures), celle-ci a été prise d'une part deś études géologiques sur la conduite spatiale des systémes de diaclases et d'autre part de l'interprétation géoestatistique.

INTRODUCTION

It is known that foundation road engineering is also supported by geology. All geological data used to inventory, analize and sinthesize natural phenomena are backgrounds on which an engineer can design, plan and build in a secure and harmonious way in accordance with the natural environment.

One of the most significant parameters provided by foundation engineering is rock structural features, especially geological discontinuities, such as: faults, fractures, bedding planes, cleavage, joints, dikes and so on. Discontinuities have a great influence on rock mechanical behaviour because they usually increase the rocky mass strain and decrease its resistivity. These two events are responsible

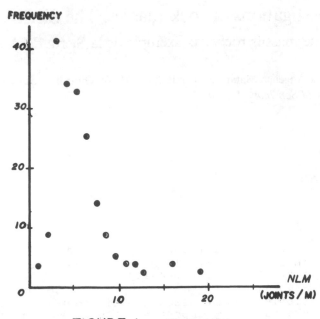

FIGURE 1 - NLM Histogram.-

for most of the faults occurred under static or dynamic charge conditions.
In order to obtain trend, dip, spacing, length, fill, wrinkle and density data, a
complete discontinuity geological mapping of the rocky massif based on a planned
study is required. Then it is possible to assess the number of joint families pre
sent, their distribution, significance, intact rock natural blocks, etc. in quantitative and qualitative terms.
Geostatistics is another approach use to accomplish these objetives.

METHODOLOGY USED

The general methodology employed in the determination of the least fractured area
of a rocky massif was as follows: Collection of data to define the final grid and
its sampling.
Twenty eight approximately isodistanced stations supplied values of "Number of
joints" (N) and their "Spacing" along 10 m lines normal to the strike of the two
main systems of joints that affect the rock mass. Then "samples"of the definitive
grid (188 stations) were obtained together with the geology and survey of the
area.
Rock quality was evaluated, from the mechanical point of view,through the N/L and
RQDS parameters, the latter being defined as the summation of spacing lengths
(more than 10 cm long) per linear meter. N/L and RQDS values were estimated taking these primary data as basis.

Determination of the definitive grid.
  Different forms and mesh densifications were proposed once the required accuracy
  in the knowledge of the variable had been established.This analysis was carried
out only with the N/L parameter.

Histography of N/L and RQSD.
  The distribution pattern to which the sampling of N/L and RQDS obey,was used in
  selecting the type of semivariography to be estimated from them.
  The chi-square criterion allowed to define the adjustment of the N/L histography to the corresponding distribution.
The correlation between the N/L and RQSD parameters is shown at this point (Figure 1).

Semi-variography of N/L and RQSD.
  To carry out the geostatistical representation of the variables employed, the
  model of variability from the corresponding experimental semivariographies were

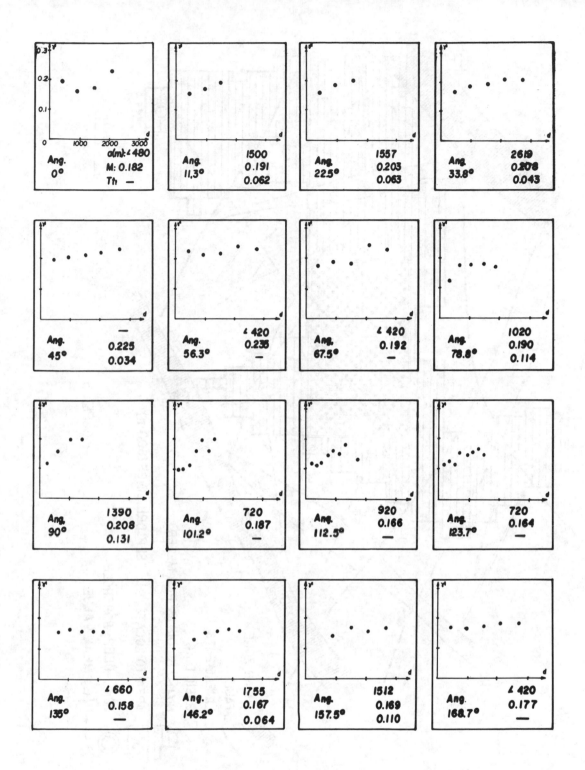

FIGURE 2 - NLM Experimental semivariograms.-

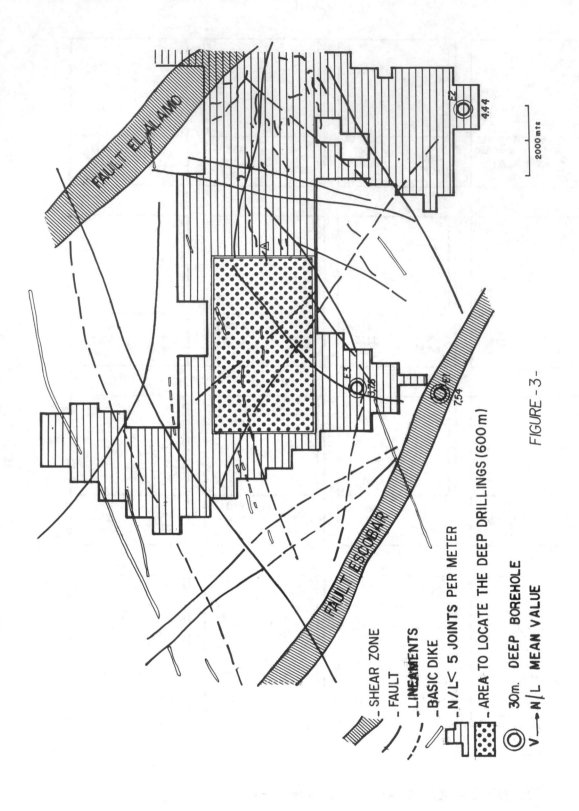

SHEAR ZONE
FAULT
LINEAMENTS
BASIC DIKE
N/L < 5 JOINTS PER METER
AREA TO LOCATE THE DEEP DRILLINGS (600 m)

30 m. DEEP BOREHOLE

V ⟶ N/L MEAN VALUE

FAULT EL ALAMO

FAULT ESCOBAR

2000 mts

FIGURE - 3 -

50

obtained. These were estimated considering N-S, NE-SW, E-W and NW-SE strikes (Figure 2).

To check the realiability of the N/L semivariography to be used in the krigeage operator, the Re-Use method was applied.

Determination of the minimum fracturation area.

It was based on the kriging estimation of the mean values of the variables corresponding to each area in which the sampled surface was assumed to be divided in.
At the same time, the geologic data were analysed in detail which allowed to establish the reliability of the geostatistical model of the investigated area (Figure 3).

DATA ON DEPTH

Once the minimum fracturation area had been selected and in order to begin with the study in depth, the location of 4 deep core drillings (450 to 750 m) was proposed. Such core drillings could be of the inclined or vertical type. Several 4 core drillings - configuration were proposed in the area taking into account the surface fracturation semivariography model. The mesh that became to be related with the least typical deviation of kriging was that one elongated east-west with a side ratio of 5/3.
It should be mentioned that at this stage of the planning, three core drillings between 200 and 300 m deep had already been performed;thus some fracturation characteristic data were available.
Four deep core drillings were made -PP1; PP2; PP3 and PP4- and from them the one called PP1 had a 60°south dip on purpose, to detect the influence of some geological structures.
All this work yielded data on the number of fractures per linear meter N/L from 3.05 m deep cores.
The N value comprises all fractures and it is independants of their characteristics. The L value is referred to the core run.
The N/L mean values, corresponding to the bearings resulting from intersecting the drillings with the horizontal bench model derived from the highest point in Sierra del Medio - i.e. Chivas Hill, some 1.350 m a.s.l. -, to be used in the three-dimensional model calculations, were computed. The benches were assumed to be 25 m high, this value resulting from studying the semivariography of N/L values for core lengths of 3.05 m.
Since the ranges determined for the vertical drillings are consistent with the bench height, inclusion of short range variabilites is avoided in this first stage of the work. Figure 4 shows, schematically, the bench model proposed.

CONCLUSIONS

This study has allowed to determine the location of a minimum fracturation area of about 600 x 1,000 meters in Sierra del Medio. Geostatistical techniques were applied to the parameters "Number of joints per linear meter (N/L)" and "Index of surface rock quality (RQSD)".
To accomplish the previously mentioned determination, N/L was employed because it is the most representative parameter of the surface structural characteristics of the rock mass.
The minimum fracturation area has maximum values of 5 joints per meter and 25% typical deviation (this error affecting the former in 1 discontinuity at the most).
The reliability of this area location was corroborated not only by the good correlation existing with data resulting from the Geology in detail, but also by the 200 m deep boreholes in the sampling zone.
It is noteworthy that the application of Geostatistics to this type of problems is associated with a time shorter than that required for the application of traditional exploration techniques.
A three dimensional model using the N/L surface mean values and the depth ones is being carried out at present.

REFERENCES

Perucca,J.C. y otros,1980: "Investigación sobre antecedentes de cuerpos intrusivos trusivos graníticos de la República Argentina". Inf. Int. IIG. UNSJ.

I.I.M., 1982-83: "Investigación detallada del intrusivo plutónico discordante del tipo batolítico denominado Sierra del Medio.

Girardi, J.P. et al, 1984: "Selection of a minimum fracturation density area by Geostatistics in crystalline rock". Geostatistics for Natural Resources Characterization, NATO ASI Series - D. Reidel Publishing Co..

Girardi, J.P, Sarquís, M.A. 1984: "Selección de un área de mínima fracturación en roca cristalina mediante Geoestadística. Funciones Aletaorias Intrínsecas de Orden k". Inf. Int. IIM. UNSJ.

Application of complex of geological and geophysical methods for evaluation of rock masses as foundations for waterworks

Emploi de l'ensemble de méthodes géologiques et géotechniques pour l'évaluation des massifs rocheux en qualité des fondations d'ouvrages hydrauliques

A.V.Moulina & L.D.Lavrova, *'Hydroproject' Institute, Moscow, USSR*

ABSTRACT: Studying the rock masses as foundations for waterworks is carried out by a complex of geological and geophysical methods, the main of which are geological survey, exploratory drilling and excavation, laboratory tests, hydrogeological and geophysical studies. Depending on the type of the proposed structure, engineering geology,design stage, etc. extent of field investigations, their objectives and approach to choosing the basic methods of investigation tend to vary as well as their combinations. Usually, at the early stages, geological survey, exploratory drilling and excavation are used in combination with laboratory tests, hydrogeological tests and geophysical areal studies (magnetic, seismic and electrical prospecting). Properties, state of the rock mass are identified by geophysical (mainly, by seismo-acoustic) methods in combination with geological, hydrogeological, geotechnical studies at various scale levels: from samples to field studies. Seismo-acoustic methods play the pacing role in distributing the comprehensive studies of the properties of the rock mass in separate points to the whole mass or elements. Described herein are two examples of studying and developing the schematic pattern of the structure on strength, deformation, permeability and other characteristics used for calculating the interaction between the foundation and the structure.

RESUME: L'étude des massifs rocheux en qualité des fondations d'ouvrages hydrauliques s'effectue à l'aide des méthodes géologiques et géophysiques, telles que: levé géologique, travaux de mine et de forage, essais de laboratoire, études hydrogéologiques et géophysiques. Selon le type de l'ouvrage projeté, les conditions géotechniques, le stade d'élaboration des projets, etc., changent les détails, les buts des études et l'approche au choix des méthodes principales d'études et leurs combinaisons. D'habitude, aux stades préliminaires du projet on utilise le levé géologique, les travaux de mine et de forage avec les essais de laboratoire, les études hydrogéologiques et les études géophysiques de surface (prospection magnétique, séismique, électrique). La définition des propriétés et de l'état du massif rocheux est faite par les méthodes géophysiques (essentiellement, par les méthodes séismoacoustiques) ensemble avec les études géologiques, hydrogéologiques et géotechniques aux différents niveaux - à partir des échantillons jusqu'à la reconnaissance in situ. Les méthodes séismoacoustiques jouent un rôle décisif dans l'extension, sur tout le massif ou sur ses éléments, de l'étude complexe des propriétés du massif dans les points isolés. Sont donnés deux exemples de l'étude et de la schématisation des fondations des ouvrages compte tenu de la résistance, déformation perméabilité et autres caractéristiques à utiliser dans les calculs de l'interaction de la fondation et de l'ouvrage.

Nowadays it is hard to imagine that studying the rock masses for hydraulic engineering purposes is done without application of a complex of geophysical methods, especially seismo-acoustic ones as mostly informative. But the practical experience proves that efficiency of application of the geophysical methods from the viewpoint of the required detailing and credibility of the information to be received as well as the cost and time to be taken by the investigations depends on their proper mix with other methods and, primarily, with geological, hydrogeological, geomechanical ones. Detailing, to certain extent, objectives of the investigations and approach to choosing the main methods of investigation and their combinations are obviously the functions of the type of the proposed structure, engineering geology and the stage of design development. In practice, application of the said complex of the methods is proved to be necessary in studying the structure and morphology of the rock mass, its properties and state. The main methods, out of the said ones, are geological survey and exploratory drilling and excavation conducted parallel with laboratory rock testing, hydrogeological tests and geophysical studies. The geological survey gives an idea about the rock mass structure while geophysical areal explorations (electrical, magnetic and seismic prospecting) in conjunction with exploration pits make it possible to detail and to reveal the typical and weakened zones of the mass. Solving these problems allows for proceeding to the zoning of the rock mass (Kayakin, V.V. 1984), i.e. to delineating of the zones which are quasi-uniform by state and properties within the mass boundaries as well as the "key" areas for performing detailed rock drilling, laboratory, hydrogeological and geophysical studies. Basing on the results of detailed studies, the rock mass is schematized by strength, deformation, permeability and other characteristics to be used for calculation of the interaction between the hydraulic structure and its foundation.

Studying the rock mass morphology, i.e. its boundaries and primarily the upper one drapped over by the mantle of loose deposits is one of the main problems. When solving the problem, the geophysical methods function consists in detail exploration of variability of the above boundary in plan. In this case the results of drilling are used for verification of the findings of geophysical studies and serve as the base for their unambiguous interpretation, while geological survey and aerospace methods allow for proper planning of the geophysical studies, focusing them on studying one or another aspects, e.g. buried valleys, etc.

Worthy to note especially the important role of integrated geological-and-geophysical explorations in complicated natural conditions (spread of perennial frozen rocks, landslide, karst-prone, etc.). Integration of geological and geophysical methods such as seismic, magnetic and electrical prospecting supplemented by logging (Savich, A.I., A.D. Mikhailov, V.I. Koptev & M.M. Iljin 1983) yields the best results.

Identification of the rock properties, its state and rock mass anisotropy by geophysical methods is carried out in combination with geological, hydrogeological and geotechnical studies at various scale levels - from samples to in-situ studies. In this case, the geophysical studies and primarily seismo-acoustic explorations play a crusial role in extending the results of integrated investigations of the rock mass properties obtained at individual points to the whole mass or its separate elements.

Important achievement of the seismo-acoustic studies is the possibility of constructing the detailed deformation models of the rock mass basing on the correlation of dynamic and static deformability indices. The data available show that all these correlations are stable and can be used for assessing the deformability of the rock masses often with the required precision without recoursing to the static tests (Savich, A.I. & L.G. Yascherko 1979).

Depending on the stage of design development and the required detailing in assessment of the rock mass properties the aforesaid problems are solved in combination with various geophysical methods. If at the initial stages of studies the emphasis is on in-line geophysical observations further they are supplemented by observations at the inner points of the rock mass, i.e. with studies to be conducted lengthwise the axis and around the perifery of the adits, by sounding between exploratory excavations and by well logging.

Given below are particular results of studying the state and properties of the rock masses served as the foundations of hydraulic structures which now are under construction on the Katun' River (Altaj mountain range), on the Irtysh River (Vostochnyj Kazakhstan).

Composition, state and properties of the rock masses and the patterns of their spatial changeability are analyzed on the basis of geological survey, exploratory drilling and excavations, geophysical studies (seismic shooting, seismic sounding, occasionally EM prospecting, well logging), mass laboratory identification of physical properties of the rocks (water absorption, density), as well as compression strength limit. These data allowed for zoning of the rock masses by the areas which are quazi-homogeneous by the state and by the properties of their elements.

The former mentioned rock mass, cut through by the Katun' River, is composed of Cambrian metamorphosed rocks (tufogenic sandstone, aleurolite, agglomeratic tuffite, spilite) and is located in the area of a large fault, which is manifested by a pronounced blocky structure of the mass with an amplitude of the relative displacement (more than one km) of the largest blocks (0.5-0.6 sq.km in area), the presence of multiple rupture dislocations and highly-intensive jointing. Alongside its characteristic feature is healing of the dislocation as a result of later-period silicification and epidolization, healing of the large fissures with quartz and calcite, closing of the joints of all systems and different magnitude outside the zones of weathering and relaxation in the absence of somewhat significant rejuvenation of discontinuity. Analysis of changeability of the rock mass state and properties by a combination of evidences shows that:

(1) Heterogeneity of the rock mass is conditioned by the processes of weathering and relaxation, the presence of tectonic ruptures dislocations. Due to a high degree of the rock metamorphism and joint closure, the strength, elastic and deformative properties of the rock mass are little affected by the differences in lithological composition. Relaxation of the rock mass caused by the valley entrechment is manifested mainly in the zones of tectonic rupture dislocations and large fissures (tens of meters long), which is recorded in the adit documents.

(2) Below the relaxation zone (20-30 m in the valley bottom and upto 40-50 m in the valley flanks) the rock mass is subjected to considerable compression, which is confirmed by high velocities of longitudinal seismic waves propagation (V_p = = 6.0-6.4 km/s), insignificant difference in the value of the parameter at various scale levels in the rock mass and lower values acquisited from the sample tests (V_p = 5.3-5.5 km/s).

The region of variations in property values, corresponding to the relaxation zone of the rock mass is clearly seen over the height of the mass beginning from the roof in the chart of variations in averaged values of properties indices (V_p km/s by the results of seismo-acoustic studies and ultrasonic logging, specific water absorption q l/min by the results of pumping-in tests, water absorption W by the results of laboratory tests on the samples from its roof downward (Fig. 1). Seen as well are the variations in thickness of the zone as applied to the delineated structural   elements of the rock mass and depending on the hypsometric setting.

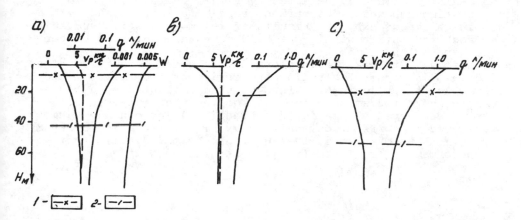

Fig. 1. Variation of rock mass properties downward from the roof (the Katun' River): a - for valley flanks; b - for valley bottom in structural-tectonic blocks, c - for valley bottom in tectonic rupture dislocation zones; 1 - lower boundary of weathering zone; 2 - ditto for relaxation zone

In the first case the governing factor is a significantly different intensity of the rock mass jointing within the boundaries of structural-tectonic blocks and tectonic rupture dislocations zone; in the second case - it is duration of the valley formation (slopes - bottom), partially its morphology (in this case steepness and height of the slopes). The latter is clearly seen in Fig. 4.from comparing the relaxation zone thickness on the right and left banks of the river.

Analyzing the meaningfulness of the various methods for engineering geology zoning of the mass it has to be said that the relaxation zone can be best identified by the results of profile shooting, sounding and permeability tests on the wells. Delineation of the zone by visual features (by bore-hole core, adit inspection) involves difficulties, because ultrasonic logging in the heavily compressed rock mass is less effective, but the lower boundary of the zone was traced with confidence following the regular variation of water-absorption ratio values. Decrease in rock mass strength within the zone is reflected also in the results of

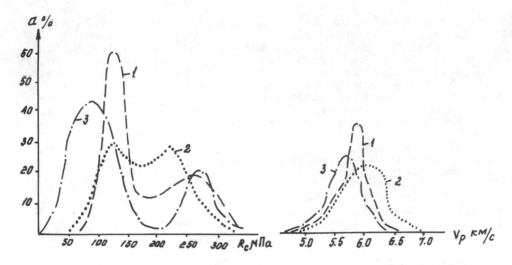

Fig. 2 Diagrams of distribution of ultimate compressive strength
and longitudinal wave velocities by ultrasonic logging (the Katun'
River): 1 - within the zone of relaxation; 2 - within the zone of
undisturbed rocks; 3 - within the zone of tectonic rupture dislocations.

uniaxial compression tests (Fig. 2). The sub-surface zone which is the most intensively destressed zone with slight manifestation of weathering and the tectonic
rupture dislocations can be confidently detected by visual evaluation of the rock
state, intensity of jointing and water absorption, bore-hole logging and with a
strata more than 3 m in thickness - by the results of seismo-acoustic studies as
well. Changes in the rock mass state within the said zones are detected by permeability test on the bore-holes.

Zoning of the rock mass and its detailed exploration by seismo-acoustic methods
made it possible to develop a model of elastic properties of the rocks, to identify the methods and areas of application of the geomechanical studies aimed at
determining the strength and deformation properties of the rock mass. Results of
rock deformability tests conducted by static (geomechanical) and dynamic (seismo-
acoustic) methods were used in correlation analysis and in plotting the diagram
of relationship between modulus of elasticity (Eg) found by the dynamic test results and modulus of deformation (D) taken from the results of static studies
(Fig. 3).

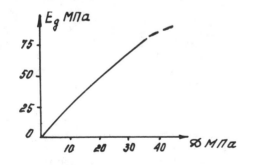

Fig. 3. Modulus of elasticity vs modulus of
deformation (the Katun' River).

On the basis of the established relationship and zoning of the rock mass, the results of the static studies were extended to the entire field of interaction between the hydraulic structure and its foundation. This made it possible to develop
eventually a model of deformative properties of the hydraulic project area (Fig.4).

Under somewhat different geological-and-morphological conditions of the Irtysh
river region in eastern Kazakhstan utilization of such a complex of geological-
-and-geophysical methods allows development of a structural scheme of the rock
mass (to single out structural-tectonic blocks and tectonic rupture dislocations
limiting these blocks, to determine their relative displacements, etc.) to characterize its state and properties.

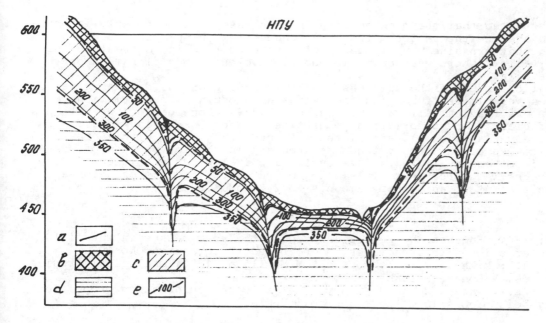

Fig. 4. Seepage and deformation pattern of the rock mass (the Katun' River): a - tectonic rupture dislocations; mass features with design specific water absorption, l/min; b - 0.5 (zone of weathering); c - 0.2 (zone of relaxation); d - 0.05 (zone of undisturbed rocks); e - isolines of modulus of deformation, 10² MPa.

The rock mass is composed of metamorphized pyritized sandstones, aleurolites, coaly-clay shales and covered on top with a strata (upto 15-20 m in thickness) of sandy-clayey and gravel-pebble deposits. The delineated structural-tectonic blocks reach in cross-sections first hundreds of meters. The rock mass properties are different in each block near the roof and depend on their relative displacement (Fig. 5): in sunken blocks (curves 3-4) the weathering zone is 30-40 m thick, in uplifted ones (curves 1-2) it comes to about 10 m.

The analysis of changes in the state and properties of the rock with depth considering the regional paleogeography allowed these changes to be related to two stages of weathering: older one occurred before neotectonic movements and the surface dissection by erosion in Miocene Quarternary period, and recent stage, associated with the formation of Miocene and modern valleys. In the first case, the pattern of changes in properties is not connected with the modern outlines of the rock roof surface (stepless increase in density below Els. 210-230 m, where slightly weathered rocks are developed, Fig. 5), in the second case, the above pattern is associated with the position of groundwater table, i.e. with the location of erosion base level during formation of the valley. In the diagrams this zone is traced along the inflection of δ and V_p curves and corresponds to the occurrence of the intensively weathered rocks.

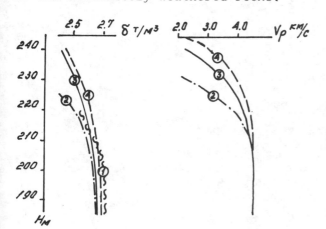

Fig. 5. Variation in rock density and longitudinal waves velocities in structural-tectonic blocks of rock mass depending on abs. elevations (the Irtysh River, eastern Kazakhstan).

It should be noted that the rock density in this case is a reasonably informative index to analyze the pattern of spatial changes in the properties of the rock mass. This is explained by abundance of weathered rock in contrast to the Katun' River valley, where weathering is poorly developed and relaxation of the rock mass prevails.

The deformation model of the area was developed basing on the diagram of correlation between velocities of longitudinal waves and modulus of deformation for metamorphic rocks (Savich, A.I. & L.G. Yaschenko 1979). Considered therewith was anisotropy of elastic and deformation properties of the rocks (sandstones, aleurolites and coaly-clay shales), associated with bedding joints. With increase in jointing the degree of anisotropy tends to grow and in the top portion of weathering zone it makes up $X_v = V_n/1 = 1.4$ by longitudinal wave velocities, in the undisturbed rocks X_v decreases to $1.1 \div 1.15$. According the modulus of deformation, the coefficient of anisotropy amounts to 2.2 and 1.3.

REFERENCES

Kayakin, V.V. 1984. Engineering-geological zoning the rock masses in the field of their interaction with structures. Hydroproject Proceedings, No 103.
Savich, A.I., A.D. Mikhailov, V.I. Koptev & M.M. Iljin 1983. Geological studies of rock masses. V Congress ISRM, Melbourn.
Savich, A.I. & L.G. Yaschenko 1979. Study of elastic and deformation properties of rocks by seismoacoustic methods. Moscow: Nedra.

Surfaces wave velocity measurement as a laboratory nondestructive control of the rocks properties

La vitesse d'une onde de surface comme essai de laboratoire de contrôle non destructif des propriétés des roches

Joanna Pinińska, *Warsaw University, Poland*

ABSTRACT: The paper deals with the investigation for the quick,nonestructiv testing of the geotechnical rocks parameters by using the new method of the surfaces waves generation. Since when the material parameters control by ultrasound became so popular and valuable in laboratory, one erised the obligation to enlarg its efficiency. Very useful is to apply the ultrasounds for defects detection as well as for structures, porosity, density and strength establishing. Very often are also utilized the relations between the ultrasonic waves velocity and the elasticity constants of the rocks. Such researches are basing on the measurements of the longitudinal and transwersal waves and are well grounded. But is also well known that in many cases the rocks are not very tractability medium for the transwersal waves propagation. Hence very promising solution may be given by the analise of the surface wave velocity. Paper presents the results of the experiments provided by those way on different rocks and the correlations with the other ultrasonic standards methods.

RESUME: Cet article traite des recherches d une prompte methode indestructible pour tester les proprietes geotechniques des roches. Dans ce but on a employe une nouvelle methode de generer les ondes de rurfaces. Depuis que le controle des proprietes de materiau par les ultrasons est devenu usuel dans les laboratoires a augmente l obligation pour amplifier son efficassite. Jusqu a present il est repandu l application des ultrasons pour la detection des defectuosites et encore pour etablir les structures, les porosites, la densite et la resistance. Souvent on use aussi la relation entre la vitesse d une onde ultrasonique et d elasticites constances des roches. Ces recherches basent sur les mesures de la vitesse d une onde longitudinale et transversale, ce qui est bien fonde. Mais, il est aussi bien repandu que les roches dans diverses cas ne sont pas un bon medium pour generer les ondes transversales. C est pourquoi comme une promise solution peut etre une analyse de la vitesse d une onde de surface. Cet article presente les resultats des experiences menees selon cette methode pour diverses roches et leurs corelations avec les autres ultrasoniques standards methodes.

DISCUSSION AND NEW OBSERVATIONS

The present geological-engineering works tend to a unified and quick identification of the rocks properties. The latter should be based or on the simple instruments for direct measurements of the needed parameter, or through a correlative connection with the indirect parameter to be used for finding the other necessary for the project calculations. For example a rock strength or a predisposition of mining galleries to falls can be evaluated by the Schmidt s hammer /1, 5/.

Methods of acoustic tests are also very useful and quite simple ones in this field /7/. They allow for a direct determination of the elasticity constants of the a material and basing on correlations, also many other physico-mechanical features. On the other hand, a non-destructive type of measurements enables an infinite recurrence of results.

Finding the elastic properties of a material by acoustic methods i.e. ultrasonic, sonic and inphrasonic ones, is particularly useful for a possible record

of very small deformations and measurements can be carried through with pieces
of various sizes - so no samples of standard sizes are necessary. All dependencies needed to determine these properties are based in a simple way on known
rules of their relation upon elastic wave velocity propagation, if a medium is
sufficiently large and isotropic.

On the other hand, much trouble arises in measuring a velocity in anisotropic
media. In this case the direction of anisotropy and the functions of the material constants changes should be known.

In the laboratory, due to the small sample sizes the ultrasonic waves should
be applied, and on the basis of detailed direct observations these features of
the material should be examined that influence in a measurable way the waves
propagation in the medium, the absorption of their velocity or the reflections
from inner defects. But such tests call for a keen and competent interpretation.
As is well known the different features are reflected in various way in the measurements and also, in different bodies a propagation velocity can be the same
in spite of varying structural properties.

Thus, the ultrasonic tests should be carried through from different points of
view and e.g. absorption or dissipation of waves or density of the media can be
amidst as the additional identification features.

A choice of measurements of a suitable parameter is therefore important, for
it should be the most sensitive in the given structural case. The difficulties
for the rocks properties arise to a number of factors that influence the propagation parameters. Therefore the measurements for a definite purpose should be
carried through with a use of the material of a known and well examined structure, or for defined earlier correlative indices, and same ambiguity of conclusions can also be eliminated by measurements of as much as possible ultrasonic
parameters in a function of e.g. frequency, temperature or time.

In a reference to a correct evaluation of the modulus of rock elasticity, in
each case the velocities of longitudinal and transversal waves should be defined
for a univocal determination of the Poisson's coefficient. In fact it makes numerous troubles due to intensive absorption of a transversal wave in rocks and
its varying evaluation dependent on a kind of medium anisotropy /4/. So particularly here, an introduction of every auxiliary parameter of identification increases considerably a reliability of interpretation.

For this reason, measurements of the surface wave were done in this test. They
formed together with measurements of velocities of longitudinal and transversal
waves the wide basis for a geotechnical observation of the rocks lithology,
strength characteristic and density etc.

Ultrasonic measurements of the surface wave velocity have not been widely used
in the evaluation of rock properties due to considerable troubles in their generation. Previous methods, based on a phenomenon of refracted longitudinal waves
with a use wedges, suitably selected to the material, have not given good results for the rocks. An application of the ultrasonic refractometer also resulted in limited possibilities.

Therefore, the tests were carried through with a use of the new Polish method
of the surface wave generation, presented in the papers 2 and 3. The wave is
generated by metal wedge-shaped elements with sharp edges. At the plane of a
such wedge the surface wave is generated by a transformation from the longitudinal wave due to refraction at the contact of two media. A time of the pass is
measured for a constant distance of about 11 cm with a use of the transmitting
/T/ or receiving /R/ head /Fig. 1/.

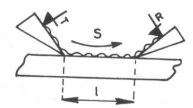

Figure 1. Schema of the surface wave
measuring /after 2/.

A record of the passing time of a surface wave was done with a use of the
ultrasonic defectoscope DI-23P, INCO made, at a room temperature and with a use
of heads of frequency equal 2.5 MHz. In similar conditions the passing times of
longitudinal and transversal waves were also recorded with a use of suitable
heads.

The usability of the applied method was the main item of considerations over
the evaluation of elasticity constants of tested rocks and especially, of the

dynamic elasticity mudulus E_d as the most important in geotechnic examinations.
In the common engineering practice, the evaluation of this parameter by the acoustic testing is done by finding the velocity of a longitudinal wave C_L, determination of the volume density / / and using the simplified calculation system for a wave propagation along the bar. Seldom a tabular value of the Poisson's coefficient is applied and a calculation formula for a non-limited medium is used. Both these simplifications result from difficulties in measurements of a transversal wave in rocks, particularly during laboratory tests. The results received in this way are much different from the real ones and do not supply with data on individual veriability of the rocks samples /6/. So a certain solution is created, by an introduction of the parameter of a surface wave velocity /C_S/. A direct record of a surface wave itself, is an important identification feature of the examined material as being the effect of a continuous measurement of the surfaces wave. In this way an approximate value of a transversal wave velocity /C_T/ can be also obtained easily for any fragment of the tested body /Fig. 2/.

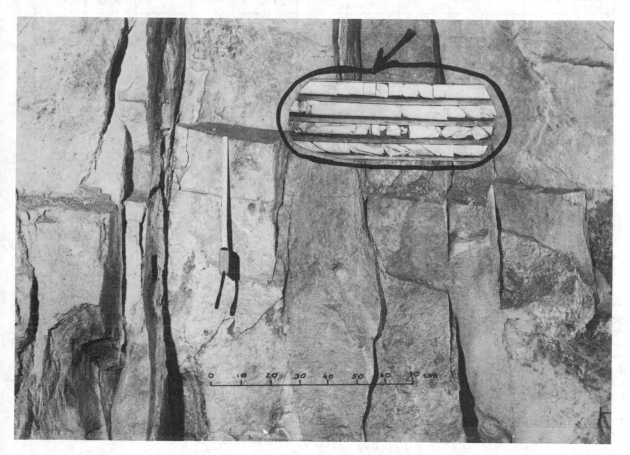

Figure 2. Variability of the rocks massif and rocks bores.

Results of determinations of the value C_S for rocks of varying lithology are presented in the Table 1, as well as the measured velocity values of C_L and C_T.
The results enabled to find the values of the Poisson's coefficient /V/ and to calculate by various methods the dynamic modulus of elasticity /E_d/. At first the value of the modulus E_{dI} was determined for the non-limited medium, with a use of direct measurements of the longitudinal and transversal waves velocities finding the value V_I for each sample, and known formulae applied:

$$V = \frac{0.5 - \left(\frac{C_T}{C_L}\right)^2}{1 - \left(\frac{C_T}{C_L}\right)^2} \quad [1] \quad \text{and} \quad E_d = C_L^2 \cdot \rho \, \frac{(1 + V)(1 - 2V)}{(1 - V)} \quad [2]$$

Table 1. Acoustic waves velocities for various types of rocks.

Lithology	fabric marks	volume density G/cm³	waves velocity [m/s] longitudinal C_L	transversal C_T	surfaces C_S	C_S/C_T index „K"	average	Transversal waves velocity calculated due to rule [4]	due to average index K
rhyolites	pelitic homogenous	2,63	5000 / 4920	3012 / 2840	2875 / 2735	0,95 / 0,96	0,96	3194 / 3042	2984 / 2852
marles	fein crysta-llinic	2,30	2800	1961	1854	0,94		2060	1961
metamorphic slates	fein crystallinic laminated	2,79	3642 / 4322 / 4545 / 4347	2475 / 2480 / 2777 / 2500	2211 / 2129 / 2395 / 2395	0,89 / 0,85 / 0,86 / 0,95	0,88	2456 / 2365 / 2661 / 2661	2475 / 2419 / 2721 / 2721
serpentinites	fein crystallinic	2,75	5250 / 5130	3281 / 3156	2875 / 2718	0,88 / 0,86	0,87	3194 / 3042	3304 / 3124
limestones (1)	fein grains		4482 / 4811	2888 / 3023	2300 / 2395	0,80 / 0,80	0,80	2555 / 2661	2875 / 2993
limestones (2)	organic debris	2,66	5019 / 5389	2849 / 3012	2053 / 2254	0,72 / 0,74	0,79	2281 / 2504	2812 / 3087
diabases	mineralised composed grains	2,78	4897 / 4782 / 5238	2750 / 2750 / 2894	2500 / 2395 / 2300	0,66 / 0,87 / 0,79	0,77	2777 / 2661 / 2555	3246 / 3110 / 2987
thysches sandstones	convoluted	2,67	5263 / 5376	3030 / 3067	2300 / 2300	0,76 / 0,76	0,76	2555 / 2555	3030 / 3026
granites	coarse crystallinic	2,62	5625 / 4909 / 5185 / 5000 / 4912	3450 / 3000 / 3333 / 3320 / 3210	2500 / 2446 / 2395 / 2500 / 2613	0,72 / 0,81 / 0,71 / 0,75 / 0,81	0,75	2777 / 2717 / 2661 / 2777 / 2903	3333 / 3261 / 3193 / 3333 / 3484

These determinations formed a background for the analysis of the modulus values, received from the formula for the bar $E_{dII} = C_L^2 \cdot \varrho$ [3], with a use of the tabular value of the Poisson's coefficient v_{III} and by an evaluation of the Poisson's coefficient in an indirect way v_{IV} and v_V, on the basis of calculations of the value C_T from the value C_S and following them the E_{dII}, E_{dIII}, E_{dIV}, E_{dV} values due to formulae [1] and [2]. In this last case two systems for a determination of the value C_T on the basis of C_S were applied, basing on the calculation according to the formula [4] and finding the individual calculation index for a tested rock. The velocity of a surface wave C_S is known to be always slightly lower than the velocity of a transversal wave C_T /about 5-10%/ and was accepted to be described by the following formula:

$$C_S \approx \frac{0.87 + 1.12 \, v}{1 + v} \cdot C_T \approx 0.9 \, C_T \qquad [4]$$

The carried tests proved that a relation of velocities C_T and C_S differs slightly, different for various rock media. This relation was preliminary found to be about $C_S \approx 0,75 \; C_T - 0.95 \, C_T$. For the rocks of distinctly heterogeneous structure is lower as noted for a coarse-crystalline granite /0.75/, sandstones of a convolute structure /0.76/ and limestones with organic debris /0.79/ and arise to about 0.95 for pelitic, homogeneous marls or compact fine-crystalline rhyolites. For this reason the values of the index $k = \frac{C_S}{C_T}$ was calculated for particular lithologic groups and then used to find the values of other parameters.

The values of elasticity parameters of the tested rocks, received on described various ways, are listed in Table 2. A confrontation of the results suggests that if an earlier recognition of the medium is possible, then a use of the index k results in particularly good effects. It does not replace in fact, a direct measurement of the velocity of a transversal wave to find the value v but

Table 2. Elasticity modulus $/E_d/$ and Poisson coefficient $/\nu/$ values due to the different methods of calculation.

| Lithology | For the bar due to formula (3) | E_d FOR NON-LIMITED MEDIUM DUE TO FORMULA (2) | | | | | | | |
| | | used formula | | used tables | | used $C_S/C_T = 0{,}9$ | | used individual C_S/C_T index $_k$ | |
	E_{dII}	ν_I	E_{dI}	ν_{III}	E_{dIII}	ν_{IV}	E_{dIV}	ν_V	E_{dV}
rhyolites	6,57	0,21	5,78	0,35	4,07	0,18	6,04	0,21	5,78
	6,36	0,26	5,13		3,54	0,20	5,72	0,26	5,19
marles	1,80	0,02	1,78	0,22	1,54	-	-	-	-
metamorphyc slates	3,70	0,10	3,58	0,15	3,47	0,10	3,58	-	-
	6,69	0,26	5,41		5,35	0,29	4,14	0,28	5,21
	5,76	0,20	5,18	0,27	4,60	0,25	4,83	0,21	4,63
	5,27	0,26	4,26		4,21	0,20	4,74	0,19	4,79
serpentinites	7,57	0,19	6,88	0,29	4,99	0,21	6,66	0,19	6,88
	7,23	0,20	6,50		5,42	0,24	6,07	0,21	6,36
limestones (1)	5,34	0,16	4,96	0,31	3,84	0,26	4,38	0,16	4,96
	6,15	0,19	5,59		4,42	0,28	4,79	0,19	5,56
limestones (2)	6,07	0,27	4,85	0,30	4,49	0,37	3,39	0,27	4,85
	7,72	0,28	5,86		5,71	0,36	4,40	0,26	6,25
diabases	6,66	0,27	5,32	0,24	5,59	0,27	5,32	0,12	6,39
	6,35	0,26	5,14		5,33	0,28	4,55	0,14	6,03
	7,62	0,28	5,48		6,40	0,35	4,72	0,26	6,24
fhysches sandstones	7,39	0,26	5,98	0,21	6,50	0,35	4,58	0,26	5,98
	7,71	0,26	6,24		6,78	0,35	3,86	0,27	6,16
granites	8,29	0,20	7,45	0,18	4,22	0,34	4,55	0,24	6,95
	6,31	0,20	5,67		3,21	0,28	4,92	0,12	6,05
	7,04	0,16	6,54		3,59	0,32	4,83	0,20	6,33
	6,55	0,12	6,28		3,34	0,28	4,89	0,12	6,28
	5,39	0,13	5,17		2,74	0,24	4,52	0,02	5,33

results in a higher accuracy. During a continuous testing of a core this method makes it possible to define the varying properties in every point of it and for example the unhomogenuity of the medium in the mineralisation bends, inserts and impregnation /Fig. 3/. It can be done even in the field and to facilitate the decision on further, more expensive examinations. Therefore if a medium has been sufficiently tested and laboratory researches defined suitable correlations then measurements of C_S velocity are simple eliminate the problems that accompany the measurements of a transversal wave. They eliminate also the error of an overstated value of the elasticity modulus, resulting from simplifications applied for a bar. This error can reach even several dozen per cent if a sample size is close to a length of the generated wave $/\lambda/$. In the case of the sample size smaller than three wave lengths, the results are overstated of about 30%/6/.

The carried experiments were done with a use of samples of a non-polished, rugged rock surface. Such surface is common for core walls and applied for standard resistance tests. But for a comparison the measurements of the velocity of wave propagation were done by the mentioned method at specially polished surface. Similar results were obtained. Thus, a varying characteristic of surfaces accepted in geology does not influence the resultant values. It can be found therefore that on the basis of direct measurements of a surface wave, all the other material constants as K, G, λ, can be defined with a sufficient accuracy.

But it should be underlined that within the ultrasonic investigations, a progres in the research technics gets immensely ahead the competence in interpretation of the results. Therefore, every result should be analyzed in detail against geologic factors. There is still much to do in the field of acoustic tests to give to carried measurements the identificatory and proper physical sense and properly to snow the relations of individual dependences.

Figure 3. The examinations of the bore
inserts, impregnations and fractures.
1,2... measurements points.

CONCLUSIONS

1. The carried examinations of the velocity of the surface wave propagation
   point out the possibilities to evaluate by a simple method the elasticity
   parameters with a high accuracy.
2. A velocity of surface wave is also another identification parameter for lithologic subdivisions of rocks. It can be also a helpful measurement to find the
   proper time needed for passing of surface waves in the anisotropic body where,

64

depending on its characteristic, several transversal waves can be generated. In the case of a strong absorption of a transversal wave in porous media, a measurement of a surface wave can be the only technical basis to define this value.
3. At a detailed record of changes of a rock along a core this methodology allows for a check-up without a special treatment of the material.

REFERENCES

/1/ Kidybinski, A. 1981. Bursting lability indicee of coal. Int. J. Rocks Mech. Min. Sc. and Geomech. vol. 18 no 1.
/2/ Pęski, Z. 1984. Nowa metoda generowania fali powierzchniowej - 13 Krajowa Konferencja Badań Nieniszczących.
/3/ Pęski, Z., Ranachowski, J. 1984. Projekt wynalazku P-245876 zgłoszony do UP. "Sposób i urządzenie do wytwarzania akustycznych fal powierzchniowych".
/4/ Pinińska, J. 1978. Anisotropy of acoustico properties of sandstones in the nothern flysch Carpathians. Proc. of III Int. Cong. I.A.E. G. Madrid.
/5/ Pinińska, J. 1982. Comparison of some dynamic nondestructiv test results for different geological conditions. Proc. of the second European Symposium on Penetration Testing /ESOPT II/ Amsterdam.
/6/ Pinińska, J., Karska, Z. 1982. Uproszczona kontrola parametrów sprężystości skał przy zastosowaniu zasady propagacji fali ultradźwiękowej wzdłuż pręta. National Symp. I.A.E. G. Badania Geol. Inż. w Górnictwie, Kraków.
/7/ Savit, C.H. 1980. Geophysics will find the elusive trap - Oil and Gas Journal - November.

Seismic reflexion for geotechnical exploration

Utilisation de la réflexion sismique en prospection géotechnique

L.F.Rodrigues, *Laboratório Nacional de Engenharia Civil (LNEC), Lisbon, Portugal*
J.D.Fonseca, *Physics Department, IST, Lisbon and Laboratório Nacional de Engenharia Civil (LNEC), Lisbon, Portugal*

ABSTRACT: The paper presents some results obtained in a research program carried out for the evaluation of the seismic reflexion method to map sub-surface structure in geotechnical investigations. Field techniques and data processing metodologies used to improve shallow seismic reflection interpretation are described and discussed. The results obtained at a particular site applying both reflection and seismic refraction profiling are also presented and compared.

RÉSUMÉ: La communication présente les résultats d'une investigation sur l'aplication de la sismique réflexion à faible profondeur, surtout à l'égard de l'exécution des éssais e du traitement des données. Les résultats obtenus avec des essais par réflexion e par réfraction sont aussi presentés et comparés.

1 INTRODUCTION

The seismic refraction method is probably the best known and widely used geophysical method in geotechnical investigations for determination of sub-surface structure.

On the other hand the seismic reflection method has a very limited application in this field of geophysical exploration but it is almost the sole seismic method applied at the great depths that are usual in the oil prospection.

The reasons for this fact can be found in the laborious processing procedures inherent to the last method, and also in the noise problems arising with its application at shallow depths. In recent years, however, with the proliferation of micro-computers, processing of digital data became cheaper and faster and new equipment was designed to account for the stringent problem of noise contamination of the data. Seismic reflection is now being regarded as an alternative in shallow depth investigations and in the early eighties the first reports appeared illustrating its applicability (Hunter et al 1984 and Singh 1984).

The seismic refraction method is currently used at the Site Exploration Division of LNEC for geotechnical investigations, mainly to obtain information about thickness of the overburden layers and total depth down to the bedrock and also to assess rock quality. In the last two years, however, some research was carried out for the implementation of shallow reflection seismics in geotechnical investigations and we shall now synthetize the acquired experience in field procedures and data processing.

2 BASIC CONCEPTS

The basic idea about data collection for reflection interpretation is that noise can be fought with data redundacy. The most popular method of data acquisition based in this idea is known by common reflection point or common depth point technique (CDP). It is illustrated in Figure 1.

The field layout provides that each point of the reflecting surface is sampled by a fixed number of seismic rays; this number we shall call multiplicity. Of course each seismic trace will present the reflected signal with a different arrival time, due to the difference in source-receiver offset. Introducing the number X for this offset and with the assumption of a constant velocity layer with horizontal interfaces, it is easy to see that the arrival time must obey the hyperbolic relation:

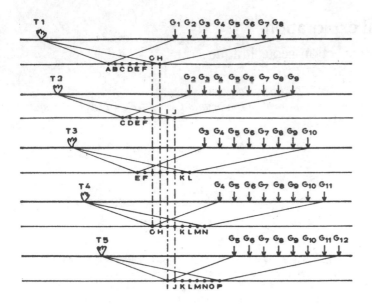

Figure 1. The CDP method

$$t_r = \left(\frac{4h^2}{V^2} + \frac{x^2}{V^2} \right)^{1/2} \qquad (1.1)$$

or

$$t_r = \left(t_o^2 + \frac{x^2}{V^2} \right)^{1/2} \qquad (1.2)$$

$t_o = \frac{2h}{V}$ being the normal incidence travel time (the time that the reflected wave would take to travel down and up along a vertical path). For the ideal model we are discussing, two seismic traces with different X would be enough to solve (1.2) for the unknowns t_o and V. Shifting the entire dispositive horizontally along a profile, another point of the reflecting interface is sampled, and this for as many different points as desired.

The difficulties that arise when reflected pulses are to be identified in a noise-contaminated seismic trace restrain the direct approach above referred (obtaining t_o and V from two traces alone) and a more efficient technique called velocity analysis (Taner & Koehler 1969) may be used to extract information from the seismic signals, using a higher multiplicity. All the other events present in the traces, like direct, refracted or surface waves must be considered noise for this purpose.

Two obvious complications of the simple model in Figure 1 are represented in figures 2 a)   and 2 b).

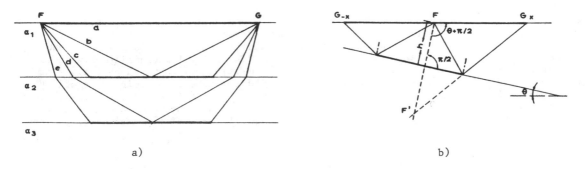

        a)   b)

Figure 2. Multi-layer model (a) and dipping interface (b)

For the multi-layer case equation (1.2) may be generalized (approximately) as

$$t_n = \left(t_{on}^2 + \frac{x^2}{\bar{v}_n^2}\right)^{1/2} \qquad (1.3)$$

where the subscript n refers to the n-th layer (n = 1,2...), and $\bar{V}_n$, the stacking velocity, may be approximated by the r.m.s velocity defined by

$$V_{rms} = \left[\frac{\sum_{i=1}^{n} V_i^2 t_i}{\sum_{i=1}^{n} t_i}\right]^{1/2} \qquad (1.4)$$

t_i being the vertical travel time inside the i-th layer (Sheriff & Geldart 1983). After the determination of $\bar{V}_n$ and t_{on} for all values of n, the seismic velocities of the layer may be computed by

$$V_n = \left[\frac{\bar{V}_n^2 t_{on} - \bar{V}_{n-1}^2 t_{o,n-1}}{t_{on} - t_{o,n-1}}\right]^{1/2} \qquad (1.5)$$

and the iterative computation of depths is straightforward. For a dipping interface, with dip θ , the time-distance becomes

$$t = \left[\frac{v^2 + 4h^2 + 4Xh\sin\theta}{v^2}\right]^{1/2} \qquad (1.6)$$

and the additional unknown θ may be found recording the same shot in simetric position along the profile, and using

$$\theta \simeq V(t_x - t_{-x})/2X \qquad (1.7)$$

3 SPECIFIC PROBLEMS AT SHALLOW DEPTHS

The field and laboratory techniques of seismic prospection were born and develloped in the oil industry where the depths of interest are one or two orders of magnitude greater than those in geotechnical problems. When brought to these shallower depths, some difficulties have to be worked out, mainly concerning noise conditions, weathering effects and source characteristics.

We have already said that other events than reflected waves were to be considered noise, and this applies specially to surface waves, which are high amplitude and low frequency signals pro-pagating from the source with weak attenuation and low velocity. For some range of distances from source to receiver, we may expect waves to be completely obscured by superimposed surface waves of greater amplitude. This can be noted on the seismogram presented in Figure 3, traces 11 to 17, which was recorded in a shallow reflection survey performed at Stº André, in the south of Portugal.

Besides the surface waves, when the distance is very small (traces 18 to 20), the direct waves travelling straightforward from source to receiver have a greater amplitude when compared with reflected waves (amplitude decays like r^{-1}, in homogeneous media, with ray-path r) and is not easy to record these last waves at this range. Automatic gain control (AGC) at the recording amplifiers may solve in part this problem, but not for very short time intervals between the two kinds of waves, as occurs with shallow reflectors and small offsets. Some criteria based on the adequate selection of the distance values were develloped under the name of optimum window techniques (Hunter et al 1983), being the upper limit imposed by taking in account the distor - tion that occurs on wide-angle reflected signals.

Stacking several records obtained with different shots at the same position is an efficient way to cancel random noise associated with traffic,wind and so on. This operation may be made in the digital memory of an enhancement seismograph. Adequate shifts in the position of the source during the stacking may be introduced to cancel surface waves, and it is a well known fact that reflected waves are not significantly affected (this procedure is usually employed in the equiva-lent way of a multi-geophone pattern recording the same shot and stacking the various records).

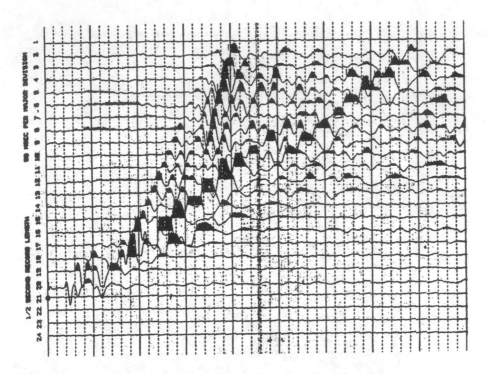

Figure 3. Seismogram recorded in a shallow reflection survey

Another problem is the influence of near-surface low-velocity anomalies due to weathering;these anomalies are irregular and usually confined to the first few meters of depth, and they have im - portant values. If not taken into account, they will introduce spurious irregularities at the interpreted interfaces, and this aspect is of great importance when the depths involved in the interpretation are small and the relative errors get higger. To avoid this, a time correction is introduced in the signals that virtually puts all shots and all receivers at some reference datum below the base of the weathered layer. These weathering corrections are usually made in two steps: the preliminary static corrections, that use waves refracted at the base of the weathered layer, and the automatic static corrections, which use computational techniques of cross-correlation analysis between cleverly grouped signals (Hileman et al 1968 and Disher & Naguin 1970). Of course, the topography of the free surface, where the geophones are placed, must be taken into account to reduce the signal to the reference datum.

Finally, the seismic source will introduce an important problem not yet considered: the fre quency content of the signals. The importance of the frequency content gets clear if we consider that the smallest geologic anomaly that a seismic wave can "report" is of the order of magnitude of its wavelength (Widess 1972). So, high frequencies will be needed to achieve the high resolu- tion necessary in shallow prospection. For a P wave velocity of 2000 ms^{-1}, a frequency of at least 100 hz would be necessary to detect a layer 20 m thick. These high values of frequency are not to be expected either in natural seismic waves or in artificial waves that have travelled a few kilometers, because it is well known that the earth behaves as a low-pass frequency filter. For shallow depths, however, 100 hz signals may be recorded using simple sources like a seismic hammer (weight of a few hundreds of kilograms that slides vertically to impact the ground), and this is exemplified with the spectral analysis of such a signal, shown in Fig. 4.

Higher frequencies may be obtained using explosives, but a few hundred meters away from the source the results obtained with the seismic hammer are almost equivalent. Clearly, a weak source like a seismic hammer would not be useful for deeper prospection. But in shallow applications it allies economy and safety with repeatability, which is important when the source is used with the enhancement seismograph. Also the vertical directivity of the seismic hammer seems to be an impor tant advantage in shallow prospection.

Along with the intrinsic dificulties of shallow depth reflection, a few advantages may be found and the main one is that the amplitude of the reflected pulse is of the order of magnitude of the direct wave, except for very small offsets or very weak reflectivities at the interface. The extraordinary recording dynamics that is necessary with conventional reflection prospection to preserve the shape of the direct wave and also to record the reflected wave (sometimes four orders of magnitude weaker) is not at all necessary with shallow reflection.

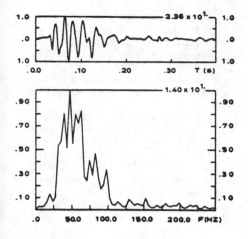

Figure 4. Spectral analysis of a hammer source signal

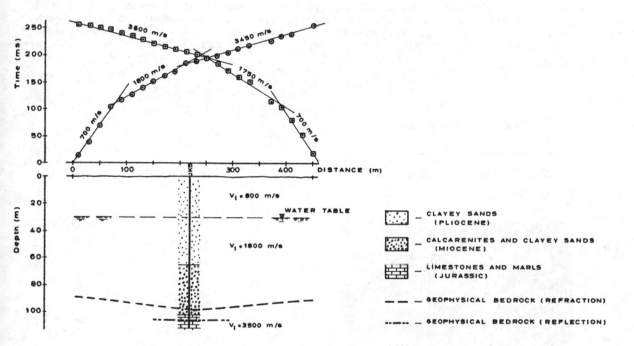

Figure 5. Stº. André - seismic refraction section

4 FIELD RESULTS

The application of the reflection method at shallow depths was tested at Santo André, in the
south of Portugal, a local previously studied with the seismic refraction method and drilled for
ground-water exploration.

The local has a smooth topography and the borehole drilled for ground-water exploration detec -
ted the next sequence of lithologic complexes (Fig. 5):

```
0 - 63 m - sands and clayey sands (Pliocene)
60- 100 m - clayey sands, clays and calcarenites  (Miocene)
 > 100 m - limestones with marl layers (Jurassic)
```

A seismic refraction profile with a total length of about 450 m was performed at the same area in the summer of 1984. As it can be seen in Fig. 5 the corresponding time-distance graphs are very regular and two seismic boundaries can be detected. The first boundary occurs at about 28 m depth and according to the results furnished by drilling it will correspond to the water table; the second boundary has a regular topography and corresponds to the bedrock, formed by the jurassic formations. The calculated depth for the seismic bedrock in the middle of the refraction profile, near borehole JK5, is of about 96 m and is in a good accordance with the depth of 100 m indicated by that borehole for the top of jurassic formations.

The field works for seismic reflexion were directed towards the coverage, with the horizontal resolution of 5 m and multiplicity 6, of a small profile 55 m long. The field crew was composed of three elements with a jeep and the work took two incomplete days to be accomplished. The equipment was composed of an enhancement seismograph Geometrics, model 2415-F with 24 channels, two cables of geophones with twelve geophones each and a seismic hammer mounted in the jeep, with a mass of 120 kg raising to a maximum height of 1,1 m. The natural frequency of the geophones was 14 hz, and the analogic filters of the seismograph were set up to select frequencies from 80 hz to 250 hz. At this particular site the surface waves (with a typical frequency of 50 hz) were efficiently removed by the filters, resulting in a considerably simplification of the field work because source shifting was not necessary.

Figure 3 is an exemple of the seismic signals recorded. The signals were grouped according to their depth-point and after being digitized, were processed in a DEC 10 computer. The search for reflected waves was made with velocity spectra which display graphycally the phase coherency of the signals in each CDP set, after an offset-dependent hyperbolic time correction to account for ray geometry (normal moveout correction). This analysis was made sistematically for each plausible pair of values $(t_{on}, \bar{V}_n)$, with reasonable increments. The result of one particular CDP-set is shown in Fig. 6, and exibits a sharp coherency anomaly for $t_o \simeq 0.150$ s and $V \simeq 1600$ m/s.

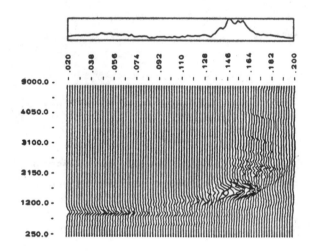

_ VELOCITY SPECTRUM _

Figure 6. Velocity spectrum

For a single horizontal reflector, these values corresponds to a depth of 120 m. A more detailed inspection of the entire set of spectrum, however, revealed the existence of a weaker reflector at a shallower depth, attributed to the water table, with the values 0.072 s and 550 m s^{-1}. Velocities and depths had to be computed following (1.5): the top of the water table was 20 m under the surface and the main interface depth was corrected to 104 m; the velocities for the dry and saturated sediments were found to be 550 ms^{-1} and 2155 ms^{-1}, respectively.

At this point it is interesting to compare the numerical results with those obtain by the refraction method. The difference in the depths determined for the water table was expectable since the field works were undertaken in different periods of the year (summer of 1984 for seismic refraction and winter of 1985 for seismic reflection). The velocities obtained for the saturated sediments are 1800 ms^{-1} with the refraction method and 2155 ms^{-1} with the reflection method. This difference is also expectable since the refraction method is sensitive to the velocity immediately under the top of the layer while the reflection method "integrates" the velocity from top to bottom in each layer. The differences in depth values for the bedrock are quite acceptable:

while the drilling indicates 100 m, the refraction method gives 96 m and the reflection method gives 104 m, i.e., each geophysical method results in a relative deviation of 4%, with different signals.

The detected reflections were checked through the analysis of the stacking of the corrected signals in each set: during this procedure, a reflected wave must grow up, since it sums in phase, while any other signal will sum randomly and tends to cancel. The stacking procedure for one of the CDP-sets is shown in Fig. 7.

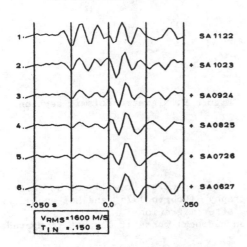

Figure 7. Stacking of corrected seismic signals

The result of the sum is a new signal representative of the mid-point of the CDP-gather. The graphic display of the stacked signals (at the vertical of the point that they represented) is the seismic section and Fig. 8 shows the one obtained with the signals recorded at Santo André. Obviously, the horizontal length of the seismic section may be extended with no limits, being proportional to the time necessary to perform the field work; with the present equipment a few hundreds of meters may be covered in one week, with a horizontal resolution as high as 5 meters.

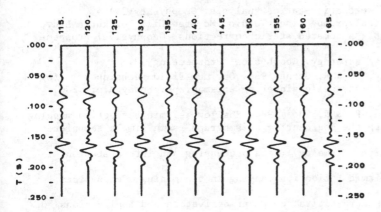

Figure 8. Seismic reflection section

Though the seismic section in Fig. 8 be very suggestive about the local structure it must be remarked that the vertical axis has units of time, and the relative importance of the layer thickness is not preserved, since velocities are variable. Also the topography of the interfaces is not preserved in this representation. The conversion of a time-section to a depth section, along with the correction of the image obtained for the interface is an important topic of research,

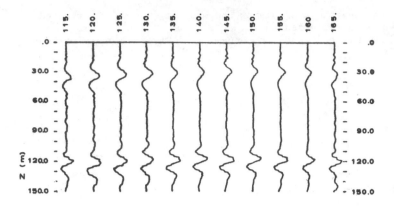

Figure 9. Migrated seismic section

falling under the designation of seismic migration. Due to its complexity it will not be dis -
cussed here,but several methods may be found in the bibliography (Dobrin 1976 and Hosken & Dere
gowsky 1985). Fig. 9 shows the same section as in Fig. 8, after migration.
   The migrated section is also plotted in Fig. 5 (reflexion bedrock) where it may be   compared
with the results obtained by refraction and by drilling.

5 CONCLUSIONS

The results presented in the paper point out that the seismic reflection method can be used   to
map geologic interfaces at shallow depths and may be a useful tool for geotechnical exploration.
Its application will be especially advantageous for high resolution studies in areas of relati -
velly thick overburden, where the seismic refraction requires much longer spreads   and   larger
source strengths. However, refraction and reflection do not exclude each other and the combined
use of both methods is probably the best way to apply seismic methods in   many   geotechnical
problems.

REFERENCES

Disher, D.& P. Naguin 1970. Statistical automatic statics analysis. Geophysics(35) 4.
Dobrin, M. 1976. Introduction to geophysical prospecting. Toquio: Mc Graw-Hill Kogakusha Ltd.
Hileman, J. Embree, P. & J. Pflneger 1968. Automated static corrections. Geophysical Prospec-
    ting 16.
Hosken, J. & S. Deregowsky 1985. Migration strategy. Geophysical Prospecting 33.
Hunter, J., Burns, R., Good, R., Mc Anley, H. & R. Gagné 1983. Optimum field techniques   for
    bedrock reflection mapping with the mutichannel engineering seismograph. Sunnyvale: EGG
    Geometrics.
Hunter, J., Pullan, S., Burnes, R., Gagné, R. & R. Good 1984. Shallow seismic reflection mapping
    of the overburden bedrock interface with the engineering seismograph - some simple techniques.
    Geophysics (49) 8.
Sheriff, R. & L. Geldart 1983. Exploration seismology: data processing and interpretation.
    Cambridge: University Press.
Singh, S. 1984 High frequency shallow seismic reflection mapping in tin mining. Geophysical
    Prospecting 32.
Taner, M. & F. Koehler 1969. Velocity spectra-digital computer derivation and applications of
    velocity functions. Geophysics (34) 6.
Widess, M. 1973. How thin is a thin bed? Geophysics (38) 6.

Characterization of the rock mass at the McLellan Mine

Caractéristiques du massif rocheux dans la mine Mac Lellan, Canada

J.L.Rotzien, *University of British Columbia, Vancouver, Canada*
H.D.S.Miller, *University of British Columbia and International Mining Services Inc., Vancouver, Canada*
I.M.Plummer, *Sherritt Gordon Mines Ltd, Lynn Lake, Manitoba, Canada*

ABSTRACT:

The characterization of a rock mass surrounding an underground opening is often difficult due either to inadequate data gathering or to massive amounts of information that cannot be readily assimilated. At the McLellan Mine at Lynn Lake, Manitoba, geotechnical logging of the diamond drill core was completed concurrently with geological logging. The data was entered directly into computer files using terminals in the core logging shed. Computer printouts of the diamond drill core logs were then provided to U.B.C. to produce rock mass ratings (Bieniawski, 1984) in a summarized form.

The rock mass classification data was then placed into computer files at U.B.C., and scanned to provide a better understanding of the spatial variation in rock mass quality. Three computer programs were written to a) provide a means for input of raw data; b) convert the raw data into a more usable form; and c) scan the data files for relevant information and complete statistical calculations on the data of interest. These programs were written in HP Basic to be used on an HP 9845A.

In addition to the rock mass classification work, uniaxial and triaxial compressive tests and specific gravity tests were completed on BQ core samples in the rock mechanics laboratory at U.B.C.

The results of the laboratory testing program and the rock mass classification work produced a clear understanding of the character and variability of the rock mass surrounding the planned stopes. The results also provided input data for numerical modelling and a basis for interpretation of the numerical models.

INTRODUCTION

In October 1983, the Department of Mining and Mineral Process Engineering of the University of British Columbia, Vancouver, B.C., signed a research contract with Sherritt Gordon Mines Limited to conduct a rock mechanics research program for the McLellan Mine. The contract was reviewed in January 1985, to complete the Phase II program. One phase of the program was to set up the geotechnical logging of the diamond drill core at the mine site and to complete laboratory tests on drillcore samples to aid in the characterization of the rock mass at the McLellan Mine. The work was completed in February 1986.

The McLellan Mine is a gold-silver mine located in the Precambrian Shield of Canada approximately 725 kilometres northeast of Winnipeg, Manitoba (Figure 1). Due to the irregular nature of the ore zone, a decision was made in October 1985 to develop the mine by cut-and-fill methods.

GEOLOGY

Regional Geology

The McLellan Mine is located in a gently undulating glacial outwash plain (Figure 2) underlain by Proterozoic rocks of the Lynn Lake Greenstone Belt which comprises part of the Churchill Structural Province. Two extensive successions of the Wasekwan Group of the Lynn Lake Greenstone Belt have been mapped in northeasterly trending belts. The McLellan Mine is located within the northern belt between Motriuk Lake and Eagle Lake. This northern belt consists primarily of mafic volcanic rocks with intercalated felsic and intermediate units.

The emplacement of small subvolcanic plutons and large mafic and felsic plutons during and subsequent to the deposition of the Wasekwan Group resulted in metamorphism and folding of the sequence.

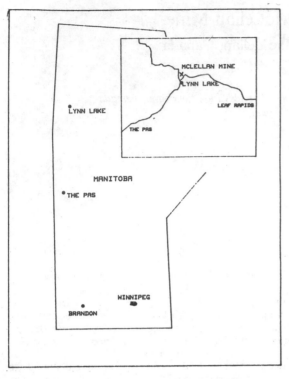

Figure 1. Location map.

Site Geology

At and immediately surrounding the McLellan Mine site, the Wasekwan Group consists of steeply inclined intermediate and mafic volcanic flows and tuffs, dacite and greywacke-siltstone. The Main Zone of the mine is composed of a series of steeply dipping, parallel, tabular zones of ore grade material closely related to beds of altered, clastic, biotitic sediments. (Field name, tuffaceous sediments.) The sediments occur in an extensive high magnesium mafic unit (Field name, Chloritized Andesite.) consisting of actinolite-chlorite schist. The mafic rocks which contain both sedimentary and volcanic features host some gold mineralization when they are silicified, carbonatized and mineralized with sulphides.

Structural Geology

At the Mine Site, the Wasekwan Group has been folded into a steeply inclined sequence dipping at approximately 85° to the north. The intrusion of plutons has resulted in regional metamorphism and a foliation which strikes regionally at approximately north 45° east and dips at 85° to the southeast. Within this sequence the zones of ore grade material dip at approximately 85° to the north.

The ore grade mineralization is terminated to the north by the North Shear. This structure consisting of a 0.3 metre to 6.0 metre thick zone of shearing is sub-parallel to the foliation, generally striking at approximately N 50° E and dipping at 70° to 85° to the northeast. It is the most significant structure at the site.

In addition to the North Shear, three sets of discontinuities are evident. The most common set of discontinuities is along the foliation, striking N 50° E and dipping at 80° to 90° to the northwest and southeast.

The second most significant joint set is near perpendicular to the foliation, striking at N 40° W and dipping at 80° to 90° to the northeast and southwest.

The third joint set consists of relatively flat-lying joints dipping at 0° to 25° with no preferred strike direction.

LABORATORY TESTING

Unconfined compressive tests were completed on 54 samples of BQ (36.5 mm) core. The results of the testing indicated that the foliated rocks are strongly anisotropic. The Ucs of the samples loaded sub-parallel to the foliation is approximately 15 times the Ucs of samples loaded sub-perpendicular to the foliation. For design purposes, the lower values, with foliation at 45°

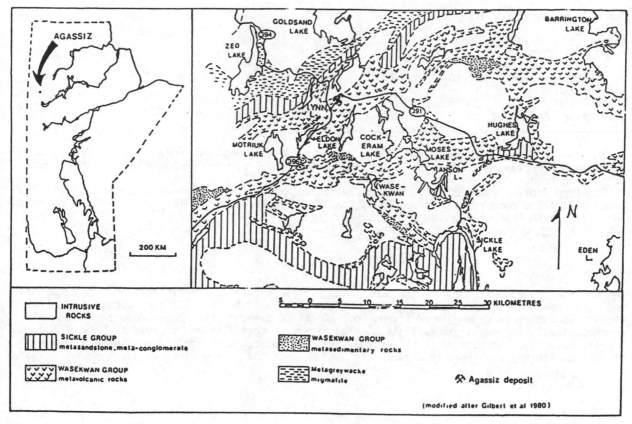

Figure 2. Regional geology.

to 90° to the direction of Loading (Table 1) were used.

Triaxial tests were completed on eleven BQ diamond drill core samples. From these tests it was estimated (Figure 3) that the internal angle of friction ranges from 55° to 21° between zero confinement and a confining stress of 20.69 MPa. The cohesion was estimated at 11 MPa.

ROCK MASS CLASSIFICATION

In Stage I of the research, a geomechanical core logging system was initiated and the data was incorporated with the standard Sherritt Gordon diamond drill core logs. After an initial training period, Sherritt Gordon personnel completed all of the core logging duties. At the McLellan Mine this entails direct entry of data into computer files. Hard copies of the core log files were then sent to U.B.C. where the rock mass rating of the core was completed.

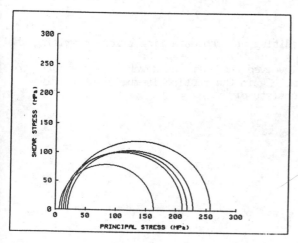

Figure 3. Mohr's circle plot.

Table I. Summary of design values.

Lithological Units	Ucs	E	ν
	(MPa)	(MPa) (x10^4)	
Tuff Sedimentary	92.83	2.61	0.29
Chloritized Amph.			
Andesite	50.45	1.74	0.20
Chloritized Andesite	99.29	2.64	0.16

Due to the extensive data base developed from the development drilling, a simple, quick classification system was required. Thus, the CSIR geomechanical system developed by Z.T. Bieniawski
(1984) was used.

Corelogs from selected holes were also classified by the N.G.I. system developed by Barton and
Lunde (1974) for comparison. It was found that the values for RMR (CSIR) and Q (N.G.I.) correlated
adequately with the relationship:

RMR = 9 log$_e$ Q + 44.

METHODOLOGY

Because the ore zone at the McLellan Mine is a relatively thin, near vertical, and irregular deposit, it was anticipated that the quality of the rock would be variable even within an individual
domain for each rock type. In addition, both AQ and BQ diamond drill holes were completed and
geomechanically logged by six geologists. The effect of core size and individual bias of the core
loggers were also anticipated to affect the variation in rock mass quality and had to be examined.
Thus, a system was developed to examine the variation of the rock quality with:

1. Location
2. Hole size
3. Domain
4. Lithology
5. Core logger.

Three computer programs were therefore written for the Hewlett Packard 9845B at the Mining
Department.

The first program written consists of a routine to store the required data in the form of string
variables. The first line of data for each hole is composed of the general hole information
including:

1. Drill hole number   (6 characters)
2. Core size   (2 characters)
3. Latitude of collar   (7 characters)
4. Departure of collar   (7 characters)
5. Elevation of collar   (7 characters)
6. Number of datra entries   (3 characters)
7. Number of downhole surveys (2 characters)
8. Initials of core logger   (2 characters).

Subsequent lines consist of:

1. The start depth of interval
2. End depth of interval
3. Domain
4. Lithology
5. RMR.

This program is fully interactive and allows for editing of each data line prior to writing the
data in the file.

The second program developed to process the data involved the entry of downhole survey data to
convert the stored field data into a more usable form. Again the routine is fully interactive
with editing capability. The output to disc files consists of:

1. Drill hole number
2. Core size
3. Start latitude
4. Start departure
5. Start elevation
6. End latitude
7. End departure
8. Domain
9. Lithology
10. Core logger
11. RMR.

The third program developed for data processing consists of a scanning and sorting routine with simple statistical capability. The program has been written to scan for intervals with appropriate parameters, save those intervals in a new file, and compute the weighted mean, standard deviation, and variance of the rock mass rating for those intervals. These scans can be completed for intervals of diamond drill core with a specified

1. Domain
2. Lithology
3. Core size
4. Core logger
5. Interval of latitude
6. Interval of departure
7. Interval of elevation.

In addition, the statistical analysis can be computed for all of the core intervals in the data file.

The output data from this routine consists of

1. An echo of the specified parameters
2. The total length of core intervals with the specified parameters
3. The weighted mean of the RMR for those intervals
4. The standard deviation of the RMR
5. If desired, a hardcopy output listing those core intervals
6. A new data file consisting only of those intervals.

ANALYSIS

The first step in the analysis was to test the consistency of the core logging and the RMR from hole to hole. Several scans were completed to determine the weighted mean of the intervals logged by each core logger for the major lithological units. In addition, two scans were completed to test the variation in RMR with core size. From this work, it was evident that the summarized Rock Mass Ratings produced were very consistent.

The next step in analyzing the data was to scan the main data file for each domain, hanging wall, footwall and North Shear, to compile smaller data files for the domain and the determine the total length logged in each domain. In addition, scans were completed to determine the total length logged and overall rockmass rating of each lithological unit within each domain (Table II).

Table II. Summary of rock mass ratings for each domain.

DOMAIN	LITHOLOGICAL UNIT	METRES SUMMARIZED	% OF TOTAL	RMR
Hanging Wall	Tuffaceous Sediments	1750.10	29.5	75.38
	Chloritized Andesite	2752.36	46.4	70.89
	Variable Andesite	766.44	13.0	69.75
	Amphib. Andesite	56.00	0.9	78.21
	Andesite	514.29	8.7	75.47
	Other	90.47	1.5	--
	Total	5929.66	100.0	72.56
Footwall	Tuffaceous Sediments	1714.36	38.5	76.79
	Chloritized Andesite	2366.04	53.1	73.62
	Variable Andesite	175.47	3.8	77.96
	Amphib. Andesite	163.54	3.7	83.80
	Andesite	7.80	0.2	75.24
	Other	30.39	0.7	--
	Total	4457.60	100.0	75.40
North Shear	Tuffaceous Sediments	21.50	11.1	43.58
	Chloritized Andesite	171.10	88.4	43.90
	Variable Andesite	0	0	--
	Amphib. Andesite	0.72	0.4	56.00
	Andesite	0	0	--
	Other	0.18	0.1	--
	Total	193.50	100.0	43.91

From the results of these scans, it is readily obvious that the tuffaceous sediments and chloritized andesite make up the bulk of the rock mass at the McLellan Mine, especially in the footwall and North Shear where the active mining will be located. The remainder of the analyses of the rock mass rating in the footwall therefore concentrated on these two lithological units. The analysis of the rock mass in the Hanging Wall was confined to intervals within 10 metres of the North Shear, the zone that was anticipated to have a major effect on dilution and stability of the openings.

The effect of the North Shear on the proposed cut-and-fill stopes was investigated in two stages. The first step consisted of determining the spatial variation of the rock mass rating within the North Shear. The second stage involved determining the location of the North Shear with respect to the stopes and the extent of any zones assaying less than 0.8 oz of gold per ton within three metres of the stopes. Thus, the areas which could cause significant dilution were estimated using drill hole data and stope plans showing the results of a block model of the gold values.

RESULTS

Within the Hanging Wall rocks, the rock mass ratings are generally lower near the North Shear. In addition, the variation in rock mass quality more than one stope diameter beyond a stope will have little effect on the reaction of the stopewalls to excavation. Thus the spatial variation of the rock mass quality in the Hanging Wall is confined to those intervals within 10 metres of the North Shear. All discussion of the Hanging Wall rocks is restricted to this area.

The rock mass rating of the Hanging Wall rocks is relatively consistent at 70% to 75% below elevation 1360 m. In the interval 1360 to 1480 m, the rock mass rating appears to decrease to approximately 65%. However, because the data points are spaced at 40 m intervals, the decrease in rock mass quality likely exists at or near 1380 m elevation. Between 1480 m elevation and the surface, the rock mass rating is approximately 55% (Figure 4).

The rock mass ratings of the Hanging Wall rocks appear to be zoned along the strike of the North Shear. Within the interval -100 mE to 100 mE, the rock mass rating averages approximately 70%. Elsewhere, the rock mass ratings are above 75% (Figure 5).

These values are conservative due to the effect of the lower rock mass ratings obtained above elevation 1380 m.

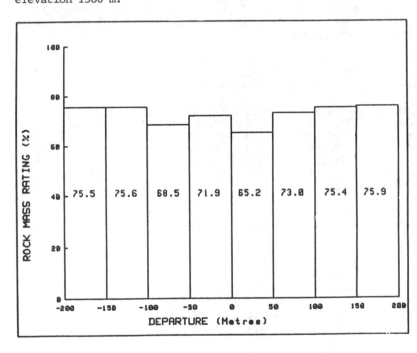

Figure 4. Rock quality vs. departure: hanging wall.

Footwall rocks

In the tuffaceous sediments, the rock mass rating is greater than 75% for 95% of the area between elevations 1160 and 1380 m.

From elevation 1380 m, the rock mass rating generally decreases towards the surface with a zone of comparatively lower rock quality between -200 mE and 0 mE. However, due to a lack of data near the surface, the trends are not as well defined. In this area the contours are the result of

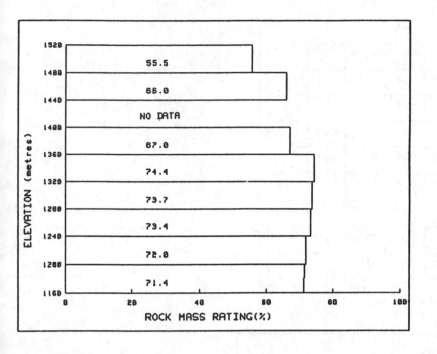

Figure 5. Rock quality vs.
elevation: hanging wall.

interpretation of general, non-geotechnical drill core logs and correlation with hydrological studies (Rotzien and Miller, 1984).

In the chloritized andesite, the rock mass rating is greater than 70% for 80% of the area between elevations 1160 and 1380 m. Within this area, the rock mass rating of the chloritized andesite does not fall below 60%.

Above elevation 1380 m, the rock mass rating generally decreases towards the surface, with zones of comparatively higher rock quality between -100 mE and 0 mE and between 100 mE and 200 mE. As with the tuffaceous sediments, this interpretation is based heavily upon correlation with previous work.

North Shear

Within the North Shear, the rock mass rating generally increases with depth. However, no data is available between elevations 1400 m and 1440 m, and only one data point is available between elevations 1380 m and 1400 m. The trend is therefore not clear in this area.

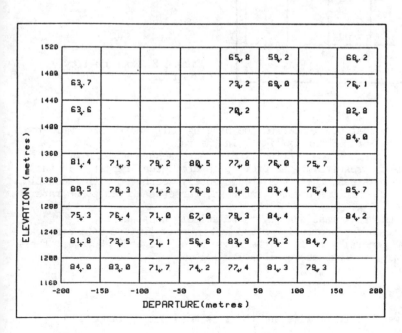

Figure 6. Rock quality of the
tuffaceous sediments.

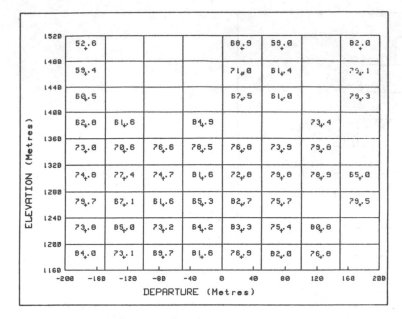

Figure 7. Rock quality of the chloritized andesite.

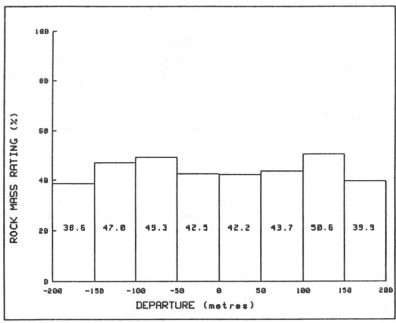

Figure 8. Rock quality vs. departure: North Shear.

By comparing the available data with that obtained from the Hanging Wall and the Footwall rocks, it appears that the rock quality in the North Shear is likely in the range of 25% to 30% above elevation 1380 m. At or near 1380 m, the rock mass rating likely increases to a range of 40% to 45%. This assessment is validated by the results of the previous hydrological report. In that investigation, it was found that in some areas, the rocks are intensely weathered with limonitic alterations within joints and faults. Within the area of interest, these weathered zones rarely extended below 1400 m elevation.

Laterally within the North Shear, the rock mass quality is relatively constant but between departures -200 m and -150 m, and between departures 100 m and 200 m, only limited data are available. Elsewhere the rock mass rating generally falls in the range of 40% to 45%. However, these values are affected by the low rock mass ratings above elevation 1380 m.

Estimated stand-up times

Using the empirical relationship between stand-up time and the rock mass rating proposed by

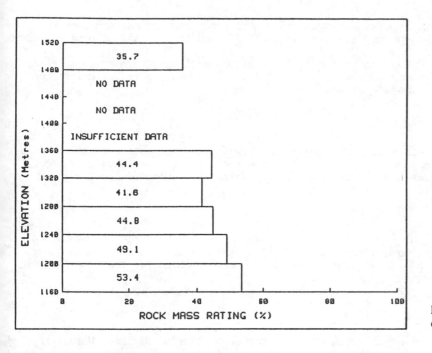

Figure 9. Rock quality vs. elevation: North Shear.

Table III. Estimated stand-up time for unsupported spans.

DOMAIN	LITH. UNITS	UNSUPPORTED SPAN	LOW RMR	STAND-UP TIME	HIGH RMR	STAND-UP TIME
HW	ALL	4 m	55%	10 days	75%	8 mos.
NS	ALL	3 m	25%	0	45%	2 days
		4 m	25%	0	45%	2 days
FW	TS	3 m			75%	9 mos.
		5 m			75%	8 mos.
		10 m			75%	6 mos.
		20 m			75%	0
FW	CA	3 m	60%	1 month	70%	5 mos.
		5 m	60%	25 days	70%	4 mos.
		10 m	60%	15 days	70%	3 mos.
		20 m	60%	0	70%	0

Bieniawski (Bieniawski, 1976), the stand-up times for unsupported spans were estimated as summarized in Table III.

Although the stand-up times are only order of magnitude estimates, they are helpful in outlining the problem areas.

Table III indicates that some immediate loosening and ravelling may occur in the stope roofs wherever the stope widths exceed 20 metres. In the footwall rocks, the stope walls and roof less than 20 metres wide will likely stand-up for the duration of each cut-and-fill cycle. Where the rock mass rating is less than 60%, the roof will likely experience some loosening and ravelling prior to the completion of each cut.

Wherever the North Shear is exposed in the stope walls and roofs, significant dilution is likely to occur due to the extremely short stand-up time.

DISCUSSION

The rock mass at the McLellan Mine consists predominantly of relatively hard, competent chloritized andesite and tuffaceous sediments. The quality of these rocks varies considerably with elevation and from domain to domain. However, below elevation 1380 m minimal dilution is generally anticipated from the Hanging Wall and Footwall Domains where the width of the cut is less than 20 metres.

Where the width of the cut is greater than 20 meters and/or the rock mass ratings are less than 70%, the dilution is anticipated to be in the order of 10% to 15% and the installation of temporary support may be required.

It is estimated that the 50% of the hanging walls of the stopes will be within or adjacent to the North Shear. Wherever this occurs, the dilution is anticipated to be 15% to 20% unless controlled blasting techniques and temporary support are used.

CONCLUSIONS

The rock mass characterization defined in this paper lends itself to initial estimates of the dilution and outlines the prime areas of concern. However, it is not intended that this work should stand alone. These initial estimates of dilution will be confirmed by numerical modelling and empirical methods.

ACKNOWLEDGEMENTS

The authors wish to thank Sherritt Gordon Mines Ltd. for their permission to publish this paper. In addition, we are indebted to the geology department at the McLellan Mine for their patience in producing a large base of geotechnical data.

REFERENCES

Augsten, B.E.K. 1985.Ore mineralogy of the Agassiz Stratiform Gold Deposit. Carleton University, B.Sc. Thesis.

Barton, N., Lien, R. and Lunde, J. 1974. Engineering classification of rock masses for the design of tunnel support. Rock Mechanics, Vol. 6, No. 4: 183-236.

Bieniawski, Z.T. 1984. Rock Mechanics Design in mining and tunnelling. A.A. Balkema, Boston, p. 97-135.

Sherritt Gordon Mines Ltd. 1984. Excerpt from Untitled, Internal Report, Sherritt Gordon Mines Ltd.

Engineering and geologic studies providing control
of a rock mass during mining operations

Etudes géologiques et techniques de surveillance d'un massif rocheux
pendant des travaux miniers

V.V.Rzevsky, *Moscow Mining Institute, USSR*

ABSTRACT: Monitoring of rock mass deformation and failure during mining operations are treated as the main principle of control over its state.

RESUMÉ: La prévention et le contrôle des processus de déformation et de
destruction du massif lars des travaux miniers sont considérés comme la base
de contrôle de son état.

Earth's crust dynamics problems stipulated by engineering activity of man are
connected to a great extent with the increase of depth and scale of mining
operations. The study of engineering and geologic conditions of a mineral
deposit has major importance for the solution of fundamental problems in underground and opencast mining: determination and monitoring of rock pressure,
stability and displacement of rock masses, evaluation of rock workability,
ensuring completeness of extraction from the entrails, environmental protection
and reclamation of areas affected by mining operations.

Engineering and geologic observations at all the stages of a deposit exploitation, as well as the analysis of engineering and geologic situation, emsure
effective structural and geomechanical typisation of rock masses for certain
objects. Here, the solution of the following problems ia necessary: study of
mechanical properties and structural and mechanical features of rock masses,
investigation and prediction of mechanical processes in massifs during mining
operations and possible manifestation of these processes (rock pressure),
monitoring of mechanical processes in rock mass or its state control, including
working out of means of this control, taking into account the type and character
of mining operations.

The study of a rock mass structural and mechanical features should be concentrated, first of all, on structural features (fissility, layering, cavitation,
presence of clumps, etc.) disturbing uniformity of rock masses, in order to
use reasonably methods of uniform medium mechanics for description of mechanical processes in rock masses; establishing regularities of anisotropy hetero-
genecity formation under the influence of above mentioned structural features;
the study of natural or initial stress condition of a massif for mechanical
processes description.

Rock pressure in a certain point of a massif as to its value, direction and
character of manifestation is a combined result of the following:

1. First stock forces effect - external static pressure and gravitation
forces (according to rock unit weight), taking place in a massif from cleat to
cleat and observing the laws of forces distribution between solid bodies, not
necessarily vertically $P \neq H$;

2. Tectonic forces effect in a massif not affected by mining operations, and
defined mainly by geologic structure of a massif.

3. Manifestation of physical and technical parametres, both of "elementary
volumes" and of a massif consisting of them; actual values of these parametres
determining the terms of spatial redistribution of forces and appearance of
local zones of rock pressure.

4. Effects produced by mining operations technology: various mining activities
on the whole and individual processes lead to different spatial redistribution

of rock pressure and, therefore, to variety of its manifestation.

The perspective way of rock pressure calculation is representation of each influence factor by the methods of entrails geometrization and, then, their account using mechanical models of rock deformation and failure for certain engineering and geologic elements of a massif.

The control over a state of a massif includes prediction, monitoring and purposeful change of intensivity of deformation and failure processes developing during underground and opencast operations.

In underground coal mining, the major problem is the elimination of methane emission, dust, instant coal and gas bursts and spontaneous combustion.

The theory and industrial testing of the coal seams properties and condition regulation were based upon thorough investigation into natural conditions: structure and texture of dangerous zones; physical and chemical features of coal seams and host rocksin dangerous zones; filtration, collecting, dense and deformation seam characteristics and regularities of their manifestation during active influences; gas dynamics of coal bearing seams, conditions for dust formation and spontaneous combustion of coals during mining operations; physical and chemical interrelations between coal and fluid environments; life conditions of methane eating, methane forming and hydrogene forming bacteria in coal bearing seams.

The following effects are widely used as active methods and means of control over state and features of a massif:

1. hydrodynamic, by means of drilling speccially directed holes from the surface and from underground workings;

2. pneumatic, by means of pneumatic treatment of dismembered coal seams through the drill holes, as well as by means of primary pneumatic dismembering of a coal seam, to increase phase and total permeability of the latter;

3. thermic, by means of dismembered coal seams treatment, to decrease coal sorption capacity;

4. by means of explosion of methane-oxygene mixture in the seam;

5. wavy, using mechanical oscillatoryeffects in acoustical frequency range;

6. physical and chemical, by means of H-Cl, etc. solutions, injecting monomer solutions, providing their solidification in a coal seams massif;

7. microbiological, by creating conditions ensuring vital activity of methane absorbing bacteria in coal bearing seams;

8. multistage, affecting the seams by different combinations of methods mentioned above.

Engineering and geologic investigations at all the stages of mineral deposits explotation by opencast method envisage engineering and geologic shematization of a massif and differenciated choice of mechanical models over a massif elements. Provision of quarry pit edges and dumps is based upon establishing interrelations between opencast mining dynamics and the following processes: massif deformation during deep falling of the water level, decrease in strength and development of rock shear strains, compaction of a dump body and foundation. The initial parametres for the dependencies are found from engineering and geologic zoning of quarry and dump areas. Establishing character and scale of engineering and geologic processes during mining operations makes it possible to reason out the complex of steps necessary for purposeful influencing rock masses, rising engineering and geologic information available during exploration of a deposit. The initial data for planning steps are determined using the system of a massif state control, and their effectiveness is evaluated.

Bringing about technological and specific measures for purposeful influencing the edges and dumps quarry massifs is the most effective way of ensuring economy and safety of opemcast operations and environmental protection. At present, a new approach has been formed to the solution of the problem of providing edges and dumps stability on the base of the theory of quarry slope state control. The recommendatinns on the control of quarry slopes should be based on the resulus of engineering and geologic zoning of quarry and dump areas.

The exploitation of water filled deposits leads to specific deformations of sand-clay and semi-rocky massifs.

The creation of large quarries and mines during exploitation of water filled deposits, implies drawdown of water-bearing horizons in adjacent areas; the size of cones of water table depression being tens of kilometres, The depression compaction of rock depths is connected with falling of waterbearing horizons head. The prediction of surface displacement is necessary both for ensuring normal work of surface structures, and for the choice of rational design and parametres of underground workings support, which are disturbed in the process of depression. During opencast mining of water filled deposits, the process of consolidation of massifs edges made of sand-clay and semi-rocky material, due to withdrawal of hydrodynamic and hydrostatic pressure, is "overlaid" by the process of reducing these rocks strength in the course of time, the effects of which are determined by the slope life term.

Determination of height and bulk of dumps and areas required for their locating is an important element in planning opencast projects. Variations in the course of time occurs, mainly, due to their consolidation, prediction of which is made by means of filtration consolidation or creep theory (depending on the rock phase composition), allowing to take the dynamics of dump progress into account. Durability characteristics of dump rocks are made depending on their compaction degree, determined by the order of dump formation. After checking up the dump formation regime from the point of view of slope stability, the necessary technological and special steps are chosen providing minimum withdrawal of land for locating dumps, and effective reclamation activities. It should be noted that the dumps present a considerable external load to the massif they are made upon.

The use of engineering and geologic analogies method, in order to determine the parametres of durability, creep and compactability, has major importance for obtaining reliable initial information during planning mining operations. To verify the initial information during construction and exploitation of mining projects, the programme of observations and field experiments is necessary, providing continuity of designing and optimal decisions about rock massif control at different stages of a deposits exploration. The massif state control should be carried out using both stationary observation network and mobile units, providing unformation on water, physical and mechanical rock features at any moment and quarry or dump areas.

True picture of a massif "life" at different stages of mineral deposits exploration may be obtained as a result of summarizing constantly flowing field information. The apparatus for the control by means of physical methods of stress and deformation measureng should be put into the massif during prospecting for the deposits, and keep functioning at all stages of mining operations, ensuring, thereby, obtaining of a wide and representable range of indeces and their variability in the course of time. The solution of reverse problems, in order to establish the role of certain factors, finding parametres and regularities of processes being studied, has considerable importance here.

The creation of effective system of stationary and mobile rock massif control together with the principles of continuity, adaptation and feed-back, provides reliable geomechanic description of geologic situation. Mining operations experience presents good material which may be used both prediction of engineering and geologic conditions on promising deposits, and for solving theoretical and methodical problems of engineering and geologic investigation as to different branches of engineering activity.

Thus, during mineral deposits exploitation, the evaluation of engineering and geologic conditions and rock massif state control has major importance, including working out of the means of prediction and monitoring of mechanical processes developing in a massif, and taking the type and character of mining operations into account.

Geodynamic phenomena induced by the opencast mining activity at Bełchatów

Phénomènes géodynamiques provoqués par les exploitations minières à ciel ouvert à Bełchatów

Lech Wysokiński, *Warsaw University, Poland*

ABSTRACT: The seismic activity phenomenon induced by the mining works at the Bełchatów open-pit mine of depth dawn to 200 m is described in the present paper. The phenomenon mechanism was discussed and the forecasted values of tremors calculated on the basis of the simplified seismic hazard method are presented.
RESUME: Phénomèmes géodynamiques induits par l'action de mine à Bełchatów. Dans le travail présent on a discuté le phénomène de séismicité induit par les travaux miniers à la mine de Bełchatów de profondeur plus que 200 m.
On a décrit le mécanisme probable du ce phénomène ainsi qu' on a donné les valeurs prognostiques des séismes et l' estimation de la menace des objets de mine et d'usine électrique.

The Bełchatów brown coal deposits are of the tectonic origin, cavity type, mostly occuring in tecton ditches. The whole area is 40 km long and 1.2 ÷ 2.5 km wide. The overburden density is equal to 130 m and the average coal layers thickness amounts to 60 m. Chiefly sands, which constitute 60% of the material, are found in the overburden. First the deposits were dried up by the depression wells method in 1975 and then one year later exploitation works were commenced. In 1986 the winding efficiency was fully reached and amounted approximately to 40 mln m^3 of coal at the overburden output of 110-120 mln m^3. The mine depth approximately equals to 200 m at present and is estimated to be equal to 250 m in the future.
Tremors induced by the exploitation began in August, 1979. Detailed studies of the tremors began after 29th November, 1980 and on this very day the strongest tremor of a magnitude equal to M_L = 4.6 took place. After the tremor a measuring system has been established, and for the past 2 years 5 recording stations to provide measurements of seismic activity within the mine area have been operating.
Up till 1st October, 1985 about 270 tremors have been detected.
The most strongest tremors recorded and the tremors frequency may be found in Tables 1 and 2 respectively.
The strongest tremors that occured in 1980 and 1985 have been analised in detail. Data concerning the 1980 tremor have been carefully studied by:
/Głazek, Gibowicz, Wysokiński 1981/ in Polish, /Gibowicz 1981/, /Gibowicz, Guterch, Lewandowska, Wysokiński 1982/.

DATE			MAGNITUDE	SEISMIC ENERGY	TIME G M T		
1979	08	17	3.6	~ 10^8	14	00	16
1980	02	26	3.5	~ 10^8	12	01	31
1980	04	01	3.5	~ 10^8	16	26	20
1980	11	29	4.6	~ 10^{11}	20	42	19
1980	12	16	3.4	~ 10^9	17	26	40
1981	02	28	2.8	~ 10^8	17	04	53
1982	03	26	3.3	$6.9 \cdot 10^8$	9	26	19
1984	03	30	2.4	$2.6 \cdot 10^7$	10	32	18
1985	01	17	4.2	$3.6 \cdot 10^9$	13	51	12

Tab 1. List of stronger tremors within the Bełchatów Mine.

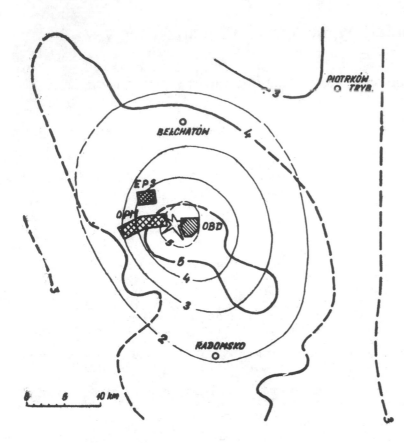

Fig 1. Map of isoseists of
the strongest tremors

EPS - Elektric power station
OPM - Open pit-mine
OBD - Over burden deposit
☆ Epicenter

Fig. 1 presents a map of isoseists of the strongest tremors of 1980 and 1985. The
tremor of November, 1980 was
detected at a quite vast area
of a radius of about 100 km.
The tremor at the mine was
detected with a maximum intensity of about $I_o = 7$ according
to the MSK-64 scale. An average radius of the isoseises
$I_o = 6$ is equal to 2 km.
Applying calculations of the
attenuation curves a depth of
the tremor focus was determi-
ned to be about 2 km, Gibowicz /1982/. However, the focus mechanism was estimated
only for the tremor of 29th November, 1980, /Gibowicz, Guterch, Lewandowsk , Wysokiński 1982/. The tremor was detected by all of the European seismic stations and
many others throughout the world. The directions of the P-wave were recorded by
28 seismic stations all around Europe ranging from the Chorzów Station, situated
108 km from the epicentre, to the Kevo Station /Finland/ situated 2096 km from the
epicentre. The model focus mechanism suggests a displacement of rock masses equal
to a reverse fault of a considerable displacement component, thus the fault of a
throw-displacement type.
For the tremor of 17th January, 1985 a modelling of the focus has been made to
enable determination of the focus depth equal to about 1 km. The problem presented
herein concerns evaluation of hazard to the power station objects determined by
the seismic risk method as well as an attempt of explanation of the repeating tremors phenomenon. Table 2 shows the number of tremors of magnitude greater than 2,
specially classified according to the tremor strength.

YEAR	NUMBER OF TREMORS	DATE	MAGNITUDE
1980	10	29.11	$M_L = 4.6$
1981	34	28.02	$M_L = 2.8$
1982	8	26.02	$M_L = 3.3$
1983	43	20.01	$M_L = 2.2$
1984	108	30.03	$M_L = 2.4$
1985	68	17.01	$M_L = 4.2$

Tab. 2. Annual fre-
quency of tre-
mors and their
maximum magni-
tude

Statistic evaluation of the tremors by /A. Kijko 1975 / enabled to determine that
the maximum magnitude should not exceed 4.8. The calculations have been made assuming that the tremor number per time unit is a random variable of the Poisson's
distribution. As the logarithm of the tremor number N of a magnitude not less than

m is indicated by log N = a - b m, where a, b are constant, then the density function of a magnitude probability m is shown as below:

$$f/m/ = \beta \; \frac{\exp\left[-\beta /m - M_{max}/\right]}{1 - \exp\left[-\beta /M_{max} - M_{min}/\right]} \quad , \text{where}$$

M_{max}, M_{min}, m - maximum, minimum, and expected magnitudes respectively, β - parameter.
The calculating methods may be found in the following papers:
/A. Kijko 1975/, whereas Table 3 shows calculations of the seismic risk. In order to provide a quite realistic seismc risk forecast one should know a geomechanic model of the phenomenon as well as the geological structure Table 4. The model is

TIME (YEARS)	2.5	3.0	3.5	4.0	4.2	4.5	4.7	4.8
1/12	0.348	0.249	0.162	0.091	0.066	0.032	0.011	0.002
3/12	0.716	0.572	0.412	0.249	0.185	0.093	0.034	0.006
6/12	0.914	0.811	0.650	0.433	0.334	0.176	0.067	0.011
9/12	0.972	0.914	0.789	0.571	0.455	0.252	0.098	0.017
1.0	0.990	0.960	0.871	0.674	0.554	0.320	0.129	0.022
1.5	0.99	0.991	0.951	0.810	0.698	0.438	0.186	0.033
2.0		0.998	0.980	0.888	0.795	0.535	0.240	0.044
2.5		0.999	0.992	0.933	0.860	0.614	0.290	0.054
3			0.997	0.960	0.903	0.680	0.337	0.065
35			0.998	0.975	0.933	0.734	0.380	0.075
4				0.985	0.953	0.778	0.431	0.086
5				0.994	0.977	0.845	0.493	0.106
6				0.998	0.988	0.891	0.557	0.126
8					0.997	0.946	0.660	0.163
10					0.999	0.972	0.738	0.200
15						0.994	0.862	0.284
20						0.999	0.926	0.358
30							0.978	0.484
50							0.998	0.655
100								0.882

MAGNITUDE M_L

Tab 3. Predicted magnitudes calculated on the basis of statistic analisis of the recent tremors.

rather considered to greater extent as a hypothesis not a shear fact. The deepest bore-hole in the trench is over 1000 m and reaches the middle Jurasic limestone layers. More deeper bore-holes /over 3000 m/ found in the distance of 3 ÷ 20 km from the trench prove that the salt layers occur at the depth of about 3 km. It has been assumed that the tremors are induced by the mining activity and the mine is situated within an aseismic area. Recent tremors of a considerable strength took place in 1932 and were detected all over the Central Poland.

MAGNITUDE M_L	NUMBER OF TREMORS	
	PREDICATIVE	REAL
2.0	50	47
3.0	9	8
3.5	6	4
4.0	3	2
4.2	2	1
4.5	1	1
4.7	0	0

Tab 4. Predicative and real number of tremors.

On the basis of the archival seismic records it has been proved that in 1971 a tremor of magnitude equal to 3.6 took place and whose epicentre may be located between Łódź and Częstochowa, therefore somewhere about the studied area. It is not entirely credible to associate the tremor source with the Bełchatów trench structure, however, it is thought to be possible.

Dams inducing tremors have been commonly described in the literature. Due to open-pit mining works the seismic activity has been observed in the South Africa /Mc Garr 1979/. It is also belived that the tremors that took place in June, 1974 in the New York State /USA/ were caused by unloading within the dolomites quarries area, /Pomeroy and others 1976/.

The strain equal to the overburden weight /lithostatic pressure/ excavated in the open-pit mine at Bełchatów /about 100 m thick in 1980/ amounted to about 2.5 MPa, thus three times smaller than strain drop observed at the strongest tremors, and many times smaller than estimated average strains within the focus area. The overburden excavation is not then cause of strain changes related to the observed seismic activity, however, can bring in unloading discharge of the tectonic strains. It is difficult to estimate values of the strain changes related to hydrological alterations caused by ground water pumping. Seismicity induced by pumping water out of the earth´s crust is usually associated with deep differential compaction, caused by drop of liquid pressure in pores and therefore rise in effective strains /Yerkes and Castle, 1976/. If the effective strain increase is presented accordingly by the relevant hydrostatic pressure changes, then one can conclude that the increase is about twice smaller than the lithostatic pressure. Both of the factors can play the role of seismity inducing factors. We may presume that the most important indirect factor is the local tectonic strain state, on which strains induced by mining activity are imposed. Tectonic investigations of the trench for exploitation purposes are carried as deep as 600 - 800 m. However, up till now the trench origin has not been well recognized. Having drawn some parallels we may state that they resemble other structures of this type observed in the central Europe, Poland, West Germany, East Germany.

In Fig. 2 one may find a geophysical picture of a similar structure in the northern Poland. It should be emphasized that in the middle of the trench the salt diapir has been found and its activity has been scietifically proved until pliocen and perhaps it is still taking place.

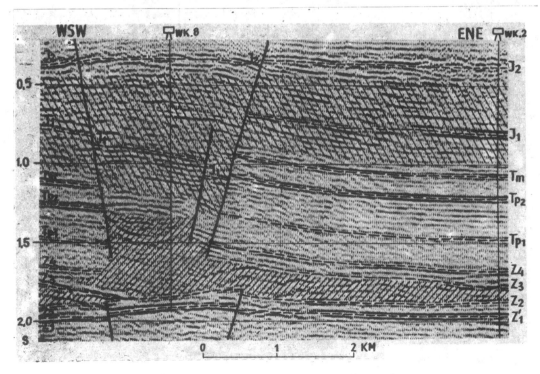

Fig 2. Time seismic section by L. Kniszner at all /1983/.
J_2, J_1 - reflection boundaries in Jurassic; T_k, T_m, T_{p2}, T_{p1} - as above, in Trissic; Z_4, Z_3, Z_2, Z_1 - as above, in Zechstein; Lower Jurassic marked with loosely spaced strokes, and Stassfurt salt- with densly spaced strokes

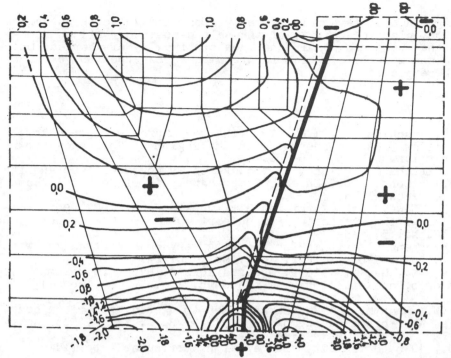

Fig 3. Isolines of horizontal displacements calculated assuming occurrence of dewatering, fault with out salt deposits.

Modelling of displacements induced by the mining activity by the finite element method on the basis of a simplified model of the trench enabled to indicate that, only when the salt shift is taken into account, the considerably great effects of compareable extent with the observed ones are gained. Having failed to consider the salt occurrence in the model one cannot explain a strain followed by displacements. Figures 3,4 present horizontal displacements. Vertical displacements of a few tens of cm have been recorded within the mine area. This sort of displacements is caused by dewatering and they are hardly distinguished from those induced by tremors. A fault appeared in the mine scarp after the tremor of February, 1980. The fault has been measured, where the vertical and horizontal displacements

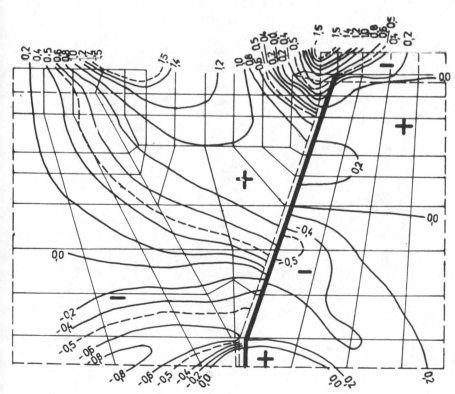

Fig 4. Isolines of horizontal displacements calculated assuming occurrence of dewatering, foult and salt deposits.

are equal to over 1.50 m and about 0.7 m respectively. For practical purposes /e.g. estimation of hazard to the power station objects/ it is proposed to accept the observed isoseists resolution of 29th November 1980 as the basis for evaluation of

the seismic hazard.
The estimated maximum magnitude differs only slightly from this value. High degree of the rock desintegration, plastic activity of the salt deposit as well as numerous tremors prove that the tectonic strain shall be continually discharged. One may conclude that during the mine life, which is presumed to be about 80 years, no serious hazards are to be expected.

REFERENCES

Gibowicz S.J., Głazek J., Wysokiński L. 1981. Zjawiska sejsmiczne w rejonie kopalni węgla brunatnego Bełchatów /Seismic phenomena in the region of Bełchatów brown coal mine/. Przegląd Geologiczny, No 5, 246-250.
Gibowicz S.J., 1981. The Bełchatów, Poland, earthquake of 29 November 1980 and its tectonic and mining implications. Proc. Intern. Symp. "Analysis of Seismicity and Seismic Hazard", Liblice, Czechosłowacja.
Gibowicz S.J., Guterch B., Lewandowska-Marciniak H., Wysokiński L. 1982. Seismicity induced by surface mining: the Bełchatów, Poland earthquake of 29 Nonember 1980. Acta Geophys. Pol., v. 30, No. 3, 193-219.
Kijko A. 1975. Some methods and algorithms for locating very near earthquakes with a digital computer. Publ. Inst. Geophys. Pol. Acad. Sc., No. 84.
Mc Garr A., Spottiswoode S.M., Gay N.C. 1979. Observations relevant to seismic driving, stress, stress drop, and efficiency. J. Geophys. Res., v. 84, No. 11, 2251-2261.
Pomeroy P.W., Simpson D.W., Sbar M.L. 1976. Earthquakes triggered by surface quarrying - Wappingers Falls, New York sequence of June, 1974, Bull. Seism. Soc. Am., 66, 685-700.
Yerkes R.F., Castle R.O., 1976. Seismicity and faulting attributable to fluid extraction, Engin. Geol., 10, 151-167.
Wysokiński L. 1984. Modelowanie przemieszczeń rowu tektonicznego dla określenia związków zjawisk geodynamicznych z odkrywkową eksploatacją górniczą /Modelling of dislacations of trench for defining relations betwen geodynamic phenomena and open-pit mining exploatation/. Technika Poszukiwań Geologicznych 2, 13-25.
Knieszner L., Połkanowa L.P., Gulińska A. 1983. Geneza struktur rowowych w kompleksie mezozoiczno-kenozoicznym niżu polskiego /On the origin of trough structures in the Mesozoic - Cenozoic complex in the Polish Lowlands/. Przegląd Geologiczny 7, 408-414.

High ground stress and mechanical properties of rock mass
Contraintes élevées et propriétés mécaniques d'un massif rocheux

Xi-cheng Xue & Da-nian Wang, *Tianjin University, China*

ABSTRACT: This paper analyzes and discusses high ground stress and problems related to rock mechanics in concerned regions by way of some engineering examples at high ground stress areas. According to the stress ratio of in situ examples, $n = \Sigma \sigma_{wi}/\Sigma \sigma_{ei}$ %, stress regions are divided into three types: high, moderate, and ordinary ground stress areas. In the formula, $\Sigma \sigma_{wi}$ represents the total sum of normal stresses of dead weight while $\Sigma \sigma_{ei}$ stands for the sum of normal stresses of ground. By means of such grouping, geological origins and states of ground stress field are explained. Under the effect of a high circumferential pressure, the founction of the structural planes of rock mass decreases (e.g. percolation, deformation and strength), and its continuity and elesticity increase; the unloading can give diskal rock core fracture and rock burst as well. The thick microfissuring in rock mass may be caused by high ground stress and when the circumferential pressure is freed, the change in the engineering properties of rock mass may be caused. High ground stress also has a historical and geological effect of deciding the engineering properties of rock mass.

SOMMAIRE: Par quelques examples de génie sur territoire de tensions de roche haute dans cet article on a analysé et discuté les tensions de roche haute et ses effects sur la propriété méchanique de masse rocheuse. Selon le taux de tensions mesurées dans l' example réel $n = \Sigma \sigma_{wi}/\Sigma \sigma_{ei}$ % on divise la région de tension en trois classes: la région de tension de roche haute, milieu et générale, où $\Sigma \sigma_{wi}$ est la somme de tension principale de gravité et $\Sigma \sigma_{ei}$ est la somme de tension principale de tension de terre, par moyen de telle classification, les origines et les conditions géologiques de champ de tension de terre sont montrées. Sous l' action de haute pression circonférenciele l' effect de plan structurel de masse rocheuse diminue (e.g. percolation, déformation et résistance), la propriété consécutive et l' élasticite augmente, le phénomene de disque fracture de carrote et d'éclatement de roche produit pendant la roche est déchargé, la microfissure épaise peut exister avec la tonsion haute dans la roche, pendant la tension est réduite, les propriétés de génie de masse rocheuse seront changées. Les propriété de génie de masse rocheuse aussi dépendent la fonction géologique d' histoire de tensions de roche haute.

1 REASONS FOR THE PROPOSAL OF ROCK MECHANICAL PROBLEMS AT HIGH GROUND STRESS REGIONS

During the latest years, many questions have been advanced regarding the matter of rock mechanics related to ground stress in large-scale hydrauelectric projects and in the survey, experiments, and construction of some mining and metallergical projects in China: a great deal of in situ ground stress measurements have been made and big amounts of theoratical analyses and studies on ground stress field have been carried out as a consequence. Along with the gradual deepening in research activities, the conception of high ground stress has been put forward. It is understood that many problems about rock mechanics are related to high ground stress. And these problems can be classified into:

1.1 Rock Burst and Fall

In the exploratory tunnel, along with some sound pieces of rock shot off from the walls of the tunnel, or in pace with the sound, rock fall followed, or when a certain depth of cutting was made in the tunnel for experimental rock specimen cubes, the cubes sprang up suddenly. The plane caused by bursting is of a new fresh rock. Rock burst is closely related to the direction of the tunnel.

* Project Sponsored by the Science Foundation of the Chinese Academy of Sciences.

1.2 Diskal Rock Core Fracture

Diskal rock core fracture found out in the drill holes had fractured planes of a new fresh rock (Photo 1). At the same place, the thickness of diskal rock core fracture is in positive proportion with the diameter of the drill holes. Different locations present various thicknesses of diskal rock core fractures. Accurate survey revealed that diskal rock core fractures take an oval shape and their broken planes are zigzagged.

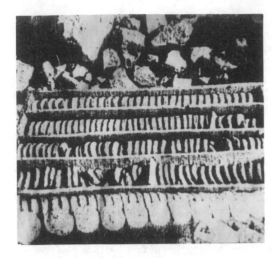

Photo 1.

1.3 Unpermeability

The result ω 0.01 liter/min.m.m., obtained through hydrogeological tests showed that the sections where rock burst and diskal rock core fracture happen have a good relationship with unpermeability.

1.4 Linear Elasticity of Rock Mass and Decrease of the Effect of Structural Plane

Experiments in elastic waves Vp m/min. were conducted toward rock mass and rock specimen cube at the same working point: 27 pieces of rock specimen cubes have an average value Vp of 5163, among these, the smallest value being 4326, the biggest 5832; 13 experimental sections of rock mass have an average value Vp of 6707, the smallest value being 4360, and the biggest 5988. The average value and range value are almost equal to one another.

The Static Elastic Modulus: The average value of the indoor rock specimen dubes stands at 53.85 x10^{9}Pa, while the average value of 9 measuring points in situ obtained by jacking is 53.43x10^{9}Pa.

1.5 Microfissuring

Fresh rock specimen cubes dug out of the tunnel, when exposed in moderate carbonic acid, would weather and break into small pieces within a few months due to the unloading of the circimferential pressure. Microscope observation identified that there is a concentrated developement of joint in rock specimen cubes.

The occurrence of the above stated physical phenomena is related to high ground stress. The discussion of such a problem, therefore might lead to the solution of problems in the rock mass engineering and it is necessary to understand rock mass and too, to conduct calculation and analyses. This paper aims to discuss: 1. What high ground stress is, how to identify it, and its formation causes and features; 2. The relationship between high ground stress and rock mass, i.e. the relationship of the rock engineering, rock mass and the state of stress.

2 ACTUALLY MEASURED PROJECT DATA

15 groups of actually measured data on 6 projects (R.H.L.W.G.LA.etc.) are listed in the table. These projects are all located in China at the regions with high and moderate ground stresses. Table 1 excludes measuring points where projects are close to the dead weight stress field. Table 2 presents a calculation contrast between the ground stress value actually measured and the dead weight stress value. In terms of Table 1 and Table 2, a brief analysis is attempted on the directions and amounts of ground stress.

2.1 Ground Stress Directions

Following general laws, the direction of the actually measured principal stress should coincides with the direction of the acting force of a regional geological structure. It is obvious from Table 1 that because of the differences of the geological terrain in practice, the following are the different factors in determining the direction of principal stress:

1. Deep River Valley Region: as measuring points lie 300 meters embedded below the ground (Table 2), the direction of principal stress is confined by terrain conditions.

Project R: The direction of the acting force of the regional geological structure stands at

Table 1. Ground Stress Actually Measured and Geological Condition

Measuring point	Stress $\times 10^5$ Pa		Angle of orientation	Dip angle	Direction of regional stress field	Geological conditions
R_1	σ_1	96.0	N7W	31.0		Syenite. trend of river valley slope, N 60 W, incline, SW, $\angle 35°$.
	σ_2	45.0	N17W	-51.0		
	σ_3	43.0	N83E	-4.0		
R_2	σ_1	222.0	N30E	20.0		
	σ_2	152.0	N72W	20.0	NNW-SSE	
	σ_3	44.0	S20E	60.0		
R_3	σ_1	264.0	N35E	23.0		
	σ_2	89.0	S39E	33.0		
	σ_3	25.0	N38E	48.0		
R_4	σ_1	600.0	N30E	0.0		Valley bottom, concentrated area of stress.
	σ_3	315.0	N60W	0.0		
H	σ_1	341.4	S87E	-15.0		Mica schist, guartz, gently sloped terrain.
	σ_2	203.3	S72E	75.0	NWW-SEE	
	σ_3	65.8	S13E	-4.0		
L_1	σ_1	170.0	N70W	-25.0		Triassic period, limestone, dolostone. Attitude of rocks, NE30, SE$\downarrow$15-30°. Trend of river valley slope, NW45, SW $\angle 70°$.
	σ_2	118.0	S15E	-50.0		
	σ_3	45.0	N40E	-27.5		
L_2	σ_1	173.0	N70W	-43.0		
	σ_2	130.0	S35E	-45.0	NW-SE	
	σ_3	52.0	N12E	-10.0		
L_3	σ_1	143.0	N50W	-15.0		Triassic period, calcisiltite. Attitude, NW10-15, NE, $\angle$ 60°. Trend of river valley slope NW70, SW, $\angle 60°$.
	σ_2	106.0	S50W	-50.0		
	σ_3	40.0	N45E	-25.0		
W	σ_1	101.9	N80E	60.0		
	σ_2	48.9	N80E	-40.0		
G_1	σ_1	42.0	N20E	0.0		Marble. Smooth terrain.
	σ_3	35.0	S70E	0.0		
G_2	σ_1	320.0	N32E	6.0		
	σ_2	214.0	S43E	67.0	NE-SW	
	σ_3	206.0	N60W	22.0		
G_3	σ_1	168.0	N28W	57.0		
	σ_2	121.0	S35W	16.0		
	σ_3	58.0	S63E	28.0		
LA_1	σ_1	228.7	N10W	-41.0		Granite. Trend of river valley slope, E-W, incline N, $\angle 60°$.
	σ_2	132.9	N69E	11.0		
	σ_3	94.7	N33W	46.0		
LA_2	σ_1	227.0	N22W	-33.0		
	σ_2	186.4	N88E	-27.0	NNE-SSW	
	σ_3	131.3	N28E	45.0		
LA_3	σ_1	205.0	N12E	39.0		
	σ_2	140.3	N82E	-22.0		
	σ_3	57.0	N29W	42.0		

NOTE: Above dip angle, positive; below dip angle, negative.

Table 2. Calculation Chart of Ground Stress Measured Actually and Stress of Gravity at Various Points

Point	Actually surveyed ground stress x 10^5 Pa								Stress of gravity x 10^5 Pa						$n = \dfrac{\Sigma \sigma_{wi}}{\Sigma \sigma_{ei}}$ %
	σ_1	σ_2	σ_3	σ_m	S_1	S_2	S_3	$\Sigma \sigma_{ei}$	deep m	σ_x	σ_y	σ_z	σ_m	$\Sigma \sigma_{wi}$	
R_1	96.0	45.0	43.0	61.3	34.7	-16.3	-18.3	184.0	50.0	1.20	2.22	11.9	5.11	15.3	8.3
R_2	222.0	152.0	44.0	139.3	82.7	12.7	-95.3	418.0	260.0	14.39	14.29	69.7	32.80	98.4	23.5
R_3	264.0	89.0	25.0	126.0	138.0	-37.0	-101.0	378.0	280.0	22.95	18.07	83.4	41.47	124.4	32.9
H	341.4	203.3	65.8	203.5	137.9	-0.2	-137.7	610.5	400.0	32.70	32.70	116.0	60.47	181.4	29.7
L_1	170.0	118.0	45.0	111.0	59.0	7.0	-66.0	333.0	269.0	35.04	29.84	100.0	55.16	165.5	49.7
L_2	173.0	130.0	52.0	118.3	54.7	11.7	-66.3	355.0	308.0	34.52	29.75	100.7	54.99	165.0	46.5
L_3	143.0	106.0	40.0	96.3	46.7	9.7	-56.3	289.0	322.0	34.59	29.71	100.6	54.91	164.7	57.0
W	101.9	48.9	/	50.3	51.6	-1.4	/	150.8	188.0	13.20	13.10	49.4	25.23	75.7	50.2
G_1	42.0	/	35.0	25.7	16.3	/	9.3	77.0	44.0	3.80	3.80	11.9	6.50	19.5	25.3
G_2	320.0	214.0	206.0	246.7	73.3	-32.7	-40.7	740.0	480.0	40.80	40.80	129.6	70.40	211.2	28.5
G_3	168.0	121.0	58.0	115.7	52.3	5.3	-57.7	347.0	120.0	10.20	10.20	32.4	17.60	52.8	15.2
LA_1	228.7	132.9	94.7	152.1	76.6	-19.2	-57.4	456.3	280.0	34.00	29.83	95.7	53.18	159.5	35.0
LA_2	227.0	186.4	131.3	181.6	45.4	4.8	-50.3	544.7	240.0	33.59	25.25	76.1	45.00	135.0	14.8
LA_3	205.0	140.3	57.0	134.1	70.9	6.2	-77.1	402.3	160.0	33.87	23.66	69.0	42.18	126.5	31.5

NNW-SSE, and the river valley slope runs in the direction of N 60°W. The surveyed direction σ_1 of points R2,R3,R4, after projected on a plane is N 30°E, vertical to the flow direction of the river. This is not the same with the direction of the acting force of the regional structure, but it tends to be in accordance with the direction of the slope.

Project LA: The direction of the acting force of the regional structure stands at NNE, the slope of the river valley runs in the direction of E-W, and the direction σ_1 also is proved vertical to that of the slope, and has a tendency of going with the slope in direction.

2. Geologically Flat Region: A general agreement is found between the actually measured direction σ_1 and the direction of the acting force of the regional structural stress field.

Project G: The direction of the acting force of the structure is NE-SW, and the direction σ_{11} actually surveyed is NE 20°-30°.

Project H: The direction of the structural acting force is NWW-SEE while the direction σ_1 measured in fact is S 87° E, agreeable with the acting force direction of the regional structure.

3. Control by the Attitude of Rock Layer

Project L: The analyzed direction of the acting force of the regional structure stands at NW-SE while the attitude of the limestone layer NE 30°, SE, $\angle$ 15°-30°. The in situ surveyed stress is NW 70°, vertical against the strike of the rock layer. The inclined direction of σ_1 agrees with the incline of the rock layer, showing a difference of 20°-30° compared with the direction of the acting force of the regional structural stress.

Project W: The attitude of the powder calcareous sandstone stands at NW 10°-15°, NE, $\angle$ 60°. The direction of σ_1 measured in fact is NE 80°, vertical to the direction of the rock layer. σ_1 agrees with the incline of the rock layer.

Therefore, the direction of principal stress, measured within a depth range of 300 to 400 meters, has a close relationship with the condition of the local geological terrain. Nevertheless, the direction of principal stress is not all decided upon by the direction of the regional stress field.

2.2 Stress Value and Stress Ratio (Table 2)

1. Calculation of the Dead Weight Stress Field

If the ground surface stands even near a half-space, $\sigma_z = \gamma H$, $\sigma_x = \sigma_y = K_o \sigma_z$, K_o stands for the parameter of the lateral static pressure. The above stated formula could not be applied at river valley regions as there is a big difference in terrain undulation. In Table 2, σ_x、σ_y、σ_z are worked out by the FEM, $\Sigma \sigma_{wi} = \sigma_x + \sigma_y + \sigma_z$. The hydrostatic pressure is $\sigma_m = \Sigma \sigma_{wi}/3$.

2. Ground Stress Value Actually Surveyed

$\Sigma \sigma_{ei} = \sigma_1 + \sigma_2 + \sigma_3$, hydrostatic pressure stands at $\sigma_m = \Sigma \sigma_{ei}/3$, deviatoric stress, $S_1 = \sigma_1 - \sigma_m$, $S_2 = \sigma_2 - \sigma_m$, $S_3 = \sigma_3 - \sigma_m$. According to traditional mechanical conceptions, hydrostatic pressure (or sphere tensor) only produces an effect of compaction on rock layer, and shear deformation is related to deviatoric stress Si.

3. Stress Ratio

$n = \Sigma \sigma_{wi}/\Sigma \sigma_{ei}$ explains the proportion the dead weight occupies in the actually surveyed ground stress. When value n is small, the gravitational factor makes little contributions to ground stress, the remaining effects are provided by the acting force of the structure. Value n explains in fact the formation causes of the ground stress field. The value of hydrostatic pressure σ_m, big or small, partially is decided by the amount of the structural stress field, the amounts of deviatoric stresses S_1, S_2, S_3, are tightly connected with the effect of structural stress. This indicates the reason why at high ground stress regions, more opportunities are afforded for the crevices of rock mass to be compacted closely and for the shear fracture and microcracks to take place.

3 CLASSIFICATION OF HIGH GROUND STRESS AND ITS FORMATION CAUSES

The standers in identifying high ground stress (high ground stress regions) need not only to explain the quantity of the value of high ground stress, but also its state (the distribution law of stress), and its formation causes. The following points need explaining: 1. The so-called "high" refers to the fact that, within a certain embedded depth, the value of the ground stress of the concerned point is much more bigger than the stress produced by dead weight, and such understanding is a relative concept; 2. Regions with violent geological structural movements (including mild movements) belong to the high ground stress. In accordance with view points on geological mechanics, the fact that the structural acting force stands at near water level determines the distribution characters of the stress field at high ground stress areas. We advance the following identification formula:

$$n = \frac{\Sigma \sigma_{wi}}{\Sigma \sigma_{ei}} \%$$

In the formula,

$\Sigma \sigma_{wi}$ -- sum of normal (principal) stresses of dead weight,

$\Sigma \sigma_{i}$ -- sum of normal (principal) stresses of ground stresses at corresponding points.

Suggested standerds are as follows according to the actually surveyed data and the ground stress field (omitted from this paper) which we have worked out (see Table 1, Table 2):

n < 50%, high ground stress region,

n= 50-90%, moderate ground stress region,

n > 90%, ordinary (close to dead weight) ground stress region.

The focus of such a classification lies on being in the light of formation causes. Variety of formation causes makes different amounts big or small, in the ground stress value and leads to differences in distribution laws. Causes for the shaping of the ground stress are: dead weight, acting force of structure, underground water, ground temperature. Among these, the dead weight and acting force of structure are of the first importance. While dead weight remaines stable, various areas under the structural acting force are greatly different, causing the varied amounts of value n. This can also make it clear that whether a high ground stress area can shape is determined by the strength of the structural acting force. Under the same condition of embedded depth, the smaller value n stands at, the higher the ground stress value will become.

3.1 High ground stress region

n < 50%, the proportion in the ground stress value occupied by dead weight stands at less than 50%, i.e. more than 50% in the percentage is taken by the structural acting force. In Table 1, in addition to Project L and Project W, all the remaining projects belong to the high ground stress area. Features of these regions are that the horizontal stress plays a leading role. Take Project R and Project LA for examples, when all the stress values at points R_1, R_2, R_3, and LA_1, LA_2, LA_3 are converted one by one onto rectangular coordinate systems, the mean of quantities R_1, R_2, and R_3 is $\sigma_x = 176.1 \times 10^5$ Pa, $\sigma_y = 185.156 \times 10^5$ Pa, $\sigma_z = 87.77 \times 10^5$ Pa, $\sigma_x/\sigma_z = 2.01$, $\sigma_y/\sigma_z = 2.11$; the mean of LA_1, LA_2, LA_3 stands at $\sigma_x = 150.03 \times 10^5$ PA, $\sigma_y = 169.48 \times 10^5$ Pa, $\sigma_z = 58.07 \times 10^5$ Pa, $\sigma_x/\sigma_z = 2.5$, $\sigma_y/\sigma_z = 2.92$. The horizontal stresses σ_x or σ_y are all more than twice as much as the vertical stress σ_z is.

3.2 Moderate ground stress region

n = 50-90%. The proportion of dead weight may reach 50-90%. Project L and Project W fall into the moderate ground stress area. The average mean of points L_1, L_2, L_3 stands at: $\sigma_x = 78.4 \times 10^5$ Pa, $\sigma_y = 139.4 \times 10^5$ Pa, $\sigma_z = 110.37 \times 10^5$ Pa, $\sigma_x/\sigma_z = 0.71$, $\sigma_y/\sigma_z = 1.26$. Its horizontal stress does not occupy a dominant position.

3.3 Ordinary (close to dead weight) ground stress region

n > 90%, and this is the dead weight stress field we are formilar with, in which the dominant position is taken by the vertical stress.

Using the foregoing standerds in dividing ground stress regions may calculate the amounts of stress value, distribution law and formation causes with one universal standerd, as the amounts of value and the law of distribution rely on formation causes. And such a type of classification is instructive in analysing the ground stress field and for the rock mass engineering as well.

4 RELATIONSHIP OF HIGH GROUND STRESS AND ROCK MASS ENGINEERING, ROCK MASS

The foregoing discussion not only leads us to a better understanding of the law of the ground stress field, but also provides us with some scientific information for our analyses of problems appearing in the rock mass engineering, especially in analyzing these problems occuring in underground buildings at high ground stress areas.

4.1 Choice of Axial Line of Tunnel

Large-scale underground houses of hydraulic power stations and tunnels at mining areas are horizontally spreading in most cases, and σ_1 of high ground stress areas stands close to level. It would be most benificial if the axial line of buildings is parallel with the orientation angle of

σ_1. The orientation angle of σ_{II}, projected on a horizontal plane, is related to the orientation of the regional structural stress field, and related, too, to the conditions of the local geological terrain. For example, at river valley areas of magmatic rock (Table 1), measured data showed that σ_{II} is often vertical to rivers and the axial line, following the suit, distributes vertically against rivers. As seen in Project R, the originally designed axial line for the houses underground is NW 60°, parallel to the river valley, vertical to σ_1. And yet, when rock burst was located inside the exploratory tunnel, corrections were made to being NE 30°, parallel to σ_1. As a result, rock burst was avoided.

If there are difficulties in the lay-out of the axial line parallel with σ_{II}, the lay-out of the axial line parallel to σ_3 also has an advantage. The most disadvantage is shown in laying out the axial line parallelly to σ_2, as the shear stress on the tunnel section is at its greatest.

4.2 Choice of Tunnel Section

The length of the section of a tunnel and the short axial ratio should be on such a basis: σ_z (vertical stress) / $\sigma_{x(y)}$ (horizontal stress) = a/b, $\sigma_z = \sigma_x$, a=b=r, a circular section $\sigma_z > \sigma_x$, a > b; $\sigma_z < \sigma_x$, a < b. In this way, the appearance of tension stress areas in the circumferential rock could be avoided. And when the contradiction appears between the demand in application and type choice in stress ratio, the above classification of stress may have the advantage of predicting tensile cracks and shear fracture areas that may possibly turn up in the circumferential rock.

4.3 Diskal Rock Core Fracture

Diskal rock core fracture develops after the circumferential pressure on the rock core is freed in most of the drill holes made in the river bed (stress concentrated regions), and this development happens all in the complete rock that has a good storage capacity. The happenening of diskal rock core fracture provides us with the information about high ground stress. To explain the relationship between ground stress and the formation of diskal rock core fracture, some researchers have applied elastic mechanics, some used fracture mechanics in their analyses of stress under the condition that the strength factors are controlled by the tension and shear strength. The condition under the control of tension strength is $\sigma > Rt/0.22$. In Table 1, measuring point R_4 (river bed), $\sigma_1 = 600 \times 10^5$ Pa, syenite tension strength Rt = 100×10^5 Pa. As $\sigma > 454.5 \times 10^5$ Pa is obtained diskal rock core fracture occurs. There are more than 7 to 8 types of such identification formulae. The most possible conception out of the analytical results for us is: while horizontal stress $\sigma_1 > 450 \times 10^5$ Pa, diskal rock core fracture appears. The research result can be a reference for the study of the ground stress field, and it has a guiding significance in the excavation depth of the basement pit of a project and for the type choice in excavation.

4.4 Relationship Between Hydrostatic Pressure (σ_m), Deviatoric Stress (S_i) and Mechanical Properties of Rock Mass

Comparatively, high ground stress regions are with a high stress value, and as a result, hydrostatic pressure (sphere tensor) become big in amount; since the shape of high ground stress depends on the structural acting force, the distribution of the principal stress, therefore, has an obvious directivity, leading to a big deviatoric stress (Si). This makes up the causes of changes in physical mechanical properties of rock mass at high ground stress locations.

The amounts of permeability value of rock mass and the amounts of deformation value (the amounts of deformation modulus) and the transmitting speed of elastic waves are all conditioned by the continuity and completeness of rock mass. Most of rock specimen cubes can be taken as an elastic continueous mass. Due to the cracks within rock mass and because of the existence of some structural planes, such as joint plane, the elastic modulus value of rock and wave speed are all far more than the relative value of rock mass. Nevertheless, the experiments at high ground stress regions in rock specimen cubes and rock mass showed that the two values are much close to each other in amounts. The rock mass under experiment in situ was unpermeable, ω 0.01 liter/min.m.m.

The compaction degree of rock mass is conditioned by the amount of hydrostatic pressure. In terms of data shown in Table 2, a distribution chart has been drawn on the hydrostatic pressure of ground stress (see Figure 1). In the chart, Line OA shows the line for the hydrostatic pressure under the effect of the gravitational field, while the hydrostatic pressure of Line OB is double that of Line OA. The hydrostatic pressure of OD is 4 times against that of OA. At high ground stress areas, the parameter of hydrostatic pressure is several times larger than that of the gravitational field, thus making cracks in rock mass compacted and closed, increasing the continuity and linear elasticity of rock mass, and causing the structural plane effect to decrease remarkably. This is a change of rock mass caused by the high circumferential pressure. So more attention should be paid to the relations between rock mass and the stress state which acts on it when analyzing the physical mechanical properties and hydrogeological properties of rock mass

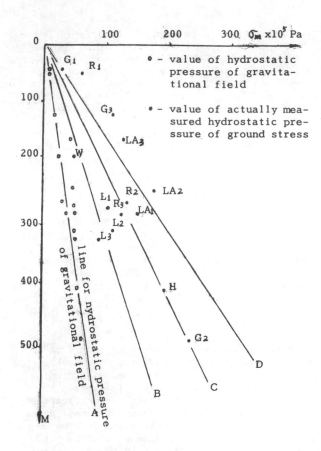

Figure 1. Distribution Chart of Hydristatic
Pressure at Various Measuring Points

The formation of high ground stress determines that high ground stress has an obvious directivity, and the directivity increases the deviatoric stress (see Table 2), causing the shear deformation of rock mass. In the granite of Project LA and the carbonatite of Project L, a developement of microcracks can be detected under microscope. After the circumferential pressure is unloaded, rock mass is more subject to weathering and cracking, and within a not quite long period of time, rock mass will fall apart when hammered. So, after excavating construction starts, care should be taken to protect the mass.

CONCLUSION

1. This paper carried out the comparative analyses of the direction of actually measured ground stress, the value of actually surveyed ground stress, and the stress value of gravitational field.

2. In accordance with the ratio of the dead weight stress field and the actually surveyed ground stress, stress regions are divided into areas of high ground stress, moderate ground stress, and ordinary ground stress. The classification indicates the formation of the ground stress field. Such a classification can use one standard in deciding the amounts of the value of stress field, its distribution laws, and its formation causes.

3. A close relation lies between the features of rock mass and the state of high ground stress. Hydrostatic pressure σ_m keeps rock in a highly compacted state, making its continuity go up, decreasing the effect of structural plane; while deviatoric stress (S_i) contributes to the existence of microcracks in rock mass.

4. Research activities in high ground stress areas present scientific information for the line choosing in underground projects, for the type selecting, and for analyzing the stability of the circumferential rock.

REFERENCES

R.Q. JIANG, X.Y.MU, 1981, Basis for Plastic Mechanics, Mechinery Industry Publishing Press, Chapter II.
S.W.BAI, G.L.LI, 1982, Study on Rock Mass Stress Field of Dam Area of Ertan Hydraulic Power Station, Chinese Journal of Rock Mechanics and Engineering, Vol.1, No.1, p. 45-55.
H.Z.GUO, and Others, 1983, Analytical Method of Original Ground Stress Field of Rock Mass, Chinese Journal of Geotechnical Engineering, Vol.5, No.3, p. 64-75.

1.2 Geotechnical classifications: Practical applications
Classifications géomécaniques: Applications pratiques

Geomechanical classifications for the lay in foundations of large dams Application to the river Pas dam, Cantabria, Spain

Classifications géomécaniques utilisées pour les fondations de grands barrages Application au barrage de la rivière Pas, Espagne

A.Foyo Marcos & C.Tomillo, *Santander University, Spain*

ABSTRACT: The geomechanical classifications in operatio nowdays are still utilitarian and precise, and their application to lithologicaly in no-homogeneous rock masses has not been solved. On the other hand, the subjetivity of great number of - characterization parameters used, in some way reduces defined value all of them. In the present work, the application of characterization and classification methodology of rock masses developed in the last year by the Department of Applied Geology to the study of the foundation lay in for the future Pas Dam in Cantabria, has been able to analyse the real value of each typical geomechanical parameters, in their application to the characterization lithologicaly heterogeneous rock masses. The results obtained, reveal the existing dependence between the true geological characteristics and the geomechanical performance in the same way as the influence of the properties of the whole rock on a rock mass unit, in this case lithologically and structurally heterogeneous.

RESUME: Les classifications géomecaniques en vigueur actuellement sont encore utilitaires et ponctuelles, et leur application pour des massifs rocheux non-homogènes litologiquement n'est pas resolue. D'autre part, la subjectivité d'un grand nombre des paramètres de caractérisation utilisés en quelque façon réduit valeur définitoire à l'ensemble d'eux. Dans ce travail, l'application de la méthodologie de caractérisation et classification de massifs rocheux, développée par le Départe-- ment de Géologie Appliquée pendant la dernière année, à l'étude de fondement du - future barrage du Pas à Cantabria, a servi á analyser la valeur réelle de chacun des paramètres géomecaniques classiques, dans leur application à la caractérisa-- tion des massifs rocheux litologiquement hétérogènes. Les résultats obténus expo- sent la dépendence qui existe entre les caractéristiques géologiques proprement di- tes et le comportement géomécanique, de même que l'influence des propriétés du ro- cher intact dans l'ensemble d'un massif rocheux, dans ce cas-lá, litologique et - structurellement hétérogène.

1 MODIFYED ROCK MASS GRADE CLASSIFICATION

Following the view of Kikuchi et alt. (1982), wiews is proposed a classification system for heterogeneous rock masses and their use as foundation of Large Dams, - which satisfays two of the objectives that the mentioned authors propose for a -- classification of this type.
 1. Objectivity and simplicity
 2. Utilization of desing parameters
 There is no doubt that the thrid objective, is the application to all of the he- terogeneous rock masses, although it is very possible, must be proved by expe-- rience.
 The parameters used in the elaboration of the R.M.G. classification, a variation of Kikuchi et alt. (1982), include in themselves the characteristics, both of the intact rock as well as rock masses, that in the clasic classification they intro- duce a big subjectivity charge in the global appreciation of the quality of the rock mass, in their possible application to foundations of Large Dams.

Table 1. Modifyed classification R.M.G. for heterogeneous rock masses. Tomillo et alt. (1985)

Grade	σ	E_t	V_l	V_f	E_d	F_c
A	2000	1,50	5000	5000	400	1,50
B	1500	1,00	3500	3500	200	1,00
C	1000	0,50	2000	2000	100	0,50
D	500	0,25	1500	1500	50	0,25
E	250	0,15	800	800	10	0,15
F	250	0,15	500	500	---	0,15

Where:
σ : Uniaxial compressive strength in Kgr/cm^2.
E_t: Tangent modulus for the 50% of the fracture charge in $Kgr/cm^2.10^5$.
V_l: Transmission velocity of longitudinal seismic wawes measure in the laboratory.
V_f: Transmission velocity of the longitudinal seismic wawes in m/s. Deduced of the field measures.
E_d: Dynamic modulus obtained of the field measures and assimilated to the analysed prooves in laboratory in Tn/cm^2.
F_c: Parameter of the hydraulic fracturing.
Likewise the grade,replays to the opinions of Kikuchi et alt.(1982)

Table 2. Utilization criterious of the rock masses as foundation of Large Dams.

Grade	Arch Dams	Gravity Dams	Embankment Dams
A	Very good	Very good	Very good
B	Very good	Very good	Very good
C	Good to regular	Good to regular	Good to regular
D	Bad.In the hard or semi-hard rocks - cam be considered included in grade C.Soft rocks are not adecuated as dam foundation	Bad.In the hard or semi-hard rocks - can be considered	Good to regular funtion of the behaviour is resistance.
E	Very bad	Bad.It must be maded treatment of the rock mass for being used as foundation dams.	Generaly,the massif of this grade are not suitable as dam foundation
F	Very bad	Very bad	bad

So:
1. σ and E_t are parameters of the intact rock that include the suitable characteristics of each litology.
2. V_f and E_d are parameters of the rock mass influenced clearly by the discontinuities,joints and stratification. Lithological and structural heterogeneity of the rock mass.
3. V_l is the parameter of the intact rock, controled by the suitable heterogeneity and which allows it contrast with V_f, similar field parameter.
4. F_c is the hydraulic fracturing parameter which controls the traction strength of the rock mass from the Lugeon permeability-test.

In conclusion,the parameters used are of obligated obtention for their use in the desing and they control the resistance characteristics,of deformability and permeability of the rock masses.

1.1. Characterization of some Spanish Dam foundations by means of Modifyed R.M.G. Classification

The result of the utilization of the modificated R.M.G. Classification in the characterization as foundation of the dam, of the rock mass of the River Pas Dam, (Cantabria,Spain) is significant.

Table 3. Modifyed   R.M.G. Classification. River Pas Dam. Spain

Grade	σ	E_t	V_l	V_f	E_d	F_c
A		s				
B	s	+ c			s	
C	+ c		+ c s	+ o c s	+ c	+ o c s
D	o	o		o		
E					o	
F						

+ Average value of the parameters along the foundation
o Shales
c Calcareous sandstones
s Siliceous sandstones

Taking into account the average value of the parameters along the foundation,the rock mass makes up of the Pas Dam, is grade C. The inclusion of some of the values of the parameters of the intact rock in others differents grades,allows in the same way, to anticipate possibles incidences and to use the classification with the necessary precautions,the juntion of the dispersión that is apreciated in the definitive parameters of each lithology.

Using the dates given by the corresponding geological information,the modificated R.M.G. classification has been applyed to the characterization of the foundation of four Spanish Dams,and the result of which is expounded as follow.

Huarte Dam

The Huarte site of foundation dam, is placed near Pamplona and is sited in the sheet 115,Gunila, of the geological map of Spain.

Geologicaly, the region is into the south-pyrenean zone inside the chain of the Alpine range of the Iberian Peninsula. The site is made up by an alternation of marls and calcarenites which belong to the formation of "blue marls of Pamplona" of tertiary age, exactly to the Eoceno.

Taking into account the average value of the parameters along the foundation we could considerer Huarte Dam as grade C, good to regular for any type of dam.

Nevertheless, the classification reveals how the characteristical parameters of intact rock give indexes of less quality than the define ones of the rocks mass.

Consecuently, as it is above mentioned in the Geological Public Work Service inform " the dam foundation will try to establish itself on the most adequate calcarenite packets or else in the alternation of calcarenites and marls" in the case of a concrete dam.

If it concerned a enbankment dam, the classification in grade C of the rock mass, it is conditioned because of the low resist nce capacity of the marls and calcarenites.

The Geological Public Work Service inform doesn't shave this conservative criterion to which considers a medium resistance of 300 Kg/cm^2 enough as an index of the carrier capacity of the blue marls.

Table 4. Modifyed R.M.G. Classification. Huarte Dam. Spain

Grade	σ	E_t	V_l	V_f	E_d	F_c
A						
B						
C				+ m c	+ c	+ c
D	+ c.	c				
E	m	+				
F		m				

+ Average value of the parameters along the foundation
m Marls
c Calcarenites

In summary, the modifyed R.M.G. classification, applied to the foundation analyse of the Huarte dam manifestes the relevance of the intact rock parameters in the behaviour of the whole.

Huesna Dam

The area of realization of the works for Huesna reservoir is situated near Villanueva del Rio y Minas, it is placed in the Geological Map of Spain in the sheet number 941, called Ventas Quemadas.
  Geologicaly the emplacement of Huesna reservoir is sited in the Osa-Morena zone of the Iberian Hercynian Massif, in its South part, close to the border with Guadalquivir depression and the supposed site of foundation is set up by a unit of siliceous shales with alternance of quarzites, the whole of it of Cambrian Age.
  The great heterogeneity of the rock mass which sets up the Huesna dam foundation, is no manifested in the observed dispersion in the application of the modificated R.M.G. classification for the characterization of the massif.
  However, a detailed analysis of the results, allows to obtain some conclusion

Tabla 5. Modifyed R.M.G. Classification. Huesna Dam. Spain

Grade	σ	E_t	V_l	V_f	E_d	F_c
A			$q\ p_s$. +		p_s	
B				p_s .	. +	q .
C	q		p	q +	q	p_s +
D	p_s + .	q	l	p	p	p
E		. +				
F	p	p p_s				

+ Average value of the parameters along the foundation
p Weathered shale
q Quarzites
p_s Shale
. Average value of the parameters without "p"

On one hand, if we bear in mind the average values of the massif parameters, the resultant great dispersion obeys fundamentaly to the undoubled influence that the superior levels of weathered shale has in the obtencion of the average.
  If we eliminate this superficial zone indicated by the S.G.O.P. at 10 meters minimum depth and 20 meters maxin depth, we find as the same as in Huarte dam, the whole of the rock mass indexs allow really an optimisian classification, grade B,

as contrasted with the intact rock parameter which do not allow to exceed grade D, this anomaly that the S.G.O.P. inform manifests and attributes to the "existence of main planes of fracture join to the microfissure or to the schistosity".There is no doubt that the complete elimination of the weathered zone is too excessive and probably, such as is noted in grades D and E characteristics of the R.M.G. - classification, the treatment of this zone of the rock massif by consolidation inyections would increase the geomechanical superficial characteristics and probably a new application of the modifyed R.M.G. classification will give more coherent results.

Mondin Dam

From the geological point of view the Mondin reservoir is placed on the top of Pre Cambrian materials of the Recumbent-Fold Core of Mondoñedo at the border between the occidental Astur-Leonesa zone and the Central Iberian zone of the Hercynian in the Iberian Peninsula.
    The Pre-Cambriam age materials are formed by biotitics schistes of the schist-grauwacke serie of Villalba.
    Considering the Mondin foundation dam as an homogeneous rock mass, the results obtained in the characterization allow to determine the grade C to the whole foundation at a first approach.

Table 6. Modifyed R.M.G. Classification. Mondin Dam. Spain

Grade	E_t	V_l	V_f	E_d	F_c
A					
B				+	
C		+	+		+
D	+	+			
E					
F					

+ Average value of the parameters along the foundation

Again,the resistant parameter of the intact rock, tends to reduce the first valuation of the whole. In this case, and following the determined opinions of each one of the R.M.G. classification grades, it is possible to reach the following conclusions:
    1. A part from the topography,the solution of a arch dam results refused by the characteristics of the soft rock, made clear by the value of the intact rock parameters.
    2. For a gravity solution the geomecanical conditions of the foundation could obviously be better by consolidation injections. The following classification to the treatment, would allow to define the viability of this type of dam.
    3. The solution of enbankment dam would be the correct one, bearing in mind that the resistance conduct may be esteemed good enough,even if it is included in grade D.

Endara-San Anton Dam

The studied zone is placed in the municipal district of Lesaca, in Navarra province and it is localized in sheet 65, Vera Bidasoa, of the geological Map of Spain.
    Geologicaly the location area of Endara-San Antón Dam, is situated in the fur--thest occidental Pyrenees, exactly in the massif of Cinco Villas, where the Paleozoic indication allow to consider it as the occidental end of the Axial-Pyrenean zone.
    The Peñas de Haya mountain where the site of foundation is placed, is characterized by the presence of a great granite formation, formed the most part by alkali granites of medium to thick grain joined to which basic rocks of diabasic nature appear.
    The R.M.G. classification gives to the foundation rock mass of Endara-San Antón

Table 7. Modifyed R.M.G. Classification. Endara-San Antón Dam. Spain

Grade	E_t	V_l	V_f	E_d	F_c
A					
B					
C	+ g B	+ g B g_a	+ g B		+ g B g_a
D	+ g B				
E	g_a	g_a			
F		g_a			

+ Average value of the parameters along the foundation
g Granite
g_a Weathered granite
B Diabasic

Dam, a grade C, good to regular for any type of dam, a part of the geomorphological considerations.
   As is observed in the classification, the influence on the weathered zones in the medium value of the classification parameters is negelible, consecuently with tne scase depth that it reaches, for what it is possible to built up a dam of an about 50 meters in hight in the site, either of concrete prepared or enbankment dam.

2. CONCLUSION

In summary, the utilization of this type of formulas for the caracterization of heterogeneous rock masses, has as it main premise the detailed knowledge of the rock masses that aims to characterize, based above all on the perfect evidence of the drilling and on conventional field and laboratory tests that in this way show their great resolution capacity

REFERENCES

Bieniawski, Z.T. 1979. The geomechanics classification in rock engineering applications.4 Int. Congrss. I.S.R.M. Montreaux. 2:41-48.
Cameron-Clarke, T.S. & Budavari, S. 1981. Correlation of rock mass classification parameters obtained from borecore and in-situ observations. Engineering Geology 17:19-53
Foyo Marcos, A. 1983. Analyse des caracteristiques geomecaniques de massifs rocheux au moyen d'essais hydrauliques de type Lugeon. Bull.I.A.E.G. 26-27:411-414
Kikuchi, K. & Saito, K. & Kusunoki, K.I. 1982. Geotechnically integrated evaluation on the stability of dam foundation rocks. 14 ICOLD. Río de Janeiro. Q.53, R.4

Engineering geological features of joint sets according to their genesis in slaty cleaved bedding slip folds

Les caractères géotechniques des groupes de discontinuités d'origine différente dans les plis schisteux du type 'bedding slip fold'

Karl-Heinz Hesse & Joachim Tiedemann, *Fachgebiet Ingenieurgeologie, Technical University, Berlin, FR Germany*

ABSTRACT

For the design of large engineering structures in bed-rock under variable geological conditions and for estimating the hydraulic situation clear divisions of rock areas are necessary. Within these regions, so called homogeneous areas, respective relevant characteristics should be equally distributed.

Geological and engineering-geological studies of slaty cleaved bedding slip folds of the palaeozoic Rhenish Massive have shown, that criteria for a practical definition of homogeneous areas can be found even for complicated rock structures.

The geological studies revealed that folds are subdivided according to their geometry and lithological structure almost exclusively by certain joint systems.

The next step taken was to examine the joint systems further as regards their engineering geologically relevant structural geological parameters as:

- spacing
- surface properties
- aperture
- filling
- degree of separation

As the result a descriptive engineering-geological discontinuity characterization was obtained. It demonstrated clearly that the development of structural geological features are determined by discontinuity-genesis and lithology. However, a point that should be taken into consideration is that primary fracture is not the last stage of the genesis of the discontinuities. The process of folding continues progressively and the discontinuities can thus be altered secondarily. The genetic discontinuity groups must therefore be subdivided further in accordance with their time of origin during folding.

Discontinuities with corresponding genesis, the same time of origin during folding and comparable lithological conditions are similar in structural-geological and engineering-geological features that determine their mechanical and hydraulic behaviour.

The third stage of investigation is to ascertain whether the regularities between the joint systems that had been formed and the structural-geological features could also apply to their friction properties. At present extensive in situ and laboratory experiments are beeing carried out on this point.

RESUME

Pour la planification de constructions dans les massifs rocheux, ainsi que pour l'estimation du comportement hydraulique des masses rocheuses, aux conditions géologiquement hétérogènes, il est nécessaire de délimiter des zones homogènes, dans lesquelles les caractères géotechniques d'importance restent uniformes.

Les recherches géologiques, réalisées dans les plis schisteux du type "bedding slip fold" du Massif Rhénan en Allemagne occidentale, ont démontrées que les plis sont décomposés par des groupes de discontinuités bien définis et d'orientation systématique, en fonction de la géometrie et du caractère lithologique des plis.

Dans la deuxième phase des travaux, on a analysé les paramètres structuraux d'importance pour la Géologie de l'Ingenieur:

- espacement interfractural
- nature des surfaces
- ouverture
- remplissage
- degré de séparation

Le resultat de cette étude était d'obtenir une caractérisation descriptive des discontinuités pour les besoins de la Géologie de l'Ingenieur.

Il s'ensuit clairement que les diaclases, qui coincident par rapport au genèse, au moment de formation pendant le plissement, et aux conditions lithologiques, sont en accord par rapport aux caractères, qui déterminent le comportement mécanique et hydraulique. Cependant, il faut noter que l'orientation ainsi que les caractères structuraux des fractures peuvent être modifiées secondairement, grâce à la continuation progressive du plissement.

La troisième phase de travail consiste à examiner, si les relations naturelles entre les groupes de discontinuités génétiques et leurs caractères structuraux existent aussi par rapport au caractère de friction. De nombreux essais en laboratoire et sur le terrain sont en train d'être effectués pour étudier les propriétés de friction des discontinuites.

1. Introduction

The characterization of the rock mass for engineering geological purposes in practice proceedes with respect to homogeneous zones. The evaluation of homogeneous zones is necessary for the solution of all engineering geological problems concerning heterogeneous rock mass conditions. The differentiation depends on the type of rock structure as well as on the relevant geotechnical conditions. Quasi-homogeneous zones are rock mass sections that show uniform geological characteristics with respect to the designated structure..

The relevance of rock mass characteristics depends on

- structural factors
- lithological factors
- exogeneous factors
- the deversity of technical requirements.

Fig. 1 shows the application of separated homogeneous zones to the underground powerhouse Waldeck II in W.-Germany. For the differentiation and definition of the homogeneous areas I - XIII

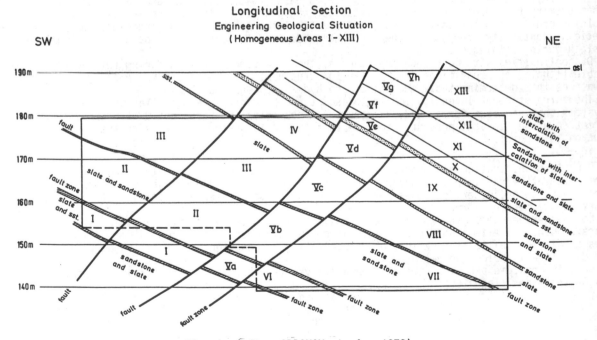

Fig. 1 (after ABRAHAM et al., 1973)

a direct impression of the lithological situation and the network of discontinuities was apre-
requisite. In engineering geology, this procedure is very often applied, but it fails if there
are large engineering structures to be designed in badly outcropped geological structures with
frequent changes. Under such conditions the best results of investigations can be expected,
if principle regularities between the general geological structure and the engineering geologically relevant features of the rock mass are known and can be taken into account.

Thus, the following report is a contribution towards the definition of engineering geological homogeneous zones within the folded Rhenish Massive in W.-Germany.

The most important question to examine was if similarity between lithology and tectonic
deformation of geological structures implied similar engineering geological features with
respect to distribution and characteristics of the discontinuities.

2. Investigated geological structures

To date, studies have been restricted to asymmetric, slaty cleaved bedding slip folds,
occuring in a monotonous alterating sequence of Lower Devonian silty-sandy shales and fine
textured graywackes within the Ahr-valley-anticline. The engineering geological survey was
concentrated on well outcropped anticlines, all belonging to the type shown in Fig. 2.

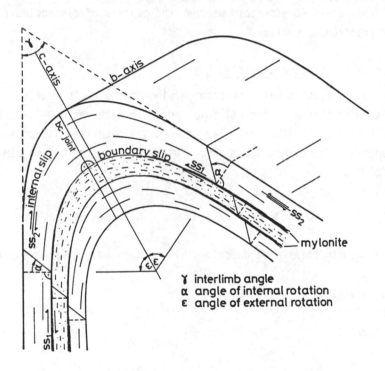

Fig. 2   Sketch of a typical anticline in the Ahr-valley
(from TIEDEMANN, 1983)

The fold formation is a result of SE - NW directed compression. Three different types of deformation, facilitated in particular by the special lithological composition, enabled folding to
occur:
- Boundary slip (ss_4) between different layers such as graywacke and shale.

- Internal slip within fine textured, plastic reacting layers; this means a simple shear on synthetic planes that does not generate any macroscopic fabric (BREDDIN, 1968). With progressing steepening of the fold limbs, the magnitudes of the normal stress components acting across the bedding planes increase until the rate of shear strain becomes zero.
- Tilt of the cleavage planes, which were originally perpendicular to the bedding planes, associated with internal deformation. This enables the folding to continue (WEBER, 1976).

The internal slip together with the tilt of the cleavage planes causes an internal rotation, which is directed towards the hinge zone of the folds.

The axial planes of all investigated bedding slip folds are dipping towards NW. The axial lines are sub-horizontal or gently dipping (5° - 15°) to SW or NE.

3. Investigations and results

The investigation program was devided into two parts:
1) Surface mapping of the outcropped anticlines on three different scales and a detailed description of discontinuity systems in selected areas.
   A quantitative or semi-quantitative interpretation of the recorded data for a classification of the structural features of the discontinuities.
2) Laboratory and field experiments to ascertain whether the genesis of discontinuities exercises an influence on their friction properties.

3.1 Structural characteristics of discontinuities

The first stage of investigation was a surface mapping of the geological structures on a scale of 1 : 10.000. Basing on this map, typical folds and fold sections were selected for a second mapping on a scale of 1 : 500. The next step taken was a detailed discontinuity survey of planes with trace lengths of more than 0,5 m and a detailled lithological mapping in particular sections of the folds as
- NW limbs of anticlines
- SE limbs of anticlines
- hinge zones.

In total, about 10.000 discontinuities were described with regard to the following features:
- Lithology
- Orientation (measured directly)
· Aperture (measured directly)
- Filling (mylonite or quartz)
- Spacing (measured directly)
- Area (by estimation)
- Surface appearance on three scales (qualitatively):
  - m² - scale (deviation from a true plane)
  - dm² - scale (macroroughness)
  - cm² - scale (microroughness)

Finitially, the interpretation of the field data concentrated on analysing the genesis of discontinuities, in order to understand the mechanism of joint formation and to evaluate the stress history. For this purpose, all investigated folds including the recorded discontinuities were once more transformed to a horizontal position by using spherical projection methods. In this manner, the internal rotation and the external rotation were reversed and nine different orientated sets of discontinuities could be distinguished. Seven of these sets must have been present since the initial stage of folding and have been involved in the continuing fold formation. Two additional groups of discontinuities appeared during a later stage of folding.

The actual position and orientation of discontinuities is a result of external and internal rotations, which change the primary positions and orientations of discontinuities in different manner (Fig. 3).

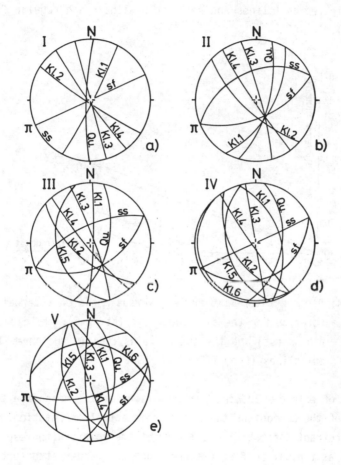

Fig. 3 a - e: Great circles of the nine discontinuity sets in the investigated NW-limbs of anticlines in the Ahr-valley after transforming the b-axis into a horizontal position; hinge zone (I), NW-limbs (II-V) (from TIEDEMANN, 1983)

External rotation

External rotation is a simple rotation of layers parallel to the orientation of tectonic compression; in this case the rotation axis is the b-axis of the folds. If the complete fold is rotated the a-axis is mostly the rotating axis.

Internal rotation

Internal flow and tilt of the cleavage planes increase with progressive fold formation. Both kinds of deformation cause internal rotation which acts mainly in plastic reacting material and distorts the complete fabric in the direction of the hinge zone. Thence, it grows up progressively up to the transition to the following syncline or anticline.

Accordingly, the angle of internal rotation α is zero in the hinge area and, starting from here, it increases until the inflexion point of the layers. The inter-limb angle γ develops in an inversely proportional manner. It is 180° in the hinge zone and comes to a minimum at the point of inflexion (Fig. 2).

The relation between the inter-limb angle γ and the angle of internal rotation α is derived from the investigated anticlines and seems to be linear and regular for all sets of discontinuities (Fig. 4).

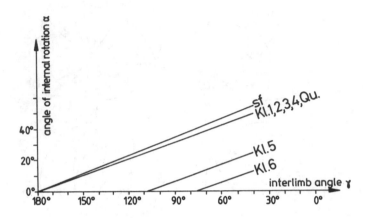

Fig. 4: Plot of inter-limb angle γ versus angle of internal rotation for the six joint sets (Kl. 1-6), the cleavage planes (sf) and the quartz veins (Qu) in the Ahr-valley anticlines (from TIEDEMANN, 1983)

Due to the effect of external and internal rotation, it is impossible to draw conclusions concerning the genesis of the discontinuities from their actual orientation. Both components of deformation have to be retracted first. The symmetry of the joints with respect to the abc-coordinate system as well as typical surface features such as feather structures, step faces and fringe faces are suited to determine the geneses.

The nine discontinuity sets were distributed within the genetic system as follows:

Open and closed bedding planes } Closed cleavage planes	- shear fractures
Open cleavage planes	- combined shear fractures and tensile fractures
Joint sets Kl. 1-6	- displacement fractures
Quartz veins	- tensile fractures by torsion

(Fissures with evident aperture are signed as open discontinuities. So called joints do not show any tracks of shear displacement and are usually closed).

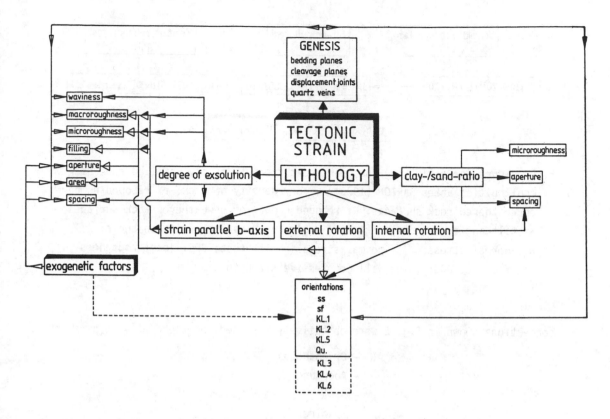

Fig. 5: Scheme of the connections between geological factors and the observed
discontinuity features (from TIEDEMANN, 1983)

After the analysis of the displacement history and the genesis of all discontinuity sets,
the recorded joint characteristics were interpreted statistically. Fig. 5 shows schematically
the joint features discovered and the geological factors that have determined the structure of
the folds.

The influence of the lithology, especially the silt/sand-ratio of the rocks and the
degree of exsolution was investigated. Furthermore, the influence of fold formation on the
engineering geological relevant characteristics was examined. For this purpose the whole process of folding was separated according to the chronological sequence into four kinds of deformation:

- Discontinuity formation (primary genesis)
- External rotation of the discontinuities
- Internal rotation of the discontinuities
- Strain parallel the b-axis

Usually, lithological composition and the rock deformation are combined inseparably
(Fig. 5). But for very large discontinuities these lithological factors become insignificant.
Each lithological and tectonic factor shown in Fig. 5 determines the characteristic of the
discontinuities very specifically. Interactions are possible.

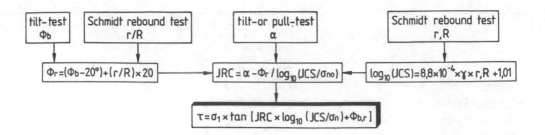

Fig. 6: Procedure for estimating the peak shear strength of dry unfilled rock discontinuities after BARTON (1973). γ: dry density of rock; r: rebound number
on weathered rock surfaces; R: rebound number on unweathered rock surfaces;
α: tilt angle; Φ_b: basic friction angle; Φ_r: residual friction angle;
σ_n: normal stress; σ_{no}: normal stress in tilt test; JRC: joint roughness
coefficient; JCS: joint wall compressive strength

The connections shown in Fig. 5 were objectively assigned for the following parameters:

- Orientation
- Aperture
- Area
- Spacing
- Filling.

The description of the joint roughness and joint undulation could only be carried out
qualitatively. Although it is impossible to derive shear parameters from this there is an important advantage having recorded all the planes in this manner; because of its application to
statistical methods.

The most important result of the whole discontinuity analysis is witnessed by the fact
that all engineering geological relevant discontinuity characteristics are determined by lithological composition and by the process of folding. Therefore, bedding slip folds can be subdivided into homogeneous areas according to lithology and the size of the inter-limb angle γ
(TIEDEMANN, 1983).

3.2 Friction properties

The interpretation of the structural geological joint features indicates a qualitative
connection between genesis, lithology and friction properties of the joints. In order to quantify
these results, appropriate geotechnical studies have been performed.

Concerning the great number of joints to be tested a method based on an empirical law of
friction was employed. It was developed by BARTON (1973) and revised by BARTON & CHOUBY (1977)
and BANDIS et al. (1981). According to BARTON, the friction law can predict the peak shear
strength τ_ρ for dry rock joints without gouge. Simple index tests are sufficient in order to
determine the joint roughness coefficient (JRC), the joint wall compressive strength (JCS) and
the basic friction angle (Φ_b). Some of the index values can be determined in different tests.
Fig. 6 shows the complete procedure schematically.

The discontinuities were tested as follows: The joint wall compressive strength was measured by Schmidt hammer tests. According to BARTON this is a suitable method for determining the material properties, especially of weathered rock planes. Since the thickness of weathered zones on joints may be a fraction of a millimeter up to a few millimeters, other mechanical experiments fail.

Tilt tests and pull tests were carried out in order to determine the joint roughness coefficients. The samples used were cement replicas of natural discontinuity surfaces. The basic friction angles Φ_b were estimated from tilt tests on dry unweathered sawn surfaces.

Parallel to the investigations according to the BARTON concept, other replicas (300 cm²) of the same discontinuities were determined by direct shear box tests. The results for open bedding planes (ss), quartz veins (Q) and displacement joints (Kl. 1-5) are presented in Table 1.

A comparison of the values shows that the friction angles determined by the index tests after BARTON are lower than the friction angles determined in direct shear box tests.

Table 1: Friction angles of the discontinuity sets ss, Q and Kl. 1-5 determined in tilt tests or pull tests after BARTON in comparison with friction angles determined in direct shear-box tests (after STRIEGEL, 1984)

arc tan (τ/σ) = Φ_r + JRC $\log_{10}$ (JCS/σ)						
location	M	D	M	D	M	D
	tilt-test		pull-test		shear-box	
σ_n	0,3 MN/m²					
ss	arc tan τ/σ	43,1°	42,4°	43,1°	40,1°	50,6°	55,4°
175/65	Δ	+7,5	+13,0	+7,5	+15,3		
Kl 5	arc tan τ/σ	46,6°	41,1°	49,2°	42,5°	57°	55,3°
350/35	Δ	10,4	+14,2	+7,8	+12,8		
Q	arc tan τ/σ	45,6°	45,6°	47,4°	47,8°	62,8°	64,2°
110/50	Δ	+17,2	+18,6	+15,4	+16,4		
Kl 1-4	arc tan τ/σ	42,5°	39,0°	43,7°	40,0°	42,6°	40,3°
70/65-260/60	Δ	+0,1	+1,3	+1,1	+0,3		

Δ = [arc tan$(\tau/\sigma)_{measured}$ - arc tan$(\tau/\sigma)_{predicted}$]

The reason for the deviation of the values is that the joint roughness coefficients are not constant for all levels of normal stress. Contrary to BARTON's opinion, the values of JRC increase with rising normal stress σ_n. An error in the BARTON concept is obvious at this point. Tilt and pull tests, performed under extremly low normal stress are very strongly influenced by microroughnesses (HEITFELD & HESSE, 1985).

Apart from this systematic fault, the Schmidt hammer test proved unsuitable for the determination of the joint wall compressive strength, as the results could not be reproduced. Therefore, the rebound tests were substituted for uniaxial compression tests (STRIEGEL, 1984).

The failure of the Schmidt hammer test is extremely inconvenient as there is thus no suitable experiment for determining the strength of weathered zones. Such a test seems even more necessary since microscopic examinations of joint sections have shown that a high quota of discontinuity walls is weathered.

The reliability of the replicas should be questioned as their upper and lower parts are always indented completely by the process of production. On the other hand the statistic evaluation of the recorded discontinuity features has shown that the walls of many fissures and open bedding planes are not indented completely due to secondary tectonic shear displacement or exogenetic factors.

4. Summary and further studies planned

The results of these investigations may be assesed in several different ways.

In particular, the investigations, dealing with structural geology showe some important rules concerning the fabric of the circular flexural slip folds in the Ahr-valley. These rules usefully describe and define homogeneous areas with similar lithology and similar joint features. The most important parameter is the inter-limb angle, γ.

A classification of the joint systems for engineering geological purposes requires the deduction of their genesis, taking into account their full strain history. Joints with corresponding lithology and genesis are similar in most structural-geological features that determine their mechanical and hydraulic behaviour.

The relationship found between the geological structure and the relevant engineering geological features of joint sets is applicable to the solution of geotechnical problems in practice. For example the locations for exploration drilling could be selected more effectively. Furthermore, the results of drilling and other direct investigations can be extended to wider areas of such folds.

The next step was to examine whether corresponding genesis and lithology of joints imply friction properties of the same order of magnitude.

This problem could not be solved unequivocally, because it became apparent that the methods according to BARTON were not adequate.

Taking the results to date, further inquiries are orientated in two directions.

1.) The principle of rock characterization for engineering geological purposes, as developed for the Ahr-valley will be applied to bedding slip folds in other regions of the Rhenish massive.

2.) The correlation between the genesis of joints and their friction properties is to be tested in shear box tests of natural specimens with an area of 300 cm². Before shearing the degree of weathering must be known for such tests.

5. REFERENCES

ABRAHAM, K.-H. & PORZIG, R.: The prestressing anchors of the Waldeck II pumped - storage scheme.- Baumaschine und Bautechnik, 6 + 7, 209-220 and 273-285, 1973.

BARTON, N.R.   : Review of a new shear-strength criterion for rock joints. - Engineering Geology, 7, 287-332, 1973.

BARTON, N.R. & CHOUBY, V. : The shear strength of rock joints in theory and practice. - Rock Mechanics, 10, 1-54, 1977.

BANDIS, S., LUMSDEN, A.C. & BARTON, N.R.: Experimental studies of scale effects on the shear behaviour of rock joints. - Int. J. Rock Mech. Min. Sci & Geomech. Abstr., 18, 1-21, 1981.

BREDDIN, H.   : Quantitative Tektonik 2. Teil, III. Faltung. - Geol. Mitt., 7, H. 4, 333-448, Aachen, 1968.

HEITFELD, K.-H. & HESSE, K.-H.: Untersuchungen über typische Verteilungsmuster von Trennflächen und deren Ausbildung unter ingenieurgeologischen Gesichtspunkten. - In Heitfeld, K.-H., 1985, Ingenieurgeologische Probleme im Grenzbereich zwischen Locker- und Festgesteinen, Springer-Verlag Berlin, Heidelberg, printed in Germany, 695 p.

STRIEGEL, K.-H.   : Untersuchungen zur Scherfestigkeit rauher Trennflächen im Ahrtalsattel. - Mitt. Ing.- u. Hydrogeol., 19, 168 p., Aachen, 1984.

TIEDEMANN, J.   : Geologisch-ingenieurgeologische Untersuchungen zur Abgrenzung von Homogenbereichen innerhalb der Mittleren Siegener Schichten (Ahrtalsattel). - Mitt. Ing.- u. Hydrogeol., 16, 202 p., Aachen, 1983.

WEBER, K.   : Gefügeuntersuchungen an transversalgeschieferten Gesteinen aus dem östlichen Rheinischen Schiefergebirge - Ein Beitrag zur Genese der transversalen Schieferung. - Geol. Jb., Reihe D, 15, Hannover, 1976.

Evaluation of the rock grade classification using RCI & RQD$_N$

Classification des roches à partir de RCI et RQD$_N$

Kokichi Kikuchi, Makoto Fujieda & Hajime Shimizu, *Tokyo Electric Power Services Co. Ltd, Japan*

ABSTRACT: When the bearing capacity of hard rocks is judged in reference to drilled cores, the distribution of fissures in the drilled cores is an important element. This study has examined the correlativity between the RQD known as a method for quantitatively evaluating rocks in reference to the distribution of fissures in drilled cores, and rock grades. As a result, RQD$_{(N)}$ and RCI have been contrived as new evaluation indicators, and their applicability has been examined. As a result, it has been found that as for RQD$_{(N)}$, the use of RQD$_{(5)}$, RQD$_{(10)}$ and RQD$_{(30)}$ in combination allows distinct rock grade classification, that the RCI is very effective as a method for quantitatively evaluating rock grade classification, and that high correlativity exists between physical values of rocks and RCI.

RESUME: C'est pourquoi lorsque la capacité portante des roches dures est jugée d'après les carottes de sondage, la répartition des fissures desdites carottes de sondage, les conclusions tirées à partir de cette analyse revêtent un caractère capital. Cette analyse a été le sujet de l'observation d'une corrélation existant entre l'appellation de la qualité des roches (RQD) en termes de procédé d'évaluation quantitative des roches et en référence à la répartition des fissures dans les carottes de sondage ainsi qu'à la qualité des roches. Il en résulte que l'appellation modifiée de la qualité des roches RQD$_{(N)}$ et le répertoire de classification des roches (RCI) ont été créés pour servir de nouveaux indices d'évaluation tandis que leur possibilité d'application ont également fait l'objet d'une analyse particulière.
Il en résulte que pour l'appellation modifiée de la qualité des roches RQD$_{(N)}$, l'utilisation de RQD$_{(5)}$, RQD$_{(10)}$ et RQD$_{(30)}$ combinés permettent d'effectuer une classification précise des roches et que d'autre part, le répertoire de classification des roches (RCI) se révèle être un procédé particulièrement efficace pour effectuer une classification de la qualité quantitative des roches (RCI) et qu'enfin, il existe une corrélation entre les indices physiques des roches et le répertoire de classification des roches (RCI).

1. INTRODUCTION

To know the bearing capacity of foundation rocks of a civil engineering structure is a prerequisite for the stability of the structure. Thus, before designing and constructing a civil engineering structure, it is required to evaluate the bearing capacity of the foundation rocks located within the range of affecting the stability of the structure.

Physical values showing the bearing capacity of foundation rocks can be obtained by various in-situ tests and laboratory tests, but the values are local. For this reason, in the investigation of geologic features for civil engineering, rocks are evaluated in reference to the rock grades established to express bearing capacities.

In Japan, Kikuchi and Saito's rock classification (Kikuchi and Saito, 1982) is widely used. The rock classification covers massive rocks and is related with the modulus of deformation (D), static modulus of elasticity (Et), shearing strength (τ), etc. obtained by in-situ rock tests. The rock classification is made based on the observation of outcrops, test adits, drilled cores, etc. and greatly depends on the experience of the geologist concerned.

The present paper examines the applicability of the Modified Rock Quality Designation (RQD$_{(N)}$) and the Rock Classification Index (RCI) for rock classification.

2. ROCK CLASSIFICATION

Drilling is generally practiced for investigating the foundation rocks of civil engineering structures. Factors of rock classification obtained from drilled cores include:

Table 1 Criterion for rock classification by
the evaluation of drilled cores
(massiv hard rocks)

Rock grade	As a rule of thumb, acceptable rocks are 80 MPa or more in the unconfined compressive strength of naturally dried test pieces of fresh rocks. Such fresh rocks generally generate metallic sound when outcropping rocks are hit by a hammer.	
	General properties of rocks	Conditions of drilled cores
A	Rocks are very fresh in lithologic character. The rock forming minerals of igneous rocks or the component grains of sedimentary rocks are not weathered or altered at all. Few joints are distributed. Rock mass is solid and compact.	Cores are like rods of 100 cm or more, very fresh in lithologic character, and very smooth on the surface. No joint is observed. (Namely, the rocks are intact with no fissures observed in a 1 m square core box.) Core recovery is very good.
B	Rocks are fresh in lithologic character. The rock forming minerals of igneous rocks or the component grains of sedimentary rocks are little weathered and altered. Joints are sparsely distributed in compact texture. Rock mass is solid and compact.	Cores are mostly long columns of about 40 to 50 cm, fresh in lithologic character, and smooth on the surface. Few joints are distributed, in compact texture. Joint surfaces are rarely contaminated. Core recovery is very good.
C_H	Rocks are almost fresh and hard in lithologic character. In the case of igneous rocks, feldspars and color minerals such as mica and amphibole among rock forming minerals may be slightly weathered and altered. In the case of sedimentary rocks, feldspars and color minerals existing secondarily as component grains may be slightly weathered and altered. Joint surfaces are mostly weathered and altered, to be colored and contaminated. Sometimes, weathered substances are thinly deposited, but texture is generally compact. Rock mass is hard.	Cores are mostly columns of about 10 to 30 cm, almost fresh in lithologic character, and almost smooth on the surface. Joints are rather developed, and joint surfaces are often weathered and altered to be colored light brown. But weathering and alteration do not progress inside. Sometimes, weathered substances may be thinly deposited on joint surfaces. Core recovery is good.
C_M	Rocks are generally rather weathered and altered in lithologic character. In the case of igneous rocks, feldspars and colored minerals excluding quartz are weathered, and often colored brown or reddish brown. In the case of sedimentary rocks, feldspars and colored minerals existing secondarily as component grains are weathered and altered and often colored brown or reddish brown as in the cse of igneous rocks. Joints are open and often contain clay or weathered substances. Rocks of this grade often contain many fine hairlike fissures. Rocks of this grade often contain many fine hair like fissures. Rocks which are fresh in lithologic character but have open joints remarkably distributed to show a cracky state are also included in this grade.	Cores are mostly short columns of about 10 cm, and even if they are fragmentary, the fragments can be combined to make columns. They are weathered and altered in lithological character. The cores are mostly rough on the surface. Joint surfaces are weathered and contaminated, and weathering progresses inside. When a core is taken from the core barrel, new fissures are generated. Core recovery is approximately more than 80%. Rocks which are fresh in lithological character but have open joints developed in short cores are also included in this grade.
C_L	Since the rock forming minerals of igneous rocks or component grains of sedimentary rocks are remarkably weathered, whole rocks are generally colored brown or reddish brown. Joints are open and contain clay and weathered substances considerably. In the rocks of this grade, fine hairlike fissures are remarkably distributed, and weathering progresses along the fissures. Rocks which are fresh in lithologic character but have open joints remarkably distributed to show a state like masonry are also included in this grade.	Cores are mostly fragmentary, and even if the fragments are combined, it is difficult to form columns. Since rocks are weathered in lithologic character, cores are rough on the surface and are generally colored brownish or liver brownish. Weathering and alteration progress not only near the joints but also generally. When a core is taken from the core barrel, it is liable to collapse. Core recovery is approximately less than 80%. Rocks with cores of short columns and cores of sand and clay repeated are also included in this grade.
D	The rock forming minerals of igneous rocks or component grains of sedimentary rocks are remarkably weathered and often contain sandy and clayey portions. In the rocks of this grade, the distribution of joints is rather unclear.	Cores are mostly sandy or clayey, and at first sight, it is difficult to distinguish them from the rock covering layer. But they are relatively compact. In ordinary fresh water drilling, core recovery is very low, even if a double core tube is used.

a) Strength and deformability of rocks themselves
b) Distributional density of joints
c) Opening degree of joints
d) Condition of joint and properties of inclusions

The criterion for rock classification based on drilled cores is shown in Table 1. Factors for the classification are shown in Tables 2 and 3. The classification is based on empirical and qualitative evaluation.

The relationships of the rock grades with the values of the modulus of deformation $(D_{(LLT)})$ and the static modulus of elasticity $(Et_{(LLT)})$ obtained by borehole lateral load tests are shown in Figs. 1 and 2. Correlativity can be seen between these physical values of rocks and rock grades.

Table 2 Classification of rock classification factors for the evaluation of drilled cores

| Symbol | Standard classification of weathered conditions | Hardness of rocks themselves | | Distribution of jonints | | | |
| | | Standard classification | Rule of thumb | Spatial density of joints | | Standard classification of opening degrees of joints | Standard classification of joint surfaces conditions |
				Standard classification	Rule of thumb Average core length		
○	Fresh (Joint surfaces are not weathered either.)	Hard	80 MPa or more in unconfined compressive strength of dry rocks	Little distributed.	60 cm or more	No gap at all.	Not weathered
△	Almost fresh (Joint portions only are weathered.)	Almost hard	80 to 40 MAa in unconfined compressive strength of dry rocks	Sparse	30 to 60 cm	Little gap	Rather weath- and may be contaminated.
▲	Weathered (Weathered along joints.	A little soft	40 to 20 MPa in unconfined compressive strength of dry rocks	Distri- buted.	10 to 30 cm	Some gap	Weathered and contaminated. Wethered sub- stances are deposited thinly.
●	Very weath- ered. (No fresh portion is observed.)	Soft	20 MPa or less in unconfined compressive strength of dry rocks	Remark- ably dis- tributed.	10 cm or less	Clear gap	Remarkably weathered and contaminated. Clay or weath- ered substances are remarkably combined.

Table 3 Factors of rock classification in the evaluation of drilled rocks

| Rock grade | Elements of classification | | | | |
| | Condition of rock forming minerals or component grains | | Condition of joints | | |
	Weathering	Hardness of rocks themselves	Spatial density of joints	Opening degree of joints	Condition of joint surface
A	○	○	○	○	○
B	○	○	△	○	○ or △
C_H	○ or △	○ or △	▲	△	△ or ▲
C_M	△ or ▲	△ or ▲	▲ or ●	▲ or ●	▲ or ●
C_L	▲ or ●	▲ or ●	●	●	●
D	●	● (very soft)	(-)	(-)	(-)

125

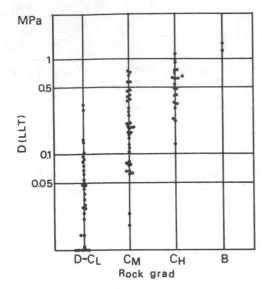

Figure 1. Relationship grad
and the modulus of in massive
rocks

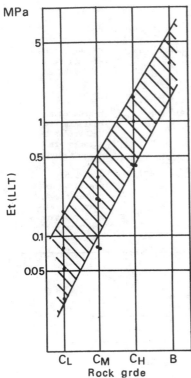

Figure 2. Relationship between the rock
grade and the static modulus of elasticity
in massive rocks (borehole load tests)

3. JUDGEMENT OF ROCK GRADES BY USE OF DRILLED CORES

3.1 Purpose and point of attention

In the judgement of rock grades, it is important to obtain objective results.
  The distribution of joints and the condition of weathering in massive hard rocks are important
factors which decide the mechanical properties of rocks, and distribution of joint spacing in
drilled cores are closely related with the weathering of rocks. For this reason, examination
was made with attention paid to the distribution of joints.

3.2 Judgement of rock grades in reference to RQD

The RQD was proposed by Deere (1973), and is generally widely used of late.
  The RQD refers to a rate of the total length of 10 cm or longer cores to a drilled length,
when the cores are not smaller than the NX size (53 mm in core diameter), and is defined as
follows:

$$RQD = \frac{L}{1} \times 100 \ (\%)$$

where L: Total length of 10 cm or longer cores in a drilled length
1: Drilled length (usually 100 cm)

Relationships between RQD values and rock grades are collectively shown in Fig. 3. It can be seen that higher rock grades show higher RQD values. Above all, a clear difference is observed between grades C_H and C_M. However, between other grades, the respective frequency distributions have large overlapped portions, and therefore it is difficult to judge rock grades in reference to RQD only.

3.3 $RQD_{(N)}$ and rock classification

Since it is difficult to judge rock grades in reference to RQD only as described in Section 3.2, RQD was modified, to examine the applicability in reference to $RQD_{(N)}$ designed to reflect the joint characteristics of respective rock grades.

1) Definition of $RQD_{(N)}$

The $RQD_{(N)}$ refers to a rate of the total length of N cm or longer cores in a unit length (usually 100 cm), and is defined as follows:

$$RQD_{(N)} = \frac{L_{(N)}}{1} \times 100 \ (\%)$$

where $L_{(N)}$: Total length of N cm or longer cores in a unit length
1: Unit length (usually 100 cm)

The drilled cores are required to have been obtained by drilling with a bore diameter of 56 mm or more (a core diameter of 40 mm or more).

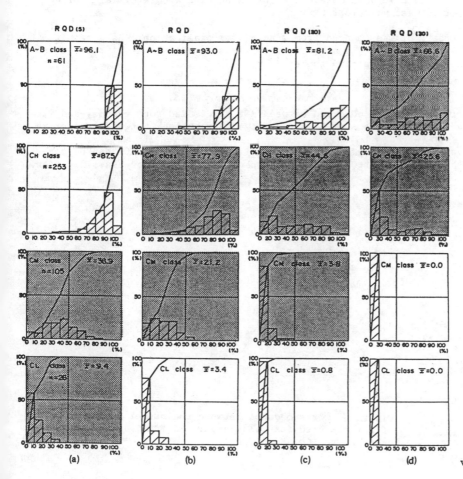

Figure 3. Relationship between RQD and $RQD_{(N)}$ values and rock grades

2) Judgement of rock grades in reference to $RQD_{(N)}$

With regard to the correlativity between $RQD_{(N)}$ values and rock grades, examination was made for cases of 5, 20 and 30 cm as N values.
   The results are shown in Fig. 3. As can be seen, grade C_L is clearly distinguished from grade C_M in reference to $RQD_{(5)}$, and grade C_H, from grade B in reference to $RQD_{(30)}$.
   Thus, it was found that $RQD_{(N)}$ is effective as a measure for rock classification, if $RQD_{(5)}$, RQD ($RQD_{(10)}$) and $RQD_{(30)}$ are used in combination.

3.4 RCI and rock classification

The bearing capacity of rock mass greatly depends on the combination between the strength of rocks themselves and the condition of fissures. The fissures observed in a drilled core include two kinds; those of rocks themselves and those generated during drilling. Of them, since the fissures generated during drilling must have been caused in relation with the strength of rocks themselves, it can be concluded that two kinds of fissures can be handled together. That is, it can be generally said that larger bearing capacity of rocks results in longer maximum and average core lengths obtained by drilling.
   With attention paid to this point, Rock Classification Index (RCI) was defined. The index includes both the factors of maximum core length and average core length. The judgement of rock grades in reference to this measure was examined.

1) Definition

The RCI is defined by the following equation:

$$RCI = \frac{Ln}{n} \ (cm)$$

   where Ln: Total length of cores from the longest one to the n-th one in a unit length, in cm
         n : A given number of 3 or less
   The drilled cores are required to have been obtained by drilling with a core diameter of 40 mm or more, using a diamond bit in massive hard rocks.

   For this definition, the number, n, was changed from 1 to 3, for examination. As a result, it was found, as shown in Fig. 4, that with n = 3 or more, the values of RCI become almost constant. Thus, n = 3 was decided as a standard number.

2) Judgement of rock grades in reference to RCI

As shown in Fig. 5, clear correlativity is observed between RCI values and rock grades, and it was especially found that overlapped portions of RCI frequency distributions between respective rock grades are small.
   The relation between RCI values and rock grades is as shown in Table 4. The RCI can be considered to be highly applicable as a measure for quantitative evaluation in rock classification.

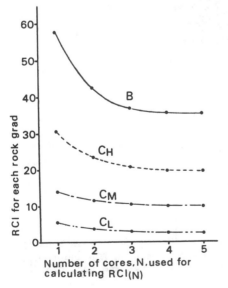

Figure 4. Corrspondence between the number of cores, N, used for calculating and RCI calculating $RQD_{(N)}$

Table 4. Relation between RCI values and rock grades

Rock grade	RCI (cm)
B	27 or more
C_H	15 to 27
C_M	8 to 15
C_L	0 to 8

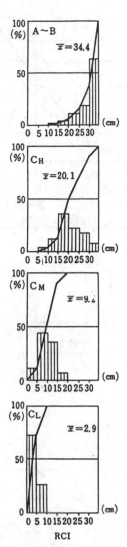

Figure 5. Relationship between RCI values and rock grades

3) Correlativity between RCI values and physical values of rocks

If the correlativity of RCI values with physical values of rocks such as the modulus of deformatin (D) and the tangent modulus of elasticity (Et) can be identified, the RCI by use of drilled cores is surmised to be an effective method for evaluating rocks.

For this reason, the correlativity of $D_{(LLT)}$ and $Et_{(LLT)}$ values obtained by borehole load test, with RCI values in the same test sections was examined.

As shown in Fig. 6, good correlativity was found between respective test values and RCI values. Larger RCI values correspond to larger $D_{(LLT)}$ and $ET_{(LLT)}$ values.

On the other hand, the relationships of RQD values with $D_{(LLT)}$ and $Et_{(LLT)}$ values obtained by a similar method are as shown in Fig. 7. Although correlativity of RQD with $D_{(LLT)}$ and $Et_{(LLT)}$ can be seen, it is far less than the correlativity of RCI.

4. CONCLUSION

This paper has examined the applicability of the rock evaluation measures obtained from drilled cores, RQD, $RQD_{(N)}$ and RCI. As a result it was found that the RQD and $RQD_{(N)}$ are effective as a measure for classification, if $RQD_{(5)}$, $RQD_{(10)}$ and $RQD_{(30)}$ are used in combination. The RCI was found to be highly correlative to rock grades and physical values of rocks.

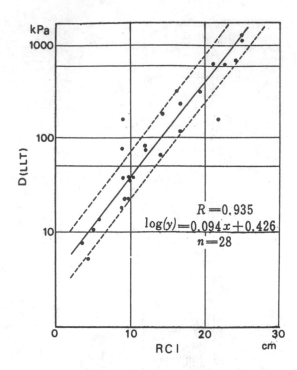

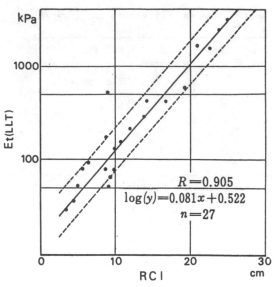

Figure 6. Relations of RCI values with the values of the modulus of deformation (D) and that static modulus of elasticity (Et)

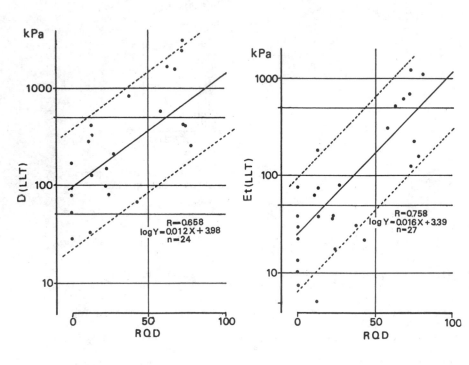

Figure 7. Relations of RQD values, with the values of the modulus of deformation (D) and the static modulus of elasticity

REFERENCES

Kikuchi, K. & K. Saito, 1982, Geotechnically integrated evaluation on the stability of dam foundation rocks. 14th International Congress on Large Dams, Rio de Janeiro: 49 - 74.
Deere, D.U., 1973, Technical description of rock cores for engineering purposes. Felsmech and Ing. - geologie 1: 16 - 22.

Discussion about engineering geologic division of rockmasses for engineering geologic mapping

Discussion sur le classement géotechnique des roches pour la cartographie géotechnique

Liu Guangrun, *Geologic Bureau of Hubei Province, Wuhan, China*

ABSTRACT: This paper expounds the threefold concepts concerning engineering geologic division of rockmasses — typologic division, unit division and gradation; presents the concret schemes of engineering geologic type division and unit division of rockmasses; proposes a use of the engineering geologic units of rockmasses to replace the common stratigraphic units for engineering geologic mapping, and gives a practical using case of these ideas in engineering geologic mapping of Yichang Area.

RÉSUMÉ: Au sujet de la division de la géologie de l'ingénieur du corps de roche, il faut differencier clairement ces trois notions: division des unités, division des types et hiérarchisations. Cette thèse présente un schéma concret sur la division des unités et des fypes. Elle propose qu'on substitue les unités de géologie de l'ingénieur à celles de couches conventionelles et offre un modéle protiqué dans la dessination géologique de l'ingénieur de la région Yichang.

INTRODUCTION

Engineering geologic division of rockmasses is the first important problem for engineering geologic survey and mapping. Although many papers have discussed this problem, but it has not been solved perfectly as yet. Now we would also like to have a try to approach it based on the practice of engineering geologic mapping of Hubei Province and Yichang Area.

THE WAYS OF ENGINEERING GEOLOGIC DIVISION OF ROCKMASSES

It is need to distinguished clearly the threefold concepts concerning engineering geologic division of rockmasses — rockmass type division, unit division and gradation. They are three ways of engineering geologic division of rockmasses related as well as different from each other. The rockmass type division considers only the rocknature, but disregards their position. Reckmasses under a common type might exist in different periods and areas. The rockmass unit division considers not only the rocknature, but also their real position. Rockmasses within a common unit have the same or similar engineering geologic features and occupy a definite and continuous geologic horizon. The rockmass gradation may be carried out by the quality and quantity. The qualitative gradation of rockmasses is generally applied to the rockmass quality evaluation of foundation, with less relations to engineering geologic mapping, therefore it is not discussed here. The quantitative gradation of rockmasses refers to the size gradation of engineering geologic units of rockmasses, it is an important object that we should deal with.
  The rockmass type division is suitable to the generalization of the similarity and difference of rockmasses, and to the study of common characters. The rockmass unit division is suitable to the discription of the single rockmasses and to the study of their special characters. The gradation of rockmasses would suit the needs of engineering geologic mapping on different scales.

ENGINEERING GEOLOGIC TYPE DIVISION OF ROCKMASSES

Engineering geologic type division of rockmasses is the scientific combination of rockmasses by differences and similarities of their engineering geologic characters, according to the principle of marging the similars and dividing the differents. Difference or similarity of engineering geologic characters is the foundamental basis for rockmass type division and the gualitative factor of rockmass unit division.

Rockmasses are generally the complexes consisted of rockblocks and separative or weak structural planes. The engineering geologic characters of rockmasses shown mainly in their strength,
deformation (in compressing and sliding) and permeability are usually determined by the development situation of structural planes and the engineering geologic nature of rockblocks, it is the
composite effect of both these factors. But the roles played by each factor in determining the
engineering geologic characters of different rockmasses are not alike. Generally speaking, the
higher the strength of rock, the more notable the action of rockmass structure; the lower the
strength of rock, the greater the role played by engineering geologic nature of rockblocks. The
outline of this situation may be shown in the following sketch:

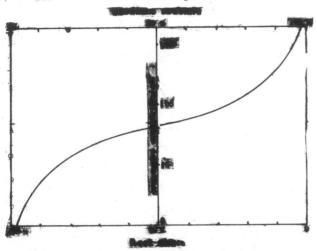

Fig.1 Sketch showing the leading factor determining the variation of engineering geologic
character of rockmasses

Since the variation of engineering geologic
characters of most of bedrocks is controlled
by rockmass structure, and the variation of
all the inconsolidant coverburden and soluble
bedrock is determined by rocknature, therefore the principle of common engineering
geologic type division for all rockmasses
might only be the combination of rockmass
structure with rocknature. In accordance with
this principle, all rockmasses may be divided
into seven engineering geologic rockkinds,
namely: A. Massive engineering geologic rock-
kind(MSRK), consisting of all igneous rocks,
massive volcanic rocks and part of metamorphic
rocks with massive structure as gneiss etc;
B. Stratiform engineering geologic rockkind
(STRK), consisting of all clastic sedimentary
bedrocks, stratified volcanic rocks and part
of metamorphic rocks with bedding structure
as slate, quartzite etc; C. Schistose engineering geologic rockkind(SCRK), consisting of
various schists, phyllites; D. Soluble (in
water) engineering geologic rockkind(SLRK),
consisting of various limestones, dolomites,
marbles, gypsums, rocksalt and other carbona-
tites; E. Kataclastic engineering geologic rockkind(KTRK), consisting of various tectonites;
F. Cohesive engineering geologic rockkind (CHRK), consisting of various clayey rocks; G. Loose
engineering geologic rockkind(LSRK), consisting of various loose sediments of sand, gravel and
other debris.

The classification (this term has usually both the meaning of type division and grade division,
here it refers to the former) of rockmasses as mentioned above suits only regional engineering
geologic investigation and brief appraisal of rockmasses. For engineering geologic investigation
in a smaller section or construction site and more detailed appraisal it is necessary to provide
the secondary and more detailed divisions of rockmasses, such as subkinds, rocktypes and subtypes,
according to the more detailed rockmass structure and rocknature. But for the secondary as well
as more detailed division of rockmasses, it is not needed to take a common criteria. It should
be carried out according to the concret situations of mapping area.

The main engineering geologic characteristics of every rockkind and the criteria for their
secondary division are shown in table 1.

ENGINEERING GEOLOGIC UNIT DIVISION OF ROCKMASSES

Engineering geologic units of rockmasses are the foundamental units for engineering geologic mapping and the concret objects for engineering geologic exposition. The dividing lines of engineering geologic units of rockmasses should be the conspicuous variational limits of engineering
geologic characteristic of rockmasses, that may be or not identical with the dividing lines of
common stratigraphic units.

To suit the needs of engineering geologic mapping on different scales, engineering geologic
rockmass units should have their own grade series.

In the gradation of engineering geologic rockmass units, both quantitative (size) and qualitative (engineering geologic character) factors should be taken into account, so as to reflect the
value of every rockmass unit (the degree of its influense upon the regional engineering geologic
conditions). Within the permissive circumstance of mapping scale, those rockmasses possessing
conspicuous engineering geologic characteristics, especially different from the adjacent units,
should be divided as detailed as possible; On the contrary, those ones lacking conspicuous
characteristics may be divided rouphly. Theoretically, there isn't any relation between the
grades of engineering geologic units and the common geologic time grades.

Relying on the principle as mentioned above, engineering geologic rockmass units may be divided

Table 1. Scheme of engineering geologic classification (typologic division) of rockmasses

Engineering geologic rockkind	Typical lithologies	principal engineering geologic characteristics	Basis of subdivision (into subrockkind, rocktype, subrocktype)
Massive rockkind (A)	Intrusions of all kinds; massive volcanics; massive metamorphic rocks,	Empty of bedding structure, with high integrity and high strength, The variation of engineering geologic characteristics is controled by fectonic and wethering fractures	Genetic conditions, lithologic nature, fracture development, strength homogenity, antiweathering capacity etc.
Stratiform rockkind (B)	All the clastic sedimentary bedrocks, stratiform volcanics, stratiform meta-rocks.	With bedding structure in common. Often assumes interbedding of soft and hard rocks. The strength and stability are usually controlled by the weak bands	Lithologic nature and its combination, bedding development
Schistose rockkind (C)	Various schists,, phyllites.	With thick schistose structure, low integrity and low isotropy. The engineering geologic characteristic is controlled by schistosity and rock-nature, poor usually in stability.	Lithologic nature and its combination, shcistosity development.
Soluble rockkind (D)	Various limestones, dolomites, marbles, carbonatite, rocksult, gypsum.	Besides the common character of stratiform rockmasses, soluble in water, with the karst phenomena usually, often appear the problem of karst leakage and karst gush.	Lithologic nature, solubility, karst development, existence of insolubbe bands.
Kataclastic rockkind (E)	Various tectonites, such as kataclasite, Breccia, Mylonite etc.	Crushed intensively. Lost the feature of original rocks, The integrity and strength are decided by the level of crushing and recement, but poor usually in both of them.	Level of crushing and recement situation
Cohesive rockkind (F)	Various clayey rocks	With higher cohesion, plasticity, compressibility, and lower permeability, The variation of engineering geologic character in a great degree is controlled by the moisture content	Granule composition, organic content. The kind of clay menerals and the practise physical and mechanic indexes.
Loose rockkind (G)	Various sands, gravels, and other loose debric sediments.	Ampty Cohesion, Poor in compressibility, with higher permeability. The variation of engineering geologic character is controlled by the granule composition and clay content.	Granule composition, grain form, clay content, and the practise physical and mechanic indexes.

into four grades, which are: 1. Engineering geologic group; 2. Engineering geologic formation; 3. Engineering geologic section; 4. Engineering geologic belt (or block).

An engineering geologic group consists of rockmasses possessing generally a continute stratigraphic horizon formed in the uniform genetic conditions, corresponding to a large lithologic formation, belonging to the same or similar engineering geologic rockkinds, with a thickness of more than 1000m. It is suitable to serve as a basic unit for the engineering geologic mapping on small scale (less than 1:1,000,000).

The divisional conditions of engineering geologic formations are similar to that of engineering geologic groups, but the former with their dimension are smaller than the latter. An engineering geologic formation may be composed of a main engineering geologic rockkind or of a specific association of several rockkinds, usually has a higher uniformity of engineering geologic character than a group. Its thickness may vary from 200m to 2000m. It is suitable to serve as a basic unit for engineering geologic mapping on the middle-small scale (1:100,000 1:500,000).

An engineering geologic section possesses generally considerable uniformity in the rocknature and rockmass structure with a thickness of 50 500m. It is suitable to serve as a basic unit for engineering geologic mapping on the middle-large scale (1:10,000 1:50,000).

An engineering geologic belt or block possesses the homogeneous rocknature and the same of similar rockmass structure of original and secondary genensis with a thickness less than 100m. It is possible to be represented by a united strength index and suitable to serve as a basic unit for engineering geologic mapping on large scale (more than 1:5000).

The boundaries of engineering geologic groups, formations and sections are usually served by the sedimentary contact lines or intrusive contact lines and by the tectonic lines along some large tectonic fractural zones in a few cases. But the boundaries of engineering geologic belt

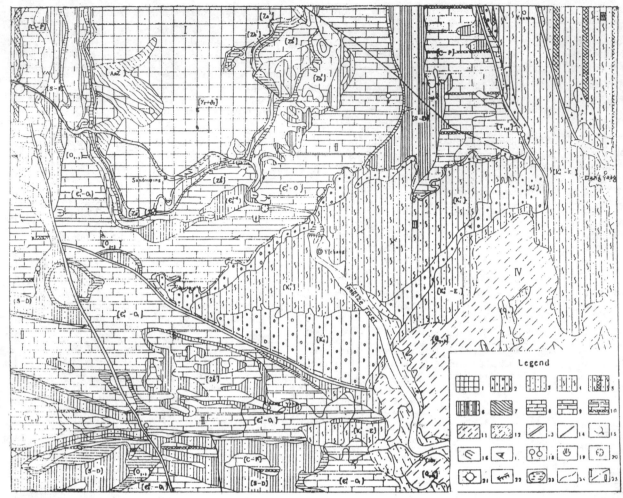

1. Massive rockkind(A); 2. Mainly conglomerite subkind of stratiform rockkind(B)$_1$; 3. Mainly sandstone subkind of stratiform rockking(B)$_2$; 4. Clayey sandstone and mudstone subkind of stratiform rockkind(B)$_3$; 5. coaly subkind of stratiform rockkind(B)$_4$; 6. Mainly shale subkind of stratiform rockkind(B)$_5$; 7. Schistose rockkind(C); 8. Limestone subkind of soluble rockkind(D)$_1$; 9. Limestone with shale subkind of soluble rockkind(D)$_2$; 10. Limestone with coal series subkind of soluble rockkind(D)$_3$; 11. Old cohesive soil subkind of cohesive rockkind(F)$_1$; 12. New cohesive soil subkind of cohesive rockkind(F)$_2$; 13. Neoid active fault; 14. Old fault; 15. Landslide; 16. Rockfall; 17. Debris flow; 18. Spring; 19. Thermcale; 20. Karst cave 21. Epicentre of earthquake; 22. Dam; 23. Reservoir; 24. Limit and numbers of engineering geologic zoning.

Fig 2. Schemetic engineering geologic map of Yichang Area   scale 1:500,000

or block may often cut the lithological boundaries due to the variation of detailed rockmass structure.

Every engineering geologic rockmass unit should have its own special symbol, and it is necessary to build a symbol series corresponding to the series of rockmass units. In order to reflect the geologic horizon and the deformational history of every rockmass unit, we suggest to take the symbols of the two grades in higher classes of engineering geologic units, group and formation, using the time symbols of strata consisted in the engineering geologic rockmass units with double and single squar brackets outside separately. Thus it would be well to show both the relation and difference between the engineering geologic rockmass units and the common stratigraphic units. With the subdivisions, engineering geologic section and belt, since their thickness becomes smaller and smaller, usually there aren't such small common stratigraphic units corresponding with them, and therefore, it is impossible to build the independent symbols form them. They might be represented by ordinary numbers under the formation which they belong to, adding outside the small brackets for sections and the circle for belts.

The gist of engineering geologic unit division of rockmasses is given in table 2.

Table 2. Scheme of engineering geologic unit division of rockmasses

Gist of division \ Grade \ unit	Engineering geologic group (e.g.group)	Engineering geologic formation (e.g.formation)	Engineering geologic section (e.g.section)	Engineering geologic belt (e.g.belt)
	1	2	3	4
mapping scale	$\dfrac{1}{1,000,000}$	$\dfrac{1}{100,000-500,000}$	$\dfrac{1}{10,000-50,000}$	$\dfrac{1}{5,000}$
size (Thickness)	>1000m	200-2000m	50-500m	<100m
Lithologic conditions	Formed in the same or similar sedimentary, metamorphose or irruptive conditions. Correspond usually to a large lithologic formation, with complex rocknature.	Formed in the same genetic conditions, correspond usually to a small litholotic formation or a large lithologic number of a large formation. Rock nature may be complex or more unitary.	May be a unitary lithologic section, a regular interbeding of two rocknatures or a main rocknature with thin bands of others.	unitary lithology, or thin bed interbedding of two kinds of rocknatures.
Rockmass structure	With a same or similar macro rockmass structure	With a same rockmass structure. Stratification thickness should be also as same as possible	Uniform basically in the detailed original structure and the main secondary structure	Original and secondary detailed structure tend to uniform
Engieering geologic character	Has a rough uniformity, belongs to the same or similar engineering geolotic rockkind	Has a considerable uniformity, belongs to the same or similar subrockkind. Might surve to a regional water bearing or water insisting complex, and the rockmass unit of suitability appraisal for regional plan.	Strength, integrity, and permeability tend to uniform, might surve to a principal unit for site selection, a local water bearing or water insisting layer, and the rockmass unit of suitability appraisal of the site selection of large projects.	Uniform rocknature. Might be given a united strength index. Might surve to the rockmass unit of suitability appraisal for smaller projects or the ground treatment.
Division boundaries	Delimited usually along the unconformity, paraconformity planes, irruptive contact limes, or regional large fault	Delimited usually along the conformity or unconformity planes, irruptive contact lines, large fault planes.	Delimited usually along the conspicuous boundaries of rockmass structure, the marginal boundaries of fault zones.	Delimited usually along the stratification planes and the conspicuous variation limits of rockmass structure
Unit symbol	Time symbol of strata contained by an e.g. group. Within the double squarbrackets such as: 〖Z-O〗	Time symbol of strata contained by an e.g.formation, within the single squar brackets such as: $[C_1^2-O_2]$ $[r_5]$	Ordinal numbers under a same e.g. formation for the common strata, fault symbol for the large fault zones making up an e.g. section alone, within the small brackets. such as: (1). (F_5)	Ordinal numbers under a same e.g. section for the common strata, fault symbol for the fault zone making up an e.g. belt alone, within the circle. such as: ②. ⓕ$_6$

135

Table 3. Engineering geologic division of rockmasses for engineering geologic mapping in Yichang Area

common stratigraphic division				Thickness (m)	columnar section	Typical lithologies	Engineering geologic unit division (eg.formation)	Engineering geologic type division	
Era/them	sys-tem	se-ries	local unit					Rock-kind	subrockkind
Cenozoic Era	Tr	E	Jianacao formation	1700		Brownish yellow, brownish red thick and very thick clayey sandstone, sandy mudstone	$[K_2^2-E]$	Stra-tiform (B)	clayey sandstone and mudstone
Mesozoic Era	K	K_2	Paomagang formation K_2^3	700		Purplish red thick mudstone, siltstone			$(B)_3$
			Honghuatao formation K_2^2	260		Brick red very thick clayey siltstone			
			Luojingtan formation K_2^1	880		light greyish red very thick conglomerite	$[K_2^1]$		Mainly conglo-merite $(B)_1$
		K_1	Wulong formation K_1^2	350-1500		Greyish red, greyish green thick clayey sandstone with thin mudstone	$[K_1^2]$		$(B)_3$
			Shimen f. K_1^1	0-200		Greyish red conglomerite	$[K_1^1]$		$(B)_1$
	J	J_{2+3}	J_{2+3}^4	600		Arkose and saudy mudstone	$[J_{2+3}]$		$(B)_3$
			J_{2+3}^3	1200		Thick saudy mudstone with sandstone			
			J_{2+3}^2	900		Mudstone, sandy mudstone with sandstone			
			J_{2+3}^1			Thick sandy mudstone with sandstone			
		J_1	Xiangxi group	520		sandstone, shale, coal series	$[J_3-J_1]$		Coaly formation $(B)_4$
	T	T_3	Badong formation	120-980		Red sandy mudstone with limestone and shale			
		T_{1-2}	Jialingjiang formation	500		Deep grey medium bedded crystaline limestone	$[T_{1+2}]$	Soluble (D)	Limestone $(D)_1$
		T_1	Daye group	900		Grey thin limestone with calc-shale			
Paleozoic Era	P	P_2	Longtan Dalong formation	200		Greyish black chert limestone and coal series	$[C-P]$		Limestone with coal series $(D)_3$
		P_1				Deep grey bituminous lime-stone,coal series			
	C	C_{2+3}		40		Thick crystaline limestone			
	D	D_{2+3}		100		Quartz sandstone, shale,	$[S-D]$	(B)	Mainly shale $(B)_5$
	S	S_2	Shamao group	200		Greyish green thin saudstone with shale			
		S_1	Luoreping formation	200-700		Yellowish brown sandy mudstone with sandstone			
			Longmaxi formation	500		Blueish green shale, mudstone with sandstone			
	O	O_{2+3}		100		Clayey limestone, shale ls	$[O_{2+3}]$	(D)	Limestone with shale $(D)_2$
		O_1		200		Grey thick crystaline			
	ϵ	ϵ_3	Sanyoudong group	300		Grey white thick crystaline dolomitic limestone	$[\epsilon_1^3-O_1]$	(D)	$(D)_1$
		ϵ_2	Qinjiamiao group	300-700		Deep grey, blueish grey medium bedded delometic limestone with calc-shale			
		ϵ_1	Shilongdong formation	100		Thick limestone with dolometic limestone			
			Tianheban f.	100		Banded dolomitic limestone			
			Shipai f.	200		Shale with sandstone	$[\epsilon_1^1-\epsilon_1^2]$	(B)	$(B)_5$
			Shuijingtuo f.	100		carboraceous shale, limestone			
Proterozoic Era	Z	Zb	Dengying formation	500		Thick siliceous dolometic limestone	$[Zb^2]$	(D)	$(D)_1$
			Doushantuo formation	200		Medium bedded dolomieic limstone with shale	$[Zb^1]$		
		Za	Nantuo glacier	0-100		Glacier	$[Za]$	(B)	Mainly sandstone $(B)_2$
			Nantuo f.	0-200		Sandstone		Schis-tose (C)	
	Pt		Kongling group	1000		Quartz-mica schist, Quartz actinolite schist, limestone	$[Pt]$	Massive	(C)
						Granite, diorite	$[r_2]$	(A)	(A)

136

THE PRACTICE OF A USE OF ENGINEERING GEOLOGIC DIVISION OF ROCKMASSES IN ENGINEERING GEOLOGIC
    MAPPING OF YICHANG AREA

The mapping area is about 10000km², where the regional geologic and hydrogeologic survey have
been carried on with the scale 1:500,000, and a lot of engineering geologic prospecting for many
important constructions have been finished. Existing data can fullfill the needs of engineering
geologic mapping on that scale.

    The mapping work is started on the engineering geologic division of rockmass units based on the
regional stratigraphic columnar section, which are divided into 19 engineering geologic formations belonging to the massive, stratiform, schistose, soluble and cohesive engineering geologic
rockkinds separatelly.

    A common geologic map is used as the base map. Drawing up all the dividing lines of engi-
neering geologic rockmass units and marking their symbols, make the engineering geologic formations to form the basic units of rockmasses of engineering geologic map. When drawing normally
in secondary step, all the common stratigraphic lines and marks without relation to engineering
geologic division of rockmasses are omitted, because their imformations on geologic period and
lithology for engineering geologic map with such scale have been reflected perfectly into engineering geologic formations.

    In the mapping, colours, the most conspicuous means, are used to show the engineering geologic
rockmass classification, correctly on our map, the engineering geologic rockkinds and subkinds.
Every engineering geologic rockkind is shown by a given colour, and their subkinds are expressed
by various tones of that same colour. On the black and white map, they might be shown by the
rocknature symbols. In a word, the colours on the engineering geologic maps are used to shown
the engineering geological character of rockmasses at all and still have not any meaning about
geologic time as that on the common geological map.

    The engineering geologic map drawn up as discribed above could shake off the unreasonable yoke
of common geologic map, and possesses more independent character.

    The concret results of engineering geologic division of rockmasses and engineering geologic
mapping of Yichang Area are shown in table 3 and figure 2.

Potential collapsibility of a rocky mountain

Potentialité d'écoulement dans la montagne rocheuse

T.F.Onodera, *Nippon Geophysical Prospecting Co. Ltd, Tokyo, Japan*

ABSTRACT

A case study was made to express the collapsibility of a mountain composed of rock.
A representative sample area was selected on a mountain region in the central part
of Japan where detailed Sabo (world-widely used technical term in the field of erosion
control works and its technology) data are available. Number of collapse in the concern
ed region, area and volume of each collapse are contained in these data.
Topographical studies were made to allocate the sample area. Geological studies
were so made as to map in a scale of 1/1000. Observations on rock joint were carefully
made on the field outcrops and on the wall in test adits of a proposed dam. Records of
core boring and geophysical prospecting were also taken up. Necessary field and laboratory rock tests were done.
The value e = 100-RQD of a rock mass or a bed can reasonably be an index of potential collapsibility of the mass. The total summed up E on the concerned extent is to
express the potential collapsibility of that part of the mountain.
Relations of E to lineament density, volume of talus deposit and to the collapsed
volume were discussed.

RÉSUMÉ

C'est une étude destinée a présenter une potentialité d'éboulement de la montagne
rocheuse.

Une endroit d'échantillonnage representatif a éte choise dans la résion montagneuse
centrale du Japon où nors trouvons facilement des donées detaillées sur le Sabo (c'est une
terme technique utilisé largement dans le monde, indiquant les tracaux et la technologie
d'antiérosion). Le nombre, la superficie et le volume d'éboulement dans cette région sont
contenues dan ces données.

Les études topographiques ont été effectuées en vue de justifier le choix de l'endroit d'échantionnage. Les é tudes géologiques ont été faites également dans le but d'.
établir une carte à l'échelle de 1/1000. La diaclase de roche était soigneusement ob-
servéepour l'affleurement et la paroi de la galerie d'essai prévue pour un barrage à
construire. Ici, sont prises également des données obtenues au cours du carrotage et de
la prospection géophysique. Des essais méchaniques, considérés comme nécessaires, ont
été organisés sur le chantier et dans le laboratoire.

La valeur e = 100-RQD d'une certaine roche ou d'un giaement, elle peut servir comme indice de la potentialité d'éboulement de terres. La somme des valeures E à l'étud-
endue concernée peut impliquer la potentialité de terres dans cette partie de la montagnes.

La valeur E par rapport à la densité de linéation, volume du dépôt de talus et au
volume d'éboulement a été examiée.

INTRODUCTION

The author has discussed the mechanical properties of rock mass relating to the progressive failure of the earth crust (1986). It is natural that the mechanical property of basement rock or a mountain land originated from tectonic force of the region, plays an intrinsic role on the mass movement together with its topography.

This reasoning means that mechanical property of the basement rock becomes the very primary factor to the potentiality of mass movement because the collapsing property of rock mass is the reciprocal of its strength property. This also infers that it will govern the thickness of talus and quantity of stream bed deposit along the upper reaches.

For the purpose to study the aspect of relation of mechanical property to mass movement, a small sample area about 5.7km^2 as shown in Fig. 1 was taken up on the upper reaches of the River Arakawa which rises near the sample area and flows to the Bay through the City of Tokyo.

The mountain region containing the sample area had sufferd from severe mass movement disaster caused by heavy rain several times in the near past and comprehensive investigation results for Sabo purpose have been published (TANIGUCHI 1961).

This paper discusses the relation between the minor geological structure and collapsibility of mountain rock mass.

1. LOCALITY AND TOPOGRAPHY

The sample area is situated between two nearly parallel tributaries abriviated here KML on the north and NKL on the south, the elevation is about 440-490m on the north and 420-500m on the south, the dividing ridge of nearly east-west direction elevates 1055-1151m.

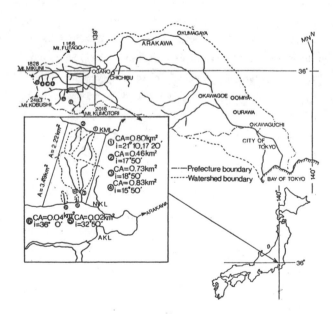

Fig. 1. Location map of sample area and control area.

Table 1 Number of spot and quantity of mass movement
in control areas

geology	river system	drainage basin	mass movement			
			total		per km²	
			no. of spot	volume (10³m³)	no. of spot	volume (10³m³)
Paleozoic	N	O G	6	4.6	1.1	0.82
		O Y	12	7.8	2.3	1.34
		O W	6	13.5	0.8	1.80
Jurassic	O	O N	19	21.2	2.9	3.21
		S K	6	308.4	0.8	42.20
	A	O Y	18	138.8	3.6	21.80

Table 2 Number of spot and quantity of mass movement
in the upper reaches including sample area

geology	drainage basin	mass movement			
		total		per km²	
		no. of spot	volume (10³m³)	no. of spot	volume (10³m³)
Paleozoic	KML	27	33.2	1.2	1.44
and	NKL	36	48.3	1.5	2.00
Jurassic	ARU	30	104.5	2.8	9.68

Six small drainage basins with similar geology and topography to the sample area
were taken up as control areas to compare and verify the appropriateness of the sample
area in inferring the results over the wide upper mountain reaches because the sample
area is very small and lacks the detailed Sabo data. Quantity of past mass movement
of the control areas are summarized in Table 1.

The mass moved volume on three tributaries of the upper reaches, abriviated here
KML, NKL and ARU containing the sample area are summarized in Table 2.

Comparing Table 2 with Table 1 it is understood that in these reaches, where the
sample area is contained, the scale of mass movement is averaged compared to the control areas.

Topography and drainage pattern of the sample area are shown by Figs. 2 and 3 respectively.

1.1. DRAINAGE DENSITY

Fig. 3 shows the distribution of mountain streams of the sample area read on the
photomap made from aerial photographs 1/12,500 scale.

Fig. 2 Aerial photomap.

Fig. 3 Drainage pattern map.

Fig. 4 Lineament map.

142

Owing to the difference of photographic tone both total length of the streams and the density per km² (HORTON 1945) are much greater on the north side slope.

Instead the erosion stage of the sample area is similar to those of control areas, stream density of sample area is from 5 to 10 times of control area.

From these points it shold be said that the stream density can not be adequately taken up   as element to show erosion stage.

1.2. LINEAMENT DENSITY

Fig. 4 shows the lineaments read similarly to the above case on the photomap. Density of total length per km² is shown in Table 5 later.

In this case though the difference of photographic tone by north and south slopes gives influence, the difference is not so serious as to obstruct expressing the minor geostructure.

Lineament is originated by erosion along shear fracture surfaces such as fault or joint in the basement rock on the mountain slope and corresponds to the very early stage of zero order streams (TSUKAMOTO 1973) in the developing course of mountain streams.

1.3. USE OF HYPSOMETRIC CURVES

Developing aspect of erosion on mountain topography appears in the change of hypsometric i.e. area-altitute curve (STRAHLER 1957). When the topography of different area is to be compared percentage hypsometric curve is effective which adopts both co-ordinates in percentage.

The shape of a hypsometric curve expresses the status of erosion and the area beneath the curve h_i (percentage hypsometric integral) expresses the stage of erosion.

On the Paleozoic north slope of the sample area two small drainage basins ① and ② are contained and on the south slope each two, ③ and ④ in the Paleozoic and ⑤ and ⑥ in the Jurassic System are contained.

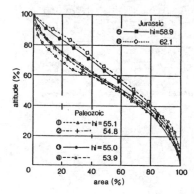

Fig. 5 Percentage hypsogram
of sample area.

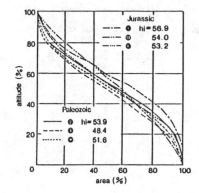

Fig. 6 Percentage hypsogram
of control area.

Fig. 5 shows the percentage hypsometric curves for the minor drainage basins in the sample area. It is shown that the erosion stage in the Paleozoic System on the north and south slopes is not so different except for the point that the erosion is more advanced on the higher part of ④ than other streams.

Though any decisive conclusion must be avoided because of the smallness of the drainage basins and the limited location of the Jurassic System, erosion on the Jurassic System seems considerably behind that of the Paleozic System.

Fig.6 shows the percentage hypsometric curves for the control areas by TANIGUCHI (1961) in which curves for each area in the Paleozoic System are similar to each other.

Though as a whole group of curves for the Paleozoic System is in the lower position than group of curves for areas in the Jurassic System, it can be said that there is no serious differnce between these groups of curves.

From the facts described above it is inferred that firstly though the general erosion stage on the control area is somewhat more advanced in the area of Paleozoic System, compared to the area of Jurassic System, there is no remarkable diffence between them, and secondly the sample area selected here, small as it is, can represent the wide mountain upper reaches on the point of erosion.

2. GEOLOGY

Sample area is made up of a System in the so-called Chichibu Paleozoic Formation and Jurassic System which comes on the south in contact with the Paleozoic System by a reverse fault striking WEW-ESE and steeply dipping to north.

Fracture system of small fault and joint is developed by this reverse faulting. On the one hand the fracture system governs the mechanical properties of basement rock to the deep inner part, on the other hand it gives directional trend on the minor physiography as lineaments on the surfaces.

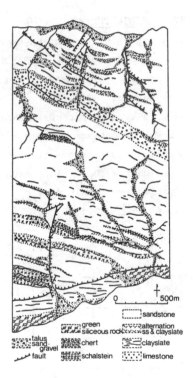

Fig. 7 Geological map.

2.1. GEOLOGICAL STRUCTURE

Fig.7 shows the geological map of the sample area. The Paleozoic System in this area is composed of sandstone, clayslate, their slternation, chert, limestone and schal stein in the order of prelalence. It shows a monoclinal structure with strike N40°-70°W and dip NE30°-70°. This System has many small intraformational foldings and shear joints are developed in this System and also in the Jurassic System.

Reverse fault which borders the Paleozoic and Jurassic Systems here is one of the main tectonic lines in this mountain region. It strikes N70°-80°W and dips NE70° or so.

The thickness of individual slip-crushed layer within this fault is about 10cm and the total width of tht fault smounts to 10m. Small fault and joint system in this area is in conjugate to this main fault.

Jurassic System in this area is mainly composed of sandstone, clayslate and sili-ceous green rock and some limestone, schalstein and chert. It shows a monoclinal structure striking nearly parallel to the main fault and dipping NW20°-80°.

In addition to the field observation detailed minor structures are observable on the walls of the test tunnels for a proposed dam close to the sample area.

Almost all of the fractures in the basement rock belong to the system of shear fractures conformable to the main fault stated above.

Besides the basement rock there distribute talus and detrital deposit on the slopes or stream beds of the sample area amounting to from several to dozens of meters thick. Sand and gravel beds also distribute locally. These clastic deposits become the direct supply source of washout sediment by heavy rain.

Though thickness of the topsoil amounts to the extent of one or two meter, seismic prospecting shows that below this depth there is so-called low velocity layer down to about 15m or so, caused by gradual weathering along the joint surfaces which leads to the opening of the surfaces.

2.2. PROPERTIES AGAINST WEATHERING

Limestone as a mass is the most resistant to mechanical weathering and it shows projecting topography in the sample area.

Chert is partially severely jointed to disintegrate into angular fragments near the earth surface. However as a whole mass it is reistant to weathering to form pro-jecting topography.

Siliceous clayslate shows resistant property to weathering close to chert.

Sandstone, except the part which intercalates much clayslate is the most resist-ant to weathering as in the limestone bed. Joint spacing is wide and opening of joint is small in sandstone bed.

2.3. DISTRIBUTION AREA OF EACH BED

Horizontal distribution areas of each bed were measured on the geological map of the scale 1/2,500. The areas were compensated according to each mean inclination of north slope, south Paleozoic slope and south Jurassic slope as shown in the location map in Fig. 1 to obtain the inclined actual areas of each bed listed in the columns of A in Table 3.

T a b l e 3 Distribution Area and Potential Collapsibility

slope direction			north			south		
geology			e (%)	A (km²)	e A (km³)	e (%)	A (km²)	e A (km³)
basement rock	Paleozoic	sandstone	27	0.90	24	32	1.78	57
		alternation of ss and clayslate	33	0.63	21	36	0.08	3
		clayslate	37	0.29	11	47	0.18	8
		limestone	20	0.24	5	28	0.31	9
		chert	16	0.03	0	37	0.35	13
		schalstein	(25)	0.02	1	—	—	—
		total		2.11	62		2.70	90
	Jurassic	sandstone				(23)	0.08	2
		clayslate				23	0.47	11
		siliceous green rock				18	0.13	2
		total					0.68	15
		total		2.11	62		3.38	105
talus			100	0.11	11	100	0.13	13
grand total				2.22	73		3.51	118

2.4. JOINT SYSTEM

Basement rock in the sample area contains systems of many small fault and joint with regular directions. Striations and minor slips are sometimes observed on the surface of joint showing the scale and character of joint are continuous to those of fault.

Frequency and direction of joint are various locality by locality. When the joint surfaces are plotted on Schmidt projection using the upper hemisphere, according to north and south slopes, to geological Systems and to rock types the predominant direct ions become very obscure as shown by Fig. 8.

However when the joint system at a locality is plotted, the plots show clear direc- ion systems as shown by Fig. 9 for example. These plots show that joint and fault in the basement rock are in conjugate shear fracture surfaces.

Therefore we have to recognize that in spite of the difference of opening of sepa- ration surface, separating character of joint surface continues down deep inside of the mountain rocks and governs the mechanical properties thereabout.

In many cases distinct two and another indistinct direction are observed. One of them often coincides with bedding plane in this area.

As a whole no special feature of joint system direction by geology, rock types or by the direction of slope is recognized and roughly speaking, general predominant direc tion is in around N30°W, NE 35° and subpredominant directions around N50°W, SW35° and around N35°E, SE65° are seen.

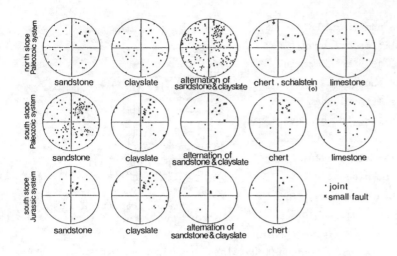

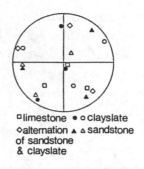

Fig. 8 Schmidt projection of fault and joint.

□ limestone ● ○ clayslate
◇ alternation ▲ △ sandstone
of sandstone
& clayslate

· joint
× small fault

Fig. 9 Example of Schmidt
projection showing
conjugate
joint system
at individual
localities.

T a b l e 4   Basic physical properties of rock in the sample area

rock / property	sandstone	clayslate	chert	limestone	schalstein
dry bulk sp. gr.	2. 55~2. 69	2. 53~2. 72	2. 62~2. 66	2. 62~2. 75	2. 68~2. 75
water absorption (%)	1. 40~0. 24	2. 71~0. 17	0. 87~0. 08	0. 10~0. 88	0. 49~0. 13

This set of joint and minor fault increases tendency of rock as a mass to fracture, lowers the resistance to mechanical weathering and becomes the source to produce
rock blocks of various size and shape such as platy or columnar. Number of joint and
minor fault are increased near the major fault to produce heavy fracturing to basement
rock.

3. STRENGTH PROPERTY OF BASEMENT ROCK MASS

3.1. BASIC PHYSICAL PROPERTIES OF CONSTITUENT ROCK

Basic physical properties of rock in the sample area are as summarized in
Table 4.

Bulk specific gravity and water absorption of a rock can be taken up as index
properties to express its basic physical properties as DUNCAN presented (1969),
and these values generally represent the mechanical properties of the rock.

The author had modified DUNCAN's classification denominating consistency division
as shown in Fig.10 (1974). The index properties of intact constituent rocks of basement rock of this area are plotted in Fig.10 which shows all specimens belong to very
high consistency division or DUNCAN's grouping I,II and III except one of clayslate.

Similar data on about the same number of Paleoaoic and Jurassic rocks from other localities of Japan (TAKATA 1933) are also plotted in Fig. 10. By comparing these plottings it can be said that though the latter distribute down to consistency C, all the intact rocks of the basement rock in this area belong to the most high consistency class and that there is no difference to treat separately between the Paleozoic and Jurassic rocks. In other words if there be no discontinuity surface such as joint in the mountain rock mass, the mass must alsmost uniformly resistant to mechanical forces such as abrasion, impact and scratching and so on.

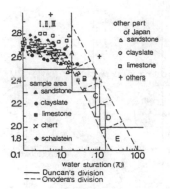

Fig. 10 Index properties of Paleozoic and Jurassic rocks in Japan.

This implication means that the mass movement of mountain composed of such highly consistent rock is mostly governed not by the properties of constituent intact rock but by the discontinuity surfaces such as fault, joint and bedding plane (ONODERA 1986). Therefore lineament desity becomes to bear relation with the potantiality of basement rock to cause mass movement. This fact is one phase of manifestation that minor geological structure govern the mechanical properties of mountain rock mass.

3.2. MECHANICAL PROPERTY OF ROCK AND ITS ANISOTROPY

From the statistical data on various types of rock it is recognized (JUDD 1961) that scleroscopic and impact strength of rock are in general increased with strength properties such as compressive strength or elastic modulus. Moreover it is recognized that they are in good linear relation when rock type is limited (ONODERA 1974).

As the mechanical properties of rock in the area is limited strength properties presented by tensile strength or Young's modulus are thought to be directly related to strength against physical erosive agencies even freezing and thawing.

Therefore tensile strength was taken up and was obtained on moist specimens by point load test in test tunnels and as its control diametral compression test on dry, moist and water saturated specimens was performed in the laboratory.

In place of elastic modulus test supersonic velocities were measured under comression up to about 60MPa on dry and water saturated specimens. From the test results which had been presented elswhere (1986) the following points are suggested.

Practically speaking, discontinuity surfaces in this area will be closed at tne overburden pressure roughly 50MPa which corresponds to rock overburden about 200m. Oppositely speaking discontinuity surface will more prevail in the shallower part than this depth and werthering process will be activated as the depth become shallower.

When the opening of discontinuity surface is wider as normally called to be joint or fault, this phenomenon will become more serious. These observable and dormant discontinuity surfaces govern endogenically the strength properties of mountain rock mass.

4. POTENTIAL COLLPAIBILITY

4.1. DIVERSION OF RQD

Principle of expression of mechanical property of a rock mass containing discontinuity surfaces is based on the combination of factor of constituent intact rock and factor of discontinuity. Then on the problem of the sample area where the strength property of constituent rock can be regarded as invariable, strength property of rock mass can be expressed by the factor of discontinuity only.

RQD is a practical method to express the soundness of rockmass including the mechanical properties of its constituent rocks (DEERE 1961). However in this case as

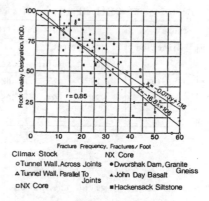

Cllmax Stock
- Tunnel Wall, Across Joints
- Tunnel Wall, Parallel To Joints
- NX Core

NX Core
- Dworshak Dam, Granite Gneiss
- John Day Basalt
- Hackensack Siltstone

Fig. 11 Relation of RQD and frequency of discontinuity surface.

has been discussed above, it can be used mainly to designate the discontinuity frequency.

Fig. 11 is the one originally presented by DEERE and retouched by the author (1974) to obtain the correlation equation. High coeffecient of correlation as 0.85 shows that RQD and discontinuity frequency are in very good linear relation and it is inferred that when the properties of rock are not so variable as the case of this area the relation will be better. From this point frequencies or spacing of discontinuity surfaces observed on exposures are simply converted to RQD and vice versa.

RQD can be interpreted to be the degree of resistance by which the rock mass does not decrease its mechanical strength property by the existing discontinuity surfaces. Then when we take 1-RQD (in ratio) or 100-RQD (in %) and denoting it e, e signifies the decrement of mechanical property due to discontinuity surfaces. It must imply the potential collapsibility of the mass.

Columns e in Table 3 show the value e converted from measurements of joint spacing on several to dozens exposures of each bed.

4.2. POTENTIAL COLLAPSIBILITY

When a bed i with uniform potential collapsibility occupies area A_i on a mountain slope, this bed is thought to share potential collapsibility e_iA_i on the surface. When depth d_i of this bed suffers mechanical erosive stress from the surface $d_ie_iA_i$ will be the potential share of bed i to mass movement of the slope.

For the area which is composed of n beds the total potentiality E to mass movement is given by $\sum_1^n d_ie_iA_i$, where d_i is quite unknown. It will be small where e is small and may be larger where e is larger.

Presently on the bold assumption that e is not so differnt by bed as a whole, by taking d as an arbitrary unit value, potential collapsibility of the total area is given by $e = \sum_1^n e_iA_i$, where the dimension of E is length to the 3rd power i.e. volume and e must be given by ratio not by per cent.

Table 3 where the values of e are expressed in percent, summarizes these values on the provision that e for talus be 100. e is assumed 25 and 23 for schalstein and Jurassic sandstone respectively for the practical convenience.

T a b l e 5 Collapsibility and its related quantities

slope direction	collapsibility				volume of talus deposit (10^{6}m^3)	lineament density (km/km^2)
	coefficient					
	basement	total	basement	total		
north	29	33	62	72	0.535	5.21
south	31	34	105	118	0.630	Pal. 6.92
						Jur. .0.85

Total eA in Table 3 gives the potential collapsibility of the sample area provided e is invariable throughout the bed. If e varies by point, it may be combined according to equal e. In the case when the distribution area of basement rock only is taken up E for basement rock is expressed and so on.

5. RESULTS AND DISCUSSION

Potential collapsibility E is presumed to have an intimate relation with mass moved or eroded quantity and with topographical elements that are taken up widely to be ralated to mass movement.

5.1. RELATION TO QUANTITY OF DETRITAL DEPOSIT

Volume of talus on the north and south slopes is shown in TAble 5 on the assumption that the mean depth of talus is 5m using teh results of seismic prospecting.

Volume of detrital deposit by past mass movement along the tributaries KML and NKL along the upper reaches including the sample area is as given in Table 5.

When the potential collapsibility of basement rock or total sample area is taken as abscissa and those values of detrital deposit are plotted on the ordinate, the results fall as shown in Fig. 12. The plottings for talus are in a positive relation with the potential collapsibility of basement rock and the plottings of the volume of detritus by near past mass movements are in another positive relations with the collapsibility of total area or basement rock of the sample area.

5.2. RELATION TO LINEAMENT

Plottings of lineament density against the potential collapsibility of basement rock are also known to be in a positive relation and it reflects an important implication that the substantial geology results in a topographical erosion index at the initial stage on the upper mountain reaches.

5.3. COEFFECIENT OF COLLAPSIBILITY

Dimension of potential collapsibility in Tables 3 and 5 and Fig. 12 are given by taking the unit of e in percent. These values must be divided by 100 to oabtain E in km unit.

Next dividing E by $\sum_{i}^{n} d_i A_i$ we get the dimensionless coefficient of potential collapsibility ß which means potential collapsibility of near surface unit volume of the mountain.

In this study case ß is obtained simply by dividing total eA by total A in Table 3 and we get the following values.

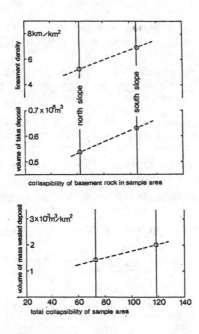

Fig. 12 Relation of potential collapsibility
versus volume of detrital deposit and
lineament density.

	basement rock	as a whole
north slope	0.30	0.33
south slope	0.31	0.34

If we know ß by any means we can estimate the potential collapsibility by multiplying it by the concerned volume.

ACKNOWLEDGEMENT

This paper presents a part of the results of the studies on natural disasters while tht author had been serving as the professor of Rock Engineering Chair, Department of Geotechnology, Saitama University, succeeded by Prof. R. Yoshinaka. The author is grateful to the subsidiaries of the Ministry of Education for this study at that time and he also is much grateful to the help of the present faculties and staffs of the Chair for arranging the manuscript of this paper.

REFERENCES

DEERE D.E., HENDRON A.N., PATTON F.D., CORDING F.M. (1967): Design of surface and near-surface construction in rock, Proc. 8th Symp. Rock Mech., Minneapolis, AIME. pp. 237-302.

DUNCAN N. (1969): Engineering geology and rock mechanics, VOL. 1, pp. 103-110, Leonard Hill, London.

HORTON R.F. (1945): erosional development of streams and their drainage basins, hydrophysical approach to quantitative morphology, Bull. Geol. Soc. Am., vol. 56, pp. 275-370.

JUDD W.R.,HUBER C. (1961): Correlation of rock properties by statistical methods, Int'l. Symp. on Mining Res., rolla, Mo.

ONODERA T.F. (1961): Directional physiography and the mechanical anisotropy of rock in the region suffering habitual disater by mass movement, Proc. 8th Symp. Nat. Disaster, pp. 183-186. (in Japanese)

ONODERA T.F. (1974):Interrelation of physical pro erties of rock, Engrg, Properties of Rock mass and their application to design and excecution, Chapt. 4, pp.229-240, JSSMFE Publication, Tokyo. (in Japanese)

ONODERA T.F. (1986): Minor physiography and mechanical properties of rock as affected by progressive failure of the earth crust (I),(II), Engineering Geology in the Construction of Asia, IAEG, Science Press in China , Beijing. (in press)

STRAHLER A.N. (1952): Hypsometric (area-altitude) analysis on erosional topography, Bull. Geol. Soc. Am.vol. 63, pp.1117-1142.

STRAHLER A.N. (1957): Quantitative analysis of watershed geomorphology. Trans AGU vol. 38, pp. 913-920.

TAKATA A.,MAZAKI S. (1933): Rock material for pavement, Rept. Civ. Engerg. Expt. Sta., vo. 24 Supplement.(in Japanese)

TANIGUCHI T., MURANO T., IZUMI I. (1961): Report of investigation distribution characteristics of mass movement area in Saitama Pref., Rept. Pub. Works Res. Inst. (in Japanese)

TSUKAMOTO R., HIRANO Y., SHINOHARA S. (1973): Study on the growth of stream channels (III)- Relationship between zero order channels and landslides, Shin-Sabo, vol. 26, pp14-20. (in Japanese)

Distribution and characteristics of rock types of Singapore

Distribution et caractéristiques de différents types de roches à Singapour

S.D.Ramaswamy, *National University of Singapore*

ABSTRACT: The Characteristics of rock types of Singapore island and their distribution are becoming clearer only in the past decade as revealed by rapid urban constructions such as highrise foundations and tunnels. The main solid formations reveal igneous and sedimentary rocks which are often deeply weathered and frequently exhibit an abrupt interface between completely weathered and slightly weathered rocks. Igneous rocks exhibit a high degree of fracturing while the sedimentary formations show up a high but variable dip. The Northern and central parts of the island consist of igneous rock while the rocks in the Western and Southern parts are of sedimentary origin. The transition between the igneous and sedimentary rocks is often abrupt. The Eastern areas the bedrocks are overlain by variable quarternary deposits consisting of soft alluvial clay and peaty clay.

In this paper, the author describes the general distribution and characteristics of rocks in Singapore based on his research findings as well as from the recent data available from construction sites and publications.

RESUME: Seulement dans les derniers dix ans ont été mis en évidence les divers types de roches qui constituent l'île de Singapore. Ceci a été possible grâce a l'évolution rapide des constructions urbaines et souterraines. Les roches solides sont distribuées de la maniere suivante: au Nord et au centre de pays il y a sourtout des roches eruptives, tandis que au Sud et a l'Ouest les formations plus répandues sont d'origine sédimentaire, la transition est fréquement abrupte. A l'Est la couverture du"bed-rock" est composée par des argilles molles et des argilles tourbeuses. L'auteur décrit la distribution générale et les caractéristiques des différents types de roches sur la base de ses rechereches, de la bibliographie existante et à partir des nouvelles informations des constructions récentes.

1. INTRODUCTION

The location of Singapore island is towards the extreme south of the Southerly projection of the Eurasian tectonic plate. Although just north of the Java Tranch which is part of the Northerly terminations of the Indian Plate, the island is free from the influence of tectonic activity.

The rocks of Singapore (PWD, 1976) primarily consist of four solid series as shown in fig. 1.

Jurong Formation	- Upper Triassic, Lower and Middle Jurassic
Gombak Norite	- Upper Palaeozoic
Bukit Timah Granite	- Lower and Middle Triassic
Sajahat Formation	- Lower Palaeozoic

Table 1 shows a more detailed succession with a brief listing of lithological characteristics. A fair conception of the distribution and characteristics of rock types of Singapore has emerged from the recent investigations by the author and others. (Ramaswamy 1975 a, 1975 b, Ramaswamy and Aziz, 1977, 1981; Ramaswamy et al 1978, Pitts, 1984 a, 1984 b; Aziz and Ramaswamy, 1978)

The Jurong formation consists of sedimentary rocks belonging to late Triassic and Lower to Middle Jurassic age. The rock types consist of conglomorates, sand stone and shale deposits. The Sajahat formation is known to exist beneath younger deposits in the eastern part of the island and extend over a small area only. Rocks of basic composition forming the Gombak group have intruded into the Sajahat formation. These intrusions are primarily gabbroic and noritic

Table 1 Lithological characteristics

PALAEOZOIC		MESOZOIC				No local rocks of later Mesozoic and early and mid Cainozoic ages exist	CAINOZOIC				Formation / Description
		Triassic			Jurassic		Pleistocene		Holocene		
Lower	Upper	Lower	Middle	Upper	Lower		Early	Late	5000 BP	Present	
										X	**KALLANG FORMATION** — Reef member: coral, unconsolidated calcareous sand, some quartz, iron-cemented sand
										X	Transitional member: unconsolidated dark mud, muddy sand or sand with peaty layers.
										X	Littoral member: well sorted unconsolidated beach quartz-sand with some laterite, shell and sandstone fragments. Iron-cemented beach rock also exists.
									XXX	XX-	Alluvial member: alluvial pebble beds, sands, muddy sands clays and peat.
								X	XX	XXX	Marine member: mainly unconsolidated blue-grey clayey mud with peat and sand horizons.
									X		**TEKONG FORMATION:** unconsolidated sand with some cobbles.
							XX				**HUAT CHOE FORMATION:** white kaolinite-rich clay and occasional quartz gravel.
							XX				**OLD ALLUVIUM:** loose quartz-feldspar sand and gravel with occasional weak sandstone and conglomerate.
		X									**JURONG FORMATION** — Murai Schist: mudrock cleaved and sheared by dynamic metamorphism on thrust faults.
		X	X								Tengah Facies: muddy sandstones with grits and conglomerates. Poorly cemented.
		X	X								St John Facies: muddy sandstones with carbonaceous laminations.
		X									Rimau Facies: well cemented quartz sandstone to quartz conglomerate.
		X	X								Ayer Chawan Facies: well bedded, mainly black sandstones and mudrocks; red conglomerates, basic lava, and volcanic ash are also included.
		X									JONG Facies: well cemented conglomerate and sandstones and occasional mudrocks and basic lavas.
		X	X								Queenstown Facies: red to purple mudrocks, sandstones and occasional conglomerates; some volcanic ash also occurs.
	X										**DYKE ROCKS.** Acid (older) and basic (younger) dykes intruded into all older rocks.
	X	X									**BUKIT TIMAH GRANITE.** Mainly acid igneous rocks, but with some less acid forms due to mixing of the granite and Gombak Intrusives e.g. a 'Hybrid' granodiorite is recognised.
X											**PALAEOZOIC VOLCANICS.** Coarse and fine grained pyroclastic rocks. Mainly on offshore islands.
X											**GOMBAK INTRUSIVES.** Basic igneous rocks of mainly noritic and gabbroic composition. Now much altered by later intrusion.
X											**SAJAHAT FORMATION.** Hard quartzite, sandstone and mudrocks.

rocks. Both these series are intruded in turn by a more acid sequence of granitie rocks. The plutonic rocks are intruded by small dykes of both doleritic and aplitic composition (pitts 1984). These plutonic series of rocks are quaried extensively for use in construction. The granitic rocks in particular produce high class aggregates for use in road construction and for making concrete.

2. DISTRIBUTION OF ROCK TYPES IN SINGAPORE

Basic rocks of Singapore, shown in Fig. 1, are composed of sedimentary and igneous rock types (Geology of the Republic of Singapore, 1976). Each group occurs fairly in well defined areas

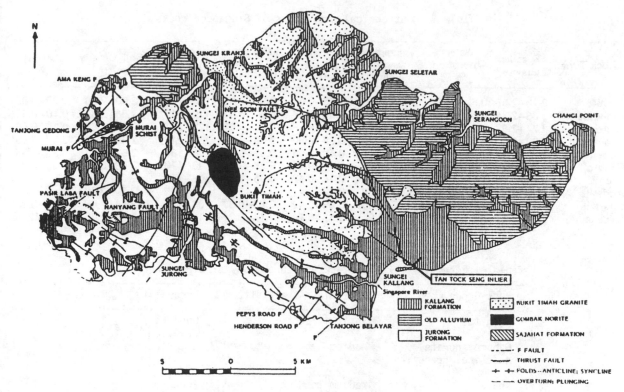

Figure 1. Rocks of Singapore

Table 2 Analysis of granitic group

Constituents	Granite PWD Quarry Mandai	Granite PWD Quarry Pulau Ubin	Granodiorite Swee Constr. Quarry Buket Timah	Microgranite Gim Huat Quarry	Microgranite Aik Hwa Quarry No. 1	Adamellite Singapore Granite Quarry Buket Timah	Gabbro Yun Onn Quarry
SiO_2	76.55	71.69	68.98	69.92	69.09	75.60	51.52
Al_2O_3	12.78	17.26	13.80	14.30	14.82	13.03	13.29
Fe_2O_3	0.38	—	0.48	—	—	0.11	0.30
FeO	0.37	--	3.44	—	—	1.44	9.91
Total iron as FeO	—	2.47	—	3.78	3.82	—	—
MgO	0.12	0.28	1.56	0.81	0.80	0.21	12.31
CaO	1.46	2.63	4.32	2.88	2.28	0.84	8.77
Na_2O	2.56	3.16	2.86	3.67	3.77	3.58	1.01
K_2O	5.04	2.01	3.26	3.32	3.29	4.17	0.39
H_2O+	0.33	--	0.07	—	—	0.57	1.54
H_2O-	0.22	—	—	—	—	0.13	—
TiO_2	trance	0.23	0.49	0.27	0.24	0.16	0.58
P_2O_5	trance	0.13	0.08	0.05	0.11	0.06	0.06
MnO	trance	0.06	0.02	0.08	0.05	—	0.29
CO_2	—	—	0.03	—	—	—	—
Total	99.81	99.92	99.82	99.08	99.27	99.90	99.97

possessing distinct type of terrain. All sedimentary rocks mainly shales occur in South, Southeastern and southwestern parts of Singapore, Pulau Tekong and a group of southern islands. The term shale is normally used for all sedimentary type weak rocks which encompass mudstone, siltstone and claystone (Ramaswamy and Aziz, 1977). Igneous rocks are divided into the central Singapore granite, lying west of the Pongol-Siglap line and Changi-Pulau Ubin granite. The term granite is used in a general sense for the entire family of acid rocks including granite, adamellite, microgranite and the acid and intermediate hybrids mainly of

Table 3 Geotechnical properties of Singapore rocks

Rock Type	Specific Gravity	Water Absorption %	Aggregate Abrasion Value	Polished Stone Coefficient	Sideways Force Coefficient	Compressive Strength N/mm^2	Bulk Density gm/cm^3	Porosity %
Granite	2.73–2.85	0.50–0.80	6.00–7.60	0.56–0.65	0.53–0.56	130–220	2.60–2.88	0.20–0.85
Granodiorite	2.68–2.74	0.65–1.10	5.00–6.40	0.54–0.63	0.50–0.53	120–215	2.56–2.85	0.10–0.60
Microgranite	2.74–2.83	0.60–0.90	6.50–7.40	0.55–0.58	0.48–0.51	125–220	2.61–2.82	0.15–0.65
Adamellite	2.69–2.82	0.80–1.34	5.80–6.20	0.52–0.59	0.49–0.52	120–210	2.58–2.90	0.20–0.90
Dolerite	2.68–2.77	0.96–1.26	4.30–6.10	0.50–0.56	0.42–0.47	100–180	2.53–2.67	0.25–0.85
Gabbro	2.65–2.72	1.35–2.35	4.40–5.70	0.42–0.46	0.36–0.43	90–170	2.58–2.90	0.2 –0.8
Norite	2.64–2.70	1.40–2.40	3.00–3.80	0.38–0.42	0.35–0.38	60–150	2.35–2.60	–
Andesite	2.63–2.68	1.30–2.25	3.50–4.80	0.40–0.45	0.36–0.39	65–165	2.50–2.60	–
Quartzite	2.62–2.67	1.80–2.42	2.4 –3.20	0.35–0.42	0.33–0.37	80–155	2.54–2.68	0.15–0.86
Shale group	2.32–2.62	2.20–5.60	1.22–4.0	0.15–0.29	0.14–0.22	15–80	2.00–2.45	8.00–12.50

Table 4 Potential alkali-attack and durability of Singapore rocks

Rock Group	Potential Alkali Attack	Durability
Granites	Insignificant	Good to excellent
Gabbro, norite, andesite, quartzite	Medium	Satisfactory
shales	Medium to High	Poor to satisfactory

granodioritic and dioritic composition, resulting from the assimilation of basic rocks within
Other types of rocks norite, gabbro and noritic gabbro are also found to be exposed at Bukit
Panjang and Bukit Gombak. Analysis of some Singapore rocks is given in Table 2 (Geology of the
Republic of Singapore, 1976.)

3. GEOTECHNICAL PROPERTIES OF IGNIOUS ROCKS

Rock samples were collected from various locations of the Republic of Singapore and tested for
various geotechnical properties. The results are presented in Table 3. It is observed that the
properties of various rocks in the granite group consisting of granite, granodiorite,
microgranite, dolerite and adamellite are almost similar. The other group comprising gabbro,
norite, andesite and quartzite distrinctly exhibits inferior qualities in comparison with those
of the granite group. The shale group obviously exhibits inferior properties. After thorough
testing, the potential alkali attack and durability of Singapore rocks were ascertained and the
results are presented in Table 4. The group of rocks comprising gabbro, norite, andesite and
quarzite are found to be moderately strong and are suitable for making medium strength concrete
having satisfactory durability. The granitic rocks can be used to produce strong and highly
durable concrete with insignificant alkali attack. Crushed granite fine aggregates can be used
where high strength concrete is essential for some special concrete structures.

4. DISTRIBUTION OF SHALES IN SINGAPORE

Shale deposits are distributed over the entire south, southeast and southwest of Singapore, and
also on Pulau Tekong and the Southern island. The deposits are considered to be of shallow
water origin and are probably estuarine. Shale deposits generally fall under three groups: (a)
micaceous, (b) argillaceous and (c) arenaceous. The thickness of beds is variable, measuring
about 800 metres deep in some places. The beds have a general dip away from the central granite
deposit but dips are variable. In some places, the dips are almost vertical or even dip towards
the granite deposits (Ramaswamy, 1975). The general composition of Singapore Shales is shown
in Table 5.

Table 5 Composition of Singapore shales

Chemical Composition*		Mineralogical Composition	
Component	%	Component	%
SiO_2	48–58	Clay minerals	26–58
Al_2O_3	12–15	Quartz	26–36
Fe_2O_3	4–6	Feldspar	7–18
FeO	2–4	Organic matter	1–2
MgO	2–3	Limonite hematite and pyrite	1–7
CaO	1–2	Other minerals	trace
Na_2O	1–2		
K_2O	1–3		
MnO	trace		
CO_2	1–3		
SO_3	0.1–0.6		
Misc.	trace		

* Individual sample showed wide variations in almost all components

Table 6. Properties of Singapore shales

Location	Natural Water Content %	Saturated Water Content %	Specific Gravity	Natural Density gm/cm^3	Crushing Strength MN/m^2	Aggregate Abrasion Value	Polished Stone Coefficient	California Bearing Ratio (CBR), %	
								Unsoaked	Soaked in Water for 4 days
Tanjong Pagar	1.46–2.69	3.86–4.32	2.41–2.56	2.10–2.54	1.32–1.65	2.20–2.64	0.24–0.26	50–60	22–28
Keppel Road	2.36–2.84	4.15–4.86	2.32–2.48	2.00–2.49	0.82–1.36	2.24–2.80	0.21–0.27	50–60	23–27
Mt. Faber	2.85–3.20	4.34–5.60	2.40–2.52	2.15–2.53	0.96–1.72	1.30–2.34	0.20–0.25	–	–
Pasir Panjang	1.82–3.45	4.25–5.74	2.43–2.56	2.35–2.60	1.22–1.73	1.95–2.73	0.23–0.28	–	–
Kent Ridge	1.26–2.34	3.46–4.23	2.49–2.62	2.40–2.61	1.43–1.76	1.86 2.30	0.24–0.29	50–00	22–30
Ulu Pandan	1.35–3.16	3.55–4.30	2.47–2.58	2.32–2.57	1.30–1.68	1.34–3.60	0.21–0.25	50–60	20–27
Faber Avenue	1.38–2.85	2.20–3.40	2.46–2.55	2.24–2.36	1.13–1.46	2.23–2.92	0.18–0.24	–	–
Jurong	2.56–2.43	3.72–4.63	2.44–2.57	2.26–2.33	0.82–1.24	1.65–3.84	0.16–0.23	50–50	20–25
Chua Chu Kang Road	3.15–4.50	3.85–5.20	2.40–2.52	2.28–2.47	0.73–1.18	1.22–4.10	0.18–0.22	–	–
Pulau Tekong	3.40–4.75	3.70–5.52	2.38–2.54	2.23–2.55	0.68–1.16	1.30–3.82	0.15–0.23	50–55	21–26

4.1 Properties of Singapore Shales

Shale samples collected from various locations of the Republic of Singapore and laboratory investigations carried out revealed results presented in Tables 6.

5. SANDSTONE BOULDER DEPOSIT

A boulder bed forms part of the Jurong Formation. The deposit consists of boulders of sandstone normally up to 2 or 3 m^3. The boulders are cemented together with stiff, overconsolidated silty clay. The proportion of silty clay to boulders varies generally from 40% to 50% at shallow depths to 90% at lower levels (Sehested, 1960). The boulders are generally quite fresh with little signs of weathering. The boulder bed forms a lozenge-shaped deposit from the Southern part of the central business district north wards and it thins quickly to the east and disapapears towards Marina Bay. Some of the highrise developments have been built on the boulder deposit utilising either deep caissons or thick mats.

CONCLUSIONS

A fairly clear picture about the distribution and characteristics of rock types of Singapore has emerged during the past decade primarily due to the enormous construction activities involving deep excavations and tunnels. The publication on the geology of Singapore by the Public Works Department contributed significantly to the existing knowledge. Several publications later appeared over the past few years enhancing the understanding of the Singapore rock types. Basically, it is clear now that the igneous acid rocks and sedimentary rocks occur in distinct areas. Although many variations in such rocks are found, from the engineering construction point of view sedimentary types are found to be of little value where as the igneous types of which granite is most widely distributed are of immense value.

REFERENCES

Aziz, M A and S D Ramaswamy, 1978. Some geotechnical properties of Singapore shale and its potential use in road construction. Proceedings of the 6th Asian Regional Conference on Soil Mechanics and Foundation Engineering, Vol. 1, p. 201-204.

Pitts, John, 1984a. A survey of engineering geology in Singapore, Geotechnical Engineering, Vol. 15, Bangkok, p. 1-20.

Pitts, John 1984b. A review of geology and engineering geology in Singapore, Quarterely Journal of Engineering Geology, Vol. 17, Londodn, p. 93-101.

Public Works Department, Singapore, 1976. Geology of the Republic of Singapore, Public Works Department, Singapore, p. 79.

Ramaswamy, S D 1975a. Some geological problems of Singapore as applied to civil engineering. Proceedings of the Regional Conference on Geology and Mineral Resources of South East Asia, Jakarta, p. 213-221.

Ramaswamy, S D 1975b. Regional deposits of Singapore. Proceedings of the Fifth Regional Conference on Soil Mechanics and Foundation Engineering, Vol. 2, Bangalore, p. 82-87.

Ramaswamy, S D and M A Aziz, 1977. Identification and excavability of weak rocks of Singapore, Journal of the Institution of Engineers of Singapore, Vol. 17, p. 461-473.

Ramaswamy, S D and M A Aziz, 1981. Importance of polished stone coefficient in the choice of stones for road surfaces in Singapore, Proceedings of the Third Conference of Road Engineering Association of Asia and Australasia, Taipei, Vol. 1, p. 461-473.

Ramaswamy, S D, M A Aziz and C K Murthy, 1978. Selection of Singapore rocks for durable concrete, proceedings of the International Conference on Materials, Bangkok, Vol. 1, p. 233-237.

Sehested K G 1960. The geology of Singapore as seen by a civil engineer, Journal of Singapore-Malaya Joint Overseas Engineering Group, p. 8-18.

Geomechanics classification for support selection in Indian coal mines – Case studies

Classification géomécanique pour le choix des appuis dans des mines de charbon en Inde – Exemples

A.Sinha & V.Venkateswarlu, *Central Mining Research Station, Dhanbad, India*

ABSTRACT: Engineering classification of rock masses has been widely applied for tunnels, and in mining. It was found that the several classification systems in vogue are not applicable to coal mine rocks in India because the ground conditions here are different. Hence a new rock mass classification, "CMRS Geomechanics Classification", has been developed for support design in coal mines. Based on extensive geotechnical studies in different coalfields of India, five parameters were selected: 1) layer thickness, 2) weatherability, 3) strength, 4) groundwater, and 5) structural features. Appropriate ratings were allocated for each parameter. Rock mass rating (RMR) is obtained as a sum of the ratings for individual parameter values. Adjustments are made, and the final value classifies the roof into five classes ranging from Very Poor to Very Good. Support guidelines have been prepared from correlations between RMR and effective support systems. The effectiveness of the new system is discussed with three case studies.

1 INTRODUCTION

Engineering geologic investigations in recent years are relying more and more on classification systems for characterization of rock masses and for a more accurate prediction of ground conditions. A wide range of classification approaches are currently in use, and many of them find application in mining excavations also. Particularly in coal mines, where multi-opening excavations are made in heterogenous bedded strata, the selection of effective supports needs a rational approach like the classification.

2 NEED FOR A NEW CLASSIFICATION

The available systems of classification, specifically Q-system (Barton, Lien & Lunde 1974), and RMR approach (Bieniawsky 1974), were applied for Indian coal mines. However, it was observed that they do not properly represent the actual roof conditions. For example, the generally known stable roof conditions (requiring light supports) in New Kenda colliery were classified as "fair" in RMR system and as "very good" in Q-system; the supports suggested in the two approaches are also not practised in India. In a similar way, the generally observed fair condition at Lachipur and bad roof condition at Bhelatand were classified as otherwise by the two systems, with support recommendations totally at variance with the normal support measures in these mines.

The reason for non-applicability of the two systems to Indian coal mines lies basically on the selection of parameters. In Q-system more importance is given to joint attributes whereas in coal measure strata bedding planes play a greater role. Joint properties like roughness and alteration have little variation from mine to mine. The SRF values are also not relevant to coal mining stress fields. In Bieniawsky's system the parameters are based on conditions existing in American coal mines. In this also due importance is not given to factors causing roof problems in India.

3 APPROACH TO CMRS GEOMECHANICS CLASSIFICATION

Because of the difficulty in applying other rock mass classifications, the need
was felt to develop a system particularly suitable for Indian coal mines. Thus
a new classification, named as CMRS Geomechanics Classification was developed
taking into consideration the existing geomining conditions in coal consideration
the existing geomining conditions in coal mines of India (Venkateswarlu, Sinha &
Raju 1985).

The first step in the development of any new system is collection of data. For
this, all the relevant field information as well as the physico-mechanical properties of the roof rocks were collected through detailed geotechnical investigations
in different coalfields of the country. Statistical analyses were carried out to
determine their relative importance. Finally, five parameters have been selected.
They are: 1) RQD, 2) rock strength, 3) groundwater seepage, 4) rock weatherability,
and 5) structural features.

As the present classification is basically derived from Bieniawsky's system,
many of the terms like RMR are common to both. It can be seen that the first
three parameters are similar to those in Bieniawsky's classification. Emphasis
has been given in the present system on weatherability of rocks and structural
features, (which were not considered in the other classification), because they
are the chief factors contributing to roof deterioration in India.

4 TESTING PROCEDURES AND RATINGS

Rock Quality Designation (RQD) is the index for layer thickness of roof rocks.
This is obtained from double barrel diamond drilling vertically into the roof.
The percent recovery of core pieces longer than or equal to 10 cm is taken as RQD.
In the absence of drilling facilities underground, the layer thickness (S, in cm)
as measured in any roof fall area can be converted to RQD by the relation:

$$RQD = 100 (0.1\lambda + 1)e^{-0.1\lambda}$$

where λ is layer frequency (=100/S)

Strength of the rock is measured in the field using point load tester on any
irregular sample. The following empirical relation was obtained between point
load index (I_{pl}) and uniaxial compressive strength (σ_c):

$$\sigma_c = 14 I_{pl}$$

Rate of groundwater seepage is measured by drilling a 1.5 m long hole in the
roof and collecting the water percolating through the hole. It is expressed in
ml per minute.

Weathering property of rocks is determined by ISRM slake durability test. First
cycle durability index is taken for the classification.

All the geological features are recorded through detailed geotechnical investigations. Anomalous sedimentary and structural features are assessed for their
unfavourability.

Depending on statistical analyses, the maximum ratings allocated for the five
parameters are:

R Q D	30%
Structural features	25%
Weatherability	20%
Rock strength	15%
Groundwater	10%

Further subdivisions in each parameter are made according to their values.

5 RMR AND ADJUSTMENTS

Ratings are given separately for each rock type in the immediate roof. In the case of structural features, they are awarded on the basis of the geologist's judgement about their influence on the roof stability. For the others, the parameter value determines the rating (Table 1). The ratings for the five parameters are added together to get the rock mass rating (RMR). The RMR for the whole roof is obtained by weighting the RMR of each bed with its thickness as follows:

$$\text{Combined RMR} = \frac{\text{(RMR x bed thickness)}}{\text{total thickness of the beds}}$$

In order to account for some special factors which are not directly considered in the calculation of RMR but which contribute to roof failure, some adjustments are made to the RMR. For example, it was found that in workings of the same seam, the roof conditions deteriorate with increasing depth. The following adjustments were introduced for depth correction:

Depth	Adjustment	Adjusted RMR
Less than 300 m	none	RMR x 1.0
300-400 m	10% reduction	RMR x 0.9
400-500 m	20% reduction	RMR x 0.8

Similarly, it was seen in the same type of roof, the introduction of continuous miners improved the roof condition substantially, while in districts with solid blasting the conditions worsened. In the former case, 10% addition is made to RMR, while in the latter case 10% reduction is made. No correction is required where blasting with undercut is the normal practice. Other adjustments, like for lateral stresses, may be introduced depending on specific conditions encountered in a mine.

6 CLASSIFICATION AND THE SUPPORT SYSTEMS

According to the value of the adjusted RMR, the roof is classified into one of the five classes ranging from Very Good to Very Poor, as follows:

RMR	Class	Description
80 - 100	I	Very Good
60 - 80	II	Good
40 - 60	III	Fair
20 - 40	IV	Poor
0 - 20	V	Very Poor

The CMRS Geomechanics Classification has been correlated with effective support measures. For each group of RMR, design guidelines have been prepared (Table 2). These are based on experiences with different support systems from various mines in India.

7 APPLICATION OF THE CLASSIFICATION

Till the time of writing this paper, the classification has been applied to 47 coal mines. The supports have been found to be satisfactory. As examples for the application of the CMRS Geomechanics Classification, three case studies are given below. They are Bhelatand ("Poor"), Lachipur ("Fair") and New Kenda ("Good") collieries.

Table 1 Geomechanics classification ratings for parameters

PARAMETER		RANGE OF VALUES				
1 Spacing of horizontal weakness planes						
a) R Q D	(%)	0 – 25	25 – 50	50 – 75	75 – 90	90 – 100
b) Layer thickness	(cm)	< 2.5	2.5–7.5	7.5–20	20 – 50	> 50
	Rating	0–4	4–12	12–20	20 – 26	26 – 30
2 Intact rock strength	(kg/cm^2)	< 100	100–300	300–600	600–900	> 900
	Rating	0–2	3–6	7–10	11–13	14 – 15
3 Groundwater flow rate	(ml/min.)	> 2000	2000–200	200–20	20–0	– dry –
	Rating	0–1	2–4	5–7	8–9	10
4 Weatherability (I Slake durability)	(%)	< 60	60–85	85–97	97–99	> 99
	Rating	0–3	4–8	9–13	14–17	18 – 20
5 Structural features	Rating	0–4	5–10	11–16	17–21	22 – 25
R M R		0–20	20–40	40–60	60–80	80 – 100
CLASS		V	IV	III	II	I
DESCRIPTION		Very poor	Poor	Fair	Good	very good

Table 2 Guidelines for supports (bolting) based on classification

R M R	Permnent roadways	Temporary roadways
90–100	none required	none required
80– 90	spot bolting	spot bolting
70 – 80	regular bolting	bolting in structurally disturbed areas
60 – 70	full-column grouted bolts in 1.2 x 1.2 m pattern	roof stitching or rope dowels
50 – 60	quick setting cement grouted bolts in 1.0 x 1.2 m pattern; wire netting	roof stitching and rope dowels
40 – 50	resin bolts, wire netting; steel straps connecting bolts	roof stitching with grouted bolts
30 – 40	resin bolts in 1.0 x 1.0 m pattern with steel channels	roof stitching at 1 m spacing supplemented by grouted bolts
20 – 30	resin bolts in 1.0 x 1.0 m pattern; wire netting and steel channels; steel props	resin anchored roof truss with grouted bolts or rope dowels
10 – 20	yielding steel arches	wooden or steel arches
0 – 20	cannot be supported	cannot be supported

7.1 Bhelatand Colliery –

In this mine, the Top section of No. 15 seam of Baraker Measures is being worked at a depth of 310 m. Sequence of the beds is as follows:

	fine grained sandstone	
Immediate roof	block/flaky shale	– 0.75 m
	coal – 0.6 m	
Working section	coal – 1.8 m	15 SEAM TOP

Since it was proposed to use dint header, the whole seam thickness will be taken, exposing the shale. So, both coal and shale beds in the roof were classified. The parameter values and the ratings for the two are shown below:

	Parameter	Coal Value	Coal Rating	Shale Value	Shale Rating
1	Layer thickness	1.5 cm	8	3.57 cm	5
2	Rock strength	89 kg/cm^2	2	375 kg/cm^2	7
3	Slaking index	99.26%	18	98.8%	16
4	Groundwater	1.5 l/min	1	1.5 l/min	1
5	Structural features	unfavourable cleats	15	close-spaced joints, slips	10
	R M R		44		39

Coal-shale combined RMR $= \dfrac{(44 \times 0.6) + (39 \times 0.75)}{(0.6 + 0.75)} = 41.2$

Depth correction is 10% reduction (depth more than 300 m). Therefore,

Adjusted RMR = 41.2 x 0.9 = 37.1

The final RMR of 37.1 classifies the roof as IV B "Poor Roof". The supports recommended for this were full-column grouted bolts at 1.2 m spacing and connected with steel straps. Wire netting shall be used if coal is to be taken down. However, as the RMR for the shale roof comes to 35, it was advised to leave 20 to 30 cm thick coal in the roof to protect the shale. This support system is being practised in the mine and it is found to be working satisfactorily.

7.2 Lachipur Colliery –

The formations in the immediate roof here are 1.0 m shale overlain by about 1.0 m thick laminated sandstone. In addition, 0.5 m of the 3.0 m thick Sonachora seam is also being left in the roof. The summarised classification for the three beds is given in the accompanying table.

	Parameter	Coal Value	Coal Rating	Shale Value	Shale Rating	Sandstone Value	Sandstone Rating
1	R Q D (%)	0	0	16	2	44	13
2	Strength(kg/cm^2)	425	8	124.4	3	464.5	9
3	Slaking Index (%)	98	15	68	5	96.4	12
4	Water seepage	damp	9	damp	9	damp	9
5	Structural features	cleats & minor slips	18	joints & minor slips	18	joints & minor slips	18

	Coal	Shale	Sandstone
R M R	50	37	61

$$\text{Combined RMR} = \frac{(50 \times 0.5) + (37 \times 1.0) + (61 \times 1.0)}{0.5 + 1.0 + 1.0} = 49.2$$

With an RMR of 49.2, the roof was classified as Class IIIA "Fair roof". Resin anchored bolts at 1.0 m interval (and a row spacing of 1.2 m) were recommended for this roof. This system of support has been working well.

7.3 New Kenda Colliery -

An 8.5 m thick seam is being worked in this mine at a shallow depth of 30 m. Top section workings were studied where 0.6 m coal is being left as the immediate roof. Following is the roof character:

Parameter	Value	Rating
1 Layer thickness	9 cm	17
2 Rock strength	350 kg/cm^2	7
3 Slakke durability index	99%	17
4 Groundwater seepage rate	dry	10
5 Structural features	random joints	21
R M R		72

The RMR value of 72 categorises the roof into Class II B ("Good roof"). For this, the only support required is spot bolting. In the mine also, bolting is being done only in structurally disturbed areas.

8 CONCLUSION

Experience with the use of CMRS Geomechanics Classification, together with the support systems, has been encouraging. As there were no proper guidelines for support section till date, this Classification will fulfil a long felt need of a rational approach. It is proposed to extend the system for estimation of support loads also through instrumentation and monitoring in select mines.

ACKNOWLEDGEMENT

The work is part of the S&T project "Geomechanical Classification of Coal Measures Roof Rocks vis-a-vis Roof Supports". Sincere thanks are due to the Department of Coal and Central Mine Planning & Design Institute for the financial support. Grateful acknowledgements are also due to Dr NM Raju for his valuable suggestions. The authors also thank Dr B Singh, Director, CMRS for his encouragement and permission to publish this part of the work.

REFERENCES

Barton, N., R. Lien & J. Lunde 1974. Engineering classification of rock masses for the design of tunnel support. Rock Mechanics, 6: 189-236

Bieniawsky, Z.T. 1974: Geomechanical classification of rock masses and its application in tunneling. Proc. 3rd Cong. ISRM, Denver. 2A: 27-32

Venkateswarlu, V., A. Sinha & N.M. Raju 1985: Classification of coal mine roof rocks in India. Workshop on Engg. Classification of Rocks, New Delhi: 173-184

Engineering geological predictions in mineral deposits exploration
Prévisions géotechniques lors de l'étude de gisements de matériaux

B.V.Smirnov & A.I.Dymna, *All-Union Geological Research Institute, Rostov-on-Don, USSR*

ABSTRACT: One of the important tasks of engineering-geological study of mineral deposits is prediction of processes able to produce harmful or useful effects when developing these resources. The investigation has proved that the character and scale of these processes are generally determined sumultaneously by many natural and anthropogenic factors, rather than by any single one. It complicates the use of traditional prediction methods based on prediction classification, extrapolation, mathematic modelling etc. The best results are achieved with the help of methods of probabilistic recognition. To a high degree these methods correspond to a random nature of engineering-geological processes and to probabilistic character of prediction. They permit to consider practically any number of significant factors and implement system approach to predictive estimate of objects, to achieve a high share of correct solutions.

RESUME: Une des tâches importantes de l'étude géotechnique des gisements de minéraux utiles, c'est la prévision des processus qui sont susceptibles d'influencer leur exploitation de façon nuisible ou favorable. Les investigations ont démontré que le caractère et l'échelle de ces processus sont déterminés non pas par des facteurs isolés mais simultanément par de nombreux facteurs naturels et techniques. Ceci rend difficile l'utilisation des méthodes classiques de prévision basées sur les principes de classification prévisionnelle, d'extrapolation de simulation mathématique, etc. Les meilleurs résultats sont obtenus par l'emploi des méthodes d'identification probabiliste. Ces méthodes correspondent le plus au caractére aléatoire des processus géotechniques et au caractére probabiliste des prévisions. Elles permettent de tenir compte pratiquement de n'importe quel nombre de facteurs substantiels, de réaliser une approche systématique à l'évaluation prévisionnelle des objets et obtenir un pourcentage grand des solutions correctes.

Human impact on geological environment in the course of development of mineral deposits gives rise to various engineering-geological processes which express themselves in phenomena able to complicate, and sometimes to favour working conditions of mining enterprises. Influence of these processes and phenomena reaches its maximum when deep stratas of deposits are developed by large, highly mechanized mines. Development of mineral deposits at greater depths is usually accompanied by significant changes in rock stress, increase of their deformation, destruction and displacement, higher frequency of dynamic and gasodynamic phenomena, including those formely unknown (sudden rises of mine floor etc.). These dangerous processes and phenomena can be avoided and their harmful effects significantly weakened by special engineering measures. However, to take effective preventive measures, it is necessary to know beforehand where, when and which particular engineering-geological complications are likely to occur in the future mine. Gaining such information - engineering-geological prediction in deposit exploration - has become no less important and responsible task of geological survey then correct quantitative and qualitative estimate of explored reserves of mineral deposit.

To provide full and reliable engineering-geological predictions in exploration of coal deposits of the USSR, the authors and their co-workers were implementing the following main tasks:

1) collection, generalization and analysis of data on engineering-geological processes and phenomena in coal mines; regularities of their localization and influence on efficiency and safety of mining;

2) determination of a complex of geological and anthropogenic factors affecting occurrence and scale of engineering-geological processes and phenomena in underground mine openings, which must necessarily be considered when making predictions;

3) elucidation and expression of natural relations between engineering-geological processes and phenomena and factors which characterize them;

4) choice of most effective methods of engineering-geological prediction, allowing to use these relations;

5) estimate of reliability of predictions based on chosen methods;

6) setting up of automated systems of engineering-geological predictions.

Investigation of concrete forms of geological environment response to mining, analysis of modern concepts of the nature and the causes of this response allowed to isolate simple and complex engineering-geological processes and phenomena. The first of the above groups was assumed to include processes (phenomena) whose occurrence, nature and scale are generally determined by one or two geological factors which can be recorded on the basis of geological exploration data. Thus, for example, the group of simple processes includes creeping of rock into underground openings, which is typical of massifs with steeply dipping coal-bearing layers, separated by plastic clayer or coaly bands serving as sliding surfaces. Collapses of so-called false roof, formed when coal seam is overlain by thin layers of rock with low mechanical strength, are regarded as simple processes. This group of processes also includes subsidence of floor composed of mechanically weak, soakable rocks under the load developed by supports and mining equipment.

Complex engineering-geological processes and phenomena are determined by three or more nearly equally combined geological factors. This group includes dangerous collapses of immediate roof rocks into near-face area of producing openings, subsidence and failure of main roof, resulting in damage of openings and mining equipment, extreemly dangerous phenomena of sudden coal or rock-and-gas outbursts, and many other significant engineering-geological complications.

Establishment of a complex of significant factors of complicated processes (phenomena) was based on statistical estimate of relations between properties of lithogenetic, tectonic, hydrogeological, geochemical and other components of geological environment in particular areas of openings and characteristics of processes (phenomena) in the same areas. With this aim in view the factual data on several thousands of producing openings in Donetsk, Kuznetsk, Pechora and Karaganda coal basins were used. To measure the strength of relations values of mutual conjugation coefficients were calculated according to A.A.Chuprov, methods of factor analysis and estimation were used. As an example, results of investigation of factors, determining nature of subsidence and failure of main roofs into mine faces in Pechora basin are given. It has been found, that lithology of rock layers overlying coal seams, as well as their thickness, mechanical strength, jointing density, ratio of immediate roof and coal seam thicknesses are essential geological factors determining these dangerous phenomena. Complexity of this and other processes and phenomena is further increased due to influence of anthropogenic and technical factors along with geological ones: dimension of openings, their orientation, rate of face advance, means of rock pressure control etc.

Multi-factor dependence of complex engineering-geological processes and phenomena prevented us from establishing strictly determined relations (expressed in the form of algebraic, differential equations and systems of equations) governing these processes. Available for prediction are mainly statistical regularities, which are manifested through mass correlation of geological and anthropogenic parameters with characteristics of processes and phenomena of interest. Probabilistic models are a most suitable form of expression of statistic regularities and relations. These models are represented by the summary of parameters $P(B_i/A_j)$ - empirical frequences of occurrence of features related to essential factors, in actual realization of processes or phenomena or of their characteristics. Summary of distribution of investigated openings in coal mines of Pechora basin according to intensity of immediate roof collapse (A_j) and density (B_i) of tectonic jointing is shown as an example (Table).

In the given example regular intensification of rock failure into the face with the increase of jointing density is rather pronounced, as the sequence of groups with increasing jointing density $(B_1-B_2-B_3-B_4)$ shows increase in the

Table. Distribution of openings according to immediate roof collapse and density of tectonic jointing

Intensity of rock collapse, A_j	Density of jointing, joint per meter, B_i			
	< 1 (B_1)	$1-1,9$ (B_2)	$2,0-2,9$ (B_3)	$> 2,9$ (B_4)
Low, A_1 (stable roofs)	0,499	0,214	0,072	0,215
Medium, A_2 (medium stable roofs)	0,025	0,300	0,300	0,375
High, A_3 (unstable roofs)	0,069	0,167	0,284	0,480

frequency of occurrence of immediate roofs with high intensity of rock failure and a decrease in the frequency of realization of stable roofs.

Engineering-geological prediction in the course of mineral deposits exploration generally includes synthesis of information about future state of object under exploration, possibility of occurrence and probable nature of engineering-geological processes and phenomena based on combined account of both direct results of exploration and engineering-geological regularities, allowing to proceed from initial data to prediction itself. In search of optimum algorithms of engineering-geological prediction, possibilities of the following principles and methods of prediction have been investigated:
- mathematical, physical, physico-chemical, cybernetic and other types of modelling of engineering-geological processes and phenomena;
  - empirical, probabilistic-statistical, determined classification;
  - analogy;
  - extrapolation;
  - interpolation;
  - symptoms;
  - expert estimates;
  - graphycal and grapho-analytical recognition;
  - probabilistic-statistical recognition.

The ennumerated methods are acceptable for prediction of majority of simple processes and phenomena. Methods of modelling, classification, analogy, extrapolation and graphycal recognition are developed and successfully used in exploration of mineral deposits.

Inadequacy of our knowledge of the nature and mechanics, multifactor dependence and, consequently – random character of complex processes and phenomena are the reason why these phenomena, as opposed to simple ones, do not lend themselves to prediction with the most of the above methods. Thus, for example, attempts to make correct predictive classifications, reflecting relation of complex processes and phenomena to all significant factors, lead to extreemly cumbersome, practically inoperative classification schemes, containing hundreds and thousands of finite dividing terms, which can not be reliably characterized on the basis of empirical data. Probabilistic-statistical recognition of objects as a method of prediction of complex engineering-geological processes and phenomena has an advantage over all the other methods, as its use practically invariably gives beneficial results. Essence of probabilistic-statistical approach to prediction lies in verification of statistic hypotheses about coal-bearing rock masses or fragments of these masses belonging to one of previously established classes with prevailing probability of occurrence of concrete engineering-geological phenomena or realization of their characteristic parameters. In our case the best results were achieved with the help of Bayes strategy of recognition by means of estimation of conditional probability $P(A_j/B_i)$ of realization of A_j hypotheses (classes) with actually registered events (features, factors) B_i. Calculation of conditional probability as applied to objects to be predictively estimated, was implemented according to the simplified modification of Bayes formula, as follows:

$$P(A_j/B_i) = \frac{P(A_j) \quad \prod\limits_{i=1}^{n} P(B_i/A_j)}{\sum\limits_{j=1}^{m} P(A_j) \quad \prod\limits_{i=1}^{n} P(B_i/A_j)} ,$$

where $P(A_j)$ - quasi-a priori probability of A_j hypothesis, estimated by empirical frequency of their realization;
$P(B_i/A_j)$ - frequency of occurence of features, calculated on the basis of retrospective information; the essence of the parameter is revealed in the above text.

Substituting into prediction formula the empirical estimates of apriori probability and probabilities of factual features, related to significant geological and anthropogenic factors, it is possible to estimate probabilities of realization of each class, recognize the class with highest probability of realization and to place the object under evaluation in correspondence with phenomena and other engineering-geological parameters most typical of a given class.

Verification of reliability of engineering-geological predictions based on probabilistic methods, was carried out on control samples of objects, which are well characterized by factual data on openings in operation. From the results of this test based on criterion of reliability of obtained solutions, complexes of significant factors involved in prediction were optimized. As a final result it was generally possible to achieve significant (75-90%) share of correct recognitions of objects having high probability of occurrence of dangerous phenomena.

Along with satisfactory reliability of solutions, probabilistic-statistical prediction has a number of other obvious advantages: it most closely corresponds to a random nature of complex engineering-geological processes and phenomena, allows to consider practically any number of significant geological, anthropogenic and even so called systemic factors, which express mechanic, chemical, physico-chemical etc. types of interaction between coal seams and overlying and underlying layers of enclosing rocks when driving openings of coal mines. The use of this particular method of prediction enables us to consider previous mining experience to a high degree; reliable predictions can be obtained even in the absence of adequate information on certain factors. Meanwhile, it should be noted, that in the case of "new" deposits, where mining operations are not yet performed on a large scale, probabilistic-statistical prediction can be used only on the basis of retrospective information gained in other, similar coal-bearing regions, which does not always allow to get satisfactory results immediately.

Large-scale implementation of probabilistic-statistical prediction requires constant renewal and processing of large number of retrospective and geological exploration data. This calls for setting up of computerized data acquisition and processing systems as well as creation of automated prediction systems. First among such systems are already in operation in the main coal basins of the USSR. The complex programs already developed or those being developed for ES-series computers will allow to solve the following interrelated problems:

a) input of retrospective information into computer, processing this information along with formation of a complex of significant natural and other factors, calculation of empirical frequences of occurrence of features related to significant factors, periodical correction of factor complex and values of empirical frequences of occurrence of features based on new data and results of test on reliability of prediction; storage of retrospective information in the form of original and pre-processed data;
b) input and pre-processing of geological-exploration data;
c) input and pre-processing of technical data, reflecting the nature and intensity of anthropogenic effect on rock masses in the course of future mining works;
d) calculation of conditional probabilities of hypotheses using Bayes formula and predictive recognition of objects;
e) test for prediction reliability;
f) printout of results;
g) plotting of engineering-geological maps with prediction elements.

Automated systems are already widely used for prediction of immediate roof collapse into near-face area of openings, and of subsidence and collapse of the main roof. Investigations carried out by scientists of All-Union Coal Geological Research Institute proved that in a number of cases processes of rock extrusion and volume of water inflow into the worked out space can be predicted with the use of automated multifactor method. Possibility of reliable prediction of sudden coal-and-gas outbursts, of spontaneous coal ingnition has also been proved. The above principles and methods of engineering-geological prediction in coal deposits exploration can also be used for other mineral deposits, development of which is accompanied with dangerous effects of geological environment on mining.

It should be emphasized that variation in the volume and nature of original geological and technical information in the process of deposit exploration, as well as alteration of structure, properties and state of rock masses associated with extraction of minerals necessitate permanent implementation of engineering-geological prediction with periodical test of reliability of former predictions and their subsequent correction. Such corrections of initial predictions are most expedient at the stage of mine construction, when developing deeper strata, and prior to mine reconstruction. Depending on the scale and the nature of tasks related to commercial development of mineral deposit, engineering-geological prediction can be performed for ore- (coal) bearing regions, mining districts or large deposits, separate mines or quarries and their large sections (wings, stratas, extraction areas) as well as for particular mine openings.

Classification of the rock mass and its application in mining engineering

Classification des massifs rocheux
Application à la géotechnique minière

Xu Bing & Li Yurui, *Institute of Geology, Academia Sinica, Beijing, China*

ABSTRACT: The basic principles of rock mass classification are to reflect objective natural characteristics of rock mass on the one hand, and to apply to engineering requirements on the other hand. The objective natural characteristics of rock mass may be presented by three conditions: engineering geological rock group; rock mass structures and existent environment of rock mass.

RESUME: Le principe fondamental de la classification des masses rocheuses est de refleter des caractéres naturelles objectives des masses rocheuses d'une part, et de s'applique à la demande d'ouvrage d'autre part. Les caractéres naturelles objectives des masses rocheuses peuvent être présentees par les trois conditions: le groupe rocheux géologique de l'ingénieur, la stracture des masses rocheuses et son environnement existant.

The mining engineering occupies an important place in the human engineering activities. It has close relationship to geological environment. The direct object of study in the mining engineering is considered to be the ore body and surrounding rock. They are referred to as rock mass. It should be noted that, to recongize and to master the nature characteristics of rock mass is one of the fundamental problems in mining engineering. However, in the early mining practice, mining engineers studied the crust pressure and support measures passively only according to mining technology, ignoring the rock mass characteristics and its changes. The complicated rock mass properties are often treated as a single rock proportics in the initial stage of rock mechanics studies. Although in the later stage the importance of rock fissures has been recognized, but the fissures are only considered as the effecting factor. However, a great number of engineering practice indicates that, the rock mass deformation and failure are mainly controlled by various surfaces within the rock mass and blocks bounded by these surfaces. The form and degree of rock mass deformation and failure are different according to the surface behavior and the figures of the block. The researchers, holding the view point of engineering geomechanics in Chinese engineering geological community draw great inspiration from this practice experience. They considered that the rock mass, including the surfaces of fault, joint, bedding plane, schistosity etc., should be treated as an intercorrelated natural body. Individual rock mass has its own internal structure. Different types of rock mass deformations and failure depend on the difference of rock mass structures. From this point of view, therefore, the rock mass having structure can only be considered as the object of study. Keeping all these abovementioned in mind, this paper will deal with the rock mass classification and its application in mining engineering.

1

The rock mass structure has been formed in the long history of its geological evolution. However, every rock mass has passed two stages in its evolution, that is the stage of formation and the stage of transformation. The formation is the material basis of rock mass. Its engineering geological characteristics may be generalized as the engineering geologic groups. And the transformation is the structural basis of the rock mass, and its engineering characteristics are generalized as the rock mass structure. As to the ground stress and the underground water, they can be regarded as the environment in which of a rock mass exists. They are also considered to be changeable factors, and are controlled by the concrete conditions. Therefore, the engineering geological rock groups and the rock mass structure types are the two basis of rock mass classification.

Engineering geological rock group: In order to facilitate the study on rock mass, from the view point of engineering geology, rock strata, having similar engineering geological behavior, can be grouped. That is the engineering geological rock group. The rock group is mainly based on the stratigraphical division, considering the lithological characters and the layer feature. Some particular layers, especially, soft and intercalation layers should be marked distinctly. The rock combination in nature is complex. At least, there are four basic types: single, double, multiple and mixed. They are depending on genesis and the concrete enviornment, in which they are. In order to recognize comprehensively the rock groups, three fundamental aspects should be studied: genesis types and lithologic characters, lithofacies; material composition and rock texture; layer-formation condition and variation of layer-thickness.

Rock mass structure and its types: The various geological boundaries in rock mass are called structure surfaces. The rock block, bounded by the structure surface, is referred to as a structure body. The so-called "rock mass structure" is a combination of the structure surfaces and the arrangement of the structure bodies in rock mass. This is a highly theoretical generalization. It pushes the study on rock mass fissures to a new level and it serves as an applicable fundamental theory for the recognition of a rock mass nature and for the rock mass stability analysis. The various types of rock mass structures can be classified according to their degree of development, the magnitude, the arrangement and combination of these structures, as well as the contact features of the structure surfaces and bodies. In the study of rock mass structure, the key problem is the study on the structure surface. For the study of the structure surface characteristics it is necessary to classify the structure surfaces, according to the scale and magnitude, into different orders. Obviously, the structure surfaces of different scales and magnitude play different role in the deformation and failure of rock mass.

The basis of study on rock mass structure is the regular pattern of the tectonic fault network.

2

This paper deals with a case history in the 2nd area, Jinchuan mine, located in north-west China. The concrete rock mass classification has been given as well.

This mining area is located in the north-eastern corner of the Hexi Corridor, province Gansu and is referred to the fault-folding zone of mountain Long Shou. Under the action of multiple tectonic movement the fault tectonic has developed, the rock mass fractured and metamorphosed to high and moderate grade; the magmatic activity was estensive and frequently, and the lithology and lithofacies changed greatly. In general, the engineering geological cindition is not very favourable.

Engineering geological rock group: The strata in mining area are composed of a crystalline formation Pre-Cambrian in age and the Quaternary unconsolidated deposits. According to the principles of the engineering geologic rock groups division there are 11 types of engineering geologic rock group, mainly considering the lithology, thickness and some original structure surface (especially the weak interbed). Because of the repeated action of multiple tectonic, the strata position of the fault, as F1 and F16, are stable. Therefore, an independent rock group can be distingusihed. (Fig.1)

The III_3 and XI rock groups are the key to study among the 11 rock groups, because not only the character of rock mass is very soft and lossened, but also they have close relationship to the engineering construction.

Type of rock mass structure: The form of the tectonic network is characterized by a set of strike faults (well developed), which is of predominant place; by a pair of shear fault (developed locally) and by the tension fault (developed weakly). (Fig.2)

The types of rock mass structure are controlled by the spatial form of fault and its difference of the degree of development. According to the character (weak and rigid), degree of development and combination form of the structure surfaces the rock mass structure in Jinchuan mine may be classified into 6 types:

1. Rhombus and blocky structure: the rigid structure surface is dominant (developed not well).

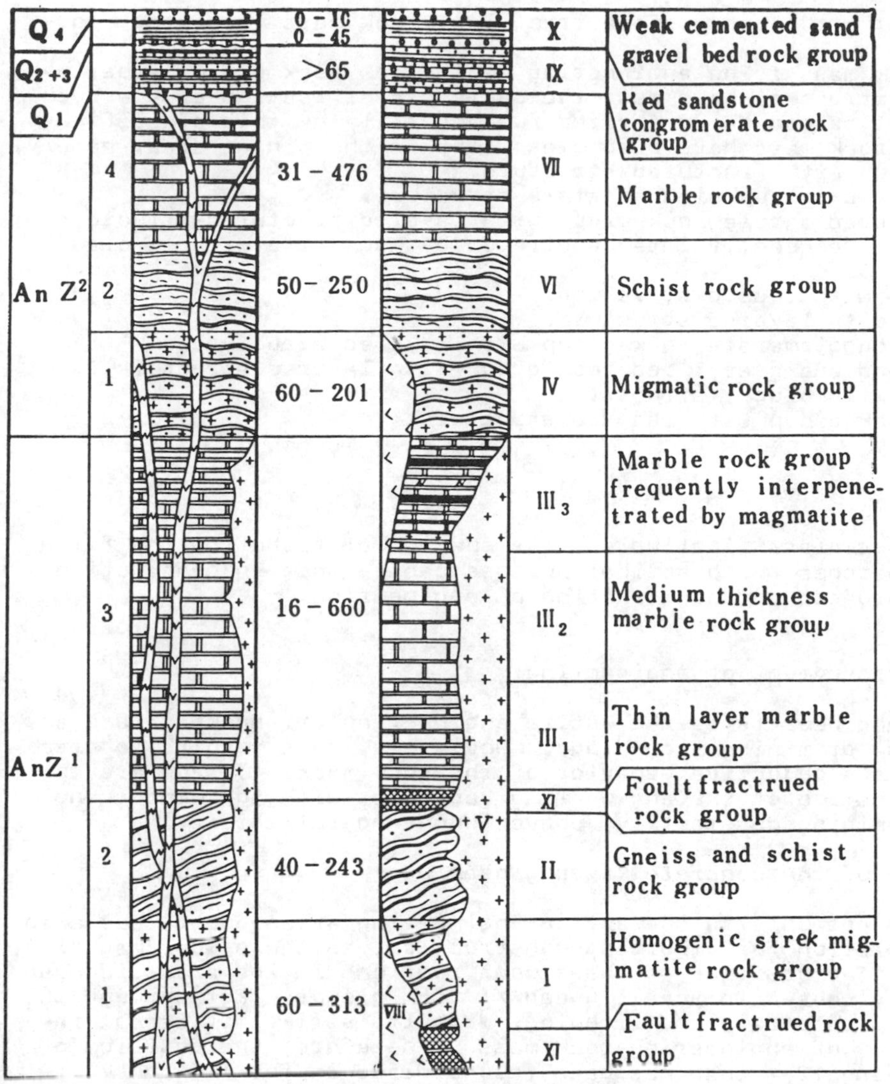

Fig.1 (stratigraphic column):

Age	Unit	Thickness	Symbol	Rock group
Q4		0–10 / 0–45	X	Weak cemented sand gravel bed rock group
Q2+3		>65	IX	
Q1				Red sandstone congromerate rock group
An Z²	4	31–476	VII	Marble rock group
	2	50–250	VI	Schist rock group
	1	60–201	IV	Migmatic rock group
An Z¹	3	16–660	III₃	Marble rock group frequently interpenetrated by magmatite
			III₂	Medium thickness marble rock group
			III₁	Thin layer marble rock group
			XI	Foult fractrued rock group
	2	40–243	II	Gneiss and schist rock group
	1	60–313	I	Homogenic strek migmatite rock group
			XI	Fault fractrued rock group

Fig.1 Schematic diagram of engineering geological rock group division.
V — Fractured granite rock group
VIII — Ore ultrabasic rock group

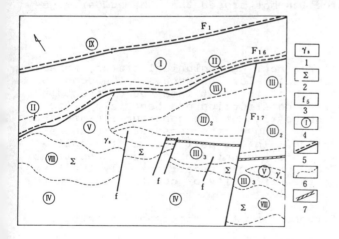

Fig.2 Distribution of fault and rock group in the mining area.
1 granite   2 ore ultrabasic
3 number of fault   4 numbers of rock group   5 fault   6 boundary of rock group   7 weak interbed

2. Layered structure: mainly a set of structure surfaces.
3. Mosaic structure: multiple rigid structure surface, developed well.
4. Layered fractured structure: mainly developed a set of weak structure surface.
5. Fractured structure: multiple weak and rigid structure surfaces, developed well.

6. Lossened structure: structure surface irregular, rock mass fractured, loosened.
  Classification of rock mass : The engineering geolgoical rock groups superimposed on the rock mass structure have made the basic classification of the rock
mass. The rock mass nature can be summarized fundamentally by this classification. The 10 types of rock mass have been classified in the mining area: namely,
  1. Migmatite rock group with fractured structure.
  2. Gneiss and schist rock group with fracture structure.
  3. Medium and thin bedded marble rock group with layered fractured structure.
  4. Marble rock group, frenquently interpenetrated by magmatite with layered
fractured structure.
  5. Granite rock group with mosaic structure.
  6. Marble rock group with layered structure.
  7. Red sandstone and conglomerate rock group with layered structure.
  8. Weakly cemented sand and gravel bed rock group with layered structure.
  9. Marble rock group with blocky structure.
  10. Ore ultrabasic rock group with rhombus structure.

3

On the basis of rock mass classification and its spacial distribution the following 5 aspects of researches for practical purpose can be made in conjunction
with the arrangement, project and construction of engineering.

1. Direct the entire arrangement of engineering.

Generally, the mining engineer takes the entire engineering arrangement into account from the view point of mining technology. Obviously, this results in disastrous effect ignoring the objective behavior of the rock mass. In 2nd area of
Jinchuan mine the combination of the engineering geologist with the mining engineer can guarantee the mining possibility, convenience and reliability.

2. Correct the position of the concrete key engineering.

We should pay attention not only to the entire engineering arrangement, but also
to the concrete demonstration for important construction, as the powder magazine,
water storehouse etc.. The position of these constructions in 2nd area Jinchuan
mine had been originally chosen in the III_3 and XI rock groups. But through the
analysis of relationship of strata and lithology with its strike, it is not favourable to the stability of engineering rock mass. Therefore, the position of
the powder magazine having 12 m span has been reconsidered in the III_2 rock group with a block structure. This construction has proved that the powder magazine is safe and stable.

3. Engineering geological prediction.

During the construction the prediction of engineering geological status, especially, of the bad rock mass became very important. For instance, in the construction of the transport bottom the relationship of the soft interbed or fractured
zone in III_3 rock group with the engineering construction has been taken into
account closely. And the prediction and suggestion of the construction are made
previously, thus it not only can promote the tempo of the construction, but also
can ensure safety and quality of construction. In the past because the engineering geological prediction was not taken into account, in the same conditions
a large rock falls had been occurred. Because of this hazard not only the time
table for a project was delayed, but also the difficulty of construction and the
engineering cost were increased.

4. Point out the possible deformation of tunnel and the corresponding measures.

In the mining construction according to the given rock mass the analysis of its
possible deformation and support forms is also an important applied field. For
instance, the granite rock mass V is very fractured and a section of tunnel would
be through it. Therefore, according to the rock mass characteristics and the high level ground stress it should be noted that the ground pressure in this tunnel
is great, the rock deformation can be developed rapidey. For this, the entire
envelope support form ―― the best support measure may be suggested. The great

deformation in a section of tunnel (the dimentions of the tunnel section has decreased and bottom distension was about 1 m) had occurred because the correct suggestion of engineering geologist is not applied. For instance again, due to
the bottom transport locates in the III_3 rock group with layered fractured structure and the tunnel strike is vertical to the max level principle stress direction, the engineering geologist has exhorted the mining engineers to that: in the
construction of this tunnel section the inclined pressure will occur; the rock
fall will occur in the weak interbed and fractured zone; and the roof rock can
fall in the tunnel crossing. Because the mining engineer accepted these suggestion and revised project, thereby this section tunnel had been excavated successfully.

5. Suggestion of the stope arrangement and the excavating sequence.

According to the occurrence of the ore ultrabasic rock mass, the direction of the
ground stress (max. principle stress is directed about NE 30°) and the well developed strike structure surfaces, it has been suggested, that the stope length
direction is best vertical to the strata strike and the mining sequence is favourable by means of the skip a chequer.
  Because the mining method in Jinchuan mine is the filling means, the above mentioned suggestions are of obvious effect.
  In brief, in the engineering geological researches on the 2nd area Jinchuan mine, the rock mass classification for practical application has been mede and
from the viewpoint of the relationship of engineering geological conditions with
the concrete engineering constructions the analysis of engineering stability and
the hazard prediction have been made as well. The engineering practice has proved that these researches are successful.

Geomechanical classification of rock masses for evaluation of strength and deformation parameters

Classification géomécanique des massifs rocheux pour l'estimation des paramètres de résistance et de déformation

Yudbhir, *Indian Institute of Technology, Kanpur*
Dacha Luangpitakchumpol, *Asian Institute of Technology, Bangkok, Thailand*

ABSTRACT: This paper describes results of rockmass classification studies conducted on the left abutment and power house area at the Khao Laem dam site in Western Thailand. Details of field data and sample computations for different classes of rockmasses are presented. During construction, the cutslopes in these critical areas experienced instability. Parameters for Geomechanic classification were evaluated and on the basis of Rock Mass Rating values of strength parameters obtained on the basis of Geomechanics classification are compared with laboratory direct shear tests data and failure criteria for rock masses. Using stereographic projection techniques, guideline to adjust the basic Rock Mass Rating, for slope has been proposed.

RESUME: Cet article contient les résultats d'étude de la classification de la masse d'ensemble des roches, fait sur la rive gauche et sur l'aire d'usine de l'électricite du barrage Khao Laem situé en Thailand de l'ouest. Les détails de donées et des calculs pour plusieurs cas de masse d'ensemble sont présentés. Pendant la construction du barrage on a trouvé que les profiles des pente dans ces régions critiques sont instables. Les paramètres de la classification géomécanique sontevalués et on a proposé les valeurs optimales de paramètres de resistance et du module de deformation basées sur lo grade de masse d'ensemble des roches. Les paramètres de résistance obtenus en considération de la classification géomécanique sont comparés avec ceux d'éssais de cisaillement et de critère de défaut de roches. En utilisant la technique de projection stéréographique on a proposé l'idée directrice pour régler le grade de masse d'ensemble selon la pente.

1 KHAO LAEM DAM

Khao Laem dam is a multipurpose rockfill dam located on the river Quae Noi river about 240 km north-west of Bangkok, Thailand. It has a maximum height of 92 m above the stream bed and a crest length of 910 m. The dam has a volume of $8 \times 10^6 \text{m}^3$ of lime stone as rockfill with an upstream concrete face.

2 TOPOGRAPHY AND GENERAL GEOLOGIC STRUCTURE

Plan layout and longitudinal sections are depicted in Figure 1. The right abutment rises majestically as a cliff in sheared limestone while the left abutment rises at a slope of 25° and steeper from elevation 100 m to 300 m and beyond. The left abutment is founded on spur S_2 which is bounded by steep gullies G_2 and G_3 which have incised 30 to 40 m into the valley slope. Similar spurs S_1 and S_3 are located upstream and downstream of spur S_2. The right abutment cliff extends downstream for about 150 m and upstream for over 500 m.

2.1 In the dam site area the strata are folded. A major fault zone lies on the right abutment near the toe of the cliff and is exposed as fault scarp. The rocks of the Thung Song group appear to be folded in a complex syncline along the river. The strike is mostly NNW-SSE, with dips, in exposures on the left bank and in the river bed, mostly between 40° to 60°, towards the right bank. Massive primary lime stone with solution cavities, belonging to the Ratburi group of rocks, comprises the right abutment. On the left bank at the power

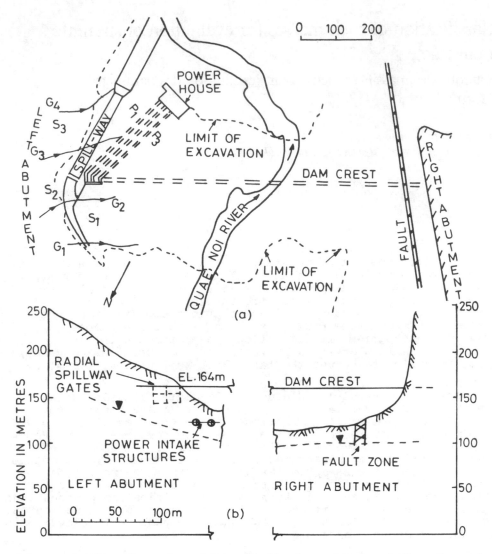

Figure 1. Plan layout and longitudinal sections of Khao Laem dam.

house site, generally moderately to highly weathered interbedded limestone and shale are exposed.

3 SLOPE INSTABILITY PROBLEMS

During construction, landslides took place in the excavations for power house area, power conduits, P_1, P_2, P_3, left abutment on spur S_2 and also along the diversion channel on the right bank. A major rockfall controlled by joints and triggered by blasting took place on the right abutment involving a rock volume of about 5000 m³. Detailed study of slope instability problems is reported elsewhere (Nattawuth 1983).

4 GEOMECHANICAL CLASSIFICATION STUDIES

In order to assess the rock mass stability, especially in the left abutment spur S_2, and excavations for the power house and power conduits, it was decided to carry out geomechanical and quality index rock mass classification studies in these areas (Bieniawski, 1973, 1976 and Barton et.al. 1974). Limited studies were also conducted on the right abutment where tunnelling was the main concern. Main objective of this study was to finally calibrate rockmmasses in different areas of the project according to various rock mass classes predicted by these classification systems (for details see Dacha 1982).

5 FIELD DATA AND ROCKMASS RATING ASSESSMENT

Field data was mainly collected from excavations in the power house, power conduit spur S_2 on left abutment, and a gallery 'C' on the right abutment.

5.1 A total of 187 locations covering a linear distance of 935 m was covered in the power house and power conduits area and three slopes at elevations 142, 167 and 207 m on spur S_2 were investigated at the left abutment. Schmidt hammer and point load strength index tests (Figure 2) and joint studies in terms of spacing, roughness and separation (Figure 3) were conducted in the left abutment and power house areas. Most of the joints in the power house area were found dry and only some seepage was observed along the bedding planes at the east slope of the power house excavation at the rate of 0.75 litre/minute. In comparison to the power house area, the spur S_2 area was less jointed and the joints were mostly dry. Instability in the spur S_2 area was mainly controlled by clay seams. The RQD values for rocks in all the areas on the left abutment were estimated from the measurement of joint volumes on the outcrops mapped (Barton 1974 in Hoek and Brown 1980). Detailed mapping of joint attitudes was carried out (Dacha 1982).

5.2 On the basis of extensive field measurements Rock Mass Ratings, RMR, and rock mass classes were evaluated (Bieniawski 1973). Table 1 gives a typical calculation sheet for the right slope of the power conduit P_1. Similar calculations were carried out for the rock masses in the power house, spur S_2 on the left abutment and gallery 'C' on the right abutment areas.

6 INTERPRETATION OF TEST DATA AND DISCUSSION

Based on these field studies the rock masses in these areas have been classified and the values of strength and deformability parameters have been assigned (Table 2) where E_m is the rock mass modulus, E_1 is the rock material modulus, and C', $\emptyset'$ are the shear strength parameters. E_m and E_m/E_1 parameters were assessed from RMR values using the available correlations (Bieniawski and Orr 1976a, Chappell and Maurice 1980, Williams 1980).

6.1 A number of direct shear tests on block samples were conducted in the laboratory (AIT 1981) and the test results (Figure 4) are compared with those obtained from RMR studies (Bieniawski 1973) and the predictions from the engineering failure criteria for rock masses (Hoek and Brown 1980). It will be seen that the value of the angle of shearing resistance, $\emptyset'$, varies from $20°$ to $40°$ with an average of $30°$. At stress level less than $100t/m^2$, the laboratory test data and field estimates show reasonable agreement. These values were used to check stability conditions.

6.2 The relationship between RQD and joint volume as proposed by Barton was checked at this site. Large number of core logs were analysed in respect of RQD and joint volume (Figure 5). Best fit to the data shows good agreement with Barton's relationship.

6.3 Rock mass rating and Q values obtained for gallery 'C' at the right abutment are compared (Figure 6) and the results fall within the limits of suggested correlations (Bieniawski 1976).

6.4 While guide lines for adjusting RMR values due to joint orientation in case of tunnels and foundations (Bieniawski 1973, 1976a) are available and were used in this study, no definite guide lines are available to decide favourable and unfavourable discontinuity attitudes for slopes. On the basis of graphical procedures (Markland 1972, Hocking 1976 in Hoek and Bray 1977) control of major joint sets on stability of slopes on the left abutment spur S_2 and in the power house area was investigated. Typical results (Figure 7) for the right slope of power conduit P_1 illustrate Hocking's method. The values of $\emptyset'=20°$ to $40°$, as evaluated earlier, were used and as shown both values produce the same results. Goodman's method (See Hendron et al 1980) gave identical results. On the basis of these graphical studies, a tentative inter-relationship between the angle of dip of the joint, δ, the slope angle, α, angle of shearing resistance, $\emptyset'$, and the angle between the joint strike and the slope line, β is formulated (Table 3). Adjustments to the RMR values due to joint orientation for slopes was applied on the basis of this guide line.

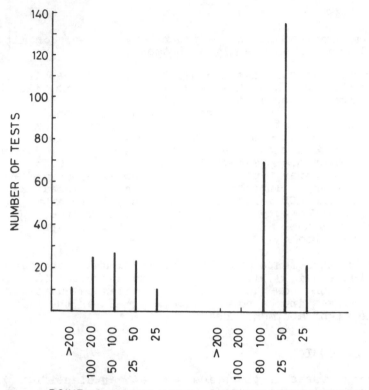

Figure 2. Point load and Schmidt hammer tests.

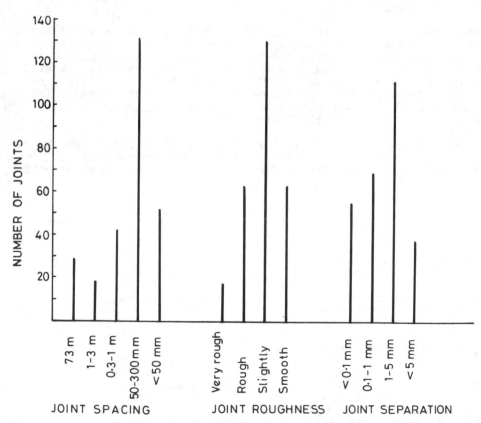

Figure 3. Joint spacing, roughness and separation data on the left abutment.

Table 1. Rock mass rating for right slope of power conduit P_1

Parameters	Value/Description		Rating
Uniaxial compressive strength	25 – 50 MPa		4
RQD	69 percent		13
Joint spacing			
Set 1	Close,	50–300 mm	10
Set 2	Close,	50–300 mm	10
Set 3	Very close,	50 mm	5
Average rating of joint spacing			8.33
Joint condition			
Set 1	Continuous, no gouge, moderately open, smooth, hard wall, planar		6
Set 2	Not continuous, no gouge, tight, smooth, hard wall, planar		25
Set 3	Not continuous, no gouge, tight, smooth, hard wall, planar		25
Average rating of joint condition			18.67
Ground water condition	dry		10
Basic or insitu rating			54
Adjustment due to joint orientation for foundation			
Set 1	N9W 50W (favourable)		-2
Set 2	N18E42E (favourable		-2
Set 3	N89W59S (favourable)		-2
Average rating of joint orientation for foundation			-2
Rock mass rating for foundation			52
Rock mass rating class			III
Adjustment due to joint orientation for slope			
Set 1	N9W 50W (very favourable)		0
Set 2	N18E42E (unfavourable)		-50
Set 3	N89W59S (unfavourable)		-50
Average rating of joint orientation for slope			-33.33
Rock mass rating for slope			20.67
Rock mass rating class			IV

Table 2. Results of geomechanics classification

Location	RMR	Class	E_m, GPa	E_m/E_1	C', KPa	ϕ'
Left abutment Power house area –						
Foundation	52–32.3	III–IV	6–20	0.2–0.25	–	–
Slope	21–<20	IV–V	–	–	0–100	25°–30°
Spur S_2	57–<40	III–IV	–	–	100–150	30°–40°
Right abutment Gallery 'C'						
Tunnel	66	II	20–60	0.22–0.50	200–300	40°–45°

Table 3. Assessment of joint orientation favourability upon stability of slopes

Dipping into the slope				Dipping away from the slope	
$\beta < \delta < \alpha$					
$0 \leq \beta \ 45^o$	$45^o \leq \beta < 90^o$	$\delta < \phi < \alpha$	$\delta > \alpha$	$0 \leq \beta \leq 45^o$	$45^o \leq \beta \leq 90^o$
Very unfavourable	Unfavourable	Fair	Very favourable	Very favourable	Favourable

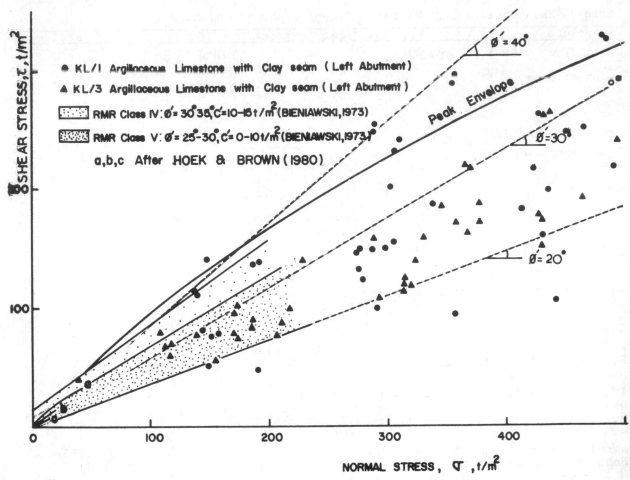

Figure 4. Shear strength characteristics of rock masses at left abutment.

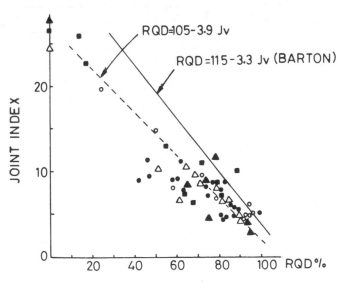

Figure 5. RQD vs joint volume relationship.

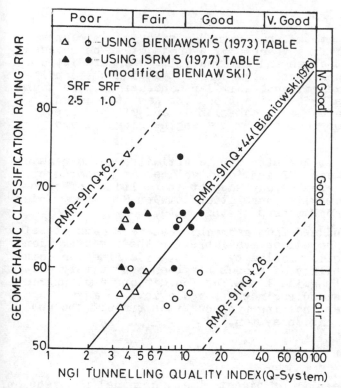

Figure 6. Relationship between RMR and Q index.

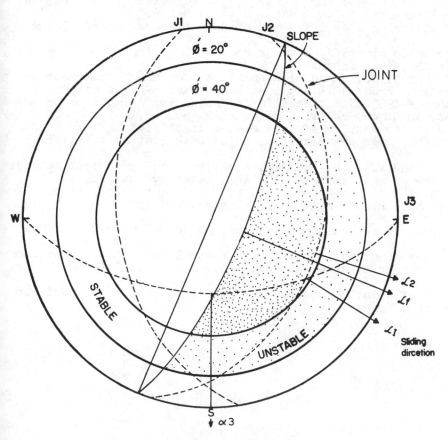

Figure 7. Hocking's test for stability of right slope of power conduit P₁

7 CONCLUSIONS

On the basis of detailed field studies at the Khao Laem dam site it has been demonstrated that the rock masses in different areas of a project can be successfully classified in respect of their strength and deformability characteristics. Rock mass classes of IV to V in the power house area correlated with the conditions of instability witnessed during excavations. Similar conclusions were found valid for the spur S_2 area on which the left abutment of the dam is located. These studies deduced shear strength parameters, cohesion C' and Ø' for these rock masses which agree well with those predicted by the engineering failure criteria (Hoek and Brown 1980).

7.1 The value of RQD for the rock mass at the site can be estimated quite reasonably by using the relationship between RQD and joint volume, as proposed by Barton. Results obtained in the present study support the validity of such a relationship. Furthermore Bieniawski's correlation between RMR and Q index is shown to be applicable for the rock masses at this site.

7.2 Based on the strength parameters deduced from geomechanical rock mass classification and the use of graphical procedures available for the investigation of conditions of slope stability as controlled by joints, guide line for joint orientation favourability upon stability of slopes has been tentatively proposed. This along with Bieniawski's guide lines for foundations and tunnels would assist geotechnical engineers and engineering geologists to make quantitative assessments of civil engineering project sites through the use of geomechanical rock mass classification system.

REFERENCES

Asian Institute of Technology. 1980. Results of direct shear, uniaxial compression, and indirect tensile tests on rock samples. Report to EGAT, Khao Laem project, Thailand.

Barton, N., R. Lien and J. Lunde 1974. Engineering classification of rock masses for the design of tunnel support. J. Rock Mech. 6(4): 189-236.

Bieniawski, Z.T. 1973. Engineering classification of jointed rock masses. Trans. S. African Instn. of Civ. Eng. 15(12):335-344.

Bieniawski, Z.T. 1976. Rockmass classification in rock engineering. Proc.Symp. on Exploration for Rock Engineering. Johansburg.

Bieniawski, Z.T. and C.M. Orr 1976a. Rapid site appraisal for dam foundations by geomechanics classification. Proc. 12th Cong. on Large Dams. Mexico City.

Chappell, B.A. and R. Maurice 1980. Classification of rock mass related to foundations. Proc. Int. Conf. on Structural Foundns.on rock. Sydney.

Dacha, L. 1982. Geomechanical evaluation of rock masses in the left abutment and power house area at the Khao Laem dam site. Master of Science thesis. A.I.T. Bangkok, Thailand.

Hendron, A.J., E.J. Cording and A.K. Aiyer 1980. Analytical and graphical methods for the analysis of slopes in rock masses, U.S. Army Waterways Experiment station, Vicksburg, Miss.

Hoek, E. and J. Bray 1977. Rock slope engineering. London. The Instn. of Min. and Met.

Hoek, E. and E.T. Brown 1980. Underground excavations. London. The Instn. of Min. and Met.

Nattawuth, U. 1983. Stability of slopes of diversion channel and excavat d slopes in the left and right abutments at Khao Laem dam site. Master of Science thesis, A.I.T. Bangkok, Thailand.

Williams, A.F. 1980. Effect of jointing on rock mass modulus. Proc. Int. Conf. on Structural Foundns. on rock. Sydney.

1.3 Surface and underground excavations of large dimensions: Problems related to stresses
Fouilles à ciel ouvert ou souterraines de grandes dimensions: Problèmes liés aux contraintes

Rock mass monitoring of the Tabuaço hydraulic circuit
Contrôle des massifs rocheux dans le complexe de Tabuaço

António P.Cunha, *Laboratório Nacional de Engenharia Civil, LNEC, Lisbon, Portugal*
A.Ferreira da Silva, *Structural Safety Division, Electricidade de Portugal*

ABSTRACT: An important water inflow was found in galleries in the neighbourhood of the surge shaft of Tabuaço, a few months after a strong earthquake. The paper deals with the geological investigations and the rock mass monitoring which assisted in the diagnosis of the hydraulic phenomenon, and in the evaluation of the drainage effectiveness, slope stability and repairing works.

RÉSUMÉ: Des débits importants firent leur apparition dans des galeries au voisinage de la cheminée d'équilibre de Tabuaço, quelques mois après un sysme important. On décrit les études géologiques et les résultats de la surveillance du massif qui ont permis d'établir l'origine des débits, et l'efficacité du drainage, la stabilité du talus et les mesures de réparation de la cheminée.

1 INTRODUCTION

The Vilar - Tabuaço hydroelectric development (figure 1) begins at the Vilar rockfill dam, the water being conveyed 15 km through a 3 m diameter pressure tunnel which ends at a 90m high surge shaft. This 3 m diameter shaft, as well as its upper and lower expansion chambers, are located near mountainous edge of the narrow river Távora valley. After the surge shaft and the valve chamber, a steel penstock leads the water down the slope to a pressure shaft, and finally, about 400 m below the surge shaft, to the 2 x 32 MW units of the underground power station of Tabuaço (figure 2).

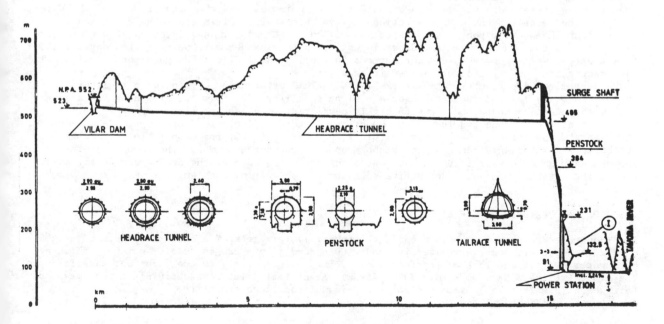

Figure 1. Longitudinal profile of the Vilar - Tabuaço hydroelectric development

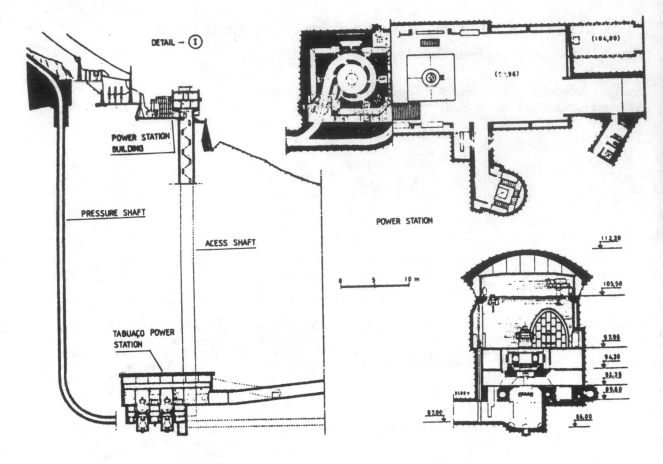

Figure 2. Pressure shaft and Tabuaço power station

The operation phase of the development started in 1965. A few months after the strong February 1969 earthquake, an important water inflow was found in the access gallery of the valve chamber, in the neighbourhood of the surge shaft.

A drainage system for the rock mass was designed, which allowed the measurement of the water flow and the establishment of a connection between the water level in the surge shaft and the infiltrations. Meanwhile, the visual inspection of the surge chamber showed the existence of several cracks in the concrete lining. Several repairs on the chamber wall were carried out, at different dates, which, however, did not succeed in stoping the infiltration phenomenon, which always reappeared some time after the repairing works.

The economical importance of the water losses (values of about 5 m^3/min. were registered in the drainage system) and the danger for the slope stability, due to water pressures in association with potential undrained flows (a highly persistant joint set parallel to the slope can be seen both at the surface and inside the mountain – figure 3),determined Electricidade de Portugal (EDP) to ask the National Laboratory for Civil Engineering (LNEC) for cooperation, in order to establish a monitoring scheme in the neighbourhood of the surge chamber. The results of the monitoring campaign would assist EDP in the control of the slope safety, the drainage effectiveness and in the establishment of a definitive solution for the inflow problem.

2 GEOLOGICAL FEATURES OF THE SURGE SHAFT ZONE

The Tabuaço surge shaft was excavated in the generally fair rock of a moderately jointed granitic rock mass. However, some very extensive and wide opened joints can be seen (figure 3), allowing both water seepage and rock alteration. These joints belong mainly to the joint set parallel to the slope (strike N30º – 50ºW, dip 30º – 45ºNE). A vertical joint sets defining a rock mass pattern which is unfavourable from the point of view of the slope stability.

Figure 3. A suggestive view of the rock slope near the surge chamber

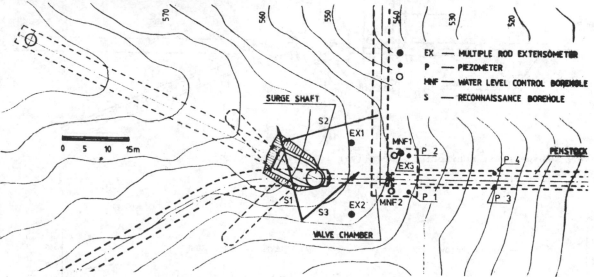

Figure 4. Surge shaft monitoring scheme (plant)

3 MONITORING SCHEME AND RESULTS

In two boreholes, drilled parallel to the shaft, which allowed also the geological study of the surrounding rock mass (see section 2), two multi-rod extensometers EX1, EX2 (figures 4 and 5) were installed, aiming at controlling any eventual sliding along the slope-parallel joint set. The borehole cores revealed the presence of some wide opened joints of this set, which, due to their location, clearly explained the direct water seepage from the cracked zone of the surge shaft lining to the drain net in the neighbourhood of the valve chamber. Another two-rod extensometer (EX3) was installed downwards from this chamber, allowing both a deeper control of the slope stability and the study of the rock mass below the chamber, which had never been sampled before.

The Lugeon tests in the extensometer boreholes showed a marked trend to flow increase with pressure increase, due to opening of joints and progressive erosion of its fillings, therefore explaining the progressive increase of infiltrations after each repair. On the other hand, as a raise of the water level in the surge chamber increased the aperture of the lining cracks and rock joints, the water level had a notorious influence on the water inflow (figure 6), before the definitive re

189

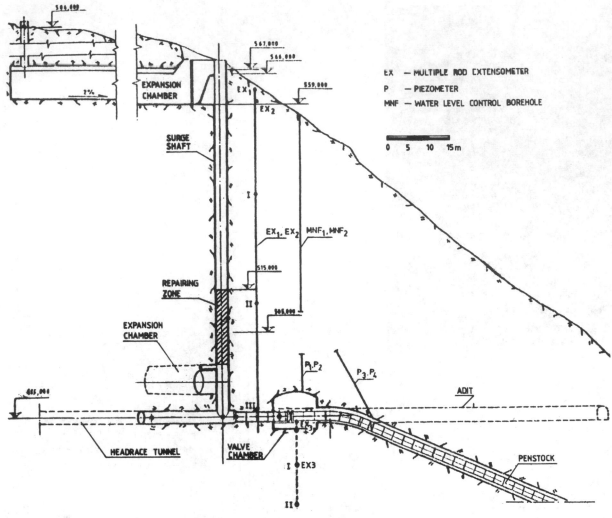

Figure 5. Surge shaft monitoring scheme (vertical cross-section)

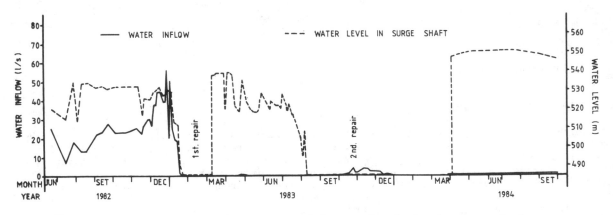

Figure 6. Water inflow in the drainage system and water levels in the surge chamber

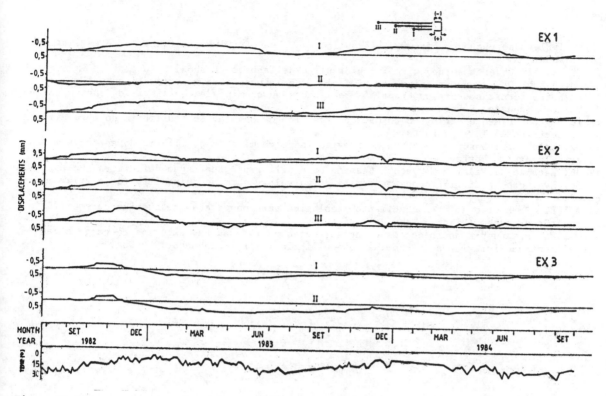

Figure 7. Extensometer measurements

pairing works. The repair involved the complete substitution of the non-reinforced concrete lining of the surge shaft between the levels 495 m and 515 m, by a reinforced concrete lining, 0.25m thick, connected by appropriate water-stop joints to the remaining support.

The analysis of the extensometer measurements (figure 7) shows a reversible and periodic evolution of the displacements, which follows the yearly thermic wave, as shown in the same figure. No major scale slope sliding could be detected during this period.

As concerns the hydraulic monitoring of the surrounding rock mass, the drain net was cleaned and the number and lenght of drains increased. Some piezometers were installed and the ground water level determined by means of appropriate boreholes. Since the repairing works no significant water pressures have been measured and a clear reduction of water inflow from 100 to 1 has been observed (figure 6).

Although the monitoring is still being carried on, bearing in mind the progressive feature of the seepage in rock masses, the rough topography of the site, the slope stability problems associated, and the economical importance of the hydraulic development, some conclusions of the monitoring program can already be drawn.

4 CONCLUSIONS

The analysis of the removed lining, and the geotechnical studies of the rock mass associated with the monitoring scheme of the Tabuaço surge shaft, allowed the establishment of a comprehensive mechanism for the genesis and evolution of the water infiltrations.

Due to the 1969 earthquake, the surge shaft lining was separated from the surrounding rock mass. As above level 495 m no reinforcement of the lining existed, the concrete ring could not stand alone the water pressure, and cracked. Water infiltrations through the fissured support, had first to find out their way through the rock mass joints, and then, by progressive erosion of the fillings, progressive flows could infiltrate through the rock mass.

The drainage system proved to be satisfactory, and the extensometer measurements showed no perspective of irreversible movements of the slope. As to the repairing works of the surge shaft lining, they so far proved their effectiveness.

REFERENCES

Cunha A.P. 1981. Observação do circuito hidráulico de Tabuaço - Plano de trabalhos e estimativa de custo. Lisboa.
Cunha, A.P. 1985a. Observação do circuito hidráulico de Tabuaço. Lisboa. LNEC.
Cunha, A.P. 1985b. Observação do circuito hidráulico de Tabuaço (2nd report). Lisboa. LNEC.
EDP. 1981a. Notas técnicas sobre os trabalhos realizados na chaminé de equilíbrio do aproveitamento de Vilar - Tabuaço. Porto.
EDP. 1981b. Reconhecimento geológico do maciço granítico da chaminé de equilíbrio do aproveitamento hidroeléctrico do Távora. Porto.
EDP. 1983a. Nota técnica sobre a reparação provisória da chaminé de equilíbrio do aproveitamento de Vilar - Tabuaço. Porto.
EDP. 1983b. Escalão de Vilar - Tabuaço - Chaminé de equilíbrio - Reparação definitiva. Porto.
EDP. 1983c. Evolução das infiltrações na câmara de válvula - chaminé de equilíbrio de Vilar - Tabuaço. Porto.
Tecnasol. 1982. Relatório do reconhecimento geológico para observação da chaminé de equilíbrio e reforço da drenagem (Vilar - Tabuaço). Lisboa.
Tecnasol. 1984. Nota sobre as injecções de consolidação executadas nas galerias de carga e câmaras de alimentação inferiores da chaminé de equilíbrio (Vilar - Tabuaço). Lisboa.

The superiority of the comprehensive evaluation method in the stability estimation of the surrounding rocks of karst caves – A practice in the expansion and reinforcement of a natural karst bridge

Superiorité de la méthode du jugement général dans l'estimation de la stabilité de la masse de roches entourant des cavités naturelles – Travaux d'agrandissement et de consolidation d'un pont naturel

Chang Shibiao & Zhang Wenqing, *Exploration Company, NORINCO, Beijing, China*

ABSTRACT: The Protodyrakonov's Theory has been replaced gradually by the Comprehensive Evaluation Method in the stability estimation of the surrounding rocks of a artificial cave since 1970's. The latter greatly promoted the development of underground construction of artificial caves for it is more practical than Protodyrakonov's Theory. It is pointed out in the book 'Engineering Geological Survey for Underground Construction' published in 1974 that comprehensive Evaluation Method should also be used in the stability estimation of the surrounding rocks of natural caves, but some people are still using the Protodyrakonov's Theory in the estimation for natural caves. Natural caves are formed through a long geological history and the deformation of the surrounding rocks almost stopped. Most of the rock mass are not loose but typically textured. Servative conclusion will be made if the Protodyrakonov's Theory is applied in the stability estimation for natured caves. The superiority of the Comprehensive Evaluation Method is evidenced by a practical example in this paper.

RESUME: En Chine, dupuis des annees 70, la méthode du jugement général dans l'estimation de la stabilite de la masse de roches entourant des cavités artificielles a graduellement remplacé la méthode de considérer la masse de roches comme la théorie de plotodyrakonov sur l'arche effondrée dans la substance desserée. plus proche de la pratique que cette théorie cette méthode a évidemment poussé le développement de la construction des travaux souterrains de la cavité artificielle. Bien que l'article sur la prospection géologique des travaux de la construction souterraine rocheuse publie en Chine en 1974 demande d'adapter la méthode du jugement général de la combinaison de l'analyse géologique des travaux avec le calcul de la dynamique, lors de l'estimation de la stabilité de la masse de roches entourant des cavités naturelles, il y en a encore qui utilisent, jusqu'a présent, la théorie de plotodyrakonov dans le travail. Comme la cavité naturelle est un produit de la longue histoire géologique et la déformation de la masse de roches touche à sa fin pour la plupart, la masse de roches n'est absolument pas la substance desserrée en ce qui concerne l'intermédiaire, mais au contraire, la plupart de roches sont des substances structurées typiques. C'est ainsi que la copie de la théorie de plotodyrakonov risque de causer la conclusion très conservatrice pour des travaux. Par la différence évidente de la rentabilité des travaux dans la pratique d'agrandissement et de consolidation d'un pont naturel en utilisant respectivement la théorie de plotodyrakonov et la méthode de jugement intégral, on essaie, dans le présent article, de fournir ses appreciations sur la supériorité de la méthode du jugement intégral dans l'estimation de la stabilité de la masse de roches entourant des cavités naturelles.

1 PROBLEMS IN EVALUATION OF THE STABILITY OF THE SURROUNDINGS ROCKS OF NATURAL KARST CAVES

There are great numbers of karst caves in the vast carbonatite areas in China. Most of them can be used for various underground projects. However, the first problem we meet before making use of them is how to evaluate the stability of the surrounding rocks of these caves.

When we began to make a programme about utilizing some natural karst caves in last 1950's, designer raised a point of order, that they could be used if surroundings are stable, with other conditions concerned are also proper. Otherwise we will ont consider them until sufficient measures are taken to reinforce them. As for what kind of measures to be used, it should be recommended by engineering geologists.

Since there had been no new methods of the stability estimation expounded we had to deal with this problem by using Protodyrakonov's Theory about artificial caves evaluation. Several contradictions were soon found, for example: the height of collapsed-arch is determined by the following formula:(See Fig.1.)

$$h_1 = a_1/f_{K\rho} = [a+h\,tg(45^\circ - \phi/2)]/tg\phi$$

Obviously, the vertical rock mass pressure $q=\gamma h_1$, is always above zero, because these parameters γ, a and h are greater than zero. thus, it must be concluded that rock mass pressure exist in any natural karst caves and therefor it would not be stable without supporting measures. Forthermore from the formula $q=\gamma h_1 = \gamma a_1/f_{K\rho}$ it is clear that rock mass pressure is directly proportional to the cave span a, so, enormous rock mass pressure would appear in many natural caves with big span.' It would be very complicated and expensive if taking measures to reinforce them.

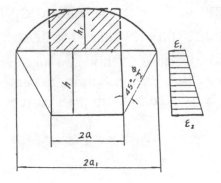

Fig.1

According to our in-situ investigation to many of the caves, it has been proved by both corrosion phenomena at the roofs and walls a and accumulation at the bottoms of caves, that they have been stable for quite a long geologic period. One of the convincing examples in mordern time is that some natural karst caves were onece used as plant during anti-Japanese war in late 1930's. Although no measures were taken for the roof at that time, there have been no droppings found after more than 40 years up to now. This example illustrates that the rock surroundings of these caves are stable with out reinforcement, therefore no rock mass pressure exist in these caves.

Since we knew little essential distinctions between natural and artificial caves, we tried to solve the problem by installing pressure cells and setting up settlment measuring sites in two caves in order to determine the strains and stresses. Through more than one year work, we found no signs of rock mass pressure and settlment at the bottoms. All showed that it was not appropriate to evaluate the surrounding's stability of a natural karst cave by measurement technigue and Protodyrakonov's Theory.

Through the study of several natural caves in utilization, we have gradually realized that it is the properer way to evaluate the rock surroundings' stability of a natural cave on the basis of engineering geology anylasis combined with the quantitative mechanical calculations ---what we call Comprehensive Evaluation Method, which first appeared in' The Measures of The Rock Underground Structures In China' published in 1974 at chapter of Geology Seney. As a matter of fact, however, there have been guite a few engineers who continued to use Protodyrakonov's Theory in recent years to evaluate the stability of the surrounding rock, and that would inevitably meet with such contradictory as those mentioned above, and cause a lot of loses and unnecessary waste due to his/her unsuitable conclusion.

In 1982, an expansion and reinforcement project of a natural karst cave in Sichun province, south China was put forward its rock surroundings' stability. We debated with Protodyrakonov's Theory holders upon the scheme and economic benefit of the project in comparation to the comprehensive method. Finally we found the great differences between them. Conclusions and suggestions resulted from comprehensive evaluation method were adopted at last. The expansion and reinforcement construction was carried on soon afterwards and was completed at early summer in 1984, this project experienced a trial of floodwater in 1984. It has shown great superioity of the construction and especially the obious economic benific of the project.

2 EXPANSION AND REINFORCEMENT

2.1 Origin of the project

A factory is located in a closed karst valley area scoured and deepened by limestone. The only outlet of the surface runoff whithin the area is a natural karst caves that forms a natural bridge on which obliquely lies the in and out road of the factory.(Fig.2.)

The cave's cross section was about 40 square meters which was not large enough to drain off 100-year frequency flood. Should the water be blocked up, the factory would be in danger of inundation. In July, 1975, the cave was partly stopped up during a flood, and that caused a disaster in various degrees to both the factory buildings in several blocks and residentia quarter, that created a direct loss of more than 480,000 yuan (RBM). Since then great attention was paid to the capacity of draining off water of the cave, and many experts had been asked to solve this problem for the factory. It was that time that the expansion and reinforcement of the cave was brought forward.

2.2 Topographical and Geological Conditions Around the Karst Cave

The factory is situated on Sinian-cambrian ilmestone rock base where karst corrosion is well developed. Up apart from the natural karst bridge is a closed valley which is long and narrow

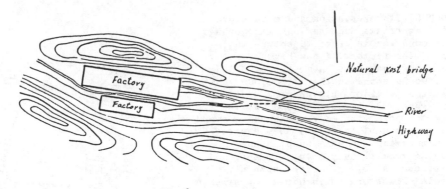

Fig.2

in width ranking from 100 meters up to 200 meters, paralleled at the sides by hills of 150 to
250 meters in height. The bridge reaches the narrowest place of the valley with only about 20
meters in width.

The coveraging area up the bridge is about 31 square kilometers, the surface runoff goes
down totally through the karst cave. Since there exist either mountains or cliffs at the sides
of the valley, there are no other choices but the bridge which could be used as a way of comm-
nication. It is the most important link to guarantee both factory safety and people's life.

The tectonic nearby the bridge is simple and there are no folds and faults which possibly
influence its stability. Outcrops of the rock strata show no differences its occurrence 290°-
315∠17°-20°, thickness of stratum being 0.3 to 1.0 meter and there are no soft layer among them
Joints are well developed in just two main sets: its occurence of the first sets 60∠83° and the
second 230∠85°, which nearly cut across the strata in right angle, eroded pits along the joints
sets can clearly be seen in the cave. Stalagmites and stalacites appear at the roof and especially whithin the range of 5 to 10 meters symmetrized the cave's axis. cross sections of the
corroded limestone show a 'V' shape which embeds into lime rocks (Fig.3) Leakage from the roof
was rather serious and it was considered that the leakage had a connection with precipitation.

Experiment shows a series of results of the rock
mechanical parameters. e.g.Protodyrakonov's sturdy
coefficient is f=1.5, limited compressive strenth
of the corroded limestone is 1429 kg/cm², mean
limited compressive strenth of the corroded limestone is 118 kg/cm², the minimum is 79 kg/cm².
angle of internal friction of corroded limestone
is ϕ=46°, cohesion is c=6kg/cm².

The karst bridge was 74 meters long and 7 to 10
meters wide, 7 to 14 meters high before being
expanded the over-laying rock was 8 to 11 meters
thick. Now it was extended to 14 meters at bottom,
8 to 14 meters high, cross section area was doubled, it would be safe to drain off 100-year frequency flood.

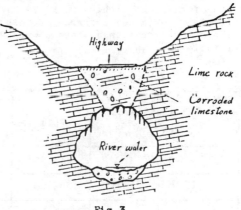

Fig.3

2.3 Evaluation of the surrounding rocks of the
      natural karst bridge

1. Expansion
The experts held several interesting discussions on whether the natural karst bridge could
be reexcavated or not. The protodyrakonov's theory holders thought, on the basis of 'railway
Engineering Geology manual', that the safety thickness of a roof arch could be determined by:
$$H=[a+H \cdot tg(90°-\phi)]/f_{K\rho}$$
where: H=safety thickness of the roof
    a=half span, which is 5 meters here
    ϕ=angle of internal friction which is assumed 60°
    H =height of cave which is 14 meters
    $f_{K\rho}$=Protodyrakonov's sturdy coefficient, here is 1.5
so H=[5+14×0.577]/1.5=13.08m>8m
It can be seen that thickness of the roof(8) is smaller than calculated safety roof(13.8),
hence the cave bridge is unstable and it is not proper to expand it but a permenant solution
was blown up rock cover into a open through, then to build a reinforced concrete highway
bridge over it. (Fig.4.)
There were two problems with this programme, first was that it would cost 900,000 yuan(RMB)

(about 300,000 USD)with excavation and move aways the cover, building bridge and dearing with garbage proposal. The second was in order to ensure the transportation of the factory, a new highway about 1.5 kilometers along the valley near the cave had to be built, it would spend another 1,000,000 yuan (RMB) on it, so the total of this project would reach 1,900,000 yuan(RMB) (about 630,000 USD).

However, from the pointview of comprehensive evaluation method, we thought there are some shortages according to engineering analysis such as the thinest overlaying rock is only about 8 meter, and the formation at central of the cover is mainly formed by corroded limestone which has a low strenth and serious leakage, however its geological structure is simple and there are no unfavourable folds and faults, so the walls are stable. Also the 'V' shape between the limestone provides the stability in its entirety though the corroded rock itself is in low strength.

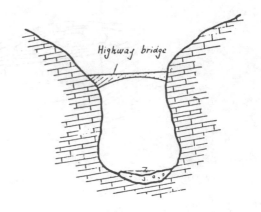

Fig.4

Furthermore we used two kinds of mechanical methods to evaluate the stability of surrounding rocks with a quanlitive calculation. At first, taking the karst bridge as a hingeless arch in calculation. (Fig.5)

where V=the vertical reactive force of the hingeless arch at the arch ends.

M=the moment of the hingeless arch to its axis.

H=the horizontal force of the hingless arch at the footings.

$R=\sqrt{H^2+V^2}$, is resultant force at each ends of the arch.

$b=2R/\delta$ is the minimum of safety thickness of the arch.

δ=allowable compressive strength of the rock mass formed the arch, here is 79 kg/cm².

At first, we obtained a serres of data from the formula mentioned above showed as below: table 1.

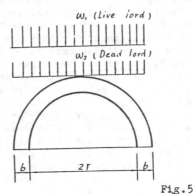 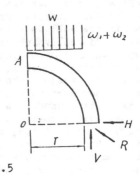

Fig.5

Tab.1.

Assumed Thickness of the Hingeless Arck (m)	V(t)	M(t)	H(t)	R(t)	B(m)
5.0	413.90	1550.35	221.48	469.40	5.87
6.0	485.60	1768.00	252.57	523.55	6.54
7.0	503.30	1799.20	258.61	528.69	7.94
8.0	548.00	2244.00	320.57	634.88	7.94

From the table we know that if the thickness of the hingeless arch is not smaller than 8, the rock mass pressure at both roof and ends of the arch would be whithin the allowable strength of the corroded limestone. In fact, the overlaying rock is little bigger than 8. So it is considered as safety.

Secondly, according to Chang Shibiao's formula $P=\delta_\theta K-2cK_1K_2$ where, P=rock mass pressure at a point on roof of a cave. δ_θ =tangential stress of a point on the roof of a cave, K_1=tg $(45°-\phi/2)$, K_2=tg$(45°+\phi/2)$, c=cohesion of surrounding rocks, θ =angle between the centre point of the arch and horizontal line, degree.

On the basis of lab experiment, several data were obtained as followings, to corroded limestone μ=0.4, γ=2.4g/cm³, ϕ=46°, c=6kg/cm², and c=2kg/cm² in calculation, while to limestone, μ=0.2, γ=2.8g/cm³, ϕ=60°, c=5kg/cm². Taking relevant values back into the formula, we get a serres of rock mass pressure values at vauious positions of the cave from the tables bellow. Data in table 2 were on the supposition that the whole surroundings would be formed by limestone, while those in table 3 by corroded limestone.

Tab.2.

(degree)	0	15	30	45	60	75	90
P(kg/cm²)	−1.95	−2.08	−2.27	−2.45	−2.59	−2.65	−2.70

Tab.3.

(degree)	0	15	30	45	60	75	90
P(kg/cm²)	−0.25	−0.42	−0.65	−0.86	−1.03	−1.13	−1.46

It is seen from both table 2 and table 3 that either corroded limestone or limestone surrounding the cave, there are no rock mass pressure at the walls of the cave. Hence, we come to the conclusion that the results by using two kind of mechanical calculation reach a unanimity with that by engineering geology analysis. Thus, we suggested that a little amount of explosive be used to extend the cave to 14 meter in diameter by keeping over-laying rock of cave bigger more than 8 meters in thickness, which can assure the stability of the cave with no reinforced support.

This proposal was adopted by the factory in 1982, and whole section excavation was soon carried out without reinforcement, and was completed whithin the same year. During the construction the karst bridge could pass through freely.

This program not only cut down 1 million (RMB) yuan invesment but also ensure the factory's normal prodrction.

2. Reinforcement

'Is it necessary to reinforce the surrounding rock and how to do it ?' was the problem followed soon after the excavation. Another argument arose. The outline of the roof of the cave looked much clearer than before. Although there were no unexpected drops from the roof of the cave, yet overmuch explosive had been used and so caused serious tensive cracks in the roof of the cave, leakage was even worse than it used to be. Protodyrakonov's theory holders advanced that whole section reinforced by building a arch lining (Fig.6) according to Protodyrakonov's collapsearch theory, they made a design that top arch should be 1.0 meter thick, 1.5 meter at the feet of the arch. Furthermore, in order to satisfy with draining off capacity, it is necessary to expend 1 meter at the cave's footings. It would cost nearly 650,000 yuan and that did not yet include the last item.

The comprehensive evaluation method holders insisted that no room available for further excavation when even first cut in the cave was not allowed accordiog to Protodyrakonov's theory. Based on comprehensive evaluation method, it is considered that there are no rock mass pressures in the walls of the cave, it is unneccessary to support the lining with whole section reinforced. However, the explosive effection should not be neglected and nor the tensive cracks and heavy leakage. In prerequisite condition of ensuring the satety of the karst cave bridge, and the draining off water we proposed the programme of top-bottom bolts lining with surry concrete.(Fig.7)

In this way, it is possible to strengthen the cave's entirety and mechanical strength of the corroded limestone, and meantime, decrease leakage, so that the cave's stability increased. All this programme would only cost less than 250,000 yuan, and it is easy to handle the construction, This was adopted by the factory at last.

After completion of the reinforcement in 1984, leakage of the cave gets greatly improved and the whole construction has stood a severe flood test in July, 1984. That shows a success

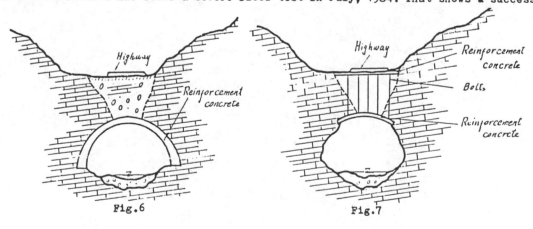

Fig.6   Fig.7

in programme which not only shortens construction period but also save an investment about 4oo,ooo yuan. (the design of top-bottom bolts lining with slurry concrete is omitted)

3 CONCLUSION

Natural karst caves are the products of long geologic process. Many of them has corroded in the walls and roofs and stalatites and stalagmites are well developed. However, those karst caves,which there are no obviously modern drops or collapses, are not different from an unsuported artificial cavern with long geologic process, collapses of the surrounding rocks have already stopped at unstable places. So it is unpractical to evaluate the stability by Protodyrakonov's collapse-arch theory. Furthermore, the rheologic and the rheopoxic deformation and settlment have almost finished, so it can be considered as a structure. Appearently the combination method of qualitives engineering geology analysis and quantaive mechanical calculation has superiority over Protodyrakonov's theory in surrounding rocks stability evaluation of natural karst caves.

Pumped-storage schemes in tectonically deteriorated rock masses

Projets de centrale à accumulation dans le massif rocheux détérioré tectoniquement

M.Matula, R.Ondrášik & P.Wagner, *Department of Engineering Geology, Comenius University, Bratislava, Czechoslovakia*
A.Matejček, *Engineering Geological & Hydrogeological Exploration (IGHP), Žilina, Czechoslovakia*

ABSTRACT: Engineering geological investigations for the construction of a pumped-storage power plant in tectonically badly deteriorated crystalline rocks are described. The obtained data made possible to recommend the optimum siting and composition of the dams and underground power station.

RESUME: On décrit les investigations géologiques de l'ingénieur pour la construction de la centrale a accumulation dans les roches cristallines fortement détériorées tectoniquement. Les données obtenues permirent de recommander le placement optimal et la composition des barrages, ainsi que la centrale électrique souterraine.

1 INTRODUCTION

Major faults and their recent activity are the principal limiting factors in siting and construction of complex pumped-storage power plants in crystalline igneous and metamorphic rocks. In the first stages of designe, when the most responsible decisions concerning the scheme conception and structural arrangement are to be made, an evaluation of tectonic conditions and of the irregular rock mass deterioration is of primary importance for a full success of the whole project.

Although the crystalline rocks in the Czechoslovak Carpathians (which are a young Alpine-type mountain system) lithologically are very variagated, in a "sound" state not depreciated by faulting and weathering, they represent a solid bedrock and good foundation grounds. For common civil engineering works the major part of granitoids and highly metamorphosed rocks offer quasi-homogeneous rock masses, characterized by a differently pronounced anisotropy (e.g. in gneisses, mica-schists). In tectonically deteriorated rock masses, however, these favourable characteristics are rather exceptional in large regions of the country and the deep weathering and hydrothermal alteration processes extensively developed in faulted rocks increase their heterogeneity and spatial physical variability. Recent tectonic movements are a common feature in this young mountain system of Carpathians.

The evaluation of both, the tectonic-structural heterogeneity and the recent faulting activity in crystalline rocks, is very difficult, because of the lack of marker beds which would indicate the original position of rock bodies. Therefore the investigation of rock mass structure and of present physical state of rocks here must be based on a reconstruction of lithogenetic conditions and postgenetic (mainly tectonical) history of igneous and metamorphic masses. Thorough petrological, tectonical and geomorphological analyses are necessary for the construction of reliable structural models at various levels of detailness serving a successful engineering geological investigation, exploration and solution of geotechnical problems.

The high efficiency of such an approach was well shown in all the previous works of authors oriented to the construction of complex hydraulic structures (Matula et al. 1985). This approach was also applied in preparing the engineering geological documentation for the new water pumped-storage scheme on the Ipel river (Carpathian Mts., Central Slovakia).

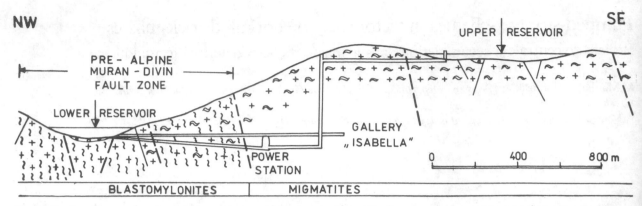

NW   SE

UPPER | RESERVOIR

PRE - ALPINE
MURAN - DIVIN
FAULT ZONE

LOWER | RESERVOIR

GALLERY
„ISABELLA"

POWER
STATION

0   400   800 m

BLASTOMYLONITES | MIGMATITES

Figure 1. Pumped-storage power scheme on the Ipel river is situated in a Paleozoic mass of granitoids and migmatites which are blastomylonitized within the
major regional fault zone (situation of this cross-section is shown in the map,
Fig. 2).

2 ORIENTATION AND PRELIMINARY INVESTIGATIONS FOR OPTIMUM SITING OF THE POWER SCHEME

For the construction of this pumped-storage power plant one favourable valley
section was proposed due to the most convenient hydrological and landscape conditions. A rockfill dam 78 m high will create the lower reservoir with a storage
volume of 16 mil. cubic meters. The rockfill dam of the upper reservoir is 60 m
high and the head is up to 400 m. The underground power station is to be excavated in the heterogeneous and tectonically affected Paleozoic mass of migmatites (Fig. 1).
   The pumped-storage power plant is to be located in a brachyanticlinal structure, the major faults being oriented along the brachyanticline axis in SW—NE and
in NW—SE directions (Fig. 2). Down to 400 to 500 m deeply eroded and weathered
remnant of an ancient Tertiary peneplain in the central part of this massif is
characterized by deep elluvial (residual) soils. The valley of the Ipel river
follows one of the oldest and principal fault zones in the Czechoslovak Carpa-

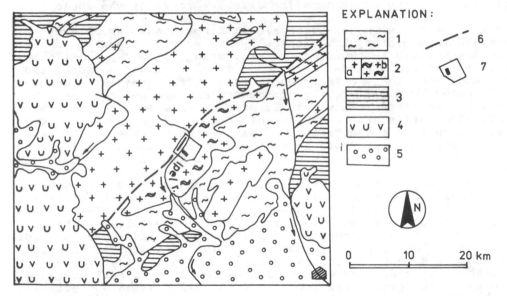

EXPLANATION:

Figure 2. Geology of the larger area of the pumped-storage
power scheme on the Ipel river (Czechoslovak Carpathians Mts.)
1 - Paleozoic (partly Pre-Cambrian?) gneisses; 2 - Paleozoic
granodiorits (a) and migmatites (b); 3 - Mesozoic sedimentary
complexes; 4 - andesitic neovolcanites; 5 - Tertiary sedimentary complexes; 6 - Muran-divinsky regional fault zone; 7 - area
of the power scheme with the exploration gallery.

thians (Muran-divinsky fault zone, SW-NE) which may interfer with the interests of building the lower reservoir dam and the underground power station beneath the left slope.

Although it was clear enough in the preliminary considerations that the scheme was situated in a major fault zone both, the designers and the geologists did believe not to meet here any serious troubles due to poor quality rocks, young tectonical movements and/or seismicity. This simplified understanding was improved soon during the preliminary engineering geological investigation. This included mapping on scale 1:50 000 and 1:10 000, geophysical measurements and exploratory works (186 boreholes with a total length of 6407 m, 23 hydrogeological wells - 592 m and 159 shafts - 1238 m). The most important was the 1070 m long gallery "Isabella" driven into the left bank slope where the real nature of rock mass was discovered in its full complexity and variability of rock characteristics and structure (Matejček 1985).

In the gallery the bedrock is represented by various types of migmatites which in the first 600 m are blastomylonitized. The rock structure is complicated by irregular fault dislocations. The rocks in the fault zones are altered to tectonic breccias, kakirites, ultramylonites. Clayey dislocation gouge, in some faults up to 2 m thick, made difficult driving the gallery in such sections due to swelling of clays and overbreaks.

The boreholes and shafts revealed the irregular cover of surficial deposits and variable depth to the bedrock all over the area of interest. Elluvial (residual) soils up to 3 to 5 m thick are beneath the designed upper reservoir, but also in the upper parts of the valley slopes they are up to 3 m thick. Stony and sandy loams of talus reach 5 m at places up to 10 m thickness. A great surprise, however, was the thickness of proluvial and fluvial deposits (up to 35 m) just in the left bank slope beneath the designed dam of the lower reservoir.

Fault phenomena and very variable thicknesses of residual soils and Quaternary deposits indicate the differentiated neotectonical movements of individualized rock mass blocks. This led to detailed mapping of the area of interest on scale 1:10 000, to extensive geo-structural studies and to additional geophysical measurements and drilling works oriented to explain the genetical conditions and tectonic history, which were decisive in the geological and geomorphological development, and responsible for the present state and quality of the local engineering geological enviroments. Results of this study may be summarized as follows.

3 EXPLANATION OF LITHOGENETIC CONDITIONS AND HISTORY OF THE LOCAL ENGINEERING GEOLOGICAL ENVIRONMENT

The origin of the granitoid rock mass is related to the Pre-Alpine anatexis of Old-Paleozoic (partly Pre-Cambrian?) paragneisses. In the proposed construction site of our power plant there occur mainly various types of migmatites created by partial anatexis (Ondrášik et al., in print). Younger, but still Pre-Alpine orogenic movements (Carboniferous-Permian) produced an extensive blastomylonitization (blastomylonites s. s.) along the major regional complex fault zone (Muran-divinsky fault zone) reaching here about 600 - 1000 m in width. During the Alpine orogeny granitoid mass subsequently emerged from the depth about 10 km, forming a brachyanticline which was accompanied by differentiated faulting and relative subsiding of the uppermost part of brachyanticlinal structure.

In analyzing the types of faults it is possible to deduce that the oldest of them originated in the Paleo- and Meso-Alpine period due to predominant minimum stresses of WNW-ESE, later NNW-SSE strikes. Rocks within fault zones of this generation were altered to partly recrystalize blastomylonites and there occured qurtz veins and lenses. Later the Neo-Alpine period (Neogene and Quaternary) was characterized by partial healing of fractures with calcite. Faults of that time were conditioned by minimum stresses of W-E, later of SW-NE strikes, the principal stress was approximately vertical and in the movements vertical components prevailed. During this period parts of the old major Muran-divinsky fault zone were intensely reactivated and a wide spectrum of tectonites (including tectonic breccias, kakirites to ultramylonites) developed, as well as tectonized rocks altered by revived hydrothermal and deep weathering processes. Tectonic clays discovered in the gallery are composed of 20 to 40 % (exceptionally 60 %) clay particles. According to chemical, RTG, DTA and SEM analyses illite-montmorillonite minerals are dominant (Ondrášik et al., in print b). Accompanying joints of the oldest generation are filled up with quartz, later with calcite, and the youngest have clayey, sandy or no infilling. Some joints have the aperture up to 10 cm and are displaying strong seepage.

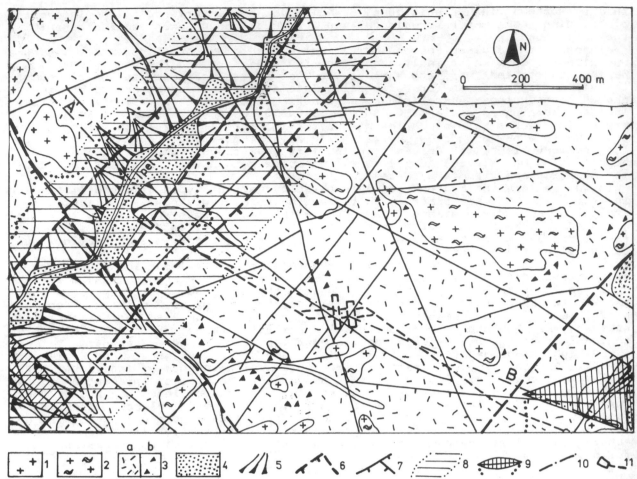

Figure 3. Geological-structural sketch derived from a detail engineering geological map 1:10 000 represents the block structure of the basement rocks and the extent of deteriorated rock masses within the regional Muran-divinsky fault zone. 1 - biotite granodiorites, with some paragneisses; 2 - migmatites and hybridized granodiorites; 3 - slope outwash (a) loams, (b) with talus blocks and fragments (more than 1 m thick); 4 - fluvial deposits; 5 - proluvial fans; 6 - regional and subregional faults; 7 - local faults; 8 - extent of the rock masses affected (mylonitized) by the Muran-divinsky fault zone; 9 - earth dams and the water level in reservoirs; 10 - cross-section along the exploration gallery (see in Fig. 1); 11 - designated power scheme.

During the neotectonical brachyanticlinal uplift of the granitoid mass, in the granitoid mass, in the arch bend its individual structural blocks subsided differently along faults parallel with the anticlinal axis, as well as crossing it perpendicularly. Morphologically the most impres ive depressions, including the Ipel river valley, developed upon the reactivated Muran-divinsky fault zone. Subsiding blocks are well indicated by accumulation of surficial materials. Recent tectonic movements of individual blocks are testified by active erosional and accumulation processes, as well as by sudden changes of the gradient curve of local water courses on different bedrock blocks. From the geomorphological analysis based on differences of old levelling surfaces and of the bases of Quaternary deposits, the rate of vertical uplift was calculated, on some blocks reaching several milimeters and on others up to 3 to 4 centimeters per hundred years.

Atmospheric precipitations saturate mainly the upper part of the rock massif where within the surficial deposits and in weathered jonted rocks a shallow water circuit does not endanger the slope stability. Part of the ground water, however, descends into deeper circuit through permeable fault and joint zones and frequent inflows to the exploration gallery are indicating the degradation of rocks. Hydrogeological conditions within the rock mass are very complicated due to

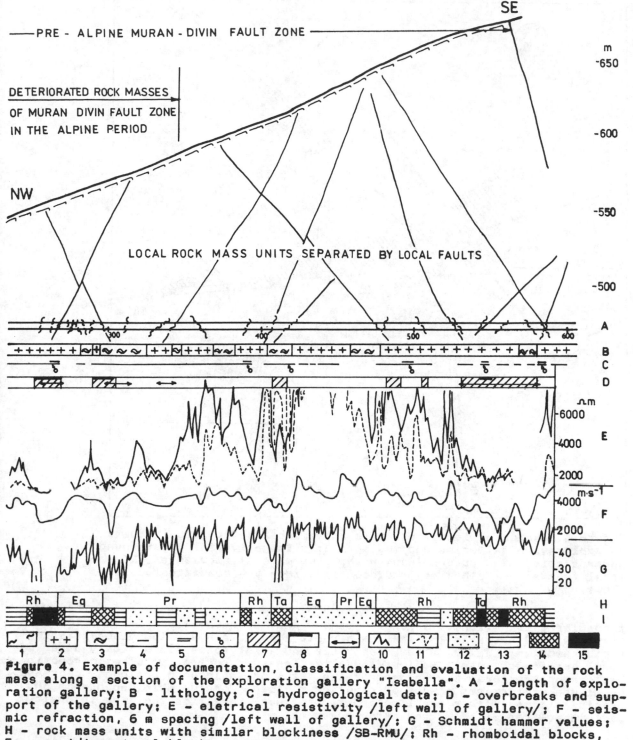

Figure 4. Example of documentation, classification and evaluation of the rock mass along a section of the exploration gallery "Isabella". A - length of exploration gallery; B - lithology; C - hydrogeological data; D - overbreaks and support of the gallery; E - eletrical resistivity /left wall of gallery/; F - seismic refraction, 6 m spacing /left wall of gallery/; G - Schmidt hammer values; H - rock mass units with similar blockiness /SB-RMU/; Rh - rhomboidal blocks, Eq - equidimensional blocks, Pr - prismatic blocks, Ta - tabular blocks; I - rock mass rating classification /by Bieniawski/ - evaluation in the direct of the gallery. 1 - mylonitized zones, kakirits; 2 - migmatitized granit, migmatit; 3 - paragneiss; 4 - moisture sectors of gallery; 5 - dripping and unconcentrated inflows; 6 - concentrated inflows more than 0,05 l.s^{-1}; 7 - support parts of gallery; 8 - overbreak sections; 9 - zones of strong rock deterioration; 10 - spacing of electrodes 14 m; 11 - spacing of electrodes 6 m; 12 - good rock /61 - 80 points/; 13 - fair rock /41 - 60/; 14 - poor rock /21 - 40/; 15 - very poor rock /0 - 20/.

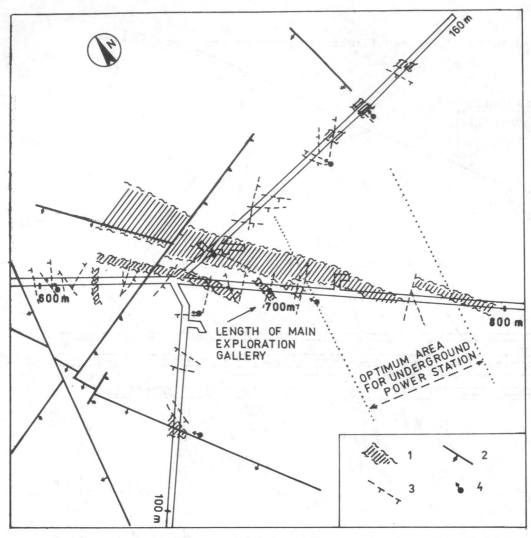

Figure 5. Sketch of local rock mass units separated by local faults
in exploration galleries and the optimum alternative for underground
power station. 1 - paragneisses within the mass of migmatites; 2 -
local faults; 3 - other major discontinuities; 4 - concentrated inflows.

irregular alternation of strongly jointed and impermeable barriers of tectonic
clays. Ground water in the shallow circuit zone displays a low mineralization
(acid reaction). Water in the deep fault and joint zones flows slowly under natural conditions, has a medium mineralization an a neutral to alcalic reaction.
Exploratory gallerie act as deep drains, causing an intense drainage, decrease
of mineralization and change of reaction to acid. Considerable dependence on
precipitations (with a small retardation effect) testifies the high intensity
of this new regime of ground water movement.

4 RECOMMENDATIONS BASED ON MODELS OF ENGINEERING GEOLOGICAL CONDITIONS

The engineering geological data derived from the enlarged Preliminary Investigation Program (including mapping on scale 1:50 000 and 1:10 000, boring and other
exploratory works, geophysical measurements, "in situ" testing, etc.) made it
possible to delimit quasi-homogeneous rock mass units (RMU) of different ranks
and to characterize them in terms of engineering geology and geotechniques.
Litho-structural and fault zone RMU is distinguished according to the classification system by Golodkovskaya, Matula and Shaumyan (1982).
   Regional rock mass units (R-RMU) with their individual lithology, structure

and history of tectonic regimes were delimited on the map on scale 1:50 000.
They are separated by regional Muran-divinsky complex fault zone (RF-RMU). These
units were divided into sub-regional (SR-RMU) characterized by a higher degree
of internal structural uniformity and behaviour. They are bordered by subregional
fault zones (SRF-RMU).

Within the suggested construction area local rock mass units (L-RMU) are delimited on the basis of mapping on scale 1:10 000, as well as additional geophysical and exploratory works. They represent individual blocks with different neotectonical activities, separated by local faults (LF-RMU). These regional, sub-
regional and local RMU are shown in Fig. 3 and in the cross-section, Fig. 4.

Detailed engineering geological observation and measurements of electrical
resistivity, seismic refraction and Schmidt hammer values in the exploration
gallery enabled a still more precise clssification of rock mass units based on
the lithostructural homogeneity and tectonic deterioration of rocks. Rock mass
units with similar lithology and blockiness (SB-RMU) were divide within local
units, as well as units with similar type and degree of tectonic deterioration
within fault zones are distinguished (Fig. 4).

Delimitation of sections in the gallery with lithologically and physically
similar rocks (i.e. engineering geological types in the sense of UNESCO-IAEG,
1976), as well as the characterization of these quasi-homogeneous rock units in
terms of electrical resistivity, microseismic waves velocity, Schmidt hammer
rebound values, ground water inflow, overbreak and other data - all of them represent the best basis for the evaluation of the physical state and behavioural
qualities of rocks and for their classification in terms of any geotechnical
rating system for underground construction. It also permits to suggest the most
convenient site for underground cavern (Fig. 5) and designate sections with less
favourable tunneling conditions.

The classification of lithogenetic conditions and tectonic history of rock
environments in the area has demonstrated its reasonability in many ways. It
enabled mainly: (a) to understand better the structural heterogeneity and geodynamic nature of local engineering geological conditions; (b) to interpret
efficiently the information data obtained by exploratory procedures; (c) to
estimate more realistically the natural requirements for optimum siting and
designing individual componental parts of the scheme (e.g. dams of both water
reservoirs were replaced after obtaining new geostructural data, as well as the
proposed position of the underground power station and tunnels; (d) to prepare
a reasonable program for the following main investigation stage for the final
design of the project.

REFERENCES

Golodkovskaya, G.A., Matula,M., Shaumyan, L.V. 1982. Classification of Rock Mas-
    ses. In Proc. IVth Congr. IAEG. Vol. II. th. 1, p.25-32, New Delhi.
Matejček, A. 1985. Výsledky inžinierskogeologického prieskumu na lokalite Ipeľ.
    In Proc. Symp. Inžinierska geológia a energetická výstavba, p.35-43. Brno.
Matula, M., Ondrášik, R., Wagner, P., Holzer, R., Hyánková, A. 1985. Regional
    Evaluation of Rock Mass Conditions for Pumped-storage Plants. In Bull. IAEG
    No 31, p.89-94. Paris.
Ondrášik, R., Hovorka D., Matejček, A. (in print - a). Prejavy muráňsko-divín-
    skej poruchovej zóny vo veporickom kryštaliniku v štôlni PVE Ipeľ. Mineralia
    Slovaca. Bratislava.
Ondrášik, R., Matejček, A., Klukanová, A. (in print - b). Zlomové poruchy vepo-
    rického kryštalinika v prieskumnej štôlni PVE Ipeľ. In: Západné Karpaty, sé-
    ria hydrogeológia a inžinierska geológia. Bratislava.
UNESCO - IAEG 1976. Engineering Geological Maps. A guide to their preparation.
    UNESCO Publishing House. Paris.

The foot wall slope stability analyses for open pit of Nanfen Iron Mine, Liaoning Province, People's Republic of China

Analyse de la stabilité des pentes dans la mine de fer à ciel ouvert de Nanfen

Shuju Zhou & Jiuming Li, *Baoding Institute of Geotechnics, Hydrogeology & Photogrammetry, MMI, China*

ABSTRACT: The foot wall slope of the Nanfen Iron Mine is cut by six joint sets, A to F, among which the F joint set extends more larger than the others and has smooth planes with the dip towards the pit, and their mean dip angle is a bit larger than the slope angle. The angle between the average dip directions of this joint set and the slope ranges from 25° up to 51°. There would not be any slide along the joint planes.

The C joint set also extends considerably large, and the angle between the dip directions of C set and the slope is 20°. The dip angle is too gentle (approximately 18°) to form plane failure. The overall slope failure pattern is likely to be the bi-plane type formed by F and C sets. This type of failure pattern may be typical of some rock slope.

RESUME: Le talus du mur iferieure de mine de fer a' Nanfen est coupé par les joints de six groupes, de A à F, dans lesquelles les joints du groupe F étendent plus large que les autre, avec plan lisse et la pente envers le quit, et leur angle moyen de la pente est un peu large que l'angle du talus. L'angle entre la direction de la pente moyenne de cette joint groupe et la direction de la pente moyenne de cette joint groupe et la direction du talus rande de 25° a' 51°. Il n'y aura pas de glissade le long le plan.

Les joints du groupe C le long le talus étendent considérablement large aussi, et l'angle entre la direction de la de C groupe et la direction du talus est 20°. L'angle de la pente est trés doux (18° approximativement), la destruction de type-plan n'est pas possible d'être formée séparément. Le modéle de destruction du talus général est probablement de type de destruction de plan de glissade doublée, composée des joints du groupe F et du groupe C.

General Condition

Nanfen Iron Mine, situated in the east of Liaoning province, China, has been mined for several decades. Its open pit now is about 2000 meters from north to south and 1600 meters from east to west. The foot wall slope is on the east side of the pit with its top level of 717 meters and the pit's bittom level of 442 meters. The temporarily existing slope near the final boundary has reached the height of 213 meters, which is mostly composed of merged benches of vertically 24 meters heigh. The final design level for the pit bottom is -2 meters, and at the end of mining the foot wall slope will be 720 meters high. Investigation and assessment of the final slope is still going on, analysis of the failure pattern and stability assessment are made here just for the 213-meters heigh slope formed so far. We think the results will be useful for a further study of the final slope stability.

Geologic Conditions

The ore body of this mine exists in Archaeozoic metamorphic rock of the Anshan group. The rock formation forms comparatively stable monoclinal structure, and dips to the west with an angle of approximately 45°. The rock mass behind the foot wall slope consists of hard bimicaceous quartz-schist and epidote-hornblendite (epidote-hornblende schist) of the Archaeozoic erathem. Since their schistosities haven't been well developed, they do not form weak discontinuities. There are intrusive veins of granite-porphyry and basalt in the slope rock mass.

Through the pit goes a reginal fault F_1 which in general extends north north-eastwards and in the mining area stretches approximately north-southwards. On the west side of the pit it goes across the end wall of the north side and the hanging wall. Because of its effect, the formation of Sinian System has been pushed upon the Archaeozoic earthem formation, thus forming the rock mass of the hanging wall. Judging from the trace of the fault F_1, the tectonic principal stress in the area of the pit should be in a direction approximately from east to west.

There are faults F_2, F_6 and F_7 distributed in the slope of the foot wall. These faults as well as veins intersect the slope at a large angle with the strike of the slope and their dip angles are approximately 70-80, so they have little influence on the stability of the foot wall slope.

The developement of jointsis quite well in the rock mass of the pit. Data have been collected from over 2000 joints in the foot wall slope after measurements by detailed line survey. Through calculation with computer, the preferred orientations for six sets of main joint planes were determined, and their attitudes are shown in the table below.

Number of joint set	A	B	C	D	E	F
Attitude of joint	214/83	108/54	288/18	266/84	314/59	295/47

The properties of each joint set are briefly described as below:

F joint set. A set of shear joints formed on a vertical section under the action of nearly east-wetward tectonic principal stress. Its attitude is almost the same as the fault F_1's. The joints of this set are widely distributed, with spacing of 50-100 cm. The joints stretch through all benches and all the bench slopes are actually formed almost by large smooth faces of the joints. It is thought that the joint planes of F set are completely throughout the slope. The joint planes are smooth and have sliding trace on them. Joint roughness coefficient, or JRC, is about 2-4. As the joint plane's strike is perpendicular to the direction of the tectonic principal stress, the plane is clearly characterized by being pressed. In some sections these planes are gently wave-shaped not only in strike but also in dip direction, and in certain sections greater undulations occur where the dip directions range from 248° to 340° and the dip angles range from 25°to 60°. Sometimes adjacent joint planes are unparallel, but are arc-crossed at small angles.

C joint set. Joints of C set are relatively widely distributed with the spacings of 50-200 cm. The joints extend very long and the traces of them on the benches reach the length of over 100 meters. The joint planes are rough, filled with quartz veins which are generally 20-50 mm thick, though the greatest thickness could be 500 mm. It is assumed from this appearance that these joint planes are formed by tension stress. According to observation of rock cores, it is known that the quartz-veins of C joint set are closely combined with their adjoining rock without opening, but the quarts-veins appeared on the benches are nearly broken; the upper benches have long been formed and the quarz-veins there are almost all broken. This may be caused by the effect of temperature and weather on materials of various colours. Unloading and shock by blasting also have great influence on the brokness of emerged quartz-veins. On the basis of field observation, the JRC value of C joint set can be 16-20. Because of its wide distribution and smaller dip angles, C joint set can catch and transport some water permeating from upper cracks so that at emerged part of this joint set trace of water can often be seen, and a flow may come out in rain season.

B joint set. A set of shear joints which conjugates with F joint set, arranged closely with spacing of 5-50 cm, tipping in reversed inclination of the slope. This joint set has been developed quite well, but not extending long, only 2-5 meters. These joints are filled with quartz-veins of 1 to 5 millimeters thick.

A joint set. Spacings are about 60 cm, extending 5 to 10 meters. Joinit planes are nearly upright, severing the slope vertically.

E joint set extending 5 to 7 meters with spacingsof about 50 cm.

D joint set extending 0.5 to 1 meters with spacings of 10 to 50 cm, and only occuring in certain sections of the slope. The nearly upright joints are filled with quartz-veins of about 1 mm thick.

Stability Condition and Failure Mode Analysis of the Existing Slope

The slope that has been formed now is 213 meters high with a overall angle of 30°. At 670-meter level there is a wide platform. Between level of 504 and 670 meters (or at the range of 166 meter vertical height) the slope angle is 37°. Now the slope as a whole is stable without great destructive deformation, but failures on a scale of 1-2 benches are quite serious, the most serious sliding can cover three benches up to 72 meters high.

After studying the positions and attitudes of various faults and veins in the foot wall slope, we think they don't do much harm to the stability of the slope. As joint set F and C are widely extending with the dip towards the pit, so their attitude conditions are unfavourable to the slope stability. It is shown from field observations that failures in existing slope occur mostly along joint planes. Analysis of slope failure mode is done by use of a stereographic projection of attitudes of six sets of joints and the slope face in which friction angle of 30° of the joint planes is taken for the friction circle, with average slope dip direction value of 268° and average dip angle of 37°. This stereographic projection is projected on a upper hemisphere with equiangularity (Fig.1).

It is a general judgement criterion that when the angle between the strickes of a discontinuity and a slope is smaller than 20°, and the former's dip angle smaller then the later's, plane sliding would occur. If the intersecting point of great circles for two discontinuities falls into the crescent dangerous area formed by the great circle of slope plane and friction circle, wedge sliding would occur. And, if the discontinuity dips in opposite direction to the slope, toppling failure would possibly occur. It can be seen from these principles that B joint set will probably result in toppling failure, and plane or wedge failures cannot   occur. .. In fact toppling failures caused by B joint set do occur now and tehn in the benches of the slope, but these joints only extend 2 to 5 meters which are relatively short, so they do not have much threat to the overall slope.

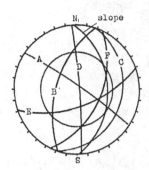

Fig.1 Stereographic projection of joint planes

In the light of their average strike and average dip angle, F joint set will not cause plane sliding which threatens the overall slope stability. As described before, F joint set with wide spread assumes a pattern of gentle waved form either in strike or in dip direction. Though the intersecting angle between their general strike and the strike of the slope is 27°, but in some sections this angle may be very small or even the strike of joints and the slope are nearly parpllel as showed in Fig.2. Much the same, the general dip angle of F joint set is steeper than the slope angle, however, dip angles of joints vary with wave-shaped undulation everywhere, so that in some sections joint planes daylight on the slope because of their less dip angle, as shown in Fig.3. The slope may be severed vertically by other joints. Combination of all these conditions will result kinematically in plane failure along planes of F joint set. Of cause, sliding face is some what a curved plane instead of a strict plane. Such failures can appear in a single bench only, and it can also cover several benches. In fact, most slidings happened on the existing slope benches just belong to this type of failure.

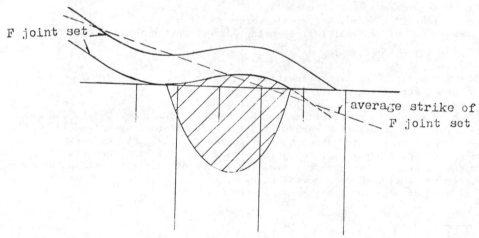

Fig.2. Plan of F joint set plane

The intersecting angle between the strike of C joint set and that of the slope is 20°, and the average dip angle of C joint set is only 18°, which is smaller than friction angle. Therefore this joint set can neither form sliding plane alone to bring about plane failure, nor form step failure as a main sliding plane. However, C joint set extends long and its combining with F joint set may produce double plane sliding to cause overall slope failure. It is thought that the most possible failure pattern in the foot wall slope of Nanfen iron mine just belongs to this type of double plane sliding. Based on this pattern, calculation and assessment for the existing slope stability is done here.

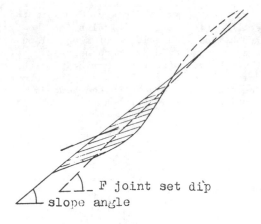

_F joint set dip

_slope angle

Fig.3 Profile of F joint set plane

Calculation for Stability

Calculation is made just for composite sliding of double plane. F set joint plane is considered as thoroughly severed, and the shear strength of discontinuity is taken as calculating strength. The part of C joint set adjacent to the slope face is open. Supposing the open part of the joints is 20 meters long, the shear strength of this part is rated as calculating strength in terms of discontinuity strength. For the other part out of the 20 meters long, its friction angle is still expressed in terms of friction angle of discontinuities, and because of the close combination of quartz vein with its adjoining rock, its cohesion may be taken in calculation in terms of cohesion of the rock mass. Condition and parameters for calculation are determined as follows.

1. Ground water level: According to the measurements of a piezometer located at the top of the foot wall and to the water depths in boreholes on the operating platform near the foot of the slope, and supposing that the water level is 32 meter deep at the top of the slope and at the middle of the 670 meter level wide platform, and ground water comes out from the foot of the slope, it is thereby possible to determine the normally falling down curve of the water table.

2. Seismic loading coefficient: Based on the result of seismic danger analysis for the mine area, seismic loading coefficient is determined as 0.15. Corresponding period of reappearance is 50 years with occurance probability of 25%.

3. Rock unit weight is $2.8t/m^3$. Water unit weight is $1t/m^3$.

4. Shear strength of discontinuities: By using Barton empiric formula (1):

$$\tau = \sigma tg\,(\phi_b + JRC\,\log\frac{\sigma_J}{\sigma}) \qquad\qquad (1)$$

where τ – shear strength of discontinuities (Kg/cm^2);
     σ – normal stress (Kg/cm^2);
     ϕ_b – basic friction angle, and $\phi_b = 30°$;
     JRC – joint roughness coefficient. For F jointset, JRC=3; for C set, JRC=18;
     σ_J – uniaxial compression strength of joint wall material, and $\sigma_J = \frac{1}{3}\sigma_c = 287\ Kg/cm^2$;
     σ_c – uniaxial compression strength of rock specimen, $\sigma_c = 860\ Kg/cm^2$.

τ – σ curves are drawn separately for joint sets F and C. Taking tangents of the curve against various normal stresses, then we may determine the values of friction angle φ and cohesion C of the joints under the action of each normal stress.

5. Shear strength of rock mass. After reference [2].

$$\frac{C_m}{C_i} = 0.114e^{-0.48(i-2)} + 0.02$$

where   i – fracture intensity, according to the measurement of fracture of cores in borehole, i=6 line/m;
     C_i – cohesion of intact rock, according to the experiment data, $C_i = 148\ Kg/cm^2$;
     C_m – cohesion of rock mass, by calculation, $C_m/C_i = 0.0367$; $C_m = 5.43\ Kg/cm^2$.

Shear strength of rock mass can be evaluated in another way. Relationship between shear strength τ of rock mass & normal stress σ can be expressed as below [1]:

$$\tau = A\sigma_c(\sigma/\sigma_c - T)^B$$

where A, T and B are empiric constants determined by rock types and quality of rock mass [1]. In Bieniawski's rock mass quality classification [3] CSIR, rating is 56; while in rock mass quality classification by Barton, Lien and Londe [3] NGI, rating is 5.94. Therefore, the rock mass shear strength envelop will be within the following two lines:

$$\tau_1 = 0.295 \times 860 \, [\sigma/860 - (-0.0003)]^{0.691}$$
$$\tau_2 = 0.525 \times 860 \, [\sigma/860 - (-0.002)]^{0.698}$$

Corresponding to each value of the normal stress, we may draw tangent to the curve and then determine the cohesion C and friction angle ϕ. We do this for the two curves respectively. The average value is taken as rock mass shear strength parameters.

For rock mass shear strength parameters in vertical strip boundaries, the second approach mentioned above is used, while for the cohesion in uncracked part of C joint set, the two ways mentioned above are summed up, having $C_m = 5.0$ Kg/cm².

Calculated Results:

Assuming that the sliding plane formed by C joint set passes through the foot of the slope, and that formed by F joint set has different positions, thereby different sliding block may appear . For each sliding block safety factor K is calculated respectively in which the smallest K value is the safety factor required, and the corresponding double plane sliding is just the most dangeous one, shown in Fig.4 where joint planes of both F set and C set are drawn in apparent dip.

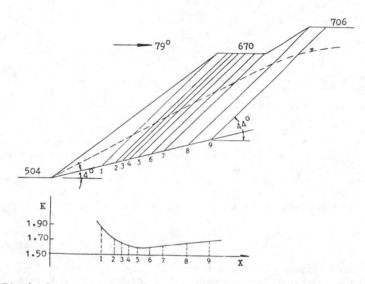

Fig.4 Determination of dangerous sliding plane position

Calculation is done by cutting the whole sliding mass vertically into several blocks. Assuming that the unbalanced force between the two blocks should be paralle to the base of the upper block. The safety factor thus calculated is 1.63. Checking computations are also carried out to get the safety factors in vertical boundaries between blocks, the smallest safety factor being 1.75 (Fig.5).

It is shown from the calculated results that the slope now is in a state of stability, which is proved by the existing stable condition of the overall slope. However, safety factor is not so great in consideration of the great mining depth designed. It may be expected that in a certain mining depth some effective remedial measures must be carried out to prevent double plane sliding failure in the slope.

Conclusion

1. Discontinuities caused by tectonic stress and under the influence of pressure is often wave-shaped. In analysis of this failure mode it is necessary to study not only the relationship between its average dip angle and general strike with that of the slope's, but also the possible daylighting of the curved plane, as well as to estimate the possible failure scale. Generally speaking, a set of wave-shaped plane with longer extension and smaller spacings is easily inter-sected by slope face, where its dip angle becomes gentler to form a sliding plane. The scale of failure is in direct proportion to wave length and undulation.
2. Planar, wedge, toppling and circular failures are the basic modes in rock slope failure. Practically, failures in engineering projects are often in a composite pattern which depends on

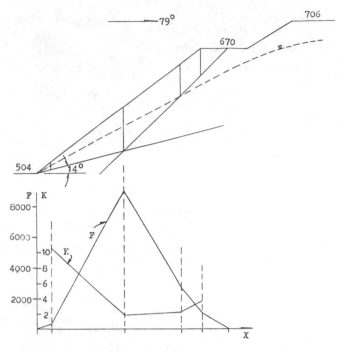

Fig.5   The internal forces between blocks

attitudes and composition of discontinuities in rock. If discontinuities are approximately parallel to a slope with its dip angle equal to or a little larger than the slope dip angle, the
discontinuities cannot form a plane failure kinematically. But they could not be thought to be
always stable. The discontinuities will form a double planes failure when the main discontinuities are intersected by another set of discontinuities parallel in strike with comparatively
gentle dip angle towards the pit. Sliding will also occur in a slope with great height where rock
material at the slope foot is broken under over great pressure stress or because of the rock mass
shear strength being exceeded. This specific kind of geological condition is typical in a foot
wall slope of sedimentary deposit or sedimentary-metamorphic deposit.

REFERENCE

[1].E.Hoek, J.W.Bray: Rock Slope Engineering 1981.
[2].G.Manev, E.Avromova - tracheva: Proceedings, 2nd Inter.Cong of Rock mech. Belgrade, 1970.
[3].R.E.Goodman: Methods of Geological Engineering in discontinuous Rocks. 1976.

An analysis for the unloading effects of the rocks around the foundation pit at the Gezhouba Project, Yangtze River with a fracture mechanical model

Analyse de l'influence sur les roches du déchargement aux alentours des fouilles de fondation du site de Gezhouba, sur le fleuve Yangtze
Utilisation d'un modèle mathématique pour l'étude de fractures

Li Zonghua, *Gezhouba Institute of Hydro-Electric Engineering, Yichang, China*

ABSTRACT: A linear elastic fracture mechanical model of unloading effects of the nick is established by using the foundation pit of the Second Channel Power Plant at Gezhouba Project as an example, while taking an artifical surface trench where unloading effects produce and a natural erosion valley as tip nick of halfspace. Also the basic hypothesis conditions of the model are presented. An elastic stress field expresion of the bottom of the nick and several boundary line equations of the failure zone are set up. Of all the equations, each one can meet its given failure criteria. As well as a horizontal displacement expression is given in the paper. The horizontal displacement is caused by the elastic recovery during which both slopes of the nick are moving inwards. The results obtained by using the model in analyzing the unloading effects of the pit coincide with presentations measured in many aspects. The theoretical results are close to the relative solutions obtained by numerical analysis or model tests.

EXTRAIT: A partir de la fouille du centrale hydro-électrique au chenal secondaire du complexe hyraulique de Gezhouba. Le présent mémoire donne, en prenant des fosses artificielles et des vallées d'érosion par la nature sous l'effet de décharge pour des brèches aux bornes du corps demi-infini, un modèle mécanique de rupture linéair de son effet de décharge. L'article a fait des conditione fondamentales de supposition pour ce modèle et fourni à la fois des formules afin de calculer le champ de contrainte élastique au pied des brèches, une équation de limite dans le domaine de rupture, à laquelle correspondent les divers critères de destruction et celle de déplacement rétabli d'élastioité en horizontale. des brèches dont les pentes des deux côtés inclinent vers la fouille.

En fonction de ce modèle, on marque une identité en plusieurs points entre les résultats de l'analyse de la fosse de fondation en exemple avec ceux de l'observation en place et pourrait les comparer avec ceux de l'analyse du modèle en valeur correspondante et de l'experience de ce dernier.

INTRODUCTION

Earlier on, some phenomena relate to the rock stress were noticed during the excavation of foundation pits of some dams in USA, such as Grand Coule'e Dam.J.L. Serafin(Portugal) pointed out that such phenomena as that of Grand Coule'e Dam were also detected in portugal at the Seventh International Conference on Dams(1961); L.Miiller (Austria) once explained the influence of residual stress on ground projects and put forward some ways to dispose at the International Crustal Stress Condition Symposium (1963) with Vajont Arched Dam (Italy) as an example; N.Hast (Sweden) once pointed out the effect of the crustal stress on a dam at the 8th International Conference on Dam (1966). He defined that the stress conditions in rocks had to be analyzed thoroughly before a dam would be built, and that the changes of the stress state ought to be inferred based on the analytical results after the dam to be built; At the 3rd International Rock Mechanics Conference (1974) some European schoolars, such as Manuel Rocha, state to excavate foundation would bring about shear fault which was heterotropic with the earth surface and separation of bedrock of a foundation pit.

Today the special topic on large-sized surface and underground excavation and its problems relative to stresses are placed on as one of main subjects to discuss.It shows that it has been attached its due importance with man's daily deepening of cognition to.

So far thousands of thousands of datum of crustal stress measured in-situ all over the world show that the examples of the horizontal stress is greater than the vertical stress or stress due to rock weight have meanly made up 75% of the total. Several large-sized hydro-electric projects in China,such as Gezhouba. The Three Gorge Projects and Ertain Hydro Power Station, all belong to such as these.Under great initial horizontal stress,another decidable factor of

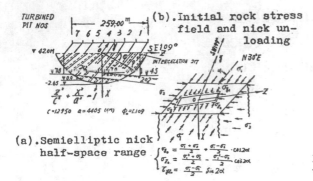

(a).Semielliptic nick half-space range

(b).Initial rock stress field and nick unloading

$\begin{cases} \sigma_{\theta_1} = \frac{\sigma_1 + \sigma_3}{2} + \frac{\sigma_1 - \sigma_3}{2} \cdot cos 2\alpha \\ \sigma_{Z_1} = \frac{\sigma_1 + \sigma_3}{2} - \frac{\sigma_1 - \sigma_3}{2} \cdot cos 2\alpha \\ \tau_{\theta_Z} = \frac{\sigma_1 - \sigma_3}{2} \cdot sin 2\alpha \end{cases}$

Figure 1.Longitudinal section simplification and unloading for the pit

stress-strain state change of rocks resulting from large-sized ground excavation is the surface geometric shape change with excavation processes. To analyze the unloading effects of the surface nick naturally suggests the linear elastic fracture mechanical model.Fracture Mechanics is of Griffith's glass microcrack analytical origin. It is later applied to the macro industrial nicks, and now the analysis for the unloading effect of rocks around a surface nick.It's certain that the geometrical dimensions have been enlarged. It's because of their major dimensions that initial stress conditions and medium characteristic indices can be counted based on presentations of multi-point measurement at different positions and depths,and initial stress conditions and medium behaviour indices and so on can be added up to study. This is a good condition to analyze the unloading effects of rock masses accurately. What of the linear fracture mechanical model? There is no harm in trying.

BASIC HYPOTHESIS

The model may be used to analyze the change of stress-strain state and failure of the rocks around the earth surface nick due to the large-sized excavation to the ground or the underground, or the unloading of natural ersion. The following conditions are suitable for the model:
  1. The rock is of homogeneous or isotropical viscoelasticity or elastic body;
  2. Much greater horizontal stress exists in the far-field relative to the depth of measuring points and rock strength. Its value is greater than the vertical stress or stress due to rock weight;
  3. The nick tip contour curvature is so great that the effects of the vertical stress and the horizontal stress running parallel to the major axes of the pit may be neglible;
  4. Plastic zone of the bottom of the nick relative to space dimensions of the nick is limited to a small scale.

SETTING UP THE MODEL AND ANALYZING THE EXAMPLE

  1. Using a Quadratic Curve Fits Nick Section Shape
  In Figture 1 (a), the length of the foundation pit of the Second Channel Power Plant at Gezhouba $2L= 259.00$m with a slope elevation of 7.80m or 4.50m at an end respectively,and 0.00m in the centre, i.e. h=42.00m which is below the mean elevation of initial bedrock. It is assumed that the major semi-axis c of equivalent semiellipse, $C = (1+\alpha)\cdot L$, and minor semi-axis $a = (1+\alpha)\cdot h$; thus, its area equals the read area of the longitudinal section of the pit to cut. The parameters were obtained:a = 4,465cm, c = 13,766cm,Type II elliptical integration $\hat{k}$ = 1.109. The endpoint of minor semi-axis of equivalent semi-ellipse is at elevation of −2.65m below the centre of the bedrock in the Turbined Pit 4.
  The traverse section was cut deepest in the pit section with 32m wide. Both side slopes rose with a slope of 1 on 0.7, reaching an elevation of 16.2m or 15m respectively, and then rose in a slope with terrances or gentle slope. Considering loose influnce on both graded shoulders by blasting, we can fit parabola into the traverse section shape. The focus is at $\frac{1}{2}$,i.e.the vertical distance between the endpoint of minor axis and the focus,$r= \ell/(1+cos\theta)$,where ℓ = 1,992cm (shown in Fig.3).

  2. Initial Horizontal Stress Measured In-situ in the Pit and the Horizontal Unloading of Its Both Side Slopes
  By means of the vertical borehole method,the measured horizontal stress in-situ at 400-meter downstream of the pit was obtained by Institute of Rock and Soil Mechanics,Academia Sinica,Wuhan (Li Guangyu and Bai Shiwei,1981.) The borehole was 39 metres deep. The initial horizontal

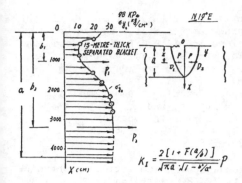

$$K_I = \frac{2[1 + F(a/b)]}{\sqrt{\pi a} \cdot \sqrt{1 - b^2/a^2}} \cdot P$$

Fig.2. Symmetric distribution or concentrated force in nick interior

Table 1. Initial horizontal stress results measured in-situ at Gezhouba dam site

Hole depth(m)	16.8	17.3	21.6	27.4	30.2	34.2	37.2	39.4
σ_1 (KPa)	196.0	1666.0	970.2	2048.2	2430.4	2930.2	3087.0	3008.6
σ_3 (KPa)	1685.6	852.6	833.0	1489.6	940.8	1705.2	2597.0	2567.0
ϕ (deg.)	NE48	NW49	NE2	NE60	NE32	NE12	NE62	NE26

stresses measured from 8 points with various depths are respectively presented in Table 1, which stand for the stresses on farfield without excavation influence.

The horizontal unloading stress component is estimated by plane static balance conditions, (see Figure 1.b). The computed results σ_{y_0} =3008.6KPa, σ_{z_0} =2283.4KPa, τ_{xy} = 147.0KPa. The error is not greater than the engineering precision when the pit is regarded as Type I nick. for the real bottom width of a pit is only 12% as long as its length of the same pit, and the value of τ_{y_0} is rather small. Since the vertical unloading is one-third less than the horizontal unloading relative to the pit weight, the bottom of the pit is so narrow that the unloading due to rock weight is almost negligible. The normal unloading stress σ_{y_0} on both vertical planes of the side slopes is determined by the horizontal stress value and its direction measured from various depths. The horizontal unloading distribution along the rock masses is shown on Figure 2. Considering the unloading effect of σ_{v_0}, we add up the unit length resultant value in two parts according to the rock depth and determine the positions of acting points. Thus, p_1 =352,800KN/m, b_1 =1,000cm, p_2 =72,118KN/m, b_2 =3,233cm (Fig.2.).

When the nick is unloaded, only the elastic deformation can restitute. Applying the imaginary waiting extensions p_1 and p_2 on the vertical surfaces of both side slopes, We can analyze the effect of the horizontal unloading of rocks, especially of the pit.

3. Elastic Stress Field of the Bottom of the Pit
(1). The stress intensity factor of every point of the nick tip

The stress intensity factor of a nick tip point is calculated by linear elastic fracture mechanical formulas. These formulas can be found in any books on stress intensity factors.

For two-dimensional surface nicks, such as trenches, valleys or ditches, Equation(1) may be used to compute their stress intensity factors, which is described as

$$\begin{Bmatrix} K_I \\ K_{II} \\ K_{III} \end{Bmatrix} = \frac{2}{\sqrt{\pi a} \cdot \sqrt{1 - b^2/a^2}} \begin{Bmatrix} F(b/a) \cdot P \\ F(b/a) \cdot Q \\ S \end{Bmatrix}$$

where P.Q.S. for the forces of unit thickness; while F(a/b) can be found in the relate tables or Figures.

For the semielliptic nick of semi-space, in consideration of the stress intensity factors of the minor semi-axis endpoint of plastic and shape corrections, the factors can be obtained from Equation (1) only multiplied by $Q^{\frac{1}{2}}$

$$Q = \left[\phi_0^2 - 0.212 \left(\frac{\sigma_{y(0)}}{\sigma_{ys}} \right)^2 \right]$$

where σ_{Y_0} for uniform stress in location, σ_{Y_S} for effective yield limit, i.e. σ_1 in the plastic zone. If plastic yield zone acturally approaches to plane strain state, the value of uniaxial compression limiting intensity of the rock σ_c is close to σ_{Y_S}. P.C.F. stands for the plastic constrained factor, P.C. f=1.43 in the plastic zone. The stress intensity factor of any Point A at the nick tip is given by Point A's eccentric angle ϕ shown on the graph in Fig.1(a), i.e., Eq.(1) multiplied by $Q^{-\frac{1}{2}} \cdot [1 - K'^2 \cos^2\phi]^{\frac{1}{4}}$ (where $K'=1- (a/c)^2$).

For the minor or major axis endpoint of equivalent semiellipse, fetching the maximum or minimum value of K_I respectively, substituting values of P_1, P_2, B_1, b_2, a, and ϕ of the above example, and assuming $\sigma_{Y_S} = \sigma_c =1241.7$KPa, the uniform equivalent stress in location $\sigma_{Y_0} = 1965.9$KPa. Hence the minor axis endpoint $K_{I max} =31,830.4$N.cm$^{-\frac{1}{2}}$; K_I (ϕ =60°) =25,695.6N.cm$^{-\frac{1}{2}}$ for nick tip point of the middle sections of the Turbined pit NO.2 or 6.

(2). Elastic stress field of the bottom of the pit

As shown in Fig.3, when using a parabola to fit cross section shape of the bottom of the nick, the elastic distrubition of the bottom may be estimated from Eq.(3) or (4) given by Greager,M. and Paris, P.C. (1967) as

$$
\begin{Bmatrix} \sigma_x \\ \sigma_y \\ \tau_{xy} \end{Bmatrix} = \frac{K_I}{\sqrt{2\pi r}} \frac{\rho}{2r} \begin{Bmatrix} -\cos\frac{3\theta}{2} \\ \cos\frac{3\theta}{2} \\ -\sin\frac{3\theta}{2} \end{Bmatrix} + \frac{K_I}{\sqrt{2\pi r}} \cos\frac{\theta}{2} \begin{Bmatrix} 1 - \sin\frac{\theta}{2}\sin\frac{3\theta}{2} \\ 1 + \sin\frac{\theta}{2}\sin\frac{3\theta}{2} \\ \sin\frac{\theta}{2}\cos\frac{3\theta}{2} \end{Bmatrix} \quad (3)
$$

$$
\begin{Bmatrix} \tau_{xz} \\ \tau_{yz} \end{Bmatrix} = \frac{K_I}{\sqrt{2\pi r}} \begin{Bmatrix} -\sin\frac{\theta}{2} \\ \cos\frac{\theta}{2} \end{Bmatrix} \quad (4)
$$

The principal stress, shearing stress and their directions can be computed based on plane static equilibrium conditions. And the computed results of elastic stress field of the bottom of the turbined pit section4 is shown in Fig.3 as well.

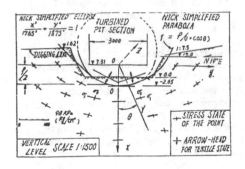

Fig.3. Graph and elastic stress field of bottom of the nick

4. Mechanical Parameters for the Rocks and Rock Masses around a Nick

Using a finite element method, the rock deformation of the pit was analyzed by Institute of Rock and Soil Mechanics.Acadenia Sinica,Wuhan.The results verified that the deformation of the rock masses around the pit depends on the whole deformation behaviour of the rocks and weak intercalations in the whole area, even if the rock masses belong to the Cretaceous sedimentary deposit with multilayers, close space. In view of the above,for the lateral isotropic body, the weight of characteristic index of deformation and strength of all sorts of rocks based on their thickness represents the characteristic index of the rock masses around the pit.

On the statistical datum measured in-situ from 4 maximum and 3 minimum dimension boreholes in various positions of the pit,the thickness ratio was determined for different rocks. After synthesizing the results of major, medium and minor triaxial tests in or out of the lab as well as other factors,the mechanical parameters for different rocks were determined by survey and design

Table 2, Strength and deformation parameter of rocks and rock masses

Rock and rockmass Parameter		clay-stone	clayish duststone	sandstone	Rock mass of the pit (Weighted average)
Rock thickness ratio (%)		6.75	45.00	48.43	100.00
Triaxial test strength	F	2.4	3.0	4.5	3.687
	σ_c	4.2	6.0	20.5	12.67
Deformation parameter	E	0.4	2.74	3.15	1.68
	μ	0.31	0.19	0.33	0.27

units together.The different mechanical parameters for rock masses around the pit were ~~determined~~ by the weighted average of the relative indices of the rocks according to their thickness.These values are given in Table 2. The triaxial intensity parameter for rock masses of the pit is described by linear equations of the Coulomb Failure Criteria, i.e. $\sigma_i = F \cdot \sigma_r + \sigma_\circ$ where F for the parameter relative to shearing strength index of the rock masses, $\sigma_\circ$ for the uniaxial compression ultimate strength. The uniaxial tensile ultimate strength of rock masses is determined by Griffith Cleavage Criteria $(\sigma_i - \sigma_3)/(\sigma_i + \sigma_3) = 8\sigma_t$ and the value of σ_t; while for that of the Gezhouba pit $\sigma_t = 155.2$KPa.

5. Failure Zone of Rock Masses around a Nick

Having studied many examples,such as the granitic foundation of Grand Coule'e Dam in USA or the foundation pit of Stagnate Water Pond to the Oahe Dam, which lies in the Gretaceous weak bedrock,or the pit at Gezhouba Project,China, we can see that foundation excavation will cause horizontal separation of the bottom bedrock, and heterotropic shear fracture with the ground.

Hence the stress on the boundary between plastic or cleavage zone and elastic zone is satisfactory to failure criteria and elastic conditions simultaneously.

(1). Boundary equation of Griffith Cleavage Zone

Substituting elastic conditional equality of the bottom of a nick into cleavage criterion equation gives:

$$\left\{ \begin{array}{l} \gamma = \left[\dfrac{1}{40.11} \left(\dfrac{k_1}{\sigma_t} \right) \right]^2 \cdot m_\theta^2 \\[2em] m_\theta = \left[\left(\dfrac{f}{\gamma} \right)^2 + \left(Sin\theta \cdot cos \dfrac{3\theta}{2} \right)^2 - \dfrac{f}{\gamma} Sin\theta \cdot Sin\, 3\theta \right] Sec\, \dfrac{\theta}{2} \end{array} \right. \qquad (5)$$

which is the boundary equation of Griffith Cleavage Zone.
Assuming $\theta = 0°$ gives:

$$R = \left[\dfrac{1}{40.11} \left(\dfrac{K_1}{\sigma_t} \right) \cdot P^2 \right]^{2/3}$$

which is the maximum depth characteristic size of the cleavage zone.
And while for $\theta = 60°$, $r_{(\theta = 60°)} = 1.059R$.

(2). Boundary equation of Coulomb Plastic Zone

Substituting elastic stress conditional equality of the bottom of a nick into Coulomb failure criterion expression, obtaining boundary equation of Coulomb Plastic Zone:

$$\left\{ \begin{array}{l} \gamma = \dfrac{1}{2\pi} \left(\dfrac{K_1}{\sigma_\circ} \right)^2 \cdot (m_\theta)^2 \\[2em] m_\theta = (1-F) cos \dfrac{\theta}{2} + \dfrac{1}{2}(1+F) \left[Sin^2\theta + \left(\dfrac{f}{\gamma} \right)^2 \right]^{1/2} \end{array} \right. \qquad (6)$$

(3). Boundary equation of shear failure zone along horizontal weak plane

In theory, for along any mechanical weak plane, the boundary equation of shear failure zone can be obtained by substituting elastic stress conditional equality of this weak plane into Coulomb plastic conditional equality or Mcolintock-Walsh failure criterion equality. There is an important but simple condition: rocks around the pit,such as that of Gezhouba pit in China or of Stagnate Water Pond to Oahe Dam,are horizontal sedimentary deposit.Therefore, the normal stresses σ_x and shear stress τ_{xy} are applied on the bedding plane or weak intercalations. The boundary equation of shear fault zone along horizontal mechanical weak plane under the bottom of the nick is described as

$$\left\{ \begin{array}{l} r = \dfrac{P \left(f \cdot cos \dfrac{3\theta}{2} + Sin \dfrac{3\theta}{2} \right) - 2B \cdot \gamma \sqrt{2\pi r}/K_1}{Sin\theta \cdot Cos \dfrac{3\theta}{2} + 2f cos \dfrac{\theta}{2} \left(1 - Sin \dfrac{\theta}{2} Sin \dfrac{3\theta}{2} \right)} \\[2em] B = \left\{ \begin{array}{l} \tau_\circ \\ 2\sigma_t \end{array} \right. \end{array} \right. \qquad (7)$$

which for Coulomb conditions.

which for Griffith correction conditions.

where $f = tg\,\varphi$, $\tau_\circ$ for shearing strength index of Coulomb rock mass; σ_t for uniaxial tensile ultimate strength of Griffith rock mass. Therefore it can be assumed that B=0 in Eq.(7), and Eq.(7) can be further simplified, for σ_i's directions along mechanical weak plane fault and the bottom of the nick are close to level, and unusual deviatoric stress. Subslituting the relate

217

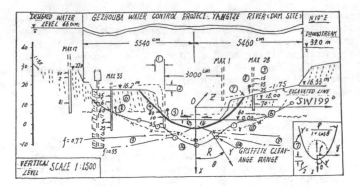

Note: 1.Pit walls moving to its interior; 2.Rising pit wall; 3.Rising bottom; 4.Intercalation fault along mudded layer (mm); 5.Measured intercalation fault along layers(mm); 6.Fracturing zone of both slopes; 7.Simplified parabola; 8.Measured absorbed water Q > $10^7 m^3$/min; 9.Horizontal fault danger zone along the intercalations; 10.Griffith Cleavage Zone boundary; 11.Coulomb plastic zone boundary.

Fig.4. Fracture mechanical model for deformation and failure of the pit

Table 3. The boundary dimensions of different theoretical failure zone around the nick foot of the second power plant at Gezhouba

Theoretical base of the boundary	Nos.for the turbined pit	θ (Degree)						Calculated parameters
		0°	30°	45°	60°	75°	90°	
Griffith Cleavage	4	2103	1925		2228	2955	4833	K_I=31830N·cm*, r=1992cm,q=155KPa;
Zone boundary	2 & 6	2781	2578		2957	3775	5633	K_I=28478N·cm*,r=3000cm,q=155KPa;
Coulomb Plastic	4	2803	1692		2414	3231	4542	K_I=31830N·cm*,F=3,687,r=1992cm,q=1241KPa;
Zone boundary	2 & 6	2198	2420		3285	4130	5286	K_I=284788N·cm*,F=3,687,r=3000cm,q=1241KPa;
Faulted Zone boundaryalong horizontal plane	4	996	1594	2436	4182			τ=0,ς=0; f=0.55,r=1992cm;
	2 & 6	1500	2392	3071	4499			τ=0,ς=0; f=0.77,r=3000cm;

parameters for Turbined Pit 4 and 2 or 6 sections into Equations (5),(6) and (7),we obtained their boundary dimensions, (listed in Table 3 and shown in Diagrammatic sketch 4).

6. Elastic Recovery Displacement of Both Side Slopes Inwards

The elastic recovery displacement Δy_F of both slopes of the nick moving inwards is solved from Eq.(8):

$$\Delta y_F = \int_b^{a'} \left[\frac{1}{E'} \left(K_I \frac{\partial K_{IF}}{\partial F} + K_{II} \frac{\partial K_{IF}}{\partial F} \right) + \frac{1}{G_e} K_{II} \frac{\partial \bar{K}_{IIF}}{\partial F} \right] da$$

where $a' = a+r_p$ for the effective depth of thenick; and $\gamma_p = \frac{1}{2} R \smile (R - \frac{r}{2})$ for the depth of stress weaken zone; for Gezhouba pit r=554cm, a'=5,019cm; E'= $E/(1-\mu)$ =1764.0KPa for plane strain state elastic modulus; and for the Turbined Pit 4, K_I=31830.4N.cm$^{-1/2}$; and the values for the horizontal elastic recovery displacement obtained from Eq.(8) are shown on the graph in Fig.5;

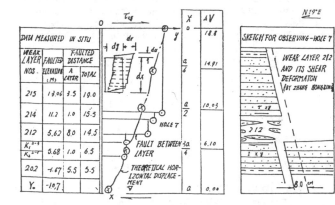

Fig.5. Value comparison between the theoretical horizontal displacement and measured layer fault of rocks of the Gezhouba pit

on which Δy_F is the displacement of different depth points below the mean bedrock of one slopeinwards

With large-sized excavation of earth surface or underground, or the natural erosion processes of the rocks around the nick, the stress-strain is under dynamic change state due to the unloading and that the rocks produce spreadwide rebound effect. After the Gezhouba pit was excavated and formed, when $r = \frac{\ell}{2}$, while θ changes from zero to $\frac{\pi}{2}$, σ_3 is decreased from zero to -1837.5KPa which is the extension stress towards the pit; with the result that σ_1 of compression stress is decreased from 8919.0KPa to 8144.8KPa. Because of the above changes, deformation and failure of the rocks take place. The deformation and failure is manifested as time-dependent cubical dilatation processes. These stress changes give: (a) Up rise of the bottom of a valley; (b) Occurrence of separation phenomenon under part rises sticking out above the bottom in the middle of the valley; (c) Raising a slope top; (d) Appearance of fracturing joint zone inside both side slopes; (e) Movement of both side slopes towards the pit axis; (f) Turning weak intercalations mylonited or mudded due to shear displacement which takes place along the intercalations sticking out above both side slopes; (g) The part thrust fault formed at the bottom of the nick.

These phynomena were observed at the Gezhouba pit. Especially, on the stop tip of the middle section of more than 100 metres long of the pit, a row of boreholes had been faulted by 7 to 8cm along weak intercalations directionally and in step. The boreholes were used to prevent the rocks from breaking (see Figures 4 & 5), which drew engineering units' and research departments' great attention. Therefore the theoretical results obtained by using the model in analyzing the deformation and failure of the pit and stress state of the built pit coincide with the presentations measured in-situ in many aspects.

1. Theoretical Stress Field and Measured Stress Field In-situ
Measured stress in-situ at Borehole 1602 (on the central line of the 3rd Turbined Pit section) and Borehole 1601 (on the dam axile, i.e. at the terrance with an elevation of 16.2m). (shown in Fig.6).

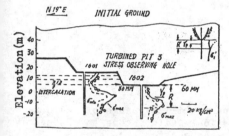

Fig.6. Measured value of horizontal stress in the pit (By Bai Shiwei)

(a). In a weaken stress zone: σ_1'=1120.1KPa was obtained at 4 measuring points in Borehole 1602 (The 4 points were between an elevation of -3.2 to 3.1 m); and $\bar{\sigma}_1'$ =1432.8KPa at 5 points in Borehole 1601, (between an elevation of -0.3 to 13.3m). The mean value for the above 9 points $\bar{\sigma}_1'$=1293.6KPa, and is approximately equals to the effective yield stress of Tresca Yield condition on plane stress state; the value of σ_N =1241.7KPa. (b). In a greater stress zone: The value for 3 measuring points $\bar{\sigma}_1'$=3998.4KPa is obtained between the elevation of -15.0 to -8.6m in Borehole 1602. P.C.F. approaches to 3. It's close to Von Mises plastic constrained conditions on plane strain state; (c). On an X axis: The theoretical failure depth R'=R-ℓ/2 =10.87m; r_P=½R' =5.44m for the depth of weaken stress zone, i.e. it's over an elevation of -8.08m. It's just coupled with the measured results in-situ.

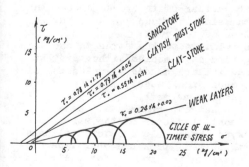

Fig.7. Coulomb-Mohr strength of rocks, rock masses and weak intercalations of the pit

2. Elastic Recovery Displacement of Both Slopes Inwards and the Measured Intercalation Fault In-situ

The rocks around the Gezhouba pit are of a level sedimentary deposit with multiintercalations and close space,i.e. lateral isotropic body.The shear strength indices of various rocks and weak intercalation are of vast difference.(see Fig.7). The elastic recovery horizontal displacement is manifested as intercalation fault moving from upblanket relative to its underlayer.The difference of elastic recovry displacement between the two layers is the faulted amount between them. Thus,adding up each intercalation faulted number along the elevation of one slope from down to up,we can obtain the value of elastic recovery displacement relative to any point. The value is relative to the point whose elastic recovery displacement equals zero.It is the very point that is taken as the starting point of calculating elastic recovery displacement.

(a). In Fig 5. the theoretical elastic recovery displacement of one slope with an elevation of -8.08 to 42.00m from top to bottom is 18.8cm;which the number of the measured intercalation fault of that slope by adding up numbers along the elevation in a large-sized borehole is 19.0 cm. (b) The theoretical displacement is coupled with the measured faulted amount in-situ at the bottom of the nick (i.e. below the elevation of 5 to 6 m).

3. Theoretical Failure Zone and Exploration In-situ, and Foundation Treatment

The deformation and failure of rocks caused by the unloading effect of the earth surface nick to decrease the rigidity and strength of rock masses greatly, conversly to increase their permeability. As a result that the stability of rocks of a dam and slopes (including permeable stability) is reduced.In resistance to sliding analysis, stresw-strain state change and failure conditions of the rocks caused by unloading must be considered.

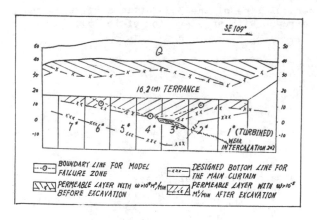

Fig.8. Water permeability change of bedrock and the theoretical failure zone before and after the entrance section excavated

(a). Fig.8 shows that the bottom margin which is permeable bedrock with $W = 10M^{-5}/m \cdot m \cdot min$ of water-pressing test after the main waterproofing curtain cut is basically coupled with the boundary line of Coulomb Plastic Zone or that of Griffith Cleavage Zone; (b) It is only part section from 3 to 6m under the foundation surface or along weak intercalation 202 had good grout absorbability in grout consolidation treatment of the main curtain, but under the above section had good water permeability, while no grout obsorbability. That can be well explained by the theoretical analysis of the model.

CONCLUSION AND SUGGESTION

Considering the universal factor of great horizontal stress in the rock masses, we put forward the model. Based on the analytical results obtained by the behaviour of the rocks around the foundation pit of the Second Channel Power Plant at Gezhouba,using the model we have put forward in this paper, some conclusions and suggestions are given as follows:

1. When there is a great enough initial horizontal stress relative to the strength and vertical stress on the rock mass, even if the rock mass is regarded as a medium without weight for a narrow and deep earth surface nick, the deformation and failure caused by unloading of the rocks around are still suitable for the linear fracture mechanical model; The theoretical solutions coincide with the presentations measured in-situ in many aspects.

2. The geometrical shapes of the nick here mainly mean slope height, grade, and space shapes of the nick, especially the shape of a nick tip.This is the control factor of the stress and displacement fields,and failure range of rocks around a nick,if there is a greater horizontal stress. Hence the designs of both geometrical shapes of earth surface nick and operating processes are matters worthy of note.The model may be used to get a better result for designers.

3. It's necessary for the medium or large-sized project to measure the initial rock mass

stresses scientifically. To develop the evaluation of the rockmass stability to quantum, the initial stress field of the rock mass must be regarded as the calculated base number.

4. The engineering problems to make geological environment ought to be highly stressed. The disastrous rockfall and landslipe must be averted, such as those had taken place at Vajont Dam in Italy and Yianchihe Phosphatic Mine in Hubei Province,China. The former caused by cutting at its slopefoot; and the latter by the underground-nine-square-mine in the shear stress concentrated zone at its slopefoot.

Acknowledgement

The auther wishes to express his gratitute to Associate Professor Ge Qida and Comrade Wang Jingwu, who showed concern and support for this paper, and Comrade Cao Mengshuai,wno gave some valuable suggestions; Comrade Chen Qian, who read and revised the English translation, and Comrade Zhou Shangxin, who drew all the graphs of this paper; Ying Yifang, who typed this paper.

References

Han Zhicheng, 1984 . Effect of excavation of the foundation pit on the ground at Gezhouba Water Control Project Yangtze River,(in Chinese). Hydrogeology and Engineering Geology, Total No.75, P13 .

Mogi, 1978 . Rock mechanics and earthquake,(in Chinese). Institute of Rock and Soil Mechanics, Acaderia Sinica,Wuhan,China.

Aeronautical Engineering Research Institute,China,1981 . Handbook of stress strength factors, in Chinese, Sicence Publishing House.

Institute of Rock and Soil Mechanics, Acaderia Sinica,1981 . To research the foundation problems for Gezhouba Water Contral Project, Yangtze River(in Chinese), Wuhan, China.

On the regularities of rockmovement at upper levels in mines of the thin tungsten lode series

Problèmes des mouvements des roches dans les niveaux supérieurs des mines exploitant du tungstène en filons minces

Wang Sijing & Ding Enbao, *Institute of Geology, Academia Sinica, Beijing, China*
Li Zhonglin, *Jiangxi Metallurgical College, China*

ABSTRACT: Since the middle of the 1960s, thirty one times of large scale rockmovements happened at upper levels in ten mining districts in the Southern of Jiangxi Province, China. It caused a serious loss in economy and natural resources.

In this paper, through engineering geological study rock structure survey, rockmovement measurement, structure model test and numerical analysis, some regularities of rockmovement in mines of thin tungsten lode series have been drawn:

1. Rockmovement is controlled by rock mass structure. All large scale rockmovement occur in the rock mass masses with wedge blocks.

2. The development of rockmovement is closely related to the excavation process.

From the obtained results, it was found that the wedgeblock model may reflect well the processes of observed phenomena

According to the proposed mechanism of rockmovement, the principles of reinforcement for stoped out area and prediction of rockmovement are suggested.

RESUME: Depuis le milieu dex années 1960, 31 fois du rocheux movement de l'échelle grande se sont passés aux niveaux en haut dans 10 mine quarties à la province Jianxi en China, Il a cousé les dommages graves en écomomie et en ressources naturelles. Dans cet article, d'après la prospection de geologie dell'ingénieur, l'analyse de la stracture rocheuse et les documents mesurés dans ces quarties, un certain nombre des regularités du rocheux movement aux niveax en haut pour les tropes des veinces minces de tungsten mine sont obtenus.

1. Le rocheux movement est controlé par la structure rocheuse. Tous rocheux movements de l'échelle grande se sent passés a la structure rocheuse avec un bloc de coin.

2. La valeur de rocheux movement est en relation avec le procédé d'excavation etroitment.

Pour vérifier la fonction de coin bloc au rocheux movement, l'expérimentation de structure modèle et la FEM analyse sont faits. Selon les résultats obtenus, it était trouvé que le modèle de coin bloc peut refléter bien la regularité du phénomène observé. D'après le mecanisme proposé du rocheux movement le principe du perfectionnement pour la région exploité et la prédiction de rocheux movement sent proposé.

1. INTRODUCTION

From the beginning of the 1950s to the middle of the 1960s, large scale rock-movements happened in some mining districts. In Jiangxi Province, there are ten mining districts where rockmovements occurred thirty one times. A typical example is rockmovement at upper levels occurred in 1966 and 1967 in P tungsten mining district. Another example is T mining district where a large scale rockmovement happened in 1979.

This paper discusses the regularities of rock movement during excavation that occurred at up levels (in upper 300m) in thin tungsten lode series in the southern part of Jiangxi Province.

The mining districts are located at a composite area of Nanling latitude tectonic system and Wuyi mountain NNE tectonic system. So, there exist several sets of fault in the area.

The topography in mining area is mauntainous. Upper level are developed with gallery and inclined shaft. Every section is about 50 m high. The blast hole stoping system and short-hole shrinkage are taken as the mining method.

2. ENGINEERING GEOLOGICAL CONDITION

2.1. Lithological characteristics and its physical mechanical properties

The lithology of surrounding rock mainly includes hard and brittle quartzite, metamorphic sandstone and a few soft slate and phyllite. Besides tungsten lode, lithology of lode rock mainly is basaltic porphyrite. The results of laboratory tests for several main rocks are given in table 1.

Table 1. Engineering geological properties of rocks

	Compressive strength (Mpa)	Tensile strength (Mpa)	Young's modulus E (Gpa)	Poisson's ratio $\sqrt{}$	Cohesion C (Mpa)	Angle of friction ϕ	Bursting index (Mpa)	Brittleness
Quartzite	261.8	13.5	60.49	0.176	27.4	57°	0.566	24.15
Ruduceous sandstone	84.7	9.5	48.66	0.3385	20.1	59°	0.0741	8.91
Silistone	228.9	14.4	68.02	0.205	24.5	56°	0.385	19.38
Basaltic Porphyrite	133.5	4.3	44.20	0.271	8.82	62°	0.202	30.96

2.2 MAIN STRUCTURE PLANES AND ROCKMASS STRUCTURE

The main structure surfaces existing in the area are as follows.
1) Bedding planes: Contact planes between two layers conjugate well.
But in partial mining districts there exist interlayer sliding plane between hard metamorphic sandstone and weaker slate and phlyllite with 0.5 to 2cm argillaceous intercalations on it. It is of low strength and long stretching.

2) Faults: Owing to the area being located in tectonically composite part the faults developed well.

The major faults in the area are compresso-shearing structure plane with steep dip angle. The fault planes are usually straight, plain and smooth with plastic gauge ranging in thickness 20 to 30 cm, such as F19, F38 in T mining district; F3, F5 in P mining district.

3) Lode rock: At later metallogenetic epoch there are basaltic porphyrite lode etc. intruding in surrounding rock. Its strength is high (Table 1) and conjugate well with surrounding rock. It was usually formed along faults which developed at later stage. In addition to this, they were usually weathered into Kaolin forming weak cutting planes, such as basaltic porphyrite lode in P mining district.

4) Joints: There exist 3 to 5 sets of joint in each mining district.

The generally density is 2 to 3 joints per meter. Its average value of intactness factor is about 0.5. They extend less than 10 m. The shear strength of joints is very high. According to the results of shear test of joint planes in P mine, taking the yielding strength as criterion, the value of $tg\phi$ is 1.52 to 2.52; C is 0.2156 to 0.9927 Mpa.

The rock mass is often cut into wedge block which sharpens downwards by two faults or weak lode rock with opposite direction of dip. It is this geological structure condition to cause rockmovement on a large scale in the area.

2.3. In-situ stress and underground water

The results of field stress measurement indicate that the stress operating at upper levels are predominantly vertical compressive stress increased with depth.

The hydrogeological condition is relatively simple. There is no water bearing stratum. The underground water in the area is fissure water in bed rock.

With the depth of excavation increased and emptied stopes grew in number, rockmovement becomes intensified with opening of joint and formation of new cracks. These make the speed and quantity of seepage increased and intensified argillitization and softening of weak structure planes. So, the rainfall season is usually a period for intensive rockmovement.

3. THE RESULTS OF ROCKMOVEMENT OBSERVATION AND THE PATTERN OF ROCK MASS DEFORMATION

According to field survey and analysis of measured data at three tungsten mining districts, the following ragularities of rockmass deformation, movement and failure can be found.

1) Rockmovement occurred in two forms of rock mass structure. First, sliding of unstable rock block to emptied stopes along a fault with steep dip angle and opposite direction of dip to emptied stopes. This pattern is shown in Fig.1. The second pattern is the wedge mentioned above (Fig.2).

2) The scope of rockmovement on a large scale is limited in the wedge cutting by boundary faults.

At the points along faults the maximun value of settlement occurred at surface and underground levels. For example, the scope of rockmovement in P mining district is limited in the wedge formed by F3 and basaltic porphyrite lode (Fig.2).

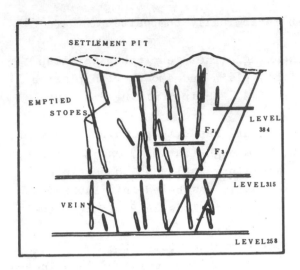

Fig.1 Geological section in Sh mining
district

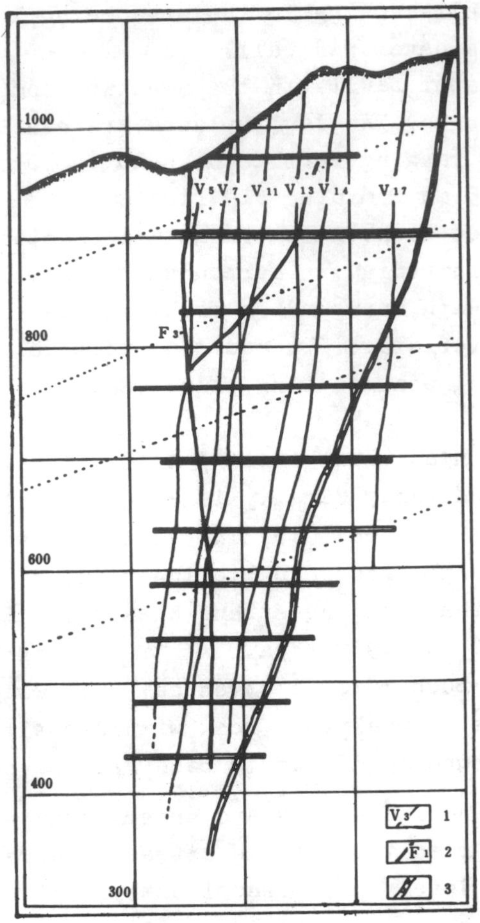

3) The trace of rockmovement at all
measuring points are oriented to concentrated emptied stopes and it does not depend upon whether they are in the wedge
or out of it, for example,the trace of
rockmovement in P mining district points
to the concentrated stopes of lode 7 zone,
and in H mining district to them of lode
39 to 62 zone.
4) The value of rockmovement is closely
related to excavation process (Fig.3),

1. TUNGSTEN LODE; 2.FAULT;
3. BASALTIC PORPHYRITE LODE
Fig.2 Wedge block in P mining
district

5) The wedge is cut by the other third class structure surface into multi-blocks.
They make the form of rockmovement to be multi-blocks movement. This conclusion
can be confirmed from the facts that the points at which larger settlement or
sudden change of it on surface and in several underground levels occurred are
located in the faults (Fig.4)
6) Side walls closed up slowly. That is the main form of deformation and failure
of emptied stopes. There are a few slides along joints or layer planes causing
the wall partialy callapse, such as the wall between 3318 and 3320 emptied stopes sliding along inclined direction of layer plane.

4. ANALYSIS OF ROCKMOVEMENT MECHANISM

4.1. Rock mass structure in H mining district
    The outline of the geological structure in H mining district is shown in Fig.5.
The strikes of major fault sets are NNE and NEE. The characteristics of these are

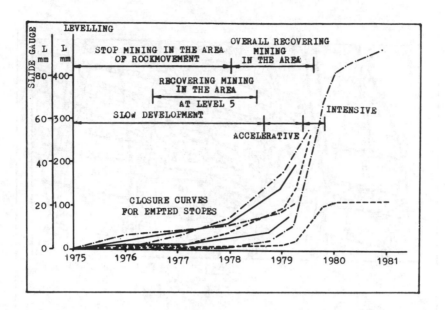

Fig.3   The relationship between rockmovement and
        excavation process in T mining district

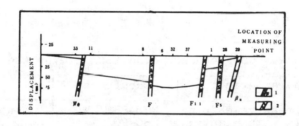

1. FAULT;   2. VIENS

Fig.4 Curves of displacement at 889 level along lode
      in P mining district

faults are the same as mentioned above. Faults F19 and F17 cut the rock mass into a wedge block. The main concentrated emptied stopes of lode zone 39-62 are
developed in the right side of the wedge which forms a compressible boundary.
4.2. Analysis of factors
1) F19: Its shearing strength and stiffness is low, its angle of dip is close
to $45+\phi/2$ and also to the direction of released load of northern emptied stopes,
so the wedge is easy to slide along it and to thrust and press northern emptied
stopes.

227

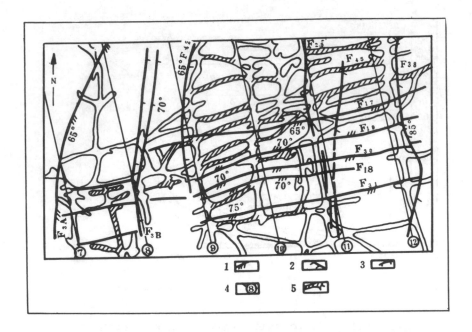

1. COMPRESSIVE FAULT; 2. TENSO-SHEARING FAULT;

3. TENSILE FAULT; 4. EXPLORATION LINE;

5. EMPTIED STOPES

Fig.5   Tectonic map in H mining district

Table 2. The boundary condition of the wedge

Name	Characteristics	Occurrence	Location
F19	compressive-sharing fault with plastic gauge 20-30	N70-85E/NW 60-70 along lode strike	southern boundary of the wedge
F17	"	N70-85E/SE 60-70 along lode strike	northern boundary of the wedge
Continuous emptied stopes of 62 lode		NEE/SE 65-75 along lode strike	northern compressive boundary
F38	"	N5-30E/SE 60-85 perpendicular to lode strike	western boundary of the wedge, lateral cutting surface
F25	"	N5-30E/SE 60-85 perpendicular to lode strike	eastern boundary of the wedge, lateral cutting surface

2) F17: The direction of its dip is opposite to F19, in combination they form
the wedge. Owing to the function of the wedge, the action of thrusting and pres-
sing emptied stopes are increased. Besides, it reduce the resistance of different block movement which is caused by the slip of the wedge along F19 and displacement of right block of F17 toward emptied stopes. That caused the wedge
settlement and rockmovement intensified.

3) The continuous emptied stopes: They not only provide compressible space but
also produce released load. With the increase of excavated depth of northern
lode zone, the space between stopes and F19 is constantly reduced and small. As
a result, the rockmovement gets serious day by day. Especially, when the sixth
level was excavated the equilibrium of the rock mass is destroyed. The settlement value of the wedge increase greatly and rockmovement on a large scale occurred.

4.3. Structure model test and FEM analysis

In order to verify the function of the wedge in rockmovement the structure model test and FEM analysis are taken. Rock mass structures are divided into three
kinds of models:

model 1: assumed intact rock mass, no fault;

model 2: there exists one fault. Its direction of dip points to emptied stopes;

model 3: there exist two faults with opposite direction of dip. They form the
wedge;

Material of model are composed of 25% gelatin, 25% glycerine and 50% water.

The size of model block is $60 \times 36 \times 4 cm^3$ which is put on model frame. Its two
lateral surface are confined. The deformation of the model is controlled by gravity stress field and any external loads have not been applied.

The test is divided into three steps to get on: 1) excavation of four emptied
stopes; 2) cuting F19; 3) cuting F17. At every step the displacement are observed. Testing results are shown in Fig.6.

Three kinds of structure model mentioned above are respectively computed by

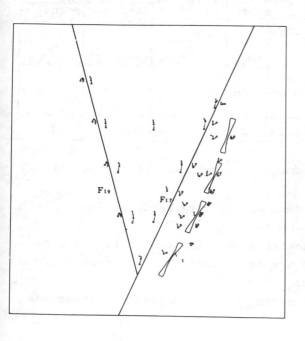

Fig.6 structure model testing result —— displacement
trace

. the position of observation
point before excavation

x after excavation

△ after cuting F19

□ after cuting F17

Fig.7 Nodal displacement for
model A

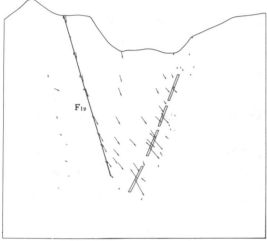

Fig.8 Nodal displacement for
model B

Table 3   Calculating parameters

| | modulus of elasticity E (GPa) | poissons ratio ν | unit weight (T/m³) | tensile strength σ_t (MPa) | shearing strength | | friction strength | | stiffness | |
					tgϕ	c (MPa)	tgϕ	c (MPa)	tangential (GPa/m)	normal (GPa/m)
meta-morphic sandstone	19.6	0.25	2.9	0.98	0.61	1.96	0.61	0		
F19,F17					0.28	0		0	0.294	2.94

using F.E.M., then to compare their displacement regularities. The computational model is simplified as following.

1) Section 11 perpendicular to lode is taken as calculating section. It is regarded as plane strain problem.

2) The emptied stopes of lode 62 are most concentrated, its continuity is considered to be best. So, in calculation they are only considered. Excavation from above to below is divided into four steps to get on.

3) Gravity is only considered to be initial stress field. Computational parameters are shown in table 2.

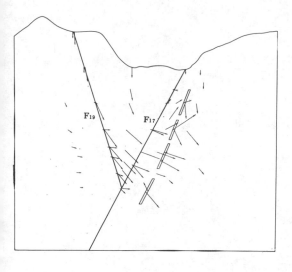

Fig.9   Nodal displacement for
        model C

Computed results are shown in Fig.7, Fig.8, Fig.9.

From the results of computation and model test two can find clearly that the results of the third test and calculating model C may reflect well the regularities of observed data and the function of the wedge is very important for rockmovement.

5. CONCLUSIONS

The following conclusions may be drawn from the preliminary analysis:
- Rockmovement is controlled by rock mass structure. Under the same conditions comprising excavated depth, surrounding rock and distribution of emptied stopes, it is more easy for rockmovement to take place when the rock mass structure is a wedge formed by intersection of two faults with opposite directions of dip. It is very interesting that all rockmovements on a large scale observed in the area have occurred in this geological condition.
- Deformation and failure of surrounding rock in mining area mainly are of forms such as settlement of the wedge along weak structure surface, closure of side walls of emptied stopes and collapse of partial pillars and side walls along joints and bedding planes.
- Weak structure surface are important factor and especially the second and the third class structure surfaces are often determinant for rockmovement.
- The observation data indicate that the value of rockmovement is closely related to excavation progress. With the increasing depth of excavation and the value of stoped out area especially with the closing of stopes to the buttom of the wedge, rockmovement becomes more intensive. Stopes out area supply not only a compressible space for rockmovement but also released load.
- Under the guidance of the conducted study the emphasis of improvement for stoped area should be taken in the places inside the wedge or near it, and monitoring and prediction can be gone on successfully.

REFERENCES

Gu Dezhen, (1979): Fundamentals of the engineering geomechanics of rockmass, p. 231-296. Science Press, Beijing (in Chinese)

Stability analysis and strengthening for high side walls of the underground power house and draft tubes at Baishan Hydropower Station

Analyse de stabilité et renforcement des murs de soutènement de la station souterraine et des conduites sur le site de Baishan

Zhou Xianjie & Feng Chongan, *Northeast Institute of Exploration & Design, NWREP, Changchun, China*

ABSTRACT: This paper expounds engineering geological condition of the Baishan underground power house, the analysis of rock stability and strengthening of the high side walls and draft tubes. Reinforcement effect is analysed from observed data. The paper suggests that stability condition of surrounding rock of underground chambers built in sound blocky rock is governed by a variety of geological structure planes cutting rock mass. Hence an appraisal of surrounding rock stability should be studied comprehensively according to the cutting condition of rock mass, the relationship between geometry and exposed surfaces of the chamber and the mechanical strength of structure planes etc. On the basis of the above conditions, a comprehensive appraisal of rock stability is given, after mechanical analysis and computations. The engineering treatments have been completed with reference to the above results. Operation proved effect good.

RÉSUMÉ: Cet article présente les conditions géologiques de travaux, l'analyse de la stabilité de roche en masse et son renforcement sur les hauts murs et les tubes d' eau d'aval des bâtiments souterrains à la Centrale Hydraulique de Bai-shan, en même temps, il analyse l'effet du renforcement à partir des resultats observés. On croit que les conditions de stabilité de roches de parois de la chambre souterraine fondée dans une roche dure en bloc sont essentiellement contrôlées par les différents plans de structure géologique qui coupent la roche en masse. De ce fait, l'évaluation de la stabilité de roches de parois doivent etre etudiee synthetiquement selon les conditions de ooupure de roche en masse, la relation entre la géometrie et les surfaces de murs de murs de la chambre, et aussi l'intensité mécanique sur les plans de structure etc. Sur la base de conditions citees plus haut, l'évaluation synthétique de la stabilité de roche en masse a été obtenue en réalisant une analyse mécanique et une verification des donnees. Les traitements de traveaux correspondant à cette evaluation ont été entrepris. Durant l'opération, un bon effet a été demontré.

INTRODUCTION: The underground power house of Baishan Hydropower Station is in the right bank, downstream side of dam. It is about 90m from dam, axis direction N45° E, buried depth 55 to 115m. The crisscross power house system consists of the main power house, accessory house, air conditioning room, air intake, bus-bar tunnel and main transformer room etc. (Fig. 1).

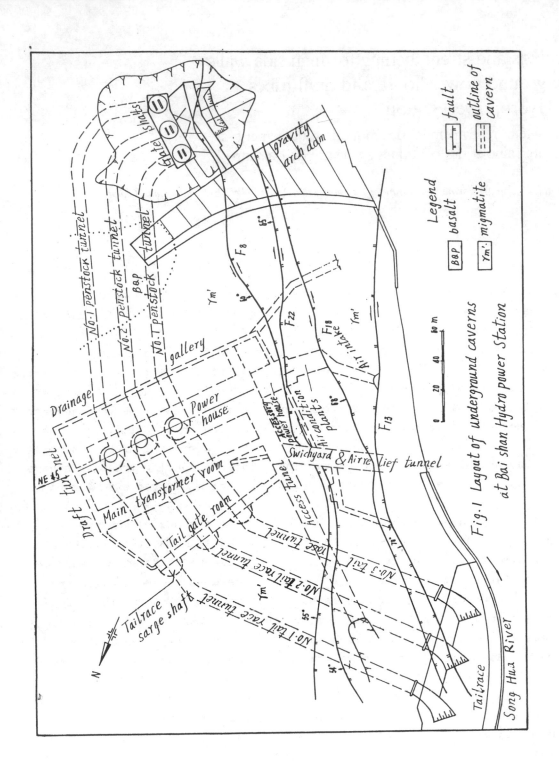

Fig.1 Layout of underground caverns
at Bai shan Hydro power Station

Legend

B&P basalt   Ym' migmatite

━━━ fault   ┄┄┄ outline of cavern

234

The dimension of the main power house is 121.5m in length, 25m in·span and 54m in height.

The stability analyses and strengthening treatments of surround rock are main design problems because of crowded and large excaved caverns. Particularly, on the downstream high side wall, due to unfavourable composition of geological structure planes, there is an instable rock mass. Stability of the three draft tubes under this wall due to thin rock wall between them were also worried, endangering the power house and neighbouring underground chamber. Therefore, problems of rock stability were analysed in some aspects from engineering geological research linking with stress and deformation conditions of the surrounding rock, then, corresponding construction methods and support measures were employed in light of particular conditions of the project, with good results.

1 Rock stability of the downstream high side wall of the power house

1.1 Main conditions

The downstream high side wall is 48m in height, three bus-bar tunnels, draft tubes and the access tunnel is perpendicular to the wall at elevations of 289.70m, 270.15m and 303.5m respectively (Fig.2). There is a main transformer room in parallel with the power house and downstream from it, the rock wall between them is only 16.5m in thickness (Fig.3).

The side wall consists of Pre-Sinian migmatite. The rock is fresh and sound with saturated compressive strength of 122.58 MPa (1250 kg/cm), the intact rock mass has a deformation module as much as 18,613.3 MPa (20 X 10 kg/cm). Three faults (F8, F22 and Fi) pass through the side wall.

The faults F8, F22, strike NW, normal to the power house axis, with thickness from 5cm to 70cm, do not affect the sidewall stability.

Fi is exposed at the middle and upper part of the downstream side wall from 0+40m to 0+82m, with a strike of N 70 E (oblique crossing the wall at angle of 25°), dip 50°SE (incline to the house). It is 10 to 80cm in thickness contents greyish white gouge of thickness 0.2 to 4cm and stretches continuously. The gouge is gravelly heavy sandy loam. Its strength is tg $\emptyset$=0.3, c=49.03 KPa (0.5 kg/cm²). According to the lab shearing tests. The fault is a possible sliding plane which endangers the stability of the downstream side wall, because it has a small angle with the house axis, dips toward the house and possesses a low shearing strength. In the rock mass, on the river side of the fault, a set of high dip angle joints, ex. T195, T49, T44, T56 etc., strike 340°- 350°(normal to the side wall), dip angle 80° to 85°, toward NE, have good continuity, smooth and planar surface. These joints are filled with clay and slightly opened locally. So the lateral shear strength is greatly weakened and near the arch foot at Fi, there is a set of low dip angle fissure (T80) which dips downstream, has good continuity and forms a horizontal cutting plane on the upper part of the side wall. A rock mass which is unfavorable to stability consists of a possible sliding plane Fi and cutting planes of T80 at the top and T195 at the side, and has a volume of 2,670m³ in this section of the side wall(as shown in Figs 2, 3 and 4).

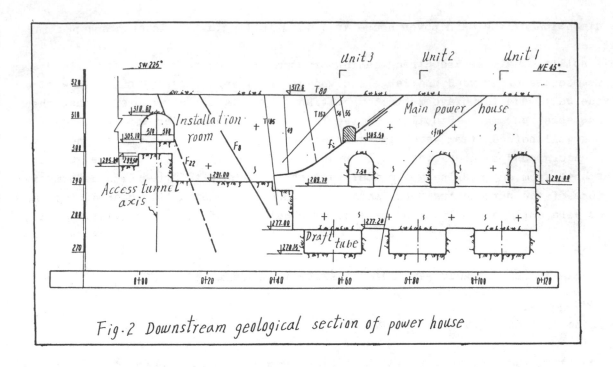

Fig.2 Downstream geological section of power house

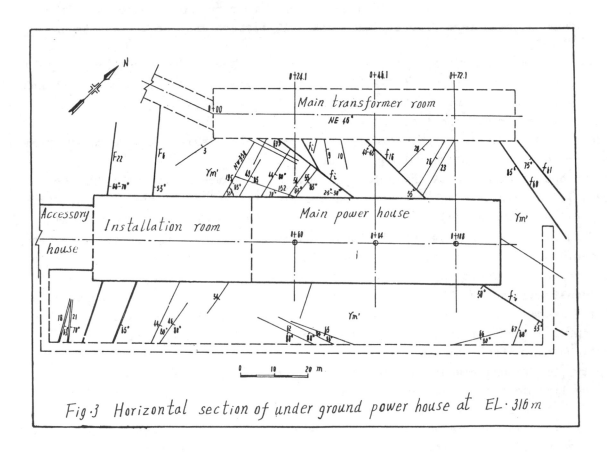

Fig.3 Horizontal section of under ground power house at EL·316m

1.2 Stability analysis

The characteristics of the rock block evaluated from stereographic projection and geometry are shown in Table 1.

According to the space composition, relationship of structure plane the rock block, which cut by Fi fault plane (main sliding plane) and T195 cutting plane, may slide along the intersection line of them. For simpleness and conveniency in calculation and slightly safety, it is assumed that the rock block will slide towards the dip direction of Fi, because the dip angle of fissure T195 and others are rather high and their strike is almost normal to Fi. As to external water pressure used in stability calculation, as the borehole information, the natural underground water table is about 15-70m higher than arch roof. The external water pressure measured in the upstream side testing tunnel is about 300 KPa. In order to prevent external water pressure built up and seepage water from penstock tunnels, consequently reduce stability of the power house side wall. Besides a drainage curtain consists of upper and lower drainage galleries and a curtain of drainage holes are laid out upstream from the power house. Drainage holes are also constructed at the roof and side walls of the power house and main transformer room. After adopting these drainage measures, the external water pressure in the side wall of the power house has been greatly lowered. The effect of drainage is very good. The water pressure of every hole is zero according to measurements on both of the upper and lower drainage galleries. So the external water pressure was not considered in the stability calculation. As to the rock stresses, on the basis of measurement results in the adits, all the vertical and horizontal stresses are about 98.066 KPa (100 kg/cm^2), the coefficient of the lateral compression is equal to 1 approximately. The stresses in the rock mass is close the state of static water pressure. The Baishan dam is situated at a stable massive that is not a high stress district and the underground power house is located in the hill side, buried shallowly. The rock mass is dissected by fault F4 and Laoling gully at the upstream side of the power house, Xiaoping gally at the downstream side, the deep dissected river bed on the left side and the fault F21 that is more than 20m in width on the right side. Therefore, the ground stresses are relieved to a certain extent. After excavating the power house, because rock mass displacement towards the cavern, rock stresses redistribution and blast shock effect, rock stress in the upfaulted block of Fi may be lowered obviously, therefore, only weight of the block is taken account in the calculation, the safety factor against sliding is given by

$$K_c = \frac{G\cos^2\alpha \, tg\phi + C_1 F_1 + C_2 F_2}{G\sin\alpha}$$

Where G is the weight of block,
$\quad\alpha$ is the inclination of the sliding plane,
$\quad tg\,\phi$ is the coefficient of friction of the fault plane Fi,
$\quad C_1$ is the cohesion of the fault plane Fi,
$\quad C_2$ is the cohesion of the fissure plane,
$\quad F_1$ is the area of the Fi fault plane,
$\quad F_2$ is the area of the fissure plane.

In the calculation, shear strength of the fault Fi is taken as $tg\phi = 0.3$, $C_1 =$

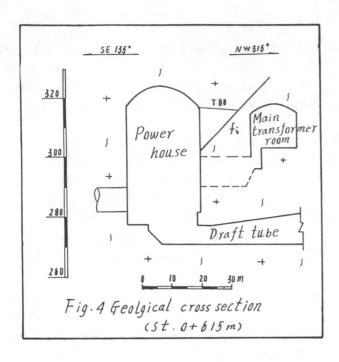

Fig.4 Geolgical cross section
(St. 0+615m)

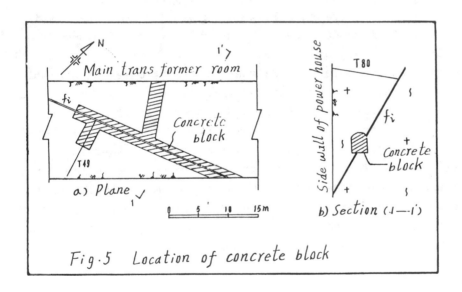

a) Plane

b) Section (1—1')

Fig.5 Location of concrete block

Table 1

Composition of structure planes	Volume of rock block (m³)	Area of sliding plane (m²)		Intersection line of sliding planes		Depth of rock block (m)	F* (m²)	L* (m)
		fissure	fault	direction	dip angle			
Fi T195 T80	2.670	204	406	158°	49° 23'	16	455	98
Fi T44 T80	977	130	277	144°	48°	13	140	57
Fi T56 T80	358	66	133	144°	48°	9.4	117	51

*: F, L are the exposed area and length of the rock block on the vertical
plane of the side wall.

49.03 KPa (0.5 kg/cm²); on the fissure plane, only C is accounted, which, considering the soil filling locally, take C_2 = 49.03 KPa (0.5 kg/cm²) for safety. The calculation results are shown in table 2.

Table 2

Composition	Kc
Fi T56 T80	1.56
Fi T44 T80	1.03
Fi T195 T80	0.82

The above calculation results indicate the stability safety factor of the rock block is not enough. If no dependable strengthening measures is applied, when the cavern is excavated below the intersection point between Fi and T195 and the rock block loses supports, the rock block may slide down.

1.3 Reinforcing measures

To ensure the stability of the downstream side wall, through calculating, checking and comparing, reinforcing measures for the rock block is as follows:

(a) Before excavation of the side wall, excavating an adit which is 23.8m long, 2.5m wide and 3.5m high, along the fault strike from the main transformer room towards the power house. The fill with concrete together with some old rails embeded in the concrete, (and grouting the gap between the concrete and rock wall). Thus, a slide resisting block was formed along the fault Fi as to increase the resisting slide force of the main slide plane. The layout of the concrete block is shown in Fig. 5.

(b) Reinforcing with prestress cables:

First, excavation depth of the power house is limited, i.e. the first stage excavation depth will be higher than the intersection point between the faults Fi and fissures T49, T195. Thus an adequate thick rock mass was retained to support the lower part of unstable rock block to assure stability in construction after excavation of the side wall 36 60T prestress cables of 16m long and penetrate the rock wall, and 18 60T, 12m long prestress cables were constructed immediately, to reinforce the rock mass (Fig 6).

(C) The 134 Ø25, 4.5m long "Joint bolts" were anchored as an auxiliary measures, along the Fi fault exposure line on the downstream side wall of the power house and the upstream side wall of the main transformer room.

The reinforcement above had made the factor of safety against sliding increased to 2.88 as to get enough safety.

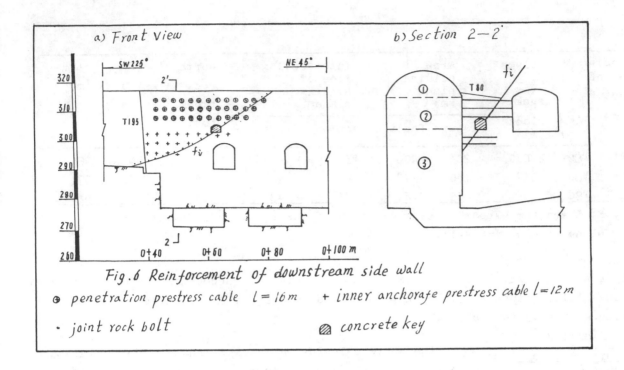

a) Front View   b) Section 2—2'

Fig.6 Reinforcement of downstream side wall

⊕ penetration prestress cable l = 16 m   + inner anchorage prestress cable l = 12 m

· joint rock bolt   ▨ concrete key

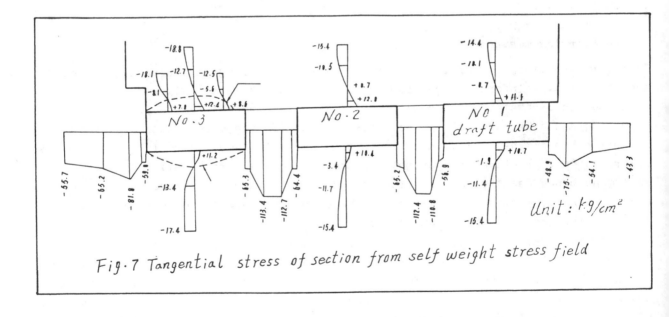

Unit : kg/cm²

Fig.7 Tangential stress of section from self weight stress field

1.4 Deformation observation and treatment effect

In order to verify the treatment effect and to acquire the surrounding rock deformation and stability information, when rock block strengthening was completed and the part below the intersection point between Fi andT195 was excavated, the surface displacement observation had been carried out at the downstream side wall, the results are shown in table 3.

The observed results indicate that although the rock block of the side wall deformed to some extent as unloading of the power house excavation, there was no trace of tension fracture along the T80 large fissure plane or shear displacement along the Fi fault plane on this rock block. Moreover the lateral displacement (20.1mm) as compared with the elastic deformation computed by finite element method is in accord basically. It is close to the measurement data of other power houses (Table 4) It may be considered that the deformation is within allowable limits and can not effect stability of the side wall.

After strengthening, the large exposed side wall surface was left without additional support for years. It was also suffered blast shock effect when its lower part, bus-bar tunnels and draft tubes were excavated. No abnormal phenomena was observed by examination. As above, the effect of the reinforcing measures used is very good.

2 Stability of the draft tube rock mass

The sections of three draft tubes that cross the downstream side wall are flat rectangles, 7m high, 17m excavation span for each. The rock wall thickness is only 6 to 7m. Because the rock walls are thin and situate at the intersect place of underground caverns, whether they are stable or not will relate to the safety of the downstream side wall of the power house and the neighbouring tunnels directly.

Table 3

Elevation	Number of the measuring point	Relative displacement of the 1st measurement (mm) Aug. 8, 1980	Relative displacement of the 2nd measurement (mm) Aug. 22, 1980	Relative displacement of the 3rd measurement (mm) Sept. 14, 1980
	D1	+3.8	+4.4	+4.4
	D2	+19.5	+20.1	+20.1
303.7	D3	+10.2	+10.8	+10.8
	D4	+11.2	+10.9	+10.9
	D1 - 1	+3.9	+3.0	+3.0
300.0	D3 - 1	+6.0	+4.9	+4.9
	D5 - 1		+5.9	+5.8

Note: The basic displacement parameters were observed on June 17, 1980.

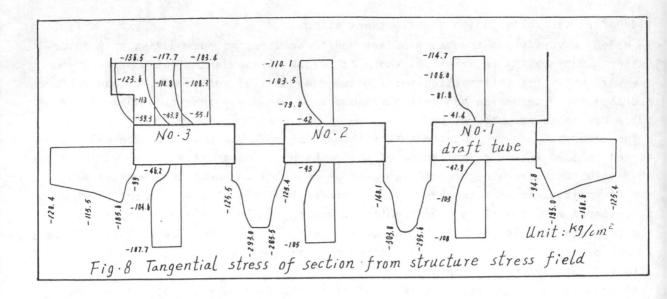

-138.5 -117.7 -103.4
-123.6 -118.8 -108.3
-113
-59.3 -43.9 -55.1

-110.1
-103.5
-79.0
-42

-114.7
-106.0
-81.8
-41.4

NO·3

NO·2

NO·1
draft tube

-46.2
-99
-104.6

-45

-42.9

-120.4 -115.5 -185.6

-126.5 -125.4

-140.1

-103

-94.8 -135.0 -160.6 -126.4

-187.7

-293.0 -285.5 -105

-303.0 -295.6 -108

Unit: kg/cm²

Fig·8 Tangential stress of section from structure stress field

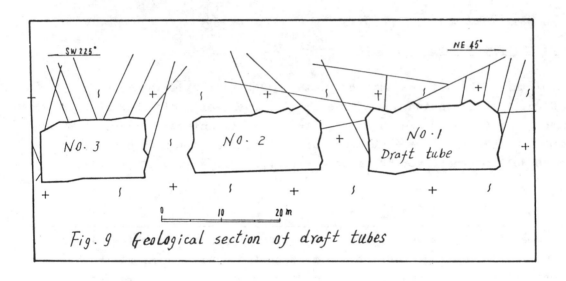

SW 225°

NE 45°

NO·3

NO·2

NO·1
Draft tube

0   10   20 m

Fig·9 Geological section of draft tubes

Table 4

Name of power station	Rock	Span of excavation (m)	Height of side wall (m)	Lateral deformation (mm)	
				upstream side wall	downstream side wall
Miranda (Portugal)	Mica-schist, migmatite, Deiterosomatic granite	20	42.7	20	
Kariba (Zambia)	Biotite gneiss	23	40	12 in general Max. 37	
Tumuf II (Australia)	Granite	17	30	Max. 12.7	max. 17.8
Jingbohu (China)	Diorite, Granite	18	27	Mrx. 24	
Baishan (China)	Migmatite	25.4	47.6	Mrx. 22	Max. 20

To evaluate the surrounding rock stability there, photoelastic analysis and finite element analysis were carried out so as to acquire knowledge of the surrounding rock stability based on analysis of the surrounding rock stress distribution along with its strength.

Mechanical charaters used in the finite element method is as follows:

Deformation modulus of the surrounding rock

Eo = 19,613.3 MPa (20 x 10 kg/cm²);

Poison's ratio μ = 0.25

Unie weight of rock r = 2.7 T/m³

Due to the blast effect, the surface surrounding rock was relaxed and regarded as "disturbed zone", 1.5m in thickness, and the deformation modulus is 0.3 E_o.

According to both of the self weight field (ground stress σ_z=2.55 MPa i.e. 26 kg/cm²) and structure field (ground strees was adopted as 9.8 MPa, i.e. 100 kg/cm²), 18 stress analysis cases were computed separately. Two stress figures of the both fields were shown in Figs 7 and 8.

The typical section stresses are shown in table 5.

The calculation results indicate that the state of the surrounding rock is complicated due to the crisscross cavern. Under the self weight field, λ=0.33, at the roof and floor of the draft tube, there exist tangential tensile stresses. Maximum tension is +1216.02 and +1098.54 KPa (+12.4 and +11.2 kg/cm²), is limited within

Table 5

Stress	Position	Roof		Side wall		Remarks
		self weight field	structure field	self weight field	structure field	
Tangential	1st layer	+12.4	−43.9	−65.3	−140.1	Roof stress is No.1 draft tube; Side wall stress is between No.1 and No.2 draft tubes; ( )-- No.3 draft tube stress; + -- tensile stress
Tangential	2nd layer	−0.2	−84.7	−113.4	−30.3	
Radial	1st layer	−0.2 (+0.5)	−1.1 (+0.6)	−2.0 (+0.6)	+2.7	
Radial	2nd layer	−2.3 (−1.1)	−7.9	−10.9 (9.0)	−34.2	
Depth of tensile zone (m)	Tangential	3.0				No.3 draft tube
Depth of tensile zone (m)	Radial	(3.0)	(1.5)	(1.5)	(1.5)	

3m from the surrounding rock surface. Under the structure stress field, $\lambda =1.0$, maximum compression stress of the side wall reached to −29714.15 KPa (−303 kg/cm²), local stresses at the corners may become even more concentrated. The stresses above have not considered the stress redistribution effect after excavation of the power house and main transformer room. If take this stress redistribution into consideration, the above stresses will increase greatly. The stress distribution regime and value, both the photoelastic test and the finite element computation under the same condition are generally consistent.

Assume the surrounding rock can not bear the tential stresses, take compression-shear strength as the result of in situ shear test. From MohrCoulomb criteria, we have:

$$[\sigma] = \frac{2C \; Sin \; \phi \; Ctg \; \phi}{1 - Sin \; \phi} + \frac{1 + Sin \; \phi}{1 - Sin \; \phi} \sigma \; min$$

let

$$[R] = \frac{2 \; C \; Sin \; \phi \; ctg \; \phi}{1 - Sin \; \phi}$$

$$F\phi = \frac{1 + Sin \; \phi}{1 - Sin \; \phi}$$

thus, $[\sigma] = [R] + F\phi \; \sigma \; min$

where $\emptyset$, $\cup$ are the internal friction angle, cohsion from the in situ shear
test of integral rock block.

σ min is lateral pressure,

[R]is allowable stress of the surrounding rock when σ min=0.

According to the value of $\emptyset$ and C from in situ tests, F$\emptyset$ =5 - 18. [R] =19.6-21.6
Mpa (200-220 kg/cm^2) were obtained; for safety, the surrounding rock compression-
shear allowable stress is $[\sigma]$ =150+4σmin (kg/cm^2).

According to above allowable stress, compression-shear failure zone may exist at roof
corners and side walls. Thus stability of the surrounding rock is unfavorable. But
when adopting a higher strength as actual rock condition, considering the self weight
field and λ = 1, the surrounding rock is still stable.

Based on expossure of the adit excavation, the three draft tube surrounding rock
is rather integral. But on the roofs of No. 1 and No.2 draft tubes, low dip angle
fissure filling with clay is developed, extended in longer distance, and intersect
with a set of high dip angle fissures forming an unfavorable composition. Rock drop
occured during the adit excavation. There are also high dip angle fissures paralled
to the adit axis on both side walls. Stability of the roofs and walls of draft tubes
may be ruther poor from the analyses of the geometrical composition between the
structure planes and tunnels (Fig.9).

Due to the unfavorable conditions above, to ensure stability of the surrounding rock
there, besides employing concrete lining of 1m in thickness, the following measures
were specified in construction.

(a) To improve the roof stress state, the roof was excavated somewhat like arch,
so as to avoid considerable tensile stress.

(b) As to the sequence of the three draft tubes, first excavating and lining No.2
draft tube in the middle, then excavate No. 1 and No. 3 draft tubes. Smooth blasting
should be employed and the excavation section should be strictly controlled in the
excavation, so as not to effect the rock stability due to the blasting shock and
excessive overbreak. But actual construction sequence is until all three draft tubes
have been excavated, then they are lined concrete.

(c) After excavation, temporary supports were applied with 3.5m long $\emptyset$25 grouted
rock bolts on the roof and its corner; at the rock wall between draft tubes, 60T
prestressed cables penetrate through the walls together with grouted rock bolts were
constructed for reinforcement, so as to increase the integral strength of the rock wall
and prevent the unfavorable joint plane from shear failure.

The three draft tubes are normal in the excavation through treatment measures mentioned above. After the reinforcement, the rock walls between the draft tubes and
the downstream side walls of the power house have been exposed for three years. Through
examination, no rock displacement and failure was seen in the surrounding rock, indicating a good effect of the reinforcement.

3 Conclusions

1) The general characteristics of underground power house of hydropower station
are: high side walls, crisscross caverns, complex section form and so on. Their sur-

rounding rock stable condition is more complicated than other caverns. It is necessary to investigate the geological condition in detail.

From engineering practice in Baishan project, deformation and stability of the surrounding rock, in a sound, blocky rock mass, are mainly controlled by structure planes of the rock mass dissected. Therefore, evaluation of surrounding rock stability should take all aspects of the dissected condition of the rock mass into consideration, such as: geometry, relation with excavation space, mechanical strength of structure planes, integral degree of rock mass, dependability of surrounding rock, hydrogeological condition and so on. On the basis of finding out the boundary conditions mentioned above, satisfy results may be gained through mechanic analyses and checking computations.

2) The ground stresses are an important factor on the rock mass stability. When considering this factor, on the one hand it is necessary to analyse the ground stress store and relief from the geological structure conditions and burial conditions of the caverns, on the other hand after excavation of caverns the ground stress may obviously decrease due to the unfavorable composition rock mass displacement towards the cavern. This situation was considered in the reinforcement of downstream side wall of Baishan power house, so as to decrease sliding force, simplify reinforcement measures successfully. Through observation, this prediction is correct.

3) Reasonable construction sequence and construction methods are an important linkage to prevent the surrounding rock of cavern from unstability. At lower stability of surrounding rock in this project, suitable construction measures were adopted in construction according to the geological condition, the results are better. The practice has proved that choosing reasonable excavation sequence and method, supporting immediately and decreasing the surrounding rock exposure time should not be ignored to keep rock stability and construction safety.

4 Acknowledgements

The authors wish to thank Chief engineers Cai Weiwu, Liu Pu and Wang Zheren for their checking, guiding and approving this paper.

1.4 Exploration and testing techniques for tunnelling
   Prospection et méthodes d'essai pour l'étude des tunnels

A critical geotechnical evaluation of Himalayan rock masses with reference to rock tunnelling projects
Etude géotechnique de massifs rocheux de l'Himalaya pour des projets de tunnels

Y.R.Dhar & V.D.Choubey, *Indian School of Mines, Dhanbad*

ABSTRACT: The engineering geological investigation in the Himalayas have revealed several significant geotechnical features. The terrain consists of diversified rock masses which show complex fold patterns and are severely disturbed. In many situations, the correct prediction of sub-surface geology from surface outcrops has been difficult problem. Other geotechnical problems which have been encountered during tunnel constructions include instances of excessive overbreaks, tunnel closures due to squeezing, abnormal rock loads, poor stand-up time, flowing ground condition in faulted and sheared gougy zones, and groundwater problems.

In order to establish a reliable geotechnical model for future planning design and construction of tunnels, an attempt has been made to study the various problems in relation to physiographic, and geotectonic set-up of the terrain. The studies have revealed that nature and type of the problems of various tunnels are broadly controlled by tectonic environments of various zones of Himalayas. Each tectonic zone, therefore, can also be defined as a geotechnical zone in which behaviour of rock mass for tunneling can be broadly identical for designing appropriate support system.

1 INTRODUCTION

During last twenty five years several kilometers of Tunnels have been constructed in Himalayas. Most of these tunnels have been constructed in outer and lesser Himalayas in connection with hydro-power generation projects. A few of these have been made along highways for vehicular transportation. Presently, many tunnelling projects are under construction stage and large number of these are under the planning and investigation stages of project feasibility stages.

Due to high relief and rugged topography of the region, it has been experienced and accepted that it is generally not possible to project accurately the observable surface geological features at the Tunnel grade and predict the same satisfactorily. There are many instances to show that initial investigations have proved to be inadequate due to complexity of geological structure. In order to solve this problem and establish a fairly reliable geotechnical model, a systematic approach,for establishing geotechnical zones in Himalayas,need to be given high priority. In an attempt to do so,a number of case studies have been evaluated on the basis of their geological physiographical, lithological and tectonic set-up. On the basis of these analytical studies, it is found that the prediction of potential hazards controlled by geological and tectonic setting will become more realistic to suit engineering requirements. The scope of this paper is rather limited to the tunnelling experience in outer and lesser Himalayan belt of India. It is in these belts where most of the tunnels have been constructed. Attempts to synthesise the geotechnical aspects of some Himalayan Tunnels have been made earlier by Hukku et al (1970), Sinha (1971), Varma et al (1971), Sinha (1971) and Singh et al (1982). In the present work various features of the tunnels have been grouped in various geotechnical zones which appear to have broadly similar rock-tunnelling behaviour.

Table : 1 Geotechnical Features of Tunnels in Sub-Himalayan Zone

Name and Type of the Tunnel	Lithology	Alignment	Tunnelling problem	Remedial measures support and
1. Nandini Traffic Tunnel 118.4 m & 122 m long 5.49 m dia	Siwalik Sandstone and clay shale	Across the formation	No major problems Minor seepage at sandstone shale contact	Brick lining in clay shale in one tube only
2. Khara head-race Tunnel 1.3 km Twin Tunnel	Upper Siwalik conglome-rate and sand rock	$15^{\circ}-20^{\circ}$ askew	Portal insta-bility poor stand up time and flowing condition when saturated with water	Heading and benching
3. Giri* Hydropower Tunnel 7.1 km 3.66 dia	Nahan claystone and silt stone	Across the formation	Mild tunnel closures	Shotcreting perfoboting flexible lining

* Major part of this tunnel passes through Nappe zone

2 PHYSIOGRAPHY AND TECTONIC SETTING

From the physiographical point of view, the Himalayas have been divided into the following lateral zones. These are, from south to north, (1) Sub-Himalayas,(2) Lower Himalayas,(3) Higher Himalayas, and (4) Tethys Himalayas. Considering the disposition of tectonostratigraphic units in nearly east west trending belts and the important dislocation zones which separate them, the above division roughly correspond to various tectonic belts (1) Foot hill tectonic belt, (2) Inner tectonic belt, (3) Axial tectonic belt and (4) Trans-axial tectonic belt respectively. The elements which control the geotechnical features of these belts are discussed below.

3 SUB HIMALAYAS

These are generally formed of low hills with gentle slopes and wide valleys and are made up of upper Tertiary rocks comprising sandstone, siltstone, shales and boulder conglomerates. Due to sharp variations in their resistance to weathering the constituent units show well marked topographical expressions with bold projecting bands of competent rocks. Structurally the rocks are folded in series of broad, low plunging, anticlines and synclines. Series of reversed strike-faults are characteristic feature in this zone. The most prominent thrust is Main Boundary Fault (Fig 1). It usually separates Siwaliks from early Tertiaries. Other important thrusts in this zone are Satillita Thrust in Punjab, Markanda Thrust in Himachal Pradesh, Kalagarh Thrust in Uttar Pradesh and Riasi Thrust in Kashmir. These thrusts show evidence of sub-recent activity with Siwaliks having been brought over the sub-recent gravels (Chaturvedi et al 1973). Besides these number of tear faults, trending NNW-SSE to NNE-SSW direction have also been recorded. Some of the Tunnels constructed in this zone with their geotechnical aspects have been given in Table-I.

3.1 GEOTECHNICAL FEATURES OF SUB-HIMALAYAS

The construction of tunnels in this belt have highlighted the following geotechnical aspects.

250

1. The whole tectonic belt of foot hill Himalayas can be divided in geotechnical zones viz (1) Upper Siwalik zone, (2) Lr and Mid Siwalik zone.

2. The Tunnels in Upper Siwalik rock shall have to be driven through weak strata of sand rock and boulder conglomerate. These rock formations are expected to have poor stand-up time.

3. Lower and Middle Siwalik tunnels composed of sandstone and shales do not pose major geotechnical problems. However, in certain situations the seepage at the contact of clay shale and sandstone can be expected.

4. Tunnels through Nahan Rocks (Lr Siwaliks) which are also similar to Murres in lithological nature comprise claystone and shale rocks. These rocks are, in general, incompetent and usually yield if overburden stresses exceed the compressive strength of these rocks. Mild squeezing stresses are therefore, expected in these rocks. It has been also found that heavy overbreaks can take place if tunnels are oriented parallel to the rock strike.

4   LESSER HIMALAYAS

This region is characterised by high mountains rising upto 4000 to 5000 m in elevation. The terrain is cut by deep narrow gorges and steep slopes.

It is in this region where most of the tunnelling activity have been carried out in India.

The area encompassing this belt is structurally complex. The rocks have been intensely folded and refolded. The folds are generally recumbent, inclined, reclined and vertical usually with axial plane. Mesoscopic structures of these rocks indicate that rocks have been subjected to polyphase folding. These rocks have also been severed by reverse faults which have, subsequently, served as surfaces along which rock slices have moved. The rock formations, in general show reverse order of succession i.e. younger formations remaining under older ones.

This belt comprises mainly a thick pile of low to medium grade rocks (slates, phyllites, schists quartzite) and high grade gneisses. These rocks are seen resting over Siwalik and Murrees perhaps, by overlapping at places, the autochthonous belt of Permo-Carboniferous and Lower Tertiary rocks. In many places they are seen to overlie the autochthon which in turn overlies Siwaliks. This situation is generally seen at lower elevations in north-western sector of Himalayas.

Various thrusts have been reported to occur in this belt. These include Murree Thrust in J & K and Nahan Thrust in H.P. and U.P. Himalayas. These thrusts have brought Lower Teriarites over upper Tertiary rocks of foot hill belt. Another thrust intervening between Permo-Carboniferous Eocene autochthonous belt and Precambrian metasediments. This has been known as Panjal Thrust in Kashmir Himalayas and Krol thrust in H.P. and U.P. Himalayas. Here Precambrian older metamorphic rocks have been brought to rest upon younger sedimentary and volcanic rocks.

4.1   GEOTECHNICAL FEATURES OF LESSER HIMALAYAS

The lesser Himalayas corresponding to inner tectonic belt can be classified into three geotechnical zones (1) Murree Autochthonous zone, (2) Subathu Autochthonous zone and (3) Nappe zone. Some of the geotechnical features of the Tunnels in these zones has been given in Table 2 and 3.

4.2   MURREE – AUTOCHTHONOUS ZONE

This zone is bounded by Main boundary fault on south and Murree Thrust in the north. The zone comprises Murree or Dharamshala Group

Table : 2   Geotechnical Features of Tunnels in Lesser Himalayan Autochthonous
Zone

Name and Type of Tunnel	Lithology	Alignment	Tunnelling problem	Remedial measures support and
1. Chenani Hydropower Tunnels 6.5 km aggregate 2.1 m dia	Murree-clay stones siltstones sandstones	Some Across the formation and some oblique	Overbreaks.More overbreaks in tunnels aligned along the strike	Controlled blasting 38 cm or more of cement concrete steel supports.
2. Yamuna* Chhibro-Khodri Hydropower Tunnel 5.6 km 7.5 m dia	Subathu formation crumpled red shales, siltstones, pockets of black plastic clays	Across the strike	Inadequacy in prediction of Tunnel grade geology Rock considerably sheared and crumpled at thrust ends. Tunnel closures and support deformation variability,overbreak at crown, water in rush and flowing condition in saturated reaches	Heading and Benching in middle and multiple drift method near thrust planes. Circular ribs at less cover rocks. Dewatering and drainage.
3. Sundernagar-Slapper Tunnel 12.28 km 8.5 m dia	Shali series shales,dolomite quartzite	--	Cavity formation water inflows and distress in steel supports	Driving of sheet channels a head of face to forestall. The moving ground fore poling, drainage steel supports
4. Salal Tailrace Tunnel 2.0 km 11.0 m dia	Sirban limestones and dolomites	--	Occasional seepage	

* Major portion of this passes through Nappe zone also

rocks comprise sandstones and purple shales which appear to have been derived from oxidized terrain along the northern border of Indian Peninsula. Construction of tunnels in these type of rocks have frequently revealed heavy overbreaks. In Chenani tunnels maximum trouble have been experienced while constructing tunnels parallel to strike of the rocks (Hukku et al 1971). The choice of Tunnel routes in these sections of tunnels had to be controlled by topographic considerations in order to avoid undesirable rock cover thickness and portal location limitation. Evidently the topographic features at times governed the choice and necessiated adversely oriented choice of Tunnel route.

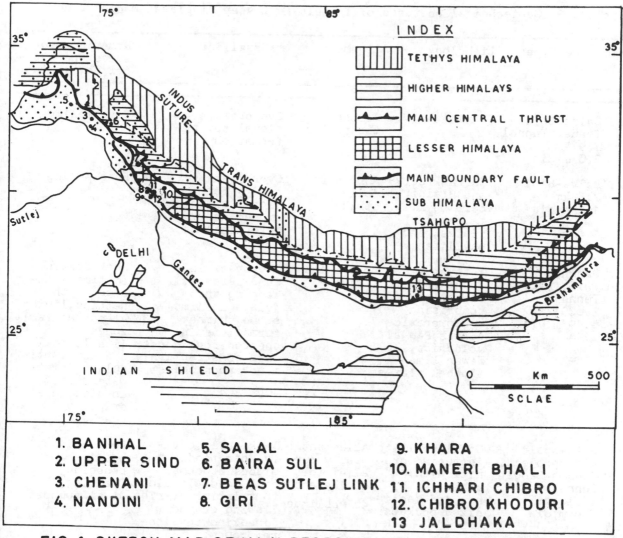

FIG·1 SKETCH MAP OF MAIN TECTONIC BELT OF HIMALAYAS
WITH TUNNEL LOCALITIES

4.3 SUBATHU-AUTOCHTHONOUS ZONE

This zone comprise Subathu Dagshai and Kasoli formation of Lower
Tertiary age. These are in general fossiliferous and comprise
predominantly greenish and purple shales, dark limestone and sandstone.
In Kashmir Himalayas this zone is caught up between two thrusts
Murree thrust in south and Panjal thrust in north. In Himachal Pradesh,
they are thrust over Siwalik rocks along Main Boundary Fault in the
south, while in the north, they are bound by Chandpur/Krol thrust.
This zone, which is believed to be tectonically autochthonous,
may contain rocks lower than Eocene age. In Kashmir Himalayas these
contain Panjal volcanic rocks with limestone impiers of Pre-Subathu
age.

The various geotechnical problems in this zone are as follows :

1. Geology of this zone is complex. The formations are traversed by
thin shear zones and faults which may have large continuity. Careful mapping
coupled with exploration need to be done in future projects in this zone to
assess the geology at tunnel grade.

253

Table : 3 Geotechnical Features of Tunnels in Lesser Himalayan Nappee Zone

Name and Type	Lithology	Alignment	Tunnelling	Remedial measures and support and lining
1	2	3	4	5
1. Baira-Baledh feeder Tunnel 7.9 km 2.5 m dia	Phyllite	Across the strike	Overbreak, occasional cavity formation	Fore poling
2. Baira-Siul 7.6 km 4.5 m dia	Infra-krol(?) and Blaini(?) carb phyllite, pebbly slates, and phyllite	Across to oblique to strike	Overbreak, cavity formation, flowing ground, heavy water inflows	Fore poling
3. Banihal Twin traffic Tunnel 2.539 km 5.49 m dia	Tethyan sediment and volcanics comprising agglomeratic slates, Panjal volcanics, Zewan Limestone & slates	Across the strike	Overbreak in some reaches, seeping through joints in limestone, chimney formations in sheared rock zones post construction minor rock falls	Rock bolting at times combined with netting, lining at portal, fault zones and lithological contacts. Articulate joints water proof insulation at dripping reaches
4. Beas-Sutlej link project, Pandoh-Baggi Tunnel 15.15 km 7.6 m dia	Chail phyllite intrusive granite with schistose bands and kaolinized pockets	Alignment nearly parallel to strike and 45° oblique	Rock fall, cavity formation. At places abnormal rock loads leading to twisting of ribs, flowing ground, Squeezing in altered zones.	Fore poling, heading and benching, Draining, Steel supports.
5. Giri* Hydropower Tunnel 7.1 km 3.66 m dia	Mandhali infrakol and Blaini slates with boulders beds	Across the strike	Overbreak, high movements in rock when aligned more parallel to strike	Shotcreting perfobolting, rock bolts, Flexible support
6. Ichari-Chhibro Hydropower Tunnel 6.2 km 7.0 m dia	Mandhali-Chandur and Nagthat - slates interbedded with quartzites, limestones	Across the strike	Poor standup time spilling, chimney formation High overbreak	Shotcreting Pre-reinforcement. Heading and benching in weathered and insufficient rock cover sections. Steel ribs with concrete blocking.

* Part of this tunnel passes through autochthonous zone

254

Table : 3 continued

1	2	3	4	5
7. Maneri-Bhali Hydro-power Tunnel 16.8 km 6.0 m dia	Janusars: quartzites, basic, rocks limestone	Across to oblique to strike	Water inflows over break and cavity formation	Drainage of rock, fore poling
8. Upper-Sindh water conductor Tunnel 4.1 km 2.59 m dia	Cambro-Silurian slates and quartzites	Oblique to strike	Minor overbreak	Steel supports only for 30% length, Roof unlined
9. Jaldhakal Head-race Tunnel 4.40 km 3.04 m dia	Daling and Darjeeling series Schists, Amphibolites gneisses and quartzite	Across foliation to oblique at high angles	Crown collapse caused by wide shear zone, chimney formation	

2. Nalas and stream channels in the vicinity of a fault may lead to sudden groundwater inflow. This is because of the fact that some faults and thrusts are intensely sheared surfaces, which, at times, are filled with impervious gougy seams and clay lenses which act as barriers for perched water bodies. Such situations have been met with tunnel crossings at Subathu/ Mandhali contact (Shome 1979).

3. The overall rock mass has been rendered weak by tectonic movement of thrust planes. The formation towards the ends of intra-thrust zones are more crushed and crumbly than towards centre. As the formation have low compressive strength, the squeezing pressure develop in the tunnel as a result of plastic response to vertical stress caused by superincumbent load. Such squeezing pressures resulting in tunnel closures were met in Chhibro-Khodri tunnel and Loktak Tunnel while through this zone. For planning finished tunnel diameters in this zone, the extent of closures have to be kept in mind. This can be achieved by initial over excavation and subsequent provision for flexible lining.

4. Lithological variations can be expected in this zone. This can make task of rendering uniform support difficult.

5. In central reaches of this zone, the rocks are relatively sound, Therefore, overburden stress can lead to only loosening pressures resulting in overbreak and cavity formation due to adverse disposition of discontinuities.

6. Since the formations of this age are expected to contain oil and gas, the presence of gas hazard can not be overlooked. In Loktak tunnel this hazard was met with during construction time which lead to multiferious problems.

7. For creating large openings in this zone, heading and benching method or multiple drift method and NATM have proved more successful in view of the low bridging capacity of the rock. In reaches with heavy overbreak forepoling with a limited pull has proved to be successful.

4.4   NAPPE ZONE

This belt comprises several sheets of older rocks lying between Panjal/Krol Thrust and Main Central Thrust. In Kashmir Himalayas Panjal Thrust is most important features of this zone. Along this

255

thrust older Precambrian rocks over ridden by Tethyan Sediments have been brought to rest upon carboniferous Tertiary rocks. In Himachal Pradesh, in between these two main thrusts separations, several structural units namely, Salkhala nappe, Jutogh Nappe, Chail Nappe, Chandpur-Krol nappe, Krol Klippe and Simla Nappes have been identified by various investigators. Similarly, in Uttar Pradesh, Jutogh Nappe, Chail crystalline Nappe, Bijni Nappe, Chandpur-Krol Nappe and Naingaon Nappe have been reported. Likewise several thrust sheets have been identified in North-eastern Himalayans. The Highest tectonic unit being either Salkhalas or Jutoghs in Western Himalayas and in Eastern Himalayas, these are represented by Dalings and Darjeeling gneiss. The rocks in this complex zone comprise metasedimentary, basic and metabasic rocks and granites with some sedimentary basins in between. These rocks have so far defied proper classification and correlation because of tectonic complexity and, therefore, the whole inner belt comprising these rocks have been described as a single geotechtonic belt.

This various Geotechnical experiences have been given in Table 3 and important feature of physiographic, tectonic and rock mass are summarised as follows :

1. In this belt the precipitous nature of topography manifested in deep narrow transverse gorges, irregular and oblique ridges make selection of tunnel routes to deviate from standard norms. In order to meet criteria for suitable rock cover the tunnel route have to be deviated at the cost of structural geological considerations. Often tunnels have to be constructed along the strike to obviate other adverse factors.

2. In many instances the rock slopes, under which tunnel is driven, is transversely drained by seasonal or perenial nalas. These nalas or brooks more oftenly indicate weakness zones as well. In Beas-Sutlej link project at Junikhand, a perenial stream passing of Pando-tunnel proved to be a weak zone and led to groundwater problems during construction. During construction the sudden in rush of water was also met during construction of Sundernagar - Slapper Tunnel, Baira siul Tunnel and Chhibro-Khodri Tunnel.

3. The anticlinal valleys delimiting synclinal hills can also lead to groundwater problems. This phenomenon was encountered in Maneri Bhali Tunnel driven through synclinally disposed quartzites.

4. Rock formation comprising of schists, phyllites, slates and quartzites and granites have proved to be good tunneling medium but due to intense tectonic activity the rocks have invariably been intensely sheared especially within quartzite and granitic reaches. For instance, in granite intrusives in Chail formation of Beas-Sutlej link project a number of such intensely sheared zones led to difficult tunnelling problems especially when saturated (Kochhar et al 1973). Flowing ground conditions and cavity formations were met in such situation leading to squeezing action on supports. Apart from these weak zones, the formations are generally competent than those in autochthonous zones.

5. This belt is characterised by thrust sheets which are, in general, synclinally folded hills. These regional synclines are locally folded in asymmetrical patterns. Such synclinally disposed thrust sheets have been reported in Giri and Yamuna tunnels. The tunnels driven within axial regions of these synclinal hills have thick rock cover leading to higher static vertical stresses which coupled with compressional nature of forces in axial regions of syncline can lead to stress problems in these portions of tunnels.

6. The base of the thrusts is usually a surface along which overlying sheets have moved long distances. The nature of contact of these sheets is variable. At places the contact seems normal to conformable but at most of the places these surfaces are highly tectonised and show belts of shearing and crushing as along Krol thrust, Tons thrust etc. The geotechnical hazards in driving tunnels through these rocks are of two categories : (1) the weak behaviour of rocks and (2) inflow of water generally at the base of sheets.

7. Away from the thrust contacts the tunnels in general pass into hard rocks which are jointed and contain some local faults and shear planes. Small magnitude water inflows, flowing ground conditions in wide fault gouges leading to cavity formation are also common problems.

8. Among the other factors, the attitude of the discontinuities in relation to tunnel axis are important for estimating overbreak. The gently dipping beds with intersection of joints have given rise to spalling action. This type of movement leading to considerable overbreak was seen in Ichari-Chhibro tunnels. These were controlled by roof bolts. But in presence of groundwater, which led to the formation of tension cracks, steel ribs 1.5 centre to centre with immediate initial concrete proved successful support measure.

9. In case where bedding planes strike across the tunnel axis with dip towards heading a series of overbreaks travelling backwards from place to the supported tunnel may take place. Such phenomenon was met with in Chhibro heading of Ichari-Chhibro tunnels. The loosening movements leading to such overbreaks were controlled by perfobolts installed through pilot drift in advance of the face. In Giri tunnel problems started suddenly in portion where tunnel was more parallel to strike of rocks (Madhvan 1982). The supports were distorted and movements could not be contained easily. In such situations it was found economic to allow some closure by providing loose packing between the rock and rib.

10. It has been ascertained that tunnelling in up-dip direction are more safer than in down dip direction. Tunnelling in down-dip direction leads to more overbreaks. This has been proved in case of Pandoh Baggi Tunnel. The tunnelling at Baggi end gave more way to overbreak than from Pandoh end (Hukku et al 1970).

11. In Tethyan rocks which overlie the Precambrian to Lower Palaeozoic slates and schists, not much problems in tunnelling has been encountered. These rocks comprising sedimentary and volcanic rocks are not highly tectonised as basal parts of sheets. The Tethyan rocks of Kashmir comprise sandstone, shales, limestones and Panjal volcanic rocks. These rocks however are prone to minor overbreaks and seasonal seepage.

CONCLUSIONS

The studies of Tunnelling projects in Himalayas have revealed that the geotechnical problems of these tunnels have been controlled by nature of topography, rock type and tectonic environment. The tunnels in a particular tectonic zone of Himalayas have broadly identical geotechnical problems. Each Physiotectonic zone of Himalaya, therefore, can also be considered as geotechnical zone in which behaviour of rockmass for tunneling can be assessed from Tunnels constructed in that particular zone.

REFERENCES

Chaturvedi, S.N. & Jalote, S.P. 1973, Geotechnical considerations in planning Hydropower houses in Northwestern Himalayas. Proc. Seminar on Foundation Problems of Powerhouses and Related Ancillary structures. Indian Soc. Engg. Geol. No. 19-20 pp 1-12.

Hukku, B.M., Chaturvedi, S.N. & Asraf, Z, 1971, Tunneling Experiences in Jammu and Kashmir Himalayas — A Geotechnical evaluation. Proc. Sem. on Engineering Geological Problems in Tunneling. New Delhi Ind. Soc. of Engg. Geol. Part I, pp 116-127.

Hukku, B.M. and Srivastava, K.N. 1970, Geotechnical problems of Tunneling at Beas Sutlej Link project. Proc. Seminar on Engg. Geol. Prob. in River Valley projects in Northern India, Roorkee University, pp 1-10.

Kochhar, D.N. and Prem, K.S. 1973, Tunneling experience through poor rock strata. Proc. Symposium on Rock Mechanics and Tunneling problems. Kurukshetra. Vol. I, pp. 103-112.

Shome, S.K., Andotra, B.S., Sondhi, S.N. 1979, Review of Tunneling problems in Subalhu Rocks of the Yamuna Project. Jour. Engg. Geol., I.S.E.G. Vol. IX, No. 1 pp. 1-13.

Singh, and Mahajan, J. L. 1980, Some Geotechnical Considerations on Tunneling in Himalayas. Proc. IV Congress Int. Assoc. of Engg. Geol. Vol. IV New Delhi pp 65-74.

Sinha, B.N. 1971, Tunnels in water conductor system of River Valley projects in Eastern India. Proc. Seminar on Engineering Geological problem in Tunneling. Ind. Soc. Engg. Geol. Part I, p. 22-39.

Madhvan, K. 1982, Tunneling in Rock with High In-situ stresses. Proc. 4th Int. Cong. IAEG, New Delhi - Vol IV pp 79-84.

Varma, R. S. & Mehta, P.N. 1971, On some Geotechnical aspects of certain tunnels in the Himalayas. Proc. Seminar on Engineering Geological Problems in Tunneling. Ind. Soc. Engg. Geol. Part I, p. 60-70.

Sen Sarma, S. B. & Choudhury, A.K. 1970, Observation of behaviour of rocks within the power tunnel of Jaldhaka Hydel project, West Bengal. Jour. Engg. Geol. Ind. Soc. of Engg. Geol. Khosla volume Number 1, p. 181-187.

Monitoring of a large portal excavated in rock
Contrôle d'un grand ouvrage creusé dans le rocher

Juan Carlos del Río & Roberto Isidro Cravero, *Agua y Energía Eléctrica, Inspección Obras Complejo Hidroeléctrico Río Grande I, Argentina*

ABSTRACT: The importance of a rock mass monitoring section during the construction of one of the most singular excavation in the Complejo Hidroeléctrico Río Grande is analized in this paper.

The Tailrace Tunnel Portal has an area of 245 m2, a semirectangular section, a relatively low overburden. All these factors determined a careful excavation methodology.

The multiple-position mechanical extensometer used are described the reading precision and their influence to control the different excavation stages, to evaluate the types of support employed are analized.

Finally, the rock mass and rock bolts interaction system to stabilize the excavation are analized.

RESUME: L'object de ce travail c'est analyser l'importance qu'a eu l'instrumentation d'une section d'auscultation dans l'ejecution d'une des excavationes plus singulaires du Complejo Hidroeléctrico Río Grande N°1 - Córdoba - Argentina.

Il s'agit de l'excavation d'un portique d'embouchure du tunnel de restitution avec une section presque rectanguliere (245, m2), avec une relative base bouchee et avec des conditions géologuiqu es défavorables qu'ont demande l'elaboration d'une méthodologie d'excavation très soigneusse.

On decrit ici les extensometres mécaniques de position multiple utilisés; la precision de leurs lectures et la signification, que elles ont eu pour la definition de une méthodologie d'excavation et des diferent types de support employés.

On analyse, en outre, les resultats obtenus de l'interaction: massit-rocheux-support dans l'esta bilitation definitive de l'excavation.

1 PROJECT LOCATION AND BRIEF DESCRIPTION

The Complejo Hidroeléctrico Río Grande is an unprecedented work in Argentina, fundamentally on account of the magnitude of its underground excavations.

It is a pumped-storage scheme with the Machine Hall in cavern. It is situated in the province of Córdoba and it is being built on the Río Grande, principal tributary of the Río III. It will have an output of 750 MW.

It is constituted by two large groups of works: Surface works (Dams and spillways) and Underground works (Conveyance and Machine Hall).

The underground conveyance joins the reservoirs: the upper one, Cerro Pelado and the lower one Arroyo Corto.

The outstanding works are the Cavern for the Machine Hall described by Dorso et al (1982) and the Tailrace Tunnel by del Río et al (1983).

The stabilization of the Portal of Tailrace Tunnel is the object of this paper (Fig. 1).

2 GEOLOGICAL CONSIDERATIONS

The Portal and the last 1000 m of the Tailrace tunnel are situated in the Contraembalse environment

It is defined geomorphologically by gentle hills covered with regolite and weathered rock sec tioned by the drainage lines.

It involved these rock types: foliated gneiss grading to schistose levels towards the east, am - phibolites and pegmatites intercalation.

Primary structures are foliation with a general bearing of 35°/235° (dip and dip direction).

The secondary structures have a great influence on the excavation.

Two main joint system are presented, one coincident with foliation lines, with oxides and some

carbonates, and the other 60°± 10°/ 280° 'less frequent with smooth and closed joint surface.
F1 (65°/025°) and F2 (60°/035°) faults have a width of 0,20 m to 0,60 m and 0,50 m to 1,0 m respectively both with fault gauge, 0,05 m to 0,10 m thick.

These faults have great influence on the excavation, because they have a subparallel bearing to the tunnel axis and dip into it (Fig. 1).

The Portal area is divided into two sectors by F1 fault. The SE one (sector A) more jointed and the NW one (sector B) where F1 y F2 fault, along with a set of shear fractures, limited a large unstable block (Fig. 2).

The groundwater seepage was less than 5 l/min.

3 EXCAVATION WITH CONTROLLED BLASTING TECHNIQUE

The coexistence of factors like low overburden, the regular quality of the rock mass, associated mainly with faults which intercept the excavation, needed special care in the excavation of the definitive section.

It must be borne in mind that there are many cases in which the greatest damage in the underground excavations was associated with the portals. This made it necessary to carry out the analysis of the excavation methodology and blasting patterns to be used, to extremes, in order to produce the least possible disturbance.

To this end the idea is to consider that the rock breakage is produced by stress waves (due to the blast) which have a characteristic propagation velocity c; similar to the velocity of sound in the rock mass.

As the waves passes, each particle in the rock mass runs through a vibrational motion, the peak particle velocity of which is V. It is conceived as a sine-wave and for an elastic material the strain ε and stress σ are related to V:

$\varepsilon = \sigma/E = V/C$; $E \sim V$

where E: elastic modulus of the rock mass.

Thus for each kind of rock mass, rock damage occurs at approximately a given critical level of particle velocity.(Maki, Holmberg, 1982).

The risk of damage produced by blastings may be evaluated by measurements of particle velocity V, which in relation to the charge weight Q, and the distance D, between the zone considered and the one where the blast is produced.

$V = K \sqrt{Q/D^{1,5}}$ (Langefors - Kihlstrom)
where
K = rock constant experimentally determined
Q = charge detonated by delay
D = distance between blast and reception point

The damage produced in tunnels due to natural vibrations or blastings, is usually manifested in sliding according faults, joint, loss of imbrication block, fenomena which are much greater in the portals. Of course its importance is relative to the type of lining of the excavation, consequently the greatest inconveniences appear in unlined tunnels..

Although, the definition of critical level of particle velocity in underground works, does not count on the precision reached in surface rock, experiences the world over were used as a reference in our case, and satisfactory were obtained.

FIG. 1 TAILRACE TUNNEL PORTAL – FRONT VIEW

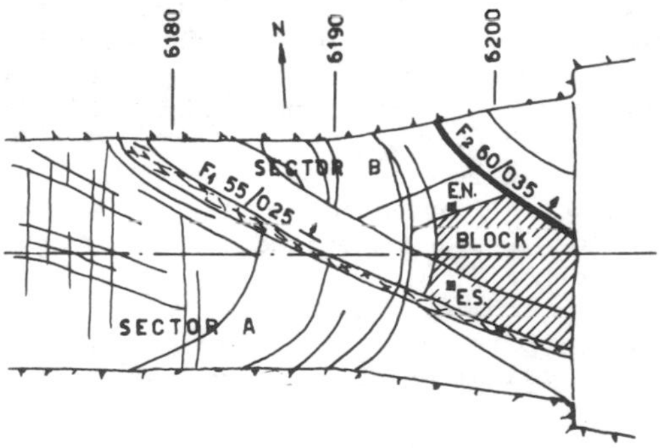

FIG. 2 ROOF GEOLOGICAL SURVEY

It must be considered that to achieve an effective blasting action, the micro-delays must be disposed in such a way as to achieve a noncooperative condition between them. For this purpose it must be considered that the maximun vibration produced is usually of short duration, not above 3 complete cycles (Langefors - Kihlstrom, 1973).

According to the above the frequency was measured in situ and the blasting patterns were designed in such a way that the velocity of particle vibration produced, in the unstable block, would always be less than 20 cm/sec.

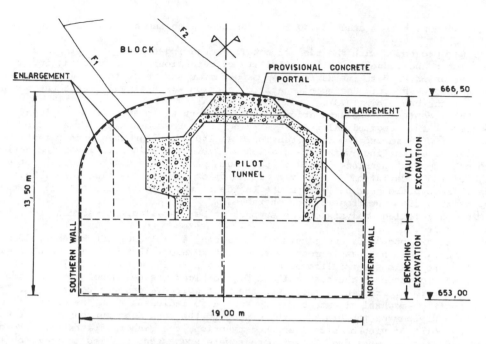

FIG.3 – TAILRACE TUNNEL – PORTAL SECTION
EXCAVATION STAGES

4 CONSTRUCTIVE PROCESS

The inlet section, with elliptical roof, and straight walls and invert, is 19,0 m wide and 13,5 m high (Fig. 3)

On account of the large size section, its semirectangular shape, and low overburden (1,5 times the maximun diameter) a small attack pilot tunnel 5,5 m wide and 4,5 m high was adopted. Because of the excavation of this tunnel was the first to be realized in Río Grande N°1 , there was not any previous experience neither knowledge of geomechanical properties of the rock mass, nor an adjusted excavation methodology. Its only counted on a geological interpretation of the front wall, and iso late data of the preliminary studies. For the reasons the pilot tunnel had the multiple purpose of investigating the principal properties of the rock mass, obtaining the first geological underground mapping, and finding and adequate excavation methodology and support design.

The pilot tunnel was begun tangentially to the tunnel vault in order to stabilize and control the excavation.

Once the excavation of the first 15 m of the pilot tunnel with the axis displaced 4,5 m towards the northern wall was completed, the overbreak was controlled and the main fault had appeared in the vault.Since there the axis direction was changed to that of the Tailrace Tunnel's. The over-break dimished and the quality of the rock improved towards the south of F1 fault (Fig. 2, sector B). Also the tunnel effect began to be noticeable.

When the pilot tunnel had a length of 23 m, the section began to be gradually widened until reach ing a width of 13 m and a height of 8,5 m, with a definitive axis (Fig. 4)

4.1. Portal Excavation

After 6 years of constructing the pilot tunnel and having completed the excavation of the Tailrace Tunnel in the 1000 m of the Contraembalse Environment, and having advanced in the knowledge of the rock mass (del Río et al, 1982) and improving the excavation methodology and type of support (del Río et al, 1983) as far as possible, the definitive construction of the Portal was undertaken.

Before widening the excavation a geological study was realized with respect to the placing and character of the principal faults and joint systems (Fig. 2), that were projected in the plant over the real vault and not over a horizontal plane, as had been done before.

This allowed the individualization of a well delimited block, of notable dimensions (7000 kN) si tuated in the southern part of the vault at the front of the entrance, of which the kinematism was impeded by the effect produced by the provisional concrete portal.

By virtue of this actualization an active support was projected in the whole of the vault already excavated (vault of the pilot tunnel plus the lateral widening), conveniently distributed and orien tated with respect to the main structures, consisting of rock bolts of Ø 25 m and between 5 and 10m with mechanical anchorage tensioned to 140 kN and grouted with cement, with a separation of 2 to 3 m.

This support was projected with the idea of totally replacing the support placed during the advance of the pilot tunnel, due to the impossibility of retensioning and grouting it, on account of the degree of corrosion, due to its having been exposed for six years to the damp and ventilated environment. In the new support design no specific function was assigned to the untensioning support installed at the begining (1976).

Previous to the widening of the excavation another support was placed from the front of the entrance to counteract the longitudinal release, trying to form a crown over the vault, with mechanical rock bolts of Ø 25 mm and 5 m long, tensioned at 140 kN and grouted with cement; placed at 1,5 m above the excavation line, with a separation of 1,0 m and a 20° dip lover the horizontal.

The need to pre-support the block formed in the south part of the vault (Fig. 3) not yet excavated, and which caused a deviation of the axis of the pilot tunnel, was also posed. This situation made it impossible to place rock bolts from the inside of the tunnel; for this reason the support had to be implemented from the front of the entrance.

This determined a detailed analysis of the support to be installed. It was concluded that 2400 kN should be installed.

Two alternatives were proposed, one consisted of placing 4 cables of 600 kN each, 15 m long and grouted with cement. The other was 16 rock bolts of 140 kN each, 12 m long and grouted with cement too. Of these two proposals the traditional one was chosen (Ø 25 mm rock bolts), for the fundamental reasons. It was foreseen in the contract and as the block was not monolithic the breakage scheme would be in the form of fragments; therefore the most suitable one was to distribute the load.

Two multiple position mechanical extensometers were also included, with the object of checking the differential behaviour of the block with respect to the rock mass surrounding the excavation.

Later, from the point of view of stress and construction, the manner of attacking the definitive excavation and widening was analysed. Two proposals were suggested: one was to approach the excavation from the outside, which pre-supposed blasting the provisional concrete portal in the first instance. The other was to begin from the inside (36 m from the portal) with controlled blasting, with 3,0 m advance. In both solutions a quasi-systematic bolting with variable lengths and inclines had been foreseen.

Excavation from the inside was chosen, on the understanding that it offered greater security and more advantages, which may be synthesized thus:

1. The excavation would be realized from lesser to greater, as the dimensions increased in that sense, 14,40 to 19,0 m wide and 8,0 to 3,50 m high.

2. The support would be installed in the zone where the "tunnel effect" predominated and would permit an increasing control of the excavation.

3. The block would be freed at the most stable part, as the direction of the kinematism had a component towards the portal.

4. An effective support could be placed on the block after each advance.

5. The provisional portal would be blasted in the last advance, maintaining the "prop effect" which controlled the kinematism of the block.

6. That being blasted, the effect would be compensated by the support placed from the front and the inside of the tunnel.

The widening of the excavation of the south part of the vault, 36 m long, was realized in 14 advances, the first 8 of 3 m and the final 6 of 2 m (Fig. 4). The diagram of the blast was a side exit and a smooth blasting with boreholes every 0,70 m. The delay-interval was designed so as not to excede the limit of 20 cm/sec of Particle Velocity, according to what was detailed before. The specific charge oscillated between 2,30 and 3,60 N/m3.

The result of the excavation was excellent. Four factors contributed decisively and they are: strict control of drilling, the amount of explosives, the delay -arrangement scheme and the systematic support placed after each advance.

In the first 25 m of excavation, Ø 25 mm rock bolts, 4 and 5 m long were placed, tensioned to 140 kN and grouted with cement, placed radially and separated from 2 to 3 m. In the final 10 m the density was increased to one rock bolt every 1,0 to 1,5 m and the length was from 6 to 8 m.

The density at the base of the block was maintained, but the length was increased to 12 m inclined and placed in the optimum direction, with respect to the direction of the sliding.

Gunite and weldmesh was placed in the vault to reinforce the very fractured surface rock, covering all the base of the block and the haunch of the section. The thickness was variable with a minimum of 100 mm and the mean value of 250 mm.

In this zone 6 "control" rock bolts were installed in the vault and walls, of varying lengths of 6 to 12 m , without tensioning or grouting. During the placing these were tensioned to 140 kN to prove the capacity of anchorage, and later discharged to an adjustment tension of 10 kN.

The end pursued was to test the behaviour of the "crust" of the excavation, as an indirect and quick measurement was obtainable of the deformation, through the control of tension that the bolts accumulated after each blasting.

Before beginning the excavation of benching to reach the invert of the project (Fig. 3), a structural support was realized in the south wall with the object of controlling the effect of toppling, caused by the principal faults, with Ø 25 mm, rock bolts, 3 to 5 m long, tensioned to 140 kN and grouted with cement.

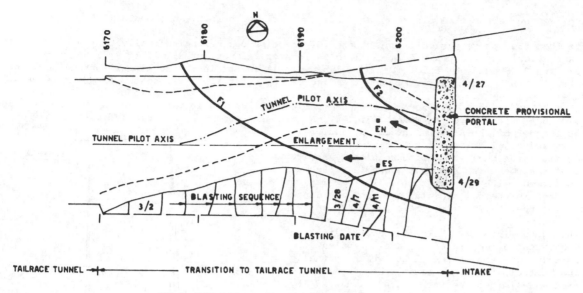

FIG. 4 EXCAVATION STAGES — PLAN VIEW

Later the benching was begun from the inside, the height of which varied from 2 to 4 m at the Portal. The different stages were concieved, considering the existing confinement, and in such a way as not to produce brusque changes in the shape of the section. Controlled blastings were used with presplitting and with the inclusion of the already mentioned criterion of damage levels. These aspects were checked through the extensometer measurements..

5 EXCAVATION MONITORING

The need to count on effective control of the excavation meant resorting to the instrumentation of the section.

As the zone was divided into two clearly differentiated areas, two multiple position rod extensometer produced in Río Grande I were installed. These were placed vertically, from the surface, by means of two drillings 27 and 28 m deep respectively.

The E.N. extensometer was placed in Sector B, while the E.S. was placed in the Sector A, specifically inside the unstable block, to measure its behaviour (Fig. 2).

Each apparatus was positioned with 3 points of reference, considering the outer head situated at the surface as the "fixed point", on elevation 695,00 (Fig. 1).

In both cases an attempt was made to place the last position as near as possible to the vault, keeping a minimum distance to ensure its protection from the effects of blasting and interpret the movement of the rock around the excavation

The precision reached in the readings was of the order of one hundredth of a millimmetre. The sensitivity of the apparatuses allowed a singularly detailed control of the effects of each blasting, making possible an inmediate evaluation of the support and, more precisely, the moment of its positioning.

6 ANALYSIS OF THE RESULTS

The definition of a monitoring section constituted an eficacious element for interpreting the behaviour of the rock mass, corresponding to the Portal, in relation to the relative dimensions of the section to be excavated, its size, its changes of form, the plan of blasting and excavation that was to be executed.

FIG. 5 PORTAL VIEW. DEFINITIVE SECTION IS MARKED IN WHITE

Displacement versus time plots of each extensometer, North (E.N.) and South (E.S.) is given in Fig. 7.

Note that the not very significant changes of shape produced longitudinally by the widening in the excavation sector, realized in the 10 blastings of a length of 25 m (Fig. 4), between the dates 3/2 and 3/20, did not originate appreciable displacements in the rock mass surroudings. The movements of points 1 and 2 started when the face of the excavation reached the extensometer station (Fig. 4).

The blasting of the 4/11 produced an appreciable modification in the section shape, as it widened 6 m (60% of the existing width) (Fig. 4). As a consequence a sudden displacement of 0,9 mm of points 1 and 2 of the E.S. was produced, and of lesser magnitude, 0,2 mm in point 1 of the E.N.

This situation began to delineate clearly the response of the rock mass since it showed a differential behaviour with respect to the block, which was appreciated in the coincidental movement of points 1 and 2 of the E.S., while 3 remained practically fixed.

In the graph (Fig. 7) the difference of displacements between curves 1-2 and 3, are equivalent to the relative descent of the block, which virtually "unlinked" itself from the massif, the kinematism of which was limited, between the 4/7 and 4/27, by the provisional concrete portal. This had a deformation equivalent to the displacement of 1,5 mm operated between these dates, as well as the deformation suffered initially during the excavation of the pilot tunnel.

This indicated that the transference mechanism of the weight of the block materialized downwards, partly, through the portal and also upwards by means of the support of the block, partially installed from the tunnel. At this moment 4/29 this situation presupposed an important change in the shape of the resistent section (Fig. 6), as the block was virtually "unlinked" from the rock mass.

After realizing the blasting of the provisional concrete portal 4/27 a sudden displacement of 1,5 mm was produced of points 1 and 2 of the E.S., which represented an increase of 74% while point 3 remained invariable at 0,8 mm, implying an increase of the relative descent of the block, which totalled 3,08-0,8 = 3,0 mm, equivalent to 48% of the displacement reached finally.

The effect shown by E.N. was less, although the increase resulted in 91%, since point 1 was displaced only 0,6 mm, totalling an absolute descent of 1,38 mm, while points 2 and 3 indicated 0,08 and 0,06 mm respectively, with the result that the relative descent of point 1 was: 1,38 - 0,08 = 1,3 mm, equivalent to 30% of the total displacement.

Fundamentally, this is caused by the loss of the rigid propping which constituted the concrete portal to that moment (compressed by the deformation), more than to that due to the change of shape of the section by the widening which was produced.

The kinematism of the block is now impeded by the support installed in it, which had to be lengthened abruptly by 1,5 mm to obtain it. Consequently a change in the mechanism of transference of the weight of the block towards the immediate upper zone was established. It followed that the response of the rock mass, at this stage, was characterized by two related aspects:
- The change of form of the excavation section (Fig. 6), from the stress point of view, by the unlinking of the block with respect to the rock mass.
- The transmission of the weight of the block to

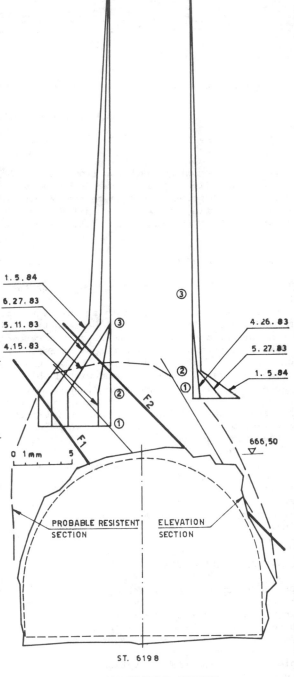

FIG. 6 MONITORING SECTION

the massif, through the support placed in same.

This explains that the deformation, by transference of weight, spread towards the inside of the massif, movilizing point 3 of the E.S. (Fig. 7) which had remained almost fixed, starting from the 4/14.

The considerable difference of incline between the coincidental curves of points 1 and 2 and that of point 3 of the E.S., between the 4/29 and 5/14, indicates that the block continues to accelerate its relative descent. This begins to lessen between the 5/14 and 7/10 given that the inclines of the curves tend to equalize. Starting from the 7/10 they equalize, therefore the relative descent of the block is annulled, which means that the block "links" itself again to the massif due to the slow deformation proper to it, plus the spread upwards and redistribution of the strain produced by the transference of the weight of the block; a fact demonstrated by the increase of the displacement of point 3 (Fig. 6).

In concordance with the date 9/10, the absolute descent equivalent to the deformation of the rock mass continues, but with less speed, given that the inclines (equal) are diminishing. Finally they tend to zero, adopting this value starting from the 9/26, which implies the definitive stabilization of the rock mass.

On the other hand the E.N. extensometer showed a different behaviour of the massif, as starting from the 4/29 the movement of the deeper points 2 and 3, described curves practically coincidental with gentle inclines, that is to say, their position was slightly modified in time, becoming stabilized after the 9/1.

On the other hand, point 1, after the blasting of the provisional concrete portal of the 4/29, it showed a sudden relative descent, later continued to increase with a lesser incline up to the 7/30. Between this date and the 8/23, the relative descent or deformation of the massif slowed down, tending to zero towards the 9/15.

After this date it stabilized definitively.

7 COMPLEMENTARY MONITORING IMPLEMENTS

With the object of increasing the area of the monitored zone, the use of "control bolts" was resorted to, consisting of mechanically anchored bolts without grouting, and of variable lengths.

At the moment of positioning, a charge of 140 kN was applied with the object of verifying the anchorage capacity, which later was reduced to 10 kN, this value being the initial measurement tension and also it constituted the adjustment of the system Nut-Plate-Rock in working position.

The purpose of these rock bolts was to test in approximate and indirect form the deformation of the surrounding rock mass, at a depth equal to the length of same, through the tension that they accumulated after each blasting.

Once the excavation was finished they were tensioned to 140 kN and grouted with cement, forming part of the definitive support.

Another auxiliary parameter was contituted by the gunite applied in correspondence with the inlet and at the base of the block, covering a surface of approximately 8 x 18 m, which was observed permanently with the object of detecting fissures or scaling produced by differential movements before the last blastings.

The absence of fissures in this case, indicated that the displacements of the surrounding rock after the blasting of the portal 4/29 were produced in a uniform manner.

8 BOUNDARY ELEMENTS METHOD

A BEM programme was used, that included Hoek-Brown's (1980) breakage criterion, which considers the resistent characteristic of rock mass, starting from the mechanical properties of the intact rock and the resultant values of the classification of the rock mass (Q and RMR).

The method operates bidimensionally in a homogeneous medium, isotropic, elastic and under a gravitational field. Due to the fact that measurements of internal tensions were not realized, and to the low overburden, two stress relations were considered (K = 0,5 and 0,8).

The results obtained for the resistent sections and the project indicated similar displacements in the surface surrounding rock, in the order of 0,5 mm for one deformation module E = 30.000MPa and for both stress relations; with breakage by shearing stress in the walls, which disappears at a radial distance of 1,5 m.

On the other hand for an E = 10.000 MPa the surface displacements values reached 1,4 mm for similar conditions, while the breakage scheme remained without variants.

A common fact for both sections and stress relations is constituted by the displacements of the interior of the rock mass (at 15 m they were equal to 70-80% of the cortical zone), which would indicate that even the surface levels were affected.

This noticeable difference between the values of displacement foreseen by the BEM, and the values obtained in the monitoring resides fundamentally in that these don't actually reflect the elastic behaviour, but include the inelastic movements due to the differential displacements between the rock blocks.

FIG.7 MONITORING OF THE TAILRACE TUNNEL PORTAL
COMPLEJO HIDROELECTRICO RIO GRANDE N° 1

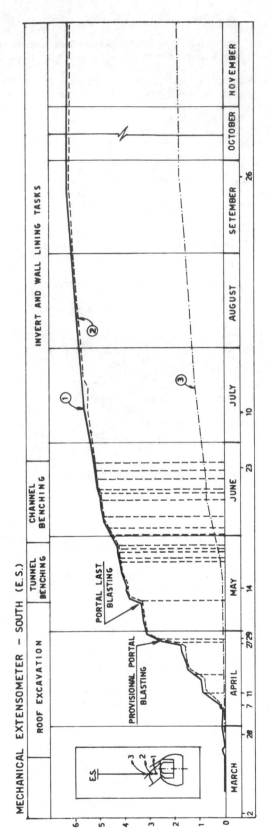

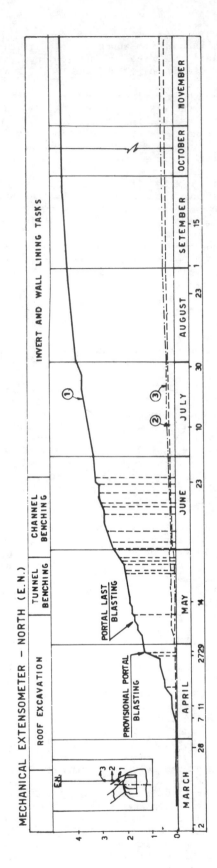

FIG.8 PORTAL. COMPLETED EXCAVATION

This type of movement is the predominant one in the area adjoining the excavation where efforts must be centered to define its predominating direction, thus permitting the adaptation of the support to be placed.

9 CONCLUSIONS

A retrospective analysis of the operative methodology employed, of the results obtained from the values indicated by the extensometers, permitted a critical evaluation of the work realized, in order to achieve a stable excavation. The results of this analysis are synthesized in the following concepts:
- The initial decision to deviate the axis of the pilot tunnel proved to be the correct one, especially considering that the operative technique for the positioning of the active support was being developed, and that the response of the rock mass to the excavations and their interaction with the support was not known.
- The operative methodology was adequate to the geotechnical characteristics of the excavation (quality of rock mass, geostructure and geometric characteristics of the section). A determining factor in the result obtained was constituted by the blasting patterns which included: precision of drilling, charge weight, delay arrangement scheme and the use of careful blasting. The direct result was that the overbreak was limited to the precision that the method permits.
- The use of 20 cm/s as critical level of Particle Velocity, in spite of the lack of defintive antecedents, proved adequate in controlling the effects of the blasting in the rock mass, although for the sectors where the discontinuities with low friction angles with clay filling, predominated, it is considered wise to lower the critical level of Particle Velocity to values of approximately 15 cm/s.
- It remains to be pointed out, with regard to the 2 excavation alternatives (from inside or outside), that the decision to use the first one proved to be right by the results obtained and datas from monitoried section. This alternative allowed that 85% of the excavation was definitively supported, including must of the unstable block, before blasting concrete provisional portal.
- The displacements measured by the extensometers, in spite of what is indicated in the above paragraph, reached important values (30% of the total) at the moment of the blasting Portal, which meant a 1,5 mm (Fig. 6) descent of the block. This descent caused an equivalent lengthning of the support placed to impede the fall of block. This indicates that if the excavation had been undertaken from the Portal, the displacements induced, without the block support, would have caused the fall of it and the consequent propagation of the deformation to the interior of the rock mass, producing alterations as far as the surface levels due to the low overburden.

10 FINAL RECOMMENDATIONS

- In the authors' opinion the construction of a portal of a tunnel is sufficiently associated with the project of a pilot tunnel, which will depend on the size and importance of the tunnel to be constructed, the transverse section, the length, the utility and character of same.
- The reach and purpose of the construction of the pilot tunnel may be diverse, but it is certain that the investment made will always yield benefits, in the measure that the object of same be precisely defined.
- It may be posed at the project and preliminary design stage; before the beginning or during the construction of the works. It may belong to the same contract as the tunnel itself or the contracted previously.
- Not wishing to lessen the scope of the subject, it may be mentioned that the original purpose of the project of the pilot tunnel is based on the achievement of an objective geological survey, which allows a geomechanics evaluation of the rock mass, and in this way to know the response of same to the excavation. Finally, in relation to these elements, to be able to define the construction process which must comprise: the method of excavation, the type of support and/or lining and an established plan of production.
- Particularly (in the case analized here), the Portal of the Tailrace Tunnel has been described in the Constructive Process, of the pilot tunnel project had a multiple purpose: "investigate the principal properties of the rock mass, obtain an underground geological survey and find an adequate excavation and support methodology" (del Río et al, 1983). Although, strictly speaking, the decision to construct the pilot tunnel was a mean to stabilize transitorily the entrance and its influence zone, which became evident in the detour from its original axis.
- It is evident that the pilot tunnel project is a decisive element with regard to the attack of an entrance, but other elements which complement the construction and contribute to the aim of the tunnel are no less important, especially to be emphasized among them are monitoring and careful blastings.
-The use of careful blastings acquires special significance in this singular zone, it being emphasized as well that is possible to include this criterion in the pattern blasting as being an integral part of the Constructive Process.

The importance of the monitoring of the singular excavation resides in the fact that the "tunnel effect" is not manifest due to its being influenced, for a correct evaluation, by means of extensometers placed on the surface in cases of low overburden, or on the other hand resort to the construction of a berm. That may be complemented with extensometers placed in the frontal slope to measure the longitudinal displacements.

This section may be confronted by another one placed at a distance where the "tunnel effect" is manifest.

In this manner a more approximate idea may be obtained of the deformation, from a tridimensional point of view, so necessary in order to interpret the structural mechanism of an entrance.

11 ACKNOWLEDGMENTS

The authors wish to thank the authorities of AGUA Y ENERGIA ELECTRICA for having permitted the publication of this paper.

12 BIBLIOGRAPHY

Agua y Energía Eléctrica, 1982. Medición de Vibraciones en el Túnel de Restitución. Jefatura de Estudios y Proyectos Región III. Informe Interno.

del Río, J.C., Sarra Pistone, R.E., Cravero, R.I., 1982. Estudio de la Estabilidad del Túnel de Restitución del Complejo Hidroeléctrico Río Grande Nº1. Actas ASAGAI Vol.II,49-67.

del Río, J.C., Sarra Pistone, R.E., Cravero, R.I., 1983. Rock Bolting used in Río Grande I. IAEG International Symposium on Engineering Geology and Underground Constructions Vol.I, II, 9-21. Lisboa.

Dorso, R.R., del Río, J.C., de la Torre, D.G., Sarra Pistone, R.E., 1982. Powerhouse Cavern of the Hydroelectric Complex Río Grande I, ISRM Symposium on Rock Mechanics. Vol. I, 221-238. Aachen. W. Germany.

Dowing Ch., Rozen A., 1978. Damage to rock tunnels from Earthquake Shaking. American Society of Civil Engineers. Vol. 104 Nº GT2, 175-191.

Hoek E., Bray J., 1980. Underground Excavations in rock. IMM 527. Londres.

Langefors V., Kihlstrom B., 1973. Técnica Moderna de Voladuras de Rocas. 276-316. Ed. Urmo. Bilbao.

Maki K., Holmerg R., 1982. The Shear Strngth of rock joints with reference to cautios blasting. ISRM Symposium on Rock Mechanics. Vol II 85-92. Aachen. W. Germany.

Monitor and forecast during construction of tunnel mount Da-Yao
Contrôle et prévisions lors de la construction du tunnel du Mont Da Yao

Deng-Yiming & Li Sheng-Tao, *Fourth Institute of Railway Exploration & Design, Wuhan, Hubei, China*

ABSTRACT: The double tracked tunnel passing through mount Da-Yao with the length of 14.295 km is being built on the Beijing-Guangzhou railway line. It is designed and constructed according to the New Austrian Tunnelling Method (NATM) and excavation is carried out mechanically by full section excavation method or half section method and it is supported by shotcrete with rock bolts.

It is incapable with simple parameters to evaluate the stability of wall rock mathematically and make a structure design which is in agreement with actual circumstances because of the complication of rock mass itself and theoretical imperfection of rock mechanics nowadays. For this reason, several means of monitor of rock mass are used during construction for mount Da-Yao Tunnel. According to information obtained and together with investigation and analysis of structure of rock mass , the stability of wall rock has been evaluated and correction to the parameters of support used in the tunnel has been offered.

I. Engineering geological condition

Tunnel through mount Da-Yao is situated in the mountain area of north Guangdong. The strata through which the tunnel passed are epizonal metamorphosed silicarenite with medium-thick layer argillite with thin layer and sandy slate (amounts to 92% of whole tunnel length), marlite and limestone (amounts to 8%).

This region has been undergone tectonic movement for many a time , the strata were severely squeezed, violently, repeatedly folded and inversed thereby. Joint, crevasse and fault well developed. (Fig. 1).

The majority of groundwater is crevice-water. The fault zone and jointy belt abound in groundwater. That in the limestone of middle part of the tunnel is chiefly karst crevice-water where karst is not well developed. Total discharge in tunnel amounts to 38,000 t/d.

IV-V classified wall rock (according to the classification by ministry of railway of China) is fairly good which amounts to 86% of whole tunnel. (Fig. 1)

II-III classified wall rock is relative poor one which amounts 14% of whole tunnel. (Fig. 1)

II. Monitor and forecast of rock mass

Test tunnel has been located to drive in the typical section according to engineering geological feature, i.e. varying region of structure (the kern or the limb of fold), varying lithological characters (limestone or sandstone with intercalated bed of slate), varying depth (800 m, 600 m 200 m) and etc. In test tunnel and its neighborhood in the main tunnel, the monitor of rock mass and investigation of rock structure have been proceeded. (Fig. 1)

On the bases of present proceeding of construction work, the two test tunnels have been excavated after two inclined shafts No.1 and No.2 at Huashipai both get to main tunnel.

The portal of test tunnel No.1 at Huashipai with its Kilmeterage DK 1996+637 sits on the left side of tunnel, its orientation of axis is SN whose angle of intersection with the axis of tunnel is 54°43'. The section area of test tunnel is 3m x 2.5m in magnitude, its length 29m and its depth 600m. Its surrounding rocks are sandstone with medium-thick layer and slate with thin layer, sometimes sandstone is the main rock, sometimes slate is the main one and sometimes their interbedding. It sits at the NW limb of an inversed anticline neighboring the kern of it where the beddings of rock have been heavily squeezed, the tracing of stagger between the layers may be found everywhere. Its main structural planes are shown in Tab. 1.

The test tunnel No.2 at Huashipai sits on the right side of tunnel, with its orientation of axis N 3°W whose angle of intersection with the axis of tunnel is 42°13'. Its kilometerage of portal is Dk 1999 + 263.5, its length 30m and its depth 200m, its section area is 3.6m x 3.6m in magnitude. The strata are sandstone and argillite and their interbeddings. The site is the kern of an anticline. (Fig. 1). The strata have been violently squeezed, fault may be found there. Joints intercrossed here and there. The spaces within the structural planes were filled and the main structural planes are shown in the Tab.2.

A. Determination of initial stress of rock mass

Stress release method is used to measure the stress. Three bore holes in varying direction which meet together at the same point would be driven each time. Adopt the 36-2 type borehole strain gauges (sensitivity 1 x 10^{-4} mm, precision 0.5%) were used to measure the radial strain several times, consequently the space stress field was calculated according to the theory of elasticity. The datum is shown in Tab.3.

The principal structure line of this region shows on the direction of SNNE base upon investigation. It is seen from Tab.3 that the direction of max. principal stress for the test tunnel No.1 at Huashipai coincides with the direction of regional stress field basically. While the direction of the max principal stress for the test tunnel No.2 at Huashipai is influenced obviously by the topographic conditions (see Fig. 2) which is acting nearly perpendicular to the direction of regional stress field. By its magnitude, the test tunnel No.1 at Huashipai with its depth 600m (as indicated above) where the difference of elevation is about 900m as considering of the topographical max. height nearby, the dead-load stress is found to be 24 Mpa approximately where is considerable residual stress at that point. The test tunnel No.2 at Huashipai with its depth 200m, where the difference of elevation is about 550m as considering of the topographical max. height nearby, the dead-load stress is found to be 14.6 Mpa approximately. Therefore, the dead-load stress is the most influential one at that point. Hence, in the deeper section such as Dk 1992 + 400 to DK 1993 + 300 and DK 1994 + 800 to DK 1997 + 100, during design of support and lining of the acting of residual structural stress should be sonsidered, in the meantime, in those section where the ratio of two principal stress is larger than 3, tensile stress would be found and local rockfall would bring about thereby. The support should be carried on as soon as possible in these regions, and reckbolts should be arranged densely, even wire mesh should be added.

B. Measurement of Displacement of rock mass

1. Displacement of sidewall

Before main tunnel excavation, from test tunnel bore a hole toward tunnel alignment with its depth 14m, six-point pole type extensometer is placed in the borehole, first point with a distance 1.5m to the sidewall and other five points with a spacing of 1m each other, the curve detected in situ is shown in Fig 3. When the tunnel is excavated to the position ofthe extensometer, the value of displacement suddenly increased, the displacement within two days amounts to about 2/3 ot total, and after eleven days the displacement shows that it is tending towards stability. Estimating by the curve in Fig.3, the limit of displacement influenced extends approximately to 1.5 width of the tunnel.

2. Convergence of displacement of the tunnel body

Within two test tunnel and neighborhood in the main tunnel body, multiple sectional observations were carried out respectively

(1). Test tunnel No.1 at Huashipai

Three sections were set up in the test tunnel with testing point setting beside side wall and arch sides, the data obtained are shown in Fig.4. The distance between I-section and tunnel body is less than that of II, III-section, its magnitude of convergence is the max, as it is influenced obviously by the tunnel excavation. While in tunnel body, two sections are set up, the data obtained are shown in Fig.5 and Fig.6 respectively. The sharp turning points of convergence value in Fig.6 are due to lower half section excavation, whereby the surrounding rock has been disturbed once again.

(2). Test tunnel No.2 at Huashipai

In the test tunnel, three sections are set up for investigation of convergence. Due to influence of the tunnel excavation, the convergence of 15 days amounts to 10mm, it denotes in the same way that the disturbance due to twice excavating produced a marked influence on the stability of surrounding rock. In tunnel body three sections are set for investigation of convergence and the data obtained are shown in Fig. 7 and Fig. 8 respectively. From them it is seen that the convergence value is inhomogeneous, horizontal convergence is larger than vertical one and that of left side is larger than that of right side. The inhomogeneity is due to chiefly distribution and combination of the structural planes and the influence of structural stress of that point as well.

By the analysis of multiple investigations related above, the following results were obtained:

① The tunnel body undertook disturbance two times or more, i.e. due to blast many times, the wall rock shocked and loosened repeatedly as it resulted, and suddenly deformed, perhaps lost its stability. Therefore, conserning tunnel excavation, it is better to take shape of opening once for it . It is suitable that full section excavation method is adopted nowadays in Mount Da-Yao Tunnel.

② The rate of deformation in general can informed that the wall rock is stable or not. So far as this tunnel is discussed, when the convergence of deformation, when 1.2mm/d, it is in the position of rapid deformation, when 1-0.2mm/d in the position of slow deformation, and 0.2mm/d appears, it shows that it is basically stable. When deformation shows that it is in the rapid deformation stage and especially, if this stage lasts for a rather long time, it is apt to lose its stability, shotcrete should have been sprayed in time and rockbolts placed or wire mesh added and other measures should have been taken into account, so as to avoid the emergence of rapid relaxation of wall rock. The first lot of shotcrete and better be sprayed several hours after excavation , namely, immediate installation of tunnel support. The second lot of shotcrete or placing rockbolts and wire mesh should take place within two days, namely, it takes place after working face 5 to 10m without delay, so as to achieve desired results of initial support of shotcrete with rockbolt.

③ The deformation of rock mass of two positions related above is relatively small in quantity, the wall rock is in stability itself, under construction for tunnel the support with shotcrete is sufficient to stand in stability.

C. Measurement of relaxation ring of wall rock

In the test tunnel or tunnel opening, detecting perforation were carried out at designated points in turn with a soniscope. In test tunnel No.1 at Huashipai the thickness of the relaxation ring is 0.4-0.8m, and in main tunnel 1.0-1.2. While in test tunnel No.2 at Huashipai the thickness of the relaxation ring is 0.4-o.6m, and in the tunnel 1-3m, By them it shows obviously that the thickness of relaxation ring of wall rock is not very large, namely the unstable block which is enclosed by the structural planes is generally not broad in scale at initial stage of loss of stability.

D. Measurement of modulus of elasticity for rock mass

By rigid bearing plate method (with plate area 2000cm^2) the moduli   of elasticity for rock mass were measured in two test tunnel repectively. The data show that in test tunnel No.1 at Huashipai E=2.30 x 10^{4}Mpa, in test tunnel No.2 at Huashipai E=1.25 x 10^{4}Mpa.

E. Sonic parameters

In test tunnel No.1 at Huashipai: longitudinal wave velocity of rock mass Vp=4800-6000 m/s; dynamic elastic modulus Ed = 6.73 x 10^{4}Mpa; Poisson's ratio (dynamic) Vd = 0.22. In test tunnel No.2. at Huashipai Vd = 3000-4000 m/s; Ed = 3.43 x 104 Mpa; Vd = 0.25.

In a word, the initial stress in Mount Da-Yao Tunnel where the depth situated larger than 200m (if consider that the topographical max. difference of elevation nearby, the depth is larger than 500m.) usually exceeds dead load stress of the overlap of that position, as the depth it situated 600m, the initial stress equals to 1.6-fold of dead load (up to 38.6Mpa), but the angle between its acting direction and the alignment of the tunnel is rather small (about 38), it influences the stability of wall rock of the tunnel not remarkably. In the section of test tunnel No.1 at Huashipai, the modulus of elasticity and modulus of deformation is rather large due to the integration and solidity of rock mass, it is in stability and it deformed relatively small in spite of rather large initial stress there. But within some areas (somewhere from the corwn to the arch springing on right side) because of the developed structural planes under action by the tensile stress, the shear deformation produces along the structural planes and loss of stability follows locally. Especially, combining with its feeble lithological characters, it has been proved just so. Therefore the analysis of stability for the tunnel body and forecast should have been undertaken upon the feeble rock mass intersected by structural planes.

F. Distinction and prediction of stability of wall rock in situ

The overwhelming parts of brocky structural rock mass of this tunnel through geological survey and monitor were found to be intersected into varying structural block by many kinds of structural planes (with different directions and different properties). Their deformation and failure chiefly were controlled by the combination of structural planes and their strength . The failure mechanism is found to be along the structural planes shearing off, sliding down and rockfall at last . Its dimension depends upon the size of structural block and limit of possible unstable block cut up by the structural planes of depends upon the range of relaxation ring found out through monitor. The form of loss stability and its dimension, the direction of displacement and its distribution along the circumference of tunnel opening can be found out by adopting stereographic projection and rock block projection. The stability analysis and calculation are proceeded according to the characteristic parameter of rock block and structural plane and monitor datum in order to offer a guide for design of the support of shotcrete with rockbolts and provide a prediction of stability for the tunnel body facing to excavated. Now take the position of DK 1900 as example to analyze below:

The condition of structural plane of the tunnel body is shown in Tab.2.

1. Determination of the dimension of possible unstable structural block and its distribution along the circumference of tunnel opening.

Fig. 9, Fig.10 shows the work done by stereographic projection and rock block projection respectively so as to determinate the distribution of possible unstable structural block along the the circumference of tunnel opening, its demension found by graphs shows in Tab.4, its comprehensive distribution shows in Fig.11 based upon above multiple graphic works.

2. Stability distinction

(1). The ortho-tetrahedral cone located upon the crown may be looked as if a suspended one, it would fall down or not depending upon the tensile strength of structural plane or its cohesion, as a rule, the strength and cohesion is so small that it is apt to fall down acting by gravity. Hence the suspended body found by graphical solution would be commonly considered as an unstable one. ABCO, DCFO, DEGO& etc. in Fig.9 are this kind of structural blocks.

(2). Except the ortho-tetrahedral cone above arch, all possible unstable structural blocks are almost clino-polyhedral cones. Their loss of stability is chiefly due to the acting of gravity which is resulting in a shearing off along the structural plane as a slip plane. The stability prediction proceeded upon possible unstable structural block found by the graphical work of stereographic projection and rock block projection, the critical one would be found out. Its stability would be analyzed base upon the theory of limit equilibrium, evaluate the results referring observed convergence value of tunnel body. Now follow above example to explain it. (Fig.9, Fig. 10, Fig.11).

Crown —— The typical structural block ABCO is an ortho-tetrahedral cone, an unstable one. The rock pressure, acting upon the support is given below:

$p = V\gamma = \frac{1}{3} w.h.\gamma = 1/3 \times 43 \times 3 \times 2.7 = 116t = 1160$ KN;

$p = 116 \times 1000/43 \times 100 \times 100 = 0.3kg/cm^2 = 30$ kpa

Side wall ——

Left side: ABCO is a clino – tetrahedral cone (Fig. 10) with single slip plane.

f (sliding force) $= P \sin\alpha = 40 \times 2.7 \times \sin 70° = 101t = 1010$ KN:

F (resisting force) $= P \cos\alpha . tg\phi + \dfrac{c.\Delta A''E''C''}{\cos\alpha}$

$= 40 \times 2.7 \times \cos 70° \times tg40° + \dfrac{3 \times 25}{\cos 20°}$

$= 111t = 1110$ KN:

K (factor of safety) $= \dfrac{F}{f} = \dfrac{1110}{1010} = 1.09$

It is alike in the position of limit equilibrium, being short of stability.

Right side: ABCD is a clino-tetrahedral cone with two sliding planes.

$f = P \sin\gamma = 54 \times \sin 50° = 41t = 410$ KN;

$F = P \dfrac{\cos\gamma}{\sin\beta_3} (\cos\beta_3 \cdot tg\phi_0 + \cos\beta_{01} \cdot tg\phi_2) + \dfrac{c.\Delta A''D''O''}{\cos\alpha_1'} + \dfrac{c.\Delta D''E''O''}{\cos\alpha_2}$

$= 54 \dfrac{\cos 50°}{\sin 51°} (\cos 11° tg 40° + \cos 40° tg 40°) + \dfrac{3 \times 13}{\cos 65°} + \dfrac{3 \times 12}{\cos 70°}$

$= 265 t = 2650$ KN;

$K = \dfrac{F}{f} = \dfrac{2650}{410} = 6.45$ Stable.

According to convergence value observed in the neighborhood, total convergence value is small (in the direction 1-3, 11.99mm, 2-3 direction, 2.4mm). The rate of deformation is not large (within a week after excavation 1.5 - 3mm/d, henceforth obviously decreases and gradually comes to be stable with a rate o.1mm/d), it denotes that the wall rock of tunnel body is basically in stability. The convergence value in the direction 1-2 is the largest, and the value in the direction 1-3 is larger than of 2-3, their difference amounts nearly to 4-fold, it shows that the deformation in the left side-wall is bigger than that of right sidewall. Therefore, it is coincident with above stability prediction.

3. Calculation for the support parameters

The data used are all obtained from above graphic work.

(1). Calculation for the thickness of shotcrete:

Crown —— According to shear off (by a push):

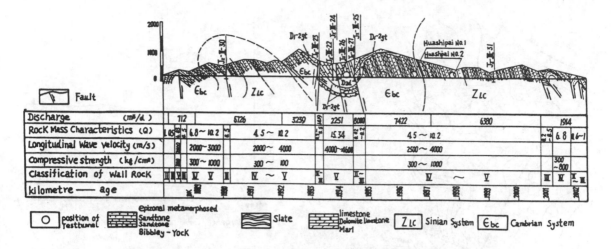

Fig.1 Engineering Geologic Longitudinal profile of Tunnel through Mount Da-Yao

1:100000

$$h = \frac{P}{uR_L} = \frac{153900}{3060 \times 6.4} = 7.8 \text{ cm};$$

According to tear down:

$$h = 3.65 \left(\frac{P}{uR_{Lu}} \right)^{4/3} \left(\frac{K}{E} \right)^{1/3}$$

$$= 3.65 \left(\frac{153900}{3060 \times 3} \right)^{4/3} \left(\frac{0.6 \times 10^4}{18 \times 10^4} \right)^{1/3} = 9.7 \text{ cm.}$$

Where: R_L —— Calculating tensile strength of 200# concrete 6.4 kg/cm^2 ,

u —— Circumference of base area of unstable structural block,

R_{Lu} —— Calculating cohesive strength between the shotcrete and rock surface 3 kg/cm^2,

k —— Coefficient of elastic elongation for the rock bedding 0.6 x 10^4 kg/cm^2,

E —— Modulus of elasticity of concrete 18 x 10^{4}kg/cm^2.

Side wall —— Basically stable. But considering the time effect of structural plane and other unfavorable conditions hardly to predict, shotcrete with a thickness of 5cm is proposed for the sake of safety.

(2). Calculation for rockbolt parameters:

① Length of rockbolt $L = l_1 + KH + l_2$

$$= 0.4 + 1.3 \times 3 + 0.1 = 4.4 \text{m.}$$

Where: L —— Total length of rockbolt,

l_1 —— Depth anchored in stable rock mass, generally 30-50cm,

l_2 —— Outcropping length if rockbolt 5-10 cm,

k —— Factor of safety, 1.3,

H —— Height of rockfall ,in this position 3m is used by calculation.

② Spacing of rockbolts:

The anchorage force of each rockbolt Q should be larger or equals to the load undertaken from max. unstable structural block, i,e.

$$Q > K_1 \gamma H S^2$$

$$S \leq \sqrt{\frac{Q}{K_1 \gamma H}}$$

$$Q = \frac{1}{4} \pi d^2 [R_p] = \frac{1}{4} \times 3.1416 \times (2.2)^2 \times 3600 = 13684.8 \text{ kg}$$

$$S = \sqrt{\frac{Q}{K_1 \gamma H}} = \frac{13684.8}{2 \times 0.0027 \times 300} = 91.9 \text{ cm .}$$

Where S —— Spacing of rockbolts (cm)

d —— Diameter of rockbolt, 2.2 cm,

R_p —— Strength against pull up, 3600 kg/cm^2,

Q —— Anchorage force of each rockbolt (kg)

K_1 —— Factor of safety, 2.

According to above calculation, when this region of tunnel is under construction, at the crown, jetting shotcrete with a thickness of 10cm was proposed and at sidewall with a thickness of 5mm. The rockbolt is placed locally with a length 4.5m and a spacing of 1m.

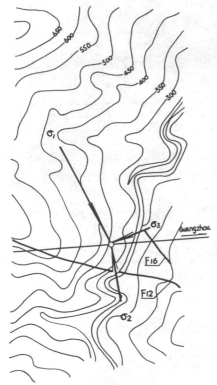

Fig. 2 Relationt between principal stress and Topography at Huashipai No.2

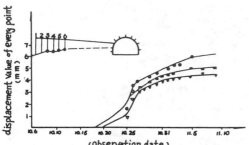

Fig.3 Displacement curve of side wall of test tunnel No.2 at Huashipai

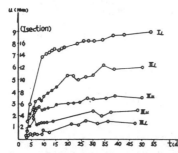

Fig.4 Convergence—Time observed Curve of Test tunnel No.1 at Huasipai

NO	strike	Dip	Dip Angle	
①	N50°—70°E	SE	60°—70°	Rock lamation or Fault
②	N30°—50°W / N30°—40°W	SW / NE	60°—70° / 60°—70°	Joint
③	N20°—40°W	SW	50°—55°	Joint
④	N 70 W / EW	SW / S	80° / 85°	Joint
⑤	N10°—30°E	SE	40°—53°	Joint

Tab 1

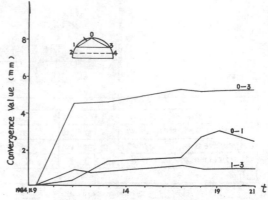

Fig.5 Convergence Curve observed at DK1997+060

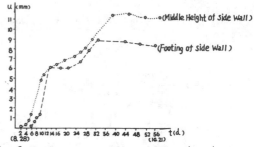

Fig.6 Sectional Convergence Value—Time Curve Observed

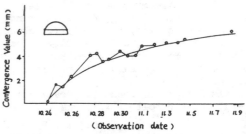

Fig.7 Convergence Curve (at DK1999+275.3) of Test tunnel No.2 at Huashipai

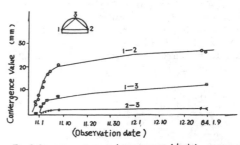

Fig.8 Convergence Curve (at DK1999+290) of test tunnel No.2 at Huashipai

Tab2

NO	strike	DIP	DipAngle	
⓪	N2°—10°E	SE	60°—70°	Formation
①	N50°—60°E	SE	80°	Joint
②	N75°—85°E	SE	70°—80°	Small fault
③	N50°—60°E	NW	70°	Joint
④	N40°—60°W	SW	50°—80°	Joint
⑤	N45°W	NE	65°	Joint

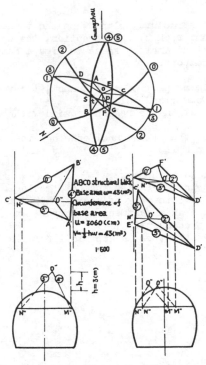

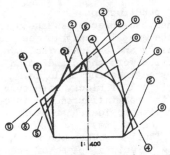

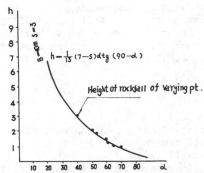

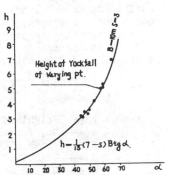

Fig.11 Circumference distribution of unstable Structural block

Fig.9 Graph of unstable Structural block at the Crown

Fig.12 Width from rockfall of side wall unto the curve drawn by empirical formula

$$h = \frac{1}{15}(7-s)\,dtg\,(90-\alpha)$$

Height of rockfall of varying pt.

Fig.13 Height of rockfall at the Crown unto the curve drawn by empirical formula

Height of rockfall of varying pt.

$$h = \frac{1}{15}(7-s)\,Btg\,\alpha.$$

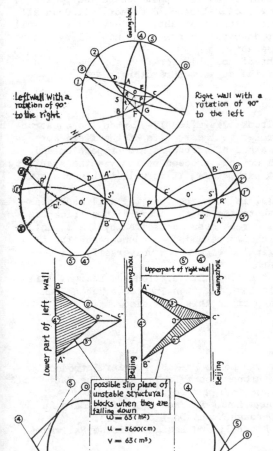

Left wall with a rotation of 90° to the right

Right wall with a rotation of 90° to the left

Upper part of right wall

lower part of left wall

possible slip plane of unstable Structural blocks when they are falling down

$$\omega = 63\,(m^2)$$
$$U = 3600\,(cm)$$
$$V = 63\,(m^3)$$

Fig.10 Graph of unstable Structural block of Side wall
1 : 400

Tab3

		principal stress (MPa)	Bearing	Dip Angle
Huashipai NO. 1	σ_1	38.6	N88°W	60°
	σ_2	21.0	N54°W	− 26°
	σ_3	13.0	S33°W	− 15°
Huashipai NO. 2	σ_1	13.3	N10°E	40°
	σ_2	5.7	S29°W	49°
	σ_3	4.1	S72°E	9°

Tab 4

Item position	Term	Main unstable structural block					Dimension			Remarks
		pattern	slip plane	Inter-secting plane	slip plane slip prism	Dip DipAngle	BaseArea (m²)	Height (m)	Volume (m³)	
Crown TOP of Crown	ABCO						43.2	3.0	43	
	CDFO						24.3	0.6	5	
	DEGO						34.5	1.0	12	
	DFCO									
	FHKO	①	①②		S30°E/80°		25.0	7.0	58	
	SBPO	⓪④	①		BO − S22°E/40°					
	tFPO	⓪②	①		FO − S60°E/60°					
	MBPO	⓪④	②		BO − S22°E/40°					
LeftArch sidesDARO	②⑤	④		DO − S65w/30-40°						
RightArch sidesEPCO	⓪⑤	⑤		CO − N40E/40°						
Side-left Wall	ABCO	④	③⓪		S40°w/10±80°		62.5	3.0	63	
Right	ABCO	⓪⑤	④		CO−N40E/40°		62.5	3.0.	63	

275

4. Forecast for construction

(1). Graphic analysis forecast : Base upon continuous survey of structural plane when the tunnel is being built, the graphic analysis goes on with stereographic and rock block projection, the calculation of stability and evaluation follows, and check the support parameter, so as to give a forecast for construction work ahead and an adjustment of design of support.

(2). Forecast by engineering analogue

① As the wall rock of blocky structure occupies the overwhelming part of this tunnel, the reason and rule of the rockfall produced in construction has been summarized below:

a. Relative to lithological charecters --- Rockfall occurs almost in the region crowded with slate and carbonaceous slate.

b. Relative to fault --- When the width of fault zone is larger than the tunnel opening, the combination of the structural plane of rock mass is unfavorable in spite of that fault is small, and in the jointy belt, rockfalls usually happened.

c. The goundwater intensifies the rockfall in the conditions related above.

By these rules, the monitor is done without interruption in construction in the analogic geologic conditions, if an abnormal condition happened, by the graphic work of stereographic and rock block projection, forecast against rockfall may be offered and an adjustment of support and other measure may be proposed for the construction work going on successfully.

② By statistics from the rockfall found in wallrock of blocky structure an empirical formula is established for estimating height of rockfall:

$$h \ (\ crown \ ) = \frac{1}{13} \ (\ 7-s \ ) \ B \ tg\alpha.$$

$$h \ (\ sidewall \ ) = \frac{1}{13} \ (\ 7-s \ ) \ d \ tg \ (\ 90° - \alpha \ )$$

Where : h --- Height of rockfall (m),

s --- No, for the classification of the wallrock,

B --- Width of tunnel opening (m),

d --- Height of sidewall (m),

α --- Dip of controlling structural plane (of slip plane).

There is 85 percent of the actual height of rockfall measured from 29 positions similar to that estimated by the empirical formula, the formula is verified. Because of that controlled by the structural plane in IV classified wallrock, the unstable structural block does not fall down all its parts entirely, the height of rockfall is therefore less than that estimated by the empirical formula, of course, the remaining part that does not drop down yet is still in unstability, therefore the estimating height is still available for the forecast.

The empirical formula can be used to forecast the height of possible rockfall for the region under construction, estimate the rock pressure and adjust the support parameters. (Fig.12, Fig.13).

III. Conclusion

A. The tunnel passed through mount Da-Yao with 86% wallrock classified IV-V, when the construction work is done in the rock mass of blocky structure, refer to the results of the field survey and monitor of rock mass structure, adopt the graphical method of stereographic projection and rock block projection and engineering geological analogue method, proceed to analyze the stability of rock mass and give a forecast and estimate the support parameters, it is well to be done simply and effectively.

B. According to field survey and comperhensive analysis, it is aware that the wall rock classified IV-V is rather good with self-stability, jetting shotcrete to the crown with a thickness 10 cm, to sidewall 5cm, stability is generally secured, the rock mass controlled by the structural plane may locally fall down where the rockbolts and mesh should be added to counter and the stability ensured. It makes thereupon an improvement on original design which universally jet shotcrete with a thickness 10cm and place rockbolts everywhere.

Remarks: The field test work is done by 4th Institute of Railway Exploration and Design. Railway Research Institute South-West Post, Chinese Scientific Investigation Institute Wuhan Rock-Soil Mechanics Research Post and other Working Units take part in this work also.

REFERENCES

1. Engineering Geologic Research Group for the Tunnel through Mount Da-Yao: A Brief Introduction to the Test and Reasearch of Rock Mechanics of Tunnel through Mount Da-Yao about the Test Shaft at DK 1999 and that Region. See " Railway Exploration and Design " Vol.2 1984.
2. Engineering Geologic Research Group for the Tunnel though Mount Da-Yao: A Brief summary in 1984 about Engineering Geologic Research of Tunnel through Mount Da-Yao. 1985.3.
3. Wang Jian-Ju: Principle for Support of shotcrete with Rockbolt and Design. Chinese Railway publication House 1984.

Experiences of the exploration technics in Talave tunnel, Spain
Etude de différentes techniques de prospection dans le tunnel de Talave, Espagne

A.García Yagüe, *Servicio Geológico de Obras Públicas and ETS de Ingenieros de Caminos, Canales y Puertos, Madrid, Spain*

ABSTRACT: The Talave tunnel, have 31,7 km lenght and circular section 5,2 m. It is 4th Section of the "Acueducto Tajo-Segura" with 255 km long.

The construction started with limited knowledge of the nature and distribution of the ground, based on surface geology, 39 boreholes and 255 electrical sounding. It were foresaw to complete the ground investigations during the works, and to adapt the tunnel trace to its results.

The total made borehole untill the end of the works were 317, with more of 66,000 m. The analysis of investigation data and the actual ground characteristic that were found during the construction allow to obtain several interesting conclussions, that are usefull to analogous problems.

RESUME: Le tunnel Talave est 31,7 km longitude it 5,2 m diametre. Il est le 4me section du "Acueducto Tajo-Segura", avee 255 km longitude.

La construction commença avee une connaissance limitée de la nature et distribution du terrain, avee une carte geologique, 39 sondages et 255 sondages electriques. La investigation du terrain on devai compléte pendant la construction, adaptant le tunnel a ses résultats.

Le total des sondages faits pour le tunnel fut 317, avec plus de 66.000 m. L'analyse des resultts de la investigation et les kéelles caracteristiques et distributions de le terrain, trouvées pendant la construction permet obtenir plussieurs conclusions, lesquels sont utiles pout problemes analogues.

1 INTRODUCTION

The "Acueducto Tajo-Segura" transfers water from the upper part of the Tajo river, in the center of the Iberian peninsula, southeast to the basin of the Segura river.
    The aqueduct begins at the Bolarque reservoir, and after more than 300 km ends at the Talave reservoir.
    The estimated maximum capacity is 1,000 hm^3/year.
    The Talave tunnel crosses the divide between the basins of the Júcar and Segura rivers and is the most difficult - and most costly to build - section of the aqueduct, this difficulty having been foreseen from initiation of the preliminary studies in the late 1920s.
    In 1964 it was decided to commence the Tajo-Segura Aqueduct, and study was intensified for analysing many alternatives. The final project was adapted to the "General Preliminary Project of the Tajo-Segura Aqueduct. D.G.O.H. (General Bureau of Hydraulic Works) - 1967". The works officially commenced on 14th May 1969 and were completed on 31st August 1980.

2 CHARACTERISTICS OF THE TALAVE TUNNEL

Length: 31,927 m (including 200 m of artificial tunnel at the tunnel's end).
Circular section: 5.05 m excavation; 4,20 m free section.
Initial floor or sill elevation: 696.50
Final floor or sill elevation: 654.03
The average depth of the aqueduct is approximately 250 m, and maximum depth approximately 310 m.

3 GEOLOGICAL AND TECTONIC CHARACTERISTICS

The terrain crossed are Marine Miocene and Continental Miocene, Cretaceous, Jurassic and Trias-sic.(Figure 1)

Four superimposed tectonic types were found in the zone:

1. Gentle folds. NNE-SSW thrust direction. few faults and rifts.
Have no effect on the Miocene (Iberian pattern).

2. Tectonic thrusts of E-W direction. Increase of a)-type folds, with thrusts and shifts which affect the tertiary. The shift planes dip between 40º and 60º to the W.

3. Tectonic thrusts of N-S to NE-W direction. Abundant in important overthrusts (Andalusian pattern). In the rest of the trace can have originated shear faults. They affect the tertiary matter.

4. Postorogenic distention phenomena which affect the basement, determining forms in the manner of grabens and horts.

The tectonic complexity of the zone was the principal obstacle to knowing the arrangement of the terrain at depth, aggravated because the zone has rather gentle morphology, recent quaternary cap-ping masses and calcareous scale common to a dry climate (average yearly rainfall of 400 mm), the existence of tertiary postorogenic capping masses, which canceal the geological structure, and a scarcity of guide levels.

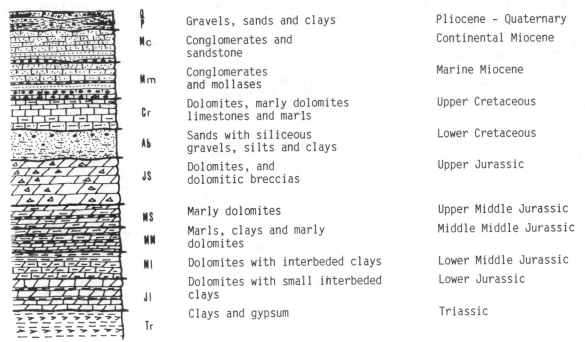

Q p	Gravels, sands and clays	Pliocene - Quaternary
Mc	Conglomerates and sandstone	Continental Miocene
Mm	Conglomerates and mollases	Marine Miocene
Cr	Dolomites, marly dolomites limestones and marls	Upper Cretaceous
Ab	Sands with siliceous gravels, silts and clays	Lower Cretaceous
JS	Dolomites, and dolomitic breccias	Upper Jurassic
MS	Marly dolomites	Upper Middle Jurassic
MM	Marls, clays and marly dolomites	Middle Middle Jurassic
Ml	Dolomites with interbeded clays	Lower Middle Jurassic
Jl	Dolomites with small interbeded clays	Lower Jurassic
Tr	Clays and gypsum	Triassic

FIG. 1   GEOLOGICAL COLUMN

4 GEOLOGICAL STUDIES REALISED

Studies of the terrain for the project were limited by the time available. For the preliminary project of 1967, general geological studies were made which permitted election of the trace and a few drill holes for stratigraphic purposes. Geological survey was made for the project, at 1/25,000 scale, of a strip 2 kms wide in which the tunnel was situated. It was agreed to continue the terrain studies without interruption during the works to thus adapt the trace of the tunnel to the conditions of the terrain, avoiding the most conflictive zones.

The beginning and end of the tunnel was fixed, the galleries and shafts and intermediate support to be specifically located by means of the studies in progress.

The geological study of the works project included:

1. Geological chart of 1/25,000 scale of a zone 2 kms wide.

2. 39 mechanical soundings, of a total length of 8,075 m.

3. 32 geoelectric profiles with a total of 225 soundings, and profile of 480 m of isorresistivities.

4. A geological profile according to the trace of the tunnel, of 1/5,000 scale.

Carried out during the construction were hydrogeological studies of the zone, geoelectric soundinsgs in two conflictive zones and above all a great number of boreholes, totalling 366 investigative soundings in all with a total length of 71,956 m. The investigative boreholes verified the terrain through which the tunnel would pass, looked for easier alternatives and investigated conflictive zones in detail. Single electrodo resistivity, spontaneous potential, neutron-gamma and thermal logging was also made, as well as gamma logging in general.

5 COMPARISON BETWEEN THE PLAN AND REALITY

It is difficult to compare the geological profile forseen in the project and the geological profile at the end of the works. The former corresponds to some unfinished studies and to a trace which underwent change, being moved up to almost 500 m at some point to avoid conflictive sectors. Nevertheless, in Table 1 we show the length of the planned strechts and of those actually traversed.

If we grade the terrain into five classes, evaluating their cost and speed of drilling according to the difficulties (1,3,5,10,40, for the cost per metre, and 1,5,7,15,60 for the time employed), we would find that the actual cost is 1.3 times that foreseen, and the actual drilling time 1.6 times that foresee. These ratios are theoretical, and in the supposition of previous knowledge of the characteristics of the terrain to foresee the changes and have adequate means for drilling and maintenance. Shutdowns for unforeseen circumstances and accidents increase the mentioned theoretical ratios 4 - 6 times for cost and 2 - 3 times for drilling time.

The difficulties of Table 1 in the terrain considered as bad or very bad would have been greater, along with the cost, without the intense work of investigation of the terrain and large number of soundings made posterior to the planning of the project, which permitted the most unfavourable sectors to be avoided.

TABLE 1   Foresseen lengths by the design and finally derilled lengths

	Lithogeologic group or tectonic accident	Design (m)	Final (m)	Variation (m)	Variation (%)
(1)	Continental Miocene	780 ...	1,000 ...	+320	+ 41
(2)	Marine Miocene	-	515	+515	-
(3)	Albian	-	135	+135	-
(4)	Upper Jurassic	4,810	3,935	-875	- 18
(5)	Upper Middle Jurassic	7,130	6,278	-852	- 12
(6)	Middle Middle Jurassic	4,590	3,042	-1,548	- 34
(7)	Lower Middle Jurassic	13,765	11,178	-2,587	- 19
(8)	Lower Jurassic	600	4,830	+4,230	+705
(9)	Keuper	-	95	+ 95	-
(10)	Mylonite of faults	-	567	+567	-
	Number of important faults	17	69	52	+306
	Number of overthrust	2	3	1	+ 50

6 KNOWLEDGE OBTAINED FROM THE STUDIES MADE

The Talave tunnel provides very valuable conclusions with respect to the possibilities and reliability of study methods, since it is possible to compare the geological logs and deduced predictions with a limited number of soundings to those obtained after drilling of the tunnel and a high
number of soundings. The conclusions are valid for like terrains and similar tunnels as regarding
to the deducible characteristics of the terrain in analogous cases, and also extrapolative in
other cases.

6.1 Surface geology

The geological map of 1/25,000 scale did not undergo substantial variation. It was difficult
and in some zones impossible, to make a more detailed map of 1/10,000 or larger scale without the
support of new soundings.
The geological profiles to a depth of 300 m deduced from the 1/25,000 map are not very reliable,
and have important local errors. The greatest difficulty lay in locating and fixing the position
of the faults.

6.2 Electrical soundings

They deduced with acceptable accuracy the distribution of materials to a depth of 100 m.
The presence of faults greatly limited their possibilities. For depths greater than 100 m, with
the existing folds and faults the interpretation was very difficult and little reliable, although
a generally instructive idea was always provided. Extrapolation at depth of the profiles obtained in the first 100 m of depth proved useful.

6.3 Boreholes

Taking into account all the drilling incidences, situation of the water level and analysing
the cores, these soundings made it possible to determine the nature of the terrain, water levels
and fracturing of rifting. Correlation of the different levels was very difficult due to the
absence of fossils and a general process of dolomitization in the zone. It was also difficult
to determine the dips of strata in the calcareous zones, and the true importance of the faults
and even their location (specifally in the dolomitic and calcareous zones and in the clays and
sands).

6.4 Well logs

The gamma logs were specially usefull, permitting the sectors to be classified stratigraphi-
cally by their general aspect. Resistivity and potential logs were also useful, although of limited application due to the need of lining or casing the drill holes. The thermal logs were of local interest, and the neutron-gamma logs were of complex interpretation due to the fracturing and
irregularity of the resulting drilling.

6.5 Hydrological test

The tests made to determine permeability of the ground for the drilling of the tunnel and
the construction of intermediate shafts gave results which were of little use, showing the difficulty of extrapolating the date from sounding to the adjacent zone.

6.6 Study methodology

The aforesaid methods are means for knowing the difficulties to overcome in construction of
the tunnel, and for estimating the machinery necessary. Therefore it is most important to indicate when and how the best results were obtained.

Geological studies, supported by mechanical soundings, obtained the best results in the Talave
tunnel when they sought, through profiles transversal to the trace of the tunnel, geological planes at the tunnel level. Tectonic complication made it difficult to achieve this objective, but
with the aid of boreholes and through successive trial and error tridimensional layouts and geological planes at different levels were achieved.

Drilling of the tunnel confirmed the correction of what was obtained, at least in the zone crossed.

In general, the geological profiles obtained with the aid of separated (200 - 400 m) boreholes were
sufficiently approximate and exact. The existence of faults made soundings necessary in each of
the blocks containing them, but even so the exact dip of the faults could not guarantee, even those
classified as subvertical.

It was very difficult to deduce the thickness of the mylonitized zone in a fault through mechanical
soundings. However, thickness estimates were generally correct when based on the characteristics
of the fault (paragraph 3) and this was already traversed by the tunnel in other zones.

Estimates of the behavious of the mylonitic zone were also quite adequate, considering all the intervening factors: types of faults, nature of the terrain and the water level situation.

The entrance of water into the tunnel, deduced through mechanical soundings and tests made in these was not generally correct, and locally was even contrary to what was expected.

In the calcareous and dolomitized masses, the permeability of the block was very high in the zone situated above the determined water level and in the first 20 m below. At depth it decreased rapidly, to the point that there was no troublesome entry of water in the tunnel in fractured dolomite and breccia, with a water head of more than 10 atmospheres. This conclusion is not valid in the 20 - 40 m adjacent to important faults.

In the studies prior to construction, 12 conflictive zones had been determined. Subsequent studies confirmed their correct qualification and location. When complementary detailed studies were made and the trace and constructive method were adjusted, they were crossed with no more incident than increased cost and longer time employed.

The incidents were vey important when it was impossible to study the zone, vary the trace or adapt the drilling method.

7 FINAL CONCLUSIONS

In the case of the Talave tunnel (31.9 kms in lenght, average depth of some 250 m and maximum of 310 m, through folded and tectonized terrain), present investigative methods permit sufficiently approximate knwoledge of the distribution of the terrain and its characteristics. The greatest difficulty and least reliability of the results is linked to the concentrated entrance of water close to important faults.

The total cost of the studies of the terrain has been 2 % of the final cost of the tunnel.

In similar cases the cost of investigation of the terrain can be considered to be about 3 %, for paradoxically if the studies are well made the percentage of their cost increases, by reason of fewer accidents and lower cost of the tunnel; if the studies are not well made or their greater part is not made before construction commences their cost percentage decreases, by reason of increased final cost due to accidents and unforeseen circumstances, although in this case not strictly necessary complementary studies are made of the terrain with the aim of finding rapid solutions and not detaining construction.

REFERENCES

Coloma J.F. & Sahun P. 1983 - Las inyescciones de cemento en el Tunel de Talave.
          Bol. de Inf. y Est. S.G.O.P. 42: 101-126

García Yagüe A. 1983. Trabajos geologicos realizados para el Tunel de Talave del trasvase Tajo-
          Segura (Albacete). Serv. Geol. de O.P. Madrid. % Volúmenes. (Report no
          published).

The Gloria fault in Talave tunnel, Spain – Used methods and consequences
La Faille Gloria dans le tunnel de Talave, Espagne – Méthodes d'étude et résultats

A.García Yagüe, *Servicio Geológico de Obras Públicas and ETS de Ingenieros de Caminos, Canales y Puertos, Madrid, Spain*

ABSTRACT: The most important difficulty to make the Talave Tunnel (31,700 m, 5.05 m , diameter) was the "Gloria" Fault ("Glory Fault). In this area the Jurassic limestones and dolomites over-thrust Cretaceous sandstones. All this area is also crossed by a system of more recent subvertical faults. To pass this tectoniced zone demand four years of works and more of 40.000.000 $ (USA) (1982).
This paper resume the investigations, accidents and final process to cross this singular zone, where the water charge were 200 m. If there had not been technical reassons, the accidents showed had been fewer and the solutions quicker and cheaper.

RESUME: La faille "Gloria" fut la plus importante difficulté pour faire le tunnel Talave (longitude 31,7 km, diamètre 5,05 m). Dans ce zone calcaires and dolomies jurassiques chevauchent grès crétaciques. Tout ce zone est aussi affecté pour un autre système de failles subverticaux plus récents. Percer ce zone très broyée demanda quatre annes de travail et plus de 40.000.000 $ (USA) (1982).
Cette communication résume les investigations faites, accidents, et méthodes utilisés pour traverser ce singulier zone, où la charge d'eau était plus de 200 m. Des circonstances ne pas techniques firent la solution très plus chère et lente.

1 INTRODUCTION

The area of "La Gloria" fault,was one of the 12 areas considered difficult.
The process of constructing the tunnel did not allow for any variation or modification in the construction system for complex but non-technical reasons.
Crossing "La Gloria" fault was the most difficult and expeensive problem for the Talave tunnel.

2 CONSTRUCTION PLAN AND GEOLOGICAL SURVEYS

In the 1968 plan, a subvertical fault was considered to exist in this area (km 13). At the level of the tunnel, this fault brought Upper Jurassic into contact with Lower-Middle Jurassic. When the work began it was considered to be one of the 12 most difficult areas to be crossed by the tunnel, but not the most difficult.
The "Servicio Geológico de Obras Públicas" began its surveys of the "Tunel de Talave" in September 1971. The construction had been started in March 1969, in accordance with a plan in which the drilling was begun with two 5,05 m "Robbins" one at the beginning (Km 0) and the other at the end (Km 31). The situation of 5 vertical shafts and another two sloping shafts were fixed simultaneously. From these it was possible to establish new work faces, either as a total section using conventional methods, or with two more smaller "Robbins" (Ø 3,8 m and Ø 3,50 m).
The work plan was fixed on general lines with the construction of the intermediate shafts.
In the draft, it was planned to adjust the design, during the construction, to the ground which was easiest at the height of the tunnel. This meant a detailed study with boreholes in front of the drilling face. Nevertheless, this flexibility was severely restricted by the situation of the intermediate shafts finished in 1971 and by the "urgent sociopolitical need" that made it impossible to both stop drilling and leave the drilled stretches. (Non-technical factors are often the cause of problems for which there is only a difficult technical solution).
For the first report made in 1972 I used the data from boreholes and reports provided by the team of geologists included in "Planning Management" (At that time there was also a Construction

Fig. 1. "Acueducto Tajo-Segura" location and Talave Tunnel Section

Management and the Construction Firm). This report pointed out that the Albian (sands, silts, and clays with gravels, (Lower Cretaceous), was the worst terrain for drilling a tunnel, especial ly when the water head reached 200 m and as a consequence the tunnel had to avoid it.

3 LA GLORIA (Stretch shaft 2 - shaft 3)

The newer studies make it possible to conclude that in the La Gloria fault area, the tunnel crossed the Albian that was overthrusted by Jurassic (Fig. 2), and with near suvertical faults, and important mylonite zone with a water head of 200 m.
   From Shaft 2 and towards La Gloria the "Robbins" Ø 3,50 m made efficient progress. The sociopo litical factors could not stop this. It was not possible to tackle the "La Gloria" fault from Shaft 3, so that water and wash load could be dislodged by gravity towards the end of the tunnel which was yet to be drilled. Neither was there time to carry out a detailed study of the area with boreholes, which could bore to a depth of 300 m that would look for a passage which was assumed to exist to the west (In 1977 this possibility was demonstrated).
   I was aware of the fact that there was very little possibility of being able to drill the zone with the "Robbins" Ø 3,50 m and that the accident was almost inevitable.
   The Management of the work took the following precautions:
   a) As the final section drilled would be 5,05 m, leave a slope towards shaft 2, to allow the water flow out by gravity to this shaft.
   b) Strengthen the means by which the water was pumped out in shaft 2 (total depth 298 m).
   c) Construct a security sluice gate at km 12,4

284

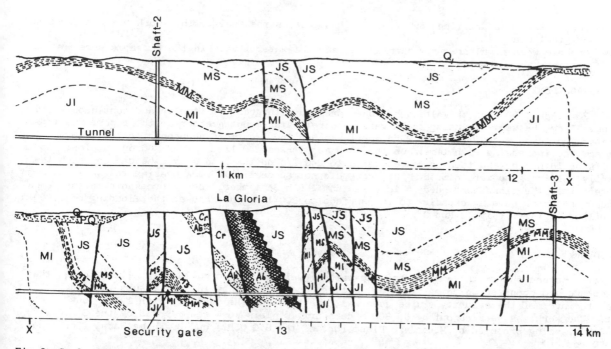

Fig.2. Geological profile shaft-2 to shaft-3 (Q = Quaternary;PQ = Pliocene-Quaternary;Cr = Upper Cretaceous;Ab = Albian; JS = Upper Jurassic; MS = Upper Middle Jurassic; MM = Middle Middle Jurassic; NI = Lower Middle Jurassic; JI = Lower Jurassic

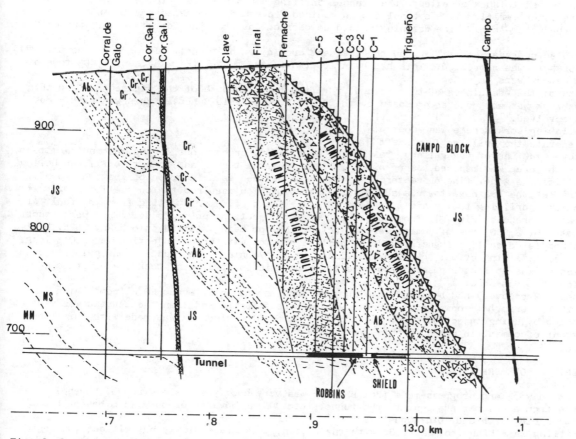

Fig. 3. La Gloria. Geological profile (Simbols as in Fig. 2).

d) Bring forward the drilling work from shaft 3 towards the fault. This work was not foreseen in the original plan.

New borehole were simultaneously carried out and the water levels in the borehole were controlled.

4 1973 ACCIDENT

By the 23rd April 90 m of the Albian had been crossed with great difficulty and continuous metal lining. From the very beginning there was a general water input greater at the floor. The inflow of water increased steadily and on 23rd April reached 150 l/s, more abundant farther from the workface and increasing with time. On the 18th May it was 240 l/s. After it dropped slowly to 190 l/s by 18th June and then increased again to 240 l/s on the 12th July, the day on which the securily sluice gate were closed. This gate did not successfully work passing the water throught the terrain. The flows continued to increase until 9th September, when a breakdown in the pumps reduced the water pumping capacity to the surface to 200 l/s, this caused the flooding of 2,800 m of drilled gallery. (Fig. 2 and 3).

5 RECOVERY

The water pumped out to the surface and the water levels in the borehole (Fig. 4) indicated that a blockage had occurred which enabled the recovery operation to begin. To ensure the success of this, the area of the security sluice gate was injected from the surface, with 1,100 tons of cement. The work was carried out by 6, 300 m deep drills. This work was carried by the S.G.O.P. (P. Sahún) being proposed by M.Lorenzo Blanc (Dr. of the S.G.O.P.) and D. Jaime Nadal (Construction Director).

As a result of this accident, about 5.10^6 m^3 of water and 40.10^3 m^3 of sand and clay had got into the La Gloria area, creating important hollows in the vicinity of the tunnel, in the Albian, of about 30.10^3 m^3 .

The study of water-levels made it possible to deduce that the water-level in the adjacent area could be lowered to such an extent that tunnel drilling could be made safe. In these estimations which were later verified it was concluded in drained flows and time necessary to obtain reductions in the water-level. The draining should be carried out from the Corral de Galo and or Campo blocks.

It was also necessary to filling up the accident area near the tunnel. In order to do this the Trigal borehole was redrilled. This had been located in the tunnel (Fig. 3), and 1,400 tons of sand and cement were injected.

In spite of the usefulness of this method, verified later in both of the sluice gate and this area, the work was not done as proposed by the S.G.O.P. (D. Pablo Sahún). This was partly due to the cost involved.

The following work scheme was adopted:

1. Accelerate the drilling from shaft 3.
2. Recover the use of the gallery, taking ever-increasing precautions on approaching the Albian.
3. Once the area was reached, carry out drillings from the tunnel and inject to fill up hollows and stabilize the zone. (The characteristics would be defined as a function of what was found).
4. Carry out the drainage towards shaft 3 by gravity, which required the completion of the tunnel.

Conventional drilling from shaft 3 discovered large flows of water starting from the fault at the Carrasco borehole (Fig. 3). From this point cement injections prior to drilling, were resorted to. A total of 2,500 tons of cement were injected, the job being started on 6th May 1975. Drilling was completed on 10th August 1976 (The drilled stretch with prior injection was 220 m long, the average, time, injection, drilling and lining is 15 m/month). On reaching the mylonite area and Albian it was no longer possible to inject. From this point digging took place with a – "Priestley" Ø 3,6 m shield.

From shaft 2 recovery had been made making the tunnel sound up to about 20 m from Albian (20th October 1975). From here onwards a concrete wall was constructed and then the terrain was made – more compact by using injections from the tunnel itself. Slow progress was made with this. The total amount of cement injected in this area was 2,500 tons.

6 ACCIDENT IN NOVEMBER 1976

The injection work and progress made from shaft 2 was very costly and slow (survey borehole, – injections from a chamber adjacent to the tunnel, etc.). An advance of only 25 m having been made between 9th March and 10th August 1976.

The drilling in Albian from shaft 3, with the shield, took place slowly but without incidents (from 10th August to 2nd November 65 m had been drilled). The fact that there had been incidents together with the difficulty in setting up drainage chambers in the "Campo" block and downstream evacuation problems incited the Management of construction to do without the drainage (the con-

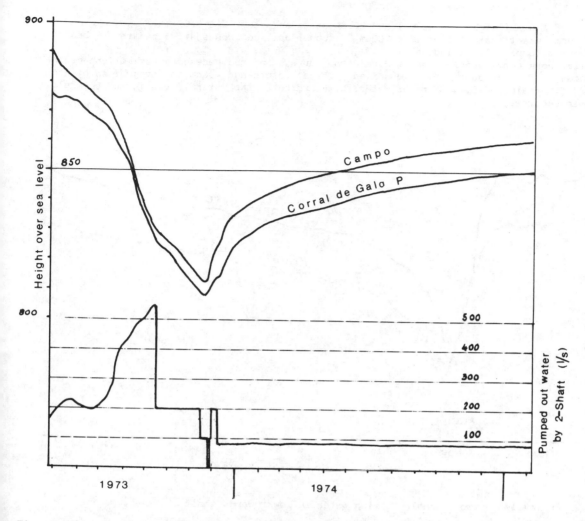

Fig. 4. Water level in "Campo" and "Corral de Galo" boreholes and pumped out water by the shaft 2 during the first accident (1973-74)

tinuity of the tunnel downstream was achieved in February 1976, and concreting was underway).

I personally considered it unlikely that it would be possible to reach the area where the Robbins was buried without drainage, as the terrain must have been very fractured, and thus would have lost its impermeability characteristics. The 30,000 m3 of hollows that had formed, even with different distribution hypotheses could not have filled up with the 2000 ton of cement injected in the vicinity of the tunnel, and the water levels had recovered.

In conclusion, the conditions of the Albian as a consequence of decompression, erosion and fracturing, and in spite of the injections carried out were worse than in 1973 and were to cause another accident.

Or 2nd November at 13.45 h water began to flow in by on exploration drill that had been done next to the left hand tunnel vall at the front of the shield. At the beginning this flow was only slight (litres/second) but it grew rapidly with the erosion process, until it reached more than 500 l/s, with intermittent pulses.

The water-levels in the Corral de Galo H and Corral de Galo P boreholes, revealed a slight drop three days before, thus warning of the phenomenon. Its hourly measurement shows perfect correlation with the increase in rate of flow, and as a whole with the 1973 phenomenon. (Fig. 5).

The sands, silts and clays with gravels of the Albian ended up blocking the tunnel and preventing the entry of water and wash load 13 days after this accident, the water levels in the adjacent boreholes rising as a consequence.

The solid material that was deposited in the gallery exceeded 13,000 m3 and that which was transported by the water (clay) 6,500 m3. A total of 220,000 m3 of water flowed in plus more than 20,000 m3 of material with the following approximate proportions: 17.5% gravel and coarse sand

($\emptyset > 2$ mm), 60% sand(2 mm>$\emptyset > 0.05$ mm) and 22.5% silts and clay ($\emptyset < 0.05$ mm).

The accident meant that at first the 700 m of tunnel drilled from shaft 3 had to be abandoned, it was in part completely filled up.

Meanwhile, work continued from shaft 2, and the surveys and the necessary estimations of the consequences of the accident were intensified. The accidents had meant the evacuation of about 50,000 m^3 of terrain and $5.2.10^6$ m^3 of water, with a drop in the water levels in the boreholes and a later recovery.

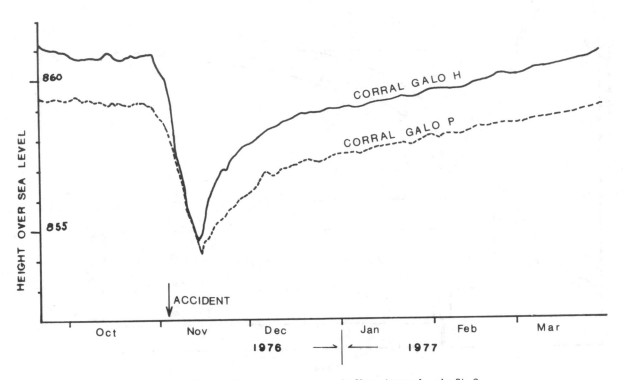

Fig. 5. Water levels in two boreholes during water inflow towards shaft 3.

7 EXPLANTION OF THE ACCIDENTS AND THE SITUATION OF THE MASSIF

When the drilling of the Albian and mylonitic area began in 1973, and in spite of the considerable mechanic lining a decompression was caused in the vicinity of the tunnel. The decompression meant an increase in porosity and permeability. The load of 200 m of water established some very steep gradient next to the tunnel, and an outflow of water began through the fissures, together with - growing wash loads as time went on. This caused the disturbance to increase in the massif and increased the permeability. The stability of the tunnel, the "gap" would not last a month, a long time, if the high tectonic pressures that the Albian had under gone are not taken into account. These were indirectly estimated as being more than 400 kg/cm^2 (Breaking of siliceous gravels and long-term consolidation tests).

The Albian was very compact but at the same time, it was tectoniced and fractures into blocks. The decompression and flow of water through the cracks and fissures opened up by the decompression turned the area into a set of blocks were falling and disintegrating and from the very beginning there was contact with the permeable calcarous Jurassic, in fact this contact already existed and was helped by the drilling of the tunnel.

The rapid communication (hours) between the "Campo" borehole and the front of the tunnel in Albian had already been tested with isotopic tracers in 1973. This had also happened with the "Corral de Galo" borehole. Both borehole were situated outside the Albian had been shown that the inflow of water at the front of the tunnel affected the water level in all the nearly boreholes, both in 1973 and 1976. This made it possible to consider that in spite of the tectonic complexity and the extremely impermeable nature of the Albian, for practical purposes, the massif behaved as though it were only one aquifer. I also concluded that the area affected by the remoral of the 50,000 m^3 of material, decompression and collapses would have a minimun volume of about 200,000 m^3. The presence of collapses was shown by the pulses in the flows and the levels observed.

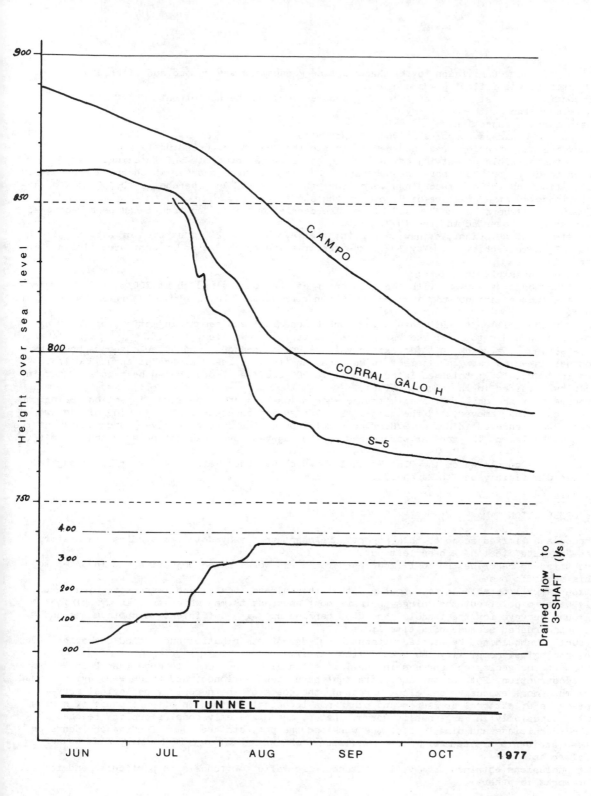

Fig. 6. Water levels in 3 boreholes and water drained towards shaft-3. Final drilling process.

8 PLAN OF ACTION

A Technical Advisory Commission to the Construction Management was set up and after studying all the observations the following was decided.
1. To continue the slow work of injecting and recovery on the northfront (Shaft 2).
2. To recover the southern gallery.
3. Begin surface injection.
4. Set up the drainage chambers in the "Campo" block.
5. Study the neighbouring area in order to locate the "path" announceding 1972.
6. Study the possible cement injection techniques, chemical products and freezing.

In order to carry out "3" above the 5 borings "C" (Fig. 3) were made in which gamma reutron-gamma, inclination and thermic recordings were done. By means of these boreholes 10,000 tons of cement were injected at a pressure of 60 atm on the surface.

The drainage chambers in the "Campo" block were carried out once the tunnel had been recovered the drainage being started in June 1977.

The existence of a variant, or new safe path, that avoided the 200 m stretch of Albian, with a length of 400 m was achieved in May 1977, the confirmation boreholes and report were finished in November 1977.

There were 4 possible solutions:
1. Construction of a variant with the abandoning of the Robbins and about 300 m of drilled tunnel.
2. To construct a variante that crossed the Albian near to the already drilled tunnel, about 200 m long.
3. Recover the drilled tunnel. This being done by chemical and cement injection, or freezing.
4. Drainage in order to lower the water-level as far as the gallery itself.

After a series of studies, not for exclusively technical reasons, and after numerous meetings,a decision was taken to carry out cement injections, and to keep the drainage at about 300 l/s, - achieved in August 1977; this was later reduced to 200-22o l/s (7 drains had been constructed in two chambers situated in the "Campo" block).

The recovery of the drilled tunnel with the shield was slow and sensitive, injections being made, lining pieces being removed, and the tunnel was widened and concreted. The joining of the two work fronts was achieved in March 1978. The completion of the stretch recovery and removal of the Robbins 3 moths later. The general water load in the adjacent area was not greater than 40 m (Fig. 6).

The tunnel was completed on the 31st March 1979 which was when water began to flow through. The work finished officialy on 7th March 1980.

9 FINAL OBSERVATION

The purpose and limited scope of this report have meant that many detailed studies, laboratory tests an "in situ" test have been left out.

I have tried to describe the most important events and works in an objective way in order to show this specific case.

The experts will be able to confirm facts that they already posess and learn something new.

I only want to point out the things that are most apparent to me, which are, in my opinion:

The ground surveys for the tunnels should be carried out with sufficient economic means and - without being rushed because of a time limit.

Construction should not be started without knowledge of distribution and ground characteristics. The more geological complexity the greater the detail.

The cost of the geological surveys in the Talave tunnel did not come to more than 2-3% of the cost of construction. Both of the accidents that have been mentioned, and which were evitable,made the work much more expensive and also increased the costs of the necessary geological surveys.

The presence of external pressures and other non-technical factors is often a cause of serious accidents, especially in large public works. Later, and mistakenly conclusions are reached in which the technicians and directors of the work and the projects are blamed for the accidents or that these accidents are caussed by the powerlessness of Applied Geology in the Engineering and Technical areas.

These conclusions obtained from "Talave" tunnel are valid for tunnels in particular and for all kinds of works in general.

REFERENCES

Coloma, J.F. & Sahún, P. 1983. Las inyecciones de cemento en el tunel de Talave. Bol. de Inf. y Est. S.G.O.P. 42: 101-126.
García Yagüe, A. 1983. Trabajos geológicos realizados para el Túnel de Talave del Trasvase Tajo- -Segura (Albacete). Serv. Geol. de O.P. Madrid. 4 volúmenes.(Report no published).

Tunnel simulation in variable geological conditions

Simulation de creusement de tunnels dans des conditions géologiques variées

D.R.Grant, *University of British Columbia, Vancouver, Canada*
H.D.S.Miller, *University of British Columbia and International Mining Services Inc., Vancouver, Canada*

ABSTRACT: Recent technological advances in tunnel excavation methods and high variability in ground conditions have rendered the task of cost estimation a complex and time-consuming procedure. The two major factors that influence the cost estimation of a tunnelling project are the uncertainties in predicting geological conditions along the proposed tunnel line and choosing the optimal construction method to match the geologic conditions. Determining the optimal construction method minimizes the time-to-completion thus reducing the total project cost.

A detailed analysis of a proposed tunnelling project becomes difficult when short time periods are allotted for project feasibility. With the large amount of information available for analysis, the computer becomes a cost-effective tool to evaluate tunnelling projects. A computer model has been developed to undertake a detailed feasibility analysis by integrating the techniques of probability, statistics, simulation, and spreadsheet manipulation. The computer program consists of two basic components -- the operational model and the cost model. The operational model simulates the advance rate of a tunnel through varying geological conditions for a conventional DRILL-BLAST tunnelling system. The Monte Carlo sampling technique is used to generate random variables from actual tunnelling activity time distributions (i.e., drill time, muck time, etc.). The output from the operational model (activity times and quantities) combined with unit price data forms the input for the cost model. The cost data is transferred through a series of databases and spreadsheets to determine the total project cost for various construction scenarios.

By varying the input data, sensitivity analyses can be performed quickly and efficiently. Computer modelling is therefore a most attractive and cost-effective method for estimation of the costs and scheduling of tunnelling projects in variable geological environments.

INTRODUCTION

Tunnelling projects are becoming increasingly prominent worldwide. The utilization of subsurface space in large urban centres in the best practical solution to cope with growing populations. Unfortunately, project costs are rising rapidly with minimal assistance from new tunnelling technologies. Tunnel excavation is regarded as one of the most complex types of construction, largely because of the variable geological conditions and their effective interaction with different construction techniques. The estimation of the time-to-completion and the cost for a particular tunnelling project is difficult owing to the large number of variables that influence the project. If the effect of these variables is examined during the project evaluation stage, the degree of project risk is reduced (see Figure 1).

The present methods for determining tunnel project feasibility are often unsophisticated because they rely on information retained by the companies' costings engineers. However, the computer can rapidly perform large amounts of data handling and simulate random processes with relative ease. Sensitivity analyses comparing construction scenarios and tunnel alignments should be considered for making effect project evaluations. These analyses incorporate a relatively inexpensive computer modelling method to analyse large amounts of data and produce realistic results. A computer model has been developed that integrates techniques from probability, statistics, simulation, and spreadsheet manipulation, to perform project feasibility analyses. Using the Monte Carlo simulation technique, random variables can be produced in an unbiased manner. These random variables simulate the operational portion of the computer model and determine the time-to-completion of the tunnelling project. The operational output is then used as partial input for the spreadsheet cost analysis.

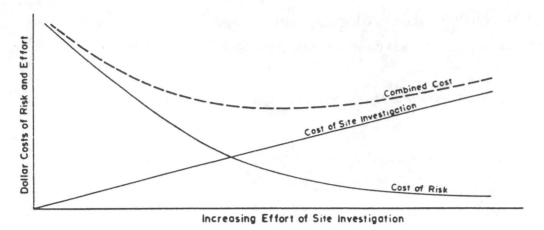

Figure 1. Cost of site investigation and cost of risk in tunnel construction
Ash et al., 1974.

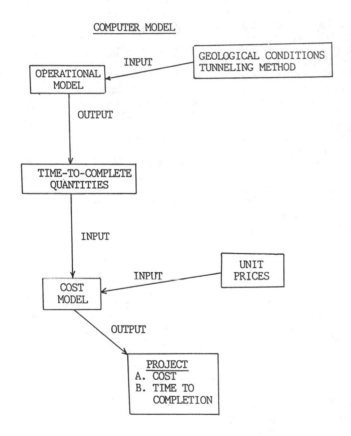

Figure 2. Structural flow of the tunnel
simulation computer model.

COMPUTER MODEL

The tunnel simulation model that has been developed simulates the construction of a DRILL-BLAST-MUCK-SUPPORT conventional tunnelling system. Expansion of the model to include Tunnel Boring Machines and Roadheader Part Face Machines is in the conceptual and data collection stages. The simulation technique used is the Monte Carlo method -- a simulation technique which consists of applying a numerical process on random numbers. The computer program consists of two major components -- the operational model and the cost model. The operational model calculates the time-to-completion for a tunnelling project and the quantities used. The output from the operational model is channelled into the cost model to determine the project cost for a particular construction scenario. If incremental changes are made to the significant project parameters, sensitivity studies can be performed. Figure 2 illustrates the structural flow of the computer model.

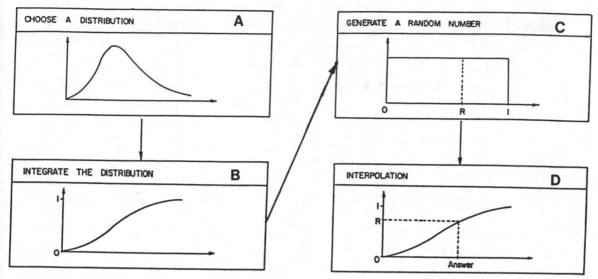

Figure 3. The Monte Carlo simulation technique.

1. Simulation

The Monte Carlo method is used to generate random variables from actual or estimated tunnelling activity time distributions (for example, drill time/cycle and muck time/cycle). A random process is first parameterized by a suitable probability distribution, which models the tunnelling activity. Through experience, it has been determined that tunnelling activities tend to be positively skewed. The application of the Monte Carlo method is best described in a four-step procedure (see Figure 3).

1.1 Choosing a probability distribution

Probability distributions are chosen to match each of the random variables (drilling time, mucking time, etc.). The probability distributions can be determined using two possible data analysis methods. First, a non-parametric approach, which uses the raw operational data obtained from field observations. Histograms are constructed using the kernel technique for maximimizing data, to determine a probability distribution function. Secondly, the data can be evaluated parametrically by fitting the distributions. These distributions are defined by unique parameters that control the shape and the skewness of the distribution. Using the sample mean and variance, the parameters that define the distributions can be estimated. See Figures 4 and 5.

1.2 Integration of the probability distribution

The probability distribution can be integrated to obtain a cumulative distribution. The limits of integration are determined by consideration of field-observed operations of the modelled activity. Judgement must be used when selecting the upper and lower limits of integration (see Figure 6).

1.3 Generate a random number

A pseudo-random number is generated to test against the integrated probability distribution. "Pseudo-random" means that there is a finite number of random numbers generated before repetition occurs. The mixed congruential method produces uniformly distributed pseudo-random numbers between zero and one. The method is defined by the formula:
$X_{n+1} = (aX_n + c) \bmod m$
and is adjusted to the interval (UL,LL) by the formula:
$r = (X_{n+1}/m)(UL-LL) + LL$

where a,c, and m are constants, UL is the upper limit, LL is the lower limit, r is the generated random number and X_n is the seed number $(0 < seed < m)$.

The uniform random number is generated when non-parametric distributions from actual observed field data are used to produce random variables. In the case of parametric distributions (theore-

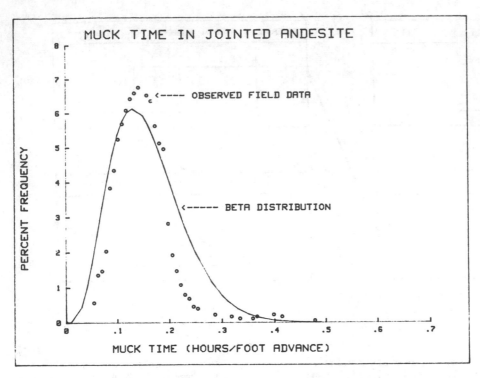

Figure 4. Percent frequency probability distribution, Beta distribution modelling observed field data (Muck Time).

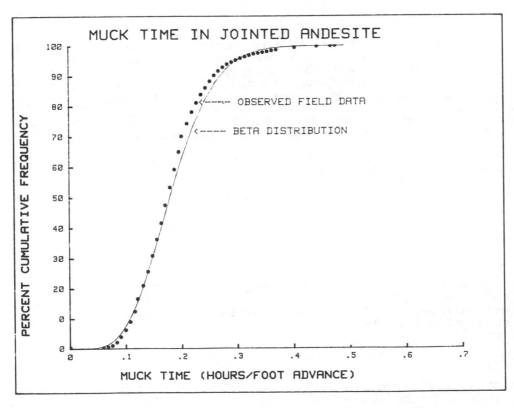

Figure 5. Percent cumulative frequency distribution, Beta distribution modelling observed field data (Muck Time).

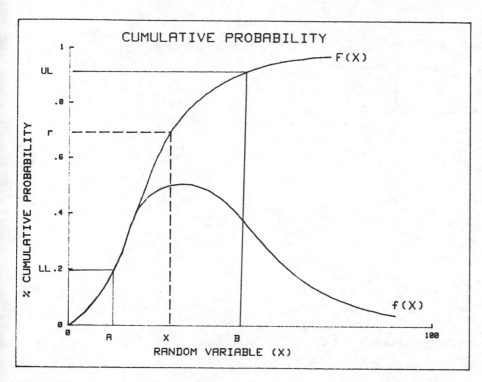

Figure 6. Integration of the probability distribution to obtain the cumulative distribution.

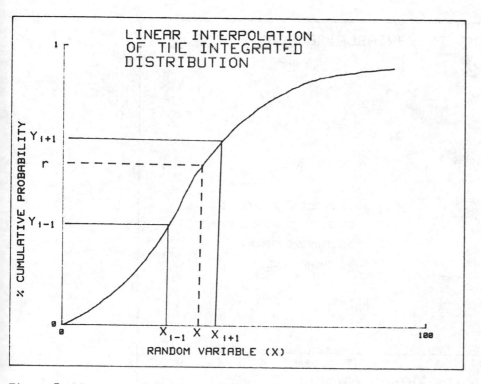

Figure 7. Linear interpolation technique.

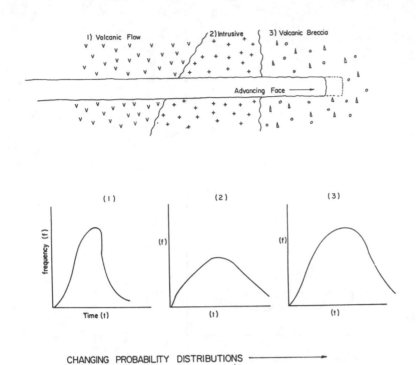

CHANGING LITHOLOGY ⟶

1) Volcanic Flow   2) Intrusive   3) Volcanic Breccia

Advancing Face ⟶

(1)   (2)   (3)

frequency (f)   (f)   (f)

Time (t)   (t)   (t)

CHANGING PROBABILITY DISTRIBUTIONS ⟶

Figure 8. Random variables as a function of geology.

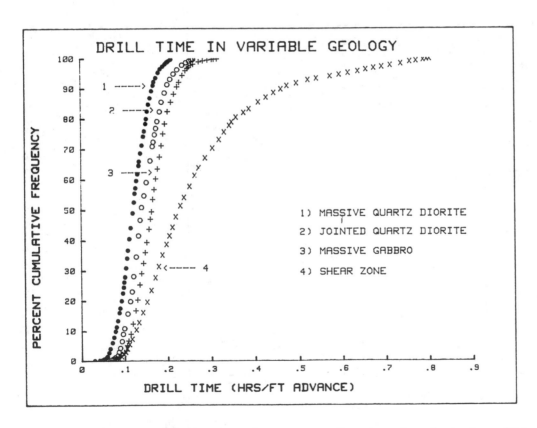

Figure 9. Changing cumulative distributions as a function of geological condition.

tical distributions modelling the actual observed field data), random number generators that produce random variables of a particular probability are used. For example, the gamma distribution can produce random variables with the following formula:

$$X = -b \log \left(\prod_{i=1}^{c} R_i \right)$$

$b = s^2/\overline{x}$ scale parameter

$c = (\overline{x}/s)^2$ shape parameter

where b, c are parameters estimated by $\overline{x}$ (sample mean) and s^2 (sample variance). R_i is a uniform random number, and X is a random variable sampled from a gamma probability distribution.

1.4 Linear interpolation

A random number, r, having been generated, linear interpolation is applied to the cumulative distribution to obtain the random variable X. The following formula describes the inverse transformation technique used to determine random variables from non-parametrically derived probability distributions (see Figure 7).

$$X = x_{i-1} + \frac{x_i - x_{i-1}}{y_i - y_{i-1}} (r - y_{i-1})$$

If a theoretical distribution (i.e. gamma or beta) is being used, step two (integration) and step four (interpolation) are not required to produce a random variable.

2. Operational model

The operational model simulates the construction of a tunnel in variable geological conditions using the conventional drill-blast-muck-support sequence of operations. The operational model forms the structural base that simulates activity times. The geology subroutine interacts with the drilling, blasting, mucking, and supporting subroutines to produce random variables. As the tunnel advances, geological conditions change, which are accounted for in the changing activity time probability distributions (see Figure 8).

As was described above, the Monte Carlo method of simulating activity times allows for uncertainties in various tunnelling operations. For example, it can be expressed as a probability distribution. The distribution may be estimated (theoretical) or obtained from actual field observations. The time to drill a round is then determined by random selection from the probability distribution.

The model currently simulates seven selected tunnelling activities. Setup, drilling, loading, smoke clearing, mucking, supporting, and downtime are treated as random variables in each simulation. Changing geological conditions are simulated by changing the cumulative distributions (construction performance curves) curves. For example, Figure 9 illustrates the adverse effect of encountering a shear zone. The probability of a time delay in a shear zone is significantly higher than in massive quartz diorite.

The drilling, blasting, mucking, and supporting subroutines simulate selected operational times for each round excavated. There are two types of activities considered during the simulation of a tunnel using the TUNSIM computer model. First, activities such as setup time and smoke clearing time, which are statistically constant and independent of geological conditions. Second, activities such as drill time and support time that have a significant variance within the same geological strata. Figure 9 illustrates the variability in drilling times between different geological strata. The figure also shows the variability of possible activity time outcomes within a geological stratum. The number of variables used in the simulation depends on the extent of the data and the degree of complexity that is desired in the project evaluation.

The predicted geology (derived from the site investigation) that lies on the tunnel line is entered into the model for each stratum that will be encountered. The input data for the operational model is mainly in the form of cumulative distribution curves (construction performance curves) when observed field data is used. If theoretical distributions are used (i.e. the gamma and/or beta distributions) then the appropriate parameters are required to define the random variable generators. The remaining input data consists of the tunnel geometry, numbers of men and equipment, materials characteristics and specifications, hours of operations, and a random seed number.

The output data generated by the operational model becomes input data for the cost model. The operational times are divided into drill, blast, muck, support, and total cycle time for each excavated round. Total weekly times for drill, blast, muck, and support are reported at the end of each week. A general statistical summary is present following the week's activity times. The last items displayed are the weekly advance, the number of rounds excavated per week, and the total advance to date. See Figures 10 and 11 for examples of the operational model's output.

```
WEEK NUMBER  13
***************
0        DRILL         BLAST         MUCK          SUPPORT        TOTAL
         ---           ---           ---           ---            ----
         0.0           0.0           2.342         0.477=         2.818
         3.743         1.882         2.955         7.194=         15.775
         6.971         1.259         2.971         6.117=         17.318
         4.817         1.267         1.150         18.153=        25.388
         3.636         2.234         4.845         9.902=         20.615
         3.833         1.982         4.696         18.018=        28.530
         6.070         2.892         1.150         11.035=        21.147
         1.300         1.226         2.290         0.490=         5.305
***************************
WEEKLY STATISTICAL REPORT.
***************************
         DRILL    BLAST    MUCK      SUPPORT     TOTAL
ACTIVITY MEANS
         4.137    2.079    3.183     11.939      21.337
ACTIVITY STANDARD DEVIATIONS
         2.298    0.864    1.401     6.831       9.047
ACTIVITY VARIANCES
         5.280    0.746    1.964     46.665      81.852
***************
WEEK NUMBER  14
***************
0        DRILL         BLAST         MUCK          SUPPORT        TOTAL
         ---           ---           ---           ---            ----
         2.261         2.151         6.138         0.0  =         10.549
         3.151         2.167         3.580         16.437=        25.336
         4.206         1.105         1.150         6.527=         12.988
         4.096         2.248         1.517         6.311=         14.172
,NEW GEOLOGICAL STRATA ENCOUNTERED
, *** SHEAR  ZONE ***
         3.165         1.574         1.263         15.761=        21.763
         3.449         1.582         1.866         25.861=        32.758
         2.926         2.584         4.346         22.729=        32.584
,NEW GEOLOGICAL STRATA ENCOUNTERED
, COPPER MINERALIZED ZONE
         2.556         0.650         6.450         0.0  =         9.656
         4.013         1.738         0.0           0.0  =         5.748
***************************
WEEKLY STATISTICAL REPORT
***************************
         DRILL    BLAST    MUCK      SUPPORT     TOTAL
ACTIVITY MEANS
         3.314    1.756    2.923     10.703      18.695
ACTIVITY STANDARD DEVIATIONS
         0.689    0.609    2.312     10.077      11.107
ACTIVITY VARIANCES
         0.475    0.370    5.346     101.552     123.358
```

Figure 10. Sample of the detailed reporting and statistics.

3. Cost model

The cost model was developed to function in tandem with the operational model to reduce the time required to prepare bids for tunnelling projects and to compare various construction scenarios. The cost model utilizes database and spreadsheet manipulation routines to systematically analyse tunnelling costs. The model consists of a large number of databases and spreadsheets that merge into a single master composite spreadsheet. At the database level, only the unit prices and corresponding units of each item are entered. The unit prices are then transferred to the intermediate spreadsheets, and quantities are assigned to each item. The total cost for each item is then determined. Costs that are in a similar category are summed into minor cost centres and subsequently into major cost centres. The minor and major cost centres are transferred to the composite spreadsheet. Indirect costs are included at this level of the model. The percentage of the total project cost that is represented by the indirect costs is then calculated. Using this percentage as a weighting factor, the indirect costs are then prorated into bid unit prices. The basic structural flow of the cost model is illustrated in Figure 12.

The input data required for the cost model are obtained from three sources. First, the activity times generated in the operational model will determine the time span of the project. All costs related to time will be affected by the operational output. Second, a database of unit prices is stored in the spreadsheet program. Unit prices are extracted and utilized as required. Finally, quantity values which are specific to the tunnelling site are entered onto the spreadsheet cost model. Cost model input is derived from three sources: operational model output, unit price database, and site specific quantities. The cost model output is presented on the composite spreadsheet as specified by the user.

ADVANCE / NUMBER OF ROUNDS PER WEEK:

WEEKLY ADVANCE	NUMBER ROUNDS	TOTAL ADVANCE	WEEK NUMBER
140.000	14	140.000	1
150.000	15	290.000	2
150.000	15	440.000	3
140.000	14	580.000	4
130.000	13	710.000	5
130.000	13	840.000	6
150.000	15	990.000	7
140.000	14	1130.000	8
140.000	14	1270.000	9
120.000	12	1390.000	10
110.000	11	1500.000	11
100.000	10	1600.000	12
80.000	8	1680.000	13
90.000	9	1770.000	14
80.000	8	1850.000	15
100.000	10	1950.000	16
110.000	11	2060.000	17
100.000	10	2160.000	18
70.000	7	2230.000	19
90.000	9	2320.000	20
80.000	8	2400.000	21
210.000	21	2610.000	22
230.000	23	2840.000	23
210.000	21	3050.000	24
210.000	21	3260.000	25
200.000	20	3460.000	26
140.000	14	3600.000	27
110.000	11	3710.000	28
130.000	13	3840.000	29
130.000	13	3970.000	30
140.000	14	4110.000	31

Figure 11. Sample of the weekly summary report of tunnel production.

STRUCTURAL FLOW OF THE
TUNNEL COST PROGRAM

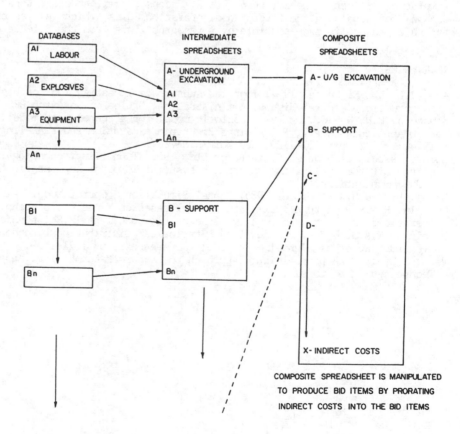

Figure 12. Structural flow of the cost model.

299

SENSITIVITY ANALYSES

The computer program that has been developed models the DRILL-BLAST tunnelling method using the FORTRAN 77 computer language and a spreadsheet software package. The model assists in evaluating project costs and analysing operating parameters. By varying the input data, sensitivity analyses can be performed to optimize the tunnel alignment, the construction method, and to determine the most efficient combination of mining equipment. The alignment of the tunnel is dependent on the geological and the geometrical constraints of the site. By balancing the cost and the time-to-completion with the given tunnel layout requirements, the most favourable geological strata can be excavated. The construction method and mining equipment combination are dependent on rock strength and structure, the length of the tunnel, the variability of geological conditions, and the availability of mining equipment and labour. The computer allows sensitivity analyses to be performed cost efficiently while minimizing project risk.

CONCLUSION

With the incredible advantages of subsurface space utilization, the methods of optimizing the feasibility, design, and construction of tunnelling projects must be improved. Recent technological advances in tunnel excavation methods and high variability in geological conditions have rendered the task of project cost estimation increasingly complex. This is largely a result of uncertainties that are involved in predicting geological conditions along the proposed tunnel line and determining the optimal combination of mining equipment and construction method. The probability of encountering adverse ground conditions cannot always be avoided, but if the adverse ground conditions are anticipated, the appropriate construction method can be employed. These variables must be accomodated in the tunnel cost analysis. This evaluation process becomes lengthy when a sensitivity study is performed to determine the most suitable construction method and tunnel line. The model can also be used as a method of estimating the consequences of specific occurrences and their repercussions on the project's progress. Computer modelling is therefore becoming more attractive as a cost effective tool for estimation of the costs and scheduling of tunnelling projects.

ACKNOWLEDGEMENTS

The authors would like to thank R. Guloglu for the initial research work on this continuing tunnel simulation project. Special thanks to G. Jacobs of Emil Anderson Construction Co. Ltd. (Hope, BC, Canada) for supplying the first tunnel operations data, which is presented in this paper. Funding for this project is made possible by a research grant from the Science Council of British Columbia.

REFERENCES

Ash, J.L. & B.E. Russel 1974. Improved subsurface investigation for highway tunnel design and construction. Federal Highway Administration, Office of Research HRS-11. Washington, D.C.

Einstein, H.H. & S.G.Vick 1974. Geologic model for a tunnel cost model. In Pattison, H.C. & E. D'Appolonia (eds.), Proceeuings 2nd rapid excavation and tunneling conference, San Francisco, 1974, Vol. 2, p. 1701-1720. New York:AIME.

Golder Associates & James F. MacLaren Ltd. 1976. Tunnelling technology: an appraisal of the state of the art for application to transit systems. Toronto: The Ontario Ministry of Transportation and Communications.

Grant, D.R. & H.D.S. Miller 1985. Tunnel Simulation Computer Model. IMM Tunnelling '85: p. 313-16.

Guloglu, R. 1979. A tunnel simulation model based on Monte Carlo techniques for analysing tunnelling operations and costs. University of Newcastle upon Tyne: M.Sc. thesis.

Hastings, N.A.J. & J.B. Peacock 1975. Statistical Distributions. London: Butterworth & Co. Ltd.

Moavenzadeh F. et al. Tunnel cost model. Reference 2, p. 1721-1739.

Vick, S.G. 1973. A probabilistic approach to hard rock tunnelling. Massachusetts Institute of Technology: M.S. thesis.

Application of comprehensive geological exploration in tunneling prospecting

Etude géologique complète pour un projet de tunnel

He Zhenning, *Third Survey & Design Institute, Ministry of Railways, Tianjin, China*

ABSTRACT: In the work of line selection for a railway tunnel and geological pioneering of the project, good results have been obtained by doing comprehensive geological prospecting adopting the method of physical exploration coupled with the methods of geological surveying and boring extensively. Comprehensive geological exploration does not merely involve the simple combination of the use of various methods of prospecting, but special exphasis is given to the comparison of various exploration data and their comprehensive analysis. Therefore, the method is effective in doing geological exploration of tunnel projects where conditions are suitable and more complicated. It also facilitates the physical methods of exploration to be used more extensively in geological prospecting of railway engineering projects.

RESUME: La grande application de la méthode de prospection physique, combinée avec autres méthodes de prospection telles que enquêtes géologiques et forages, à la prospection synthétique de géologie, obtient un bon résultat pour choisir des voies de tunnels ferroviaires et la prospection géologique de leurs travaux. La méthode synthétique de prospection géologique est non seulement une simple utilisation combinatoire de plusieures méthodes de prospection, mais aussi, et en particulier, la mise en accent sur la comparaison et l'analyse synthétique de différentes informations de prospection.

In route selection for railway tunnel and engineering geological prospecting, the extensive use of physical exploration in conjunction with the methods of geological survey, mapping and drilling in carrying out a comprehensive geological exploration can often achieve good results.

1 LIVING EXAMPLES

Case I: Tunnel "H", situated in a low-lying mountaneous region consisting of granite, has a full length of about 3 kilometer. A one hundred meter wide gully develops in the middle section of the tunnel whose buried depth is 40 -45 m. It is inferred from geological survey that the fault develops along the gully, with more groundwater accumulated. Without careful handling, not only is it possible to cause subsidence during construction but also possible to lead to such serious environmental changes as withering up of fruit trees and drying up of wells for civil use in consequence of lowered groundwater level due to huge quantities of water leakage. Physical exploration is adopted since geological survey is made difficult by coverage of a thin earth layer over the surface and intensive weathering of the bedrock's upper most layer. Drill holes are laid out on the basis of physical exploration and physical logging and hydrogeological tests are made in them. By using various explorationary methods, location, occurrence and hydrogeological parameters of the main fault are identified, providing relatively comprehensive data for tunnel construction and designing, and its grouting in advance from above ground via drill holes followed by shutting off water. (Fig. 1)

Case II: Tunnel "B", situated at the fringe of a coal-series basin in siliceous limestone area, measures 5 km in overall length. (Fig. 2)
The engineering geological conditions of the coal-series basin are quite poor, not suitable for building tunnels. Its stratum consists mainly of shale with a spongy texture, accompanied by ground water development and is susceptible to hazards of expanded incompetent

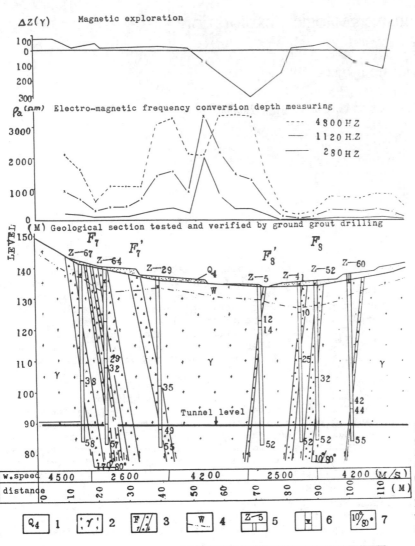

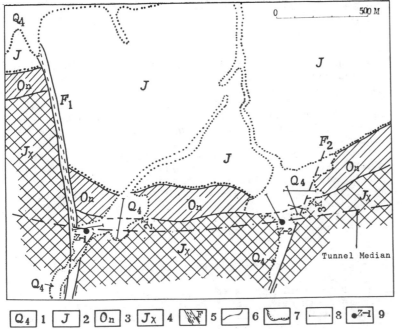

Fig.1. Longitudinal profile of comprehensive exploration of fault gully in tunnel "H". 1. Quaternary sandy strata; 2. Granite; 3. Fault fractured zone; 4. Weathered zone; 5. Drill hole; 6. Underground water level; 7. Inclination and inclined angle.

Fig. 2. Geological scheme of tunnel "B". 1. Loess; 2. Coal-measures stratum; 3. Shale; 4. Siliceous limestone; 5. Fault and serial number; 6. Rock boundary; 7. Non-conformity boundary; 8. Profile of electric detection and serial number; 9. Drill hole and serial number.

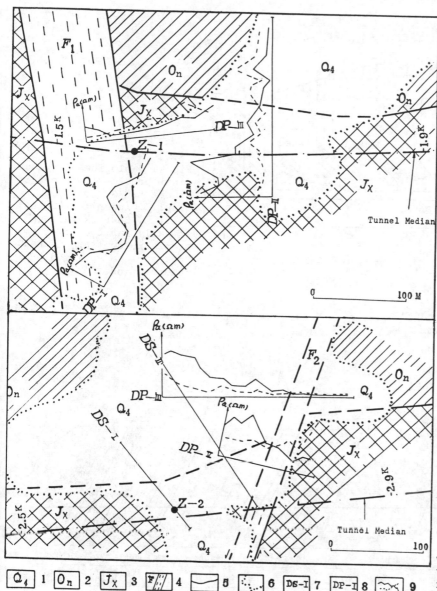

Fig.3. Plan of comprehensive exploration of tunnel "B". 1. Loess; 2. Shale; 3. Siliceous limestone; 4. Fault fractured zone; 5. Rock boundary; 6. Earth and stone boundary; 7. Depth measuring line of electric surveying; 8. Combined profile of electro metry; 9. Curves of combined profile (full line stands for electrode A; dotted line, electrode B; Intersection, rock interface or location of fault.)

rock and released coal-seam gas. For this reason, a site for driving a curved tunnel is chosen inside a siliceous limestone stratum beyond the fringe of the coal-series basin. However, in the section where two larger gullies traverse points at 1.5 and 2.5 km in Fig. 2 , there might be a developing fault stretching across the tunnel. Since the overlying earth layer prevents finding out whether the lithological interface be staggered off by the fault, physical exploration is applied to survey the location of interfacing (Fig. 3). The original plan was to lay out the drilling profile. Now, as physical exploration is properly made use of in conjunction with geological survey, a modified is implemented which involves verification by only one single drill hole, thus reducing exploratory expenses and cutting down exploratory cycle.

Case III: Tunnel "Z" has an overall length of about 2 km, with a thick layer of loess overlying its floor. The design level of the tunnel is in the neighborhood of the contact plane of the bottom of loess layer and the bedrock. By applying d.c. electrical prospecting and seismic exploration, as well as drilling a hole aimed at the position of the gully in lower terrain, the buried depth and relief of the bedrock are found out and the design level of tunnel is adjusted correspondingly exploratory data prove to be accurate, as verified by construction work. If, based only on drill data, the bedrock plane were inferred to be horizontal, with the low-lying section overlooked, it might turn out that the top of the tunnel edge in the loess, making troubles for tunnel design and construction (Fig. 4).

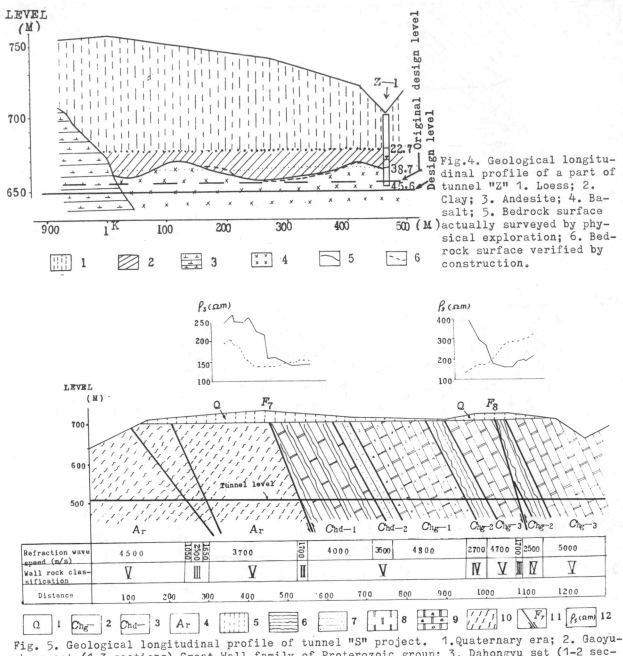

Fig.4. Geological longitudinal profile of a part of tunnel "Z" 1. Loess; 2. Clay; 3. Andesite; 4. Basalt; 5. Bedrock surface actually surveyed by physical exploration; 6. Bedrock surface verified by construction.

Fig. 5. Geological longitudinal profile of tunnel "S" project. 1.Quaternary era; 2. Gaoyuzhuan set (1-3 sections) Great Wall family of Proterozoic group; 3. Dahongyu set (1-2 sections), Great Wall family of Proterozoic group; 4. Gneiss of Proterozoic group; 5. Loess; 6. Shale; 7. Sandstone; 8. Dolomite; 9. Siliceous dolomite; 10. Gneiss; 11. Fault; 12. Combined profile with d.c. method (ρ_s - apparent resistivity).

Case IV: Tunnel "S" measures altogether 3 km long with a strike perpendicular to the fault strike, cutting across gnesis of achaean group, siliceous limestone of proterozoic group, sandstone, shale and intrusive dyke of mesozoic era. Due to the lithological variety, it is necessary to find out lithological character, width and occurrence for each stratum. Still, there remains an one km long section in the middle of the tunnel, which becomes a blank in the geological survey of bedrock. D.C. electrical survey and seismic exploration are therefore adopted to find out, on the basis of the geological survey results with the adjacent sections, the main lithological interface and the position of fault, and, at the same time, to determine the velocity of elastic waves through each stratum so as to provide basis for identification of geological characteristics and classification of various dyke sections (Fig. 5).

2 PROBLEMS IN ACCOMPLISHING COMPREHENSIVE EXPLORATION

The basic way of comprehensive exploration is choosing an appropriate physical prospecting method, on the basis of geological mapping, with geological problems and multiplex conditions requiring investigation in view, followed by laying out drill holes at critical points in accordance with engineering requirements. Apart from core drilling, physical logging and hydrogeological tests are carried out in drill holes. Finally, a comprehensive analysis of various data of prospecting is made and a report of achievements produced. The following three points in connection with proper accomplishment of comprehensive exploration are discussed.

2.1 Choice of physical prospecting method

What physical prospecting method is to be chosen by a work site depends on the geological conditions, engineering requirements, and functions of a particular method in itself. In practice, it is usually decided on through analysis of the data of physical prospecting obtained in the work site to find out whether are "multi-explanatory" or "uni-explanatory".

"Uni-explanatory" implies that the geological problem of the work site is relatively unitary, for which an exact geological explanation may be obtained by using one certain physical prospecting method. Hence one method is sufficient for the purpose. For instance, the geological problem of tunnel "B" that needs to be solved is to prospect the interface of siliceous limestone and shale. Their resistivities differ greatly from each other: the resistivity of limestone is higher ($\rho = 1000-2000 \, \Omega M$), whereas that of shale is low ($\rho = 40-100 \, \Omega M$). As far as this particular work site is concerned, by using electrical prospecting method to distinguish between these two lithological characters in conjunction with geological surveying a geological explanation can be arrived at to clarify whether it is the interface of limestone and shale or it is the fault, hence "uni-explanatory" (Fig. 3).

"Multi-explanatory" implies that the geological problem, with many factors interfering, is more complicated, for which several explanations may be arrived at by using only one physical prospecting method. On account of this, many a physical prospecting method is used and the tenability of some explanations are ruled out as a result of comprehensive analysis, thus leading to an exact, unitary geological conclusion.

For instance, the fault gully of tunnel "H" has a well-developed fracture structure. As both upper and lying walls consist of the same granite, no apparent difference in physical property is shown. In addition, its topographical condition takes the form of a "U" valley, which shows bigger relief in shape of a steep slope at each side. This condition affects to some extent the precision of explaining the data of physical prospecting. Under such circumstances, it would be difficult for any physical prospecting method alone to yield an exact explanation. Therefore, the following four physical prospecring methods are used:

Seismic prospecting —— the position of the fault is inferred from the measurements of wave velocity variation and dynamic characteristics of seismic events. The fractured rock formation at the position of the fault reflects not only low wave velocity (Fig. 1), but also variations in amplitude, phase and cycle of seismic waves.

Electromagnetic frequency-conversion depth-sounding —— a method recently introduced in engineering geological exploration. This method is least interfered by the topography and responds more definitely to geological interfaces (Fig. 1).

In applying the above-mentioned two physical prospecting methods, particular emphasis is laid on determination of the actual position of main fault.

As the intact granite in tunnel "H" has weak magnetism which attenuates and even becomes negative at the position of the fault fractured zone, the magnetic method is also used in prospecting serving as a basis of marking the limits of fault gully (Fig. 1).

Direct current exploration —— the resistivity of granite itself should be higher ($\rho > 1000 \, \Omega M$), but it decreases obviously ($\rho = 200-500 \, \Omega M$) due to the earth and water contents in fault fractured zones. Therefore, it is still necessary to use the direct current method as one of the basis for determining the location and the water-bearing feature of the fault fractured zone.

To sum up the data of four physical prospecting methods —— it is known that low resistivity, low wave velocity and negative or weak magnetism and the physical properties brought forth by the fault gully in tunnel "H" are the characteristics of fractured and ground water development. At the position of the main fault, mutations and abnomalities occur in all four physical prospecting curves. As for the complicated geological conditions like these, use of the comprehensive physical prospecting method can help bring out the more comprehensive geological explanation and is of great significance in guiding the layout of ground survey drilling and grout-injection drilling (Fig. 1).

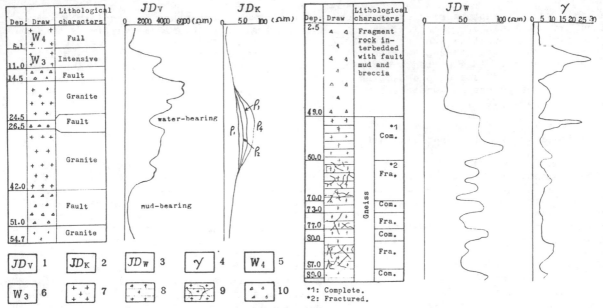

Fig. 6. Comprehensive column diagram of tunnel drill hole physical logging. 1. Resistivity logging curve; 2. Well liquid resistivity logging curve (thrown salt diffusion method); 3. Micro electric machine series logging curve; 4. Natural radioactivity logging curve; 5. Full weathered zone; 6. Intensive weathered zone; 7. Granite; 8. Gneiss; 9. Joint development and rock stratum fractured section; 10. Fault fractured zone or fragment rock.

2.2 The necessity of developing physical logging in tunnel drill holes

The physical logging is carried out on the drill holes of the above-mentioned tunnels "H", "B" and "Z", and results have been achieved in the following three aspects (Fig. 6).
  a) The accuracy of the drilling depth with respect to the lithological characters and the stratification of fractured zone is examined and modified.
  b) The location of water-bearing layer in drill hole is determined by using the thrown salt diffusion method (the curve JDK of the tunnel "H" in Fig. 6). The flow direction of ground water and coefficient of seepage of tunnel "H" are determined by the joint loggings of three drill holes arranged in ".·." form.
  c) The seismic logging method is carried out in part of drill holes and fissure coefficient is determined as :

$$\S = \left(\frac{V_p'}{V_p}\right)^2$$

where V_p' is the mean value of longitudinal wave velocity of the determined rock body in drill hole seismic logging (km/sec), and V_p is the longitudinal wave velocity of the rock core sample in the typical section of logging (km/sec).
  Before the ground grout-injection for stopping water is practised in tunnel "H", the mean value of fissure is 0.683 and the mean longitudinal wave velocity is 4.48 km/sec; while after the ground grout-injection, they become 0.810 and 4.90 km/sec respectively. This shows that after the grouting the integrality of the rock bedding is getting better and the velocity of transmitting seismic wave grows faster.

2.3 Comprehensive analysis of exploration data

It would be the key link of doing well the comprehensive exploration to have all the data, obtained in geological surveying and mapping, drilling and physical prospecting, collected and synthetically analyzed. In the practical work, the explanation of data of physical prospecting is often divorced from analyses of other exploration data, this being one of the important reasons of why the wide application of physical prospecting methods is restricted in engineering geological survey.
  To strengthen the comprehensive analysis of physical prospecting may be started with the two   aspects as follows:
  a) The analysis of strata anf structure should be integrated into the explanation of physical prospecting data.

306

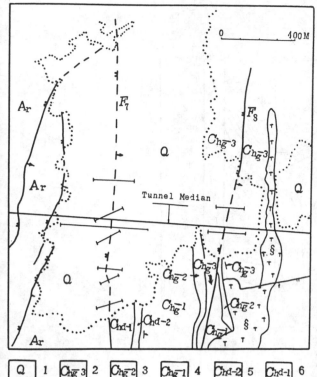

Fig.7. Comprehensive exploration plan of tunnel "S". 1. Loess; 2. Dolomite; 3. Shale; 4. Siliceous limestone; 5. Sandstone; 6. Dolomite and shale interbedding; 7. Syenite wall; 8. Gneiss; 9. Fault; 10. Earth and stone boundary line; 11. Generation pattern of rock stratum; 12. Location of measuring line of physical exploration.

For example, the location and occurrence of the faults F7 and F8 of tunnel "S" should be thoroughly investigated (Fig. 5). Faults are usually characterized by the low resistivity and low seismic wave velocity. Apart from faults, so are the shales. Sandstones and the location of dikes will also have changes in their seismic wave velocity and resistivity. Having this in mind, we can determine many sections with "low velocity", "low resistivity" and change points. It is hard to determine if there are faults or not if we are only to make analysis on the physical prospecting data and curves alone.

Due to the outcrop of bed rock around areas covered by loess, the physical prospecting work may be started at the outcrop of faults and the trace exploration can be carried out without large deviations of location (Fig. 7). But it is not enough in so doing, because the lithological characters on the different positions of faults may vary, thus bringing about the difficulty to explain the physical exploration data. Such being the case, to make the most of the conversion data, such as layer sequence, thickness and occurrence of tunnel sections and to accord with the geological structural pattern and combination relation of the area, it is possible to infer roughly the lithological characters and combination relation that would appear under the loess layer of the tunnel site (Chg 1-3, Chd 1-2 and Ar in Fig. 5, 8). This will be of very important role in putting in order all the interrelationships between physical prospecting curves and data and in bringing out the reasonable geological explanations.

In addition, tunnel "S", through the seismic exploration, will provide data of the elastic wave velocity for the classifying and segmenting of the wall rock. The seismic wave velocity actually measured are not the wave velocity of location of tunnel level directly measured, but that of the location of refraction interfaces under the weathered zones of the bed rock. Next, the section value of the wave velocity actually measured will usually be mixed and confused under the influences of various accidents or interference factors. The reasonable segmentation of the data of wave velocity and the inferring of the segmentation of wave velocity of the location   of refraction plane to the location of tunnel level are solved with the integration with the layer sequence, thickness, occurrence and structural relations (Fig. 5).

b) The comprehensive analysis of physical prospecting data and drill data

To lay out the drill holes on the basis of physical prospecting may reduce the blindness drilling. In contrast, if drilling data are available for use as given parameters, can the explanation of physical prospecting data have a clear and unamleiguous basis. For example, in the case of the above-mentioned tunnel "Z", the relief of the bed rock plane measured by electric and seismic exploration is just on the basis of the stratification data of drill holes in gullies and side-hole parameters. (Fig. 4).

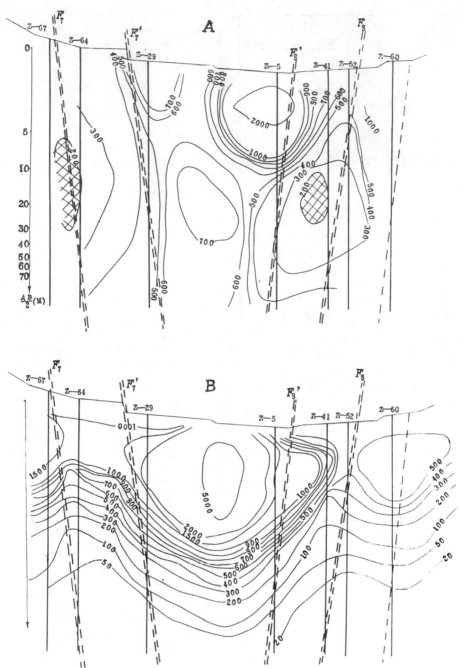

Fig. 8. Apparent resistivity isopleths diagram of fault gully of
tunnel "H" (Same location as in Fig. 1)
A. Apparent resistivity isopleths with d.c. method;
B. Apparent resistivity isopleths of electro-magnetic frequency
   conversion depth measuring.

The comprehensive analysis of drilling data and physical prospecting data may often bring
forth more audio-visual and abundant data. For instance, in the fault gully of the above-
mentioned tunnel "H", there occurs low-resistence closed circle ($\rho_s < 400 \Omega$ M equivalent
line in Fig. 8) on the equivalent line profile of apparent resistivity transending the gully
by d.c. exploration method. It has been found out by drilling, well logging and hydrogeological tests that its low resistant closed equivalent line is located on the fractured
zone of faults and near the side of hanging is resulted by the large earth content of the
fractured zone of faults, the development of fissures near the hanging wall and rich content of ground water; while the ground water is less and resistivity is high near the foot

wall. Thus, from the shape of equivalent line of apparent resistivity, we may have a more complete understanding about the fracture development of the whole fault gully and the distribution of the ground water. The shape will be clearer when it is supplemented with the equivalent line diagram of depth metering by electro-magnetic frequency conversion.

In short, the comprehensive exploration is not only to apply jointly various kinds of prospecting methods, but, as it is especially stressed, to collect and make comparative analysis of all the exploration data and to have these carried out through the whole process so as to make the most use of the physical prospecting and its data.

3 THE TUNNEL ENGINEERING GEOLOGICAL PROBLEMS SUITABLE TO BE SOLVED BY THE COMPREHENSIVE GEOLOGICAL EXPLORATION

1.) Tunnels with complicated geological structure, especially those on the section of an area where the outcrop on surface of the bed rock is less, the development of geological surveying and mapping is inconvenient and the ground water is much developed.

2.) Tunnels with thick weathered zones and large depth relief changes or developed tafoni;

3.) Tunnels with ground water which is of clearly seasonable changes. These changes can be investigated if conditions are provided for carrying out observations and analyses at a relatively long period of time through the exploration with electric method.

4.) Tunnels which provide data of elastic wave velocity are necessary for classification of tunnel wall rock.

With the above-mentioned examples and analyses, it is believed that the comprehensive exploration is an effective method for carrying out the engineering geological survey of tunnels with appropriate but rather complicated conditions, thereby facilitating the wide application of the physical prospecting method.

Monitoring of a tunnel face through the dynamic responses of the rock mass observed from the ground surface

Surveillance de la foration d'un tunnel à travers le comportement dynamique du massif rocheux observé depuis la surface

J.Moura Esteves, *Site Exploration Division, LNEC, Lisbon, Portugal*
J.Vieira de Lemos, *Dam Department, LNEC, Lisbon, Portugal*
J.D.Fonseca, *Physics Department, IST and LNEC, Lisbon, Portugal*

ABSTRACT: During the excavation of a tunnel by means of blasting, under the right bank abutment of the Castelo do Bode dam (Portugal), several sets of observation were made on the vibration levels that reached the buildings of the overlying residential area.

In addition to the purpose of controlling these vibration levels both in the buildings and in the dam itself, the study tried to evaluate the dynamic response of the rock mass through the same monitoring means, as singularities were crossed, particularly faults.

On basis of field results, an attempt was made to develop automatic computation techniques that simulated situations of that type. Results already achieved and herein presented show that significant variation occur in the dynamic response of rock masses when singularities exist in the face of tunnels.

It seems thus possible to establish a monitoring technique and methodology to detect abnormalities in the front of tunnels.

RÉSUMÉ: Pendant le percement d'un tunnel à l'aide d'explosifs sous l'appui de la rive droite du barrage de Castelo do Bode (Portugal), on réalisa plusieurs observations des niveaux de vibration qui ont atteint les bâtiments du quartier résidentiel sus-jacent.

Outre la préocupation de contrôler ces niveau, de vibration soit dans les bâtiments, soit dans le barrage lui-même, on tâcha d'évaluer à travers ce même disposi-tif d'auscultation quelle était la réponse dynamique du massif lors de la traversée de singularités, particulièrement de failles.

A partir de résultats sur le terrain, on cherche à développer des techniques de calcul automatique qui simulent des situations de ce type. Les résultats déjà obtenus et présentés dans ce travail montrent qu'il y a des variations significatives de la réponse dynamique des massifs lorsque des singularités existent au front des tunnels.

Il semble donc possible d'établir une technique et méthodologie d'auscultation du massif de façon à détecter d'éventuelles anomalies au front d'un tunnel.

1 INTRODUCTION

During the excavation of a tunnel by means of blasting, under the right bank abutment of the Castelo do Bode dam (Portugal), several measurements of vibrations were made in buildings lined up with the tunnel, wich were intended to control vibrations as prescribed in the Portuguese standard NP-2074 (1983) "Avaliação da influência emconstruções de vibrações provocadas por explosões ou solicitações si-miraes" (Assessment of the influence over buildings of vibrations produced by blasting or similar actions).

In this observations programme several abnormalities were detected in the values measured, wich were attributed to changes in the geologic structure of the rock mass, leading to different modes of energy propagation. This situation seemed more relevant when the tunnel face was close to the shear zones of significant thickness.

This work was intended for a summary study of this kind of situations, with a view to possible further development to set forth a technique and a methodology for the detection of abnormalities at the tunnel face, on basis of surface moni -

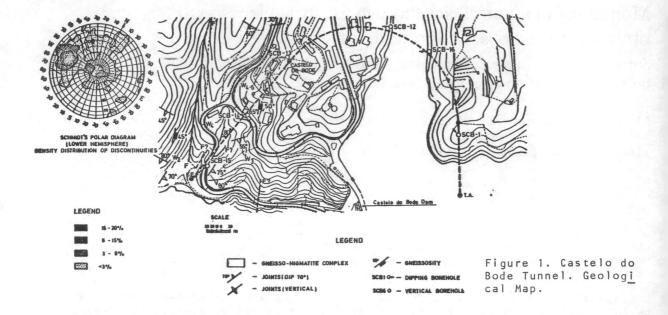

SCHMIDT'S POLAR DIAGRAM
(LOWER HEMISPHERE)
DENSITY DISTRIBUTION OF DISCONTINUITIES

LEGEND

■ 15 - 20%
■ 8 - 15%
▨ 3 - 8%
▒ <3%

SCALE
20 10 0 30
m

LEGEND

☐ – GNEISSO-MIGMATITE COMPLEX   ⚡ – GNEISSOSITY
70° ↗ – JOINTS(DIP 70°)   SCB1○– DIPPING BOREHOLE
✗ – JOINTS(VERTICAL)   SCB6○ – VERTICAL BOREHOLE

Figure 1. Castelo do Bode Tunnel. Geologi_cal Map.

toring of dynamics phenomena caused by tunnel excavation with blasting.

With this aim we established a plane model for a real situation found in that tunnel, and carried out the mathematical processing relative to different situa - tions with a generalized distinct element method (Lemos, J.V. et al.).

The tunnel was excavated in a very tectonized pre-Cambrim metamorphic complex consisting of gneisses and migmatites very similar to those which form the Caste- lo do Bode dam foundation. Some pegmatite, aptite and quartz veins with a maximum thickness of 10 cm cross the entire rock mass (Oliveira et al.).

The rock mass outcrops generally highly weathered (W_4) and some thin colluvial deposits occur at several locations.

The gneisse formations were subjected to high tectonic stresses throughout their geologic history. Nevertheless the attitude of the gneissose fabric is very much the same along the tunnel, striking around N20°W to N5°E and dipping 70° to 80° in E direction. Some tectonic features like faults can be seen.

A major fault could be detected 30 m eastside of the downstream portal of the tunnel striking N30°W, subvertical.

2 CARACTERISTICS OF THE MODEL

The model used for this study is schematically shown in Fig. 2. It is a two- -dimensional model in plane state of deformation with the following elastic cha - racteristics:

$$E = 50\ 000\ MP_a$$
$$\nu = 0,25$$
$$c_p = 4758\ m/sec$$
$$c_s = 2747\ m/sec$$

$$E' = \begin{cases} E'_1 = E/10 \\ \\ E'_2 = E/20 \end{cases}$$

The action was simulated by a pressure impulse with the form $P(t) = \frac{1}{2} P_o (1-\cos 2\pi t/T)$, $t \leq T$, with $T = 0,02$ sec. (Figure 3).

The fault (or shear zone) was simulated by a joint, and two cases were consi - dered:

- stiff joint, i.e. elastic homogeneous medium
- deformable joint corresponding to materials with $E'_1 = E/10$ and $E'_2 = E/20$.

The thickness of this joint (value d) is equal to the diameter d of the tunnel.

Based on this model we studied the hyppoteses corresponding to the following arrangements, through the determination of the vibration components Y (vertical) and X (horizontal) in stations 1 to 6 (Fig. 2):

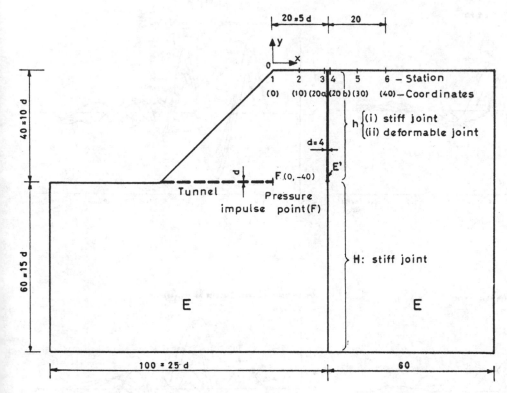

Figure 2. The model used

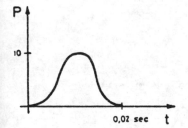

Figure 3. Impulse form

Pressure point	Type of joint	Stiff joint	Deformable joint	Very defor - mable joint
A (0,-40)		A/SJ	A/DJ	-
B (10,-40)		B/SJ	B/DJ	B/VDJ
C (18,-40)		-	C/DJ	-

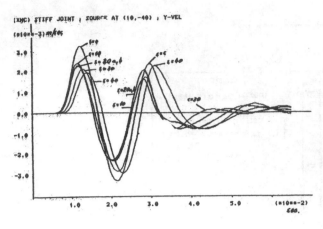

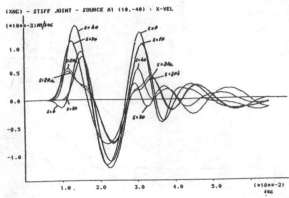

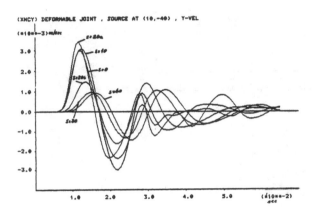

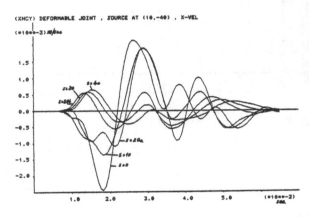

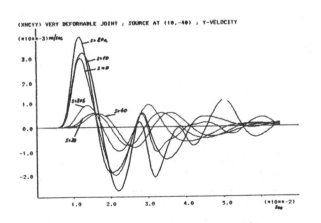

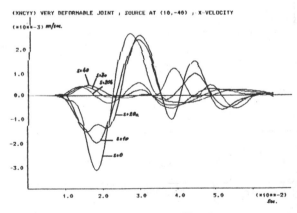

Figure 4. Vibration amplitudes Y and X

AMPLITUDE SPECTRUM
(MAXIMUM ENTROPY)

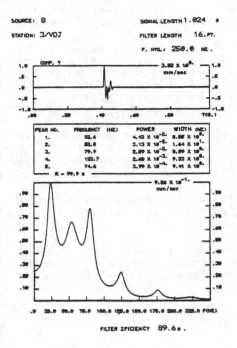

AMPLITUDE SPECTRUM
(MAXIMUM ENTROPY)

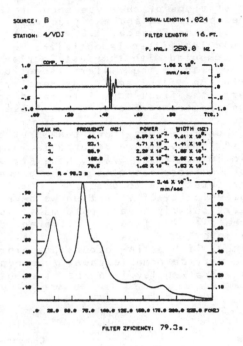

AMPLITUDE SPECTRUM
(MAXIMUM ENTROPY)

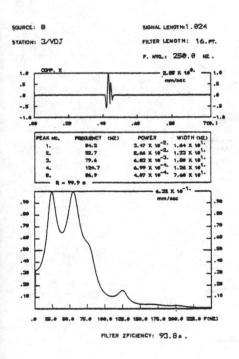

AMPLITUDE SPECTRUM
(MAXIMUM ENTROPY)

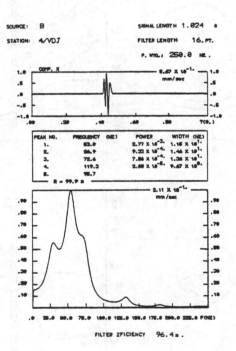

Figure 5. Distribution of the amplitude density with the frequency

3 RESULTS OBTAINED

Once the distributions of the vibration amplitudes Y and X have been obtained in the different stations of the type shown in Figure 4 the corresponding amplitude spectra were determined and analysed following the maximum entropy method (Robinson, E. et al. and Fonseca, J.D. et al.)

An error prediction filter is optimized for each signal, and the power spectrum of the filter is computed with the fast Fourier transform; the spectrum is then obtained as the inverse of the filter's spectrum. The graphycal display (Figure 5) shows the distribution of the amplitude density with the frequency and in the upper part of the figure, the frequency, integrated power and band-width of the five main peaks of the spectrum are also shown.These peaks are ordered by decreasing integrated power.

As could be expected it was found that pratically no significant abnormalities exist with a stiff joint. With a deformable joint ($E'_1 = E/10$) and chiefly with a very deformable joint ($E'_2 = E/20$), however, strong abnormalities or differences in the vibration mode were detected at the ground surface before the joint(station 1 to 3) and after (station 4 to 6).

As regards the velocity of vibration we have found, for instance, that for a deformable and particularly a very deformable joint (VDJ) the observation, from the surface, reveals a significantly different dynamic behaviour of the two blocks separated by the joint.

As Figure 6 show, if blasting occurs at point B, about 2.5 d from the joint, the presence of a very deformable shear zone near the excavation face produces the raise of the vibration level in the first block and lowering with strong con-

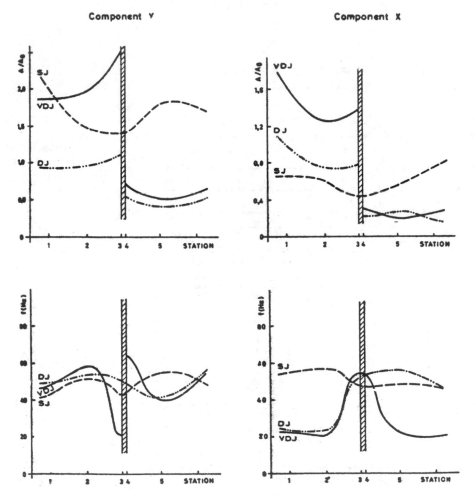

Figure 6. Level of vibration and frequency in two blocks

316

trast of levels in the second block (thus beyond the joint), and this for both components Y and X of the vibration velocity.

The Ao Is the value of the vibration amplitude at station 1 (s=0) for the impulse action at point A and for deformable joint case.

With reference to the values of frequency of the signals, there are strong contrasts and even discontinuities when very deformable shear zones are present. The lower Fig. 6 shows the trend of the frequency values that correspond to the main peak taken from diagrams identical to those presented in Fig. 5.

Given the large number of variables involved, the programmes mentioned in this work are to be developed with a view to achieving a more comprehensive analysis of the dynamic behaviour of rock masses so that even indirectly it may be possible to detect important shear zones at ahead the tunnel front face.

4 CONCLUSIONS

Based on the ground surface control of vibrations produce by blasting for rock excavation of tunnels, it will be possible to detect low-strength zones at the face of works.

A more thorough analysis of the models and techniques presented in this paper may contribute to setting down a methodology of external monitoring of the pre - sence of shear zones in the front of tunnels.

REFERENCES

Oliveira, R., Carlos Costa and José David - 1983 - Engineering Geological Studies and Design of Castelo do Bode Tunnel. Int. Symp. on Eng. Geol. and Underg. Constr. LNEC. Lisbon, Portugal.

Lemos, J.V., Roger D. Hart and Peter A. Cundall, 1985 - A Generalized distinct element program for modelling jointed rock mass. Int. Symp. on Fundamentals of Rock Joints. Bjorklinden, Sweden, September.

Robinson, E. and S. Treitel. 1980 - Geophysical Signal Analysis. Prentice-Hall, Inc. Englewood Cliffs. Usa.

Fonseca, J.D. and L.F. Rodrigues - Geological Structure Determination with Shallow -Depth Seismic Prospecting (in print.) 8th European Conference on Earthquake Engineering. 1986, September, LNEC, Lisbon, Portugal.

Development of the techniques of measurement and interpretation of the state of stress in rock masses
Application to Castelo do Bode tunnel
Techniques de mesure et d'interprétation des contraintes dans les massifs rocheux
Application au tunnel de Castelo do Bode

L.Ribeiro e Sousa, C.Souza Martins & L.Nolasco Lamas, *Laboratório Nacional de Engenharia Civil (LNEC), Lisbon, Portugal*

ABSTRACT: The actions resulting from the relief of the state of stress set up in rock masses are essentially related to the evaluation of the rock mass state of stress in situ. This paper presents a methodology to assess the state of stress around underground works, and its application to the Castelo do Bode tunnel by means of a tridimensional model by boundary elements.

RÉSUMÉ: Les sollicitations découlant de la libération de l'état de contrainte installé dans les massifs rocheux sont essentiellement liées à l'évaluation de l'état de contrainte in situ installé dans le massif. Dans cette communication on présente une méthodologie pour l'évaluation de l'état de contrainte autour d'ouvrages souterrains et son application au tunnel de Castelo do Bode, au moyen d'un modèle tridimensionnel par éléments de frontière.

1 INTRODUCTION

In the analysis of underground structures in rock masses it is necessary to predict actions on these structures, particularly those regarding initial stress relief along the boundaries created by the excavations. These actions, which, in the rock mass, bring about a redistribution of the stress field and a deformational movement towards the openings, depend on the initial state of stress prior to the construction of the work and also on the adopted construction sequence.

Accurate forecasts of the state of stress must be made by means of in situ measurements. Among the various methods used for the purpose, emphasis is laid on those based on stress relief. LNEC has been developing its own test methods in this domain, particularly the SFJ (small flat jack) method and the STT (stress tensor tube) method, both based on stress relief, the first by the opening of slots, the second by overcoring in the zone of measurements.

The tests, intended for determining the state of stress, must be carried out by phases, accompanying the design, the construction and the operation of the works. A special test methodology was established, adequate to the different phases, so as to obtain the state of stress inside the rock mass and on the excavation surfaces, using test methods and numerical models. Finally, the proposed methodology is applied to Castelo do Bode tunnel, a part of the scheme for conveying water to the Lisbon area (Portugal), using a tridimensional model by boundary elements.

2 STATE OF STRESS IN ROCK MASSES

2.1 Preliminary considerations

Due to some factors, among which the weight of overlying materials and the forces resulting from influences associated with the geological history of the formations, initial stresses are present inside the rock masses before excavation for underground works. Some other influences contribute to the setting up of stresses in situ which usually are not considered in rock mass studies, namely thermal actions due to heat from depths or to hot waters, actions by erosive agents, stresses of microscopic nature due to non-heterogeneity of the rock material, and besides factors that affect the direction and magnitude of stresses, which are related to the topography of ground surface, non-homogeneity of the rock mass and geologic structure (Goodman 1980).

Thus in order to analyse the behaviour of underground structures one has to know the distribution of initial stresses in the rock mass under consideration, and define stresses existing in the domain and along the corresponding boundaries. Stress conditions in the excavated boundaries are sometimes simulated by distributed forces, equal and with opposite signs as to the stresses prior to the construction. From the structural point of view, excavation will thus originate a defor mational movement towards the opening created, which will govern stresses in the supports, quantified by the indeterminate nature of the rock mass-support interaction, and which will provoke a complete change in the state of stress in a zone of the rock mass around the opening. This change will particularly depend on the mechanical characteristics of the rock mass and tunnel supports, on the way the latter are placed, and also on the nature of the state of stress in situ, its magnitude and principal directions. As regards underground structures, knowing the state of stress is also a basic requirement in the conception phase, as happens for instance in choosing the best location in the site, for which high anisotropy zones with reference to the state of stress should be avoided, in the orientation of caverns where the largest dimension of the cavern should not be perpendicular to the greatest principal stress, and in the definition of shapes, for which the geometry of excavations should be defined in such a way as to minimize stress concentrations.

Nevertheless stresses cannot be theoretically determined on basis of analytical models owing to the varied topographic conditions, complexity of rock mass structures and their non-linear behaviour, and lack of knowledge about tectonic forces. Hence more reliable previsions cannot be achieved but through in situ measurements. To carry out these measurements, a number of techniques have been devised, each having its own field of application and its own limitations. Measurements of in situ stresses are valid only in the vicinity of the point of measurement if the rock mass can be idealized as a continuous equivalent homogeneous medium. Nevertheless the concept of continuous medium loses meaning for volumes smaller than the elementary physical volume, and so the assumption as to uniform state of stress throughout a given zone of the rock mass is no more valid. It is thus essential to determine the minimum dimension of the rock mass volume required so that the idealization as continuous equivalent medium may hold. Possibly the choice of a suitable measurement scale is one of the greatest difficulties in this kind of tests.

It is usual practice to estimate states of stress in rock mass either on basis of in situ measurements or not. As a first approximation, the value of the state of stress is roughly calculated, as a rule, from the weight of the overlying rock, with lateral confinement and assuming linear elastic behaviour of the materials. Vertical stresses are assumed to vary linearly with depth. In relatively plane rock masses, the principal directions considered are vertical and horizontal. Nevertheless in steep zones one of the principal directions is usually normal to the slope, whereas the others are on the same plane. These gravitic conditions do not hold when there are beds folded in anticlines and synclines, or other and more complex deformation forms, being valid only in average terms. With reference to horizontal stresses, they are assumed to vary from a fraction up to a multiple of the vertical stresses (Goodman 1980), (Sousa 1983), (Martins 1985).

2.2 Procedures to determine the state of stress

The measurement of the in situ state of stress can be made by means of several techniques now available, the following procedures being mostly used: i) complete strain relief; ii) partial strain relief; iii) rock flow or fracture. Nevertheless other procedures have been developed that correlate the properties of the rock mass to stresses.

The estimation of the state of stress can also be made during and after construction, by using results of structural observation, namely displacements. Then by means of calculation models it will be possible to estimate the most probable state of stress set up and, therefore, to validate results obtained in previous tests.

Among the procedures mentioned the best known and widespread ones are overcoring, the flat-jack method and hydraulic fracturing. All stress-measurement techniques disturb the rock mass to prompt a response which can thus be measured and analysed by using a theoretical model.

Methods used to determine the state of stress on basis of complete strain relief by overcoring basically divide into two groups. In the first are those using overcoring of a section where a strain gauge is located (U.S. Bureau of Mines method, STT method of LNEC, Leeman method); in the second group one finds the doorstopper method following which a section with a gauge rosette adherent to the bottom is overcored.

As regards techniques for complete strain relief, the STT method developed by LNEC should be mentioned. This method makes it possible to determine the initial state of stress in a rock mass at test points inside boreholes, by using a stress tensor tube whose description is given in 3.1.

The doorstopper method consists of the following operations: once a borehole has been drilled to the site of measurement the doorstopper is installed at the centre of the borehole bottom. This essentially consists of a strain gauge rosette protected by a silicone cylinder shaped as a door leaf. As the hole is deepened, the surrounding rock mass zone is decompressed, following which a final reading of the strain gages is obtained. On basis of those readings it will now be possible to calculate the state of stress on the plane containing the centre of the borehole. Complete measurement of the state of stress demands at least two further boreholes in different orientations (Leeman 1970).

These techniques in general are used at small depths and although variable, the precision of results is often acceptable. Notice that the values of strains were obtained on the assumption of an elastic behaviour of the rock mass, and so only comparatively low values of stresses can be attained to avoid non-linear behaviour of the rock mass. On the other hand, almost all methods assume a fairly simplified conceptual model of the rock mass behaviour, i.e., isotropic continuous, elastic, linear. The determination of the elastic constants is considered to be one of the difficulties of these methods.

As regards methods to determine stresses by partial strain relief, the flat jack method should be singled out. It consists of the execution of a slot in an exposed wall and of restoring the initial position of measurement points by applying flat jack pressures. Pressures required for restoring those positions correspond to normal stresses acting on the surface of the slot. These methods essentially use the so-called "flat jacks" which are formed of two steel plates with a given shape, contour-welded one another, to which is conveyed a given pressure through oil inside the jacks.

With reference to techniques developed that use those methods, mention should be made of the LNEC technology with small flat jacks, called SFJ system (Rocha et al. 1966).

These methods are very useful owing to their simplicity and to the fact that it is not necessary to know the mechanical characteristics of the rock mass in the formulas used. Nevertheless these tests, as a rule, only permit stress measurement if the rock mass behaves as an elastic medium If the strength of the zone concerned by a test does not withstand the stresses set up by the excavation, plastic deformations may be appear. In these cases, the measurements with the flat jack do not allow predictions as to the initial state of stress. To interpret test suitably,it is thus necessary to carry out adequate geomechanical studies, such as for instance measurement of the virgin state of stress by other means.

Reference should also be made to the techniques using the hydraulic fracturing method which permits to estimate the state of stress at large depths, in boreholes generally vertical (Haimson 1978). The method consists in water injections in a zone of the borehole isolated by means of packers until tensile failure is produced by the increase of the injected water pressure.The existing vertical stress is assumed to equal the weight of the overlying materials. On basis of this assumption and once the orientation of the fractures has been determined, it is possible to determine the remaining principal stresses and their orientations. The difficulty encountered in the experimental determination of the rock tensile strength in general is overcome on basis of test conditions themselves, once the initial fracture has been produced. This method has also some limitations that are due to the assumptions made, i.e. elastic linear and isotropic behaviour, and requires the borehole to be lined up with the direction of one of the principal stresses. Some innovations to the hydraulic fracturing method were introduced recently (Cornet 1983).

Other procedures of stress measurement by fracturing have been proposed, namely that by Cruz (1977) which measures strains with a borehole jack, and that used in LNEC and based on dilatometer tests (Sousa & Barroso 1982).

3 TECHNIQUES FOR MEASUREMENT AND INTERPRETATION OF THE STATE OF STRESS

3.1 Measuring methods developed at LNEC

LNEC has developed its own methods for measuring the state of stress in a rock mass of which should be singled out the STT (stress tensor tube) method using complete strain relief, and the SFJ (small flat jack) method with partial strain relief.

The STT method allows complete determination of the state of stress in situ and uses a stress tensor tube which is a non-stiff plastic hollow cylinder provided with three small-base strain gage rosettes (Rocha et al. 1974). The method essentially consists in (Fig. 1): i) drilling a large diameter (~140 mm) borehole down to about the depth at which the state of stress is to be determined; ii) drilling a small diameter (37 mm) borehole, 80 cm deep, from the bottom of the precedent drilling; iii) cementing the stress tensor tube, duly oriented, to the wall of the small-diameter hole; iv) initial reading of the strain gages contained in the STT; v) overcoring of the zone containing the STT down to a depth ensuring complete relief of the strains that it undergoes, reading

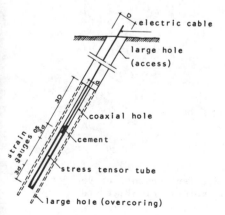

Figure 1. Stress tensor tube STT

the strain gages throughout the test up to apparent stabilization; vi) removal of the core for laboratory tests with a view to determining the elastic characteristics of the rock mass.

The readings obtained with the strain gages before and after overcoring make it possible to cal culate strains induced by the operation.

Based on formulas given by analytical and numerical finite element solutions (Rocha et al.1974) one obtains, from the strains measured and through a linear system of 9 equations in 6 unknowns, the most probable values of the components of the state of stress by means of the least-square me thod. The interpretation of the test assumes an elastic, isotropic, homogeneous medium with li - near behaviour, and whenever possible the elastic characteristics are determined through biaxial laboratory tests on the core with the STT extracted from the borehole. Besides the need to use large-diameter boreholes, a disadvantage of these tests lies in that the interpretation of the re sults depends on elastic constants.

As to the SFJ method, it can be briefly described as follows (Fig. 2):

i) Once suitably smoothened, the surface to be studied is provided with pairs of measuring ba ses, between which distances are measured.

ii) With a diamond-disk saw, open a slot between the bases; thus the normal stress on the plane is released. As a result the distance between the measuring bases varies.

iii) A circular flat jack is inserted, filling the slot, and pressure oil is injected in the flat jack. The distance between the bases is measured again until the initial position is restor- ed. Pressure introduced in the jack to restore the initial position (cancelling pressure) corres- ponds to the normal stress in the surface corresponding to the slot opened, excepting for small correction factors.

To determine the state of stress at a point, the method described is applied to three orthogo - nal surfaces, or to a further supplementary surface, to confirm results, three slots under the form of a rosette being made in each plane.

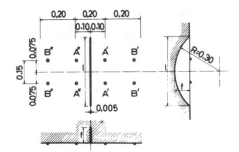

Figure 2. SFJ System - geometry of a slot

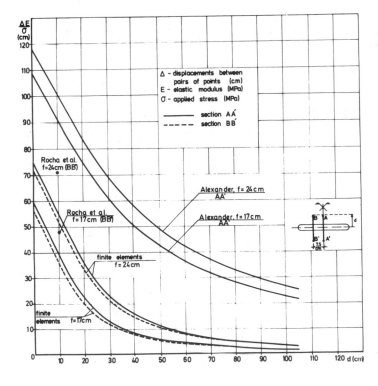

Figure 3. Displacement curves between pairs of points for SFJ tests

The SFJ method permits also to evaluate the deformability of the rock mass in the test site. The model for the interpretation of the test results uses charts deduced from a finite element tridimensional model. There the rock mass is assumed with mechanical characteristics of isotropy or such as an isotropic-transverse body. One also considers the hypothesis of a crack being generated in the plane of the slot when the flat jack applies the pressure, for the usual test depths (Martins 1985). In Fig. 3 are indicated the curves of displacements as a function of the distance between pairs of points at the same distance to the slot. The results obtained are compared with the analytical solution of the equivalent plane stress problem (Alexander's solution) and with the values obtained by an experimental model used by Rocha et al. (1966).

3.2 Methodology proposed for the evaluation of the overall state of stress

Testing for the measurement of the in-situ stresses should be programmed in stages following the design and construction of the works. In the design stage tests are carried out in boreholes; they can be performed by the STT method as there is no access yet to the future location of the main underground structures. Nevertheless as far as deep works are concerned, the use of this method should be restricted to depths smaller than that of the works, due to some technical and economic hindrances. In the construction stage one can simultaneously use complete strain relief and partial strain relief test methods inside the underground structures.

Sometimes tests are made inside test chambers and so stresses measured do not correspond in general to those existing at the surfaces of works; besides there is a dispersion of test results, mainly those of the STT method, which derives from difficulties in test techniques and from the hypoteses usually adopted for the test interpretation model. On the other hand, test places chosen generally depend on the characteristics of the rock mass in the test chamber.

The aim is to define a test methodology suitable for the construction stage, in order to as - certain the state of stress inside the rock mass and in the excavation surfaces, by means of results obtained in test chambers and by using numerical models.

Thus by tests at various points one may obtain components of the state of stress installed at those different points, grouped in a vector M_i. A matrix A_{ij} is thus determined for the structural model containing the test places, which represents the components of the state of stress at the different test points for the unit states of stress acting in the rock mass ($\sigma^o{}_x, \sigma^o{}_y, \sigma^o{}_z, \tau^o{}_{xy}, \tau^o{}_{yz}$ and $\tau^o{}_{zx}$). By representing by the vector A_j the componentes of the initial state of stress to be calculated, the following expression is obtained:

$$A_{ij} \, S_j = M_i \qquad \begin{array}{l} (i = 1,2,...N) \\ (j = 1,2,...6) \end{array} \qquad (1)$$

where N represents the total number of the components of the state of stress determined by the tests.

The matrix A_{ij} can be calculated numerically, in the case of a simple cavity shape, namely a circular gallery, by using analytical solutions (Pinto & Graça 1983) or, in the most general case, by using tridimensional numerical models by boundary elements or by finite elements (Martins 1985).

The system of equations referred to in (1) generally presents a high redundant number of equations. Thus to determine the most probable in-situ state of stress use should be made of the least-square method:

$$X_{ij} \, S_j = U_i \qquad (i,j = 1,2,...6) \qquad (2)$$

being

$$X_{ij} = A_{kj} \, A_{ki} \qquad (3)$$
$$U_i = A_{ki} \, M_k \qquad (k = 1,2,...N) \qquad (4)$$

The methodology proposed was applied to the Castelo do Bode tunnel, which is part of the water conveyance system for the Lisbon region, Portugal.

4 APPLICATION TO THE CASTELO DO BODE TUNNEL

4.1 General characteristics of the work

The new water supply system for the Lisbon region starts from the Castelo do Bode reservoir through a 125 km long conduit. Most of the circuit is constructed by cut and cover and some small tunnels were needed (Oliveira et al. 1983). The so-called Castelo do Bode sub-system begins at the water intake from the reservoir, followed by the Castelo do Bode tunnel which conveys water

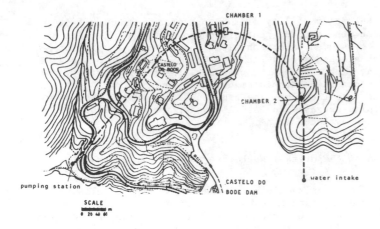

Figure 4. Castelo do Bode tunnel

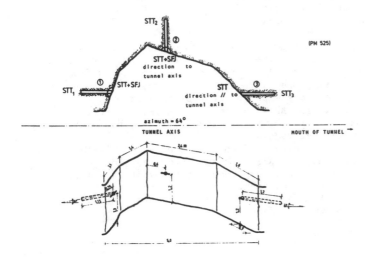

Figure 5. Tests carried out in test chamber 1

to the corresponding pumping station. This tunnel is about 1 km long, the inner diameter of its circular cross-section is 3.0 m, and has a reinforced concrete support about 0.35 m thick (Fig.4).
    LNEC gave a considerable contribution to the geotechnical characterization of the rock mass cros sed by the tunnel and to the determination of the state of stress in situ, as well as to the implementation and execution of the corresponding observation system. Prior to construction,   work included geotechnical studies of the rock mass comprising an integral-sampling exploration program me and dilatometer tests to evaluate the deformability and the in-situ stresses by means of special loading/unloading tests (Sousa & Barroso 1982). During construction, a set of observations were performed, which included systematic measurements of roof and side-wall convergences in cross-sections and testing for the determination of the in-situ state of stress in test chambers prepared for this purpose (Sousa & Martins 1983). Observations were also programmed, by using strain gages embedded in concrete and multiple position borehole extensometers which together with other complementary provisions are intended for checking the safety and operationality of the work in service.

4.2 Tests for the determination of the state of stress

The geotechnical study of the rock mass included a test programme carried out before and during the construction of the tunnel, aiming at the quantification of the in-situ state of stress. Thus in the course of the preliminary study special dilatometer tests were conducted under successive and increasing load-unload cycles (Sousa & Barroso 1982). Results obtained which cannot yet be con sidered very conclusive due to LNEC's scarce experience in the use of this technique, led to relations between horizontal and vertical stresses ranging from 0.2 to 0.8. Nevertheless one should notice that tests were made at small depths, far from reaching the depth of the tunnel.During cons truction SFJ and STT tests were carried out in two test chambers whose location is shown in Fig.4.
    In the first test chamber, located at 525 m from the tunnel mouth, two SFJ tests were carried out in horizontal slots and at half-height of side walls, as well as three STT tests in nearly parallel and normal holes to the axis of the tunnel (Fig. 5). In the second test chamber, located

Table 1. Results of SFJ and STT tests

SFJ tests		Slot	Stresses (MPa)	Elastic modulus (GPa)
Chamber 1	SFJ 1	horizontal	-2.2	22.0
	SFJ 2	horizontal	-7.8	7.6
Chamber 2	SFJ 1	horizontal	-1.2	8.2
	SFJ 2	horizontal	-0.5	17.1
	SFJ 3	45° to horizontal	-1.2	9.7
	SFJ 4	horizontal	-1.8	14.3

STT tests Chamber 1	Stresses (MPa)						Elastic modulus (GPa)	Poisson's ratio
	σ_x	σ_y	σ_z	τ_{xy}	τ_{yz}	τ_{zx}		
STT 1	-8.4	-3.4	-6.0	-2.1	0.9	+1.0	59.8	0.20
STT 2	-8.8	0.5	-7.8	0.8	1.8	0.3	18.5	0.14
STT 3	-3.1	-1.4	-1.7	0.5	1.9	-1.0	40.5	0.23

at 790 m from the tunnel mouth, four SFJ tests were conducted in three horizontal slots and a further slot inclined at 45°, at about half-height of side walls (Sousa & Martins 1985).
 In Table 1 are summed up the results obtained in SFJ and STT tests, regarding stresses and deformability characteristics.

4.3 Tridimensional model by boundary elements

For assessing the state of stress in situ in the light of test results obtained in the construction stage, a tridimensional model by boundary elements was prepared for the tunnel zone around the first test chamber, in order to apply the proposed methodology for the assessment of the most probable state of stress both in the tunnel walls and inwards the rock mass.
 The tridimensional model by boundary elements uses a numerical approximation method (boundary element method) whose approximation functions satisfy the governative equations of the problem under analysis in its domain, being approximate at the boundaries. The discretization is introduced along the boundary surfaces, following the usual techniques of approximation by finite elements (Banerjee & Butterfield 1981), (Crouch & Starfield 1983).
 Models by boundary elements have been increasingly employed in the analysis of underground works because they sometimes partly overcome several difficulties involved in the use of numerical models by finite elements, mainly as regards the handling of data and the analysis of results,which is clearly shown in the case of tridimensional equilibria. They also made it possible to solve problems dealing with infinite domains, reason why with deep underground works it is general only necessary to discretize the excavation surfaces.
 LNEC has been developing numerical models by boundary elements to predict the structural behaviour of underground works, by assuming in a first stage the simplifying hypotheses that idealize a continuous, homogeneous, isotropic medium and elastic linear behaviour. In the tridimensional model developed, a direct formulation was adopted and use is made of boundary elements of eight nodal points, curve-sided. Interpolation functions of the parabolic type are used, which are identical for displacements and for pressures (Lamas 1984). To calculate the different integrals involved in the formulation of the model, a computing algorithm was developed which makes it possible to select automatically the order of integration to be used case by case, so that one may obtain uniform precision in all integrations. To solve the system of equations whose matrix is full and non-symmetric, was adopted the Gauss algorithm by using partition techniques.
 For a region enveloping the test chamber 1 a calculation model by boundary elements was prepa-

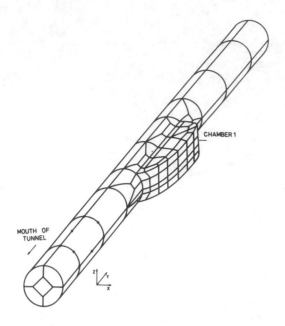

CHAMBER 1

MOUTH OF
TUNNEL

Figure 6. View of the boundary
element mesh

red considering a zone of influence of 5R (R being the tunnel radius equal to 1.75 m)   upstream
and downstream of the test chamber.The mesh used which is shown in Fig. 6 consisted of 170 boundary
elements and 612 nodal points.
    The rock mass was assumed to be linear, elastic, homogeneous and isotropic, with a modulus of
elasticity equal to the unit, and a Poisson's ratio of 0.2. Six calculations were carried out by
varying the action corresponding to relief of unit states of stress related to the six components
of the stress tensor assumed to be uniform and constant in the region analysed. The aim was thus
to obtain, for points on the excavation surfaces and inside the rock mass, the displacements   and
stresses due to relief of the unit states of stress already mentioned.

4.4 Analysis of the results obtained

By means of the calculations referred to, the coefficients A_{ij} (i = 1,2,...20; j = 1,2,...6) were
obtained which represent stresses calculated for points corresponding to the test places for the
different unit actions.
    Fig. 7 shows results in terms of displacements and stresses at the boundaries, for the actions
due to relief of the unit states of stress following two coordinate axes and for the two cross-sections indicated on the figure.

Table 2. Initial state of stress calculated

Stress (MPa)	Principal stresses (MPa)		
	$\sigma_I = 1.0$; $\sigma_{II} = -1.2$; $\sigma_{III} = -4.0$		
$\sigma_x^o = -2.4$ $\tau_{xy}^o = 1.4$	Principal axis directions		
$\sigma_y^o = 0.2$ $\tau_{yz}^o = -0.4$	Axis	x	y	z
$\sigma_z^o = -2.1$ $\tau_{zx}^o = 1.3$	I	$139°$	$72°$	$55°$
	II	$61°$	$107°$	$35°$
(–) Compression	III	$115°$	$154°$	$94°$

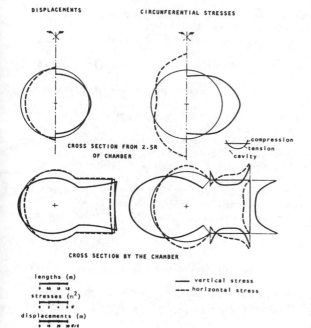

DISPLACEMENTS   CIRCUNFERENTIAL STRESSES

CROSS SECTION FROM 2.5R
OF CHAMBER

compression
tension
cavity

CROSS SECTION BY THE CHAMBER

lengths (m)

stresses (n²)

displacements (m)

—— vertical stress
--- horizontal stress

Figure 7. Displacements at the boundaries and circumferential stresses along the cross-sections

By means of the methodology proposed and based on calculated values of A_{ij} and on values given by some tests, the most probable initial state of stress in the rock mass was thus obtained (Table 2).

As regards the vertical stress, the value calculated is close to the weight of the overlying material; nevertheless there appears some horizontal tensile stress, which is not very credible.

5 FINAL COMMENTS

Emphasis is laid upon the main aspects involved in the analysis of the actions resulting from relief of the initial state of stress, which produce a redistribution of the field of stresses and a deformational movement towards the openings, in the rock masses enveloping an underground work. Problems related to aspects regarding stresses usually set up in the rock masses were discussed as well as the test methods used for their determination in situ. Special attention was given to methods developed at LNEC, particularly the STT method and the SFJ method.

A methodology was established for conducting tests and for their analysis, with a view to determining the state of stress in underground works, which was later applied to Castelo do Bode as part of a water conveyance system for the Lisbon region. With this aim use was made of a tridimensional model by boundary elements concerning a test zone. The results obtained basically express uncertainities of the values obtained in tests, as well as the idealization behind the calculation model. Nevertheless as a whole the potentialities of the methodology presented are considered of interest.

ACKNOWLEDGEMENTS

The authors are indebted to Empresa Pública das Águas Livres, EPAL, for their having allowed some data of tests of the Castelo do Bode tunnel to be included in this work. The authors also thank Ms. Fátima Bravo for the English translation of this paper.

REFERENCES

Banerjee, P. & R. Butterfield 1981. Boundary element methods in engineering science. McGraw-Hill, London.
Cornet, F. 1983. Special topics in Rock Mechanics. 5th Congress of ISRM, Melbourne.
Crouch, S. & A. Starfield 1983. Boundary element methods in solid mechanics. George Allen & Unwin, London.

Cruz, R. 1977. Jack fracturing technique of stress measurement. Rock Mechanics, V. 9, nº 1.

Goodman, R. 1980. Introduction to Rock Mechanics. John Wiley & Sons, New York.

Haimson, B. 1978. The hydrofracturing stress measuring method and recent field results. J. Rock Mech. Min. Sci., V. 15, nº 4.

Lamas, N. 1984. Three-dimensional boundary element model for underground structures (in Portuguese). LNEC, Internal Report, Lisbon.

Leeman, E. 1970. Experience throughout the world with the CSIR "Doorstopper" rock stress measurement equipment. 2nd Congress of ISRM, Beograd, 1970.

Martins, S. 1985. Contribution to the study of underground structures associated with hydroelectric development (in Portuguese). LNEC, Research Officer Thesis, Lisbon.

Oliveira, R., C. Costa & J. David. 1983. Engineering geological studies and design of Castelo do Bode tunnel. Int. Symp. on Engineering Geology and Underground Construction, Lisbon.

Pinto, L. & C. Graça 1983. Determination of the state of stress of rock masses by the small flat jack (SFJ) method. 5th Int. Cong. of ISRM, Melbourne.

Rocha, M., B. Lopes & N. Silva 1966. A new technique for applying the method of the flat jack in the determination of stresses inside rock masses. 1st Int. Cong. of ISRM, Lisbon.

Rocha, M., A. Silvério, J. Pedro & S. Delgado 1974. A new development of the LNEC stress tensor gauge. 3rd Int. Cong. of ISRM, Denver.

Sousa, R. 1983. Conception and design of large underground structures (in Portuguese). LNEC Internal Report, Lisbon.

Sousa, R. & M. Barroso 1982. Contribution to geotechnical study of Castelo do Bode tunnel (in Portuguese). LNEC Internal Report, Lisbon.

Sousa, R. & S. Martins 1985. Observed behaviour of Castelo do Bode tunnel. Geotechnical characterization of the rock mass and in situ stress (in Portuguese). LNEC Internal Report, Lisbon.

Assessment of rock strength by visual inspection point load and uniaxial testing – A comparison using data from Chinnor Tunnel

Détermination de la résistance des roches par examen visuel et par essais uniaxiaux et au pénétromètre – Comparaisons effectuées en utilisant des données du tunnel de Chinnor

J.A.Rodrigues-Carvalho, *Universidade Nova de Lisboa and CÊGÊ Lda, Lisbon, Portugal*

ABSTRACT: The assessment of strength by means of visual inspection is compared with that obtained from point load and uniaxial testing by using the results obtained from the site investigation carried out on the chalk for Chinnor Tunnel. The relative merits of those methods are pointed out in terms of reliability, cost and time consumption.

RESUME: L'auteur compare l'estimation de la resistance des roches par inspection visuelle avec les resistances obtenues par les essais de charge punctuelle (point load) et de compression uniaxial en se baseant sur les resultats des investigations realisées pour le "Chinnor Tunnel". Les mérits relatifs de ces methodes sont presentés en considerant la confiance des resultats, le coût et le temp de realization.

1 INTRODUCTION

In order to provide adequate geotechnical information about the chalk to be tunnelled at Chinnor, a two phase site investigation was performed (Priest et al, 1976; Soil Mechanics Ltd, 1974, 1975). For assessing the strength of the rock three independent methods were used:
  i   - Visual inspection
  ii - Point load testing
  iii - Uniaxial compressive testing

Visual inspection was carried out on the rock cores during systematic core logging by using a strength scale based upon that recommended by the London Geological Society (1970).

The uniaxial compressive tests were performed on 100 mm diameter rock cores except for a small number of cases were a 54 mm diameter was used. The tests were performed by following the procedure suggested by Hawkes and Mellor (1970) in that respects size and tolerances.

The point load tests were also performed on 100 mm diameter rock cores both diametrically and axially, the results indicating a relatively isotropic strength for this type of rock material (Soil Mechanics Ltd, op.cit).

The point load tests results refered to above were presented in the site investigation report corrected to an equivalent value for 50 mm diameter samples in the way suggested by Broch and Franklin (1972). The index to strength correlation factor obtained and presented in the S.I. report was 1:20.

As the majority of the uniaxial tests were performed on 100 mm diameter samples, it seems to the author that a better comparison between the uniaxial and point load test results could be achieved if the values of Is 100 for these last ones had been used. Nevertheless it seems also that this possible scale effect problem would not introduce relevant differences in that respects the matter discussed in this paper.

For the purpose of this paper only the sections of the rock core where uniaxial compressive tests, point load tests and visual strength classification have been carried out systematically are considered (see Table 1).

Quantities and assumed costs for the assessment of strength by the three different methods used are presented in Table 2 (costs are given in portuguese escudos).

Table 1 - Sections of boreholes and tests considered in this study

BOREHOLE NR.		1st PHASE OF S.I.					2nd PHASE OF S.I.							TOTAL 1st + 2nd PHASES OF S.I.
		10	11	12	14	Sub total	100	101	102	103	104	105	Sub total	
LENGTH OF BOREHOLE (m)		24.6	32.7	29.9	27.9	115.1	30.3	24.9	26.0	25.0	36.7	29.3	172.3	287.4
SECTION OF CORE CONSIDERED (m)		1.0 to 24.6	19.4 to 32.7	12.4 to 29.9	12.3 to 27.9		1.5 to 30.3	4 to 24.9	12 to 26.0	3.8 to 25.0	7.1 to 36.7	9.7 to 29.3		
TOTAL LENGTH (m)		23.6	13.3	17.5	15.6	70	28.8	20.9	14.0	21.2	29.6	19.6	134.1	204.1
NR. OF POINT LOAD TESTS ALONG ABOVE SECTION	Performed	83	49	76	48	256	37	18	14	32	35	33	169	425
	Considered in this study	73(a)	47(b)	67(c)	48	235	34(e)	18	13(f)	31(f)	35	30(e)	161	396
NR. OF UNCONF. COMP. TESTS ALONG ABOVE LENGTH		6	7	7	5	25	11	14	8	7	11 10(d)	7	57	82

(a) 5 tests not considered (failure along pre-existing fracture) and 5 anormally high

(b) 1 test not considered (failure along pre-existing fracture) and 1 anormally high

(c) 5 tests not considered (failure along pre-existing fracture) and 4 anormally high

(d) 1 test not considered (anormally high)

(e) 3 tests not considered (anormally high)

(f) 1 test not considered (failure along pre-existing fracture)

Table 2 - Quantities and costs involved with the strength assessment (see text)

Method	1st Phase S.I.		2nd Phase S.I.		1st+2nd Phase S.I.	
	Quantit.	Cost *	Quantit.	Cost *	Quantit.	Cost *
Visual Inspection	70m	5	134 m	10	204	15
Point load testing	235un.	118	161un.	81	396	198
Uniax.Comp. testing	25un.	110	57un.	251	82	361

(*) × 10^3 portuguese escudos

The costs in Table 2 has assessed based upon current LNEC-Lisbon 1985 unit prices: 4400$ and 500$ portuguese escudos for the uniaxial and point load tests respectively. The cost for the visual classification has been estimated on the basis of labour requirement, assuming that an engineering geologist would classify 40 m of rock core for strength assessment in one hour and this would cost 3000$, also considering the LNEC 1985 current prices. These costs may allow a direct comparison between the money spent in the assessment of strength by the three different methods and the effectiveness of the results obtained in each case.

2 RESULTS FROM UNIAXIAL AND POINT LOAD TESTS AND FROM VISUAL INSPECTION

Tables 3 and 4 present the average and standard deviation values for the strength of the chalk obtained by uniaxial and point load tests.

For the estimation of the strength by visual inspection the logger used a scale based on that recommended by the London Geological Society (op.cit) together with some intermediate subclasses (see Table 5).

To enable a comparison with the assessment of rock strength by using the uniaxial and point load testing methods presented in Tables 3 and 4, and attempt to quantify the estimation of strength by visual inspection has been made. For this, a value of strength has been attributed to each class and subclass used

Table 3 - Results from unconfined
          compressive tests

	1st Phase	2nd Phase
Nr. of tests	25	57
σ_{ult} MEAN (MPa)	6.2	6.7
σ_{ult} ST.DEV (MPa)	3.1	3.2

Table 4 - Results from point load tests

	1st Phase	2nd Phase
Nr. of tests	235	161
Is × 20 MEAN (MPa)	6.0	6.5
Is × 20 ST. DEV (MPa)	4.2	4.6

Table 5 - Results assessed from visual classification

SCALE USED IN VISUAL CLASSIFICATION	STRENGTH VALUES (MPa) (G.S. Scale)	ATTRIBUTED VALUE (MPa)	1st PHASE OF S.I.			2nd PHASE OF S.I.		
			NO OF METERS	ACCUMULATED PERCENTAGE		NO OF METERS	ACCUMULATED PERCENTAGE	
VERY WEAK	< 1.25	.6	1.7	2.4	100.0	—	—	—
VERY WEAK TO WEAK	(*)	1.25	2.6	3.7	97.6	—	—	—
WEAK	1.25-5	3.1	24.3	40.9	93.9	8.7	6.5	100.0
WEAK TO MOD WEAK	(*)	5.0	—	—	—	6.0	11.0	93.5
MOD WEAK	5-12.5	8.8	33.1	88.1	59.1	27.3	31.3	89.0
MOD WEAK TO MOD STRONG	(*)	12.5	1.8	90.7	18.9	40.3	61.4	68.7
MOD STRONG	12.5-50	31.2	5.7	98.7	9.3	29.2	83.1	38.6
MOD STRONG TO STRONG	(*)	50.0	—	—	—	21.7	99.3	16.9
STRONG	50-100	75.0	0.8	100.0	1.1	0.9	100.0	0.7
STRONG TO VERY STRONG	(*)	—	—	—	—	—	—	—
VERY STRONG	100-200	—	—	—	—	—	—	—
EXTREM STRONG	> 200	—	—	—	—	—	—	—

Average value for the strength derived from the visual classification:

    1st PHASE: 9.0 MPa

    2nd PHASE: 21.4 MPa

(*) Subclasses introduced by the logger

by the logger as shown in the third column of Table 5. The values used correspond to the mean of the range covered by each class and subclass and then an a-
verage value for the strength of the chalk has been evaluated taking in consideration the length of core classified in each class or subclass. The so obtained
results are also presented in Table 5.

Table 6 - Differences in the strength of the chalk from the
         three different methods

	1st Phase S.I.	2nd Phase S.I.
σ_{ult} (MPa)	6.2	6.7
Is × 20 (MPa)	6.0	6.5
$\dfrac{Is \times 20 - \sigma_{ult}}{\sigma_{ult}} \times 100$	- 3.2%	- 3.0%
(vis.insp.) (MPa)	9.0	21.4
$\dfrac{(vis.insp.) - \sigma_{ult}}{\sigma_{ult}} \times 100$	+ 45%	+ 219%

3 DISCUSSION

Table 6 presents the differences between the average of the strength calculated
from the uniaxial tests and both a) the strength calculated from the point load
tests and b) the strength evaluated by visual inspection.

3.1 The uniaxial compressive vs. point load testing

It can be considered that there is no difference between the strength obtained
either by the uniaxial tests or by the point load tests in both phases of S.I..
The average strength assessed by the uniaxial tests performed during the 2nd
phase shown a very slight increase in strength (only 8%) with respect to the
1st phase and this difference was also detected by the point load tests.
  Commenting the point load testing, Bieniawski 1975, for example, states that
"the usefulness of the test is limited for rock materials with a uniaxial compressive strength below 25 MPa. Nevertheless, the results obtained for the
chalk at Chinnor do not support this at least for the type of rock dealt with
in this paper - the chalk which present an average uniaxial compressive strength
between 6 and 7 MPa.
  The scatter of the results shown by the standard deviation is higher for the
point load than for the uniaxial tests (see tables 3 and 4). Standard deviation
is about 50% of the mean for the uniaxial tests and about 70% for the point
load tests. The differences being not very important this do not agrees with
the sated by Broch and Franklin (1972) but is in accordance with Bieniawski (op.
cit.) who also encountered greater scatter for the point load test results
although refering to harder rocks.

3.2 The visual inspection

The way adopted to assess the average value for the strength of chalk by visual
inspection may probably be discussed. Other ways could probably be selected. Ne
vertheless, it seems good enough to give an idea about the reliability of the
method by comparing the results with the ones obtained by the two other methods
used.
  Considering the results from visual inspection and the ones of the unconfined
compressive and point load tests the following can be sorted out (see tables 5
and 6).

3.2.1 1st Phase of S.I.

The logger's judgment has overassessed the strength of the rock in 45%. This
seems to be good enough for a visual classification and the result obtained
from the uniaxial tests and from the logger's judgment fall in the same range
of strength - moderately weak - of the scale used (see table 5).

3.2.2 2nd Phase of S.I.

In the 2nd phase of S.I. the logger's judgment has overassed the strength of the chalk in 220% and this seems not to be an adequate appraisal. In this case the visual classification includes the rock in the moderately strong class of the scale. It may be point out that:

i - About 70% of the core has been classified in classes stronger than the one correspondent to the average of the uniaxial and point load testing results.

ii - Within the uniaxial test results only 2 of them in a total of 57 gave strength values over the moderately weak class and these results were 14.0 MPa in both cases.

4 CONCLUSIONS

The effectiveness of the strength assessment by visual inspection depend upon the ability and experience of the estimator and this tends to be very subjective. In the case dealt with in this paper it seems that the task of core logging in the two phases of S.I. was not commited to the same engineering geologist but the author is not aware of this. The method is unreliable and this must be beared in mind chiefly in the case of low strength rocks were more accuracy in the assessment of strength is required.

In that concerns the point load testing it has been shown that this method provides a very good index to investigate the strength of rock even for low strength rocks (6 to 20 MPa according to the ISRM classification) as proved by the results from the investigations at Chinnor tunnel.

The benefits of selecting a good strength index which can be correlated with the values obtained by laboratory tests enables one to get adequate information on a site with a minimum of more sophisticated and time consuming tests. Furthermore this reduces very considerably the cost to obtain the required information. This approach is especially useful in the first stages of a site investigation. In further stages this usefulness continues by providing a check of material variability.

The geotechnical indexes that have been determined for the chalk at Chinnor included moisture content, porosity, specific gravity, bulk density and calcium carbonate content in addition to the strength tests discussed earlier. Within these, only the point load and uniaxial tests are closely correlated and this supports the stated by D'Andrea et al. (1965) who tried to correlated several different indexes with the strength determined by uniaxial compressive tests.

The point load is a reasonable quick and cheap test (less than 15% of the cost of equivalent uniaxial compressive test) and can be performed in the field on a variety of roughly prepared samples. These are good reasons to justify its use as an adjunct to uniaxial strength testing and to adopt its systematic utilization during the rock core logging.

AKNOWLEDGMENTS

The author wishes to aknowledge Dr. M P O'Reilly, head of Ground Engineering Division at the Transport and Road Research Laboratory for allowing the use of the site investigation reports for the Chinnor Tunnel which provided the basic data for this paper.

BIBLIOGRAPHY

Bieniawski,Z.T. 1975. The point-load test in geotechnical practice. Eng.Geology, 9(1), p. 1-11.
Broch,E. and Franklin,J.A. 1972. The point-load strength test. Int.Journ. Rock Mech.Min.Sciences, 9, p. 669-697.
D'Andrea,D.V.; Fisher,R.L. and Fogelson,D.E. 1965. Prediction of compressive strength from other rock properties. U.S. Depart. of the Interior - Bureau of Mines. Report of Investigation 6702.
Geological Society Engineering Group Working Party Report 1970. The logging of rock cores for engineering purposes. Q.Journ.Eng.Geology, 3, p. 1-24.
Hawkes,I. and Mellor,M. 1970. Uniaxial testing in rock mechanics laboratory. Eng.Geology, 4, p. 177-285.
International Society for Rock Mechanics 1972. Suggested method for determining the uniaxial compressive strength of rock materials and the point load strength index. Lisbon.
Priest,S.D.; Hudson,J.A. and Hurning,J.E. 1976. Site investigation for tunnelling trials in chalk. TRRL report Nr. 730.

Rocha,M. 1975. Alguns problemas relativos à mecânica das rochas dos materiais de baixa resistência. 5th Cong.Panamericano de Mecânica dos Solos e Engenharia de Fundações. Buenos Aires.

Soil Mechanics Ltd. 1974. Chinnor tunnelling trials - Geotechnical investigation Vol. I - Preliminary investigation (Report Nr. 6139/1).

Soil Mechanics Ltd. 1975. Chinnor tunnelling trials - Geotechnical investigation Vol. II - Main investigation (Report Nr. 6139/2).

1.5 Rock mass monitoring: New techniques and results
Surveillance des massifs rocheux: Nouvelles techniques
et résultats obtenus

Direct measurement of recent movements along seismic faults and creep slope deformations

Mesures in situ des déplacements actuels sur des failles actives et des déformations de pentes induites

Elka Avramova Tacheva & B.Kostak, *Geotechnical Laboratory, Academy of Sciences, Sofía, Bulgaria*

ABSTRACT: Three observation points with threedimensional dilatometers were set up in a seismically active area in Bulgaria, where two main tectonic faults cross each other and the greatest neotectonic uplift is recorded. The area is morphologically carved by complicated tectonics and slope deformations. The points are included into a larger geotechnical polygon.

RESUME: Dans la région seismoactive de la Bulgarie sud-ouest ou deux ruptures principales tectoniques se coupent et le plus haut dérangement vertical néotectonique est déterminé, deux points d'observation avec les dilatometres spatiales on été fondus. Les glissements de terrain sont fréquent ici, participant effectivement dans la formation de la morphologie locale. Les points d'observation sont inclus dans une polygon géodynamique qui est en construction dans la région.

GENERALLY

Geotechnical projects often offer far reaching conclusions only on the basis of indirect signs. There is a need for direct measurements in active tectonic zones to find relations between seismic data, and the present dynamics of a region. This may give a chance to assess prognostic data for dangerous seismic actions and, possibly, even such effects like slope deformations. In-situ observations of displacements must be carefully planned so that contributions of individual factors to resulting deformations could be recognized. In practice, gravitatonal factors on slopes are very effective, and must be considered in view of the fact that active tectonic regions are disposed to form young morphology with steep slopes. Therefore, tectonic and slope movements are in close relation, both having their specific importance.

Successful detection of large scale slope movements due to creep in Bulgaria promoted the idea to use the instrumentation that was developed (Kostak and Rybar 1978) for slope movement detection to be applied in seismoactive fault dynamics evaluation. Besides, earlier experience can be used to recognize the interference of the slope dynamics in the deep seated tectonic deformation process. Contact measurements on faults, though suggesting specific methodical problems, may be helpful in this trial.

FIELD SITUATION

A seismically active region in   Bulgaria was chosen to establish an observation system which is expected to add some information concerning general views about contemporary movements along active faults. The region represents a junction of two deep faults, the longitudinal being oriented generally at 155° and the transvers at 45°. The junction (Fig. 1) induces very serious neotectonic and present activity and high seismic intensity.

Geological and tectonic conditions of the region and its neotectonic evolution have been described by different authors, chiefly by B. Vrabljanski, and also by E. Avramova-Taceva and B. Kostak, G. Milev et al. During the neotectonic period the longitudinal fault is active at the beginning of the Upper Miocene and after

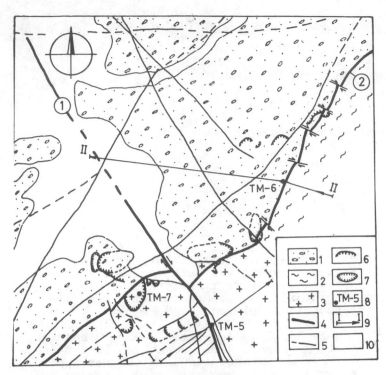

Figure 1. Investigated region with two main tectonic faults crossing. 1 - Pliocene sediments of
the graben; 2 - Proterozoic metamorphites; 3 -
South Bulgarian granitoids; 4 - the main tectonic faults; 5 - other faults; 6 - slope deformations: rockfalls, slides, creeping rock masses; 7 - accumulation of an old rockfall; 8 -
observation points with a target-meter TM; 9 -
cross-sections; 10 - Quaternary sediments.

Pliocene. Evolution of the transvers fault is continual from the Upper Miocene,
with the vertical movement reaching 3400 m.
   At the end of the Upper Miocene an inextensive tectonic graben was formed
here, filled up with rough sediments, molasses, up to 1400 m thick in the central part. A river take advantage of the longitudinal fault dewatering the basin
and cutting the S marginal slopes of the graben. A narrow canyon valley has
been cut through by the river.
   Granitoids and amphibolites, at the S slope of the basin (graben), are highly
fractured and disintegrated. They are affected by tectonic shearing along the
faults, the structure of the massif became granular with a chaotic orientation
of fractures. Slopes exhibit a wide spectrum of deformation forms. A number of
slides which occured here is evidently confined to the two main faults, and the
fault zone itself is coincident with the zone of gravitational deformations.
Other slope deformations in granitoids are confined to subparallel fault zones
and tectonic fractures.
   In granits of the main transvers fault in the area, one finds a rockfall the
volume of which can be estimated in 10^4 m^3. It originated in 1904 due to an
earthquake, and it appears to be a part of an older extensive slope deformation, the scars of which can be traced up to the highest levels of the slope.
Two rocky ribs step out of this slope with a number of horizontal depressions
(Fig. 2). The earthquake of 1904 was very strong, and the S extremity of the
basin was heavily affected.

OBSERVATION POINTS

Local survey of the region resulted in the positioning of three observation
points of the contemporary movements.
   The first (TM-5) point is situated in the canyon close to its entrance, i.e.

Figure 2. A rockfall near the canyon entrance at the place of
an older huge slope deformation; observation point TM-7.

to the fault junction (Fig. 1). This point is supposed to monitor movements on
the tectonic zone of the main longitudinal fault (Fig. 3). The zone, oriented
150° on the left river bank, belongs to a set of zones well morphologically expressed as depressions in granite over a long distance in the canyon (Fig. 4-a).
    The monitoring system of target-meters used here (Košťák 1969), has been se-
veral times applied to slope stability problems (Cacoń and Košťák 1976). It is
able to monitor any relative spatial movement across a tectonic zone, between
two faces of a depression or crack (Fig. 4-b), i.e. to register shearing and
opening of it. Results must be interpreted regarding all the different factors,
like gravitational and thermal displacements that may be superposed upon displacements of the tectonic or seismotectonic origin. At the first point the

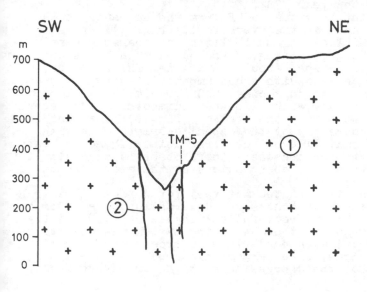

Figure 3. A cross-section I-I of
the longitudinal fault zone
through the observation point
TM-5. 1 - South Bulgarian granitoids; 2 - faults.

a)

b)

Figure 4. a) Parallel fracture
zones of the main longitudinal
fault in the canyon, well de-
tectable in morphology; b) Observation point TM-5.

gravitational process was not expected by the survey to be dominant. The observation started at the beginning of the year 1982.

The second (TM-6) observation point is situated NE from the entrance to the
canyon and is supposed to monitor the main transvers fault (Fig. 1). It is
founded on the contact between Pre-Paleozoic amphibolites and coarse grained sediments of Pliocene that fill up the graben (Fig. 5 and 6). A trench was dug
across the contact through the tectonic zone about 3,4 m wide. The zone contains: grey-green tectonic clays in contact with sound amphibolites (0,7 m), amphibolite breccia (1,5 m), grey-black tectonic clay (0,3 m), highly remolded
granitic gravel (0,9 m) in direct contact with Pliocene sediments. A long steel
tube was laid down into the trench to bridge the contact (Fig. 7), then cut into
two pieces supposed to be driven separately by the opposits brims. Otherwise,
the same monitoring equipment was used, like at the first point. There are many
details of such an installation that were necessary to be observed carefully to
avoid interference of unwanted factors. The observations at the second point
started in August 1982.

The third (TM-7) observation point is situated not very far W from the entrance of the canyon, on a slope modelled by recent slope deformations, as described
earlier (Figs 1 and 2). The slope exhibits rock-falls triggered by 1904 earthquakes. Again, the same instrumentation of the monitoring system was used to register displacements between two rock faces bordering a narrow depression with
shattered rock. There are several signs, like warm exhalations from the depth,
that may manifest close connection of the depression with the transvers fault.

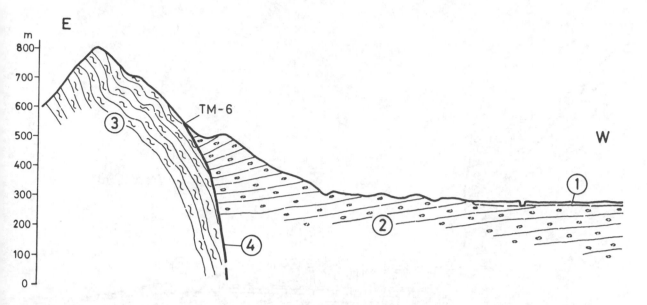

Figure 5. Cross-section II-II of the transvers fault zone. 1 - Alluvium sediments; 2 - Pliocene sediments; 3 - Proterozoic metamorphites; 4 - fault zone.

Figure 6. A contact between amphibolites and Pliocene sediments near TM-6. The contact can be traced at a long distance, marking the transvers fault.

Accordingly, the third observation point is supposed to monitor tectonic, seismotectonic, as well as gravitational displacements along the related section of the main transversal fault. A chance to register dangerous gravitational deformations cannot be avoided here, regarding the positioning of the point on a very steep and shattered mountaneous slope. However, such deformations are not independent from the seismic factor, and there is a practical reason to combine the two standpoints of the observation. Monitoring at the third point started at the end of 1983.

Three observation points described above represent two considerably different situations that must be respected in the technical details of the monitoring

ESE 110°

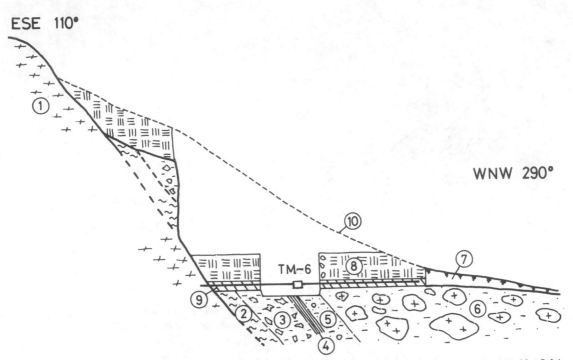

WNW 290°

TM-6

Figure 7. Observation point TM-6 on the transvers fault. 1 - amphibolites; 2 - tectonic clay; 3 - amphibolite breccia; 4 - tectonic clay; 5 - granitic gravel; 6 - Pliocene sediments; 7 - Quaternary slope sediments; 8 - filling; 9 - concrete; 10 - original surface.

system. Points one and three are designed to register displacements of faults in hard rocks, while point two designed to monitor a contact between hard, solid, and incompetent, loose sedimentary rock. The first situation is simpler than the other from the technical point of view. However, the interpretation of results obtained in a simple situation may have special difficultes, as secondary cracks may become a factor, too.

OBSERVATIONS

Displacements registered at the three observation points can be found in Figs 8 and 9. Each point has orientation of spatial rectangular coordinates x, y, z in respect to the local situation, i.e. x perpendicular, y parallel with the orientation of the section of the investigated fault, and z vertical. General orientation and interpretation of displacements can be found in Tab. 1. It should be noted that y and z has always the meaning of relative shearing, while x defines a change in width.

Table 1

Observation point	Orientation and interpretation of displacements		
	+x	+y	+z
	Horizontal zone opening	Horizontal zone shearing	Vertical zone shearing
TM-5	225°	downslope SW block to 135°	downslope SW block down
TM-6	290°	basin to 200°	basin down
TM-7	350°	downslope N block–260°	downslope N block down

Earlier observations on slopes show a significant correlation between dilatations and temperature, variation in respect to the seasonal volumetric changes in rock. The reaction can be considered parasitic, however, it is a normal reaction of rock, that may be effective in a slope deformation process also. The same reaction can be traced in the diagrams of Figs 8 and 9, it is not of constant character, it is masked often by other factors, and only some periods and some components reflect it clarly, while in other periods and other components it seem almost absent.

It is the TM-5 point, and the x component in it, where the simple sinusoidal seasonal variation can be followed from the beginning of the observation untill the end of 1983, when the seasonal variation reached temperatures from 0° to 32°C. The seasonal amplitude of x is about 1,0-1,2 mm at the TM-5 point, and it represents closing of the depression during the period of thermal energy increase, and vice versa, which is in a perfect agreement with the volumetric change of the rock that forms the opposite faces of the depression. This is a common observation, and usually, a certain delay is observed. It can be said that the point reacts in a normal way.

An ispection of TM-7 point, which is situated analogically, does not show, however, any clear temperature effect. An inspection of TM-6 shows a clear variation in x from the beginning of the observation untill the summer 1983, with amplitude of about 3,8 mm, however, the reaction opposite, i.e. after a period of closure of coming winter the fault opens during the spring 1983. A close inspection of the diagram TM-6 shows that the same variation is continual all over the observation period untill now. One can see that starting from the fall 1983 another factor interferes.

The reaction of the instruments is not very simple, therefore, and the complexity of reaction increases with the complexity of the situation in the observation point as well as with time. It may be reasonable to argue, that the reaction of TM-5 is normal, while TM-6 reflects an extremal delay of the volumetric effect or possibly seasonal swelling and shrinkage of the tectonic zone, and in case of TM-7, a high shattering of the rock at the site cuts down any clear seasonal effect. Yet, in any case one conclusion can be made that the instruments react with consistency, and second, that other factors than those of the volumetric reactions interfere with their reactions.

Since summer 1983 the components that untill then were completely regular, lose in all the diagrams their regularity, which can be traced again during winter 1984-85 for several months. One can find that the period of irregularity coinsides with a period of increased seismic activity in the central region of the Mediterranian Seismic Belt, which is indicated by series of earthquakes with epicentres in Aegean Sea, Northern Greece, Italy and Černa Gora.

Since August 1983 there are two parallel and irreversible skips in the y component at the longitudinal fault (TM-5), 0,5 mm each, and the x-component shows a series of reversible pulses of about 0,3-0,4 mm. There are another series of pulses with a skip of 0,7 mm in z. Analogical pulses are found also at the transvers fault (TM-6). Two prominent alternating pulses in x of about 2,2 and 3,2 mm, respectively, in February and May 1984 are evidently connected with earthquakes near Thessalonica and Aegean Sea with the intensity between 5-6 degrees, according to Richter. Since summer 1984, after the series of pulses, a period of systematic movement can be seen. The total displacement from June 1984 till September 1985 reached 10,0 mm in x (compression), and 12,0 mm in z component (relative along the fault zone, with the sediments of the basin in the contact sliding up). The y-component at the same time indicated relative sliding of the basin to SSW by 2 mm at the contact. The movement can be related to earthquakes with epicentres outside of the Bulgarian territory (S and SW), since the x and z, after slowing down by spring 1985, has been accelerated again in May 1985.

It seems to find evidence, therefore, that the longitudinal fault where the TM-5 point is situated, appears as a conductor for seismic energy from the south. Another earthquake (Vrancea January 1984) in the north show only a limited reaction at TM-5, while TM-6 did not react at all. The same reaction has been found for nearly earthquake located at the fault on the N of the graben. The reactions appear as sudden skip in the registered displacements of components x, y, z. Therefore, it seems to find evidence for damping effect of a tectonic fault which is found in the limits of the graben in the north or in the basin itself.

The movement since summer 1984 with its simultaneous reactions in x and z (TM-6), can be well interpreted by compressive action in the contact between the rock massif and the basin along this section of the main transvers fault.

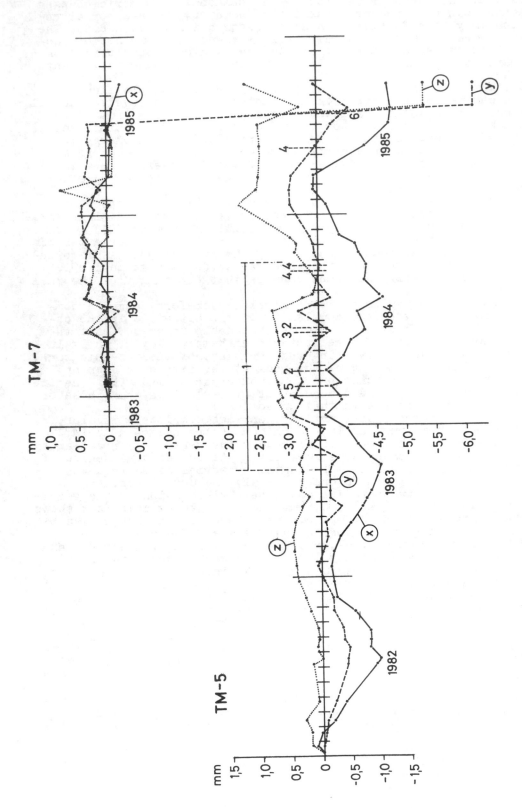

Figure 8. Displacements measured at TM-5 and TM-7 observation points. x – zone opening; y – horizontal shearing; z – vertical shearing. 1 – period of higher seismic activity in the central region of the Mediterranean Seismic Belt; earthquake epicentres: 2 – near Thessalonica: 19.2.1984, M_B 4,8 (5 Richter); 16.5.1984, M_L 3,2–3,3 (ATH) (4,2 Richter); 3 – 20 km N from the Psara Island in Aegean Sea: 6.5.1984, M_B 5,1 (5,5 Richter); 4 – Yugoslavia: 15.11.1983, M_B 4,3 (5,6 MSK); 3.3.1984, M 4,9 (7,5 MSK); 8.9.1984, $M_L M_B$ 3,5 (KBA) (8,5 MSK); 22.9.1984, M_L 2,5 (TTG) (6,5 MSK); 12.5.1985, >5,2 Richter; 5 – Vrancea: 20.1.1984, M_B 4,7 (5 Richter); 6 – NE Bulgaria: 27.7.1985, 4–5 MSK.

TM-6

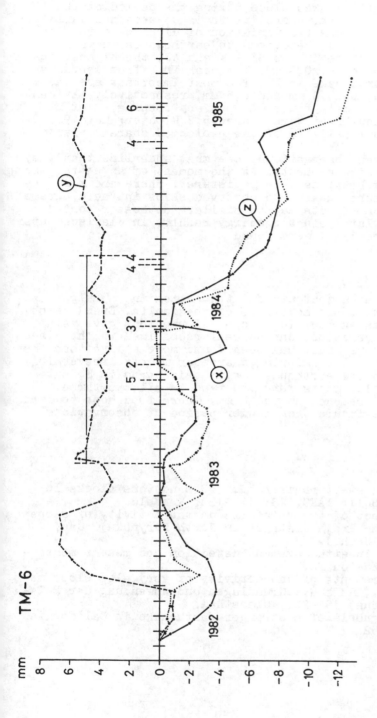

Figure 9. Displacements measured at the observation point TM-6. **x** – zone opening; **y** – horizontal shearing;
z – vertical shearing. 1 – period of higher seismic activity etc. see Fig. 8.

At the same time y-component seems to show slowing down of the movement with compression in x, and vice versa. This indicates variation of contact conditions on the fault with stress and time (Košťák, Kozák and Lokajíček 1985). No such compressional reaction was observed at the point TM-5 on the longitudinal fault.

As for the SW section of the transvers fault, some reactions of the TM-7 point can be analysed. The observation takes place during the period of the said accelerated seismic activity, and therefore, there is no seasonal variation, and all over only signs of high, chaotic variation of displacements. A closer inspection finds that there are no reactions to earthquakes in the S direction but there are responses to those in the SE, SW and NW, the highest individual response being in z-component by 0,3 mm. The peak in z from February 1985 can be related to a nearly earthquake in NW. The most important are the z and y displacements from July 1985, by 5,3 mm and 6,5 mm, respectively. Evidently, it is a slip along the transvers main fault without any opening   of it. It took place after a strong earthquake at Černá Gora (5,2 Richter; May 1985), and another one in NE Bulgaria, in the limits of the prolonged characteristic diagonal structure of Bulgaria.

Speaking generally about the given observations, one must underline that they represent permanent, irreversible displacements. At the moment of an earthquake even higher displacements due to vibrations were registered. There are more details and correlations regarding earthquakes and observed skips in the diagrams that cannot be discussed here. Such effects may be similar to those reported from San Andreas Fault, where cumulated elastic energy results in displacements of several cm to m (Wesson, Wallace 1985) were found.

CONCLUSIONS

Surface displacement measurements at a tectonic fault junction in   Bulgaria give reasonable data about seismotectonic movements on the faults. Even the observation points situated in places suspect for slope deformations give valuable data, and the seismotectonic reactions can be well recognized in the diagrams of the movements. The transvers fault has been found more active than the longitudinal one. The movements are of a skipping character mostly irreversible, and they react more sensitively to the earthquakes with epicentres in S, outside the Bulgarian territory. Displacements were observed in all the three spatial components separately, and reached up to 12 mm. There is a hope to find some direct prognostic data in the future when longer period of observations will be achieved.

REFERENCES

Cacoń, S. & B. Košťák 1976. Displacement registration of sandstone blocks in the Stolowe Gory Mts., Poland. Bull. IAEG, 13: 117-122, Krefeld.
Johnsen, G. & B. Košťák 1980. Effect of the rainfalls on the activity of slope deformations (Measurements in the Kraja district in North Durynsku). Čas. min. geol., 25, 2: 151-164. (in Čech.).
Košťák, B. 1969. A new device for in-situ movement detection and measurement. - Exp. Mechanics, 9, 8: 374-379. Easton Pa.
Košťák, B. & J. Rybař 1978. Measurements of the activity of very slow slope movements. In: Natau O., Fecker E., Reik G.: Grundlagen und Anwendung der Felsmechanik. Trans-Tech. Publications, 191-205, Clausthal.
Wesson, R. & R. Wallace 1985. Prognosis of a stronger earthquake in California. Scientific American, Feb., 2: 252.

Applications of elastic wave in engineering geological investigations

Application des ondes élastiques aux études géotechniques

Feng Guang Di, Zhang Bao Shan, Qian Bing Shen & Zhao Li Zhong, *Shenyang Engineering Exploration Corporation, MMI, China*

ABSTRACT: This paper presents the correlations between shear wave velocity and depth, and certain mechanical properties of soils. By using these correlations, the site soil classification for aseismatic design, dynamic analysis of dam stability, determination of sturdiness coefficient of surrounding rock and rigidity of natural foundations have been discussed.
All these results are in good conformity with each other and provide useful information for engineering geological evaluation.

RESUME: L'article a donné les relations entre la vitesse, la profondeur d'ondes de cisaille et certaines caractéristiques dynamiques du terrain; a expliqué comment,en utilisant ces relations, on peut résoudre les problèmes comme pour classifier la résistance sismique de l'emplacement, analyser par méthode dynamique la stabilité du corps de barrage, déterminer le coéfficient de dureté des roches d'alentour et la rigidite de la foundation naturelle etc.
Les résultats obtenus dans les differentes expérimentations sont generalement en accord les uns les autres, fournissant ainsi d'utile information pour l'évaluation géologique des travaux de construction.

PREFACE

Elastic wave velocity, especially shear wave velocity, has been used more and more in soil mechanics, rock mechanics, earthquake engineering and civil engineering etc. in recent years. A large number of theses involving the relationships between elastic wave velocity and soil characteristics (1),(2) and (3), correlations between elastic wave velocity and depth (4), and evaluation of foundation soil shear modulus and determination of stone bridge safety by using elastic wave velocity have been published (5),(6).
Since 1976, a large amount of work on elastic wave velocity testing has been done in Shenyang, Anshan, Benxi, Nanfen, Yingkou, Haicheng, Dashiqiao, Shuangyashan,Tangshan and Laoting respectively by our corporation. On the basis of these tests with downhole method as the major one, crosshole, refraction and surface wave methods as supplements, this paper presents the correlations between shear wave velocity and depth, correlations between shear wave velocity and some geotechnical engineering characteristics for various soils. It also offers ideas about how to use these correlations for site soil classification in aseismatic design, dynamic analysis of dam dam, determination of compressive rigidity coefficient of natural foundations and sturdiness coefficient of the surrounding rock of caverns.

1 CORRELATIONS BETWEEN SHEAR WAVE VELOCITY V_s AND ROCK DEPTH AND SOME ENGINEERING PROPERTIES

The elastic wave velocity of foundation soil is determined mainly by soil elastic properties, density and depth. The differences of elastic wave velocities for various soils are attributed to different constituents, textures and structures. Some wave velocity values are listed in table 1. Table 1 indicates that even for the same type of soil, the range of wave velocity V_s varies considerably because of the differences in density, depth, geological age and water content etc. But statistics of large amount of Vp and Vs which are made in accordance with the genesis and geological conditions of soil and its geological characteristics, i.e. based on the engineering geological unit, shows they are closely related to depth n and some engineering characteristics, especially the shear wave velocity. Table 2 and 3 show the regression equations of shear wave velocity of tailings materials of different sizes and natural loose layers with some geotechnical characteristics. Figs 1 and 2 represent the relationships.

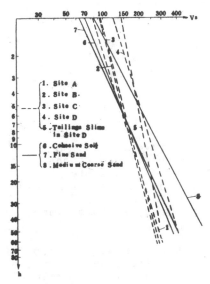

Fig.1 Correlation between Vs and h

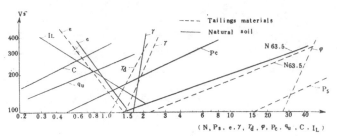

Fig 2   Correlation   between Vs and geotechnical engineering properties

TABLE 2 CORRELATIONS BETWEEN SHEAR WAVE VELOCITY Vs AND DEPTH h OF DIFFERENT SOILS

Description		Correlation	Coefficient R	Number of Statistic data	Location
Average grain size of tailings Sand	$d_{50}=0.062$ mm	$Vs=83.56 h^{0.3094}$	0.898	48	Site A
"	$d_{50}=0.083$ mm	$Vs=93.45 h^{0.2566}$	0.878	126	Site B
"	$d_{50}=0.083$ mm	$Vs=95.59 h^{0.2602}$	0.862	70	Site C
"	$d_{50}=0.173$ mm	$Vs=121.81 h^{0.2997}$	0.930	80	Site D
Tailings slime		$Vs=143.57 h^{0.1802}$	0.730	11	Site D
Cohesive soil		$Vs=70.27 h^{0.41197}$	0.969	165	Shenyang,Anshan, Yingkou and east Hepei Province
Fine Sand		$Vs=64.24 h^{0.45956}$	0.928	47	Yingkou, Anshan
Medium coarse Sand		$Vs=83.32 h^{0.51385}$	0.911	30	Shenyang, Anshan

TABLE 1 APPROXIMATE COMPRESSIVE AND SHEAR WAVE VELOCITIES OF TYPICAL MATERIALS IN THE CRUST

Item	compressive wave Vp (m/sec)	shear wave Vs (m/sec)	ρV_p (g/sec(cm 2×10 4))
weathered layer	100—500		1.2—9
Dry sand	200—800	120—500	2.0—16
Wet sand	300—800	120—500	3.8—16
Clay	1200—2500	100—400	1.8—55
Water	1430—1590		1.1—16
Soft sandstone	1500—2500	1050—1800	2.7—60
Intact sandstone	1500—4000	1300—3000	4.0—112
mudstone	2700—4800	1900—3400	6.5—130
Limestone	2500—6000	1800—4400	5.8—180
marl	2000—3500	1500—2500	4.5—100
Granite	4500—6500	3200—4600	11.0—200
Metamorphic rock	3500—6500	2500—4600	8.5—215
Basalt	4500—7500	3200—5500	12.0—250

TABLE 3 CORRELATIONS BETWEEN SHEAR WAVE VELOCITY Vs AND GEOTECHNICAL ENGINEERING PROPERTIES

Soil type	Regression equation	Coefficient R	number of statistical data	Location
Tailings materials	$Vs=73.2 N_{63.5}^{0.3909}$	0.850	252	Site A , Site B , Site C , Site D
	$Vs=29.7 Ps^{0.4439}$	0.881	301	"
	$Vs=154.6 e^{-1.866}$	−0.734	249	"
	$Vs=46.48 \gamma^{2.2045}$	0.784	253	"
	$Vs=86.99 \gamma_d^{1.8378}$	0.653	253	"
	$Vs=\varphi^{2.7427}/85.42$	0.724	193	"
natural soils	$Vs=91.35 N_{63.5}^{0.3477}$	0.833	269	Shenyang, Anshan, Yingkou and east Hepei province
	$Vs=182.89 e^{-1.340}$	−0.690	105	"
	$Vs=5.417 \gamma^{5.8801}$	0.700	105	"
	$Vs=142.39 Pc^{0.4676}$	0.769	27	Anshan
	$Vs=236.88 q_u^{0.4044}$	0.788	46	"
	$Vs=344.19 C^{0.4987}$	0.748	41	"
	$Vs=183.39 I_L^{-0.6192}$	−0.6474	122	Anshan, Yingkou and east Hepei province

Where Vs, $N_{63.5}$, Ps, e, γ, γ_d, φ, Pc, q_u, C, I_L, represent shear wave velocity, SPT value, ratio of cone penetration resistance, natural void ratio, natural unit weight, dry unit weight, internal friction angle, preconsolidation pressure, unconfined compressive strength, cohesion and liquid index respectively.

Above data show that shear wave velocities of the foundation soil are in close relation with depth and geotechnical engineering properties. The following empirical equation is obtained considering the effect of both factor on shear wave velocity through binary correlation.

For tailings dam deposit materials

$$V_s=70.21 \ N_{63.5}^{0.3312} \cdot h^{0.0761} \qquad (n=252 \quad R=0.860) \quad \dots 1$$

$$V_s=31.91 \ P_s^{0.3746} \cdot h^{0.0868} \qquad (n=301 \quad R=0.894) \quad \dots 2$$

$$V_s=100.38 \ e^{-0.9746} \cdot h^{0.1941} \qquad (n=249 \quad R=0.844) \quad \dots 3$$

$$V_s=49.71 \ \gamma^{1.6494} \cdot h^{0.1151} \qquad (n=253 \quad R=0.815) \quad \dots 4$$

$$V_s=54.44 \ \gamma_d^{1.5323} \cdot h^{0.2279} \qquad (n=253 \quad R=0.871) \quad \dots 5$$

$$V_s=\frac{\varphi^{2.1733} \cdot h^{0.1428}}{16.17} \qquad (n=193 \quad R=0.800) \quad \dots 6$$

Natural soil

$$V_s=75.67 \ N_{63.5}^{0.1436} \cdot h^{0.2563} \qquad (n=269 \quad R=0.913) \quad \dots 7$$

$$V_s=94.99 \ e^{-0.7497} \cdot h^{0.2546} \qquad (n=105 \quad R=0.872) \quad \dots 8$$

$$V_s=14.78 \ \gamma^{3.1617} \cdot h^{0.2460} \qquad (n=105 \quad R=0.861) \quad \dots 9$$

$$V_s=86.83 \ I_L^{-0.2129} \cdot h^{0.3030} \qquad (n=122 \quad R=0.882) \quad \dots 10$$

$$V_s=105.41 \ q_u^{0.0635} \cdot h^{0.2810} \qquad (n=46 \quad R=0.879) \quad \dots 11$$

$$V_s=100.08 \ c^{0.0979} \cdot h^{0.3119} \qquad (n=41 \quad R=0.897) \quad \dots 12$$

$$V_s=97.73 \ P_c^{0.0285} \cdot h^{0.2914} \qquad (n=27 \quad R=0.036) \quad \dots 13$$

Shear wave velocities can be estimated by using the above mentioned empirical formula on the basis of knowing the foundation soil depth and geotechnical engineering properties and vice versa.

2 DYNAMIC STABILITY ANALYSIS OF SITE AND DAM BY USING SHEAR WAVE VELOCITY

Dynamic stability analysis requires calculation of the stresses from earthquake to the site and dam. Dynamic elastic modulus E_d and dynamic shear modulus G_d are indispensable in the calculation. Distribution laws of dynamic elastic modulus can be obtained by using $G_d=(\mu/g)\cdot V_s^2$ and $E_d=2(1+\mu)/g\cdot V_s^2$ on the basis of knowing the relationships between shear wave velocity and depth h. Fig.3 presents the curve of shear modulus G_d with depth for tailings deposits and natural soils.

Fig.4 shows the evaluation results of liquefaction potential by the use of shear wave velocities and other necessary data in the dynamic stability analyses of Nanfen Tailings Dam.

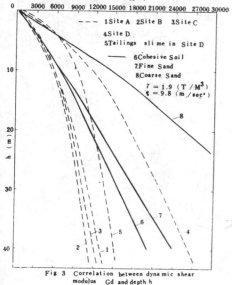

Fig 3 Correlation between dynamic shear modulus Gd and depth h

Fig 4 Basic data earthquake shear stress and liquefaction shear stress of tailings materials

Depth h MM	Layer thickness	Log	Soil type	No	Hi M	ρ	Vs m/sec	Gi T/M²
5	5		Fine Sand	1	5	0.18	97	1700
10	5			2	5	0.18	125	2800
15	5			3	5	0.19	136	3500
20	5			4	5	0.19	141	3800
25	5		"	5	5	0.19	149	4200
30	5		"	6	5	0.19	154	4500
35	5		"	7	5	0.19	164	5100
40	5		"	8	5	0.19	170	5500
45	5		"	9	5	0.19	176	5900
50	5		"	10	5	0.19	181	6200
55	5		"	11	5	0.19	186	6600
60	5		''	12	5	0.19	192	7000
65	5		Tailings	13	5	0.19	197	7400
70	5		Slime	14	5	0.19	201	7700
74			"	15	4	0.19	205	8000
				16		0.21	2185	10⁶

Shear Stress (kg /cm²)

Depth (M)

× Earthquake Shear Stress τxzd when g =0.1
○ Earthquake Shear Stress τxzd when g =0.2
▲ Shear Stress τd of Coarse Tailings Sand when γd=1.46
c Shear Stress τd of Coarse Tailings Sand when γd=1.54
△ Shear stress τd of Fine Tailings sand when γd=1.46
◇ Shear stress τd of Fine Tailings sand when γd=1.43

3 SITE SOIL CLASSIFICATION ON THE LOWER LIAO RIVER PLAIN BY MEANS OF SHEAR WAVE VELOCITIES

The classification of foundation soil in aseismatic design code is quite different in various countries. The lithological property of soil is generally considered as a factor in the classification. Location of groundwater table, state of frozen soils, thickness and structure of strata are considered in some cases. The classification seems to be only a qualitative description. A few countries adopt physico-mechanical properties to define the range of classification. In the Chinese aseismatic design code laid down in 1964, 1974 and 1978, the foundation soil classification is based on macro earthquake damage, macroseismatic observation data and theoretical analysis without classification ranges. We had made an earthquake damage investigation after the Haicheng earthquake of Feb.4, 1975, studied the macroseismic data and conducted detailed exploration and testing on the lower Liao River Plain(see table 4). Site soil classification on the lower Liao River Plain was proposed on the basis of the above work(see table 5). This classification is made mainly on the basis of shear wave velocity. This is because shear wave velocity is sensitive to types of soil with comparatively consistent values and in rather good correspondance with physico-mechanical properties of soils. Therefore, aseismatic classification of site soils is feasible from shear wave velocity.

TABLE 4 EXPLORATORY AND TESTING WORK

Locations		Number of borings	Max boring depth m	Elastic velocity	Lab. Physical & mechanical tests	SPT	CPT	Dynamic triaxial tests	Electrical resistivity	Load tests	Remarks
Yingkou district	Yingkou	4	47.90	V	V	V		V			1. V represents the tests performed 2. V/15 denominator represents number of test holes
	Dashichiao	1	35.90	V	V	V					
	Haicheng	1	34.00	V	V	V					
Anshan		15	95.00	V	V	V	V/30	V	V/3	V/15	
Shenyang	Compression Gasifying Plant	3	49.50	V	V	V					
	Northeast Electricity Bureau	4	48.00	V	V	V			V/3		

Table 5. Soil site classification on the lower Liao River Plain(proposal)

Classification		Description	V_s (m/sec)	SPT $N_{63.5}$	D_r (%)	(R) (kg/cm^2)	q_u (kg/cm^2)	Typical location
I		I-1 Stable bedrock overburden(ordinary soil) bedrock < 3m moderately-weathered slightly weathered bedrock	> 700					East hills Huaziyu
		I-2 Large crushed stone over bedrock with thickness less than 10m ground water level greater than 15m	> 500					East hills mostly at the outlet of the valley
		I-3 Hard dense old clay with thickness > 10m	> 500	> 30		> 4		Blast furnace area in Anshan I & S Co.
II		II-1 Medium dense large crushed sandy soil stone groundwater level from 6 - 10m	200-500 200-400	10-30	60-80			Planning building new plant site Shenyang
		II-2 Ordinary cohesive soil groundwater level > 5m	200-400	10-25		1.5-2.5	1.5-?	Dashiqiao Haicheng Anshan
III		III-1 Loose slightly dense sand layer groundwater level < 4m	100-200	< 10	30-50	< 1.4		Most area in Yingkou Panjin
		III-2 Muck and mucky soil groundwater level < 4m	100-200	< 10		< 1.0	0.5-1.0	Partial area in Yingkou Panjin
IV		IV-1 Liquefiable saturated silty fine sand & silty clay groundwater level < 4m	Determination of liquefaction potential according to SPT blows see article 10 clause (2) in TJ 11-78 in addition make reference to mean grain size 0.02-1.0mm uniformity coefficient < 10 relative density < 75% cohesive particle content < 10%					Certain area in Yingkou Panjin
		IV-2 Rock and soil with karst & caverns between 3-8m						Both banks of Ximu River in Haicheng County
		IV-3 Back fill (recent deposit and by dranlic fill loose tailings sand & slag)						Chemical plant of Yingkou county No.5 sinter plant of Anshan I & S Co.

Remarks: 1. The earthquake effect coefficient of site soil I II III can be obtained per the response spreotrum in Fig.2 of article 14 in TJ 11-78.
2. Usually no per manent buildings should be constructed on site IV unless detailed geological investigations and proper foundation treatments have been performed.

4 DETERMINATION OF ALLOWABLE BEARING CAPACITIES (R) OF FOUNDATION SOILS ACCORDING TO SHEAR WAVE VELOCITY

The allowable bearing capacities (R) of various foundation soils can be determined by their physico-mechanical properties, field identification results or in situ testing. Here, according to the relationship between shear wave velocity V_s and SPT $N_{63.5}$ of foundation soil,

$$V_s \text{ (cohesive soil)} = 85.46 \ N_{63.5}^{0.388}$$

$$V_s \text{ (sandy soil)} = 67.91 \ N_{63.5}^{0.4117}$$

with reference to shear wave velocity V_s, the allowable bearing capacities of ordinary cohesive soil and sandy soil is recommended in table 6.

TABLE 6 ALLOWABLE BEARING CAPACITIES (R) OF ORDINARY COHESIVE SOIL AND SANDY SOIL

	V_s (m/sec)	130	160	182	200	217	231	244	257	266	278	288
Ordinary cohesive soil	SPT blows $N_{63.5}$	3	5	7	9	11	13	15	17	19	21	23
	Allowable bearing Capacity (R T/M³)	12	16	20	24	28	32	38	42	50	58	66
Sandy soil	V_s (m/sec)	175—210			210—275			275—340				
	SPT blows $N_{63.5}$	10—15			15—30			30—50				
	Allowable bearing capacity (R)(T/M³)	14—18			18—54			34—50				

5 DETERMINATION OF COMPRESSIVE RIGIDITY COEFFICIENT C_z OF NATURAL FOUNDATIONS FROM SHEAR WAVE VELOCITY

The method of determining the dynamic parameters of natural foundations has been arousing concern. In the "DESIGN CODE OF DYNAMIC MACHINE FOUNDATIONS"(China, 1983) in effect, the compressive rigidity coefficient C_z is determined by allowable bearing capacities (R) of foundation soil. In practice, it has shown that this method has certain limitations. Therefore, the "DESIGN CODE" lays further emphasis on using in situ measuring techniques to determine C_z, but fails to point out the more reasonable methods. Of course, the results from large model foundation forced vibration test are comparatively reliable, but this method is complex, costly and time consuming. Our corporation has accumulated valuable measured data on large model foundation forced vibration tests and shear wave velocity at same places in recent 7 or 8 years. In data analysis we found that the determination of compressive rigidity coefficient of natural foundations from shear wave velocity measurement is proved as a simple and convenient method. Table 7 gives the measured data at various locations. According to table 7, the relationship between C_z and V_s can be expressed by Fig.5.

TABLE 7 MEASURED DATA AT VARIOUS LOCATIONS

Location	Soil type	V_s (m/sec)	C_z (T/M³)
Shenyang	Medium sand medium dense slightly dense	100—110	4000—5000
Fushuan	Medium—coarse sand medium dense slightly dense saturated	110—120	5000—6000
Fengrun	Fine sand dense	120—130	6000—7000
Shenyang	medium—coarse sand medium dense	130—140	6500—7500
Baotou	Gravelly sand dense	200—240	10000—11000
Fangshan	Highly weathered granite layer	300—320	11000—13000
Beijing	Silty clay plastic	120—130	4000—5000
Benxi	sandy clay plastic	140—160	6000—7000
Xian	Loessial loam plastic	160—180	7000—8000
Maanshan	Xiashu clay hard plastic	380	12000

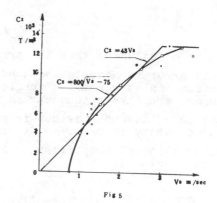

Fig 5

The relationship between C_z and V_s of various soils is mainly a parabolic type when

351

V_s increases to a certain value, C_z varies little and approaches constant. If a parabola is being used, the relationships between C_z and V_s could be approximately expressed as follows:

$$C_z = 800 \sqrt{V_s - 75} \quad \cdots \cdot 14$$

in which, when $V_s = 75$, $C_z = 0$.

This is apparently not in comformity with reality. The cause of this phenomenon is due to the lack of measured data of actual foundations having $V_s \leqslant 75$ m/sec. But , strata with small V_s naturally have small C_z, and not suitable for natural foundations. Therefore, although the theory and practice are conflicting to some extent, the above equation is still of practical value.

The relationships between C_z and V_s can also be expressed by a linear relationship. In this case, only the section with $V_s \leqslant 300$ m/sec is taken into consideration, i.e:

$$C_z = 43 V_s \quad \cdots \cdot 15$$

In fact, according to Fig.5, when $V_s \geqslant 300$ m/sec, C_z can be assumed as constant, i.e. $C_z = 13000$ t/m^3. Tests have proven that strata with $V_s \geqslant 300$ m/sec are in good soil condition, Xiashu clay in Maanshan and highly weathered granite in Fangshan,Peijing for example, their foundation rigidity is easy to meet the control of vibration amplitude in dynamic calculations. But to avoid resonance and control the vibration acceleration is of most importance.

C_z can be determined by using V_s from field measurement according to equations 14 or 15. This value represents the compressive rigidity coefficients for foundations having a base area greater or equal to 20 m^2. C_z as the actual machine foundation base area is less than 20 m^2, C_z should be corrected according to the following equation.

$$\psi = \frac{1}{0.4 + 0.6 \sqrt[3]{F/F_o}} \quad \cdots \cdot 16$$

in which:

ψ= correction factor. when correcting, multiply C_z determined from V_s by $1/\psi$.

$F = 20$ m^2

F_o= actual machine foundation base area (m^2)

Table 8 gives the comparison of C_z determined by forced vibration tests and calculation results by the use of equations 14 and 15.

TABLE 8 COMPARISON BETWEEN Cz DETERMINED BY FORCED VIBRATION TESTS AND Cz DETERMINED FROM SHEAR WAVE VELOCITY

Location	Vs (m/sec)	1 Cz determined by forced vibration Tests	2 Cz determined from formula 14	3 Cz determined from formula 15	4 Cz determined by table 1 in dynamic design code	$\frac{2-1}{1}$	$\frac{3-1}{1}$	$\frac{4-1}{1}$
Shenyang Medium sand	100	4000	4000	4300	3400	0%	+7.5%	−15%
Fushuan medium coarse sand	110	5000	47000	4700	3400	−6%	−6%	−32%
Baotou Gravelly sand	200	10000	8800	8600	4800	−12%	−14%	−52%
Fang shan Highly weathered granite	300	11000	12000	13000	10000	+9%	+18%	−10%

Table 8 indicates that the error is not more than 20% by using V_s to determine C_z.

This error is acceptable because its influence on the calculation of amplitude limitations is very small. Dynamic machine foundations are usually shallow embedded and errors are likely to occur in measuring V_s in shallow soils. Therefore, the same method -- downhole method is preferable and a proper increase in measuring numbers is necessary in order to determine the upper and lower limits of V_s rationally.

Other physical properties of soils should also be considered in determining the limiting value of V_s.

6 DETERMINATION OF STURDINESS COEFFICIENT f OF ROCK FROM ELASTIC WAVE VELOCITY

The sturdiness coefficient is a comprehensive index for the evaluation of rock stability and obtained from comprehensive geological evaluation in the past. It has brought certain difficulties to quantitative evaluation because of lack of quantitative data, and also, resulting in different values from various people as well. To solve the problem, some organizations at home and abroad have tried to use elastic wave method to study the rock stability. Fig.6 gives the correlation between f and the measured compressive wave velocity V_p in Xiguoyuan Tunnel in Anshan, Gongchangling Tunnel, Nanfeng Tunnel and Shuangyashan Tunnel. The fourteen pairs of points in Fig.4 from the data obtained from Waitoushan and Gongchangling are used

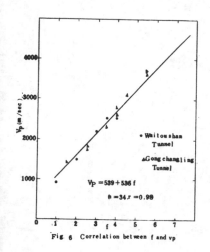

Fig. 6   Correlation between f and vp

to prove the reliability of the correlation curve in Fig.6. The comparison shows that compressive wave velocity V_p is in close relation with f. Furthermore, the above results were used in Nanfen iron mine vertical shaft project and rather good results were obtained as well. It should be pointed out that in areas with complicated structures and special geological conditions, such as hot water metamorphic zones mixed with clayey zones and siliconized zones and serpentinite zones, structural factors and other physico-mechanical properties should also be considered rather than using only the compressive wave velocity in the evaluation of rock mass stability. If the angle between the cavern axis and main fissure direction is less than 25° or groundwater occurs in rock mass, the f value determined from wave velocity should be minus 1. If the rock mass is cut and separated into rhombi by open continuous fissures, special studies have to be made to determine f. For non-softening soft rock where fissures are not developed, the f value determined from wave velocity should be plus 1.

CONCLUSION

Elastic wave velocity of soil is closely related to depth and certain engineering properties. All these relationships are in good conformity with each other and provide useful information for engineering evaluation.
Among these relationships, some of them have been used in our site investigations and good results have been achieved, such as estimating soil physico-mechanical properties; earthquake stability analysis of tailings dam, determination of sturdiness coefficient of tunnels and shaft, determination of the compressive rigidity coefficient of natural foundations etc. And some other relationship may be considered as proposals, such as classification of site foundation soil.
Downhole method is independent of geotechnical properties and conditions of strata. In cases which undisturbed samples of some soils(saturated loose sand, very soft clay and silt etc.) could not be taken for lab. testing, the downhole method can be used without difficulty. As mentioned above the elastic wave velocity method will gradually become necessary and play more and more important role in engineering project investigations.

REFERENCES

(1) IMAI Tsuneo and YOSHIMURA Masayoshi, (1972): The relation of mechanical properties of soil to P- and S- wave velocities, Geophysical Exploration, Vol.25, No.6.
(2) OHTA Yutaka and GOTO Noritoshi, (1976): Estimation of S-wave velocity in terms of characteristic indices of soil, Geophysical Exploration, Vol.29, No.4.
(3) Zhang Bao-shan and Lin Cho-pin, (1983): Shear Wave Velocity and Geotechnical Properties of Tailings Deposits, Bulletin of the International Association of ENGINEERING GEOLOGY, No.26-27 Paris.
(4) LEW Marshall et al., (1981): Correlations of seismic velocity with depth.
(5) Zhang Guo-xia and Zhang Nai-rui, (1983): A Statistical Study of Shear Modulus Determination by In-situ Shear Wave Velocity and Dynamic/Static Triaxial Tests, Chinese Journal of Geotechnical Engineering, Vol.4, No.2.
(6) OHMI MICHITO, KANEKO Katsuhiko and INOUE Masayasu, (1981): Study on the inspection of Stone Bridge by Means of Acoustic Method, Geophysical Exploation, Vol.34, No.2.

New aspects of rock state of stress study
Nouveaux aspects dans l'étude des contraintes

G.A.Golodkovskaya, L.L.Panasiyan, M.A.Petrovsky & S.A.Kolegov, *Moscow State University, USSR*

ABSTRACT: The state of stress of the rock mass is conditioned by the effect of gravitational, tectonic, hydrodynamic, geothermal and crystallization forces. It is recommended to integrate several methods: measuring stress in underground excavations and boreholes, performing special geophysical investigations, determining stresses through the use of acoustic emission and mathematical simulation. The prospect of such an integrated approach is illustrated on the example of Eastern Siberia. The conducted studies made it possible to explain the stress inversion in the subsurface zone, to locate the areas in the rock mass, where the state of stress is close to the hydrostatic one, to determine the size of these areas and their depth of occurrence and to single out from the general measured stress tensor a part which corresponds either to the local stress increase caused by the rock mass structure, or to a certain level of tectonic stress, or to the effect of the stationary thermal field.

RÉSUMÉ: L'état des contraintes d'un massif des roches est conditionnè par les forces tectoniques, hydrodynamiques, géothermiques et de crystallisation. Il est recommendé l'application d'un complexe des méthodes à mesurer des contraintes: le mesurage des contraintes dans les excavations souterraines et dans les trous de forage, les méthodes géophysiques spéciales, la détermination des contraintes à l'aide de l'émission acoustique, la simulation mathématique. La perspective de cette complexion des méthodes est illustrée sur l'exemple des gisements en Siberie de l'Est. L'information obtenue à la base d'un complexe des méthodes à permi d'expliquer l'inversion des contraintes dans la partie superficielle du massif; localiser les zones caráctérisées par l'état semihydrostatique des contraintes, determiner leurs dimensions et profondeurs; séparer du tenseur géneral le composant correspondant soit à l'augmentation locale des contraintes, liée avec la structure du massif, soit à un certain niveau des contraintes tectoniques, soit à l'influence d'un champ thermique stationaire.

The state of stress of the rock masses in the upper part of the lithosphere is the subject of much investigation by the specialists in different fields: tectonophysics, seismology, mining geology and mathematics. Their efforts are bent towards elucidating the structure of the stress field in the rock mass in order to assess the value and direction of the maximum stresses which knowledge is essential for different types of construction activities and in mining.

The study of the stressed-strained state of the rock masses is particularly topical when working deep-lying useful mineral deposits, when high stresses are the cause of rock outburst. Meanwhile, the data of the stress field structure at a considerable depth can be obtained by the existing in-situ test methods only in the process of exploitation of the deposit, while the forecast of the maximum stress values is required already at the stages of prospecting mineral deposits and designing a mining enterprise. To solve the above problem an integration of a number of methods is very promising. The results of the work done in this field at the Moscow State University are set forth below.

The studies of the state of stress of rocks in-situ under different conditions have indicated that the stresses can both correspond to the values derived by the hypotheses of A. Heim ($\sigma_1 = \sigma_2 = \sigma_3 = \gamma H$) and A.N. Dinnik ($\sigma_1 = \gamma H, \sigma_2 = \sigma_3 = \frac{\mu}{1-\mu} H$, where μ is the Poisson's ratio, γ -density and H - depth of rock occurrence) and greatly exceed these values. This is associated with a fact that the formation of the state of stress of the rock mass is going on with a great many factors participating, the most important of which appear to be the composition and structure of the rock mass, depth of occurrence, physical and mechanical properties of rocks, their density and deformability, earth surface relief, hydrodynamic conditions of the mass, geophysical and geochemical processes taking place in the mass, distribution of temperatures and man engineering activities. The most important of them from the viewpoint of most of the researchers are the gravitational, tectonic, hydrodynamic, geothermal and crystallization forces. Their effect on the mass is found to be not always unique, which in every particular case calls for tackling new methodical and practical problems.

It is customary to assume that the data of stresses measured in underground excavations or boreholes applying different methods are the most reliable ones. However, this method is not always practicable, since to obtain the statistically grounded data it requires a great many definitions to be made and the extrapolation of the obtained data is often difficult because of the lack of knowledge of the general structure of the stress field in the rock mass involved. Therefore, indirect methods of determining the state of stress of rocks are gaining recognition, in particular, geophysical studies, measuring of acoustic emission parameters and mathematical simulation. The most hopeful data can apparently be obtained using a complex of methods. Such a comprehensive investigation of the structure of the stress field of the rock mass has been conducted by the authors at several mineral deposits in the Soviet Union to forecast mining and geological phenomena at deep levels.

The most important stage in the analysis of the stress field is the creation of the model of the geological structure of the rock mass under investigation. The geological information essential to build up such a model can be obtained already at the stage of mineral prospecting. The geological structure model is required to take account of the rock structure non-uniformity when choosing the design scheme in order to be able to assign properly the areas of in-situ stress measurements for purposeful taking samples to be subjected to acoustic emission tests. It is also required to elucidate the stress anomalous values and to ascertain the major factors defining these values.

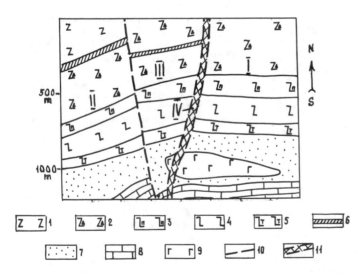

Figure 1. Geological structure model. 1-porphyric basalt; 2 -polyphyric basalt; 3-picrite-basalt, 4-tholeiite basalt; 5-titanaugite basalt; 6-tuffites; 7-sandstones, aleurolites; 8-marl, limestone, dolomite; 9-gabbro; 10-fracture lines; 11-crush zone

Let us consider the proposed approach on the example of one of the mineral deposits in Eastern Siberia, USSR, the geological structure model of which is illustrated in Fig. 1. As for the geological structure of the area, it is formed by effusive (basalt) and intrusive (gabbro-dolerites) rocks, sedimentary terrigene and carbonate deposits and can be subdivided as follows: I - eastern block with practically gentle dip of rocks with gabbro-dolerites intrusion; II - western block with monoclinal dip of rocks (westward dip at angles to 20°); III - central block, composed of highly fissured rocks and IV - fracture zone.

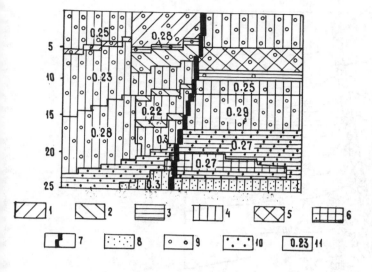

Figure 2. Model of rock properties distribution in the design
rock mass.
Elastic modulus variation range
(E·10^4 MPa): 3.5-3.6(1), 4.2-
4.5(2), 5-5.3(3), 6.1-6.5(4),
7.6-7.9(5), 9.4(6), 0.1(7).
Coefficients of linear expansion
($\alpha·10^{-5}$ degrees^{-1}): up to 0.5
(8), 0.54(9), 1.0(10).
Poisson's ratio values (11)

The geological structure model served as the basis for the engineering geological model, which schematically shows the elements alike by their composition and
by rock properties and condition, as well as their expansion limits. The given
model determines the position of a certain material point in the net area of the
design rock mass and magnitudes of the design parameters within it (Fig. 2). For
the above rock mass several calculations have been made using different schemes
taking into consideration the geological structure non-uniformity and peculiarities of rocks occurrence, combination of indices of properties of different types
of rocks and roughness of relief. Forces taken as those forming the stress field
were the dead weight of rocks which defines the main gravitational stress field,
tectonic forces and the stationary heat flow which causes additional stresses
(Golodkovskaya, G.A. & L.L. Panasiyan 1979). Rigid cohesion between the layers
was supposed to exist at their interface. The calculation was made for the pon-
derable elastic medium under the conditions of plane deformation in the section
oriented across the strike of the basic geological structures of the area.

GRAVITATIONAL STRESS FIELD STRUCTURE

When defining the gravitational stresses the Young's modulus and the Poisson's
ratio were adopted as the variable properties, since the variations of these parameters in the rock mass under study appear to exceed the variations of rock density, which in the above example was taken equal for the whole geological section.
The results of the stress calculations were obtained in the form of the verti-
cal (σ_y), horizontal (σ_x) and tangential (τ_{xy}) components in every elementary
cell. In Fig. 3 a,b the results of calculations are given in the form of equal
stress isolines.
The western block features a more quiet gravitational stress field structure.
Vertical stresses are found to increase evenly with depth and at the level of
1100 m they reach 26 MPa. In this block vertical stresses appear to concentrate
on the boundary with the Devonian basement rocks being characterized by higher
values of the modulus of elasticity. In the eastern block the distribution of
vertical stresses is much more complicated: down to a depth of 700 m they evenly
grow up and reach 13 MPa, while within the depth interval of 900-1200 m (intrusive body and terrigene rocks enclosing it) the structure of the stress field
gets markedly complicated - vertical stresses increase to about 35 MPa and drop
to 20 MPa in the rocks underlying the intrusion. The distribution of horizontal
stresses in the western block is also rather even. In the eastern block the horizontal stress field is considerably complicated by the availability of a rigid
body of intrusion rocks, in separate parts of which the magnitudes of horizontal stresses are 1.5-2 times greater than those of horizontal stresses in the enclosing rocks. The concentration of stresses is observed also in the Devonian carbonate rocks roof, the modulus of elasticity of which is 25-30 per cent higher
than that of the overlying rocks of the Tungus series.

357

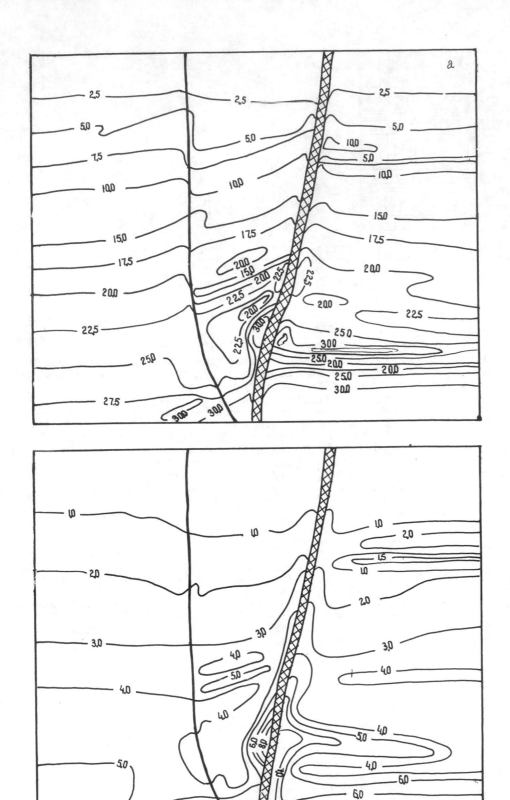

Figure 3. Gravitational stress isolines (MPa).
a) vertical (σ_y), b) horizontal (σ_x)

Especially non-uniform is the stress field in the central graben-shaped structure. In the whole the structure is characterized by higher and non-uniform jointing and, therefore, by lower and extremely different elasticity indices of rocks. So, the values of the modulus of elasticity of the basalts in the block vary between $3.5 \cdot 10^4$ and $6.3 \cdot 10^4$ MPa, i.e. almost twice as much. Therefore, within the limits of the described structure the zones of relative concentration of stresses change into the zones of lower stresses with vertical stress drop to 3 MPa for the first tens of meters.

The regularities of the stress field structure enumerated above - the concentration of stresses in more rigid rocks and decrease of the stress level in the weakened zones - are well promounced within the given structure. In particular, the zones of stress concentration are clearly defined in the near vicinity of the fracture zone. Over certain depth intervals vertical stresses in such a zone increase by 5-10 MPa as compared to the background ones. Their magnitude in the intrusive body is 1.5-2 times greater than in the enclosing rocks, and in the near-fracture zone it is 25-30 per cent greater than in the central part of the block. Just in the fracture zone σ_y does not exceed 1.5-2 MPa even at a depth over 1000 m, while the values of the background stresses at such a depth make up more than 30 MPa.

Thus, the horizontal and vertical stresses obtained as a result of calculations have not only common distribution features mentioned above, but also considerable differences stipulated by the wedging action of the central block: the degree of concentration of horizontal stresses near the fracture zone is higher than that of vertical stresses (for instance, at the level of 600 m for σ_x - up to 30 per cent and for σ_y - not more than 10-15 per cent); the magnitudes of vertical stresses at any depth on both sides (western and eastern) of the fracture zone are roughly the same, while the horizontal stresses in the fracture zone boundary part of the central block are 20-30 per cent greater than those at the same depth in the near-fracture part of the eastern block.

The studies conducted by many researchers (Non-linear geophysical problems 1981) prove, that the magnitudes of horizontal stresses almost always exceed those obtained by simulation with allowance made only for the gravitational forces. Investigations carried out by a large group of researchers (Kola superdeep borehole, 1984) have indicated that to build a tectonophysical model of the rock mass requires taking into account the thermodynamic conditions of rock occurrence, since any change in their condition especially at deep levels entails the redistribution of stresses.

STRESS FIELD WITH REGARD TO THERMAL AND PHYSICAL CHARACTERISTICS OF THE ROCK MASS

The next set of calculations has taken account of the availability of the stationary thermal field which is likely to exert a tangible effect on the state of stress. For this purpose, the use was made of the data of measurements of temperatures in deep boreholes, which helped to disclose the salient features of the thermal field of the area. In this case the design model has been developed from the model taken for defining the gravitational stresses with due regard for the distribution of temperatures with depth and thermal and physical properties of the mass-forming rocks. When analysing the state of stress with regard to the temperature the generalized values of the coefficient of thermal expansion were used, which for the rocks of effusive complex and intrusive formations was found to be equal to 0.54, for terrigene rocks it was equal to 1 and for the Devonian carbonate rocks it did not exceed 0.5.

Thus, the total values of vertical stresses obtained as a result of simulation with regard to the gravitational and thermal fields slightly differ from those calculated for the gravitational field, which complies with the theoretical prerequisites of calculation of the stressed-strained state under the given conditions. Temperature makes a material contribution to the change of horizontal stresses. In Fig. 4 are given the epures of vertical and horizontal stresses, which characterize the field stress within every tectonic block in its relatively quiet parts, while Fig. 5 gives the horizontal components of the gravitational and total stress tensor at different depth levels.

Figure 4 well illustrates the general tendency of an even increase of the total stresses with depth. Depending on the petrographic composition and state of rocks the increment of temperature stresses slightly differs, which well agrees with the accepted values of the thermal and physical properties of different types of rocks. Account must be taken of the fact, that when calculating the total stresses

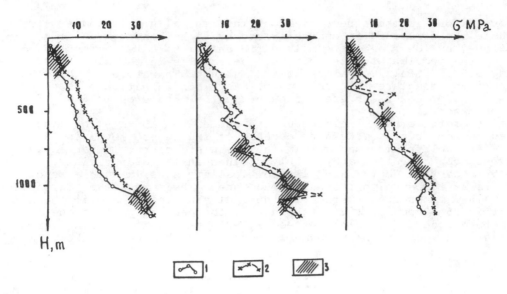

Figure 4. The epures of the total stresses.
1 - horizontal; 2 - vertical; 3 - equal-component stress zones

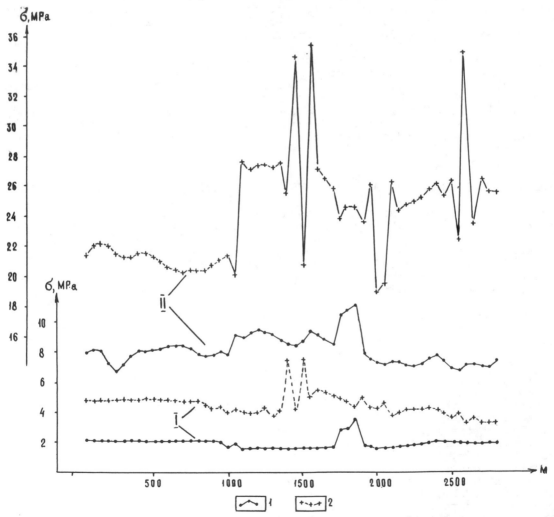

Figure 5. Distribution of horizontal stresses at the depth:
I-500m; II-1000m; 1-under conditions of gravitation; 2-with regard to the temperature field.

the moduli of elasticity of rocks within the temperature ranges available in the given area were taken constant.

As appears from Fig. 5, the main part of horizontal stresses can develop due to the effect of the stationary thermal field. In this case on separate sections the horizontal gravitational stresses make up not more than 25 per cent of the total horizontal stresses. So, at a depth of about 1000 m in the intrusive body their mean values reach 23-26 MPa, while the calculated gravitational stresses do not exceed 5 MPa.

The total stress tensor in different structural elements turns out to be equal-component (hydrostatic compression conditions) at different depth. The values σ_x and σ_y in these cases differ from each other by not more than 10-15 per cent. This factor is of particular interest, since zones with a round stress tensor have been determined by the authors before with due consideration for tectonic forces (Golodkovskaya, G.A. & L.L. Panasiyan 1979) and by other researchers when measuring stresses.

In the previous papers of the authors (Golodkovskaya, G.A. & L.L. Panasiyan 1979; Panasiyan, L.L. 1983) it was shown that the effect of tectonic forces led not only to the appearance of the said zones with "hydrostatic" distribution of stresses, but also to the inversion of stresses. The latter is manifested in the predominance of the horizontal component of the stress tensor over the vertical one in the near-surface parts of rock masses, as well as in reorientation of the axes of principal stresses and change of their angles of deviation. The effect of the tectonic forces on the distribution of stresses in the upper horizons of the earth's crust is rather complicated and depends both on the absolute value of these forces and their increment with depth and also on the state and internal structure of the rock mass, while the zones with the indicated particular conditions as well as in case of existence of the thermal field are liable to occur at different depth levels and to be of great thickness.

The data of the mathematical simulation have been correlated with the data obtained through measuring the stresses by the load relaxation method in underground excavations (Trofimov, I.M. 1983). The above measurements have indicated that the vertical components of the stress tensor of the rock mass under study are commensurable with the value of gravitational stresses, while the values of horizontal stresses are many times greater than the value which can be derived by A.N. Dinnik's hypothesis. The value σ_x, in particular, at a distance of 4-5 m away from the steeply dipping fracture of meridional orientation (Fig. 1) was equal to 60 MPa at a depth of 800 m. The mean value of the total horizontal (gravitational and thermal) stresses in the calculations for similar geological conditions is 10-12 MPa, while near the fracture the stress tends to rise by 30-40 per cent. In this case the approximate value of additional stress, which is usually called a tectonic component of the stress tensor, can be assessed at 30-35 MPA.

Special geophysical studies in underground excavations have also disclosed a complex structure of the stress field. In individual cases the obtained values of vertical stresses considerably exceeded those of the gravitational component. The horizontal stresses often exceeded the vertical ones and their orientation in the ores and in the enclosing rocks was different; at a depth of about 400 m areas with a round stress tensor were discovered.

The data of the state of stress of the mass obtained by the above in-situ methods and characterizing the stresses at individual points were found to be inadequate to build the stress field and sometimes even contradictory. So, new methods of obtaining the information of stresses in rocks are required. A search for such a new method resulted in an attempt to use acoustic emission, which had been for a long time used for flaw detection in metals.

DETERMINING STRESSES BY THE USE OF ACOUSTIC EMISSION (AE)

As is generally known, acoustic emission develops in the process of reconstruction of the internal structure of solid bodies and shows as low energy elastic waves radiation (10^{-9} - 10^{-5} J) at plastic deformations of solid materials, at phase transformations associated with changes in the crystal lattice, in the process of internal friction among structural elements at microfracture formation etc. (Greshnikov, A.A. & Yu.B. Drobot 1978; Vinogradov, S.D. 1964; Dunegan, H.L. 1973; Sholtz, C.H. 1968). The regularities of acoustic emission show in rocks are as yet imperfectly understood and it is used in mining mainly to study the kinetics of concentration of microfractures and to forecast the microdistruction, when acoustic signal duration is used at joints incipience (Physics and mechanics... 1983).

When working out the stress determining technique by the use of acoustic emission we applied the Kaiser effect (Panasiyan, L.L. & M.A. Petrovski 1984), which means the non-reproduction of acoustic signals at repeated loading cycles until the stress reaches the value relieved in the first cycle. It is well known that the effect is observed not only in metals, where it was first discovered, but also in plastic, concrete, composite materials and rocks. The Japanese researchers K. Kurita and M. Fujii (Kurita, K. & M. Fujii 1979) observed the similar effect in granites, diabases and tuffs. They called this phenomenon "stress memory". We have studied the Kaiser effect in different types of rocks (Petrovski, M.A. & L.L. Panasiyan 1983) and stated some peculiarities of its manifestation both during the uniaxial cyclic loadings and after the samples were subjected to triaxial compression tests ($\sigma_1 > \sigma_2 = \sigma_3$). In this case based on the acoustic emission kinetics (N = f(σ)), where N is the total activity) and using some diagnostic criteria (nature of the curve, angles and direction of its deviation, steps, points of divergence of several curves and so on) it is possible to recover the earlier acting stresses. In the general case, based on the acoustic emission behaviour we can judge of the history of rocks loading.

The fact that the rocks are able to store the information of the loads applied to them before, opens way for study of the state of stress of rocks with the help of acoustic emission (Borsh-Komponiets, V.I. & Yu.Ya. Katarguin 1983; Panasiyan, L.L. & M.A. Petrovski 1984; Rzhevski, V.V., Yamshikov, V.S. & V.L. Skuratnik 1983). K. Kurita and M. Fujii showed that in granites the "memory" was retained during one month after load relaxation. In our experiments we managed to restore the loads within the next two-three days practically in all types of rocks studied, while in andesites, basalts and granites such a memory was retained during 10-18 months.

Our technique was used to study over 40 samples, taken from the mineral deposits in different regions and conditions and at various depths (from 100 m to 2400 m). Under test were porphyrites from the North Urals, metasomatites from the Central Urals, granites from the Zabaikal'je, albitites from the Ukrainian shield, gabbro and basalts from the north-western outskirts of Eastern Siberia, etc. In a number of cases the data of determining stresses by acoustic emission have been correlated with those of the in-situ experiments. So, in 1981 in one of the mines of the Zabaikal'je the state of stress of rock was studied by the load relaxation method. Stresses (σ_{min}, σ_{max}) were determined and their absolute values were 51 and 96 MPa respectively. The same core samples were used in 1983 to define the acoustic emission parameters. Proceeding from the anomalous behaviour of acoustic emission during the loading process we managed to restore the stresses which the rock samples had experienced in the rock mass. By the diagrams $\Sigma AE = f(\sigma)$ (Fgi. 6a) one can define both stress values obtained by the load relaxation method to within 1.5-3 MPa. The minimum stress (σ_{min} = 51 MPa) can be determined by a slight change of the curve angle ($\alpha_1 < \alpha_2$) during the initial loading cycle (curve I), while the maximum stress is well defined by the divergence of curves at repeated (second and third) loading cycles (curves II and III).

The hopeful data obtained for the granites from the Zabaikal'je allowed us to use the same technique at the Eastern Siberia mineral deposit described hereinbefore.

Tests have been conducted to measure acoustic emission in samples taken from several boreholes drilled in different structural blocks. Experiments have been performed on 25 samples of two rock types - amygdaloidal basalts taken from the depths of 200 m and 550 m and gabbro-dolerites take from the depths between 900 and 2400 m. As diagnostic criteria we used acoustic emission curve bends during the first loading cycle and curve divergence points at repeated (up to four cycles) loadings (Fig. 6b). The tests indicate that a relatively reliable information on stresses in rocks can be obtained with the help of acoustic emission in case, if in the geological history of the area there has been a stage of stress relieve. In case, when at some point of the rock mass the state of stress has changed many times, its maximum and minimum values are recorded, the first one being recorded more clearly. There are undoubtedly great difficulties in interpreting the obtained data, but the first step towards tackling the problem is apparently the comparison of the stress values determined with the help of acoustic emission with the stress values estimated by other methods. In the tested samples the memory of several stress levels has been restored, the values being 15-20, 30-33, 40-43, 50-59 and 80-86 MPa respectively.

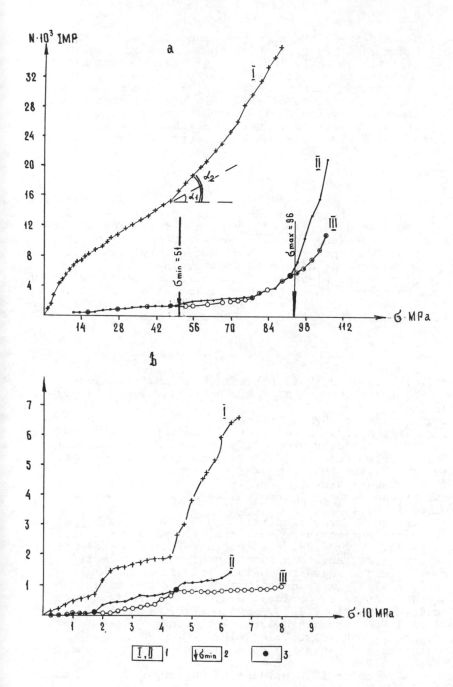

Figure 6. Determining of stresses with the help of acoustic emission.
1 - loading cycles; 2 - stress values obtained by load relaxation method; 3 -
points of curves divergence during repeated loading cycles; a) granite, the
Zabaikal'je; b) gabbro-dolerite, depth - 1250 m, Eastern Siberia.

Let us make an analysis of the test data. Stresses equal to 30-33 MPa were stated more often. They were recorded in the samples taken from different depths:
150-200 m to 1200-1300 m. Such a constancy cannot be explained by the effect of
the gravitational forces or of the horizontal component of the thermal field,
since both of them regularly alter with depth. At the same time, when comparing
the design models with the in-situ experiments it was noted, as indicated above,
that it is just 30-35 MPa which the tectonic component of the total stresses in
the area under study can be equal to. Therefore, it is rightful to consider
σ = 30-33 MPa, estimated with the help of acoustic emission, as the "memory" of
tectonic paleostresses.

The stress values of 40-43 MPa and 50-59 MPa have been recorded in the samples taken from the depth of 900-1300 m and they agree with the values of principal stresses estimated for these depths. The analysis of the geological-tectonic situation hereat has indicated that the values of 40-43 MPa are recorded in the samples taken from the boreholes located near the fracture, i.e. from the zone of the greater mass disturbance, while the values of 50-59 MPa are recorded in places where according to the estimations by the design methods and geophysical data a concentration of stresses exists. The stress values of 80-86 MPa recorded in the samples taken from the depth of 2300-2400 m can be interpreted in a similar way.

CONCLUSION

The conducted studies have shown that the general structure of the stress field for the model under study gets formed under the effect of the three most important forces: gravitational, temperature and tectonic forces. Depending on their relation the "anomalous" values of σ_x and σ_y (which do not comply with Heim's or Dinnik's hypotheses) can be recorded at different depths. The peculiarities of the geological structures of the area play a decisive part in the distribution of similar zones. It was also found that the structure of the stress field in the rock mass can be defined most reliably using a complex of methods and among them the method of acoustic emission should be considered a promising one.

REFERENCES

Borsh-Komponiets, V.I. & Yu.Ya. Katarguin 1983. Estimation of the stressed-strained state of the rock by the method of acoustic emission. Izvestia vuzov, Gorni zhurnal. 8:7-9 (Russian).
Dunegan, H.L. 1973. Emissions acusticas. Dyna. 1:28-33.
Golodkovskaya, G.A. & L.L. Panasiyan 1979. Study of the stress field structure by the design methods. Trudi Instituta "Gidroproekt", p. 45-56. Moscow (Russian).
Greshnikov, V.A. & Yu.B. Drobot 1978. Acoustic emission. Application for material test. Moscow: Izdatel'stvo standartov (Russian).
Kola superdeep borehole. 1984. Moscow: Nedra (Russian).
Kurita, K. & Fujii, M. 1979. Stress memory of crystalline rocks in acoustic emission. J. Geophys. Res. v. 6, 1:9-19.
Non-linear geophysical problems.1981. Moscow:Izdatel'stvo ONTI, VNIIYaTT (Russian).
Panasiyan, L.L. 1983. Stress distribution in non-uniform rock masses. Voprosi inzhenernoi geologii. Vypusk IV, p. 277-288. Moscow:Izdatel'stvo MGU (Russian).
Panasiyan, L.L. & M.A. Petrovski 1984. About application of Kaiser effect for estimating stresses in rocks. Inzhenernaya geologia. 2:123-128 (Russian).
Petrovski, M.A. & L.L. Panasiyan 1983. Experimental study of Kaiser effect in rocks. Vestnik Moscovskogo Universiteta, seria 4, Geologia 3:98-101. Moscow(Rus.)
Physics and mechanics of rocks destruction. 1983. Frunze:Ilim (Russian).
Rzhevski, V.V., V.S. Yamshikov & V.L. Skuratnik 1983. Emission effects of "memory" in rocks, Dokladi Akademii Nauk, USSR. 273. 5:1094-1097 (Russian).
Sholtz, C.H. 1968. Microfracturing and inelastic deformation of rock in compression, J.Geophys.Res. v. 73, 4:1417-1432.
Trofimov, I.M. 1983. Rock pressure control in productive workings of burst-hazard ores. Gorni zhurnal. 8:55-57 (Russian).
Vinogradov, S.D. 1964. Acoustic observations of the rock loading process. Moscow: Nauka (Russian).

Experience of geological-engineering investigations of andesite-basaltic lavas of Armenia by seismoacoustic and radiation logging methods

Expérience de l'étude géotechnique de laves andésite-basaltiques au moyen de la méthode séismoacoustique et du carottage radioactif

Y.M.Gorshkov, P.F.Kochetkov & T.M.Nikolaeva, *'Hydroproject' Institute, Moscow, USSR*

ABSTRACT: The paper describes the methods of integrated use of seismoacoustic and radiation logs for studying the effusive rock of Armenia.

RESUME: L'article présente l'utilisation complexe de la méthode séismoacoustique et du carottage radioactif aux fins d'étudier les roches effusives d'Arménie.

INTRODUCTION

The field investigations involving the geophysical explorations described herein below have been carried out for designing a large engineering structure, which was supposed to be constructed on the rock foundation composed of andesite-basaltic lavas formed as a result of the Aragats volcano eruption (the Armenian SSR) in the Upper-Neogene and Lower Quaternary periods. By the results of preliminary geological investigations it has been found that several different-aged lava flows took part in the geological structure of the foundation, with interstadial clastic formations occurring between these lava flows. The thickness of separate lava flows within the construction site varies from 3 to 20 m. In different parts of the same flow the state of the rock and its physical-and-mechanical properties were found to differ considerably. The porosity of andesite-basaltic rock ranges from 7 to 30% and the size of voids in this rock varies from fractions of millimetres to tens of centimetres, with the void configurations changing to a great extent. The modulus of deformation in different points of the rock mass in question, by the results of the field geomechanical tests, varies from 600 to 7000 MPa. A high degree of heterogeneity of the rock mass in the foundation of the structure presented certain difficulties in development of a design deformability model.

Application of a complex of geophysical methods, whose brief characteristic is given in Table 1, assisted in overcoming these difficulties.

Table 1

Description of explorations	Applied frequency, fr, Hz	Gauge length, L, m	Volume of sampling, W, m³
Seismic profiling	30-100	100-200	10^3-10^4
Seismoacoustic sounding	300	20-30	10^2
Seismoacoustic profiling	250	0.5	1.0-10.0
Ultrasonic prospecting in short holes and in samples	50000	0.3	10^{-3}
Density (gamma-gamma) log	-	0.5	10^{-3}
Gamma-ray log	-	-	10^{-2}

The problems to be solved by the geophysical explorations were as follows:
1. Schematizing of the rock mass in the foundation of the structure from the viewpoint of engineering geology; development of its geostructural model delineating its quasi-homogeneous (by their physical-and-mechanical properties) structural elements affecting the deformability of the foundation as a whole (the area of such structural elements should be by an order of magnitude smaller than separate structural elements of the structure, i.e. it should be equal approximately to n x 10÷100 sq.m.).

2. Determination of physical and mechanical properties of the rock in the delineated elements and their comparison with evaluation of deformation properties of
the rock mass made by conventional static tests.
3. Determination of generalized deformability indices for distinguished quasi-
homogeneous elements of the rock mass as well as for the foundation as a whole
(the procedure is described in (Recommendations... 1984)).
The first problem was solved by applying various logging methods (gamma-ray and
density logs) and seismoacoustic sounding of rock pillars between exploratory holes.
The second problem was solved by applying different-scale seismoacoustic (seismic,
acoustic and ultrasonic) investigations (Savich et al.1969) and density log.
The third problem was solved by using the data of seismic sounding and seismic
profiling.
Given below is a brief description of the physical backgrounds of the applied
techniques developed in (Savich et al.1969; Recommendations... 1984) and the main
results of the investigations.

1 SCHEMATIZATION OF ROCK MASS BY PHYSICAL-AND-MECHANICAL PROPERTIES

Identification of boundaries of separate lava flows of different composition was
made using the data of the gamma-ray logging. Potentialities inherent in the gamma-
ray logging as applied to studying the volcanic rock were often discussed in literature. Examples of successful application of this method for studying Icelandic
basalts are described in ("The Log Analyst" 1981). The physical background for zoning of andesite-basalts by the results of gamma-ray logging is the relationship
between their natural radioactivity and the content of silicon earth (SiO_2). It is
known that the content of SiO_2 in lava flows determines the lava viscosity and tex-
tural-structural peculiarities of andesite-basalts (Physical...1976):   the silicon earth-enriched lava is characterized by higher viscosity and higher gas content as compared with the lava featuring low content of SiO_2. Due to this fact,
effusive rock formed of acid lava has porous and hyaline texture. Radioactivity of
magmatic rock varies depending on the SiO_2 content: it has been established that
the amount of radioactive elements, such as uranium-238, thorium-232 and potassium-
40 tends to increase in the lavas rich in silicon earth (Physical... 1976). For
effusive rock, especially of cenotypal origin, characteristic is a strong correlation between the content of uranium, thorium and principal petrogenic components:
SiO_2 and K. This is explained by the fact that, since the time of consolidation of
effusive rocks depends on their viscosity and, hence, is very limited, there is no
possibility for the radioactive elements to migrate for a long time and concentrate
and deconcentrate in different minerals. As a result, distribution of radioactive
elements in effusive rocks depends on the viscosity of the lava flows, and the relationship between their radiactivity and SiO_2 content is close to the functional
one. The up-to-date logging radiometers allow for estimation of the rock radioactivity at an error ±5% which is sufficient for estimation of SiO_2 content with an accuracy not lower than 1%.
Table 2 based on the data of the "Hydroproject" Institute gives an average content of radioactive isotopes of uranium (U), thorium (Th), potassium (K) and silicon earth (SiO_2) in effusive rock of different composition as well as average values of their radioactivity.

Table 2

Description of rock	Content, %				Radioactivity by gamma-ray log, μR/h
	SiO_2	K	$U \cdot 10^{-4}$	$Th \cdot 10^{-4}$	
Basalt, diobase	49.0	1.0	0.7	2.3	2-4
Andesite, andesitic porphyrite	59.6	1.7	1.2	4.0	6-16
Dacite, dacitic porphyrite	66.3	2.3	2.5	10.0	16-20
Liparite, quartz porphyry	72.9	3.7	4.7	19.0	40-50

The relationship between SiO_2 content in andesitic-basaltic lavas of Armenia and
their radioactivity may be expressed by the following equation (correlation coefficient is about 0.93):

$$(SiO_2) = 38.9 + 20.7 lg(I_\gamma) \tag{1}$$

where (SiO_2) - silicon earth content, %;
I_γ - natural radioactivity, μR/h.

With regard to the above data the results of the gamma-ray log (GR) were used to identify the boundaries of different lava flows on the area of the structure. -Division of separate lava flows into quasi-homogeneous elements was made by the results of gamma-gamma density log (DL). Application of DL was dictated by the mode of occurrence of the rock stratum in question (dry rock) and its porous texture which caused great density variability of the studied rock.

The range of the density values (2.02 to 2.68 t/m³, obtained by this method) is as much as 20 times higher than the accuracy obtained using the geophysical methods .

The vertical boundaries of structural elements of the rock mass are identified by the results of multipoint seismoacoustic shooting of the pillars between the boreholes, basing on the relationship between the elastic characteristics of rocks and their texture peculiarities (Physical... 1976). The sections were plotted by the results of shooting applying the methods of seismoacoustic tomography and the results of special program computations (Yakubov et al. 1984).

2 TECHNIQUES FOR DETERMINATION OF PHYSICAL-AND-MECHANICAL PROPERTIES OF ROCK MASS ELEMENTS

The physical-and-mechanical properties of selected structural elements of the rock mass were determined by applying special techniques intended for studying density, elasticity and deformation properties of heterogeneous rock masses developed in the "Hydroproject" Institute (Physical... 1976; Recommendations 1984; Savich et al. 1969), which is based on combination of geophysical and geomechanical investigations. The geomechanical (static) methods here are aimed at obtaining the basic deformation characteristics in the most representative points of the rock mass. They consist in determining reference deformability indices in a given measurement scale and studying the main regularities of the medium deformability with the loading pattern close to operational conditions.

The geophysical methods were applied for interpolation of the results of the static determinations to rock masses in question and to the required scale levels. This problem is to be solved basing on the information obtained by geophysical methods about structural peculiarities of the rock mass, about the regularities of distribution of dynamic elastic parameters and other physical properties in the rock mass and selected blocks as well as by establishing correlation between static and dynamic deformability indices (Savich et al.1969; Recommendations... 1984).

Determination of dynamic indices of the modulus of elasticity E_d and Poisson's ratio μ_d was based on the formulas of the theory of elasticity:

$$E_d = v_p^2 \gamma \frac{(1 + \mu_d)(1 - 2\mu_d)}{1 - \mu_d} \qquad (2)$$

$$\mu_d = \frac{v_p - 2v_s^2}{2(v_p^2 - v_s^2)} \qquad (3)$$

where: V_p and V_s - longitudinal and transverse wave velocity (m/s) by the results of seismoacoustic and ultrasonic explorations; γ - rock density (kg/m³) by the results of DL.

These formulas are valid exceptionally for quasi-homogeneous and isotropic rocks. In heterogeneous isotropic media, the E_d and μ_d values calculated using the said formulas depend on the range of elastic vibrations frequency f_r, which were used for determination of the velocities, and on the measurement length Δl. As is known, values Δl and f_r determine the rock volume W, which controls the effective vibration parameters and, hence, the results of single measurements. The W value, according to (Recommendations... 1984), may be assessed by the following approximate relationship:

$$W \approx 0.2 \, \Delta l \cdot \lambda^2 \qquad (4)$$

where λ is the elastic wave length; $\lambda = \frac{V}{f_r}$.

The use of formulas (2) and (3) is possible only in those cases when prevailing linear dimensions of heterogeneity elements $L \gg \Delta l$.

In other cases E_d and M_d have the meaning of the certain effective parameters. In anisotropic rocks V_p, V_s and E_d, M_d values depend on the direction of measuring. With moderate anisotropy ($K = \dfrac{V_p \max}{V_p \min} < 2$), however, the use of formulas (2) and (3) gives an error of no greater than 10% in E_d determination.

Determination of the static modulus of deformation for different parts and elements of the rock mass using the measured value V_p and E_d is based on correlation between the static and dynamic deformability indices. The physical background of the existence of these regularities stems from their close nature conditioned by the fact that the static and dynamic indices characterize, in principle, one and the same deformation process, though its different stages.

As is shown in (Savich et al.1969; Recommendations ... 1984), in general case the correlation between the static modulus of deformation D and dynamic modulus of elasticity E_d is expressed as follows:

$$\lg D = a \ \lg E_d + b \tag{5}$$

where a and b are the constant coefficients for the given conditions of determination of moduli and type of studied rock which, according to (Recommendations... 1984), can be found from the following equations:

$$a = 1.141 + A \times e^{-\alpha P} \tag{6}$$

$$-b = 0.875 + B \times e^{-\beta P} \tag{7}$$

where: A, α, B and β are the constants for this type of rock, P is the maximum value of the load acting in static tests.

The average values of these constants for igneous and clastic rock are given in Table 3.

Table 3

Type of rock	A	α	B	β
Igneous rock	0.944	0.0188	5.730	0.0148
Clastic rock	0.460	0.0140	3.000	0.0100

Using the tabulated data (at design load p = 0.5 MPa), we obtain the following most probable correlations between the modulus of deformation and E_d for different rocks in the foundation of the structure: substituting in equation (5) the values from (6) and (7) we obtain:

for igneous rock

$$D = 6.36 E_d^{2.0} \times 10^{-7} \tag{8}$$

for clastic rock

$$D = 1.38 E_d^{1.6} \times 10^{-4} \tag{9}$$

In equations (6), (7), (8) and (9) the D, E_d and P values are in given in bar (10^2 kPa).

Since V_p, V_s and γ values used for estimation of E_d by formulas (2) and (3) are usually reciprocally correlated, formulas (8) and (9) may be often reduced to simplified relationships of $D = f(V_p)$ or $D = f(\gamma)$ type which are valid for the rocks only of the given area, but suitable for mass estimation of D value by the result of determination of either V_p or γ. Similar correlative relationships obtained for the rock stratum in question and presented in Fig. 1 were widely used for solving the above problems.

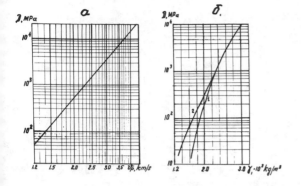

Fig. 1. Relationships between modulus of deformation D and a) longitudinal seismic wave velocity V_D in basalt and andesite-basalt; b) effusive rock density (1) and interstadial deposits (2).

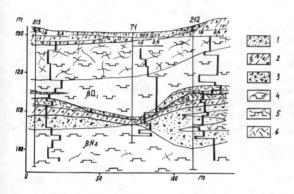

Fig. 2. Geological-geophysical section (by the geological documentation and the results of density log): 1 - debris-gruss soil; 2 - basalt; 2 - sandy loam and argillaceous sand; 6 - volcanic tuff; 5 - andesite-basalt; 3 - sand of scorificated basalt composition

3 THE MAIN RESULTS OF INTEGRATED STUDY OF DEFORMATION PROPERTIES OF ROCK MASSES IN STRUCTURE FOUNDATION

3.1 Identification of geological structure and texture

The results of the GR log were used to develop a more accurate model of the geological structure of the foundation. It was found that within the depth range studied (up to 50 m) two different lava flows with the natural radioactivity two times differing from each other take part in composing the foundation. The vertical boundaries of these flows were determined by the borehole sections as well as the thickness of clastic deposits formed in the period between two stages of volcanic activities.

3.2 Study of deformation properties and state of rock in the foundation

The results of the gamma-gamma density log made it possible to obtain detailed knowledge of the vertical and horizontal variability of the density of the rocks studied. The density of andesite-basalts was found to be essentially decreased in the perypheral parts of the lava flow. In different parts of the lava flows the voids were detected whose vertical size reached 10-20 cm. The highest concentration of these voids was found to be confined to the zones of lower density.

The detailed models of density sections of the foundation were developed for separate elements of the structure (Fig. 2).

The results of seismo-acoustic log were used to obtain the elastic properties variation diagrams of the rocks in the main structural elements of the studied rock mass (Fig. 3). The dynamic and static deformability indices of these elements have been determined. The different-scale seismoacoustic explorations

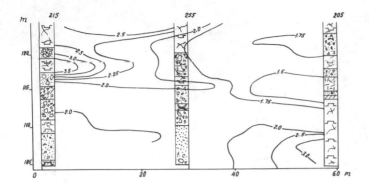

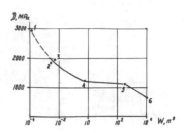

Fig. 3. Velocity profile (by the results of seismo-acoustic sounding)

Fig. 4. Relationship between the modal values of the modulus of deformation D in basalt and andesite-basalt and the scale of measurments

showed that there is a great difference in estimation of the elastic and deformation properties of the foundation rock for different scales (volumes) of a single test, W. With the increase of W the deformation and elastic properties decrease respectively. The greatest decrease in the modulus of deformation occurs in going from W 10^2 m^3 in a single test (Fig. 4). In accordance with (Recommendations... 1984), this proves the fact that the foundation is composed of the blocks of prevailing volumes from 10^2 m^3 to 10^3 m^3 and linear dimensions from 5 m to 10 m which is well correlated with the data of the other engineering-and-geological investigations.

Comparison of independent estimation of the moduli of deformation obtained for separate zones of the rock mass based on the results of the static field tests (by dilatometers and loading plates) and geophysical methods showed satisfactory agreement for respective scales of the investigations: the divergence in the estimation of the average values does not exceed 10%.

CONCLUSION

In summarizing the above said the following conclusion shall be drawn: the geophysical explorations which were conducted using the methods of seismoacoustic and radiation logs gave more profound knowledge of the geological structure of the rock mass represented by andesite-basaltic lavas in the foundation of the structure and allowed for more reliable determination of the physical-and-mechanical properties and the state of the main types of rocks and structural elements of the rock mass in the foundation.

REFERENCES

Physical properties of rock and mineral resources. 1976. Spravochnik geofizika. Moscow: Nedra (Russian).
Recommendations on application of engineering geophysics to study deformation properties of rock masses. Edited by A.I. Savich & B.D. Kuyundzhich. 1984. Trudy "Gidroproekta". Moscow (Russian).
Savich, A.I. et al. 1969. Seismoacoustic methods of studying rock masses. Moscow: Nedra (Russian).
Savich, A.I. & L.G. Yashenko 1967. Application of seismicacoustic methods for evaluation of deformation properties of rock foundations of hydraulic structures. "Gidrotekhnicheskoye stroitel'stvo": 12 (Russian).
"The Log Analyst" 1981, 22: 2.
Yakubov, V.A. et al. 1984. Application of numerical methods for processing data of seismic measurements at the Inguri and Khudoni hydroelectric stations. VIII-th Conference of surveyors and investigators of "Hydroproject" Institute. Reports and communications. Moscow (Russian).

Scale effect during compressive strength tests of rocks
Effet d'échelle dans les essais de résistance à la compression des roches

R.R.Kaczyński, *Institute of Hydrogeology & Engineering Geology, Warsaw University, Poland*

ABSTRACT: The paper presents the investigations of uniaxial compressive strength of rock samples of different sizes.The most typical rocks in Polish construction works were analyzed that is:limestone,sandstone,granite and basalt.The uniaxial compressive strength was defined for cubic samples,with their edges: 1.0,2.2,3.2,5.0 and 6.3 cm long.The uniaxial compressive stress decreases with a larger sample size.An intensity of a decrease of these functions is diversified.At first it is higher,then decreases with a lower size of a sample.

RÉSUMÉ: Dans cet article on a présenté les essais de compression simple des éprouvettes aux differents dimensions.On a examiné roches typiques appliqués dans la construction polonaise notamment: le calcaire,le grès,le granite et le bazalte.On a déterminé la résistance a la compression simple utilisant les éprouvettes cubiques aux dimensions:1.0,2.2,3.2,5.0 et 6.3 cm.Les résultats d'essais montrent que la résistance à la compression simple diminue pendant que de la dimension de l'éprouvette augmente.L'intensité de la diminution ces fonctions est variée étant plus grande dans la première étage au cours du croissement de la dimension des éprouvettes la travail de la rupture decroit.

A strength to uniaxial compression forms a principal mechanical characteristic of rocks.The problem of influence of sample size on rock strength has been already known for a long time but it is not solved in full.It is however of high theoretical and practical significance.
A determination of strength to compression of limestone,sandstone,granite and basalt on the basis of samples of varying sizes is the main purpose of this paper.Limestone and sandstone samples come from the Holy Cross Mts whereas granite and basalt samples from the Sudetes.A sandstone represents the oldest analyzed sample /Cambrian/.The other rocks are younger,in turn: granite /Carboniferous/Permian/,limestone /Jurassic/ and basalt /Tertiary/.
A limestone matrix is formed of a limestone micrite.Putroids that occur with a varying,frequency are the specific component and make the limestone look spotty.A sandstone is a monomictic and fine-grained quartzic rock with a recrystalized siliceous cement.Quartz grains are 0.15-0.20 mm in size,well sorted and angular.Granites I and II are coarse-grained rocks and in some fragments close to a granodiorite.A mineral composition is /from the most abundant components/ potassium feldspar /microcline/,plagioclase,quartz and accessory minerals.Microcline and plagioclase are kaolinized or affected by kaolinization.A basalt is an aphanitic rock.Augite and olivine are the phenocrysts,up to 1-2 mm in size.Inter-crystalline spaces are filled with nepheline.The analyzed rocks have various heterogeneities and discontinuities.
In a laboratory,cubes were cut /dimensions of 10,22,32,50 and 63 mm/ from large monoliths.The latter were oriented /top-bottom/ in relation to bedding /limestone,sandstone/,joint "L" /granites/,pillar axis /basalt/.Before a compression of samples,the non-destructive ultrasonic analyses were done.Investigations of a compressive strength were carried through in agreement with instructions of the International Bur.of Rocks Mechanics on 220 samples.The results are presented in Table 1 and Figure 1.25 series of analyses were done /5 types of rocks x 5 sample sizes x 6 samples/ and in 11 series the variation coefficient was over 26%.For this reason,a number of compressed samples was increa-

Table I. Uniaxial compressive strength dependent on sample size

ROCK	PHYSICAL PROPERTIES specific density ϱ_{s} $[\times 10^{3}$ kg/m$^3]$ volume density ϱ_{0} $[\times 10^{3}$ kg/m$^3]$ porosity n water content w [%]	MEAN VELO-CITY OF WAVE PROPAGATION Vp [m/s]	LENGTH OF CUBE EDGE a [cm]	MEAN COMPRESSIVE STRENGTH R_c [kG/cm^2], $[\times 10^{-1}$ MPa]
LIMESTONE	ϱ_s = 2,70 ϱ_0 = 2,43 n = 0,10 w = 0,13	5227 ± 260	1,0 2,2 3,2 5,0 6,3	849 ± 194 743[*] ± 243 827 ± 211 686 ± 168 538 ± 79
SANDSTONE	ϱ_s = 2,68 ϱ_0 = 2,58 n = 0,35 w = 0,18	5427 ± 256	1,0 2,2 3,2 5,0 6,3	1580 ± 244 1762 ± 438 1613 ± 156 1495 ± 299 1537 ± 471
GRANITE I	ϱ_s = 2,67 ϱ_0 = 2,59 n = 0,030 w = 0,25	4865 ± 372	1,0 2,2 3,2 5,0 6,3	1028[*] ± 434 840 ± 216 811[*] ± 383 793[*] ± 376 603 ± 117
GRANITE II	ϱ_s = 2,67 ϱ_0 = 2,60 n = 0,027 w = 0,25	4646 ±474	1,0 2,2 3,2 5,0 6,3	788 ± 145 706[*] ± 342 645 ± 136 597[*] ± 287 961 ± 129
BASALT	ϱ_s = 3,08 ϱ_0 = 2,99 n = 0,04 w = 0,29	6200 ± 311	1,0 2,2 3,2 5,0 6,3	1885 ± 359 1672 ± 426 1710 ± 561 1595 ± 387 1982 ± 543

[*] variation coefficient > 30 %

sed to 9.But still for 6 series,the variation coefficient was over 30%.Granites were found to have the largest strength variation /the coefficient to 52 %/. From a strength point of view the analyzed rocks form two principal groups /independent on sample sizes/:
1. Sandstones and basalts of mean copressive strength of 149.5-198.2 MPa,
2. Granites and limestones of mean compressive strength of 53.8-102.8 MPa.
Values of compressive strength dependent on sample sizes are presented in Table 1 and Figure 1. The length of a cube edge is from 1 to 6.3 cm, the section area from 1 to 40 cm^2 and sample volumes from 1 to 250 cm^3.In spite of a high strength variation of analyzed rocks,a dependence of compressive strength from sample sizes is noted. As samples are larger,a compressive strength gets lower. The strength fall is varying. Smaller samples indicate lower strength. The larger samples,the strength drop is more uniform. In carried tests the influence of sample sizes is disturbed in the case of a limestone /for samples of the edge length of 3.2 cm/,sandstone /for samples of the edge 1 cm/,granite II and basalt

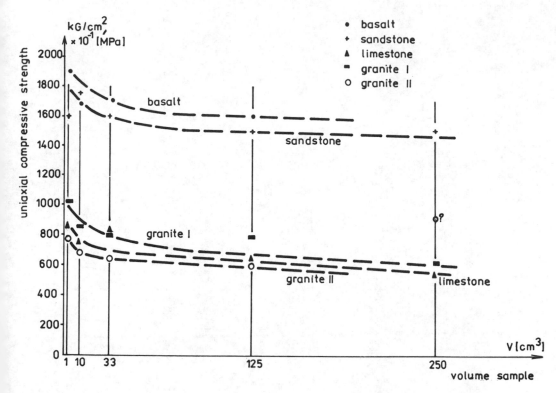

Figure 1. Relationships beetwen uniaxial compressive strength on volume samples.

/for samples of the edge 6.3 cm/. On the hand most analyzed series support the opinion on a decreasing dependency of copressive strength with larger sample sizes. A size effect is observed independently on origin of rocks: in sedimentary /limestonee, sandstone / as well as magmatic plutonic /granite/ and volcanic / basalt / ones.

Parametric back-analysis of shear strength of joints from rock slides

Analyse a posteriori de la résistance au cisaillement de joints dans des glissements rocheux

M.Maugeri & E.Motta, *Faculty of Engineering, University of Catania, Italy*
S.Wang, *Institute of Geology, Academia Sinica, China*

ABSTRACT: Geotechnical properties of joints in rock mass are the most important parametres for stability analysis and design of rock engineering. The laboratory and in situ tests can only yield some informations for small area of sliding surface. One of potential approaches to study this problem and to gain more informations may be served by parametric back-analysis from rock slides. The obtained results seem to be useful to the rock engineering problems in the same region or in the similar geological conditions. This paper describes an attempt to carry on a such analysis for joint system in carbonate rocks from the numerous rock slides during the Friuli earthquake, Italy, in 1976. The results obtained show that the friction angle of joints may be comparable with rock tests, however the values of cohesion in natural failures seem to be very low in comparison with usual results from rock mechanics tests.

RESUME: Le propriétés géothecniques des joints dans les masses rocheuses sont les plus importants parametrès pour l'analyse de stabilité et le projet de la génie des roches. Les tests de laboratoire et les tests sur la place peuvent seulement donner quelques renseignements pour de petites aires de glissement de surface. L'un des approches possibles pour étudier ce problème et obtenir les plus de renseignements possibles pourrait être une analyse paramétrique posterieure du glissement des roches. Les resultats obtenus semblent être utiles pour les problémes de la génie des roches dans les mêmes regions· ou dans les mêmes conditions géologiques. Le travail décrit une tentative de mener ce type d'analyse pour des systémes des joints dans les roches calcaires à travers l'observation de nombreux glissements de roches pendant le tremblement de terre du Friuli, Italie, en 1976. Les resultats obtenus montrent que l'angle de frottement des joints peut-être comparable à celui obtenu à travers les tests sur les roches; de toute façon les valeurs de cohésion dans les ruptures naturelles semblent être bas, en comparaison aux resultats courants obtenus dans les tests de laboratoire.

1 INTRODUCTION

The Friuli earthquake of May 6 1976 (M = 6.4) triggered off numerous rockslides in the montaneous slopes facing to the Tagliamento river valley. Most of these rockslides took place in zones which already in the past were affected by sliding phenomena, nevertheless the influence of the seismic action played an important role since the response of the rock slopes to the ground shocks was greater around the epicentral area. The rockslides occurred mainly in steep slopes and very close to the top of the mountain walls. The earthquake caused also many rockfalls of single blocks which rolled to the valley and stopped behind their usual accumulation area, in sites never reached in the past.
The rock weakness due to the intense tectonic fracturing and the steepness of the slopes, played an important role in the process of rocksliding, even if there was

not found a direct relationship between lithological condition and areal density of rockslides (Govi and Sorzana, 1977).

Landsliding of highly fractured and jointed rocks may cause severe damage to villages and lifelines thus risk evaluation methods (Civita et al, 1983) seem to be useful in forecasting and preventing damage. Among these, statistical processing of data and number of landlides and mapping of potential landslides are more used than deterministic methods. This because of the uncertainties in the shear strength evaluation along rock joint system. Nevertheless when the joint shear strength can be investigated, a deterministic procedure may produce more detailed informations in the evaluation of individual slope stability to protect buildings and infrastructures of particular importance.

This paper deals with a simple procedure for evaluating shear strength along joints by performing a parametric back analysis of the rockslidings which occurred during the Friuli earthquake.

2 GEOLOGICAL FEATURES OF ROCK SLIDES

The outcropping formations in Friuli area are mainly of sedimentary horigin and most of them indicate marine and/or transitional environment.

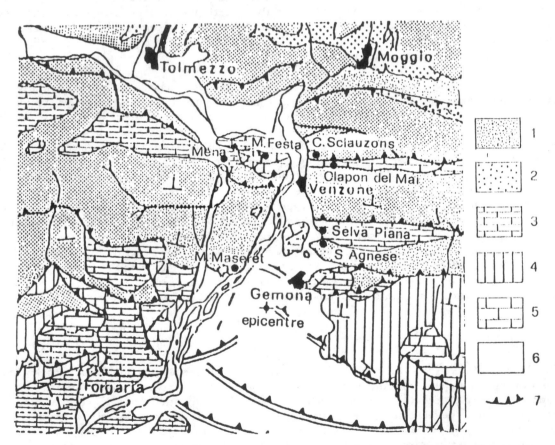

Fig.1-Geological scheme and location of the rockslides analysed in the area severely damaged by the Friuli earthquake (1976). Legend: 1 Triassic: Dolomite and dolomitic limestones; 2 Triassic: Sandstones, marls and evaporites; 3 Jurassic: chert and oolitic limestones; 4 Eocene-Paleocene: Flysch; 5 Cretaceous: platform limestones; 6 Quaternary; 7 Overthrust.

The Friulian succession, deposited during a long period from the Ordovician up to Miocene, is over 15.000 meters thick (Martinis and Cavallin, 1977). The area primarily involved in the 1976 earthquake (fig.1) is represented mainly by Julian and Carnic Prealps separated along the N-S direction by the Tagliamento river valley. Triassic deposits are represented by Dolomite and Dolomitic limestones as well as sandstones, marls and evaporites. Jurassic deposits are not uniform, nevertheless chert and oolitic limestones are the most common.

Rockslides were mainly characterized by failures along rockwalls and, according to the mechanics of the sliding process (Carson and Kirby, 1972) can be classified into three main sets (Govi and Sorzana, 1977). The first one corresponds to slab failures in which the sliding is generally translational or is due to overturnings of the entire rockwalls, depending on the slope steepness. The second group is typical of rock slopes affected by bedding planes dipping towards the slope and with a random crossed joint pattern; in this case the slides are conditioned by the direction of the fracturing planes with respect to the slope surface. This sliding process may regard single large blocks (rockfalls) as well as noticeable amount of rock material (rock avalances).

Finally the third group is characterized by failures along bedding joints in limestone sequences dipping towards the valley. This kind of failure did not modify significantly the average steepness of the slopes.

The rockslides analysed are shown in fig.2. The average steepness of the rock slopes are ranging between 50° and 70°, and the volumes involved in the failure process vary from 5000 to 80.000 cubic meters.

3 PROCEDURE OF PARAMETRIC BACK-ANALYSIS

From the field investigation it is clear that the rock slidings were controlled by joint system in the carbonate rock mass. According to the joint combination, the rock slides may be divided into three groups:

a) Rock slides with sliding surface of one moderately inclined joint or bedding plane. In this case the dipping angle of sliding surface ranges between 45°-50° (fig.3a).

b) Rock slides with sliding surface consisting of one set of steep dipping joint and another set of moderately inclined joint (fig.3b).

c) Rock slides with sliding surface consisting of one set of moderately or steep dipping joint and another set of flat dipping joint (fig.3c).

The principle of back-analysis is to determine the stress condition in the limit equilibrium state and from this to find the shear strength parametres according to the Coulomb failure criterion that even for rocks can be assumed valid (Krahn and Mongenstern, 1979). For rock slides characterized by block movement the method of limit equilibrium of rigid body may be employed.

For the rock slides of group a, the following calculation has been made to determine the normal and shear stresses on the sliding surface.

$$\sigma = \frac{W \cos \beta}{A}$$

$$\tau = \frac{W F_s \sin \beta}{A}$$

(1)

where:
W = weight of rock mass
ß = dipping angle of sliding surface
A = lenght of sliding surface in section
F_s = assumed safety factor in static condition.

In this paper the assumed safety factor is needed to be introduced, because rock slides occurred during earthquake under the action of some dynamic force and therefore, the safety factor of rock slopes in static condition would be higher than one. This problem will be dis-

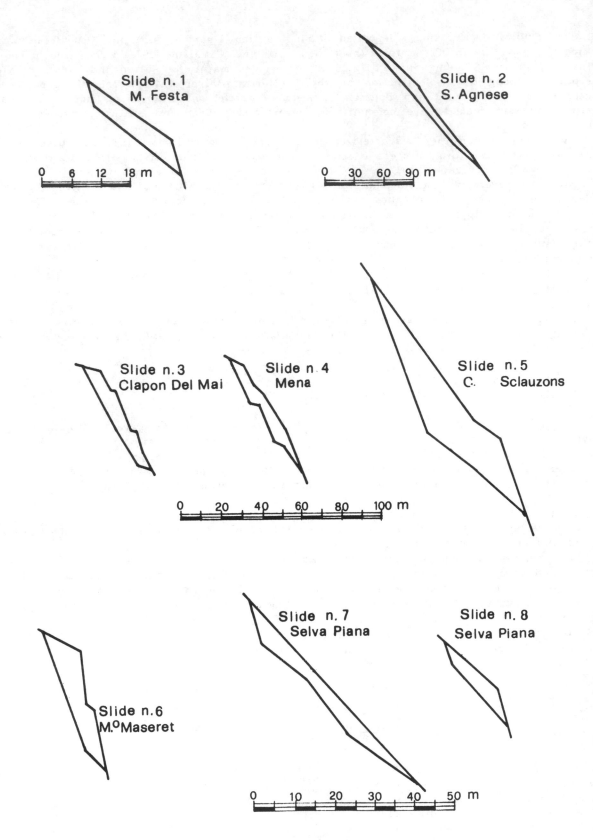

Slide n. 1
M. Festa

0   6   12   18 m

Slide n. 2
S. Agnese

0   30   60   90 m

Slide n. 3
Clapon Del Mai

Slide n. 4
Mena

Slide n. 5
C.   Sclauzons

0   20   40   60   80   100 m

Slide n. 6
M.º Maseret

Slide n. 7
Selva Piana

Slide n. 8
Selva Piana

0   10   20   30   40   50 m

Fig. 2 – Schematic rockslides profiles

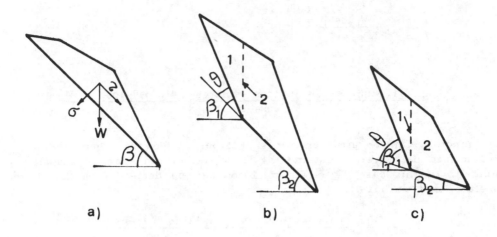

Fig.3 – Sliding models used in the back analysis

cussed elsewere in this paper lateron.

If a seismic coefficient is introduced to take the seismic force into consideration we have the pseudo–static state of stresses.

$$\sigma = \frac{W \ (\cos\beta \ - \ k\sin\beta)}{A}$$

$$\tau = \frac{W \ (\sin\beta \ - \ k\cos\beta)}{A}$$

(2)

where k is the seismic coefficient. For simplicity the seismic force is assumed to act horizontally on the sliding body.

For the slides of group b, the parametres of the moderately inclined joint have been determined in the abovementioned calculation for slides of group a, and we have to study the parametres of the steep dipping joint. As shown in fig.3b the slide may be imaginally divided into two parts. One part is resting on the steep part of sliding surface and another on the moderately inclined surface. There exists a certain interaction between them. The stress condition of the steep part of sliding will be determined not only by its own gravitational force, but also by the resisting force from the inclined part, and the system will gain the uniform state of stability. The equilibrium condition may be described as follows:

$$[(W_1 F_s \sin\beta_1 - W_1\cos\beta_1 \, tg\phi_1 - c_1 A_1)\sin\theta + W_2 \cos\beta_2] tg\phi_2 + c_2 A_2 = W_2 F_s \sin\beta_2 + (W_1 F_s \sin\beta_1 - W_1 \cos\beta_1 \, tg\phi_1 - c_1 A_1)\cos\theta$$

where:

c = cohesion

ϕ = friction angle

$\theta = \beta_1 - \beta_2$

and we obtain:

$$\frac{W_1 \cos\beta_1}{A} \, tg\phi_1 + c_1 = \frac{W_2 F_s \sin\beta_2 + W_1 F_s \sin\beta_1 (\cos\theta - \sin\theta \, tg\phi_2) - W_2\cos\beta_2 \, tg\phi_2 - c_2 A_2}{A_1(\cos\theta - \sin\theta \, tg\phi_2)}$$

Therefore, we have the stress condition in the upper steep part of sliding surface.

$$\sigma_1 = \frac{W_1 \cos\beta_1}{A_1}$$

(3)

$$\tau_1 = \frac{W_2 F_s \sin\beta_2 + W_1 F_s \sin\beta_1 (\cos\theta - \sin\theta \, tg\phi_2) - W_2 \cos\beta_2 \, tg\phi_2 - c_2 A_2}{A_1 (\cos\theta - \sin\theta \, tg\phi_2)}$$

The same principle can be used for slides of group c. However, here the parametres of the steep or inclined part are known, and the lower flat dipping part of the sliding surface is to be studied. In this case the stress condition can be derived from the abovementioned equilibrium state, and we have:

$$\sigma_2 = \frac{(W_1 F_s \sin\beta_1 - W_1 \cos\beta_1 \, tg\phi_1 - c_1 A_1) \sin\theta + W_2 F_s \sin\beta_2}{A_2}$$

(4)

$$\tau_2 = \frac{W_2 F_s \sin\beta_2 + (W_1 F_s \sin\beta_1 - W_1 \cos\beta_1 \, tg\phi_1 - c_1 A_1) \cos\theta}{A_2}$$

In order to determine the limit stress condition on the sliding surface a basic assumption to be made is the value of the safety factor. In the case of static failure a safety factor equal to one usually could be adopted. In the case of rock slides due to earthquake the safety factor of static stability will be higher than 1.0. Moreover, if a seismic coefficient is introduced into calculation, the actual dynamic stability during sliding may be less than 1.0. Therefore, an assumed static factor of safety is adopted in connection with seismic action.

For the rock slides of group a, we determined the parametres of shear strength of joint dipping in about 45° taking into account the safety factor equal to 1.0 and 1.2, and the seismic coefficient equal to 0.2. The comparison of these cases shows that F_s = 1.2 is quite reasonable to be used. In this case the determined parametres are close to the range of parametres determined with k = 0.2. This is shown in fig.4. The studied rock slides are those occurred at S.Agnese, M.Festa and Selva Piana No.8. Thus, in all the calculations the assumed safety factor equal to 1.2 in static condition is adopted.

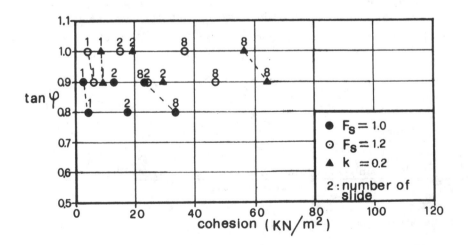

Fig.4 - Parametric back analysis for slides of group a

4 RESULTS AND DISCUSSION

The parametres of inclined joints are determined using formula (1) for S.Agnese, M.Festa and Selva Piana No.8. In the condition of $F_s = 1.2$, the value of tg ϕ reaches 0.9, and cohesion less than 50 KN/m².

The calculation of slides of group b was conducted taking into account the parametres of lower part of sliding surface which have been determined in the slides of group a. The mean values are taken for use, tg $\phi_2 = 0.9$ and $c_2 = 30$ KN/m². The slides in C.Sclauzons, M°.Maseret and Selva Piana No.7 are studied. The values of parametres of steep dipping joints seem to be somewhat lower than that of the inclined one. The mean value of tg ϕ is about 0.8 and values of cohesion are less than 90 KN/m².

The parametres of flat dipping joints are determined for slides of group c in slides at Clapon del Mai and Mena. The parametres are quite similar. The value tg ϕ reaches 0.9 and cohesion values are generally of the same order than that of inclined and steep dipping joints (Table 1).

Table 1: Shear strength parametres assesment
for the three groups analysed

GROUP	ß	tg φ	COHESION KN/m²	ROCK SLIDES
b	60°-70°	0.8	16	C.SCLAUZONS
			87	M°MASERET
			12	SELVA PIANA No.7
a	45°-50°	0.9	46	S. AGNESE
			6	M. FESTA
			23	SELVA PIANA No.8
c	20°-30°	0.9	35	CLAPON DEL MAI
			10	MENA

The limit stress conditions for all the slides are shown in fig.5. Although the results are quite scattered, the general characteristic is obvious. The value tg ϕ is ranging between 0.8 - 0.9, and the values of cohesion are more scattered. However they are generally low ranging from 0 to 90 KN/m². Similar results were obtained in previous studies by performing different numerical procedures (Maugeri and Motta, 1980; Maugeri, 1981).

In comparison with laboratory and in situ shear tests we can see that the value of tg ϕ obtained in back-analysis is consistent with peak shear strength in rock tests. This tells us that the failure process seems to overcome the peak strength. The lower cohesion value, in comparison with rock tests, may be explained by scale effect.

For large scale slides the values of cohesion have to be reduced in the stability analysis.

5 CONCLUSIONS

1. Parametric back-analysis is a potential approach to study the realistic properties of rock mass.
2. In spite of its simplicity, the calculation procedure used gives quite reasonable results in agreement with the results obtained in previous works using different calculation procedures.
3. The shear strength of joints in carbonate rock mass is generally uniform. There is not a significative variation of shear strength parametres c and tg ϕ for different sets of joints.

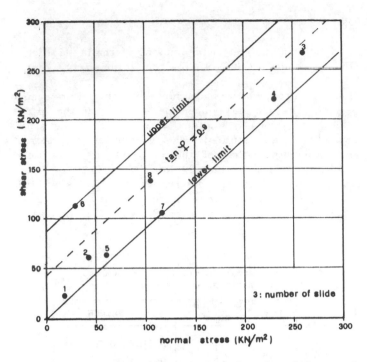

Fig. 5 - Limit stress conditions for the slides analysed

4. Parametric back-analysis shows that the shear strength along joints can be represented by the Coulomb failure criterion. According to this criterion the friction angle of 40 degrees has been evaluated and a cohesion ranging between 6 and 87 KN/m² was found.
5. In comparison with laboratory rock tests results the friction angle obtained by back-analysis seems to be in good agreement while the cohesion values seem to be much less. This fact should be considered in the stability analysis of jointed rock slopes.

REFERENCES

Carson, M.A. & Kirby, M.J. 1972. Hillslope form and process. Cambridge University press, 475 p., Cambridge.
Cavallin, A. & Martinis, B. 1977. Inquadramento geologico. In studio geologico dell'area maggiormente colpita dal terremoto friulano del 1976. Rivista Italiana di Paleontologia e Stratigrafia, 83, pp.219-235.
Civita, M., Gov, M. & Maugeri, M. 1983. La franosità dei versanti nella valutazione del rischio sismico globale. Indagini sul terremoto del Friuli (1976). Geologia Applicata e Idrogeologia, vol.18, parte 3, pp.479-506.
Govi, M. & Sorzana, P.F. 1977. Effetti geologici del terremoto: frane. In studio geologico dell'area maggiormente colpita dal terremoto friulano del 1976. Rivista Italiana di Paleontologia e Stratigrafia, 83, pp.329-368.
Krahn, J. & Morgenstern, N.R. 1979. The ultimate frictional resistence of rock discontinuities. Int. Jl Rock Mechanics and Mining Sciences, 16, No.2, pp.127-133.
Maugeri, M. 1981. Caratteri geotecnici delle frane causate dal terremoto del 1976 in Friuli. Atti della Riunione C.N.R. del Gruppo Ingegneria Geotecnica. Roma, 30-31 Marzo 1981.
Maugeri, M. & Motta, E. 1980. Determinazione della coesione di formazioni calcaree dall'osservazione di frane causate da sismi. XVI Convegno Nazionale di Geotecnica, vol.2, pp. 133-142.

On the statistical analysis of data and strength expression in the rock point load tests

Analyse statistique des données et expression de la résistance dans les essais Franklin

Xiang Guifu & Liang Hong, *Research Division of Engineering Geology, Chengdu College of Geology, Sichuan, China*

ABSTRACT: This paper deals with the influence of size and shape factors of samples on the results of rock point load tests by means of statistical analysis. In order to find out the relationship of point load test results with all of the influencing factors to establish a more rational expression, the progressive regression analysis was undertaken by use of a microcomputer. On this basis, a new expression of point load strength including all of the main influencing factors is suggested as $PLS=P/A_f$, where PLS is the point load strength, P is the failure load, A_f is the area of the surface of fracture. The expression has a definite physical meaning, and it will not be necessary to make size correction and to limit the size and the shape of samples when it is used, so it will enable the test and data processing to be simplified.

RESUME: Ce papier traite l'influence de la dimension et la forme des échantillons sur le resultat des tests de charge de pointe des roches par l'analyse statistique. Pour trouver la relation des resultats des tests de charge de pointe et tous les facteurs de l'influence afin d'établir une expression plus raisonnable, on a entrepris l'analyse de régression par le microcomputer. Sur cette base, une nouvelle expression comprenant tous les facteurs d'influence est proposée par la relation au desous, $PLS=P/A_f$, PLS étant l'intensité de la charge de pointe, P étant la charge, A_f étant la superficie de la surface de fracture. Il ne sera pas nécessaire de corriger la dimension d'échantillon et limiter aussi la dimension et la forme d'échantillon quant il est utilisé, par conséquent, il peut simplifier le test et le traitement des donnés.

INTRODUCTION

In the practices of rock point load tests of late years, we found some open questions which should be solved, such as: (1) When the point load tests are applied to the irregular specimens, the dispersity of the test results are higher than that in the diametral test of core and regular block specimens (Xiang 1981). It is shown that the size and shape of specimens have an effect on the dispersions of test results. What is the degree of the effect? How can we solve the problem? these should be studied. (2) By the original method the point load strength, $I_s=P/D^2$, which is got by the actual distance between loading points, must be corrected to the standard point load strength index $I_s(50)$ which is the I_s corresponding to the referrence distance D=50 mm in the diametral test (Kong(ed) 1979, Xiang 1981). However the standard specimen, a core 50 mm in diameter is not easily obtainable. To solve the problem, it is very troublesome to make the size correction by charts or nomograms (ISRM Commission on Laboratory Tests 1972, Kong (ed) 1979). Simplifying the data processing and adapting it to the simple technique of point load test thus is an urgent problem to deal with.

Therefore we selected a variety of block and irregular lump specimens which were in different sizes and made of different rocks for comparison tests and observed carefully. We analysed the determined results of three rocks (Limestone, Granite, Siltstone) with the method of mathematical statistics. The problems posed above are discussed and solved by the results of statistical analysis in this paper. Supposing the failure load P of the same kind of homogeneous rocks in the same size and of the same shape is considered as the same at all, then the P would not be the same for the same kind of rocks in different size and shape. Thence, we can regard the failure load P as the function of size and shape factors of specimens. We will determine the relationship of the failure load P with the related size and shape factors of specimen (D=distance between two loading points, I_{min}= mean length perpendicular to the loading axis on the minimum cross sectional area of the specimen through the loading points, L_f=mean length perpendicular to the loading axis on fracture plane of the specimen, K= shape factor which is equal to D/L_{min}, see Fig. 1) so as to take the major influencing factors into account in the

(a)

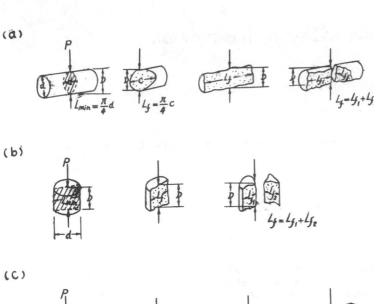

(b)

(c)

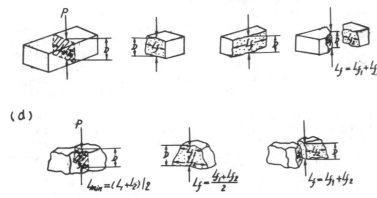

Fig. 1. Specimens with
various shapes and typical
modes of failure plane,
and determination of their
D, L_{min} and L_f. (a) The
diametral test, (b) The
axial test, (c) The block
and prism test, (d) The
irregular lump test. d:
core dia, c: The major
axis on circular and elliptical failure plane. (1)
The minimum cross-sectional
area of specimen through
loading points, (2) failure
plane

(d)

formula of point load strength, and give the point load strength a more precise physical definition. At the same time, the data processings and the size corrections are simplified.

1 ANALYSIS OF THE FACTORS INFLUENCING THE RESULTS OF POINT LOAD TEST

1.1 The size and shape factors affecting results of point load testing

D: Based on the original information (ISRM Commission on Laboratory Tests1972), the point load
strength index is defined as:
$$I_s = P/D^2 \quad \ldots\ldots\ldots \quad (1)$$
where I_s--- Point load strength index, in MN/m^2 or Mpa;
     P --- The failure load, in N or kN;
     D --- The distance between two platen contact points, in cm or mm.
  By this definition, the point load strength index of the specimen depends on the failure
load P and the square of distance D between two loading points. The exponential function relation between I_s and D has been confirmed by many workers (Kong (ed) 1979). But according to
the analysis of P vs D graph, it was found that there exists a better linear correlation between P and D (see, Fig. 2). However, when D remains constant, the dispersion of P is still
high. So we should consider some other size and shape factors which have effects on the testing
results.
K: The shape factor K is the ratio of D to L_{min}, here, L_{min} is the minimum length normal to the
D of the specimen (Kong (ed) 1979). The K is equal to or less than 1 for the flat specimens
loaded along minor axis. The K is greater than 1 for the tall or long specimens loaded along
their long axis. As for the same rock when D are same but K are different, the P values are
different either. When K is greater than 1, the P values are much less than the P when K is
equal to or less than 1, and the dispersion of P is higher when K > 1 (see Fig. 2). At the same

384

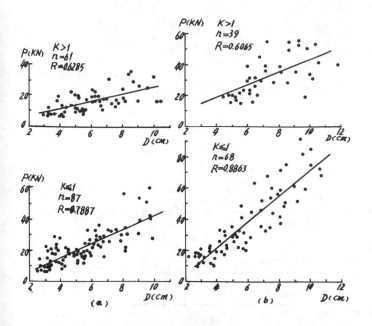

Fig. 2. The D-P relation
(a) Limestone (b) Granite

time it can be seen that there is an evident relation of inverse correlation between I_s and K in the I_s - K graph. In case K is equal to or less than 1, I_s values do not vary with K regularly for the same D value, then the results of statistical analysis show that I_s is not correlative with K (see Fig. 3).

In order to decrease the effect of the shape of specimen on testing results, the load should be applied along the minor axis of the specimen according to the discussion made above. The loading direction not noted specially is along minor axis of specimen in following tables of this paper.

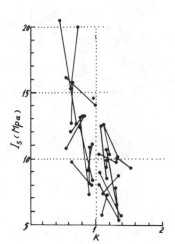

Correlation test	
K > 1	K ≤ 1
n=12	n=11
n+=2	n+=5
n-=9	n-=5
n0=1	n0=1
r=-0.793	r=0
$r_{0.05}$ =0.576	
inverse	
correlation	uncorrelated

Fig. 3. Analytic graph of correlation between I_s and K. Note: The D values are the same in both end of the lines.

L_{min}: From above, load must be applied along the minor axis of specimens. In such a case, the L_{min} is equal to or greater than D. From the relation graph of L_{min} vs P (see Fig. 4), we can know that the relationship between P and L_{min} tend to be linear. the L_{min} is then one of the influencing factors on point load test results too.

L_f: In testing, we found that P and I_s increase with L_f for the majority of specimens which D are the same (see Table 1). The I_f-P relation alike shows clearly that there is a linear relation between them (see Fig. 5).

The above analysis of mathematical statistics shows that D, K, L_{min} and L_f all have an effect on P. The effect of K may be overcomed by the method of restricting loading direction, that is, loading along minor axis. However, how do the other factors affect the P? Is each or are all nonnegligible and important? It must be tested and verified by multivariate statistical analysis.

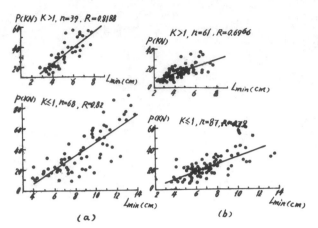

Fig. 4. L_{min}- P relation (a) Granite
(b) Limestone

Table 1.Correlation between L_f and P, I_s (specimen: granite.)

No	D(cm)	L_f(cm)	P(kN)	I_s(MPa)	Correlation of P and I_s with L_f	results
1	10.66 10.64	11.60 12.20	68 72.5	5.99 6.41	+	n=17
2	10.07 10.02	10.00 12.20	75 85.55	7.40 8.52	+	n+=12 n-=5 r=0.6026
3	9.34 9.32	13.90 18.00	35.20 91.43	4.03 10.53	+	$r_{0.05}$ =0.4821 then P and I_s
4	8.12 8.10	13.10 9.15	71.34 36.95	10.82 5.63	+	with L_f show positive correlation.
5	8.10 8.08	9.15 9.75	36.95 58.0	5.63 8.89	+	
6	7.48 7.47	9.70 10.77	58.8 53.21	10.51 9.54	−	
7	5.71 5.70	7.54 6.90	31.75 20.38	9.74 6.28	+	
8	5.48 5.43	12.05 7.35	32.34 21.36	10.77 7.25	+	
9	5.19 5.15	7.01 5.90	30.0 19.8	11.14 7.47	+	
10	4.58 4.55	9.20 4.45	22.74 30.38	10.84 14.68	−	
11	4.48 4.45	9.50 9.10	31.56 17.05	15.73 8.61	+	
12	3.73 3.70	6.55 3.95	10.78 11.27	7.75 8.24	−	
13	3.57 3.55	4.05 6.90	22.54 21.56	17.69 17.11	−	
14	3.0 3.0	5.05 5.55	16.66 20.58	18.52 22.88	+	
15	2.86 2.85	5.30 5.50	15.68 11.27	19.78 13.88	−	
16	2.8 2.8	7.6 10.0	9.8 17.64	12.5 22.51	+	
17	2.70 2.70	6.10 5.60	18.62 15.68	25.55 21.52	+	

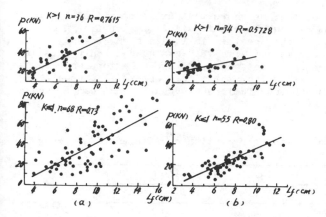

Fig. 5. The I_f-P relation (a) Granite (b) Limestone

Table 2. Significance test of trinary linear regressive analysis

Rock type	Specimen number	Significance test			Multiple correlation coefficient test			Conclusions
		F	$F_{0.05}$	result	R	$R_{0.05}$	result	
Limestone	55	36.64	2.28	$F \gg F_{0.05}$	0.826	0.377	$R \gg R_{0.05}$	Significant correlation
Siltstone	37	115.39	2.92	"	0.955	0.485	"	"
Granite	68	112.34	3.15	"	0.917	0.358	"	"
Granite (long axis)	36	23.54	2.92	"	0.836	0.464	"	"

Table 3. Results of the factors importance test

Rock type (spec.numb)	Comparison of partial regressive square sum P_i	Comparison of T_i values			Conclusions
		T_D	T_{Lmin}	T_{Lf}	
Limestone(55)	$P_{Lf} > P_D > P_{Lmin}$	1.22	0.64	2.44	Lmin is negligible
Granite(68)	$P_D > P_{Lmin} > P_{Lf}$	8.22	2.39	0.54	Lf is negligible
Siltstone(37)	$P_D > P_{Lmin} > P_{Lf}$	10.42	2.06	1.20	All have an effect
Granite(36) (long axis)	$P_{Lmin} > P_{Lf} > P_D$	0.55	3.38	1.86	D is negligible

1.2 Importance analysis for factors

First of all we begin with the analysis of trinary linear regression of D, Lmin and L_f to P. The results are given in table 2.

Table 2. shows that the trinary linear correlation between D, L_{min}, L_f and P are significant, and that the regression equation is of actual value. But of these factors which have an effect on P the major factors should be distinguished from the minor ones. Those minor factors should be deleted as much as possible, but those being important or playing the major roles can be retained. Only the new regression equation thus obtained could have more actual meaning. Therefore, we selected the partial regressive sum of squares "P_i" to judge the role that factors play in regression. The factors with great "P_i" must have an important effect on dependent variable (here, it is failure load P, see table 3). For the minor factors it should be still determined whether they are negligible. If one factor is verified to be negligible, it might not taken into calculation. Therefore we choice the statistical variable "T_i" value. The greater the "T_i" value is, the more important this factor might be. By experience, if $T_i > 1$, it has been shown that the factor has some influences on dependent variable; if $T_i > 2$, the "i-th" factor then becomes an important one; if $T_i < 1$, the factor might not be considered to have a remarkable influence. The analytical results are shown in table 3.

The conclusion drawn from the importance test of factors shows that at least two of the three factors are non-negligible and important and the negligible one is either L_{min} or L_f in case of loading along the minor axis of specimen. But if the specimen is loaded along the major axis, D is negligible and not important. However the test method with "P_i" and "T_i" is only suitable for the case in which the relations among arguments are not close. If the relations are close, it is very complicated to distinguish the major factors from the minor factors, and usually a very absurd conclusion might be obtained. Whether the conclusions above are right can be known by the correlation analysis among factors (see table 4).

Table 4. Simple correlation coefficient (R_s) and partial correlation coefficient (R_p) among factors

Rock type	R_s		Correlation factors				R_p	Correlation factors			
			D	L_{min}	L_f	P		D	L_{min}	L_f	P
Limestone	Correlation factors	D	1	0.93	0.86	0.79	D	1	0.73	-0.34	0.303
		L_{min}		1	0.96	0.79	L_{min}		1	0.81	-0.11
		L_f			1	0.80	L_f			1	0.332
		P				1	P				1
Granite		D	1	0.75	0.66	0.89	D	1	-0.43	0.44	1
		L_{min}		1	0.99	0.82	L_{min}		1	-1	-0.45
		L_f			1	0.73	L_f			1	0.55
		P				1	P				1
Siltstone		D	1	0.60	0.62	0.90	D	1	0.11	-0.25	0.86
		L_{min}		1	1	0.79	L_{min}		1	1	0.03
		L_f			1	0.79	L_f			1	0.29
		P				1	P				1

Based on the theory of mathematical statistics, the correlation among variables is very complex in multivariate regressive analysis. That is because the correlation between any two variables might exist. For two variables between which no correlation exists originally, as the correlations between them and other variables may be in existence, the simple correlation coeffecient "R_s" of them is even very great. So the actual case can not be reflected correctly only by the simple correlation coefficient "R_s" of two variables. This is also proved by the analytical results of the test (see table 4). In order to obtain real correlation coefficient between two variables, it is necessary to reject the effects of other variables on it, that is, to calculate the correlation between two variables by keeping other variables unchanged. That is why we use the partial correlation coefficient "R_p" for determination. Thereby, the "R_p " between each two among D, L_{min}, L_f and P are calculated respectively, and the results are given in table 4.

In the table 4. it is shown that the R_p between arguments L_{min} and L_f are very high, being close to or equal to 1. This is in accord with the observation in the experiment. That is, L_f approximatly equal to L_{min} in most cases. It shows that the failure planes of specimens are under the control of minimum cross sectional area of the specimen through two loading points. The L_f and L_{min} might basically be regarded as the same factor, though they are not the same. And the R_p between L_{min} and P is the smallest among the R_p between P and all arguments. The testing results of all kinds of rocks show that R_p between D and P as well as between L_f and P are greater, and the correlations are then the closest. Thus the L_{min} might be neglected, and only the effects of D and L_f on P are considered.

For further examine the correlation of D and L_f to P after the L_{min} has been rejected, we again go on with the binary regression analysis between D and L_f both and P, the analytical results are shown in table 5, 6, 7.

We compare table 5. with table 2. After the L_{min} was rejected, the multiple correlation coefficients hadn't changed much. That shows L_{min} does not play an important part in regression.

The table 6. shows that D and L_f are both important and non-negligible factors to P. The table 7. shows that R_p of D and L_f to P are greater than critical correlation coefficient, $R_{0.05}$, thereby the correlations of D and L_f with P are close.

So it is feasible for us only to consider D and L_f in all factors of size and shape which have the effects on fracture load P. The binary linear regression equations obtained by the above progressive regression analysis are shown as follows:

Table 5. Significance test of the results of binary linear regression analysis

Rock type	Spec. numb.	Significance test			Test of mulfiple corre. coeff.			Conclusions
		F	$F_{0.05}$	results	R	$R_{0.05}$	results	
Limestone	55	52.34	3.21	$F \gg F_{0.05}$	0.82	0.33	$R \gg R_{0.05}$	Correlation is significant
Granite	48	93.03	3.23	"	0.92	0.35	"	"
Siltstone	37	156.14	3.29	"	0.95	0.40	"	"

Table 6. Factors importance test.

Rock type	Comparison of P_i	Comparison of T_i		Conclusions
		T_D	T_{L_f}	
Limestone	$P_D < P_{L_f}$	1.91	3.34	All are important
Granite	$P_D > P_{L_f}$	8.52	2.88	"
Siltstone	$B_D > P_{L_f}$	9.82	5.55	"

Table 7. Simple correlation coefficient (R_s) and partial correlation coefficient (R_p) among factors

Rock type	R_s	Correlation factors			R_p	Correlation factors			$R_{0.05}$
		D	L_f	P		D	L_f	P	
Limestone	D	1	0.88	0.79	D	1	0.64	0.26	
	L_f		1	0.80	L_f		1	0.44	0.26
	P			1	P			1	
Granite	D	1	0.63	0.88	D	1	0.05	0.79	
	L_f		1	0.70	L_f		1	0.40	0.29
	P			1	P			1	
Siltstone	D	1	0.62	0.90	D	1	0.37	0.86	
	L_f		1	0.79	L_f		1	0.96	0.33
	P			1	P			1	

$$P = -9.804 + 1.768D + 3.104L_f \quad \text{for Limestone}$$
$$P = -18.72 + 6.515D + 1.716L_f \quad \text{for Granite} \quad (2)$$
$$P = -3.75 + 1.504D + 0.632L_f \quad \text{for Siltstone}$$

It is shown that the size and shape factor which has an important effect on fracture load P is not only the D; and that we should also take the L_f into consideration.

2 ON FORMULA OF POINT LOAD STRENGTH

After the main factors which have an effect on point load strength have been determined by the above analysis, we can go on to the discussion of the formula of point-load strength to formulate the best one. In this formula, the major factors P, D and L_f must be included, and the measure of strength must be obtained finally in the right of the formula. As the product of D and L_f may approximately represents the area of the failure face of specimen, it is defined as A_f. We analysed the relation of P to A_f to find out the best fitting of the regression equation between them, so that we could establish the best formula. For this purpose, we drew A_f-P relation graph, and used microcomputer for correlation analysis. The best types of regressive curves that obtained were straight line and parabola, as shown in Fig. 6. It shows that linear, parabolical and even exponential regressive curves are very proximate, and the linear and parabolical curves almost coincide with each other. Especially in the range of ordinary size the fitting regressive curves are basically straight line. The correlation of P with A_f is then linear and direct correlation. For the same group of rocks the ratios of P to A_f are very close and approach to a constant. (see Table 8 and Fig. 7).

Because ratio P/A_f has the unit of strength, it might be defined as the point load strength---PLS. It is maximum load on unit fracture area of specimens in point loading, shown as:

$$PLS = P/A_f \dots \dots \dots (3)$$

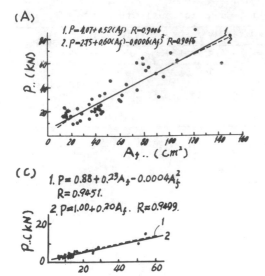

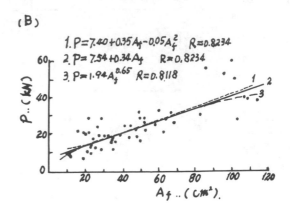

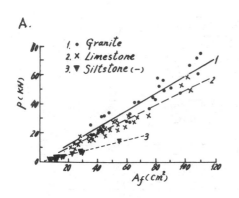

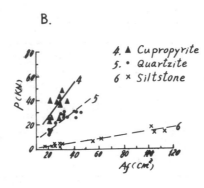

Fig. 6. Type of fitting regressive curves between A_f and P. (A) Granite; (B) Limestone; (C) siltstone.

Fig. 7. A. The P vs A_f curves. B. Curves of P vs A_f obtained by other measured data.

Table 8. The linear correlation between P and A_f and the mean of point-load strength calculated by formula (3)

Rock type	Spec. numb.	Linear corr. coef. of P with A_f r	Critical correlation coefficient $r_{0.05}$	Calcu. numb.	The mean of PLS (MPa)	Standard deviation	Coefficient of variation
Granite	68	0.9006	0.24	31	6.04	0.8154	0.1350
Limestone	54	0.8218	0.27	27	5.03	0.5588	0.1111
Siltstone(1)	29	0.9499	0.36	25	2.74	0.4283	0.1563
Cupropyrite	19	0.5529	0.46	11	13.16	2.2057	0.1980
Quartzite	20	0.6950	0.44	14	7.50	1.1662	0.1555
Siltstone(2)	23	0.8981	0.41	17	1.39	0.2655	0.1910

where, PLS in MN/m^2 or MPa; P in N or kN; A_f in cm^2 or mm^2. The formula has a strength measure and definite physical meaning, and is simple for the testing and data processing.

3 DISCUSSION FOR SEVERAL FORMULAS OF POINT LOAD STRENGTH

3.1 Several available formulas of point load strength

In the documents of suggested methods for determining point load strength, which were published successively in 1972 and 1985 by the "ISRM Commission on Testing Method", the point load strength index was defined as follows:

$I_S = P/D^2$ (in 1972) ...... (1)

$I_S = P/D_e^2$ (in 1985) ...... (4)

where, I_S is the uncorrected point load strength index;

D_e is the "equivalent core diameter", is given by (see Fig. 8):

$D_e^2 = D^2$ for diametral tests;

     $= 4A/\pi$ for axial, block and lump tests;

and A = DW = minimum cross sectional area of a plane through the platen contact points, here, W = L_{min}.

The requirments for the size and shape of specimen are also shown in Fig. 8.

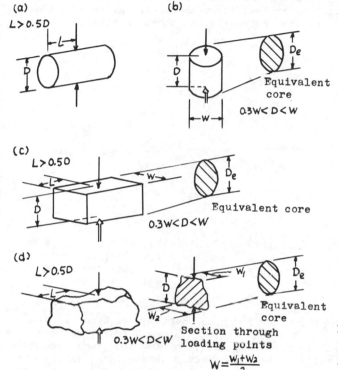

Fig. 8. The "equivalent core diameter" and specimen shape requirements for (a) the diameteral test, (b) the axial test, (c) the block test, and (d) the irregular lump test

In several papers Brook had proposed the following formula (Brook 1984, 1985):

$T^*_{500} = 2.115P/A^{0.75}$ .... (4)

where, P is the applied load in kN; A is the minimum cross sectional area in mm^2; T_{500} is the strength index in MPa, and the asterisk * indicates a value obtained by this size correction formula. The strength index is then the load for a standard cross-sectional area of 500 mm^2.

Based on a lot of point load tests with long prism samples of rectangular section, Yang Rui-lin had proposed a new formula by regression analysis as follows (Yang 1984):

$I_S = P/A^{0.75}$ .... (6)

3.2 Discussion for formulas

In the formula (1) only two factors, P and D were taken into consideration, which is evidently not enough. Because the formula was based on the stress analysis of elastic ball, it cannot meet the various complex conditions of specimens, and some factors might be omitted. From the above analysis it is known that the P is affected by size and shape factors of specimens for the same rock material besides depening on the material composition and structure characteristic of rock itself. Though D is a very important factor, however it is not the only important factor. Because in the formula (1) the important and nonnegligible factors are not taken into account, the extant size and shape effects are still existential. These have been confirmed by many workers. To obtain typical point load strength value, as mentioned in the introduction, a troublesome size correction must be carried out. If a standard shape and size is used to avoid the effects of size and shape, the work for preparing specimens would in turn increase. Therefore this method is much restricted in use.

Compared with the method in 1972 the method suggested by ISRM in 1985 has a great progress. It considered the effect of minimum cross sectional area of specimen through loading points. The great part of failure faces are under the control of minimum cross sectional area and coincident with it. However the subsequent analysis will give out that minimum cross sectional area cannot substitute the area of fracture plane at all. Because any rock mass has experienced a complex geological history, it may now show an anisotropic distribution of tensile strength. In point load testing the stress field in rock samples is highly heterogeneous (Lajtai 1980), therefore in progressive procedure of tensile fracture the anisotropy pointed out above might make minimum tensile strength not distribute on the minimum cross sectional plane of specimens, and that the fracture occurs along the plane of mimimum tensile strength. Broch once reported (Broch 1983) that diametral point load strength index, Is, is a function of the angle, α , between the foliation plane and the core axis, as shown in Fig. 9. It is

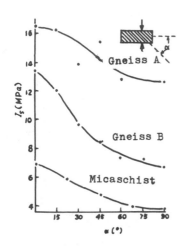

Fig. 9. Diametral point-load strength index, I_s, as a function of the angle, α, between the foliation plane and the core axis for two gneisses and a micaschist, from Aagaard.

evident that this conclusion results from the above two formulas' deficiency. Since in diametral point load test of the core the core dia D is a constant, the minimum cross sectional area of the core is then a constant. As from the above formulas, the I_s of the same rock should be approximate to a constant, too. However, in reality the point load strength index varies with the angle between the core axis and the foliation planes. The fact shows that the point load strength is not dependent on the minimum cross sectional area, but is determined by fracture area of specimen. Thus, when calculating the point load strength, it is unreasonable to still use the same D_e^2 or D^2. Therewith in the formula (4) the using of the "equivalent core diameter" instead of the area directly, increases the calculation work load, and because of the necessary size correction and the physical idea is not much clear.

In the formula (5) and (6), as compared with in the formula (4), the directly using of the minimum cross sectional area A of specimen has made a progress. However, as there are still some deficiency showed above, they need to be improved.

The formula, PLS = P/A$_f$, suggested in the present paper, is obtained from the best regression equation of a rational curve fitting in which the size and shape factors affecting P slightly are rejected and those affecting P greatly are retained by means of the progressive regression analysis based on the comparative testing data of a great variety of block and irregular lump specimens in different size. Because the A$_f$ vs P curves found in regression analysis by us are all linear, and because the actually measured P/A$_f$ ratios approach to a constant for the same type of rock (see table 8 and Fig. 7) and could simply be expected to be greater for stronger rocks and smaller for weaker rocks, the data processing is simplified greatly. Thereby a size correction, such as in the methods suggested by ISRM in 1972 and in 1985, is not necessary. According to the present method, no matter what size and shape   the rock samples are in point load testing, only by loading along the minor axis and measuring P, D, and L$_f$ values of them, we can calculate the PLS values of any rock specimens with the formula (3) directly. As least ten samples should be tested in each group. More are needed if the rock is heterogeneous or anisotropic. The mean value of PLS is to be calculated after deleting the two highest and lowest values from the 10 or more valid tests, and calculating the mean of the remaining values (ISRM Commission on Testing Method 1985), this is the point load strength of this group of rocks.

To verify the conclusions and the calculating method above, we plotted A_f vs PLS relation graph (see Fig. 10), and the verified results are given in table 9. The fig. 10 and table 9 shows that the PLS is independent of A_f. It has overcome the problem of the influence of size and shape on the point load test results reasonably. Thereby the method is feasible and reliable.

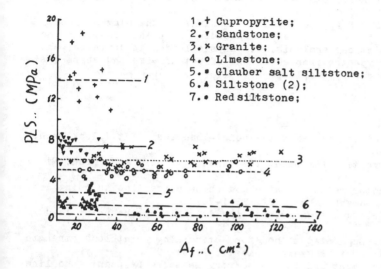

1. + Cupropyrite;
2. ▾ Sandstone;
3. × Granite;
4. ○ Limestone;
5. ▪ Glauber salt siltstone;
6. ▲ Siltstone (2);
7. • Red siltstone;

Fig. 10. Relationship between PLS and A_f.

Table 9. Test for correlation between PLS and A_f.

Rock type	Calcu. numb.	The dispersion of A_f			The mean of PLS. (MPa)	Corr. coef. of PLS and A_f; r	Critical corr.coef. $r_{0.05}$	Conclusions
		Mean (cm^2)	Standard devistion	Coef. of variation				
Cupropyrite	13	24.79	6.56	0.26	13.91	−0.1650	0.55	$|r| < r_{0.05}$, The PLS bear no relation to A_f
Sandstone	21	16.23	5.14	0.32	7.37	−0.1882	0.43	"
Granite	31	64.10	24.86	0.39	6.04	0.0382	0.35	"
Limestone	28	49.83	22.21	0.45	5.00	0.1059	0.38	"
Glauber salt siltstone	14	25.58	8.17	0.32	2.77	−0.1456	0.53	"
Siltstone(2)	28	43.57	37.23	0.79	1.49	−0.1566	0.38	"
Red siltstone	17	83.89	23.58	0.28	0.61	−0.2963	0.48	"
Siltstone(1)	25	17.48	13.16	0.75	2.74	−0.3115	0.40	"
Iron ore	11	29.82	8.36	0.28	18.58	−0.2787	0.60	"
Quartzite	14	31.78	9.02	0.28	7.50	−0.0786	0.53	"

4 CONCLUSION

1. The correlation between fracture load P and D, L_{min}, L_f —the size and shape factors of specimen was analysed by means of mathematical statistics. It is shown that D and L_f are important and nonnegligible factors which have an effect on P.
2. The minimum cross sectional area A of specimen through loading points plays a control part in the fracture of specimen, which was shown by the fact that most of the failure planes appeare along the minimum cross sectional plane. However, since rock materials are mostly heterogeneous, and characterized by anisotropy, the anisotropy of samples would make that the fracture occurs along the plane of minimum tensile strength which not coincide with the minimum

cross sectional plane of specimen in the progressive procedure of tensile fracture in point load tests.
3. The best fitted A_f vs P curves are linear in the range of usual size of specimen. The ratios of P to A_f approach to constant for the same group of rocks, Thereby no matter how size the A_f is, the size correction is unnecessary.
4. If the point load strength, PLS, is calculated by the formula (3), the area of failure plane A_f of specimen should be measured as correctly as possible. In the tests with weak, fragmental and weathered rocks, the fractured specimens should be protected, like being caught carefully by hands, to prevent from fracturing once more. Otherwise we would not be able to determine the A_f correctly.
5. The point load strength defined by PLS = P/A_f, in physical meaning, is the biggest load which the samples can bear on unit area of the fracture planes under the point load. For the formula is obtained by mathematical statistical analysis, its physical idea is definite, and it is in strength measure. Because it solves the problem of the effect of size and shape basically, the testing and data processing are very simple.

REFERENCES

Broch, E. 1983. Estimation of strength anisotropy using the point-load test. Int.J.Rock Mech. Min.Sci. & Geomech. Abstr. 20, 4:181-187.
Brook, N. 1984. Technical note, Size correction for point load testing. Int.J.Rock Mech. Min. Sci. & Geomech. Abstr. 17:231-235.
Brook, N. 1985. The equivalent core diameter method of size and shape correction in point load testing. Int.J.Rock Mech. Min. Sci. & Geomech. Abstr. 22, 2:61-70.
ISRM Commission on laboratory tests. 1972. Suggested methods for determining the point load strength index. Document. 1.
ISRM Commission on testing methods. 1985. Suggested method for determining point load strength (revised version). Int.J.Rock Mech. Min. Sci. & Geomech. Abstr. 22:51-60.
Lajtai, E.Z. 1980. Tensile strength measurement and its anisotropy measured by point--and line --loading of sandstone. Eng.Geol. 15:163-171.
Kong, D.F.(ed). 1979. On rock point load test. Selected works on Hydro. & Eng. Geo. 12. Beijing: Geological publishing house.
Xiang, G.F. 1981. The comparison studies for rock point load test. J.Hydrogeo. & Eng. Geo. 1:41-44.
Yang, R.L. 1984. Rock point load test analysis. J.Underground Eng. 6.

Application of drilling method to estimation of stresses in loosened and broken rock

Utilisation d'une méthode de forage pour apprécier les contraintes dans des roches dissociées et fracturées

Meng-tao Zhang, Ben-jun Zhao, Zeng-he Xu & Yi-shan Pan, *Research Institute of Fu-Xin Mining Institute, China*

ABSTRACT:The principles and measuring method are dealt with for fast estimation of stresses in loosened and broken rock in the paper on the basis of field measurement,lab modeling and theoretical analysis.The equation is put forward expressing relationship between amount of drillings and stresses in coal mass,with strain-softening property of unelastic coal area and volume expansion(dilatation) considered.Means of choice of calculation of parameters and practical application are discussed as well,having practically important value.

INTRODUCTION: So called drilling method is to drill small diameter holes (42-50mm) in coal mass. Stress distribution and location of peak-stress can be obtained according to the regular pattern of variation in drillings output per unit of depth of bore hole in drilling process and in the line with the dynamic phenomenon.In the early sixties this method was tested in the Federal Republic of Germany and other countries,and since then it has found worldly wide application because of its high adaptability and convenience.It has been one of the common methods to determine the danger degree of rock burst.The author hasn't seen special reports about its practical application and detail research as fast means ofor estimation of stresses in adjointing rock.

When ordinary rock stress measuring methods used(as stress relief method,hydraulic fracture test) special equipment and meters are needed,taking much time and interrupting processes of work,and its condition ofoperation are strict.It is difficult to apply this method to underground conditions. The ordinary methods will not work when it is hard to extract rock core (or coal core) and to keep the walls of bore hole to be complete.So far manyyears the authors have done some theoretical analysis and experiments based on practice of predicting danger degree of (coal) rock burst.Initial application was carried out in order to make the small diameter hole drilling method become a new way of fast estimation of stresses in   rock mass similar to or with coal mass properties.

1 BASIC PRINCIPLES

On drilling holes in coal mass the drillings extracted consist of two parts:coal flour broken from coal cylinder of the same diameter as of the drilled hole;coal flour formed by contracting deformation of hole walls caused by stresses redistribution in adjointing rock.The former is only related to the diameter of the drill bit;the latter is related to the coal machanic properties and the distribution of stress around the drilled hole.Therefore,in order to research in the relationship of drillings amount to distribution of coal mass stress it is,first,required to study the distribution of stresses and deformation in coal mass around the drilled hole.In this regard many scientists supposed the coal mass to be isotropic and homogeneous elastic body beforedrilling and to be solved as a problem of infinite square with a round hole.They applied Mohr-Coulumb criteria as the yield condition of unelastic deformation appeared after drilling,and considered the load as an axis of symmetry in hydrostatic condition.The properties of strain-softening and volume expansion (dilatation) have not been taken into account.Therefore,difference between calculated and measured values were great.An equation was built up to describe the relationship between stresses in coal mass and drillings amount,with strain-softening and volume expansion (dilatation) after appearence of unelastic area in coal considered.

In accordance with the whole course stress-strain curve in coal body (pic.1) the equation considering the strain softening is

$$\sigma = E\varepsilon \qquad (\varepsilon < \varepsilon_c) \qquad\qquad (1)$$

$$\sigma = \sigma_c\left(\frac{\varepsilon}{\varepsilon_c}\right)^m \qquad (\varepsilon \geq \varepsilon_c) \qquad\qquad (2)$$

where, σ_c, ε_c—uniaxial compression strength and correspondent strain in coal mass; σ, ε—stress and strain of unelastic deformation area;m—plasticity coefficient.

It is known from (2) that with ideal plasticity m=0,with staih hardening m>0,with strain soften-ing m<0.

Elastic plasticity is analysed with Mohr-Coylumb criteria taken as yield condition(pic.2).Radius R of unelastic strain softening area is obtained.See (Zhao Yan-sheng 1983)

$$R = a\left\{1+ \frac{(2m+q-1)(2p-\sigma_c)}{q^{1-m}(\sigma_c+(q-1))^m(q-1)}\right\}^{\frac{1}{2m+q-1}}$$ (3)

with m=0,above equation would be Kastner equation.

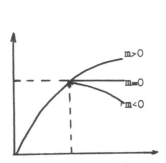

Pic. 1. Relationship of str-esses in coal-body.

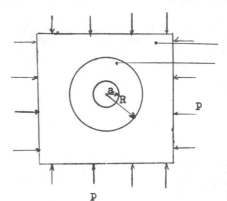

elastic area

unelastic area

Pic. 2. Analysis of ela-stic plasticity around hole

The radial displacement U_R at the boundary between unelastic area and elastic area around the hole:

$$U_R = \frac{1+\mu}{2E} R\left[\sigma_c + \frac{q-1}{q+1}(2P-\sigma_c)\right]$$ (4)

where,E—coefficient of elasticity of coal;μ—Poission's ratio;p—stress in coal body before drill-ing;a—hole radius;φ—angle of friction; $q = \frac{1+\sin\varphi}{1-\sin\varphi}$.

At present the analytic equation has not been built to describe volume expansion(dilatation). The dilatation(volume expansion)effect is calculated with average dilatation coefficient n for sh-ort.With coal-mass considered to be stable the radial displacement Ua of hole walls can be obtained with dilatation included:

$$U_a = \frac{R}{a}U_p + \frac{n-1}{2a}(R^2 - a^2)$$ (5)

where the second term n is 1.1—1.2,considering the influence caused by dilatation,in accordance with the suggestion in reference(Paituhov 1980).Therefore ,the equation has been obtained consider-ing the relationship between drillings amount G under influence of dilatation and stress P in coal mass:

$$G = \gamma(\pi a^2 + 2\pi a U_a)$$ (6)

where r—coal volume weight.

2 CHOICE OF PARAMETERS

It is very important to choice the parametes used in the equation in particular the hole radius a-nd plasticity coefficient in calculation of coal stresses by means of the extracted amount of drill-ing inthe process with small hole drilling method.

The hole diameter has greater influence on the extracted amount of drillings.Calculation shows, amount of drillings will be doubled with increase in hole diameter per 10mm,when the other codition is not changed.Note that a is not radius of drill bit but the radius of drilled hole in equation (6).Practice and theory also prove that when the same drill bits are used in drilling,the diameters of drilled holes will be increased with increase in stresses of coal mass.Therefore,while calcula-ting it is required to apply the average hole diameter of drilled holes measured on worksite as ac-curate as possible.On the basis of our experience lasted many years the diameters of drilled holes usually vary between 50—60mm.With drill bits 42mm.Used to drill in coal mass,i.e.the expansion coefficient of diameter of hole is 1.2—1.5.When the diameter of drill bits is determined,the amount of drillings must be takeninto account.small diameter is adopted when amount of drillings is less; but large diameter is adopted when amount of drillings is more.Collapse of hole wall occurs only in particular case with amount of drillings increased about 10 times.In this regard Mr.Cook has st-ated and put forward the occurence condition in reference(Jaeger & Cook 1981):

$$\left[\frac{R}{a}-1\right] \geq \frac{E}{(1+\mu)B}$$ (7)

where,B—slope rate of stress of coal unelastic area strain curve,near to constant.Above-mentioned equation was obtained on the basis of result of rock sample tested on tester without destabilization of rigidity criteria.

Choice of plasticity coefficient m is very important.It should be obtained by mediating of whole range stress-strain curves get in lab,but it is hard to get it.Because the whole range stress-strain curves are not regular on the post-peak strength limit(i.e. in the period of strain-softening), it is difficult to be mediated on the basis of exponential curve in equation (2).It will be certain that if the whole-range stress-strain curve is made smooth curve by idealizition of mediation can be done.But different idealizing methods lead to appearence of different values of plasticity coefficients.The plasticity coefficient m has been chosen by inverse analysis and test in lab,using special confined compression equipment.The authors have made pressure and drilling tests of many parties of rock samples(310×310×310mm) on 500t.press.The amount of drillings extracted under various pressure was converted into amount of drillings from unit of hole depth with certain stresses. In accordance with inverse analysis of parameters the value m is given and amount of drillings is calculated.After comparing the calculated value and measure value if the errors are greater, the value m will be optimized with optimum seeking method(usually 0.618 is adopted),then the amount of drillings is calculated with new optimized value m.After several recalculations when difference between the calculated amount of drillings and amount of drillings obtained from tests is rested within limmits(often 5%) the value of plasticity coefficient will be obtained finnally.Practice for many years shows that the value of coal plasticity coefficient m usually varies between 0.2— 0.4.

Inverse analysis can be adopted for calculation of the other parameters,but it is common to get them from lab tests directly.

3 PRACTICAL APPLICATION

Since absolute value of rock stress and distribution can obtained with small hole drilling method, this method can be applied to calculation of ditribution principles of support pressure in mining and heading faces,and can also be applied to estimation of its values.Therefore,it will play an important part not only in improvement of face-roof control and maintaining working supports,but also can be widely applied to estimation of danger degree of rock(coal) burst.

3.1 Estimation of support pressure

Support pressure means absolute value and distribution of stress in adjointing rock of mining faces. The important characteristic parameters appear to be peak support stress, its position and the area on which the support pressure acts.

The amount of drillings extracted is concerned with coal mass pressure.Distribution curve of measured amount of drillings expresses oharacristic of support pressure area.Curve of relationship between drillings amount Q and coal-bed pressure P can be obtained with above-mentioned equation, or the curve of relationship between Q and P can be got by lab. modeling.When directly measuring real amount of drillings coal bed pressure can be estimated according to the curve G—P with properties of support pressure area obtained.In table 1,there are calculation results concerned with coal beds in some coal mines.Pic.3,pic.4 correspondent curves G—p.Table 2—estimation of characteristic values of coal bed support pressure area.

3.2 Estimation of danger degree of rock(coal) burst

Small hole drilling method has found wide application to estimation and prediction of danger degree of rock(coal) burst (Zhao Ben-jun & Zhang Mend-tao 1985).The danger degree is determined according to peak value of support pressure in coal bed and its position,as well as,dynamic phenomena in drilling process.

In table 3 and Table 4 there are measured data in Coal Mines "Long-Feng" and "Men Tao-gou".

Prompt measures must be taken for saving danger area inclined to rock (coal) burst.After having taken prompt saving measures the small hole drilling method has to be adopted to examine the effect of the method.Research shows that the peak value of support pressure would be decreased when some saving measures have been taken.

Position of peak value transfers into depth of coal mass.There is no dynamic phenomenon in drilling process.It shows the coal mass is unloaded.Pic.5 and pic.6 are respectively typical curves of drillings amount in mining faces before and after the saving measures taken in Coal Mines "long-Feng" and "Meng-tao-gou".they show that peak value of support pressure transfered into depth at distance of 2—3m,with decrease in peak value by about 20%.

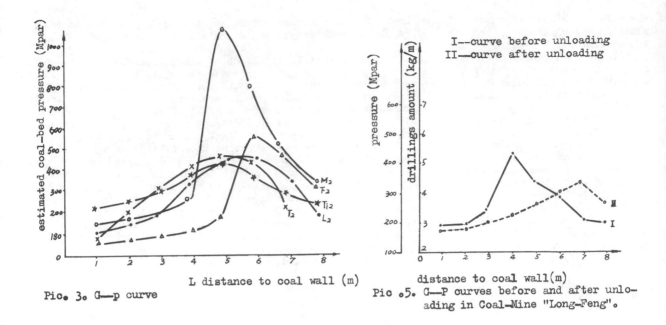

Pic. 3. G—p curve

Pic. 5. G—P curves before and after unloading in Coal-Mine "Long-Feng".

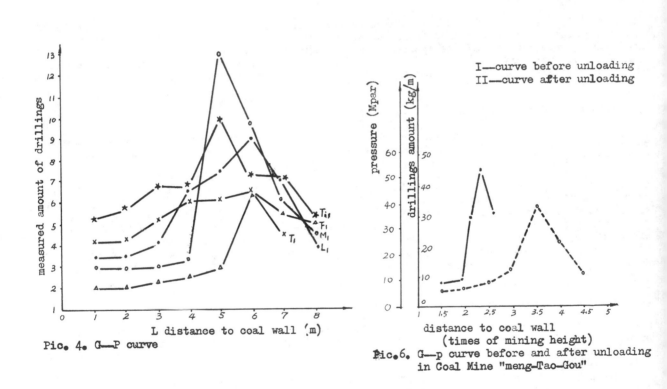

Pic. 4. G—P curve

Pic. 6. G—p curve before and after unloading in Coal Mine "meng-Tao-Gou"

REFERENCES

Jaeger J.C & N.G.W Cook 1981 .Fundamental of rock mechanics.Publishing House "Science",China.(translated from English)

Paituhov I.M. 1980.Coal burst.Publishing House of coal-industry,China.(translated from Russian)

Zhao Ben-jun & Zhang Meng-tao 1985.Research and application of small hole drilling method.Coal scientific Journal,China.(in chinese).

Zhao Yang-sheng 1983.Analysis of elastic plasticity of deep round chamber.Journal of Fu-Xin Mining Institute,China.(in chinese)

Table 1. Measured amount of drillings per meter (kg/m) and estimation of coal-bed pressure (Mpar)

amount of drilling (G) and correspondent stress (P)

Coal Mine	coal-bed	1m		2m		3m		4m		5m		6m		7m		8m	
		G	P	G	P	G	P	G	P	G	P	G	P	G	P	G	P
Fu-Shun Mine "Long-Feng"	4-5th layers	3.2	11.0	3.5	14.2	4.1	19.2	6.6	34.5	7.5	39.	9.1	46.1	7.1	37.4	4.1	19.7
Bei-Jing Mine "Moun Tao-gou"	2d bed	3.0	15.0	3.0	15.0	3.1	19.0	3.3	26.0	13.2	108.0	9.7	77.0	6.1	46.8	4.6	32.9
Bei-Jing Mine "Fang-Shan"	4th bed	2.0	6.0	2.0	6.0	2.3	9.6	2.5	12.0	2.9	18.2	6.5	56.5	5.5	48.5	5.0	44.5
Kai-lan Mine "Tang-Shan"	8-9 beds	2.35	8.7	4.25	19.3	5.2	31.4	6.1	41.4	6.2	42.5	6.6	46.1	4.5	23.0	/	/
Shi-Chuan Mine "Tian-Chi"	single bed	5.3	22.5	5.7	24.6	6.8	30.5	6.7	30.0	10.1	44.0	7.4	32.5	7.1	31.9	5.3	22.5

Table 2.

Mine	mined coal-bed	depth of mining (m)	properties of support pressure in faces			
			(m)×10 area by distinctly influenced	position of peak value (m)	peak value (Mpar)	coefficient of concentrated stresses
FU-Shun "Long-Feng"	4-5 layers	700	1.5 —2.0	3—6	30—70	2—3
Bei-Jing "Moun Tao-gou"	2d bed	500	2—3	4—7	30—100	3—6
Bei-Jing "Fang-Shan"	4th bed	500	2—3	5—10	25—80	2-5
Kai-Lan "Tang-Shan"	8-9 beds	650	1.5—2.5	3—8	25—50	2—3
Shi-Chuan "Tian-Chi"	single bed	620	2—3	4—7	25—50	2—3

Table 3.

danger degree of rock burst	Maximum amount of drillings (kg/m)			drillings amount grain more than 3mm (%)	dynamic phenomena
	l < 4m	l=4—6m	l 6m		
without burst danger	3.5	4	5.5	30	nono
with burst danger	3.5—5.0	4—5.5	6	30	burst sound stricking of drill-rod
with serious burst danger	5.0	5.5		30	hard-stricking of drill-rod

Note: l—distance of peak amount of drillings to coal wall; diameter of drill bit is 42mm.

Table 4.

depth examined B (times of mining height)	1—1.5	2	2.5	3
dangerous amount of drillings (kg/m)	4	8	16	25
dynamic effect	hard sticking of drill-rod			

Note: 1. ffectless examined depth Lo=1m.
2. Diameter of drill-bits is 42mm.

2 Engineering geological problems related to foundations
and excavations in weak rocks
Les études de géologie de l'ingénieur pour les fondations
et les fouilles dans des roches tendres

2.1 Geological and geophysical investigations of weak rocks:
Characteristics and classification
Etudes géologiques et géophysiques de roches tendres:
Caractéristiques et classifications

Cemented preconsolidated soils as very weak rocks

Sols cimentés préconsolidés considérés comme des roches à très faible résistance

E.Núñez & C.A.Micucci, *Argentina*

ABSTRACT: Within the Argentine Republic, on the right banks of the la Plata and Paraná rivers, overconsolidated by desiccation and cemented soils named "toscas" are located. They exhibit mechanical properties corresponding to weak and very weak rocks. In this paper the geotechnical characteristics of these materials are presented.
Shear and compressibility parameters related to consolidation pressures are estimated through approximate equations. Stress strain relations for shearing stage in triaxial tests are established

RESUME: En Argentine, sur la marge droite des rivières de la Plata et Paraná, on trouve des sols préconsolidés par dessèchement et cémentés, appelés "toscas". Ils ont les propriétés mécaniques qui correspondent à des roches faibles et très faibles. Dans ce travail, on étudie les caractéris tiques géotechniques de ces matériaux.
On propose des équations qui permettent de relier les paramètres de cisaillement et deformabilité avec les pressions de consolidation. On propose, aussi, rapports tension-déformation pour l'étape de cisaillement dans des essais triaxiaux.

INTRODUCTION

The geotechnic characteristics of pre-consolidated and variably cemented soils bearing the proper ties corresponding to soft and very soft rocks are investigated in this work. These soils are part of the sediments that constitute the "Pampeano formation" of the Quaternary Age formed over the so-called "Puelches sands" of the Terciary Age. In the Argentine Republic they are found over the right-bank of the Paraná river and the río de la Plata. The cities of Rosario, Buenos Aires and La Plata are situated within this area. A typical stratigraphic profile of this formation is shown in Figure 1.

The lower part of the "Pampeano formation" is likely to have a fluvial origin, while the upper part of prevailing thickness may have an eolic origin.

The particles exhibit plasticity, its range is wide enough so as to be able to classify the soils in silts or clays of high or low plasticity, i.e. MH, ML, CH and/or CL. These deposits have been formed in successive periods of extreme wet and dry climates bringing about its previous consolidation due to desiccation (Bolognesi, 1975). Therefore their consistency varies from compact to hard.

In certain locations the existence of banks of cemented soils forming the locally named "toscas" are due to the precipitation of carbonates or the presence of calcium and magnesium oxides. The rest of the profile, however, presents weaker or null degrees of cementation and the shear resistance of these soils can only be explained by the preconsolidation due to desiccation. The locally so-called "toscas" or "toscosos" soils, are silts or clays generally bearing a macro porous structure, the particles of which have a stable cementation made up by carbonates and calcium and magnesium oxides.

According to the history of the deposit formation, the cementing is quite homogeneously distri buted in the soil mass, or forming calcareous nodules appearing either in important groups or more or less scattered in the silty-clayey matrix. A gradation according to the degree of cementa tion can be therefore established:

a) "Toscas" which can be subdivided into hard and weak, when the macro porous structure and calcareous cementation is quite well differenciated.

b) The "toscosos" soils where the structures can be seen, the matrix is less enriched by calcareous.

c) "Pampeano soils" with some cementation, they are preconsolidated due to desiccation.

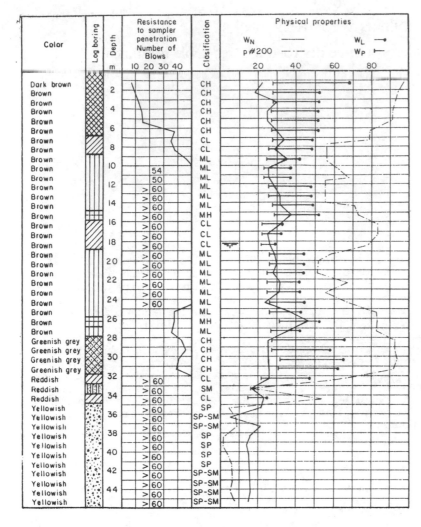

Figure 1: Typical profile

The toscas are highly resistent to unconfined compression, reaching 1000 or 2000 KPa, the "toscosos soils sometimes called "tosquillas" being the contents of small differenciated calcareous outstanding and, on the other hand the cementation of the matrix poor, are quite less resistent. In the weakest border of the formation we can find silts or clays, the cementation of which is null, but compact or very compact due to the desiccation.

According to the parameters Deere & Miller used to classify rocks, the modified graphic of Figure 2 shows the position corresponding to the studied materials; σ_{cf} corresponds to the failure on triaxial compression tests, the confining pressure of which is equal to the half of the effective overburden.

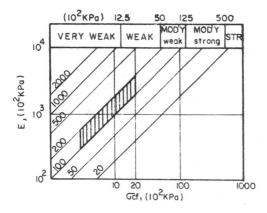

Figure 2: Modulus ratio ranges for "toscas"

PHYSICAL AND MECHANICAL CHARACTERISTICS

The physical properties have been studied through identification tests, the mechanical properties through triaxial and oedometric compression tests.

Taking into account the characteristics typical to the cemented soils or very weak rocks, the confining pressure levels in triaxial tests, and vertical pressure levels in the oedometric tests reached the highest values of 10000 KPa.

Figures 3, 4 and 5 show the typical results of triaxial tests; unconsolidated-undrained (Q); consolidated-undrained with pore pressure measurement (R); and consolidated-drained (S). An important variation of the behaviour of the different samples can be seen, due to the different cementation each of the soils bear.

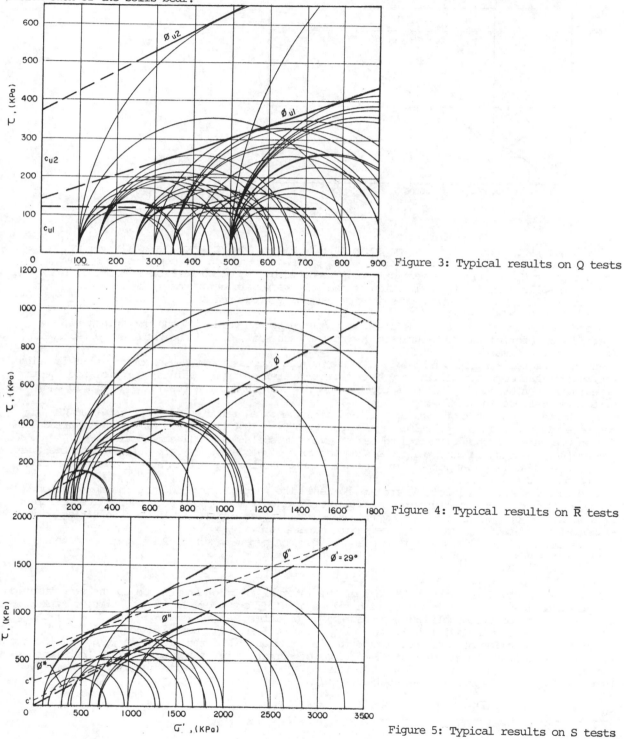

Figure 3: Typical results on Q tests

Figure 4: Typical results on R̄ tests

Figure 5: Typical results on S tests

405

Taking into consideration the results of the high pressure tests, Figure 6, the value of Ø', friction angle can be defined, which to soils with $W_L \sim 45$ is approximately of 29°. As the whole formation is pre-consolidated due to desiccation, the envelopes of Mohr circles will always bear non-null values of ordinate to origin for $\sigma'3 = 0$. This value, here is named c", depends on the structures of the clayey materials due to the action of the preconsolidation, even for null cementation values.

When the materials have own stable cementation, independently of the contents or of the water flow, the tests show an ordinate to the origin additional, which we named c"; the increase of resistance due to cementation also decreases with the increase of the effective confining pressure and generally with the increase of effective stresses. See Figure 5.

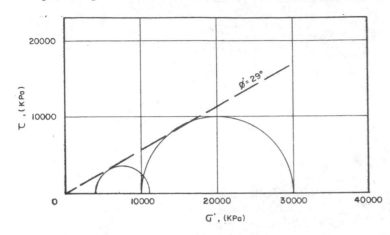

Figure 6: Results of high pressure triaxial tests.

The higher the preconsolidation pressure is, the higher the value c' generally is. For the soils of this formation, there seems to exist proved evidence that upon the deposits there have not acted important significative vertical loads produced by ice or soils later erosioned. Therefore the preconsolidation due to desiccation is the fact responsible for the value c' ≠ 0.

If we consider Bishop's old expression for partially saturated soils, we have:

$\sigma' = \sigma - [Ua - \chi (Ua-Uw)]$

Where Ua denotes pressure in the air partly occupying the voids, Uw the pore pressure, χ a coefficient refering to the saturation degree for each soil, σ, σ' the total and effective stresses respectively.

For these materials, it can be said that during the successive processes of drying-moisturing the suction has corresponded to very high negative water pressures values; for instance the air appears in open structure for low saturation degrees, Ua = 0 and the resulting isotropic equivalent effective pressure is of χ Uw.

With combinations of χ of 0,05 to 0,1, and negative pore pressures of 10000 to 20000 KPa, it can be explained that the pressures equivalent to 500 to 1000 KPa appear as effective preconsolidation pressures in the corresponding tests, even for soils with null cementation.

When the cementing is very reduced or when the time of hardening is very different to the time of drying -moisturing developped "in situ", the preconsolidation action appears in an effective reduction of the void ratio.

When the cementing is important and has hardened the particle group during the desiccation process, the resulting structure shows a high void ratio, a macro porous through quite resistent structure, not compatible with a conventional consolidation process as the one resulting for example for non-salty soils without cementing materials.

c' values of 10 to 20 KPa have been measured for non-cemented soils. Bearing in mind that:

$$\frac{c'}{\sigma'c} = \frac{N\emptyset' - N\emptyset*}{2 N\emptyset' \sqrt{N\emptyset*}} \quad (1)$$

It can be observed that the variations of the internal friction angles due to the structuration for preconsolidation is of 1° to 2°, with values of $c'/\sigma'c \cong 0,02$. In the equation, c' intercept cohesion, $\sigma'c$ for preconsolidation pressure, Ø' for internal friction angle and Ø* for true friction angle $N\emptyset = tg^2 (45 + \emptyset/2)$. Naturally for a soil normally consolidated $c'/\sigma'c = 0$.

The combined action of preconsolidation for desiccation and of the effective weight of the supra lying deposits leads to give rise to conditions of effective pressures with values Ko between 1 and Kon = η (1 - sen Ø) with $0,95 < \eta < 1$. For soils bearing null cementation it can be observed that the following expressions an applicable with enough approximation:

$$\frac{c_u}{\sigma'c} = \frac{(2c' \sqrt{N\emptyset*}/\sigma'c) + (N\emptyset* - 1)}{2 [Af (N\emptyset* - 1) + 1]} \quad (2)$$

$$\left(\frac{c_u}{\sigma'c}\right)_n = \frac{N\emptyset - 1}{2\ N\emptyset'} \qquad (3)$$

$$\frac{c_u}{\sigma'zc} = \frac{(2c'\ \sqrt{N\emptyset*}/\sigma'zc) + [Ko + (1-Ko)\ Af]\ (N\emptyset* - 1)}{2\ [Af\ (N\emptyset* - 1) + 1]} \qquad (4)$$

$$\left(\frac{c_u}{\sigma'zc}\right)_n = \frac{(2 - Ko)\ (N\emptyset' - 1)}{2\ [2\ (N\emptyset' -1) + 1]} \qquad (5)$$

Being $c_u = 1/2\ (\sigma1 - \sigma3)$ max; $\sigma'c$ isotropic pressure of consolidation; $\sigma'zc$ vertical pressure of consolidation; Af Skempton's coefficient; Ko ratio between the horizontal and vertical effective pressures, approximately equal to:

$$Ko \cong Kon + \propto \log OCR \qquad (6)$$

Where $\propto$ comes to between 0,4 and 0,5 λ; coefficient varying between 2/3 to 1, which allows to estimate the ratio:

$$\frac{c_u}{\sigma'zo} = \left(\frac{c_u}{\sigma'z_o}\right)_n \cdot OCR^\lambda \qquad (7) \text{ valid for preconsolidated soils.}$$

$\sigma'z$ is the value of the effective vertical pressure of the considered soil and $OCR = \sigma'c/\sigma'zo$ or $\sigma'zc/\sigma'zo$.

For cemented soils values of $c' + c''$ of 500 KPa have been measured. The cementation influence can be observed in the form of the curves of Figure 7, where the results of triaxial tests drained under a 100 KPa confining pressure.

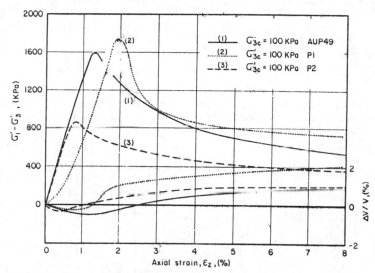

Figure 7: Results on S showing high cementation

In Figure 6, it can be seen that even an approximately $\sigma''c$ confining pressure of 5000 KPa, cementation influences upon the values of envelope of peak resistance under drained conditions; therefore, the value of the equivalent angle of $\emptyset''$ friction can be estimated with the expression (1) for each c" value.

For cemented soils an additional term appears in the numerator of the equations (2) and (4), being equal to 2c" $\sqrt{N\emptyset''}/\sigma''c$. It can be observed that the ratio c"/$\sigma''c$ can reach values of 0,1 to 0,3.

In Figure 8 Af values obtained from $\bar{R}$ tests are located in a dotted band for structured and cemented "toscas". Comparing to the typical curves of non-cemented soils, the band is moved to the right for OCR > 2.

For equal OCR, [Af = Ud/($\sigma1-\sigma3$) max], Af values are higher in cemented soils. A similar situation was found for saprolytes and saprolytic soils; both for soil with acquired cementation -"toscas" - and for soils with remaining cementation - saprolytes - the effect of cementation produces an equivalent virtual preconsolidation higher than pressure due only to previous effective stresses i.e, a higher OCR.

Being Af a measure for the development of pore pressure when failure begins, for this condition the material modifies abruptly its structure.

For example for an Af $\sim$0, the non-cementing material will show an OCR $\sim$4, but cemented "tosca" will have an OCR $\sim$10 due to its virtual preconsolidation.

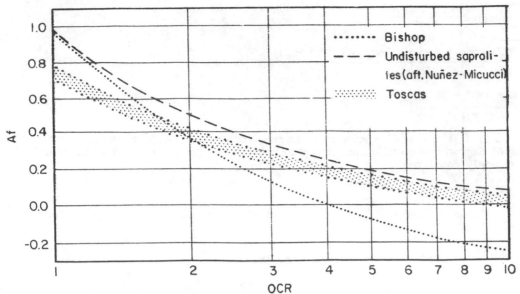

Figure 8: Af values obtained from differents structured and cemented toscas.

The Af value, as it shows (2) and (4) will be contened between:

$$Af = \frac{1}{(N\emptyset -1)} \left[\frac{(2c' \sqrt{N\emptyset}^* / \sigma'_c + 2c'' \sqrt{N\emptyset}'' / \sigma_c'') + (N\emptyset' - 1)}{2 c_u / '_{zo}} - 1 \right] \quad (2')$$

$$Af = \frac{(2c' \sqrt{N\emptyset}^* / \sigma'_{zc} + 2c'' \sqrt{N\emptyset}'' / \sigma''_{zc}) + K_o (N\emptyset - 1) - 2\frac{c_u}{\sigma_{zo}}}{(N\emptyset - 1)(2\frac{c_u}{\sigma'_{zo}} - 1 + K_o)} \quad (4')$$

$c'/\sigma'_c \sim 0{,}02; \quad c''/\sigma''_c \sim 0{,}01 \text{ to } 0{,}03$

According to Figure 9 and 10 the unidimensional consolidation tests have consistently shown over consolidation values similar to those forseen for the desiccation action. For the cemented soils as the cementation magnitude increases, the virtual preconsolidation also rises.

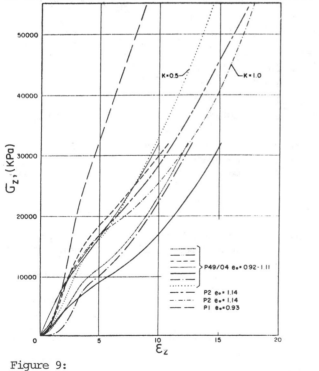

Figure 9:

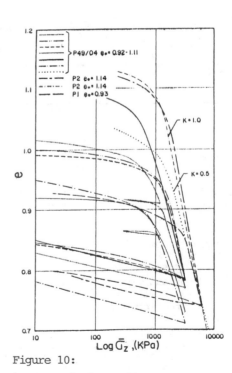

Figure 10:

Oedometric and K = cte. tests. Typical results

All the consolidations curves in graphic $\text{Log}\,\sigma'z\text{-}e$ (Fig.10) corresponding to cemented materials move towards the right side of the basic curve resulting from carrying on a test with the sample remolded at the liquid limit.

The results of the tests (Fig.9) also show that in the soils weakly cemented, for stresses higher to those of the preconsolidation, the stress-strain ratio is well represented by an expression such as $\sigma'zc = a.m\,\varepsilon z$, having as a result tangent deformation modules equal to $E_t = K_1.\sigma'zc$

On the other hand, for $\sigma'z$ values lower than those preconsolidation, the ratio is better expressed by an equation like $\sigma'z = b.\varepsilon_z{}^n$, having as a result tangent deformation modules $E_t = \left(K_2.\sigma'z\right)^{\frac{n-1}{n}}$

Generally for non-cemented soils $n < 1$ and for cemented soils $n > 1$. In the latter the increase of deformations brings about the progressive destruction of cemented bonds among the particles, when the void ratio is high and the structure is metastable at the increase of stresses, a kind of internal collapse and a reorder of particles is produced.

In fact, it is a crumbling of the material structure different to a conventional failure when the void ratio is constant and it is called critical in this case, the failure is progressive within a certain range of pressures and presenting a large variation in the void ratio.

When the triaxial tests are executed, during the first stage of which the isotropic or aniso-tropic consolidation is produced, the stress-strain ratio may be expressed through equations such as those seen before.

For the stage of application of the deviator stress, the following approximate equations are quite satisfactory:

$$\sigma_d = \sigma_R [1 - (1 - \frac{\varepsilon_d}{\varepsilon_R})^n \,] \qquad (8)$$

$$n = \frac{\lg\,(1 - \sigma_d/\sigma_R)}{\lg\,(1 - \varepsilon_d/\varepsilon_R)} \qquad (9)$$

$$n = \frac{1}{\frac{1}{n_o} + d\,[1 - (1 - \frac{\sigma_d}{\sigma_R})\,\frac{1}{n}} \qquad (10)$$

$$E_i = n_o\,\frac{\sigma_R}{\varepsilon_R} \qquad (11)$$

$$E_t = E_i.\,\frac{n}{n_o}\,(1 - \frac{\sigma_d}{\sigma_R})^{\frac{n-1}{n}}.\,\eta \qquad (12)$$

$$n = [d\,(1 - \frac{\sigma_d}{\sigma_R})\,\frac{1}{n}.\,\ln\,(1 - \frac{\sigma_d}{\sigma_R}) + 1 \qquad (13)$$

Where $\sigma_d = (\sigma_1 - \sigma_3)$; $\sigma_R = (\sigma_1 - \sigma_3)\max$; ε_d = unit axial strain; ε_R = unit axial strain at failure; E_i = initial tangent module.

If we represent $1/n = f\,(\varepsilon_d/\varepsilon_R)$, $1/n_o$ is the ordinate to the origin and d, is the slope.

HYDRAULIC PROPERTIES

It can be proved in the exposed profiles of "toscas" masses that the water flow is produced through two associated preferential systems.

A first system formed by macro fissures and little channels through which the water emerges as springs.

A second system of macro pores where the flow emerges as a free surface.

Therefore, in large masses, the permeability that the material may present is negligible with respect to these system.

Field tests carried out on "toscas" of macro porous and fissured structures, in wells partially penetrating in these free aquifers, show:

K = (permeability) : 3×10^{-3} cm/s to $1,20 \times 10^{-2}$ cm/s

T = (transmissivity) $2,15\ m^2/h$ to $6,38\ m^2/h$

S = (storativity) : $0,03$ to $0,14$

In this case the W_L values are between 25 to 30%. Obviously the permeability tests carried out in laboratories can not show the phenomenon values they show here.

The K values measured through pumping tests give values of $K \sim 5 \times 10^{-4}$ cm/s to 10^{-4} cm/s in locations similar to the results appearing in Figure 1.

CONCLUSIONS

1.-The soils preconsolidated by desiccation when bearing a null cementation at high degree of saturation, are not very sensitive as regards their resistance, though quite sensitive as regards deformations.

2.-In a great number of locations of this formation, the soils are cemented, the magnitude of this cementation is variable. The materials are sensitive as regards resistance and deformation.

3.-The laboratory investigation requires the use of undisturbed samples obtained, as far as possible from open pits. The samples obtained by means samplers set in the field, don't allow us to classify directly the mechanic properties of the soil profile.

4.-To determine the mechanic behaviour at high pressures of the soils preconsolidated by desiccation and cemented which act as weak or very weak rocks, it is necessary to carry out triaxial tests or tests of unidimensional consolidation with pressures values not less than 5000 KPa.

BEHAVIOUR OF THIS SOIL AS CONSTRUCTION MATERIAL

These preconsolidated soils, and weak to very weak rocks are locally used for construction of embankments, compacted fills, generally stabilized and mixed with cement and asphalt. The mechanic and hydraulic characteristics of the resulting materials depend on the process carried out during the working and placing of the material and on its original characteristics. These aspects are to be dealt with in another report.

REFERENCES

Bolognesi, A.J.L. 1975. Compresibilidad de los suelos de la formación Pampeano. V Panamerican Comference on Soil Mechanics and Foundation Engineering. Volume V. Pages 255-302.

Núñez, E. & Micucci C. 1985. Engineering parameters in residual soils. First International Conference on Geomechanics in tropical lateritic and saprolitic soils. Volume I. Pages 383-396.

Site investigations on weak sandstones
Etude de grès à faible résistance

Lorenz Dobereiner, *ENGE-RIO, Engenharia e Consultoria SA, Rio de Janeiro, Brazil*
Ricardo Oliveira, *LNEC and New University of Lisbon, Portugal*

ABSTRACT: Weak rocks, especially poorly cemented sandstones, are difficult to study both in-situ and in laboratory. However, the range of their strength (between 0.5 to 20 MPa) calls for quite accurate assessment of the most relevant geotechnical parameters when used as foundation of important structures and for underground construction. This paper briefly describes the results of a research on several types of weak sandstones, highlighting the advantages and drawbacks of different methods of site exploration, sampling and laboratory and in-situ testing.

RÉSUMÉ: Les roches friables spécialement les grès pauvrement cimentés sont difficiles d'étudier aussi bien seu place que au laboratoire. D'autre coté sa variation de resistance a compression (fréquemment entre 5 a 20 MPa) demande une bonne évaluation des plus importantes paramètres géochimiques quand elles sont utilisées comme fondation de structures importantes et en constructions sousterraine. Ce travail décrit de manière succinte les résultats d'une recherche sur plusieurs types de grès friables mettant en évidence les avantages et désavantages des differentes mèthodes d'exploration des sites, d'échantillonage et des testes sur place ou au laboratoire.

1 INTRODUCTION

Weak rocks are frequently avoided wherever possible, therefore they have not been studied as extensively as their stronger counterparts. They are difficult materials to sample and test, and laboratory tests of strength and deformability generally underestimate their in-situ performance. In-situ tests within them are also sometimes difficult to perform. In this paper, some methods for laboratory investigations, methods and techniques that are used in the field to asses the in-situ characteristics of some weak sandstones are discussed.

2 DESCRIPTION OF THE MATERIALS CONSIDERED

This research has investigated the properties of sediments that have a saturated uniaxial compressive strength of no more than about 20 MPa.

The youngest samples tested were sands and sandstones of Pliocene age from Turkey and Portugal; sandstones of Lower Cretaceous and Jurassic age from Portugal, and the Kidderminster sandstone and waterstones of Permo-Triassic age from Britain were also studied.

The petrography of these materials is described in table 1; their matrix accounts for 5 to 35 per cent of their total content and is in some cases calcareous (e.g. samples from Turkey) and in other cases rich in iron hydroxides and quartz overgrowths (e.g. samples from Britain).

3 LABORATORY TESTS

The laboratory testing techniques and methods used for weak sandstones include standard soil and rock mechanics procedures. Difficulties in testing due to unsuitable equipment are common. The rock mechanics equipment tends to be very strong and therefore insensitive at low stress levels. On the other hand the soil mechanics test equipment is far too weak to withstand the stress levels necessary to take weak rocks to failure.

The most relevant difficulty arises from the lack of good samples of the rock no matter the method used in their extraction.

Table 1. Composition of some samples considered

Material	Porosity (%)	Matrix (%)	Clasts (%)	Clast Mineralogy (%)					Matrix Mineralogy (%)									Observations
				Q	F	M	CP	RF	Q	F	K	I	S	IH	C	CL	A	
Waterstones (UK)	25	5	70	90	4	1	1	4	>30	10-30		>30		+		±10		1
Kidderminster Sst (UK)	31	5	64	80	4	1	5	10	<10	<10	>30	10-30	>30	+				1
Lahti Sst (Turkey)	28	25	47	30	10	5		55	<10	<10		10	>30		+	>30	>30	
Ferrel Sand (Portugal)	33	32	35	83	12	3		2	<10		>30			<10				2
Castanheira Sand (Portugal)	35	35	30	54	10	1		35	<10		>30	10-30		+				
Coina Sand (Portugal)	32	23	45	93	3	2		2	<10		>30	10-39		+				
Bauru Sst (Brazil)	27-35	5-25	40-68	90	2-5	1		4-8	10-30	<10	10-30	10-30	30	+				

1 Feldspar or quartz overgrowth observed.
2 Muscovite (30%) haematite and goethite also present
+ Quantative estimates not obtained

Q – Quartz
F – Feldspar
M – Mica
CP – Clay particles
RF – Rock fragments
K – Kaolinite

I – Illite
S – Smectite
IH – Iron hydroxides
C – Carbonates
CL – Chloride
A – Attapulgite

3.1 An index test for weak sandstones

Because weak sandstones are difficult to sample and difficult to test, and because their in-situ strength is difficult to assess, it is desirable to have a simple and economic index test of strength. Several different index test methods commonly used were evaluated.

The point load strength test described by Broch and Franklin (1972) was tested, showing a large scatter in the results. This scatter is probably caused by the friction of the ram which could account for an error of 30%, or the penetration of the platens into the rock as load increases.

The slake durability test described by Franklin and Chandra (1972) was also used, showing not to be sensitive for weak sandstones. Most rock types disintegrated completely after the first cycle (e.g. Kidderminster sandstone) and others showed a high slake durability index (e.g. waterstones).

Other index tests such as density, wave velocity and saturated moisture content were also performed. Each of these testing methods is separately evaluated in the following sections.

3.1.1 Density

Several methods for the determination of the density of soils and rocks were tested, including those described by BS 1377 (1975) and the ISRM recommendations described by Brown (1981). Due to the fragility of the samples of weak sandstones, which tend to disturb when immersed in water, it is easier to use the linear measurement or caliper technique (Brown 1981), if good sample preparation is possible, or the water displacement (BS 1377 test 15(F)) and the weighing in water method (BS 1377 test (E)), which require that the samples are coated in wax.

Density was measured in the dry and saturated state. Both the correlations between strength and density showed poor correlations, which mean that density is related to the rock's intrinsic properties (e.g. grain size, shape of grains, mineralogy) and that these are not the main controlling factors of the strength of weak sandstones.

3.1.2 Ultrasonic wave velocity

Figure 1 illustrates the variation in propagation of compressional wave velocities through samples of sandstones. Velocity of propagation varies with the moisture content of the sample. This tendency has been observed in all the materials that were tested and no constant pattern is observed for these changes. It was also observed that on samples where the waves were transmitting parallel to bedding (waterstones and Kidderminster sandstone) the velocities were higher than propagation at right angles to bedding, even though this is the direction of lowest strength.

The compressional wave velocity has also a poor correlation with the uniaxial compression strength. As for density the wave velocities depend on the intrinsic properties of the rock, showing therefore different trends of variation with moisture content for each type of weak sandstone.

The experience of the authors with the use of shear waves in weak sandstones indicates that their measurement is quite difficult; however, they are less sensitive to changes in the moisture content of the samples, as it should be expected, and they should allow for a better classification of the strength of the sandstones.

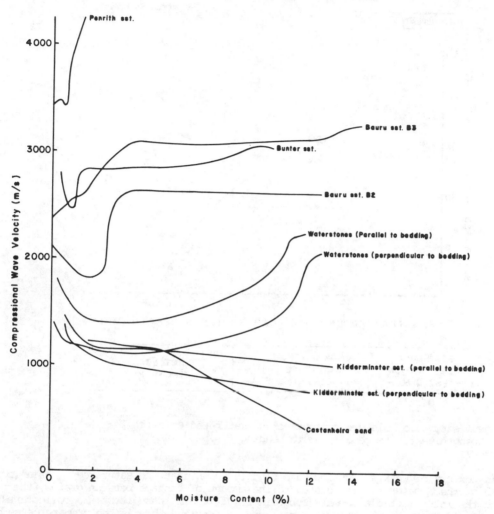

Figure 1. Variations in compressional wave velocity with changes in moisture content

3.1.3 Saturated moisture content

The correlation between the vacuum saturated moisture content and the saturated uniaxial compression strength is shown in figure 2. Saturated moisture content provides a sensitive index of uniaxial compression strength between approximately 0.5MPa and 20MPa. Over this range a change of 10% in saturated moisture content is observed (Dobereiner and de Freitas 1983). The data shown in figure 2 represents materials whose characteristics range from those of a soil to those of a strong rock.

As an index, vacuum saturated moisture content lies conveniently between the accepted working limits of the standard penetration test and the point load strength test. Saturated moisture content is a non-destructive index and can be measured on irregular samples collected from outcrops or from pieces of borehole cores.

The method of measuring this index is described by Dobereiner and de Freitas (1983) and consists of saturating samples by immersion in water under vacuum. Simple immersion alone is not sufficient to produce consistent results and samples immersed for 48 hours accepted only 85% of the volume they would accept in less than 2 hours under vacuum. Once saturated, the moisture content of the sample may be calculated in the usual way and this value compared with the relationship illustrated in figure 2.

If the samples disintegrate as they are immersed in water, they are classified as a sand and not a sandstone.

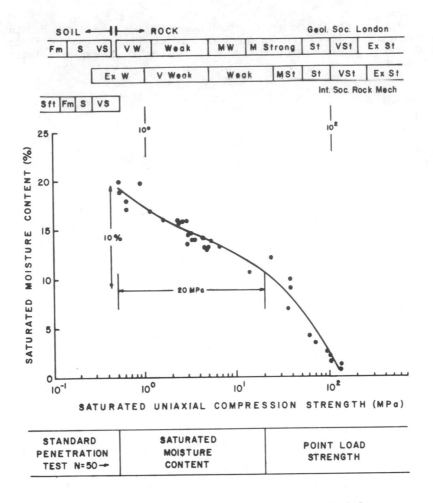

Figure 2. Vacuum saturated moisture content versus saturated uniaxial compression strength tested perpendicular to bedding

Microscopic texture of weak sandstones have clearly shown that the saturated uniaxial compression strength of these materials is related to the amount of contact between grains, the packing of grains and the inter-particle cement (Dobereiner 1984). All these items are collectively quantified in weak sandstones by the saturated moisture content and this is the primary cause of the good relationship between saturated uniaxial compression strength and saturated moisture content.

4.2 Strength and Deformability

Typical curves of lateral, axial and volumetric strain versus stress of a weak sandstone, tested in uniaxial compression, are illustrated in figure 3. The variation of volumetric strain records a decrease in volume followed by an increase in volume and this increase in volume commences at low levels of stress, often below one third of peak compressive strength (Dobereiner and de Freitas 1984).

The typical behaviour of a weak sandstone tested in triaxial compression is shown in figure 4. The pore water pressure within the samples increases with increasing stress and then decreases. This reduction in pore pressure is believed to result from dilatancy within the sample, which means the start of the process of failure (Dobereiner and de Freitas 1984).

These characteristics on the behaviour of weak sandstones indicate that certain precautions should be taken if a correct interpretation of the test results is to be made.

Firstly, it is important to emphasise that weak sandstones are remarkably sensitive to changes in moisture content and because of the difficulties of reproducing the natural moisture content, it is advisable to test all samples in a saturated state.

Weak sandstones are also markedly anisotropic in their strength even when they appear to be homogeneous in texture. The waterstones, for example, show a 30% difference in the saturated uniaxial compression strength between samples tested parallel or perpendicular to bedding. Other weak sandstones showed even greater differences up to 100%.

414

The determination of the modulus of deformation of weak sandstones should be made considering the stress-strain behaviour illustrated in figures 3 and 4. It is recommended by the International Society of Rock Mechanics (Brown 1981) to measure a tangent modulus from axial strain at 50% of peak strength. However, in weak sandstones this will be a stress level at which non-recoverable strain is occurring. It is advisable to obtain the modulus at strain levels that occur before the start of dilatancy. Therefore, it is recommended to observe the volumetric strain, which could be calculated from the lateral and axial strains in the uniaxial compression test, or the change in pore water pressure, measured in undrained triaxial compression tests, to determine the range of axial strain that should be used in calculation.

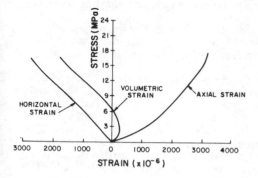

Figure 3. Stress-strain curves of uniaxial compression tests of waterstones tested parallel to bedding

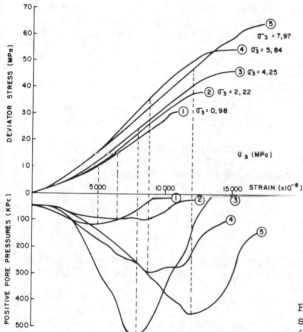

Figure 4. Stress-strain curves of triaxial compression tests with the measurement of pore pressures in waterstones tested parallel to bedding

4 SITE INVESTIGATION

Based on the experience gained from seven site investigations on weak sandstones, studied in this research, general guidelines on the most appropriate field investigation techniques to use on these materials are given. The sites on which the research was based were the Pereira Barreto canal in southern Brazil, the Castanheira do Ribatejo and the Coina tunnels, and the Ferrel nuclear power plant in Portugal, the Czire dam in eastern Turkey and the M42 motorway and Tarpsby by-pass in northern England.

4.1 Drilling and sampling

There are no hard and fast rules for a drilling and sampling technique to be used in a site investigation programme on weak sandstones.

Most specifications and drilling textbooks outline that the light cable percussion method is very versatile and that there is virtually no soil or weak rock in which it cannot be employed success-fully. This is true for boreholes in soils. However, on weak rocks it seems to be a gross general-isation. The method was used, or tried, on all the sites where weak sandstones were studied. How-ever, it was not considered satisfactory for sampling and "in-situ" testing by the designers and contractors, except for the M42 motorway (England) site investigation on the Kidderminster sand-stone. However, the information required from these boreholes was only related to the identifica-tion of the rock type.

In the Bauru sandstone in Brazil and in the waterstones of England, this method was used until the recommended penetration limit of the standard penetration test (N=50) was reached. For design purposes a SPT blow count of 50, for 30 cm penetration, obtained by the light percussion method, is sometimes considered the boundary between a soil and weak rock.

In the case histories studied in this research the quality of rotary drilling was very variable, depending on the material and type of equipment used. All the 129 boreholes drilled by rotary meth-ods in the sites studied, were performed with double tube core barrels and water as the circulation fluid.

Relatively good core recoveries were obtained within the waterstones (average near 80%), which was the strongest material studied. However, due to the equipment used (hydraulic units attached to shell and auger rig) which induced vibration within the rods, the cores recovered contained several mechanical fractures parallel to bedding, making the testing of the samples almost impos-sible.

The core recoveries from the Bauru (Brazil) and Lahti (Turkey) sandstones were considered satis-factory (70 to 100%). Some of the samples obtained were appropriate for strength and deformability testing. However, conventional rotary drilling techniques within the Bauru sandstone did not allow adequate study of the fracture pattern within the rock mass. Therefore the integral sampling tech-nique developed by Rocha (1971) was used. In all the kinds of the Bauru sandstone at all depths and even in inclined boring holes, the core recovery using the integral sampling technique was al-ways very high (Kaji et al 1981). Some problems with this technique related to the orientation of the recovered cores were encountered. However, it is thought that the problem could be solved with the use of more rigid rods (Sanchez et al 1981) or the use of a Pajari-Tropary orientation device.

The major disadvantage of using the integral sampling method, besides cost, is that the samples recovered are not suitable for any type of testing, due to impregnation.

The boreholes drilled for the investigation on the Castanheira tunnel and Ferrel nuclear power plant, showed core recoveries below 30% in the sands and sandstones. The few cores recovered are part of the better quality materials, which are not representative of the rock mass. These low re-coveries were mainly caused by the very friable characteristics of both materials and also the presence of hard cobbles in some areas.

These observations reveal the need for using more sophisticated techniques and equipment to ob-tain good sampling of weak sandstones. An improvement in core recovery and less disturbance of core during drilling can be obtained by drilling an appropriately large diameter (larger than 76 mm cores) and by the use of triple tube core barrels with air foam, as recommended by Phillipson and Chipp (1982) for residual soils in Hong Kong.

4.2 Trial pitting and sampling

The excavation of trial pits, trenches and adits in weak sandstones is important because good qual-ity sampling from boreholes is difficult to achieve and therefore a complete picture of the rock mass is rarely obtained.

On the sites studied in this research, few trial pits, trenches and adits within weak sandstones were excavated for the major projects (e.g. Pereira Barreto canal, Ferrel nuclear power plant, Czire dam and Coina tunnel). Trial pits within the superficial materials (alluvium and colluvium) were excavated more often.

The five trial pits excavated with pickaxes and pneumatic tools on the Bauru sandstone (Pereira Barreto canal) showed their usefulness. Manually excavated samples of weak sandstones obtained from depths up to 20 metres were successfully tested for materials with uniaxial compressive strength up to 5MPa.

In the site investigation on the Lahti sandstones (Czire dam), the adits excavated (approximately 40 metres long) for "in-situ" testing proved to be a very important source of information for the design. Shear zones within the rock mass, which had not been previously detected in the boreholes, were a determining factor in dictating the geotechnical design parameters to use.

The trial pits and trenches excavated on the Ferrel sand were also used for sampling and testing (Oliveira & Fialho Rodrigues 1976). Except on the Coina sand, in all other sites the use of soil samplers was not successful. Large block samples had to be hand excavated with pickaxes or chain saws, and immediately protected for later transportation to the laboratory. From these blocks, cylinders and prisms were carefully sculptured for strength and deformability testing.

4.3 Seismic surveys

Seismic refraction surveys were carried out on three of the seven sites studied in this research. A total of more than 10,000 m of seismic refraction profiling was performed in the Castanheira do Ribatejo tunnel, Ferrel nuclear power plant and Pereira Barreto canal. These profiles revealed that the seismic refraction method is imprecise for the determination of the rock head of weak sandstones and that it is not possible to obtain a detailed zoning of the rock masses concerned. The contrast in wave velocity propagation between a dense sand and a weak sandstone or between two sandstones of different strength, is usually too small to be recorded, and it may be influenced by the change in water content of the rock mass. Shear wave velocities are still not commonly used in seismic refraction but their potential as an investigation tool appears to be substantial. The seismic refraction method can be useful in the determination of the dynamic elastic constants (Young's modulus or Poisson's ratio) of weak sandstones or dense sands. However, the correlation between the static and dynamic elasticity moduli calls for many tests at each site and is not always easy to achieve.

Direct seismic methods (compressional and shear waves) have been extensively used lately (Oliveira / Graça 1986) and their results are more easily correlated with the static moduli obtained in the same locations (Fialho Rodrigues 1979). Cross-hole methods generally give a better correlation than up-hole due to the anisotropy of most weak sandstone rock masses which have a horizontal structure.

4.4 Plate loading tests

Plate loading tests were carried out on two of the sites studied in this research.

In the Coina sands 8 plate loading tests (900 cm² plates) were performed during the excavation of the trial pits. The reaction load was obtained for vertical tests from two boxes filled with soil.

Reaction for horizontal tests was given by a plate of larger dimensions on the opposite wall of the trial pit. In this material which behaved as a soil (E=40MPa), it was possible to increase the loading up to failure in most tests.

In the Ferrel sand 16 plate loading tests (1000 and 2500 cm² plates) were performed in the trial pits and trenches. The loads were vertically applied and were increased in two cycles up to failure or maximum reaction load. The results of these "in situ" tests were also of the order of three times larger than the values of the deformability modulus in the laboratory (Seco e Pinto et al 1976).

In both sites, the results obtained from the plate loading tests were considered satisfactory. The possibility of ground disturbance during the excavations of the trial pits or trenches was considered relatively small especially on the walls. Dense sands or weak sandstones are in most cases easy to excavate with pickaxes or light pneumatic tools, which cause very little disturbance. The main limitation of this testing method is considered to be the depth of the trial pit necessary to gain access to the test position. For example, in the site investigation on the Castanheira sand, the material to be excavated in the tunnel could not be tested, because of the great cover (approx. 50 metres) above the alignment of the proposed tunnel. In the Ferrel site, however, pits down to about 25 metres were excavated. It is also important to remember that the presence of discontinuities in a weak sandstone rock mass could require the use of larger plates than used in the above examples. In this case the limitation will be the maximum load attainable at the reaction point.

4.5 Pressuremeter

Pressuremeter tests were carried out during two of the site investigations studied, for the Czire dam (Lahti sandstone) and the Coina tunnel (the Coina sand). The average deformability·index obtained from the pressuremeter test results performed in the Lahti sandstone (37 tests) were considerably lower (approx. 300%) than the average modulus obtained from vertical plate bearing tests. Values of the same magnitude, for the correlation between plate bearing and pressuremeter tests, were also obtained by Meigh and Greenland (1965) on weak Bunter sandstone (England) and by Mori et al (1977) on weak sandstones and mudstones from the Kobe layers in Japan. These different values obtained for the deformability modulus are mainly caused by its dependence on the direction and level of stress at which it is measured. A discussion of these and other reasons for the variability of results is given in Mori et al (1977), Marsland & Eason (1973) and Wroth (1982).

By contrast to the Lahti sandstone, on the Coina sands (39 tests) the average vertical modulus obtained in the plate loading tests was lower (approx. 250%) than the pressuremeter modulus. No specific reason was found to explain these results, other than imprecisions in the calibration of the pressuremeter.

Pressuremeters have been developing for more than 20 years. Several manufacturers are assembling different models and the range has been extended to include rock pressuremeters, which can apply stresses up to 100MPa (Meigh & Wolski 1979). The recently developed self boring pressuremeter,

Table 2. Schedule for site investigation on weak sandstones

Category	Type	Relevant observations
Index parameters	Moisture content	Recommended for all samples tested, due to the great sensitivity of weak sandstones to changes in moisture content.
	Vacuum saturated moisture content	Very good index for assessing strength and deformability.
	Density (dry and saturated)	Poor index for geotechnical properties, useful for calculations in design.
	Ultrasonic wave velocity	Poor index for geotechnical properties, due to its sensitivity to changes in moisture content.
	Uniaxial compression strength	Most widely used test in rock mechanics. Useful for comparing one rock with another.
Design parameters determined in laboratory	Triaxial compression strength: Drained	Simple and quick method for obtaining c' and Ø'.
	Undrained with measurement of pore pressure	Very important for understanding the behaviour of weak sandstones. Onset of dilatancy can be obtained by measuring pore pressure.
	Deformability: Uniaxial compression	Most widely used method. Volumetric strain can be obtained if horizontal and vertical strains are measured.
	Triaxial compression	More representative of the in-situ conditions.
	Creep tests	Deformations of weak sandstones are time-dependent. Creep properties not well known.
Sampling	Light cable percussion boring	Suitable for sands. Very poor sampling, in weak sandstones useful for identification of rock type only. Does not consider discontinuities.
	Rotary drilling	Double tube core barrels can produce poor sampling. Triple tube barrels or internal sleeving in double tube core barrels, used with air foam produces good sampling.
	Trial pits, trenches, adits	Very useful in weak sandstones. Are limited by depth and/or length or presence of water table. Good quality sampling may be obtained.
Design parameters determined in-situ	Seismic refraction	Very low velocity contrast in weak sandstones. Useful only for determining the dynamic elastic constants.
	Direct seismic methods	Compressional and shear wave velocities. Better correlations than seismic refraction.
	Standard penetration test	Very large scatter of results if used on weak sandstones. Can be used as an index only, and in very poor sandstones.
	Deformability: Pressuremeter	Most suitable for soils. Requires more extensive application to prove suitability for weak sandstones. Difficulties in interpenetration.
	Dilatometer	Presents problems in weak sandstone mainly related to difficulties in obtaining an accurate diameter borehole.
	Plate loading	Very successful in weak sandstones. Main limitation is the access to the test location and the obtaining of a reaction load.
	Direct shear strength	Rarely used in weak sandstones. Should be very useful in site investigations for large structures.
	Permeability: Injection test	Suitable for a large permeability range. The main limitations are related to leakage due to malfunction of the packers and to hydraulic fracturing.
	Pumping test	Very useful in weak sandstones. Main limitation is high cost and the depth to be used.
	Large scale field trial	Very useful in weak sandstones. The main limitation is the high cost, making it only suitable for large structures.

which greatly minimizes the disturbance in the sides of the borehole, has been used in a very weak schist (UCS=0.7MPa) by Baguelin (1980). However, their use in weak rocks is still very limited and considerable improvements in this technique, mainly in increasing the capacity for self boring in slightly stronger material, has to be done.

4.6 Dilatometer

Borehole dilatometer tests were included in the site investigation programme for the Castanheira tunnel and Ferrel nuclear power plant. The type used was developed at INEC and is described by Rocha et al (1970) and Charrua Graça (1974). However, in both cases it was impossible to obtain any meaningful results for the deformability characteristics of the dense sands, because of irregularities on the borehole sides. Several tests were performed in the Ferrel sandstones, but no results were obtained, due to difficulties encountered in drilling boreholes with a correct diameter range (75 mm to 81 mm). These difficulties were related to the coarse grain texture and friability of both sandstones.

Better results were obtained with the borehole dilatometer on the site investigation for the canal lock of Leerstetten, part of the Europa Kanal, South of Nuremberg in Germany. The tests were performed by the LNEC on the weak Keuper sandstones. The drilling procedure to obtain a circular borehole of the desired diameter presented some difficulties. Initially a borehole of 54 mm was drilled to the entire depth. After that, a cutting shoe of larger diameter was introduced without rotation, but just with vertical pressure to enlarge the borehole in 2 m stages. After each test the procedure was repeated. However, from the 60 tests to be performed only 20 were considered satisfactory (deformability modulus obtained was between 140 and 218MPa). The lower limit of the applicability of the dilatometer test is for rocks with deformability modulus around 100MPa (Charrua Graça 1982). However, the main problem encountered in using the dilatometer for weak rocks is, as discussed above, the difficulty in obtaining a constant diameter borehole.

4.7 Permeability

"In-situ" permeability tests were carried out in two of the sites studied in this research. Pumping tests were performed at the Pereira Barreto canal (Bauru sandstone) and Lugeon tests (injection type) at the site investigation for the Czire dam (Lahti sandstone).

The eight pumping tests on the Bauru sandstones were performed in 4 trial pits (1.7 to 2.0 metre diameter) at depths of up to 20 metres and are described in detail by Ferreira et al (1981). Steady and unsteady flow were analysed by the Theis and Jacob methods. Samples were also collected in the trial pits and laboratory permeability tests were performed. The results of the laboratory tests revealed permeabilities of the order of 100 times smaller than from the pumping tests. This illustrates that the pumping tests influence a larger volume of rock, thereby testing a representative element of a discontinuous rock mass.

Lugeon tests were performed on the Lahti sandstone in most rotary drilling boreholes, at 3 metre intervals. Very low absorption (0 to 2.8 Lugeons) were observed characterising the few discontinuities present in the rock mass.

In terms of testing a large permeability range, the Lugeon tests are the most versatile. However, in weak rocks problems can be encountered with leakage, due to malfunction of the packers. This is caused in most cases by the difficulties in obtaining a constant diameter borehole in very friable materials. Lugeon tests can also be conducted with hydraulic gradients that produce a velocity of flow capable of causing erosion of friable material on joint boundaries or hydraulic fracturing of the rock. They must therefore be used with caution, and especially at reasonably low water pressures.

The pumping tests are best for higher transmissivity formations, but could have a limited application to discontinuous rock masses if the discontinuity spacing is very large. The same problem could occur with the Lugeon tests. However, the versatility of changing the length of stage to be tested can minimise these difficulties.

4.8 Large scale field trials

A large scale field trial is essentially a full scale test, enabling the ground to be tested on a scale and in conditions similar to those to be encountered during the project under investigation. There is no fixed technique or methodology which can generally be applied, each test has to be performed in such a way so as to simulate the actual site conditions.

An example of a large scale trial performed at the Pereira Barreto canal, is described by Pimenta et al (1981). A trial excavation on the Bauru sandstone was undertaken to investigate the dewatering characteristics of the rock mass, the best excavation methods to be used, the behaviour of the slopes and the suitability of the slope protection measures. The productivity of several different types of equipment for the excavation (e.g. rippers, scrapers), was also studied. With the sandstone excavated, a trial embankment for the evaluation of the degree of compaction obtainable and the trafficability of the embankment was executed. The results were very useful in defining and confirming the design parameters to be used for the large excavation. The best construction method as well as the drainage system and protection method were also obtained. The advantages and usefulness of these large scale trials is evident in weak sandstones as the results obtained are very reliable in most cases. Many times, phenomena are created in these tests (e.g. slope failure), which enable back analysis to be carried out. The main limitation which is responsible for the small number of large scale trials to be performed is the high costs involved in these tests. In most cases they are only feasible for very large projects, and only at the beginning of the construction stage.

4.9 Site observation and performance

There are two main observational site scale procedures, useful in predicting the performance of an engineering structure: the acquisition of design parameters from previously constructed sites and from natural features, as for example, the stability of natural slopes.

Observational design parameters obtained during the early stages of a project are particularly useful for very large structures and in slope stability problems, where large volumes of rock are involved. Parameters obtained can include erodability and displacement (used to calculate deformability), pressures and stresses including pore pressures and shear strength, back calculated from slide data (Meigh & Wolski 1979). None of the projects studied in this research is at present concluded, therefore no data on site performance was obtained.

The observation of natural geological or man-made structures in the vicinity of a site investigation can provide very useful parameters for design. A good example is described by Dusseault & Morgenstern (1979). The high strength of the locked sands in an intact state was obtained from back analysis of a series of steep and high natural exposed slopes.

Although the material disintegrates readily when placed unsupported in water, it is strong when in intact state and subjected to a confining pressure.

The same observations could also be made for the Castanheira and Ferrel sandstones, on the sites studied for this research. On the Kidderminster sandstone a sand quarry is in use near the M42 motorway.

The observation of these high (up to 30 metres) and steep ($>70°$) stable slopes, which were constructed several years ago, indicate that the parameters obtained from laboratory shear strength tests c'=2.6MPa and $\emptyset$ = 36° probably underestimate the real strength of the rock "in-situ". This is particularly true for quite porous rock masses easily draining the underground water.

These examples clearly show that careful observations of site performance can be of great use. The results indicate that in many cases laboratory tests underestimate the "in-situ" strength of weak sandstones. It also highlights the importance of using large trials as an accepted part of an investigation of weak materials. They will confirm or help in the assessment of the real properties of the rock mass.

5 CONCLUSIONS

A general summary of the recommended investigation techniques for weak sandstones is given in table 2. These recommendations are only guidelines and are based on the experience obtained during this research and should not be strictly regarded as a rule. A different approach may be more suitable for a particular site, depending on the requirements and/or the resources available.

At present little is known about weak rocks of the type described in this paper, and it is hoped that their characters as well as more appropriate investigation techniques, are more studied in the future.

6 ACKNOWLEDGEMENTS

The authors acknowledge the financial support given to L. Dobereiner by the "Conselho Nacional de Desenvolvimento Científico e Tecnológico-CNPq" (Brazil) which made this work possible. In addition to the "Laboratório Nacional de Engenharia Civil-LNEC" (Portugal) and to the "Laboratório Central de Engenharia Civil-CESP" (Brazil) for furnishing rock samples and their results to be used in this research.

REFERENCES

Baguelin, F. 1980. Discussion: design parameters in geotechnical engineering. V4. Proc. 7th European conference on soil mechanics. Brighton, England.
Broch, E. & J.A.Franklin 1972. The point load strength test. Int.J.Rock Mech. Min.Sci. 9:669-697.
Brown, E.T. (editor) 1981. Rock characterization, testing and monitoring. ISRM suggested methods. Oxford: Pergamon Press, 211.
BS 1377 1975. Methods of test for soils for civil engineering purposes. British Standards Institution. 143.
Charrua-Graça, J.G. 1981. Dilatometer tests in the study of the deformability of rock masses. Proc. 4th Congress of the International Society for rock mechanics. Montreux II:73-76.
Charrua-Graça, J.G. 1982. Personal communication.
Dobereiner, L. 1984. Engineering geology of weak sandstones. PhD.Thesis. Imperial College, London, England. 471.
Dobereiner, L. & M.H.de Freitas 1983. Saturated moisture content as an index for assessing strength Proc. international symposium on engineering geology and underground construction. V3. Lisbon.
Dobereiner, L. & M.H.de Freitas 1984. Investigation of weak sandstones. Geological society regional meeting. Assessing BS 5930. Guildford, England.

Dusseault, M.B. & N.R.Morgenstern 1979. Locked sands. Q.J.Engng. Geol. 12:117-131.

Ferreira, R.C., L.B.Monteiro, J.E.E.Peres & F.A.A.Prado 1981. Considerações sobre alguns modelos para análise de ensaios de permeabilidade em estratos de arenito Bauru. Anais do 3º Congresso de ABGE. 2:71-88.

Franklin, J.A. & R.C.Chandra 1972. The slake-durability test. Int.J. Rock Mech. Min.Sci.9.3:325-341.

Fialho Rodrigues, L. 1979. Métodos de projeção sísmica em geologia de engenharia. A importância de onda de corte. Tese. LNEC. Lisbon 1977.

Kaji, N., M.L.Vasconselos & M.G.Guedes 1981. Aspectos metodológicos das investigações geológicas e geotécnicas no arenito Bauru. Anais do 3º Congresso da ABGE. 2:257-270.

Marsland, A. & B.T.Eason 1973. Measurement of displacements in the ground below loaded plates in boreholes. Proc. Conf. on field instrumentation. Butterworths. 304-317.

Meigh, A.C. & S.W.Greenland 1965. In-situ testing of soft rock. Proc. 6th Int. Conf. Soil Mech. Found. Eng. Montreal. VI:73-76.

Meigh, A.C. & W.Wolski 1979. Design parameters for weak rocks. Design parameters in geotechnical engineering. 7th Euro. Conf. Soil Mech. Brighton. VS:59-79.

Mori, H., K.Takahashi & T.Noto 1977. Field measurement of deformation characteristics of soft rocks. Proc. International symposium on field measurements in rock mechanics. Zurich. VI:401-414. Publ. Rotterdam.

Oliveira, R. & L.Fialho Rodrigues 1976. Colaboração nos estudos geotécnicos e sismológicos para implantação da central nuclear do Ferrel - colheita de amostras e ensaios de laboratório. Relatório do LNEC. Proc. 54/1/4928. Lisbon.

Oliveira, R. & J.G.Charrua-Graça 1986. In-situ testing of rocks. Chapter 27 of the book Ground Engineer's Reference Book. Butterworths, London (under publication).

Pells, P.J.N. & M.J.Ferry 1983. Needless stringency in sample preparation standards for laboratory testing of weak rocks. 5th Congress of the Int. Soc. Rock Mech. Melbourne. A203-A207.

Phillipson, H.B. & P.N.Chipp 1982. Air foam sampling of residual soils in Hong Kong. Proc. ASCE geotech. enging. div. Special conference on construction in tropical and residual soil. Honolulu 17.

Pimenta, C., J.C.F.Bertolucci & M.H.Lozano 1981. Excavação experimental em arenito Bauru. Anais do 3º Congresso da ABGE. 3:255-274.

Rocha, M. 1971. Método para amostragem integral de maciços rochosos. LNEC Memoir 374, Lisbon.

Rocha, M., A.F.da Silveira, F.P.Rodrigues, A.Silveno & A.Ferreira 1970. Characterisation of the deformability of rock masses by dilatometer tests. LNEC Memoirs 360, Lisbon.

Sanchez, D.L., M.G.Guedes & J.C.F.Bertolucci 1981. Apreciação dos métodos tradicionais de orientação de testemunhos de sondagem em arenito Bauru. Anais do 3º Congresso da ABGE. 3:275-288.

Seco e Pinto, P., E.Maranho das Neves & G.de Castro 1976. Caracterização geotécnica dos terrenos de fundação da futura central nuclear do Ferrel. Relatório do LNEC. Obra 53/53/291, Lisbon.

Wroth, C.P. 1982. British experience with the self boring pressuremeter. Proc. symposium on the pressuremeter and its marine applications. Paris.

Adaptation d'un barrage voûte à une fondation formée de roches mécaniquement très contrastées

Design adaptation for arch dam on rock of contrasting mechanical properties

F.Isambert, A.Szendroi & J.Hugonin, *Coyne & Bellier, France*

RESUME : Un barrage voûte de 160 m de hauteur sera prochainement construit dans le centre de Java, à MAUNG, sur la rivière Merawu, au droit d'une gorge formée de tuf bréchique volcanique armé par deux lames sub-verticales d'andésite. Les propriétés mécaniques de ces deux matériaux sont très contrastées, encore que l'andésite soit très fracturée.

Cet article montre comment les reconnaissances ont permis d'apprécier la géologie du site, les paramètres géotechniques des deux formations en présence et comment le projet a été adapté à la présence du matériau plus tendre.

ABSTRACT : A 160 m high arch dam is soon to be built at MAUNG site on the Merawu river in Central Java. It will be built in a gorge formed of volcanic tuff breccia reinforced by two subvertical andesite bodies. Although the andesite is highly fractured, the mechanical properties of the two types of rock are significantly different.

The article details how site geology and geotechnical features of the two formations were appreciated from investigations and how the design was adapted to allow for the softer foundation material.

1. INTRODUCTION

Le centre de l'Ile de Java est montagneux et bien arrosé. Après une recherche de sites hydroélectriques, dans le bassin de la rivière Serayu, le Perusahaan Umum Listrik Negara (P.L.N.) a concentré ses efforts notamment sur le site de MAUNG, situé sur la rivière Merawu, affluent rive droite de la Serayu.

Les études de Maung sont très avancées puisque les dossiers d'appel d'offres sont achevés.

Au cours de deux phases principales d'étude : Faisabilité en 1978-1980 et Avant-Projet Détaillé entre 1982 et 1985, de nombreuses reconnaissances géologiques et géotechniques ont été effectuées.

L'aménagement (fig. 1) comprend un barrage voûte de 166 m de hauteur et de 1,2 million de mètres cube de béton, une retenue de 208 millions de mètres cube représentant une surface de 5 km², une prise d'eau en rive droite, une galerie d'amenée et une galerie de fuite, toutes deux de plus de 7 m de diamètre, respectivement de 1,630 et 2,710 km de longueur, séparées par une centrale souterraine située à plus de 200 m de profondeur et équipé de trois groupes de 120 MW, et enfin, un barrage de 20 m de hauteur régularisant et dérivant pour l'irrigation les eaux turbinées.

En amont, le rendement de l'aménagement est accru par la dérivation de la rivière Tulis vers la retenue de MAUNG. La dérivation comporte un barrage de prise d'une vingtaine de mètres de hauteur, un tunnel et une conduite forcée de 2 800 et 600 m de longueur et une centrale aérienne de 14 MW.

2. CARACTERISTIQUES GEOLOGIQUES DU SITE DE MAUNG

Le site du barrage est localisé dans une gorge profonde et sinueuse, orientée Nord-Sud d'environ 700 m de longueur. Elle est creusée perpendiculairement à l'éperon de tuf bréchique (formation Bodas) constituant le flanc aval, méridional, d'un grand géanticlinal orienté Est-Ouest datant du Zanclien (Pliocène basal).

La topographie très raide de cet éperon a pour origine d'une part le très fort pendage des couches : 70 à 85° vers le Sud-Sud-Ouest, c'est-à-dire vers l'aval, d'autre part l'armature créée par deux sills d'andésite syngénétiques avec les tufs brèchiques d'âge messinien (Miocène terminal) et plus ou moins concordants avec leur stratification.

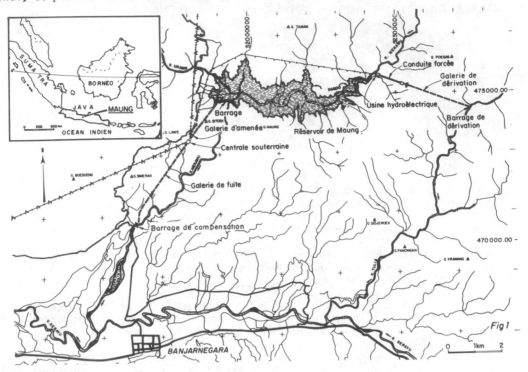

Fig 1

2.1. Reconnaissances effectuées

La structure des appuis et la continuité des deux sills d'andésite ont été précisées (fig. 2) par des sondages, galeries et grattages ainsi que par les affleurements dans le lit de la Merawu. Une quarantaine de sondages forés depuis la verticale jusqu'à l'horizontale, de la surface comme depuis les galeries, représentent au droit du site environ 2 500 m de longueur dont 1 700 m avec essais de perméabilités de type Lugeon. Six galeries ont été percées, représentant avec leurs antennes, 1 145 m, dont 622 m en rive droite et 523 m en rive gauche (trois galeries dans chaque appui).

2.2. Lithologie

La zone du site est constituée par un tuf brèchique continental (graywacke feldspathique épiclastique, d'après les nombreuses plaques minces) formé d'éléments sub-anguleux d'andésite à pyroxène gris, noirs ou rosés, sains et durs (1 à 20 cm) cimentés dans une matrice grèseuse tuffacée grossière (microlithes d'andésine et de labrador, clinopyroxènes, minéraux opaques) liée par une pâte de minéraux argileux, déterminant une roche moyennement dure à tendre et friable. Interstratification de fins lits marno-argileux et de siltstone et même de galets cimentés dans une matrice silteuse ou sableuse. Des débris végétaux sont présents.
   A proximité de la surface ou en profondeur de part et d'autre des discontinuités, la matrice tuffeuse est altérée et oxydée, donnant un matériau tendre et friable, voire sableux. Autour des sills, les solutions hydrothermales dont elles sont issues ont plus ou moins envahi la matrice, ce qui a eu à terme pour effet de l'altérer. Ceci est particulièrement net dans la moitié supérieure de l'appui rive gauche.
   Deux sills d'andésite arment ces tuffs. Le sill amont a une épaisseur augmentant de 25 m en rive droite à 75 m en rive gauche tandis que le sill aval, épais de 60 m en rive droite, ne se prolonge guère sur l'autre rive.
   Le sill amont est gris foncé à noir, sain, très dur et sa pâte finement à moyennement cristallisée tandis que le sill aval, sain et dur également, est gris et sa pâte plutôt grossière. Par contre, du point de vue pétrographique, il n'y a pas de différences significatives dans cette andésite à pyroxène : phénocristaux de plagioclases et de pyroxène, pâte de microlithes de plagicolases et de minéraux opaques.
   L'altération reste un phénomène purement superficiel de quelques mètres d'épaisseur.

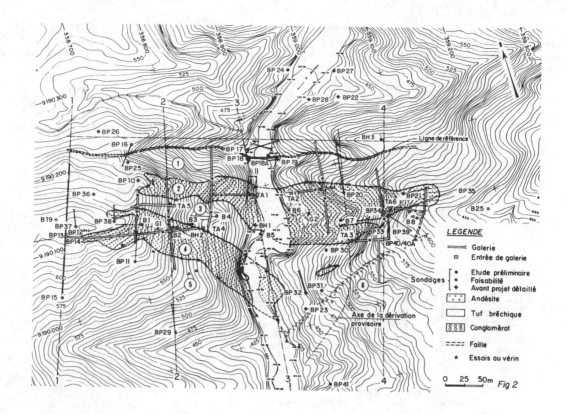

Fig 2

Une des caractéristiques principales de ces sills est leur intense fracturation par un système de joints transversaux orientés N. 110° E plongeant de 60 à 80° vers le sud, c'est-à-dire vers l'aval. Légèrement moins vertical que les sills, ce système découpe des plaques de 1 à 15 cm d'épaisseur. Dénué de remplissage, aspect essentiel pour la stabilité des versants, il constitue un excellent réseau naturel de drainage. Ce système de joint mémorisé dans la roche (sondages profonds) n'est pas en rapport avec le desserrage des versants auxquels il est d'ailleurs orienté perpendiculairement.

2.3. Description des appuis

Les sills d'andésite ont une extension géographique limitée tant vers le haut où ils ne dépassent pas les cotes (600) pour le sill amont et (630) pour le sill aval que latéralement pour le sill aval en rive gauche et le sill amont en rive droite. Ils se terminent parfois par des apophyses (sill amont en rive gauche, sill aval en rive droite).

La présence des sills détermine la succession suivante, de l'amont vers l'aval :
tuf brèchique amont (1),
sill amont d'andésite (2),
tuf brèchique intermédiaire (3),
sill aval d'andésite (4),
tuf brèchique aval (5) dans lequel on note localement un troisième sill, mince et étiré,
conglomérats (6) à l'aval rive gauche.

La structure des appuis appelle plusieurs remarques (coupes 1 à 4, fig. 3) :

1. l'éperon rive droite constitue un barrage naturel,

2. les sills d'andésite sont pratiquement verticaux alors que la stratification, indiquée par des lentilles de graviers ou de rares intercalations argileuses dans le tuf amont (faille F par exemple) est moins verticale.

3. le tuf amont (1) est massif, compact, très peu fissuré. Sous la zone d'altération superficielle dont l'épaisseur de 5 à 20 m est inversement proportionnelle à la raideur des versants, il est moyennement dur, s'écrasant sous le choc du marteau,

4. le sill d'andésite amont (2) s'épaissit du large rive droite (Ouest) vers la rive gauche (Est) et du haut du versant vers le bas : 5 m dans la coupe 1, 30 m dans la coupe 2, 60 m dans le lit de la rivière (coupe 3), 80 m en rive gauche dans la moitié inférieure de l'appui (coupe 4). Dans la moitié supérieure, au-dessus de la cote (510), sa partie aval se termine en apophyses de sorte que son épaisseur "utile" se réduit à 40 m. On peut avancer comme explication

que le sill amont est en fait formé par deux sills superposés et accolés, l'un d'une quarantaine de mètres d'épaisseur à l'amont montant jusqu'à la cote (600), l'autre à l'aval, de 40 m d'épaisseur également, ne dépassant pas la cote (530). Saine, très dure, l'andésite est très fracturée en petites dalles,

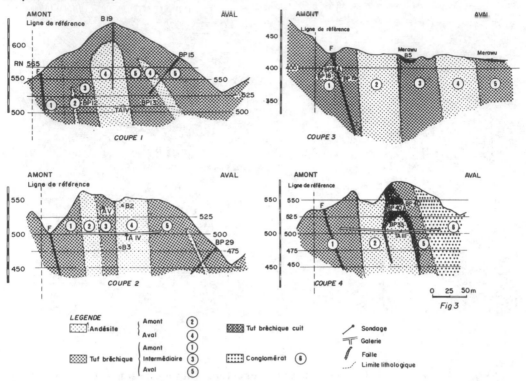

Fig 3

5. le tuf brèchique intermédiaire (3) est similaire au tuf amont. Mais la mise en place des deux sills a eu pour conséquence d'en cuire la zone de contact sur quelques centimètres ou d'en imprégner et modifier chimiquement la matrice sur plusieurs mètres d'épaisseur par les solutions hydrothermales. Cette modification chimique est particulièrement nette dans la moitié supérieure de l'appui rive gauche, autour des apophyses situées au-dessus de la cote (510). De grise à brune ou rosée, la matrice devient franchement rouge brique. Près de l'andésite aval en rive droite, une zone intermédiaire de brèche grossière andésitique à matrice rouge associée à des plans de cisaillement est rattachée à ce tuf intermédiaire. Ce tuf intermédiaire a une épaisseur croissante du haut vers le bas des rives, passant de 20-25 m à une cinquantaine de mètres,

6. le sill aval (4) est formée par une andésite grossièrement cristallisée toujours très saine et très dure, toujours aussi fissurée en dallettes de quelques centimètres d'épaisseur selon la "stratification". Mais, à l'inverse du sill amont, les dalles sont souvent tapissées d'un très fin dépôt argileux. Quelques cisaillements accompagnés de micromylonite argileuse de 1 à 3 cm d'épaisseur empruntent la "stratification". L'épaisseur du sill varie d'une cinquantaine de mètres dans toute la hauteur de l'appui droit à une vingtaine de mètres dans la rive gauche jusqu'à la cote (530),

7. le tuf brèchique aval (5) est similaire au tuf amont. Bien cimenté, moyennement dur, il est altéré superficiellement dans le versant aval de l'éperon rive droite. Il est injecté par plusieurs sills dont le plus épais ne dépasse pas 15 m,

8. le conglomérat (6) est formé de galets hétérogranulométriques pris dans un liant sableux grossier friable. Il ne participe pas aux formations concernées par le barrage.

2.4. Fracturation

Autant l'andésite des deux sills est fracturée en petites dalles de quelques centimètres d'épaisseur, autant le tuf brèchique est massif (7075 mesures de pendages dans les andésites contre 145 seulement dans les tufs brèchiques ont été effectuées dans les 6 galeries). 90 à 95 % des mesures faites dans les andésites et les tufs se rapportent à la "stratification" orientée N.110° E, inclinée de 70 à 80° sur l'horizontale vers le S.SO c'est-à-dire vers l'aval. Les rares autres joints présentent soit la même orientation transversale à la vallée mais plongeant vers l'amont, soit une orientation amont aval avec inclinaison vers l'une ou l'autre rive.

2.5. Perméabilité

Les résultats des essais Lugeon et l'observation des piézomètres montrent que les tuffs brèchiques et les andésites sont peu perméables (respectivement 7 U.L. et 10 U.L.). L'eau circule
par les nombreux joints de l'andésite alors que la circulation dans les brèches se fait par
quelques joints ouverts et par la matrice rocheuse.

3. CARACTERISTIQUES GEOTECHNIQUES

Au cours des études de Faisabilité et d'Avant-projet Détaillé ont été effectués des essais in
situ (profils de sismique-réfraction, profils de "petite sismique", essais à la plaque et essais
de cisaillement béton/rocher) ainsi que des essais en laboratoire.

3.1. Essais "in situ"

Ce chapitre ne décrit que les résultats des essais qui ont été utilisés pour la conception du
projet : petit sismique et essais au vérin.

Vingt six profils de "petite sismique" ont été effectués selon la méthode de B. SCHNEIDER dans
683 m de galerie. La majorité d'entre eux sont faits loin du versant (21) et dans les sills
d'andésite (18).

Une analyse détaillée des résultats par formation, par rive, par altitude dans le versant,
fait ressortir les points suivants :
1. la cimentation relativement médiocre des tufs brèchiques ainsi que l'intense fissuration
des andésites en dallettes de quelques centimètres se caractérisent par de faibles vitesses sis-
miques, de faibles modules dynamiques et modules statiques obtenus par corrélation à partir de
la fréquence des ondes de cisaillement, par de médiocres valeurs sclérométriques. Même si les
andésites présentent des caractéristiques supérieures d'au moins 50 % à celles des tufs brèchiques, ces valeurs restent au moins 30 % en-dessous de celles d'un béton de masse ;

TABLEAU 1 - PETITE SISMIQUE - RESULTATS PAR FORMATION

	Nbre de profils	Longueur en galerie	Onde de pression		Onde de Cisaillement		Module électrique dynamique E_D	Module correlé statique E_s	Scléro-mètre
			Vitesse V_p	Fréquence F_p	Vitesse V_s	Fréquence F_s			
	N	m	m/s	Hz	m/s	Hz	MPa	MPa	
And. (2)	11	314	3.260	533	1.660	467	18.220	7.615	31
And. (4)	7	175	4.060	532	1.920	447	24.410	7.775	30
Tuf (1)	2	30	3.430	548	1.400	349	11.650	4.280	20
Tuf (3)	5	133	2.730	463	1.425	374	12.180	5.485	22
Tuf (5)	0	8	-	-	-	-	-	-	
Congl. (6)	1	23	2.460	498	1.280	388	10.000	5.750	22
Site	26	683	3.360	459	1.650	424	18.080	6.135	28

2. les profils effectués au voisinage du versant donnent des résultats similaires à ceux
effectués plus en profondeur ;
3. la rive gauche, andésitique, est en moyenne moins bonne que la rive droite, considérée glo-
balement ;
4. la fracturation très intense influe très peu sur les caractéristiques des andésites. Ces
caractéristiques sont très légèrement meilleures dans les profils amont-aval effectués perpendiculairement aux dallettes c'est-à-dire à la "stratification",
5. les caractéritiques mécaniques des andésites diminuent sensiblement dans le tiers supérieur
des appuis du barrage.

En conclusion, la petite sismique, méthode d'investigation globale du massif, met en évidence
un contraste marqué entre les propriétés mécaniques des andésites et du tuf brèchique.

Deux essais au vérin hydraulique appuyant sur une paroi de galerie par l'intermédiaire d'une plaque rigide Ø 30 cm ont été effectués lors des études de faisabilité et vingt six autres essais en A.P.D. dans toutes les galeries à l'aide d'un vérin d'une force de 1 000 tonnes (10 000 Kn), appuyant simultanément sur deux parois de galerie par deux plaques rigides Ø 35 cm, tant sur les andésites que les tufs bréchiques.

La répartition des essais au vérin dans les galeries a visé à mettre en évidence l'influence de divers paramètres : nature du terrain, cote, orientation par rapport à la "stratification" (fig. 4). Les résultats sont donnés dans le tableau 2 où :

E est le module d'élasticité déterminé graphiquement par la pente de la tangente au point d'inflexion de la courbe de chargement du cinquième cycle, dit cycle élastique,

Γ est le module de déformation correspondant à la ligne enveloppe tangente au sommet de la courbe des cycles de chargement, Cp est le coefficient de déformation permanent, c'est-à-dire le rapport de la déformation permanente à la pression maximale appliquée.

Le tableau 2 montre que les modules E et Γ de l'andésite sont plus élevés parallèlement à la "stratification", c'est-à-dire transversalement à la vallée ; que contrairement à la petite sismique ils sont influencés par la fracturation en dallettes ; enfin que le tuf bréchique est 20 % à 100 % plus déformable que l'andésite.

TABLEAU 2 - ESSAIS A LA PLAQUE - RESULTATS

Rocher	Tuf bréchique	Andésite	
Direction de l'essai	Toutes directions	Perpendiculaire à "stratification"	Parallèle à "stratification"
Nombre d'essais	24	8	18
Module d'élasticité (MPa) E			
. Moyenne	3 500	4 450	6 420
. Ecart-type	1 280	2 030	2 020
. Coefficient de variation	0,36	0,45	0,31
. Minimum - maximum	1 360 - 6 610	2 680 - 7 650	3 720 - 10 220
Module de déformation (MPa): Γ			
. Moyenne	2 220	2 400	3 800
. Ecart-type	970	1 150	1 740
. Coefficient de variation	0,44	0,48	0,46
. Minimum - maximum	670 - 4 400	1 090 - 4 070	1 560 - 8 770
E/Γ			
. Moyenne	1,63	1,85	1,82
. Minimum - maximum	1,25 - 2,52	1,36 - 2,65	13,8 - 29,9
Coefficient de déformation permanente Cp (10^{-2} mm/bar)			
. Moyenne	0,50	0,63	0,37
. Minimum - maximum	0,19 - 1,70	0,24 - 1,17	0,12 - 0,91

La figure 4 montre qu'en rive droite les modules E et Γ des tufs bréchiques et des andésites ont tendance à se maintenir, voire s'améliorer, du bas vers le haut de l'appui alors qu'en rive gauche la tendance est à la dégradation, surtout pour l'andésite (2).

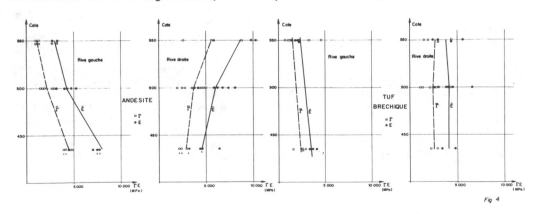

Fig 4

En conclusion, les divers essais permettent d'établir les valeurs de module à retenir pour les calculs. Ils traduisent, tant pour les andésites que pour le tuf brèchique, un rocher compact ayant un comportement effort-déformation presqu'élastique.

L'aptitude au fluage des tufs brèchiques et des andésites a été étudiée en fonction de la direction de l'essai pour détecter une éventuellement influence de la fissuration en dallettes. L'extrapolation logarithmique des résultats indique un fluage maximal inférieur à 0,5 mm (à une pression constante de 20 MPa) au bout de 50 ans pour l'andésite et de l'ordre de 0,5 mm pour les tufs brèchiques.

3.2. Essais en laboratoire

Effectués en Indonésie et en France, ils ont permis de déterminer les propriétés suivantes :
1. caractéristiques physiques : densité, porosité,
2. caractéristiques mécaniques : vitesse sismique, résistance en compression uniaxiale et par essais Franklin, résistance à la traction (essais brésiliens) et au cisaillement,
3. déformabilité : modules de déformation et d'élasticité en milieu naturel et après plusieurs jours d'immersion dans l'eau (10 à 15 jours).

Les résultats montrent que les caractéristiques physiques, mécaniques et de déformabilité de l'andésite sont bien meilleures que celles du tuf brèchique. Il faut toutefois ne pas perdre de vue que ces caractéristiques proviennent de carottes prélevées dans de l'andésite moins fissurée et dans du tuf à matrice plus indurée. On notera que l'eau n'a aucun effet sur les caracté-ristiques de l'andésite, alors qu'elle diminue fortement celles du tuf brèchique.

TABLEAU 3 - ESSAIS EN LABORATOIRE - RESULTATS

Essai	Unité	Tuf brèchique					Andésite				
		Nombre	Moyenne	Ecart-type	Coeff. de variation	Mini-Maxi	Nombre	Moyenne	Ecart-type	Coeff. de variation	Mini-Maxi
Densité	KN/m³	32	22,7	1	0,04	20,6 - 24,2	9	27,1	0,3	0,3	26,7 - 27,7
Porosité	%	30	18,4	4,8	0,26	6,7 - 27,9	2				0,6 - 1,0
Vit. sismique	m/s	14	2 470	630	0,25	1 030 - 4 120	9	5 410	220	0,04	5 040 - 5 730
Compression	MPa										
- état naturel		29	14	7	0,50	1,5 - 32,5	2				52 - 186
- après inhibition		22	5,1	2,1	0,41	1,5 - 8,3	8	58,1	29,2	0,50	8 - 92
Essais Franklin*	MPa										
- Etat naturel											
RD		205	41,3	52,8	1,28	0,72 - 384	120	189	129	0,68	1,7 - 509
RG		127	38,4	51,1	1,33	0,72 - 280	50	341	150	0,44	20,6 - 641
- après inhibition		46	32,9	27,4	0,83	4,32 - 118	2				199 - 220
Traction	MPa	4				2,8 - 4,9	4				8,0 - 16,9
Cisaillement	MPa	5					3				
Module Γ	MPa										
- Etat naturel											
4ème cycle		20	3 000	1 110	0,37		15	11 710	1 040	0,09	
Moyenne		20	2 140	930	0,43		15	5 440	1 080	0,20	
Après inhibition											
4ème cycle		8	2 660	910			13	12 410			
Moyenne		13	1 765	1 125	0,64		13	5 970	1 230	0,21	
Module E											
- Etat naturel											
4ème cycle		20	3 945	740	0,19		15	10 145	1 940	0,19	
Moyenne		20	3 560				15	8 200			
- Après inhibition											
4ème cycle		8	3 415				13	11 325	1 630	0,14	
Moyenne		13	2 975				13	8 985			

* Résultats calculés à partir de l'indice Franklin Is selon la formule Rc = 24 Is.

4. ADAPTATION DU PROJET

La conception du barrage a dû être adaptée aux conditions suivantes dictées par la topographie du site et les résultats des reconnaissances géologiques et géotechniques :
1. recherche de l'andésite en tant que matériau de fondation principale, sachant d'une part qu'en rive droite un banc de brèche intermédiaire est présent et que d'autre part sur les deux rives les sills andésitiques s'amincissent en partie supérieure et voient leurs caractéristiques diminuer.
2. recherche d'une implantation amont tenant compte de l'épaisseur limitée des appuis, notam-ment en rive droite dont la stabilité devait être garantie.

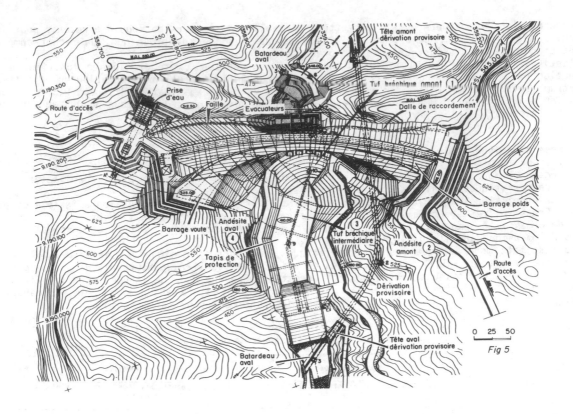

Fig 5

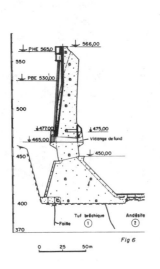

Fig 6

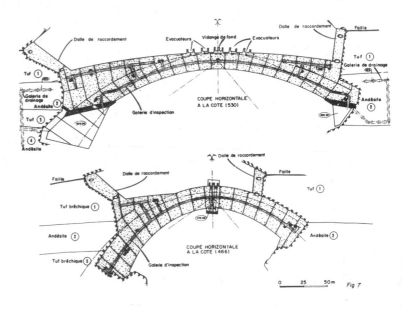

Fig 7

Compte tenu de la hauteur de l'ouvrage, le respect de ces conditions a constitué le véritable défi du projet et a conduit à essayer de nombreuses formes de barrage. Chaque essai a consisté, après dessin et examen du respect des contraintes précédentes, à faire un calcul tridimensionnel de la structure projetée.

Les formes classiques de voûte, même épaissies aux appuis pour abaisser le taux de travail de la fondation, ont été rejetées car elles faisaient apparaître notamment de fortes tractions dans la voûte et de trop fortes contraintes sur les appuis, particulièrement en partie haute.

Une structure mixte fut étudiée, comprenant :

1. tout d'abord un bouchon en fond de vallée de façon à augmenter l'effet de console et à réduire l'effet d'arc pour faire plonger les contraintes dans une direction comparable à celle des sills d'andésite. Ce bouchon, par son emprise, s'accomode également d'une fondation à la fois sur les tufs brèchiques et les andésites bien que leurs caractéristiques géotechniques soient fortement contrastées,

2. puis un profil poids en partie supérieure permettant de rejeter le parement amont du barrage vers l'amont sans trop tendre la voûte, de réduire les contraintes sur la partie supérieure des rives et de faire plonger les efforts dans les rives.

Un calcul complet de cette forme fut fait aux éléments finis tridimensionnels. La géologie et la topographie de la vallée furent modélisées. Les modules adoptés pour la fondation ont été de 4000 MPa dans les andésites et de 2 000 MPa dans les tufs brèchiques.

Les calculs ont montré des contraintes sur la fondation répondant aux critères de conception imposés. Dans la partie supérieure des appuis, les contraintes restent faibles (inférieures à 4 MPa en fonctionnement normal). Pour la majeure partie du barrage fondée sur les andésites entre les cotes 425 et 450, les contraintes sont voisines de 5,5 MPa en fonctionnement normal. Elle se réduisent à 5,2 MPa sous le bouchon de fond de vallée qui repose partiellement sur le tuf brèchique amont.

A l'issue de ce calcul, ce parti a été retenu. Il comprend donc un barrage voûte avec en partie supérieure des ailes poids et, en fond de vallée, un bouchon.

REFERENCES

I.S.R.M., (1981) : Rock characterization testing and monitoring, suggested methods, E.T. Brown editor, Pergamon Press.

Broch E., Franklin J.A., (1972) : The point-load strength test. Trans, Inst. Min. Metall.

Coyne et Bellier (1985) : Maung hydroelectric power project, detailed design report, Paris.

Bieniawski Z.T., Franklin J.A., (1972) : Suggested methods for determining the uniaxial compressive strength of rock materials and the point load strength index, I.S.R.M. Committee on laboratory tests, document N° 1.

Schneider B., (1967) : Moyens nouveaux de reconnaissances des massifs rocheux, Annales I.T.B.T.P., 20è année, n° 235-236.

Use of index tests for engineering assessment of weathered rocks
Essais pour la caractérisation géotechnique de roches altérées

R.P.Martin, *Geotechnical Control Office, Hong Kong*

ABSTRACT: Many different index tests have been used in engineering studies of weathered rocks, primarily for estimating material design properties and characterising the degree of weathering. Literature examples drawn from both these study areas are reviewed and summarised. Two case histories from Hong Kong are presented to illustrate some applications of simple index tests in engineering design assessments. There is considerable scope for practical research and development, particularly with regard to strength assessment of extensively-weathered (soil-like) materials, quantifying the degree of weathering and improving the reliability of routine descriptive procedures.

RESUME: L'étude des roches altérées, en particulier la définition des caractéristiques physiques nécessaires à un projet de génie civil et l'estimation du degré d'altération, est basée sur une multitude d'essais simples et rapides. Nombre d'exemples ont été trouvés dans des publications techniques dans ces deux domaines d'étude, et ceuxci, sont décrits et resumés. Deux études faites a Hong Kong sont présentées dans le but d'illustrer l'application de certains essais de classification d'usage courant dans le cadre des projects de génie civil. Il est démontré la necessité de faire progresser la recherche et le développement des aspects pratiques du sujet, notamment la résistance des matériaux fortement altérés ayant l'apparence de sols, la définition du degré d'altération et l'amélioration de la fidélité des systèmes descriptifs d'usage courant.

1 INTRODUCTION

Assessment of the effects of weathering is of great importance to engineering design in rock foundations and excavations. Index tests are widely used in this respect, not only for characterising the degree of weathering but also as indicators of engineering properties for use in design. These two aspects are commonly examined at both the small (material) and large (mass) scales.

The aim of this paper is to make a brief review of the use of index tests for engineering appraisal of weathered rocks, and to propose some directions for practical research and development. Emphasis is given to rapidly applied, simple tests which can be adopted for both field and laboratory use. Two case histories from Hong Kong are described to illustrate the application of selected tests.

The main concern of the paper is with index tests which can assess weathering effects over the full range of rock weathering states (or 'degrees' of weathering). For this reason applications of tests for the assessment of predominantly fresh or only slightly weathered rock (e.g. as used in the aggregate industry) are not considered. Similarly the use of routine laboratory soil classification and index tests (which are commonly applied to in situ completely weathered rocks and residual soils) are not examined in detail. The majority of the literature cited is in connection with open excavations, slopes and foundations, and no attention is given to those index tests developed exclusively for underground works (e.g. 'cuttability' indices for tunnelling machine performance).

2 INDEX TESTS

2.1 Definitions

It is not easy to make a simple, precise definition of 'index' for the purpose of weathered rock assessment. Previous engineering geological usage has tended to consider index tests as quantitative tests for classification, but this is inadequate as rock classification is not the sole purpose of index testing, and some index tests cannot be regarded as wholly quantitative. A simple dictionary definition would be 'a quantity which expresses or indicates a physical property in terms of a standard.' For weathered rocks it is helpful to regard the 'standard' as one of two things - an engineering design property of the rock, or one of the boundary states of the weathering process (i.e. fresh rock or residual soil). This basic sub-division is retained for the two review sections below (sections 3 and 4).

Although the dictionary definition implies wholly quantitative index measurements, in practice several simple rock index tests are semi-quantitative, in that the results are expressed as a number of ordered groups or classes, as opposed to a continuous numbered scale in a quantitative test. Thus quantitative test indices are typically related to engineering design properties by correlation-type statistics, while semi-quantitative tests are typically used to classify the design property into a number of fairly broad ranges. A similar distinction exists between quantitative weathering indices and semi-quantitative tests which are helpful in setting up weathering classifications. The number of classes used in weathered rock classifications is typically four to seven.

For present purposes an index test is defined as 'a quantitative or semi-quantitative test which may be related to rock engineering design properties or to the boundary states of rock weathering.' Under this definition any assessment procedure applied to weathered rocks which relies solely on observation, or is wholly qualitative, is not considered as an index test.

2.2 Applications at different scales

Table 1 shows the test categories adopted by the Commission on Standardisation of Laboratory and Field Tests of the International Society for Rock Mechanics (Brown, 1981). The twelve headings in category I of Table 1 should all be considered as potential areas for development of index tests, which can be related to the engineering design tests in category II, or used to characterise weathering. The additional column in category I of Table 1 makes a three-fold appraisal of the relative usefulness of these twelve test areas for assessing rock weathering effects.

For the laboratory tests, with the exception of permeability, most of the commonly-measured properties (e.g. strength, deformability, elasticity, porosity) of the non-soluble rocks tend to change unidirectionally and more or less progressively over most or all of the weathering scale from fresh rock to residual soil. Thus there is considerable scope for developing weathered rock index tests for intact materials.

In the field (Table 1, category I), weathering affects the rock mass most notably with respect to alteration along rock joints (aperture, infilling, roughness etc.), changes in RQD and other measures of fracture frequency, and seismic wave velocities. All these areas provide opportunities for development of rock mass index tests, but the scope is not so great as in the laboratory.

In contrast, the relevant engineering design test areas (Table 1, category II) are apparently more concerned with in situ performance rather than laboratory test properties, implying a greater need for knowledge of mass rather than material behaviour. Although this conventional approach has been queried for certain design problems by Farmer (1983), particularly with regard to rock mass strength, it still holds for most design problems in open excavations and foundations. The variable concentration of items in the four sub-categories of Table 1 is a reminder of the ever-present difficulty of extrapolating laboratory test results to the typical scale of engineering projects (Pratt & Voegele, 1984). This applies as much to index tests as any other types of test. In view of the importance of this basic distinction on scale, material and mass scales are considered separately in the following review sections.

2.3 Desirable features

Cottiss et al (1971) and Irfan & Dearman (1978a) examined the use of index tests for rock material classification and considered that they should be:
1. Rapid and simple, involving a minimum of specimen preparation.
2. Relevant to rock properties.
3. Relevant to engineering problems.
4. Capable of discriminating between grades of engineering significance.

These are all desirable criteria, but points 3 and 4 may only be satisfied after considerable experience is gained by relating laboratory tests to field tests and descriptions, and to rock performance during and after construction.

Other desirable features of index tests are:
1. Easy repeatability.
2. Lightweight, robust equipment for field tests.

Rapid and simple tests are of little use if they are not easily repeated to produce consistent results. This is an important factor in view of the great amount of routine testing which is carried out by non-specialist technical staff. The suitability of some commonly-used field equipment is often dependent on the stage at which it is used. For example, a point load test machine is an invaluable aid for assessing potential aggregate suitability once a provisional quarry site is identified, but may be too unwieldy for initial reconnaissance mapping which sets out to locate possible quarry sites.

Table 1. Rock test categories recommended by the International Society for Rock Mechanics (modified from Brown, 1981).

Category I : Classification and characterisation tests

Rock material (laboratory tests)	*Index test Potential	Rock mass (field observations)	*Index test Potential
1) Density; moisture content; porosity; absorption	3	9) Discontinuity orientation, spacing; roughness geometry; filling; alteration	1
2) Uniaxial compressive and tensile strength and deformation characteristics	3	10) Core recovery; RQD; fracture frequency	2
3) Anisotropy indices	1	11) Seismic tests for mapping and as a rock quality index	2
4) Hardness; abrasiveness; attrition	2	12) Geophysical borehole logging	1
5) Permeability	1		
6) Swelling and slake durability	3		
7) Sonic velocity	3		
8) Micro-petrographic descriptions	2		

Category II : Engineering design tests

Laboratory	Field
1) Triaxial compressive strength and deformation characteristics	4) Plate and borehole deformability tests
2) Direct shear tests	5) Direct shear tests
3) Time-dependent and plastic properties	6) Field permeability measurement
	7) In situ rock stress determination
	8) Post-construction monitoring of rock movements
	9) Insitu uniaxial, biaxial and triaxial compressive strength

*Judgement of potential for development of index tests to assess weathering effects :
1 = low, 2 = moderate, 3 = high

There is a need to develop a range of suitable tests whose results are easily inter-related when assessing rock properties and performance. Simple tests which can be easily performed in both the field and the laboratory have the greatest potential application.

3 ENGINEERING DESIGN ASSESSMENTS OF WEATHERED ROCKS

The application of index tests in making design assessments of weathered rocks is examined in this section. A literature review shows that most of the published data concerns igneous, and to a lesser extent, metamorphic rocks. Clearly this reflects the more significant influence of weathering on engineering properties in rocks formed in environments very different from those typified by present-day hydrospheric and atmospheric conditions. Reference is made throughout this section to material weathering grades, which, following the symbols and terminology recommended by Little (1969), Wakeling (1970) and others, are defined as: I = fresh rock, II = slighty weathered, III = moderately weathered, IV = highly weathered, V = completely weathered, VI = residual soil.

3.1 Design properties of rock materials

Referring to Table 1, generally the most important engineering design tests on rock materials are those concerned with strength and deformation characteristics. For the relatively less-weathered materials (i.e. 'hard' rocks), most practical engineering design is based on elastic properties and strength measured in uniaxial compression. Shear strength tests are usually of practical relevance only to discontinuities at the mass scale, while time-dependent and plastic flow characteristics are of concern for a relatively restricted range of design problems, chiefly associated with sustained heavy loading in underground excavations, or with weak sedimentary rocks (e.g. shales and evaporites).
Broadly this trend reverses through the weathering sequence so that, for the relatively more weathered 'soil-like' materials, shear strength and time-dependent deformation are generally of greater concern for most practical engineering. The choice of appropriate rock or soil mechanics

methods to investigate material behaviour is one of the most challenging aspects of weathered rock studies in the middle range of the weathering scale.

In a strict sense all mechanical rock properties measured from laboratory specimens are index properties (Table 1, Category I), because none of the tests yield fundamental material constants for particular rock types. Measured rock properties are dependent on test conditions and often vary with specimen size or shape. However, it is common practice, for example, to regard the results of uniaxial compression tests as yielding engineering design parameters.

With the above comments in mind, the material design properties considered here are static and dynamic elastic moduli, uniaxial compressive strength and tensile strength in hard rock materials, and shear strength in engineering soils.

A variety of index tests used to assess elastic moduli, compressive strength and tensile strength of rock materials are summarised in Tables 2 to 5. The majority of these tests are laboratory procedures involving determination of physical properties or petrographic studies of rock mineralogy. For rapid field use, only the point load test, quick absorption test and Schmidt hammer test are suitable. All are generally applicable to weathering grades I-IV. Of the three, the Schmidt hammer is the only truly portable piece of equipment, owing to the weight of the point-load testing machine and the need for a weighing balance and field oven (or silica gel container) for the absorption test.

The point load test has been widely used for a period of nearly 15 years. During this time a number of improvements have been made (see Turk & Dearman, 1985b; Norbury, 1985), to the point where the test can now be expected to yield reasonably consistent and repeatable results provided standardised procedures are followed regarding specimen preparation and use of size and shape correction factors (ISRM, 1985; Brook, 1985). The relationship between point load strength and compressive strength has been studied for a variety of rock types and the linear coefficient of 24 originally suggested by Broch & Franklin (1972) has been found to vary widely (i.e. 8 to > 40 , see Norbury, 1985).

The quick absorption or 'void index' test is easy to perform (Hamrol, 1961; ISRM, 1979) and has been applied in general age/strength classifications of sedimentary rocks (Duncan et al, 1968), as well as for assessment of engineering properties. The index is strongly correlated with porosity, which is commonly used as a laboratory-measured index to assess weathering effects (Tables 2 to 5). In very weathered materials the test may reveal a tendency for slaking to occur during soaking, in which case a weatherability test would be more suitable (see section 4.1).

The Schmidt hammer is probably the simplest and quickest index test to use on weathered rock materials (Hucka, 1965). However, great care needs to be taken during sampling, particularly where the surface to be tested is rough, or is noticeably affected by microfabric defects in the rock such as cracks and fissures. Poole & Farmer (1980) examined the consistency and repeatability of the test and concluded that reliable values could be obtained provided that artificially low readings were eliminated by selecting a realistic cut-off point, and peak values were chosen from a minimum of five consecutive impacts at a point. Disadvantages of the Schmidt hammer test are that it is relatively insensitive on very weak rocks which yield rebound values below 10, and it cannot readily be carried out on core, although some workers have used it on core held in a heavy vice (Taylor & Spears, 1981). ISRM (1978a) recommended the use of a simple steel cradle which holds small lengths of 'confined' core diametrically. Although such equipment is available commercially (e.g. ELE, 1985), very few examples of hammer tests on core can be found in the literature. One problem may be the difficulty of ensuring equivalent hammer contact area and degree of confinement for cores of varying size and curvature. A series of cradles and modified hammer heads (of varying curvature) to fit the common core sizes would be a useful development. In view of the simplicity and convenience of the Schmidt hammer, it is surprising that other simple rebound tests such as the Shore scleroscope (ISRM, 1978a) do not seem to have been widely used as a rock index (see Taylor & Spears, 1981).

Of the other rock index tests listed in Tables 2 to 5, laboratory-measured sonic wave velocity is also quick and fairly simple to perform (ISRM, 1978b). It is particularly valuable for assessing the effects of microstructural defects on engineering properties; the study by Berry et al (1978) of a suite of altered granitic rocks in a complex geological area of Sardinia is a good example. In principle the test can also be used in the field, but it is often difficult to ensure a good coupling of transmitters and receivers to a rough rock specimen (Dearman & Irfan, 1978b).

Shear strength testing of rock material weathered to an engineering soil traditionally incorporates a number of standard laboratory index tests which were originally developed and applied to superficial deposits and weak sedimentary rocks in temperate environments (e.g. BSI, 1975). The relative success experienced in using these tests to explain strength variability due to weathering seems to vary widely.

For tropical laterites (grade VI residual soils rich in secondary oxides of iron or aluminium), whose properties can vary significantly and irreversibly on drying or reworking, Mitchell & Sitar (1982), Lohnes & Demirel (1983) and many others have cautioned against using conventional laboratory index tests. Investigations of strength variability in some basalt-derived lateritic soils have tended to place more emphasis on detailed microfabric studies (Tuncer & Lohnes, 1977). The effect of pre-treatment on samples may have a pronounced effect on normal plasticity and grading parameters. Recent research on this topic has been reviewed and reported in some detail by the Committee on Tropical Soils of the ISSMFE (1985).

Table 2. Examples of index tests used to assess static tangent elastic modulus (E_t)

Index test	Rock type and location	ΔRelationship of E_t to test value	§Approximate weathering range	Reference
Point load strength	Granite, S.W. England	*+Pos. linear	I-IV	Irfan & Dearman, 1978a
Schmidt hammer	"	*+Pos. power	"	"
Quick absorption	"	*+Neg. logarithmic	"	"
"	Granite, Portugal	Neg. non-linear	I-IV?	Hamrol, 1961
Effective porosity	Granite, S.W. England	*+Neg. logarithmic	I-IV	Irfan & Dearman, 1978a
"	Granite, Japan	Neg. power	I-III?	Onodera et al, 1974
"	Other granites, S.W. England	*+Neg. linear	"	Dearman & Irfan, 1978a
Sonic wave velocity	Monzonite, Bulgaria	Pos. non-linear	I-IV?	Iliev, 1967
"	Granitic rocks, Sardinia, Italy	ØPos. non-linear	I-III?	Berry et al, 1978
"	Granite, S.W. England	*+Pos. power	I-IV	Irfan & Dearman, 1978a
Air porosity	Sandstone, U.K.	Neg. non-linear	?	Morgenstern & Phukan, 1967
Micropetrographic quality index	Granite, granite-gneiss, Portugal	*Pos. linear	I-III?	Mendes et al, 1966
Lixiviation index (see Table 6)	Gneiss, Brazil	*Pos. linear	IV-VI	Rocha Filho et al, 1985
SPT 'N' value	Chalk, U.K.	Pos. non-linear	I-V	Wakeling, 1970

ΔRelationships : pos. = positive, neg. = negative, power function $E_t = ax^{\pm b}$, exponential function $E_t = ae^{\pm bx}$, logarithmic function $E_t = a \pm b\log x$, where x = index test value, a & b = constants, * = explicit relationship defined, + = E_t measured at 50% σ_{ult}, Ø = E_t measured at $\sigma = 10$ MPa
§Material weathering grades : I = fresh rock, II = slightly weathered, III = moderately weathered, IV = highly weathered, V = completely weathered, VI = residual soil

Table 3. Examples of index tests used to assess dynamic elastic modulus (E_d)

Index test	Rock type and location	ΔRelationship of E_d to test value	Approximate weathering range	Reference
Point load strength	Granite, S.W. England	*Pos. linear	I-IV	Irfan & Dearman, 1978a
Effective porosity	"	*Neg. exponential	"	"
"	Granite, Japan	*Neg. power	I-III?	Onodera et al 1974
Saturation moisture content	Granites, S.W. England	Neg. non-linear	I-IV?	Duncan & Dunne, 1967
"	Granite, S.W. England	*Neg. logarithmic	I-III?	Dearman & Irfan, 1978a
Air porosity	Sandstone, U.K.	Neg. non-linear	?	Morgenstern & Phukan, 1967
% Altered minerals	Charnockite (granitic), India	*Neg. linear	I-V	Ramana & Gogte, 1982

ΔFor notes on relationships and weathering grades see Table 2.

437

Table 4. Examples of index tests used to assess compressive strength (σ_C)

Index Test	Rock type and location	ΔRelationship of σ_C to test value	Approximate weathering range	Reference
Point load strength	(General)	*Pos. linear	?	Franklin et al, 1971
"	Granite, S.W. England	*Pos. linear D&S	I-IV	Irfan & Dearman, 1978a
"	Granite, acid volcanics, dolerite, Hong Kong	"	"	Lumb, 1983
"	Limestone, dolomite, Guatemala	* "	?	Jenni & Balissat, 1979
"	Limestone, sandstone, mudstone, U.K.	* "	?	Carter & Sneddon, 1977
Schmidt hammer	Granite, S.W. England	*Pos. linear D, Pos. power S	I-IV	Irfan & Dearman, 1978a
"	Andesite, Turkey	Pos. linear	"	Pasamehmetoglu et al, 1981
"	Quartz diorite, andesite, basalt, dacite, Japan	Pos. power ?	"	Saito, 1981
Quick absorption ('void' index)	Granite, S.W. England	*Neg. non-linear D&S	"	Irfan & Dearman, 1978a
"	Andesite, Turkey	Neg. non-linear	"	Pasamehmetoglu et al, 1981
"	Basalt, India	Neg. non-linear	"	Ghosh, 1980
Saturation moisture content	Weak sandstones, U.K., Brazil, Portugal & Turkey	Neg. non-linear? S	?	Dobereiner & de Freitas, 1984
Effective porosity	Granite, S.W. England	*Neg. non-linear D&S	I-IV	Irfan & Dearman, 1978a
"	Andesite, Turkey	Neg. non-linear	"	Pasamehmetoglu et al, 1981
"	Quartz diorite, andesite, basalt, dacite, Japan	Qz diorite: Neg. linear? Rest: Neg. non-linear?	"	Saito, 1981
"	Granite, acid volcanics, dolerite, Hong Kong	Neg. exponential for porosity > 5%	"	Lumb, 1983
Porosity (unspecified)	Porphyrites, andesites, U.K. & Turkey	*Neg. power D&S	"	Turk & Dearman, 1985a
"	Basalt, India	Neg. non-linear	"	Ghosh, 1980
"	Carbonates, quartzites, quartz sandstones, USSR	*Neg. exponential	I-III?	Smorodinov et al, 1970
Density	Granite, S.W. England	*Pos. exponential D&S	I-IV	Irfan & Dearman, 1978a
"	Porphyrites, andesites, U.K. & Turkey	*Pos. power D&S	I-IV	Turk & Dearman, 1985a
"	Carbonates, USSR	*Pos. exponential	I-III?	Smorodinov et al, 1970
"	Charnockite (granitic), India	*Pos. linear	I-V	Ramana & Gogte, 1982
Dry density	Granite, Spain	Pos. linear?	I-III?	Uriel & Dapena, 1978
Sonic wave velocity	Granodiorite, western U.S.A.	Pos. linear	I-IV	Krank & Watters, 1983
"	Porphyrites, andesites, U.K. & Turkey	*Pos. power D&S	"	Turk & Dearman, 1985a
"	Quartz diorite, andesite, basalt, dacite, Japan	Pos. non-linear	"	Saito, 1981
Bulk specific gravity	Granodiorite, western U.S.A.	Pos. non-linear	"	Krank & Watters, 1983
% decomposed minerals	Charnockite (granitic), India	*Neg. exponential	I-V	Ramana & Gogte, 1982
% altered feldspars	"	* "	"	"
Microfracture intensity	Granodiorite, Australia	Neg. linear	I-III?	Dixon, 1969

ΔFor notes on relationships and weathering grades see Table 2. D = dry, S = saturated

Table 5. Examples of index tests used to assess tensile strength (σ_t)

Index test	Rock type and location	ΔRelationship of σ_t to test value	Approximate weathering range	Reference
Quartz content	Granitic rocks, California, U.S.A.	Neg. linear	?	Merriam et al, 1970
Yamanaka hardness penetrometer	'Shirasu'(welded tuff & pumice), Japan	Pos. non-linear	I-IV?	Yamanouchi et al, 1980
% altered minerals	Charnockite (granitic), India	*Neg. exponential	I-V	Ramana & Gogte, 1982
Sand blast volume loss	Basalt, gneiss, granite & dolerite, various European locations	Neg. non-linear	I-II	Verhoef et al, 1984
Dry density	Granite, Spain	Pos. linear	I-III?	Uriel & Dapena, 1978
Quick absorption	Granite, gneiss, Portugal	Neg. non-linear	I-III	Serafim & Lopes (1962) in Lama & Vutukuri (1978)

ΔFor notes on relationships and weathering grades see Table 2

In non-lateritic grade VI residual soils, many studies report apparently successful interpretation of strength variability using standard laboratory indices (e.g. Lumb (1965) for volcanic and granitic soils in Hong Kong, Chandler (1969) for Keuper marl in U.K., Pender (1971) for greywacke in New Zealand and Brenner et al (1978) for granitic soils in Thailand). However, in a review of published shear strength data for just grade VI granitic soils, Dearman et al (1978) found there was a wide variation in reported strengths (∅' varying from 20°-40°). Research by Baynes & Dearman (1978a, b) on the relationship of microfabric to engineering properties in such soils, together with theoretical considerations by Vaughan & Kwan (1984), suggests that this variability may reflect radically different microfabrics produced during the more advanced stages of weathering, which are largely dependent on the relative amounts of leaching and compaction. The inference from this work is that a good understanding of strength variability is unlikely to be obtained unless individual soil microfabrics are carefully examined. For materials with appreciable clay content this requires detailed laboratory work using the electron microscope. It seems unlikely that conventional laboratory index tests are sufficiently sensitive to allow detailed interpretation of strength variability in this type of soil.

Highly and completely weathered (grade IV and V) granitic materials appear to be more uniform than the grade VI residual soils and are characterised by granular framework microfabrics which yield consistently high angles of shearing resistance, ∅'> 30°-40°, (Dearman et al, 1978). Baynes & Dearman (1978a) studied some granites from South-West England and found there was no apparent decrease in shear strength in this weathering range, despite a progressive decrease in density.

The evidence is conflicting, but it seems reasonable to conclude that the validity of conventional laboratory index tests for assessing strength variability in in situ weathered soils should at least be questioned. Few examples of possible alternative index tests can be found in the literature, with the exception of some work done in Japan. Again working mainly in granitic soils, index tests employed by Japanese researchers include a 'consistency' test similar to the concrete slump test (Matsuo et al, 1970), specific gravity of feldspar grains (Matsuo & Nishida, 1968), soil hardness (Haruyama, 1979), grain crushability (Matsuo & Sawa, 1975), ignition loss or "high temperature drying" (Onodera et al, 1967; Sueoka et al, 1985) and feldspar X-ray diffraction intensity (Nishida & Aoyama, 1985). The moisture condition value (MCV) test (Parsons & Boden, 1979), originally developed for earthworks classification in U.K., has recently been applied in studies of lateritic and saprolitic soils in Brazil (Committee on Tropical Soils of the ISSMFE, 1985). There appears to be considerable scope for experimenting with non-standard index tests in an attempt to improve understanding of strength characteristics of in situ weathered soils.

Discussion in this section has concentrated on the direct relationships between quantitative indices and design properties. Considerable research has also been carried out on the interrelationships between different index tests (e.g. Dearman & Irfan, 1978a, b) and many of the references cited in Tables 2 to 5 contain useful tables of weathering grade classifications showing typical ranges of both engineering design test and index test values.

3.2 Engineering behaviour of rock masses

The opportunities for developing index tests which can be related to in situ design parameters at the mass scale (Table 1) are far less than in the laboratory. This reflects both the extremely varied mass weathering profiles that can develop and the difficulty of devising simple tests using reasonably portable equipment which can be applied to representative volumes of rock.

The one main exception is the use of seismic wave velocities to assess dynamic mass deformation moduli and Poisson's ratio (Evison, 1956). Bieniawski (1978) and Aikas et al (1983) discuss the use of a portable hammer seismograph for this purpose in a variety of rock types from South Africa and Finland respectively. The test is attractive because of its simplicity, but after considering the detailed study by Coon & Merritt (1970), Bieniawski concluded that there were too many pitfalls associated with simple seismic tests for them to be recommended for general use in estimating rock mass deformability.

The major use of index tests at this scale is to assist in characterising rock mass quality. Specific tests are sometimes used to assess single rock mass features; for example, the Schmidt hammer test is recommended for assessment of joint wall compressive strength by Barton & Choubey (1977). More commonly, individual tests are combined with other tests and field observations in order to assess overall rock quality. The following index tests should be mentioned:

1. Fracture spacing indices. Although not strictly 'tests', in that nothing more than linear measurement of fracture frequency is required, both rock quality designation, RQD, (Deere, 1968) and fracture spacing index, FSI, (Franklin et al, 1971) have been very widely used as mechanical indices of rock quality. When used singly on rock core, both indices can be criticised on the grounds that the values may be significantly affected by the quality of drilling. They are more commonly combined with other factors to derive classification systems for general rock quality assessment. For example, FSI is often combined with point load strength to give a two-parameter classification chart which is found useful for assessing ease of excavation (Franklin et al, 1971; Fookes et al, 1971; Baynes et al, 1978). RQD is an integral part of both the NGI Rock Mass Quality Index (Barton et al, 1974) and the South African Geomechanics Rock Mass Rating System (Bieniawski, 1974) developed for underground rock quality assessment. Abdullatif & Cruden (1983) present an interesting comparison of these three schemes as related to ease of excavation in limestone, granite, dolerite, shale and clays from South-West England. Higgs (1984) has recently proposed additional graphical measures of fracture spacing to supplement RQD values.

2. Seismic velocity index. This can be defined as the ratio of the in situ seismic velocity to the sonic velocity in a laboratory specimen. The index is sometimes squared since the dynamic elastic modulus is proportional to the square of wave velocity. Knill & Jones (1965) used this index as a measure of the intactness of the rock mass in foundation studies of dams in Iran and Sudan, while Knill (1969) used it to assess grout take at dam sites in the U.K. The field seismic velocity is also used as a rough indication of rock rippability (Attewell & Farmer, 1976). Despite these examples, Bieniawski (1978) considered both RQD and velocity index to be generally insufficient for describing overall rock mass quality. It would appear that the traditional value of geophysical techniques (i.e. helping to interpret geological structure) continues to be a much more important engineering application than the development of specific indices to assess rock mass quality.

In addition to these direct characterisations, ranges of weathered rock mass index values are often quoted for different zones in weathering classifications. For example, Dearman et al (1978) quote typical RQD values, percentage core recoveries, drilling rates, seismic velocities and resistivities in their general classification of granitic and gneissic rocks.

4 WEATHERING ASSESSMENT

This section reviews the use of index tests incorporated in weathering assessment schemes. Applications range from precise, quantitative petrographic indices for assessing degree of weathering in the laboratory to very simple semi-quantitative field tests used in weathering classifications.

4.1 Rock material weathering

Descriptive methods which rely solely on observation still form the basis of most weathering assessment schemes for rock materials, but the use of index tests to supplement basic descriptions has become more popular in recent years. The main advantage of index tests in this context is that they encourage consistency and objectivity in materials description, which is particularly helpful to the non-specialist user. Improved reliability in descriptions leads to greater confidence when making comparative studies of weathering effects for similar parent materials.

Quantitative weathering indices have been developed to assess both the degree of weathering (as produced by processes acting on a geological time scale) and weatherability, or the capacity of materials to be affected significantly by weathering effects in engineering time.

Some quantitative indices which have been proposed to evaluate degree of weathering are listed in Table 6. Probably the best-known is Lumb's (1962) degree of decomposition index, Xd. This is calculated on the basis of hand separation of quartz and feldspar minerals under the microscope. It has been applied in research studies of weathered granite behaviour (e.g. Baynes & Dearman, 1978a; Vaughan & Kwan, 1984), but is too time-consuming to be recommended for general use as a rapid index test for practical purposes. This also applies to the other petrographically-based indices in Table 6. However, several other indices such as the velocity index, quick absorption test and consistency index are much quicker and simpler to perform, and could have application to

Table 6. Examples of quantitative indices used to assess degree of weathering

Index	Rock type & location	Reference
Degree of decomposition (X_d) $X_d = (N_q - N_{qo})/(1 - N_{qo})$, where N_q = weight ratio of quartz : feldspar in weathered specimen, N_{qo} = corresponding ratio in fresh rock $(N_{qo} \sim 0.33)$	Granite, Hong Kong	Lumb, 1962
Micropetrographic quality index (K) $K = \%$ 'sound' constituents/% 'unsound' constituents	Granite, granite-gneiss Portugal Granite, S.W. England	Mendes et al, 1967 Irfan & Dearman, 1978b
% secondary minerals (degree of decomposition)	Basic igneous, South Africa	Weinert, 1964
% altered minerals	Charnockite (granitic), India	Ramana & Gogte, 1982
% altered feldspars	Charnockite (granitic), India	Ramana & Gogte, 1982
Silica : alumina ratio	Free-draining acid igneous & some basic igneous rocks, New Guinea and elsewhere	Ruxton, 1968
Abrasive pH of feldspar	(General) Granite, gneiss, Nigeria	Grant, 1969 Malomo, 1969
'Lixiviation' index (β) $\beta = (A_{weathered})/(A_{fresh} + CaO/MgO)$ where $A = (K_2O + Na_2O)/Al_2O_3$ in fresh or weathered state	Gneiss, Brazil	Rocha Filho et al, 1985
Velocity index (K) $K = V_O - V_W/V_O$, where V_W = longitudinal wave velocity in weathered state, V_O = corresponding velocity in unweathered state	Monzonite, Bulgaria	Iliev, 1967
Quick absorption index (for grade I-IV materials)	Granite, Portugal Greywacke, New Zealand	Hamrol, 1961 Pender, 1971
Specific gravity of feldspar (for grade IV-VI materials)	4 granites, Japan	Matsuo & Nishida, 1968
'Consistency' index = water content for a slump of 3 mm in a slump test (for grade IV-VI materials)	Granite, Japan	Matsuo et al, 1970
Dry density	Granite, Spain Shales, Spain	Uriel & Dapena, 1978 Dapena et al, 1978
pH development pattern	Dolerite, South Africa	Clauss, 1967

a wide range of rock types. For the coarse-grained igneous rocks, feldspar acidity and specific gravity tests are also easy to perform in grade IV and V materials which are easily separated in the laboratory.

In principle a simple quantitative degree of weathering scale can be established for any rock property which changes unidirectionally throughout the weathering scale and whose value can be readily established in the unweathered state. The scale may be given by (property of fresh rock – property of weathered sample)/property of fresh rock, e.g. Iliev's (1967) velocity index in Table 6.

A number of quantitative indices have also been used to assess weatherability, or its inverse, durability. These are most applicable to the weaker argillaceous sedimentary rocks which are noticeably affected by wetting and drying, but can also be relevant to weathered igneous and metamorphic rocks, particularly if they contain appreciable proportions of ferro-magnesian minerals (e.g. Duncan et al (1968) found that 40% of the weathered granites they tested showed some signs of swelling).

The best-known weatherability tests are the slake durability test (Franklin & Chandra, 1972) and the swelling strain index, measured in confined or unconfined conditions (ISRM, 1979). Olivier (1973), quoted in Lama & Vutukuri (1978), related compressive strength and swelling-strain index to give a strength-swell ratio for some South African mudstones. Detailed discussions of slake durability test values in various soft sedimentary rocks are given by Fookes et al (1971) and Taylor & Spears (1981). Other indices which have been used to assess weatherability are: a) elastic wave velocity, in a study of weathering in cut slopes in 10 rock types in Japan (Yamazaki et al, 1980) and in some dolerite paving stones from Portugal (Delgado Rodrigues, 1978), and b) quick absorption tests in basalt from Brazil (Farjallat & Nery de Oliviera, 1972). Fookes et al (1971) provide an interesting historical summary of various weatherability studies.

The third main application of index tests to weathering assessments in rock materials is in devising weathering classifications. Some examples are given in Table 7. Many of these tests are performed by very simple field procedures and the results are often assessable only as a series of comparative codes, e.g. the mineralogical indices developed by Weinert (1964), relative ease of core breakage, relative strength of soil grains. Other tests yield results on a continuous numbered scale (e.g. hand penetrometer, 10% fines test), but they are typically quite variable and are best grouped into classes to fit the weathering classification scale as otherwise defined.

In Hong Kong, simple index tests are proving to be particularly useful for description and classification of highly to completely weathered materials (grades IV-V) and an example will be given in section 5. It appears there is considerable scope for development of further tests of this type.

Table 7. Examples of index tests used in material weathering classifications

Index test	Rock type and location	ΔApproximate weathering range	Reference
Breakage of core or lumps by hand or finger pressure	Granodiorite, Western USA	III-IV	Krank & Watters, 1983
	Granite, Australia	III-IV	Moye, 1955
	Greywacke, New Zealand	III-VI	Pender, 1971
	Tropical rocks/soils (general)	III-V	Little, 1969
	Granitic & volcanic rocks, Hong Kong	III-IV	Hencher & Martin, 1982
Relative strength/ consistency of feldspars	Granites, S.W. England	I-V	Dearman & Irfan, 1978a
	Granitic rocks, Sardinia, Italy	III-V?	Berry et al, 1978
	Granodiorite, Hong Kong	I-V	Irfan & Powell, 1985
Schmidt rebound hammer	Granite, S.W. England	I-V	Dearman & Irfan, 1978a
	Granitic & volcanic rocks, Hong Kong	II-V	Hencher & Martin, 1982
Hand penetrometer	Granitic & volcanic rocks, Hong Kong	IV-VI	Hencher & Martin, 1982
Slakeability	Granite, Australia	IV-V	Moye, 1955
	Granitic & volcanic rocks, Hong Kong	IV-V	Hencher & Martin, 1982
	Granite, S.W. England	V-VI	Irfan & Dearman, 1978a
Mineral colour/ lustre/hardness/ crystalline state (decomposition index)	Basic igneous rocks, S. Africa	I-VI	Weinert, 1964
10% fines test (disintegration index)	Basic igneous rocks, S. Africa	I-VI	Weinert, 1964
Micropetrographic index (I_p)	Granites, U.K.	I-V	Dearman & Irfan, 1978b
Micropetrographic fracture index (I_f)	Granites, U.K.	I-V	Dearman & Irfan, 1978b

ΔFor notes on weathering grades see Table 2

4.2 Rock mass weathering

There are few examples in the literature of specific index tests being applied in rock mass weathering classifications. Martin & Hencher (1984) reviewed a number of mass weathering classifications by various authoritative bodies (Geol. Soc. 1970, 1972, 1977; IAEG, 1981; ISRM, 1981) and ten schemes drawn up by individual authors in a variety of rock types. They found that the criteria used to set up these classifications are almost exclusively descriptive, the most popular being relative proportion of 'rock' and 'soil' material within each weathering zone, presence or absence of rock mass structure and material fabric, and degree of discolouration of joint planes.

There are a few exceptions to this general rule. Knill & Jones (1965) used slakeability as a criterion to distinguish between the two most weathered zones of gneissic rocks in the Sudan, but this was presumably done on the basis of tests performed on small samples. Duncan & Dunne (1967) proposed the use of a field seismic test to establish zone boundaries in granites from South-West England. Simple relative strength indices were used by Neilson (1970), to separate five different weathering zones in some Australian mudstones, and Saito (1981), to help classify weathering zones in some basic and intermediate lavas and quartz diorites from Japan.

The difficulty of devising simple field index tests which are representative at the mass scale is obvious and there is relatively little scope for development in this area. Geophysically-based index tests are a possible exception, but the complexity and heterogeneity of typical weathering profiles in many different rocks (Deere & Patton, 1971) usually makes precise interpretation difficult. Tests based on penetration rates (e.g. by drilling or dynamic probing) are a second exception and an example is described in the next section.

5 CASE STUDIES FROM HONG KONG

Most of the territory of Hong Kong is underlain by granitic and volcanic rocks of Jurassic age, which are characterised by deep weathering profiles developed under conditions of rapid chemical decomposition. General accounts of geotechnical engineering in these rocks are given by Brand & Phillipson (1984) and Brand (1985). Two case studies are briefly described to illustrate the practical applications of some simple index tests.

5.1 Verification of pile-founding depths in granodiorite

A study of weathered granodiorite was made by Irfan & Powell (1985) for the construction of some large-diameter piled bridge foundations in the Tai Po area of Hong Kong. They used point load strength, Schmidt hammer and percussion air drill rate as index tests to assist in characterising the weathered condition of the rock mass for determination of suitable founding depths of hand-dug piles (caissons), up to depths of 60 m.

Engineering geological logging of the caisson walls was carried out during construction. A type N Schmidt hammer was used as a routine index test at pre-determined locations in each 1 m ring of the caisson as part of the assessment of rock material weathering grade. An example is shown in Figure 1 and clearly indicates the high degree of local variability in rock material weathering. This variability was confirmed by comprehensive point load testing.

Subsequently the N Schmidt hammer results were averaged for each caisson ring. These figures were then used to assess characteristic ranges of mean N rebound values to correspond with a descriptive classification of rock mass weathering, similar to that recommended by the British Standards Institution (BSI, 1981), as shown in Figure 2a. Further guidance on the assessment of suitable founding depths was provided by the results of percussive drilling through the base of the pile using a portable (25 kg) air hammer drill. Drill rate was assessed as the time required for a penetration of 300 mm. Although the air drilling rate could not identify local variations in material strength, it was sufficiently sensitive to overall rock quality that a satisfactory correlation with the mass weathering classification could be defined (Figure 2b).

Founding depths for end-bearing piles on rock in Hong Kong are traditionally based on the results of pile-specific RQD, percentage core recovery or probing tests, and are generally conservative and costly. Irfan & Powell (1985) concluded that the use of a rock mass weathering classification system, in conjunction with simple index tests, was superior to these traditional methods, and enabled limited engineering data to be applied successfully over a large project area.

Further field studies in other local rock types need to be undertaken to assess whether this approach to rock mass characterisation is of more general application to piled foundations, but the results of this study are very promising. It would be of particular interest to compare quantitative field index test parameters with the results of in-situ load tests or full-scale pile load tests in similar rocks.

5.2 Assessment of degree of decomposition and shear strength of weathered granite

Research into the shear strength of some typical Hong Kong in-situ weathered soils has been under-way in the Geotechnical Control Office (GCO) since 1978. One of the materials being investigated is a fine-medium grained (<6 mm) equigranular granite of decomposition grade IV (highly decomposed rock) to grade VI (residual soil) from the North Kowloon area of Hong Kong. (Because of the predominance of chemical weathering over other types of weathering in Hong Kong rocks, material weathering grades are traditionally described in decomposition terms as opposed to general

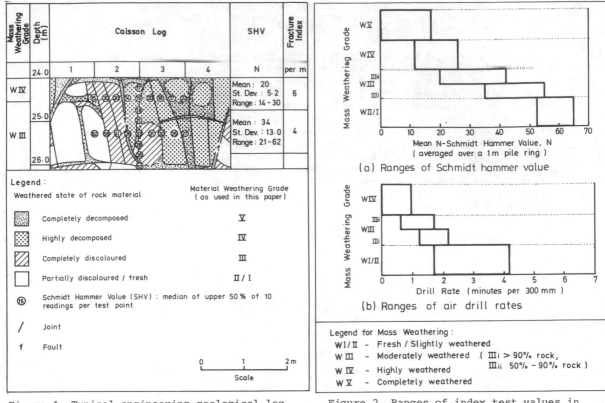

Figure 1. Typical engineering geological log of caisson wall (from Irfan & Powell, 1985)

Figure 2. Ranges of index test values in weathered granodiorite (from Irfan & Powell, 1985)

weathering terms).

Field description and classification of this granite was carried out using the methods discussed by Hencher & Martin (1982). This scheme is based on the use of several simple field index tests, of which those relevant to the material under discussion are summarised in Table 8. These tests were supplemented by visual assessment of the degree of microfracturing of hand samples.

The granite was investigated at seven different sampling sites within the weathering profile as exposed in a number of large cut slopes. In order to assess the effects of moisture content variation on index test results, trial pits were dug and index tests were performed in the sides and base of the pits under three moisture conditions viz: a) as dug ('natural' condition), b) after flooding the trial pit with water for 24 hours ('wet' condition), c) after allowing the trial pit to dry out for 2 to 3 days ('intermediate' condition).

Two of the seven materials sampled were grade VI residual soils, easily recognisable by absence of original rock texture. Two others were completely decomposed rock (grade V) which crumbled easily in the hand and slaked completely on immersion in water. The three other materials were less decomposed and appeared to be borderline between grades IV and V (highly and completely decomposed). Classification for these cases was greatly assisted by plotting the index test results on grade-index diagrams (Hencher, 1984), as shown in Figure 3. No single index test can be considered as definitive for determining grade boundaries, but combining a number of tests in such diagrams makes judgement of decomposition grade more reliable for borderline cases, and allows the effect of moisture content to be assessed.

Much of the research effort in the GCO programme has been directed to investigating unsaturated shear strength, but for comparative purposes a series of high-quality consolidated undrained (CU) triaxial tests with pore pressure measurement have also been carried out on back-saturated soil specimens taken from the same samples. Failure points from these CU tests for the materials discussed above are summarised in Figure 4. Multistage testing was adopted on three specimens from samples taken at each of the three moisture conditions, giving a total of nine test results for each material. With the exception of one of the grade VI soils (number 81/1), stress paths were typically dilatant (Figure 4) and all seven soils are characterised by linear strength envelopes. There is very little scatter in the data and strength parameters by statistical regression can be derived with some confidence.

Sorting and classifying such shear strength results is a common design problem in Hong Kong. It is often the case that the use of conventional laboratory soil index tests do not permit a fully

Table 8. Some field index tests used to describe and classify weathered granites in Hong Kong (modified from Hencher & Martin, 1982)

Index test	Procedure	Classes/Terms
Hand penetrometer strength test	Using a standard hand penetrometer, take an average of 10 values of 'undrained' shear strength (instrument reading divided by 2), avoiding disturbed or friable areas	0-250 kPa
Field strength estimate test	1. Easily penetrated by thumb 2. Penetrated by thumb with effort 3. Indented with thumb 4. Readily indented with thumbnail 5. Indented with thumbnail with difficulty 6. Not indented with thumbnail. Easily peeled with knife 7. Crumbles under firm blows with geological pick 8. Shallow indendations with geological pick 9. Not peeled with knife. No indentation with pick	1. Soft 2. Firm 3. Stiff 4. Very stiff 5. Hard 6. Extremely soft rock 7. Very soft rock 8. Soft rock 9. Hard rock
Feldspar strength test	Scratch feldspar with a pin. Where feldspars are in various states of decomposition record nature of dominant type	1. Not scratched 2. Just scratched 3. Easily scratched
Slake test	Immerse small sample (e.g. 75 mm diameter) in water. If sample does not slake in a few minutes, agitate container gently	1. Does not slake 2. On agitation breaks down to discrete fragments 3. On agitation breaks down to a slurry 4. Slakes completely

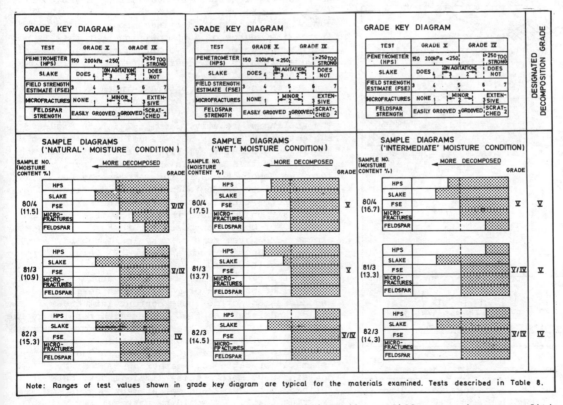

Note: Ranges of test values shown in grade key diagram are typical for the materials examined. Tests described in Table 8.

Figure 3. Grade-index diagrams for three granitic soils at different moisture conditions.

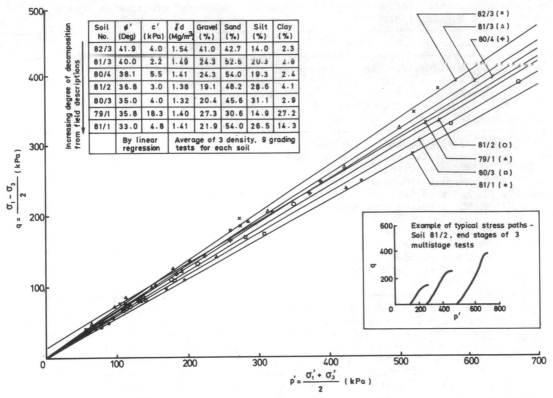

Soil No.	ϕ' (Deg)	c' (kPa)	γ_d (Mg/m³)	Gravel (%)	Sand (%)	Silt (%)	Clay (%)
82/3	41.9	4.0	1.54	41.0	42.7	14.0	2.3
81/3	40.0	2.2	1.49	24.3	52.6	20.3	2.8
80/4	38.1	5.5	1.41	24.3	54.0	19.3	2.4
81/2	36.8	3.0	1.38	19.1	48.2	28.6	4.1
80/3	35.0	4.0	1.32	20.4	45.6	31.1	2.9
79/1	35.8	18.3	1.40	27.3	30.6	14.9	27.2
81/1	33.0	4.8	1.41	21.9	54.0	26.5	14.3
	By linear regression		Average of 3 density, 9 grading tests for each soil				

Figure 4. Summary of $\overline{CU}$ triaxial tests on seven granitic soils (data from Shen, 1985)

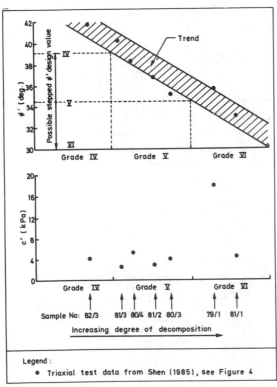

Figure 5. Strength parameters related to degree of decomposition

satisfactory interpretation of strength variability. The data in Figure 4 suggest that the frictional component of strength declines progressively with increasing degree of decomposition, yet this is not clearly indicated by the other laboratory index measures such as dry density, or summary fractions from the grading curves.

Based on the results of grade-index diagrams such as Figure 3, the strength parameters are shown plotted against degree of decomposition in Figure 5. The scaling of relative decomposition in Figure 5 is subjective and simply gives equal weight to each of the three grades(IV, V, VI). The location of each material on this scale is approximate, but the relative location of all the samples may be justified from grade-index plots (e.g. Figure 3 demonstrates that the sequence of increasing decomposition for these three soils is 82/3→81/3→ 80/4). For design purposes Figure 5 would suggest appropriate parameters of very low (probably zero) c', in conjunction with a stepped Ø' value according to decomposition grade.

Use of index tests in this study assisted in accurate description of material weathering grades and assessment of moisture content influences in the field, and interpretation of strength results in the laboratory. There is great scope for more practical research in this area. One of the most desirable aims would be to derive better, rapid quantitative estimates of degree of weathering, either by devising new, more precise simple index tests or by refining procedures and measurement scales for existing tests. It would be particularly valuable to examine the relationships between quantitative weathering indices and the microfabric structure of weathered in situ soils developed from a range of common rock types.

6 CONCLUSIONS

Index tests have been applied in two main areas to the engineering assessment of weathered rocks, i.e. evaluating weathered rock properties by empirical correlation and characterising the degree of weathering. A wide range of different tests have been employed, mostly on rocks of igneous origin. Due to practical limitations on size of equipment, and representativity of rock volumes to be tested, most rapid and simple index tests are applied at the materials scale.

For engineering design assessment, there are a number of useful simple tests, such as the point load test, quick absorption test and Schmidt hammer test, which can provide an initial assessment of rock material properties in weathering grades I to IV. In the more weathered engineering soils (grade V and VI), the success of traditional soil laboratory index tests to assess engineering properties is questionable and there is a need for research into new index procedures. Work done in Japan on various non-standard soil index tests may provide a useful starting point. For appraisal of mass engineering behaviour, fracture spacing indices and geophysical index methods are helpful in general rock quality assessment, but the scope for development of new index tests is much less than in the laboratory, and no index test procedures can be expected to replace the need for careful observation and measurement of rock mass structure.

For weathering assessment, a range of quantitative indices for evaluating degree of weathering are available, but many are based on detailed petrographic studies. There is a need to develop simpler, more rapid measures of degree of weathering which are more refined than the common six-fold, semi-quantitative classification scale. Such measures need to be related to detailed microfabric studies in order to improve understanding of engineering properties of weathered in situ soils. For improving the reliability of routine descriptive procedures, there are opportunities for incorporating a variety of simple index tests into weathering classifications. Simple material strength/hardness/durability-type tests on small exposures, hand samples or pieces of core are probably the most attractive, but there is also scope for development of penetration and probe tests at the mass scale.

ACKNOWLEDGEMENTS: The author is grateful to Dr. S.R. Hencher (Department of Earth Sciences, Leeds University, U.K.), Dr. T.Y. Irfan and Mr. D.R. Greenway (both of the Geotechnical Control Office, Hong Kong) for helpful comments on an earlier draft. This paper is published by permission of the Director of Engineering Development of the Hong Kong Government.

REFERENCES

Abdullatif, O.M. & D.M. Cruden 1983. The relationship between rock mass quality and ease of excavation. Bull. Int. Assoc. Engng. Geol. 28:183-187.
Aikas, K., P. Loven & P. Sarkka 1983. Determination of rock mass modulus of deformation by hammer seismograph. Bull. Int. Assoc. Engng. Geol. 26-27:131-133.
Attewell, P.B. & I.W. Farmer 1976. Principles of engineering geology. London: Chapman and Hall.
Barton, N. & V. Choubey 1977. The shear strength of rock joints in theory and practice. Rock. Mech. 10:1-54.
Barton, N., R. Lien & J. Lunde 1974. Engineering classification of rock masses for the design of tunnel support. Rock Mech. 6:189-236.
Baynes, F.J. & W.R. Dearman 1978a. The relationship between the microfabric and the engineering properties of weathered granite. Bull. Int. Assoc. Engng. Geol. 18:191-197.
Baynes, F.J. & W.R. Dearman 1978b. The microfabric of a chemically weathered granite. Bull. Int. Assoc. Engng. Geol. 18:91-100.
Baynes, F.J., W.R. Dearman & T.Y. Irfan 1978. Practical assessment of grade in a weathered granite. Bull. Int. Assoc. Engng. Geol. 18:101-109.
Berry, P., S.F. Martinetti, R. Ribacchi & M. Sciotti 1978. Cataclasis and alteration of a granitic rock and their influence on the geomechanical characteristics. Proc. 3rd Int. Congress, Int. Assoc. Engng. Geol., Madrid, Section II, 1:123-128.
Bieniawski, Z.T. 1974. Geomechanics classification of rock masses and its application in tunnelling. Proc. 3rd Int. Congress, Int. Soc. Rock Mech., Denver, 2A:27-32.
Bieniawski, Z.T. 1978. Determining rock mass deformability: experience from case histories. Int. J. Rock Mech. Min. Sci. 15:237-247.
Brand, E.W. 1985. Geotechnical engineering in tropical residual soils. Proc. 1st Int. Conf. on Geomechanics in Tropical Lateritic and Saprolitic Soils, Brasilia. 3:23-91.
Brand, E.W. & H.B. Phillipson 1984. Site investigation and geotechnical engineering practice in Hong Kong. Geotech. Engng. 15:97-153.
Brenner, R.P., P. Nutalaya & D.T. Bergado 1978. Weathering effects on some engineering properties of a granite residual soil in northern Thailand. Proc. 3rd Int. Congress, Int. Assoc. Engng.

Geol., Madrid. Section II. 1:23-36.

British Standards Institution 1975. Methods of test for soil for civil engineering purposes. BS 1377. London: British Standards Institution.

British Standards Institution 1981. Code of practice for site investigations. BS 5930. London: British Standards Institution.

Broch, E. & J.A. Franklin 1972a. The point-load strength test. Int. J. Rock Mech. Min. Sci. 9: 669-697.

Brook, N. 1985. The equivalent core diameter method of size and shape correction in point load testing. Int. J. Rock Mech. Min. Sci. 22:61-70.

Brown, E.T. (ed.) 1981. Rock characterisation, testing and monitoring. Oxford: Pergamon.

Carter, P.G. & M. Sneddon 1977. Comparison of Schmidt hammer, point load and unconfined compression tests in Carboniferous strata. Proc. Conf. on Rock Engng., Newcastle, U.K. 197-210.

Chandler, R.J. 1969. The effect of weathering on the shear strength properties of Keuper marl. Géotechnique 19:321-334.

Clauss, K.A. 1967. The pH of fresh and weathered dolerite as an indicator of decomposition and of stabilisation requirements. Proc. 4th African Reg. Conf. Soil Mech. Found. Engng., Cape Town. 101-108.

Coon, R.F. & A.H. Merritt 1970. Predicting in-situ modulus of deformation using rock quality indexes. ASTM Spec. Tech. Pub. 477:154-173. Philadelphia: Am. Soc. Testing & Materials.

Committee on Tropical Soils of the ISSMFE 1985. Peculiarities of geotechnical behaviour of tropical lateritic and saprolitic soils: progress report (1982-1985). Sao Paulo: Brazilian Soc. Soil Mech.

Cottiss G.I., R.W. Dowell & J.A. Franklin 1971. A rock classification system applied in civil engineering. Civl Engng. & Public Works Rev. 1:611-614.

Dapena, E., V. Escario, S. Uriel & J. Martin Vinas 1978. The influence of weathering on the characteristics of shales. Proc. 3rd Int. Congress, Int. Assoc. Engng. Geol., Madrid. Section II. 1: 57-64.

Dearman, W.R. & T.Y. Irfan 1978a. Classification and index properties of weathered coarse-grained granites from South-west England. Proc. 3rd Int. Congress, Int. Assoc. Engng. Geol., Madrid. Section II. 2:119-130.

Dearman, W.R. & T.Y. Irfan 1978b. Assessment of the degree of weathering in granite using petrographic and physical index tests. Proc. UNESCO Int. Symp. on Deterioration and Protection of Stone Monuments, Paris. Paper 2.3.

Dearman, W.R., F.J. Baynes & T.Y. Irfan 1978. Engineering grading of weathered granite. Engng. Geol. 12:345-374.

Deere, D.U. 1968. Geologic considerations. In K.G. Stagg & O.C. Zienkiewicz (eds.), Rock mechanics in engineering practice, p. 1-20. London: John Wiley & Sons.

Deere, D.U. & F.D. Patton 1971. Slope stability in residual soils. Proc. 4th Panam. Conf. Soil Mech. Found. Engng., Puerto Rico. 1:87-170.

Delgado Rodrigues, J. 1978. About the quantitative determination of rock weatherability: a case history. Proc. 3rd Int. Congress, Int. Assoc. Engng. Geol., Madrid. Section II. 1:65-71.

Dixon, H.W. 1969. Decomposition products of rock substances: proposed engineering geological classification. Symp. on Rock Mech., Sydney Univ. 39-44.

Dobereiner, L. & M.H. de Freitas 1984. Investigation of weak sandstones. In A.B. Hawkins (ed.), Site investigation practice: assessing BS 5930, p.143-154. London: Eng. Group, Geol. Soc.

Duncan, N. & M.H. Dunne 1967. A regional study of the development of residual soils. Proc. 4th African Reg. Conf. Soil Mech. Found. Engng., Cape Town. 109-119.

Duncan, N., M.H. Dunne & S. Petty 1968. Swelling characteristics of rock. Water Power. 20:185-192.

ELE International Limited (1985). Materials testing division catalogue. Hemel Hempstead, UK:ELE Int. Ltd.

Evison, F.F. 1956. The seismic determination of Young's modulus and Poisson's ratio for rocks in situ. Géotechnique 6:118-123.

Farjallat, J.E.S. & J.A. Nery de Oliviera 1972. Experimental studies on the weatherability of the Capivara dam basalts, Rio Paranapanema, Brazil. Bull. Int. Assoc. Engng. Geol. 6:83-96.

Farmer, I.W. 1983. Engineering behaviour of rocks. London: Chapman and Hall.

Fookes, P.G., W.R. Dearman & J.A. Franklin 1971. Some engineering aspects of rock weathering with field examples from Dartmoor and elsewhere. Ql. J. Engng. Geol. 4:139-185.

Franklin, J.A., E. Broch & G. Walton 1971. Logging the mechanical character of rock. Trans. Inst. Min. Met. 80:A1-A9.

Franklin, J.A. & A. Chandra 1972. The slake-durability test. Int. J. Rock Mech. Min. Sci. 9:325-341.

Geological Society Engineering Group Working Party 1970. The logging of rock cores for engineering purposes. Ql. J. Engng. Geol. 3:1-24.

Geological Society Engineering Group Working Party 1972. The preparation of maps and plans in terms of engineering geology. Ql. J. Engng. Geol. 5:295-382.

Geological Society Engineering Group Working Party 1977. The description of rock masses for engineering purposes. Ql. J. Engng. Geol. 10:355-388.

Ghosh, D.K. 1980. Relationship between petrological, chemical and geomechanical properties of Deccan basalt, India. Bull. Int. Assoc. Engng. Geol. 22:287-292.

Grant, W.H. 1969. Abrasion pH, an index of weathering. Clays and Clay Minerals. 17:151-155.

Hamrol, A. 1961. A quantitative classification of the weathering and weatherability of rocks. Proc. 5th Int. Conf. Soil Mech. Found. Engng., Paris. 2:771-774.

Haruyama, M. 1979. A method of the identifcation and classification of pyroclastic flow deposits due to hardness. Soils and Foundations. 19(1):81-92.

Hencher, S.R. & R.P. Martin 1982. The description and classification of weathered rocks in Hong Kong for engineering purposes. Proc. 7th Southeast Asian Geotech. Conf., Hong Kong. 1:125-142.

Hencher, S.R. 1984. Discussion on rock descriptions. In A.B. Hawkins (ed.) Site Investigation practice: assessing BS 5930, 2, p.121-123. London: Eng. Group, Geol. Soc.

Higgs, N.B. 1984. The profile-area and fracture frequency methods: two quantitative procedures for evaluating fracture spacing in drill core. Bull. Assoc. Engng. Geologists. 21:377-386.

Hucka, V. 1965. A rapid method for determining the strength of rocks in-situ. Int. J. Rock Mech. Min. Sci. 2:127-134.

Iliev, I.G. 1967. An attempt to measure the degree of weathering of intrusive rocks from their physico-mechanical properties. Proc. 1st Int. Congress, Int. Soc. Rock Mech., Lisbon. 1:109-114.

International Association of Engineering Geology 1981. Rock and soil description and classification for engineering geological mapping. Bull. Int. Assoc. Engng. Geol. 24:235-274.

International Society for Rock Mechanics 1978a. Suggested methods for determining hardness and abrasiveness of rocks. Int. J. Rock Mech. Min. Sci. 15:89-97.

International Society for Rock Mechanics 1978b. Suggested methods for determining sound velocity. Int. J. Rock Mech. Min. Sci. 15:53-58.

International Society for Rock Mechanics 1979. Suggested methods for determining water content, porosity, density, absorption and related properties and swelling and slake-durability index properties. Int. J. Rock Mech. Min. Sci. 16:141-156.

International Society for Rock Mechanics 1981. Basic geotechnical description of rock masses. Int. J. Rock Mech. Min. Sci. 18:85-110.

International Society for Rock Mechanics 1985. Suggested methods for determining point load strength. Int. J. Rock Mech. Min. Sci. 22:51-60.

Irfan, T.Y. & W.R. Dearman 1978a. Engineering classification and index properties of a weathered granite. Bull. Int. Assoc. Engng. Geol. 17:79-90.

Irfan, T.Y. & W.R. Dearman 1978b. The engineering petrography of a weathered granite in Cornwall, England. Ql. J. Engng. Geol. 11:233-244.

Irfan, T.Y. & G.E. Powell 1985. Engineering geological investigations for foundations on a deeply weathered granitic rock in Hong Kong. Bull. Int. Assoc. Engng. Geol., in press.

Jenni, J.P. & M. Balissat 1979. Rock testing methods performed to predict the utilization possibilities of a tunnel boring machine. Proc. 4th Int. Congress, Int. Soc. Rock Mech., Montreux. 2:267-273.

Knill, J.L. 1969. The application of seismic methods in the prediction of grout take in rock. Proc. Conf. on In-situ Investigations in Soils and Rocks, Brit. Geotech. Soc. 93-100.

Knill, J.L. & K.S. Jones 1965. The recording and interpretation of geological conditions in the foundations of the Roseires, Kariba and Latiyan dams. Géotechnique 15:94-124.

Krank, K.D. & R.J. Watters 1983. Geotechnical properties of weathered Sierra Nevada granodiorite. Bull. Assoc. Engng. Geologists. 20:173-184.

Lama, R.D. & V.S. Vutukuri 1978. Handbook on mechanical properties of rocks: volume IV. Aedermannsdorf, Switzerland: Trans Tech. Publications.

Little, A.L. 1969. The engineering classification of residual tropical soils. Proc. 7th Int. Conf. Soil Mech. Found. Engng., Mexico. 1:1-10.

Lohnes, R.A. & T. Demirel 1983. Geotechnical properties of residual tropical soils. In R.N. Yong (ed.) Geological environment and soil properties, p. 150-166. New York: Am. Soc. Civ. Engrs.

Lumb, P. 1962. The properties of decomposed granite. Géotechnique. 12:226-243.

Lumb, P. 1965. The residual soils of Hong Kong. Géotechnique. 15:180-194.

Lumb, P. 1983. Engineering properties of fresh and decomposed igneous rocks from Hong Kong. Engng. Geol. 19:81-94.

Malomo, S. 1980. Abrasive pH of feldspars as an engineering index for weathered granite. Bull. Int. Assoc. Engng. Geol. 22:207-211.

Martin, R.P. & S.R. Hencher 1984. Principles for description and classification of weathered rocks for engineering purposes. In A.B. Hawkins (ed.) Site investigation practice: assessing BS 5930, p.304-318. London: Eng. Group, Geol. Soc.

Matsuo, S. & K. Nishida 1968. Physical and chemical properties of decomposed granite soil grains. Soils and Foundations. 8(4):10-20.

Matsuo, S., M. Fukuta & K. Nishida 1970. Consistency of decomposed granite soils and its relation to engineering properties. Soils and Foundations 10(4):1-9.

Matsuo, S. & K. Sawa 1975. Studies of the crushability of decomposed granite soil grains. Bull. Int. Assoc. Engng. Geol. 11:71-76.

Mendes, F.M., L. Aires-Barros & F.P. Rodrigues 1967. The use of modal analysis in the mechanical characterization of rock masses. Proc. 1st Int. Congress, Int. Soc. Rock Mech., Lisbon. 1:217-223.

Merriam, R., H.H. Rieke III & Y.C. Kim 1970. Tensile strength related to mineralogy and texture of some granitic rocks. Engng. Geol. 4:155-160.

Mitchell, J.K. & N. Sitar 1982. Engineering properties of tropical residual soils. Proc. ASCE Specialty Conf. on Engng. and Construction in Tropical and Residual Soils, Honolulu. 30-57.

Morgenstern, N.R. & A.L.T. Phukan 1967. Non-linear deformation of a sandstone. Proc. 1st Int. Congress, Int. Soc. Rock Mech., Lisbon. 1:543-548.

Moye, D.G. 1955. Engineering geology for the Snowy Mountains scheme. J. Inst. Engrs. Australia. 27:281-299.

Neilson, J.L. 1970. Notes on weathering of the Silurian rocks of the Melbourne district. J. Inst. Engrs. Australia. 42:9-12.

Nishida, K. & C. Aoyama 1985. Physical properties and shear strength of decomposed granite soil. Proc. 1st Int. Conf. on Geomechanics in Tropical Lateritic and Saprolitic Soils, Brasilia. 1:371-382.

Norbury, D.R. 1985. The point load test. In A.B. Hawkins (ed.) Site investigation practice: assessing BS 5930, p. 344-352. London: Eng. Group, Geol. Soc.

Onodera, T.F. 1962. Dynamic investigation of formation rocks in situ. Proc. 5th Symp. on Rock Mech., Minnesota. 517-533.

Onodera, T.F., R. Yashinaka & M. Oda 1974. Weathering and its relation to mechanical properties of granite. Proc. 3rd Int. Congress, Int. Soc. Rock Mech., Denver. 2:71-78.

Onodera, T.F., M. Oda & K. Minami 1976. Shear strength of undisturbed sample of decomposed granite soil. Soils and Foundations. 16(1):17-26.

Parsons, A.W. & J.B. Boden 1979. The moisture condition test and its potential application in earthworks. Report SR-522. Crowthorne, U.K.: Transport and Road Research Laboratory.

Pasamehmetoglu, A.G., C. Karpuz & T.Y. Irfan 1981. The weathering characteristics of Ankara andesites from the rock mechanics point of view. Proc. Int. Symp. on Weak Rock, Tokyo. 1:185-191.

Pender, M.J. 1971. Some properties of weathered greywacke. Proc. 1st Australia-New Zealand Conf. on Geomechanics. 1:423-429.

Poole, R.W. & I.W. Farmer 1980. Consistency and repeatability of Schmidt hammer rebound data during field testing. Int. J. Rock Mech. Min. Sci. 17:167-171.

Pratt, H.R. & M.D. Voegele 1984. In situ tests for site characterization, evaluation and design. Bull. Assoc. Engng. Geologists. 21:3-22.

Ramana, Y.V. & B.S. Gogte 1982. Quantitative studies of weathering in saprolitized charnockites associated with a landslip zone at the Porthimund dam, India. Bull. Int. Assoc. Engng. Geol. 19:29-46.

Rocha Filho, P., F.S. Antunes & M.F.G. Falcao 1985. Qualitative influence of the weathering degree upon the mechanical properties of a young gneiss residual soil. Proc. 1st Int. Conf. on Geomechanics in Tropical Lateritic and Saprolitic Soils, Brasilia. 1:281-294.

Ruxton, B.P. 1968. Measures of the degree of chemical weathering of rocks. J. Geol. 76:518-527.

Shen, J.M. 1985. GCO research into unsaturated shear strength 1978-1982. Research Report RR 1/85. Hong Kong: Geotechnical Control Office, unpublished.

Smorodinov, M.I., E.A. Motovilov & V.A. Volkov 1970. Determinations of correlation relationships between strength and some physical characteristics of rocks. Proc. 2nd Int. Congress, Int. Soc. Rock Mech., Belgrade. 2:Paper 3-6.

Sueoka, T., I.K. Lee, M. Muramatsu & S. Imamura 1985. Geomechanical properties and engineering classification for decomposed granite soils in Kadura district, Nigeria. Proc. 1st Int. Conf. on Geomechanics in Tropical Lateritic and Saprolitic Soils, Brasilia. 1:175-186.

Taylor, R.K. & D.A. Spears 1981. Laboratory investigation of mudrocks. Ql. J. Engng. Geol. 14: 291-310.

Tuncer, R.E. & R.A. Lohnes 1977. An engineering classification of certain basalt-derived lateritic soils. Engng. Geol. 11:319-339.

Turk, N. & W.R. Dearman 1985a. Influence of water on engineering properties of weathered rocks. Proc. 21st Reg. Conf., Engng. Group of the Geol. Soc. on Groundwater in engineering geology, Sheffield, U.K. 109-121.

Turk, N. & W.R. Dearman 1985b. Improvements in the determination of point load strength. Bull. Int. Assoc. Engng. Geol. 31:137-142.

Vaughan, P.R. & C.W. Kwan 1984. Weathering, structure and in situ stress in residual soils. Géotechnique. 34:43-59.

Verhoef, P.N.W., T.J. Kuipers & W. Verwaal 1984. The use of the sand-blast test to determine rock durability. Bull. Int. Assoc. Engng. Geol. 29:457-461.

Wakeling, T.R.M. 1970. A comparison of the results of standard site investigation methods against the results of a detailed geotechnical investigation in the Middle Chalk at Mundford. Proc. Conf. on In-Situ Investigations in Soils and Rocks, Brit. Geotech. Soc. 17-23.

Weinert, H.H. 1964. Basic igneous rocks in road foundations. S. African Council Sci. Ind. Res., 218 - Natl. Inst. Road Res. Bull. 5:1-47.

Yamanouchi, T., K. Gotoh & H. Murata 1980. Stability of cut slopes in a pumice soil deposit with particular reference to tensile failure. Proc. 3rd Australia-New Zealand Conf. on Geomechanics. 2:115-120.

Yamazaki, S., S. Okuzono & S. Okubo 1980. On the stability and weathering of cut slopes. Proc. Int. Symp. on Landslides, New Delhi. 1:195-198.

The nature of foliation of the Sarmation clays in the outcrops

Nature de la foliation des argiles du Samartien à l'affleurement

A.M.Monyushko, *Institute of Geophysics & Geology, Academy of Sciences of the Moldavian SSR, USSR*

ABSTRACT: Clay soils of Sarmatian stage (N_1^3S) are widely spread in Eastern Europa. The characteristic feature of the Sarmatian clays is their disintegration in the outcrops into sheets and plates as much as some fractions of a centimeter or several centimeters thick. This disturbs the stability of slopes and complicates use of the territory. On the basis of the Sarmatian deposits of the Fore-Caucasus we could establish the reason of such clay decay. The reason lies in their microlaminated structure caused by the thinnest streaks of aleurite.

The explanation was given as to the formation of thin aleurite streaks in the Sarmatian clays against background of the pelitic mass. Microlaminated nature is explained by the activity of seasonal climatic mechanism of sedimentation.

Some recommendations it were given for projection and construction on the clays which disintegrate quickly.

RESUME: Des terrains argileux d'une couche géologique sarmate(N_1^3S) sont largement répandus an Europe Orientale. La perticularité caractéristique des argiles sarmates consiste en leur désintégration dans des affleurements en dalles et en plaques d'une épaisseur différente: dès parties d'un centimètre jusqu'aux quelques centimètres. Cela rompe la stabilité des versants et complique la mise en valeur de construction du territoire. A l'example des dépôts sarmates de l'avant-pays du Caucase nous avons constanté la cause de cette désintégration des argiles: c'est leure structure microstratiforme qui est conditionée par des minces couches intermédiaires d'alevrite.

On a donné l'explication à la formation dans les argiles sarmates des minces couches intermédiaires, d'alevrite sur la masse pélite. La microstratification s'explique par action du mécanisme des saisons et du climat de la sèdimentation.

Clay soils of Sarmatian stage (N_1^3S) are widely spread in Eastern Europe. These soils are very often foundation or excavation subject. The characteristic feature of the Sarmatian clays is their disintegration in the outcrops into sheets and plates as much as some fractions of a centimeter or several centimeters thick. Outwardly homogeneous soil foliates quickly because of surface weathering (Fig.1). This disturbs stability of slopes and complicates construction use of the territory. On the basis of the Sarmatian deposits of the Fore-Caucasus we could establish such clay decay. The reason lies in their microlaminated structure caused by the thinnest streaks of aleurite and fine sand.

According granulometric composition among the considered deposits prevail clays which clay fraction content reaches 91% and the arithmetical mean out of 92 attributes makes 68%. Dust fraction is from 6 to 81% particles using the arithmetical mean of 25%. Sand fractions take subordinate situation, their average portion makes up 6% while their particular meanings fluctuate from 0.2 to 48%. Significant mean quadratic deviation of various fraction contents of 6-15% and great variation coefficients, which are more than 20-40% give evidence about ununiformity of the Sarmatian deposits. The greatest variability characterizes sand particles and big dust content for which variation coefficient reaches 60-100%. Variation coefficients of clay particles content are significantly smaller (22-44%). The marked ununiformity must be ascribed on the account of the sublayers of grob dispersed material in the clays. Mainly fine dispersed character of the deposits states about deepwater conditions of sedimentation.

Figure 1. Foliation of the Sarmatian clays in the outcrops.

In the non-weathered condition the Sarmatian clays are mainly characterized by dark-grey, sometimes black, rarely light-grey with bluish, olive or yellow tints. The rock is compact, visually uniform, poorly lamellar or lamellar. There are thin streaks of aleurite with sparkles of mica fine and thin-dispersed sand of grey or dark-grey colour with thicks of 2-3 mm, streaks and crystals of gypsum fine-crystal thickness from 3 mm to 1-4 cm. Clays are of carbonate nature (3-14% $CaCO_3$, mean 11%), not rarely streaks of marl brown-grey colours thickness 2 cm and more lying parallel to the plain of stratification.

Under the microscope the Sarmatian clays can be seen as pelitic mass of filemot tints containing small quantity of aleurite material. In some places clays are enriched in small fine fragment material, represented by quartz, fieldspar, muscovite, sericite, glauconite and ore minerals. There are grains of carbonates, ferric hydroxides and organic matters. Texture is mainly microlaminar optically oriented (Fig.2). Lamination is caused by the alteration of dark bands

Figure 2. Microlamellar structure of the Sarmatian clays, 45^x.

of pelite mass with more light ones and marked with streched elements of rock. The width of bands is measured with tenth fractions of millimeters. Fine dispersed pelite fraction of clays according to mineralogical investigations consists mainly of minerals montmorillonite-hydromica groups admixed with caolinite, galloysite, glauconite, palygorskite, calcite, siderite, ferric hydroxides. Light bands are represented mainly by sericite and cryptocrystallic carbonate, against background of which can be met small fractions of muscovite and other minerals. Mainly along light sericite bands fractures are developed which are stretched along lamination. The width of fractures is 0.02-0.04 mm.

According to lithology of deposits and the history of the development of the area in the Sarmatian times (Hain 1964), it can be supposed that sedimentation throughout the whole period took place in the deep parts of the non-deep sea or in the transition zone from deep-water of the basin to the shallow-water with alteration of transregressions and regressions caused by oscillatory motions of the earth crust. The process of sedimentation was also effected by the seasonal climatic phenomena. Replacement and covering of tectonic and seasonal climatic phenomena gave rise to the presence in the clay fraction sand-aleurite streaks and lamination. If the role of oscillatory motions in the earth crust in the shift of the line of the basin and formation of the laminar thickness is good enough investigated, then the mechanism of the influence of seasonal climatic factors on formation of microlamination of Sarmatian clays was not clear. That's why it must be discussed more in detail.

Palinologic analysis of the investigated clays which was made for making paleoclimatic situation clear, has revealed that plant cover on the shores of the Sarmatian sea at the beginning of this period was of the same kind - forest--like. Forests were represented by various associations with coniferous and broad-leaved arts. Great content in the pollen spectre of the conifers (particularly pine-trees), variety of arts gives evidence that coniferous forests played an important role. Broad-leaved forests composed an essential part as a component of forests, represented by more various arts. Among herbaceous plants and shrubs there are arts in which forests cenozes abundant: ferns, lycopodiums, ericaceae, some cerials, green and bog moss. The presence of pollen in water and shore plants gives evidence of the existance of desalinated basins. This also characterizes plant groups along the sea shore.

In the ecological respect the majority of arboreal arts and shrubs estimated by palinological analysis is characteristic of areas with mild climate. Nevertheless arts characteristic of subtropic sarea grew there. According to the finds in the Crimea and Caucasus region that is the traces of clove trees (Mensbir 1934) we have every reason to conclude that the average annual temperature in this region in the Sarmatian age was higher than 15°C. In any case one can deny the probability of winter colds. This conclusion can be supported by the investigations (Larskaya 1965) about connections of carbonate content of Miocene (N1) clays of Fore-Caucasus with climatic conditions of the region. Accumulation of significant amounts of carbonates in the investigated clays gives evidence of half-arid climatic situation in the Sarmatian age. Thys states also the salinity characteristics of clays and its components. The analysis of 183 determinations of acqueous extracts shows that the Middle and Low Sarmatian clays are characterized by the same mean values of soil salinity (0.65 g/100 g rock). In the Low Sarmatian clays salinity variation is insignificant from 0.38 to 0.84 g/100 g rock (variation coefficient 22%); this gives evidence of its sufficient uniformity. The distribution of salinity of acqueous extracts of the Middle Sarmatian clays is characterized by the high variation coefficient (71%) and significant range of particular boundary value from 0.12 to 2.90 g/100 g rock in the condition of the high salinity of particular samples. Among cations prevail potassium-sodium ions, which average content 8.0 me per 100 g rock for Low Sarmatian clays and 7.3 me for Middle Sarmatian clays. Average content of magnesium and calcium is nearly equal; that is about 0,5 me for Low Sarmatian and 1,5 me for Middle Sarmatian clays. Sulphate ions (average content is about 5.0 me) prevail in the anion composition of water extracts of the Sarmatian clays. There are not many hydrocarbonates and chlorides: the average content of the first ones is 0.7 me for Low Sarmatian and 2.8 me for Middle Sarmatian clays; the average content of the second ones in the Low Sarmatian clays is near to the content of sulphate ions in them (4.4 me) and is lower in the Middle Sarmatian deposits (1.0 me). So the considered composition of water soluble of the Sarmatian clays of the Fore-Caucasus represents (Strakhov 1963) the conditions of physical-chemical sedimentation in the basins of some-arid zones. The degree of aridity is not so significant for our problem. The important thing is additional confirmation of the above stated fact that the effect of really cold climate in the Sarmatian time can be excluded. It can be recalled that this was a closed basin divided into the system of inner seas. By using the idea (Hough 1958) of the mechanism of deposit of banded iron formations in large comparatively deep lakes in subtropical and warm regions of temperate zones the explanation was given as to the formation of extremely thin streaks in the Sarmatian clays. There is seasonal cycle of water stirring in the isolated water basins. Stratification of the Sarmatian clays is considered to be connected with the art of particle deposition at the deep part of the basin depending on stirring. Stirring in the temperate zone and northwards where effect of changes of seasons and freezing appears to be particularly strong takes place twice a year. In subtropical and warm regions of the temperate zone stirring period lasts one year. The latter conditions as shown above existed in the Sarmatian basin. Under such conditions in summer water mass stratification takes place more evidently, that means it is divided into the upper layer of warm and light water and lower cold layer. In the cold layer the temperature can be about 4°C at which water takes higher density. The currents in summer as an effect of wind and waves stir the waters of the upper comparatively light layer while lower water mass remains slow creating the situation for line dispersed particle deposition. In the short winter period the surface water after being cooled gets as compact as deep water and arising currents stir the basin waters throughout the whole depth. As a result of the circulation of the whole water mass of the basin appear short-term conditions for deposition of more coarse, aleurite material which last 2-3 months that means the sedimentation of coarsedispearsed particles will be of seasonal character. The effect of the seasonal climatic factors we considered in the isolation from the natural situation of sedimentation in the Sarmatian time. Naturally this situation was determined not only by seasonal climatic phenomena but also by their combination with series of other factors which determine the art of deposition of the material of various coarseness. Nevertheless at the deep parts of the basin the process of sedimentation was determined by the activity of the namely seasonal according to the above considered mechanism. The thinnest aleurite streaks against background of pelite mass of the studied clays give evidence of it.

In the described data one can find explanation to the important problem for the theory and praxis of construction about the reason of the disintegration of visually uniform Sarmatian clays in situ in the weathered zone into sheets and plates. The reason of it is in the microlamellar structure of the Sarmatian clays. It is conditioned by the sedimentation with determined change of periodical natural phenomena in particular tectonic and seasonal climatic and also diagenetic and postdiagenetic changes. As is known (Logvinenko 1968) in the alkaline medium such disintegrated terrigenous components as plagioclasses and some others under definite conditions rather often turn into the aggregates of sericite. In this case usually appear also secondary carbonates. Such carbonate-sericite streaks are observed in the studied clays. In the weathered zone the connection between particles is easily lost in these sericite streaks where particular crystals are more elastic in comparison with the crystalls of hydromica and montmorillonite having content less potassium and more water. That's why the fractures observed under the microscope are related to sericite streaks. These fractures cause foliation of the outwardly homogeneous clay in the outcrops.

In the projection and construction on the clay soils which disintegrate quickly it is recommended that the foundation pit should be excavated to the design marks just before foundation with the obligatory clean-up of the fresh-weathered layer. Excavated pits should be filled up in the nearest future. The soil should be protected from the effects insolation and precipitation if it is impossible to fill up pits quickly. Cementation or other protection against weathering of clays after excavating must be foreseen. While protecting important constructions on the slopes one should take into consideration the possibility of landslips including activization of ancient landslips and foresee corresponding arrangements against sliding processes.

REFERENCES

Hain V.E. 1964. Neogenic period. In: The History of the Geological Development of the Russian Platform and its Frame, p.170-186. Moscow: Nedra.
Houch J.L. 1958. Fresh-water Environment of Deposition of Precambrian Banded Iron Formations. J. Sediment. Petrol. V.28, 3: 414-430.
Larskava E.S. 1965. Carbonate Character of the Mesozoic and Cenozoic Clays of of Western Fore-Caucasus and Paleoclimatic Conditions of their Accumulation: Lithology and Minerals. 2: 123-133.
Logvinenko N.V. 1968. Postdiagenetic Changes of Sedimentary Rocks, 92 p. Leningrad: Nauka.
Mensbir M.A. 1934. Sketches on the History of the Fauna of the European Part of the USSR (from the Beginning of the Tertiary Period), 223 p. Leningrad: State Publishing House of Biological and Medical Literature.
Strakhov N.M. 1963. Types of Lithogenesis and their Evolution in the History of the Earth, 535 p. Moscow: Cosgeoltekhizdat.

Geological and geotechnical characteristics of volcanic tuffs of Central and Southern Italy

Caractéristiques géologiques et géotechniques de tufs volcaniques du Centre et du Sud de l'Italie

Giovanni Nappi, *Centro di Studio per la Geologia Tecnica, CNR, Rome and Università di Urbino, Italy*
Mario Ottaviani, *Dipartimento di Idraulica, Trasporti e Strade, Università di Roma 'La Sapienza', Italy*

ABSTRACT

The geological and geotechnical characteristics of volcanic tuffs located in Central and Southern Italy have been studied. The tuffs, used as construction material, are the products of the Quaternary explosive activity of volcanoes such as the Vulsini, Sabatini, Cimino, Latium and Campi Flegrei complexes. The mechanical properties of the tuffs have been measured in the laboratory to determine compressive and tensile strength. Cyclic loadings have also been carried out. The values of compressive strength vary widely from about 30 to almost 300 kg/cm^2. Such a scattering of values seems to depend mainly from the eruptive mechanism, from the temperature and the size of the clasts deposited, and from the lithification processes after deposition. A linear relation between compressive strength and porosity has been determined.

RESUMEE

On a étudié les caracteristiques géologiques et géotechniques des tufs volcaniques de l'Italie centrale et méridionale. Ces tufs, utilisés comme matériel de construction, sont le produit de l'activité explosive des groupes volcaniques Vulsini, Sabatini, Cimino, Latium c des Campi Flegrei. On a determiné les caracteristiques mécaniques des tufs en laboratoire, résistance à la compression e résistance à la traction. Dans certaines cas on a méme effectué des essais à compression cycliques. Les valeurs de la résistance à la compression des tufs considerés varient de moins de 30 jusqu'à 300 kg/cm^2 environ. Cette étude a mis en évidence que cette variabilité se rapport essentiellment aux mécanismes eruptifs, à la température et à la granulometrie des clastes lors de l'eruption et des processus de litification des tufs aprés la deposition. On a determiné une relation lineaire entre la résistance à la compression et la porosité.

INTRODUCTION

Tuffs, which are lithified pyroclastic products of various origin, are commonly used as construction material where available. These rocks are largely present in the italian country and many data are available on their physical and mechanical properties. The authors of the present paper have recently investigated several types of tuff belonging to the western Vulsini volcanic complex (Nappi and Ottaviani, 1985) and have found that the widely different geotechnical characteristics shown by these tuffs could be explained taking into consideration the genetic and diagenetic processs undergone by these rocks. In fact, it has been observed that their mechanical properties depend on the eruption mechanism, on the emplacement conditions, on the temperature and on the size of the clasts deposited, and on the type of lithification process undergone by the tuffs.

In the present paper, on the basis of the findings mentioned above, a number of other tuff formations, representing a wide range of situations present in central and southern Italy, have been investigated in order to establish a more general validity of the relationship between geological and volcanological aspects and mechanical properties of this type of rock. In what follows a brief outline of the genetic and diagenetic characteristics of the tuff formations considered in this study is given together with the corresponding mechanical properties obtained by several authors.

GEOLOGICAL AND VOLCANOLOGICAL CHARACTERISTICS OF THE TUFFS

The areal distribution, the eruptive mechanisms, the emplacement conditions and the petrographic and textural characteristics of various lithological units in central and southern Italy, where active quarries are located, have been studied. These volcanites are the products of the activity of the Quaternary alcalin-potassic province except for the Cimino volcano which belongs to the acid province. The volcanic complexes considered have different age varying from 1.000.000 years (Cimino) to present (Phlaegrean volcanic complex). Each complex consists of several central volcanoes, caldera and monogenetic apparatus. The evolution of each complex is characterized by the emplacement of large pyroclastic flows due to the emptying of shallow magmatic chambers. After such parossistic eruptions, vulcanotectonic collapses lead to the formation of caldera. During collapse deep groundwater interacted with the magma causing hydromagmatic eruptions. Therefore each complex consists of basal mainly magmatic products and of less extensive younger hydromagmatic products located along the borders of the collapsed areas. The eruptive mechanisms that gave rise to the products studied are all related to high energy explosions which formed high eruptive columns of pyroclastic ejecta. Such explosions can be of magmatic (Plinian) or hydromagmatic (phreatoplinian, subplinian) type. The mechanism can evolve from magmatic to hydromagmatic or viceversa according to the water magma interaction. Generally the magmatic explosive activity changes to hydromagmatic when the flux of the water inside the open vent is prevalent on the flux of the uprising fragmented melt. The emplacement temperature of the magmatic products is, of course, higher than the temperature of the water cooled hydroclasts.

The volcanic products described in the present paper are mainly due either to pyroclastic flows or to turbulent pyroclastic flows and, more rarely, to pyroclastic surges. It has been recently recognized that ignimbrites are rocks formed by the deposit of pyroclastic flows during a single eruption often consisting of more flow units (Spark, 1975). Such deposits infill existing valleys and topographic lows and are usually poorly sorted. The pyroclastic flows have high particle concentration and move by laminar flowage whereas pyroclastic surges are mainly gaseous turbulent flows with low particle content (Walker, 1983). Products with intermediate characteristics have been called turbulent pyroclastic flows (Nappi et al., 1986).

The following formations have been studied in the present paper (see fig. 1):

Vulsini Volcanic Complex:
1) Nenfro, VW1, Western Vulsini.
2) Red tuff with black scoria, VW2, Western Vulsini.
3) Yellow tuff, VW3a and VW3b, Western Vulsini.
4) Orvieto ignimbrite, VE1, VE2, VE3, VE4, Eastern Vulsini.

Cimino Volcanic Complex
5) Cimino ignimbrite, CP1, CP2.

Sabatini Volcanic Complex:
6) Yellow tuff of Via Tiberina, ST1, ST2, ST3.
7) Magliano tuff, SM.

Latial Volcanic Complex:
8) Albano peperino, LP.

Phlaegrean Fields Volcanic Complex: 9) Campanian ignimbrite, FC.
10) Yellow Neapolitan tuff, FN1a, FN1b, FN2.

1) The products of the Vulsini volcanic complex are the northernmost among the products studied. They cover an area of about 1500 km^2 between the Tyrrhenian sea and the Tiber river. Such complex includes two large volcanotectonic depressions: the Latera and the Bolsena caldera. The eastern part, located between the Tiber river and the Bolsena lake, includes the oldest products of the complex which consist of ignimbritic plateaux, lava and other pyrocalstic products. These rocks are covered on the western part by the very thick ignimbritic series of the Latera vulcano. The following tuffs, used as construction materials, belong to this serie: "Nenfro", VW1; "Red tuff with black scoria", VW2; and "Yellow tuff", VW3. The Nenfro tuff, which is of magmatic origin, outcrops on a limited area because, being the oldest products, is found at the

456

base of the pyroclastic series. The lower part of the formation shows a micropumiceous light gray matrix including elongated dark gray pumice clasts which form a series of almost parallel fiamme (Nappi and Ottaviani, 1985). This facies shows a pseudofluidal texture with mainly unbroken phenochrysts of, in order of decreasing amount, sanidine, plagioclase and pyroxenes. The glassy groundmass includes sanidine microlites and is characterised by the presence of elongated subparallel vesicles of subelliptical shape. The upper part of the formation consists of a dark gray non welded mainly ash facies.

2) The "red tuff with black scoria", also of magmatic origin, has a considerable areal distribution around the Latera caldera except on the eastern part which is covered by the Bolsena Lake (220 km^2). This formation is found even at distances from the eruptive center of about 20 km where its maximum thickness of about 20 m is also found. This tuff presents a matrix consisting of very small pumice clasts and phenoclasts of sanidine, leucite, biotite and pyroxenes. The matrix includes large dark gray pumice clasts and lithic and lava fragments. The groundmass has a vesicular texture with glassy fragments often zeolitized into phillipsite and, to a lesser amount, into chabassite (Nappi and Ottaviani, 1985).

3) The "yellow tuff", mainly of hydromagmatic origin, outcrops uniformily all around the Latera caldera. The maximum thickness (about 70 m) is found in the North West area outside the caldera rim near Sorano. Away from the rim, the formation thins out to a few meters. The small pumices enclosed in the ash size matrix show a rounded shape and include spherical vesicles indicating a regular growth of each vesicle. The amount of vesicles appears at the microscope to be about 5-10% for VW3a and about 20% for VW3b. Zeolites are present in the groundmass: more abundant in VW3a where only chabassite is present and less abundant in VW3b where also phillipsite is present.

4) The Orvieto ignimbrite, as mentioned earlier, is the product of the older activity of the Vulsini volcanic complex. The average thickness of this tuff, which is very similar to VW2, is of about 10m but the distal deposits show a maximum thickness of about 60m under the town of

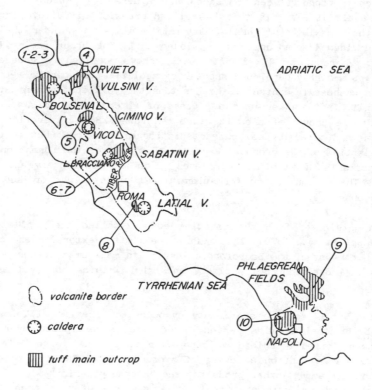

Figure 1. Map of the tuffs studied.

457

Orvieto where the VE1, VE2 and VE4 samples were taken. The Orvieto tuff (Nappi et al.,1982) shows a micropumiceous matrix which includes lithic fragments of about 10 cm size and black and gray pumices, whose size varies from a few centimeters up to a few meters in the proximal deposits. Phenoclasts of leucite and sanidine are also present.

The lithification of the Vulsini tuffs appears to be due to different causes. The Nenfro tuff was formed at high temperature as its pseudofluidal structure and the presence of sanidine microlites clearly show and therefore a welding primary process is to be indicated. For the yellow and the red tuffs, lithification is due to zeolitization. Since this process is more developed where the matrix is fine grained, the zeolites content is larger in the yellow tuff than in the red tuff. In fact the yellow tuff, being of hydromagmatic origin, was formed by fine size clasts. Moreover, it seems that the type of zeolite present in the tuff can be related to the deposition temperature of the clasts as, for instance, phillipsite is formed at higher temperature than chabassite (De Gennaro, 1984). This shows that the pyroclastic flow which originated VW3b, where phillipsite is present, had a higher temperature than the flow which originated VW3a. In fact the distance from the eruptive center is larger for VW3a than for VW3b.

5) The Cimino ignimbrite covers a large area, about 300 Km^2, around the Cimino volcano (Nappi and Lardini, 1985). Its thickness reaches about 200m the feeding fissure and decreases to about 20m in the distal area. This deposit is characterized by a fine grained mainly welded, gray and red facies which is locally called "Viterbo peperino". A large amount of biotite and sanidine clasts, less than 0,5cm long, is present in the groundmass. Also the presence of fiamme formed by glassy fragments parallel to the deposition morphology is typical of this facies. Clasts of plagioclase and pyroxenes can also be observed. The matrix shows fibrous glassy fragments with tubular, subparallel collapsed vesicles typical of pyroclastic flow emplacement.

6) The products of the Sabatini volcanic complex cover a large area between the Tyrrhenian sea and the Tiber river just north of Rome. The complex is formed by a series of eruptive centers, by the Baccano and Sacrofano caldera and by the volcanotectonic depression of the Bracciano lake. One of the oldest products of the volcanic activity of such complex is the " yellow tuff of via Tiberina" which is found between the Sacrofano caldera, from which it originated, and the Tiber river (Nappi et al, 1979). This tuff has filled preexisting valleys forming a series of tongues which reach their maximum thickness (70m) in the distal area near the Tiber river. The formation can be subdivided into an upper and a lower unit. The upper facies (ST1,ST2) shows a fine ash size matrix whereas the lower unit (ST3) shows a slightly grosser micropumiceous texture. In both facies a highly developed process of zeolitization can be recognized but while in the upper part the chabassite content is higher than the phillipsite content the opposite is true for the lower unit. In the lower unit a process of alteration into clay minerals has also accurred since a moderate amount of smectite is often found. Such alteration does not seem to depend on the chemical and mineralogical characteristics of this rock but more probably on the chemico-physical properties of the groundwater since this part of the formation is located below the water table.Phenoclasts of sanidine, pyroxenes, mica and sedimentary and lava lithics are found in the matrix. The colour of the formation is yellow but light or dark gray facies can also be seen in some quarries.

7) Another formation originated from the Sabatini complex,called the Magliano tuff (SM), has also been studied (Nappi et al., 1979). This tuff has a limited extension and can be found in an area of about 5 Km^2 Northwest of the Sacrofano caldera; its maximum thickness is about 10m. The ash sized groundmass includes some pumice clasts and other lithics of small size; the color is yellow.

8) The products of the Latial Volcano are mainly pyroclastic. The Latial Volcano is a vast and complex structure which consists of a series of edifices superimposed over each other. The pyroclastic flows have formed a shield volcano over which the large Artemisio Caldera is located. After the caldera formation a series of hydromagmatic eruptions took place from many large centers originating, among other products, the "Albano peperino" which is widely used as construction material. Such products are located over an area of more than 70 Km^2 around the Nemi and the Albano lakes just South of Rome. The "Albano peperino", studied by Berry and

Sciotti (1974), has a thickness of about 100m near the eruptive center and of a few meters in the distal area. A common feature of these products is the presence of lava and sedimentary lithics enclosed in the groundmass. In some places at the base it may contain a large amount of lithics and can be then defined as a volcanic breccia. Inversely and directly graded ashes levels showing wave structure with impact sags are also found in the basal part of the deposit. Generally, except for the upper part, the formation is lithified by a highly developed process of zeolitization. Cementation by secondary calcite is also recognized.

9) The Phlaegraean Fields volcanic complex consists of a submarine volcanic structure over which a subaerial caldera is presently found. The caldera was formed by collapse after the eruption of the Campanian igmimbrite which occurred about 30.000 years ago. Later a series of hydromagmatic eruptions inside the caldera originated first the so called "Neapolitan yellow tuff" about 10.000 years ago and then monogenetic volcanic structures (Nappi and Ardanese,1981). The Campanian ignimbrite covers an area of at least $70.000 Km^2$ of which only an area of $500 Km^2$ outcrops (Barberi et al., 1978). The accessible maximum thickness of the formation is about 60 m in the distal area near Sorrento. In the proximal area the formation shows a vertical zoning with a basal well welded "piperno" facies, and an upper facies of "gray or red and yellow tuff". The piperno facies, not considered in the present paper, is a dark gray lava-like rock which includes collapsed scoria oriented in the flowage direction. It seems that in the piperno facies the high temperature of deposition gave rise to a devitrification process causing a secondary formation of alkali felspar. In the upper levels the clasts of the groundmass are only slightly elongate and the vesicles have usually an almost spherical shape; here the variation from the gray to the yellow and red facies is due to the zeolitization processes which have lithified the tuffs. The zeolites are represented by phillipsite and chabassite in varying ratio.

10)The "Neapolitan yellow tuff" after the Campanian ignimbrite is the most widespread pyroclastic formation of the region covering an area of $180 Km^2$; its thickness is about 150m near the feeding fissures and decreases with distance. The basal levels are pyroclastic surge deposits with planar or sand wave structures. The upper levels are represented by many flow cooling units. It is generally lithified with a micropumiceous to ash sized groundmass including small rounded pumice clasts and few lithic fragments. Lithification is due to a well developed zeolitization process shown by the abundant presence of phillipsite and chabassite (Lirer and Munno, 1975).

PHYSICAL AND MECHANICAL CHARACTERISTICS OF THE TUFFS

The volcanic products described above have been studied by several authors; the summary of the results is shown in Table 1. Uniaxial compression tests have been carried out in all cases on cylindrical samples with H = 2D. Tensile strength has been measured using the indirect Brasilian test.

Nappi and Ottaviani (1985) have studied the Western Vulsini products taken from four very active quarries representing different materials. Compression and tensile tests were carried out on naturally dry and on saturated samples. The Sorano yellow tuff, VW3a, shows much higher strength than that of the other tuffs in the area. Such high values of both tensile and compression strength are essentially due to the fine grain size and to the advanced zeolitisation process wich is also shown by the low value of specific gravity γ_s (γ_s of chabassite = $2.10 \ g/cm^3$) and of porosity,n, indicating the presence of zeolites also inside the pores. The average values of σ_f for the "Montarone yellow tuff", VW3b, are lower than for VW3a because of larger grain size and of a less developed zeolitisation process shown by the higher values of γ_s and n. The σ_f values show a larger scattering for VW3b than for VW3a. The "red tuff with black scoria", VW2, has the lowest strength characteristics. This is clearly due to the high values of porosity and to the large size of the pumice clasts present in the matrix where an initial stage of alteration into clay minerals can be also observed. For the same reason a large scattering of the σ_f values has been found (±50%). The "Nenfro gray tuff" shows medium strength characteristics even though its porosity is rather high. This can be explained, as mentioned earlier, by the primary lithification process which occurred in this rock due to the

TABLE 1

Tuff	γ_d (g/cm³)	γ_s	n%	σ_f	σ_t (kg/cm²)	$\bar{\sigma}_f$	$\bar{\sigma}_t$	$E \cdot 10^3$	c	φ	Reference
VW1	1.19	2.52	53	44	7.2	40	6.5	20	-	-	Nenfro, (Nappi and Ottaviani, 1985)
VW2	1.07	2.38	55	25	4.1	22	-	18	-	-	red tuff with black scoria
VW3a	1.36	2.25	39	105	11.8	78	9.5	39	-	-	yellow tuff, Sorano
VW3b	1.32	2.48	47	52	7.8	42	-	27	-	-	yellow tuff, Monterone
VE1	1.34	2.45	45	68	9.0	46	5.6	-	-	-	Orvieto A, (Nappi et al. 1982)
VE2	1.18	2.45	51	36	5.6	23	3.6	-	-	-	Orvieto, average B and C
VE3	1.11	2.46	54	25	4.1	18	2.2	-	-	-	Orvieto, average two quarries
VE4	1.07	2.52	57	28	4.5	22	4.5	17	-	-	Orvieto, (Manfredini et al. 1980)
ST1	1.20	2.50	52	29	4.7	19	1.5	-	-	-	V. Tiberina average lower levels (Nappi et al., 1979)
ST2	1.34	2.51	46	99	9.3	55	3.3	-	-	-	V. Tiberina average medium levels
ST3	1.25	2.44	48	72	8.0	36	3.5	-	-	-	V.Tiberina average upper levels
SM	1.37	2.49	44	91	12.0	58	5.7	-	-	-	Magliano tuff
CP1	2.11	2.63	20	214	21.6	-	-	93	95	29°	Viterbo gray peperino (Berry and Sciotti, 1974)
CP2	1.93	2.59	25	220	25.8	-	-	92	101	34°	Viterbo red peperino
LP	1.95	2.67	27	267	31.6	-	-	165	78	28°	Albano peperino
FN1a	1.00	2.49	59	35	-	-	-	8	10	27°	Yellow neapolitan tuff, T1a Pellegrino (1970)
FN1b	1.10	2.44	55	58	-	-	-	15	18	27°	yellow neapolitan tuff, T1b
FN2	1.40	2.45	43	130	-	-	-	30	35	28°	yellow neapolitan tuff, T2
FC	1.18	2.44	52	53	6.7	-	-	25	-	-	Campanian ignimbrite, present work

high emplacement temperature. For all the tuffs the strength values decrease by about 20% when the samples are saturated. The average value of the elastic modulus E measured in the central part of the σ-ε curves is highest for VW1a but the ratio E/σ_f is lower for VW1a than for the low strength VW2. Fig. 2 shows stress-strain curves for the Western Vulsini products. The curves are essentially linear except at low σ_c and near failure. The E values in the initial part of the curves are lower than at higher σ_c. This behaviour is commonly found for highly porous rocks and is due to the closing of the smaller pores and to the breakdown of the weakest particle bounds. This phenomenon can be clearly recognized in the loading-unloading curves of fig. 3. In fact, only during the first cycle permanent deformations take place whereas during the second cycle the strains due to loading are fully recovered in the unloading stage. Several samples tested for about 20 cycles to 70% σ_f did not show appreciable strength reduction due to fatigue phenomena. The tensile strength, as expected, is always much lower than the compressive strength as the ratio σ_f/σ_t varies between 6 and 9.

Nappi et al. (1982) have studied the Orvieto ignimbrite. Samples have been taken at different levels on the cliffs of the Orvieto hill (VE1 and VE2) and in two quarries nearby (VE3). From the results shown in Table 1, it can be observed that quite different physical and mechanical characteristics have been found within the same lithological unit. A considerable reduction of compressive and tensile strength and of elastic modulus has been observed for samples tested in

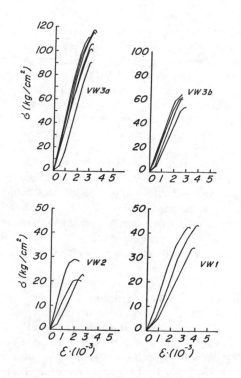

Figure 2. Stress strain curves obtained
for western Vulsini tuffs.
(from Nappi and Ottaviani, 1985)

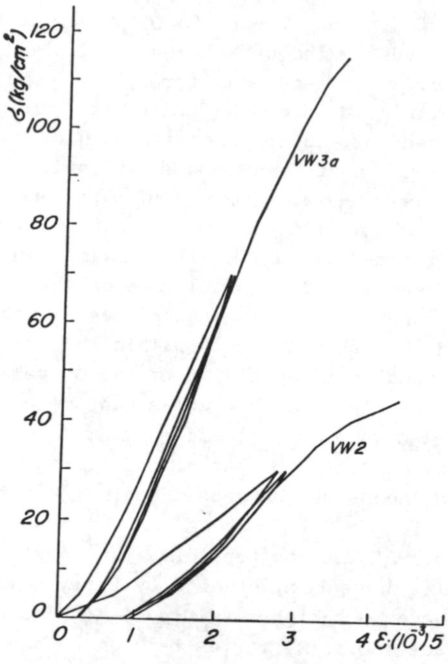

Fig. 3. Loading unloading cycles for
western Vulsini tuffs.
(from Nappi and Ottaviani, 1985)

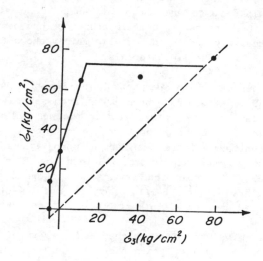

Fig. 4. Limit strength curve for Orvieto
tuff VE4.
(from Manfredini et al. 1980)

saturated conditions. The ratio σ_f/σ_t varies between 5 and 8. The tuffs of the Orvieto hill have been also investigated by Manfredini et al. (1980) who found for VE4 characteristics similar to VE2 and VE3. On more general terms, the peculiar behaviour of these pyroclastic rocks has also been shown. Fig. 4 shows the limit strength curve obtained for the Orvieto tuff. The curve can be subdivided into three parts for each of which the rock presents a different behaviour. The first vertical part corresponds to failure due to tensile stresses. In the second part the strength increases with the confining pressure according to the Coulomb criterion. As the stresses reach a critical value, failure occurs for structural collapse and the strength is independent from the confining pressure; in this stage distruction of the original texture of the rock takes place. The influence of the loading velocity has also been studied (Manfredini et al. 1980) carrying out a series of tests with load applied at a rate of 0.3 and of 0.003 kg/cm^2 per second. It was found that within this range there is no appreciable influence of the loading rate. A reduction of about 30% of the σ_f value was found for saturated samples. The values of E reported in Table 1 for VE4 are secant moduli computed at 50% σ_f. The tensile strength has been obtained also by direct tensile tests. As expected, a rather high scattering of results was found. But, contrary to what is usually found for most rocks, a good agreement with the results obtained by means of the Brasilian indirect tests was found in this case.

Two types of Viterbo peperino, the gray one CP1 and the red one CP2, together with the Albano peperino LP, have been studied by Berry and Sciotti (1974). These rocks, due to their genetic process, have much higher values of γ_d and much lower values of n than the tuffs described so far. In fact, porosity varies between 20 and 30% and γ_d between 1.9 and 2.2 g/cm^3. Samples CP1 and CP2 have been tested in two directions one parallel, $\alpha = 90°$, and the other normal, $\alpha = 0°$, to the attitude of the typical black fiamme present inside this rock, in order to find whether this characteristic texture could lead to anisotropic mechanical properties. The results of the tests have shown that indeed σ_f for the vertical samples, $\alpha = 90°$, is 20% higher than for the horizontal samples. Similar results have been obtained for tensile strength from brasilian tests. Also the Young modulus shows anisotropy with values 50-70% higher for the vertical samples than for the horizontal samples. This behaviour is due to the fact that such rocks have never sustained high lithostatic loads. The LP samples show the highest values of strength and modulus, probably due to secondary cementation by calcite. For CP1 and CP2 a slight increase of strength has been observed at increasing depth. Triaxial compression tests were also carried out for the three types of peperino tuffs at increasing values of confining pressure σ_3 (25, 50 and 100 kg/cm^2). It was found that at low confining pressure the internal angle of friction is high but decreases as σ_3 increases until a constant value is reached for $\sigma_3 \geq 25$ kg/cm^2. The values of φ and c obtained in this range are shown in Table 1. No significant anisotropic behaviour has been observed from these tests. No brittle behaviour has been observed at failure; in fact the samples were not extensively fractured but have mainly shown failure along a single inclined plane as it usually occurs for ductile materials.

The yellow tuffs of the Via Tiberina, which are part of the Sabatini Vulcanic products, have been studied by Nappi et al. (1979). Samples have been taken from several quarries, considering the areal distribution of the volcanite, in the lower and in the upper levels of the formation. In the lower level quarries (ST1) the values of unit volume weight $_d$ are lower than in the upper level (ST3) whereas the contrary is true as far as porosity values are concerned. As expected, the average value of σ_f is much higher for ST3 than for ST1 but the highest values were found for the middle level ST2. The saturated samples have shown a sharp strength drop of more than 30%. The tensile strength is about 1/6-1/10 of σ_f and the lowest values were found for ST1. The values of $\bar{\sigma}_t$ for saturated samples are almost 50% of σ_t for dry samples.

The Magliano tuffs (SM) present values of γ_d higher and values of n lower than the via Tiberina tuffs. This is because of the smaller grain size of the rock, the fewer pumice clasts included, the more homogeneous texture and the lower degree of alteration of the minerals and of the glass of the rock. The strength values shown in Table 1 reflect clearly such characteristics.

Pellegrino (1968,1970) has measured the physical and mechanical characteristics of the "yellow neapolitan tuff" in several sites. He found that at least three types of tuffs with different properties could be recognized (see table 1). Samples FN1a and FN1b show a porosity of 55-60% and a micropomiceous matrix with pumice clasts inclusion of about 1 cm size. Samples FN2 show instead lower values of porosity, smaller values of pore size, and less and smaller inclusions. The mechanical properties of these products show, consequently, as already seen for VW1a and VW1b, higher values of σ_f for FN2 than for FN1. Pellegrino (1970) has also carried out hydrostatic compression tests showing that at high pressure (100 kg/cm^2) a complete breakdown of interparticle bounds occurs and the material behaves then as a soil. Triaxial tests show that the angle of friction φ does not vary appreciably for the different types of tuff but the cohesion c is much higher for FN2 than for the other samples. It seems reasonable to explain this finding assuming that the cementation process due to formation of zeolites was much more developed for FN2 than for the other tuffs.

The authors of the present paper have studied the Campanian ignimbrite (FC) taking samples in a wide area around the town of Caserta. The tests show a considerably homogeneous behaviour of this formation which has properties very similar to those of the Orvieto tuff (VE1 and VE2) for which the same emplacement conditions have been recognized.

DISCUSSION OF THE RESULTS

It has been shown that the geotechnical characteristics of the tuffs taken into consideration in the present paper vary within a wide range. This fact is not surprising given the quite different types of genesis and diagenesis undergone by the volcanic products considered. On the basis of what has been observed above, it seems reasonable to state that in fact eruptive mechanism, emplacement conditions and lithification processes are the main factors which determine the mechanical behaviour of the tuffs. In fact, two types of eruptive mechanism have been recognized for the pyroclastic products considered. The magmatic type for which large size clasts are generated by pulverization of a high temperature molten magma. In this case the clast temperature is usually high but can vary considerably decreasing, for istance as the distance from the eruptive center increases. When their temperature is very high, the clasts are deposited still in an almost liquid state and can therefore be welded to each other causing a primary lithification process practically contemporary to deposition. This is the process undergone by the Nenfro tuff and by the Cimino ignimbrite. When the temperature is low the clasts are deposited prevalently as solid fragments and lithification is due to secondary processes such as formation of zeolites. This process is typical of the red tuff with black scoria, the Orvieto ignimbrite and the Campanian ignimbrite. The other eruptive mechanism is of the hydromagmatic type for which because of the interaction with water the clasts generated by pulverization are deposited at low temperature and are usually very fine grained. The clasts are

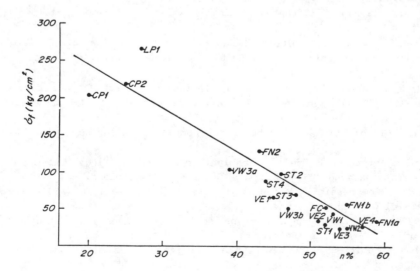

Figure 5. Plot of average porosity vs. average compressive strength for the tuff studied.

always deposited in the solid state and thus the lithification processes are secondary and very well developed, due to the fine size of the clasts which favors a high degree of zeolitization. To this group belong the Vulsini yellow tuff, the Magliano tuff, the yellow Neapolitan tuff and the Albano peperino which also shows a highly developed process of cementation by calcite. The yellow tuffs of Via Tiberina show instead an eruptive mechanism of intermediate characteristics. However, since emplacement temperature and grain size can vary with the distance from the eruptive centers it is possible to find considerable variations of geotechnical characteristics even within the same lithological unit.

From the previous observations, the one physical parameter which can be assumed to represent the various processes undergone by the tuffs seems to be the porosity which depends both from the original grain size of the products and from the secondary lithification processes. Fig. 5 shows that there is a clear linear relationship between porosity and compressive strength for the tuff studied. The relation can be expressed as: $\sigma_f = 360 - 5.8\, n\%$ which is of course suggested only in the observed range of n. The regression coefficient has been found to have in this case a value of 0.94.

A final observation can be made on the influence of the saturation condition on the values of σ_f. A considerable average reduction of about 30% of σ_f has been in fact observed for saturated samples with respect to dry samples for the samé tuff. Such behaviour seems to be due either to "stress corrosion" phenomena by the water which can ease the propagation of microfissures during the failure process and/or to capillary tension effects in the naturally dried samples.

REFERENCES

Barberi F., Innocenti F., Lirer L., Munno R., Pescatore T., Santacroce T. (1978): "The Campanian Ignimbrite: a Major Prehistoric Eruption in the Neapolitan Area (Italy)". Bull. Volcanologique, 41.

Berry P., Sciotti M. (1974) "I peperini del Lazio" Proc. 1st Intern. Congress on Exploitation of Industrial Minerals and Rocks, Torino.

De Gennaro M., Franco E., Colella C., Aiello R. (1984): "Estimation of zeolite content in Neapolitan yellow tuff. II phillipsite and chabassite in typical Phlaegraean Deposits", Rend.Soc.It.Min.Petr., 39.

Lirer L., Munno R. (1975): "Il tufo giallo napoletano (Campi Flegrei)". Period. Mineral., 44.

Manfredini G., Martinetti S., Ribacchi T., Sciotti M. (1980): "Problemi di stabilità della rupe di Orvieto", XIV Convegno di Geotecnica, Firenze.

Nappi G., De Casa G., Volponi E. (1979): "Geologia e caratteristiche del tufo giallo della Via Tiberina", Boll.Soc.Geol.It., 98.

Nappi G., Ardanese L.R. (1981): "Il freatomagmatismo nei Campi Flegrei e nell'isola di Lipari". Rend.Soc.Geol.It., 4.

Nappi G., Chiodi M., Rossi S., Volponi E. (1982): "L'ignimbrite di Orvieto nel quadro dell'evoluzione Vulcano-Tettonica dei Vulsini orientali. Caratteristiche geologiche e tecniche", Boll.Soc.Geol.It., 101.

Nappi G., Ottaviani M. (1985): "Caratteristiche geologiche e geotecniche dei tufi coltivati nei Vulsini occidentali" II Convegno Nazionale Attività estrattiva dei minerali di seconda categoria, Bari.

Nappi G., Lardini D. (1985): "I cicli eruttivi del complesso vulcanico Cimino" Rend. Soc. It. Min.Petr., 40.

Nappi G.,Martini M.,Marini I. (1986): "Turbulent pyroclastic flows: specific emplacement characteristics". International Volcanological Congress. New Zealand.

Pellegrino A. (1968): "Compressibilità e resistenza a rottura del tufo giallo napoletano", IX Convegno Nazionale di Geotecnica, Genova.

Pellegrino A. (1970): "Mechanical behaviour of soft rocks under high stresses". Proc. 2nd Intern.Congress of Rock Mechanics, Belgrado.

Sparks R.J. (1975). "Stratigraphy and geology of the ignimbrites of Vulsini Volcano, Central Italy. Geol. Rundsch, 64.

Walker G.P.L. (1983): "Ignimbrite types and ignimbrite problems". Journ.Volc.Geoth.Res., 17.

Anisotropic geotechnical properties of a residual mica schist saprolite

Les caractéristiques géotechniques d'anisotropie d'un micaschiste saprolite résiduel

O.Ogunsanwo, *Department of Geology, University of Ife, Ile-Ife, Nigeria*

ABSTRACT: The consolidation and shear strength characteristics of a residual schist saprolite for different inclinations of the planes of schistosity are compared by considerations of their coefficients of anisotropy. The coefficients of anisotropy for the consolidation and shear strength parameters determined for each locality varied between 0.3 and 2.3; 0.5 and 1.4 respectively. The geotechnical implications of these coefficients are assessed in a terrain where the inclinations of the planes of schistosity vary rather unpredictably.

RÉSUMÉ: Les caracteristiques (la consolidation et la résistance tondeuse) d'une mica schiste saprolite résiduelle pour des inclinations differentes des plans de schistosité sont étudiés selon leurs coéfficients d'anisotropie. Pour chaque endroit, les coefficients d'anisotropie pour les parametres de la consolidation et de la résistance tondeuse correspondants, varient entre 0.3 et 2.3 pour la consolidation et entre 0.5 et 1.4 pour la resistance tondeuse. La signification géotechnique de ces coéfficients sont évaluées dans un terrain ou les inclinations des plans de schistosité varient assez irregulierement.

1 INTRODUCTION

According to Arthur and Menzies (1972), Casagrande and Carillo (1944) were probably the first to model strength anisotropy in soils. In their model, they distinguished between inherent and induced anisotropy, suggesting that anisotropy may be present before the soil is strained or may be induced by the straining process. Inherent anisotropy was therefore defined as a physical characteristic inherent in the material and entirely independent of the applied strains. Induced anisotropy was defined as a physical characteristic due exclusively to the strain associated with an applied stress.

Although these definitions were intended for strength, they are equally useful for packing descriptions and also for stress-strain relationships (Arthur and Menzies 1972). A theoretical consideration of the fundamentals of anisotropy in soils was proposed by Maini (1979). In it, the limitations of the elastic theory with regards to the stress-deformation behaviour in soils are examined.

1.1 SOIL STRUCTURE

Mica schists, forming part of an amphibolite complex, occur extensively in the Ife-Ilesha district of Southwestern Nigeria. Where these rocks do not outcrop, they are overlain by lateritic soils which have been derived from the schistose rocks. The depth of weathering is often great (greater than 10m in some places) and the lateritic soils bear schistose structures to considerable heights. It is for this reason that the soils are called residual schist saprolite. The anisotropy exhibited by these lateritic soils may be termed 'residual' since it has been formed by the alteration of the anisotropic parent schistose rocks.

Generally, the dips of the planes of schistosity of the residual schist saprolites are highly variable in an unpredictable manner. It is under these conditions that constructions such as roads and bridges which require careful extrapolations of soil test results have been carried out. In this paper, samples of un-

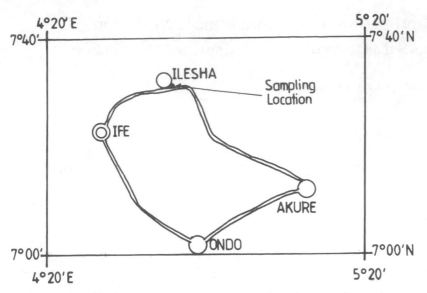

Figure 1. Sampling location of the residual mica schist saprolite.

Figure 2.   The residual mica schist saprolite showing planes of schistosity.

disturbed residual schist saprolite have been tested with varied angles of the planes of schistosity and the results thus obtained are discussed.

2   MATERIALS AND METHODS

The undisturbed soil samples were taken from a nearly vertical cut on the left side of the road on Ife-Akure Road (30 km from Ife) just on the outskirts of Ilesha in Nigeria (Figure 1). Depth of the roadcut is 8.2m and the depth of sampling is 6.5m. In the sampling area, the planes of schistosity within the residual schist were nearly horizontal (Figure 2).

The oedometer, shear box and the classification tests were carried out following the procedures laid down in the British Standards B.S. 1377 (1975). Undisturbed soil samples in which the planes of schistosity were perpendicular to the axis of compression were designated V, those in which the planes were parrallel to the axis of compression: H, and those in which the planes were inclined (45%) to the axis of compression: I (Figure 3).

The residual schist saprolite is sandy silt, the fractional percentages being Gravel: 0; Sand: 35; Silt: 60; and Clay: 5 (Figure 4).

466

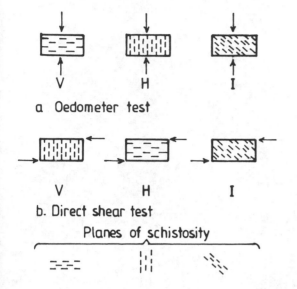

a Oedometer test

b. Direct shear test

Planes of schistosity

Figure 3. Oriented sample designations.

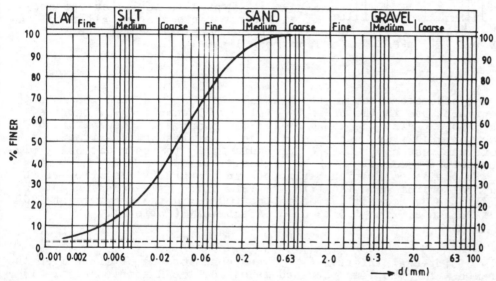

Figure 4. Grain size distribution of the residual mica schist saprolite.

Liquid limit   = 42.2%   Plastic limit = 35.7%   Index of Plasticity =6.5%
Linear Shrinkage = 2.3%   Specific Gravity 2.76   Bulk Density   =1881kg/m3

3 RESULTS AND DISCUSSIONS

3.1 CONSOLIDATION

Figure 5 shows the e-log $\bar{\sigma}_v$ curves for oriented samples from locality 1. This is
typical for the localities. Results of the consolidation parameters obtained for
all the localities are on Table 1.
   All the soil samples, irrespective of their orientations exhibit "critical pressures" (Wallace 1973) in the shapes of their e-log $\bar{\sigma}_v$ curves. Since the samples
are undisturbed, the "critical pressures" are synonymous to the past maximum consolidation pressures ($\bar{\sigma}_{v_m}$) experienced by the samples. The estimated natural
overburden (in-situ) pressure ($\bar{\sigma}_{v_o}$) on the soil samples = 122 kPa (depth of sampling x bulk density). The past maximum consolidation pressures obtained (Table 1)
show that for each locality, the V samples have the least values with the H samples
having the greatest. All the samples are however lightly overconsolidated.
   De and Furdas (1973) showed that the dislocation of the weak bonds in soils is
responsible for the shapes of the e-log $\bar{\sigma}_v$ curves and hence the "critical press-

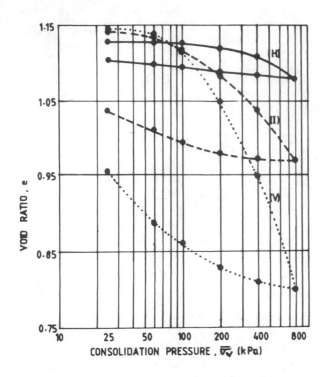

Figure 5. Typical e-log$\bar{\sigma}_v$ curves obtained for the various orientations of samples of the residual **schist** saprolite from near Ilesha.

ures" obtained in the oedometer tests. It then follows that the weak bonds in the residual schist saprolite are in the order H samples > I samples > V samples. This bond approach also explains the trend of the compression indices obtained (Table 1). Compression index (Cc) was least in the H samples and greatest in the V samples. The soil was most compressible perpendicularly to the schistosity planes and least compressible parallel to the schistosity planes.

The differences in the consolidation parameters in the oriented samples of the residual schist saprolite from each locality are very large. The coefficients of anisotropy (C.A.) for the overconsolidation ratios (OCR) and the compression indices (Cc) vary from 0.6 to 2.3 and 0.3 to 2.3 respectively (Table 1).

3.2 SHEAR STRENGTH

The values of angle of internal friction ($\varphi°$) and cohesion (c) obtained from the shear tests are greatest for the V samples and least for the H samples (Table 1) for each locality. This is due to the fact that in the V samples, shearing was across the planes of schistosity while in the H samples, shearing was along the planes of schistosity. The differences in the values of angle of internal friction for each locality are small (coefficients of anisotropy vary between 0.8 and 1.3) but the differences in the cohesion values are appreciable (coefficients of variation vary between 0.5 and 1.4) (Table 1).

Considering that this residual schistose soil consists of 35% sand, 60% silt and 5% clay and that all tests were performed on the undisturbed soil, the cohesion values obtained are quite appreciable (14 – 40 kPa). This appreciable cohesion is attributed to the presence of sesquioxides of iron in the schistose soil. They are responsible for its colour (Hue 10YR 8/6 Yellow to Hue 10 YR 5/6 Red on the Munsell Colour chart). It is on this basis, that the soil has been termed lateritic. Lohnes and Demirel (1973) have proposed a conceptual model of how the presence of sesquioxides in lateritic soils enhances their cohesion.

4   CONCLUSION

In this terrain, the soil is schistose and the angles of dips of the planes of schistosity vary unpredictably. For engineering construction work, it will be necessary to determine the consolidation and shear strength parameter values of the soil. Quite often, samples are taken from selected locations and tested. The results obtained are then extrapolated to cover some area.

This study shows that the results of consolidation and shear tests on the undis-

Table 1. Consolidation and shear strength parameter values and coefficients of anisotropy obtained for the various orientations of samples of the residual mica schist saprolite.

Localities	Orientations	σ_{vm} (x 10^2 kPa)	O.C.R. (C.A.)	Cc (C.A.)	c(kPa) (C.A.)	$\varphi°$ (C.A.)
1	V	1.3	1.1 (0.6)	0.50 (2.3)	40 (1.3)	9 (1.3)
	I	2.1	1.7 (1.0)	0.22 (1.0)	30 (1.0)	7 (1.0)
	H	4.7	3.9 (2.3)	0.10 (0.5)	16 (0.5)	7 (1.0)
2	V	3.4	2.8 (0.9)	0.23 (1.6)	35 (1.4)	9 (1.1)
	H	3.6	3.0 (1.0)	0.14 (1.0)	25.5 (1.0)	8 (1.0)
3	V	1.8	1.5 (0.6)	0.25 (1.6)	20 (1.4)	12 (1.0)
	H	3.2	2.6 (1.0)	0.16 (1.0)	14 (1.0)	11.5 (1.0)
4	V	2.4	2.0 (1.0)	0.39 (1.6)	34 (1.3)	16 (1.1)
	I	2.5	2.0 (1.0)	0.25 (1.0)	26 (1.0)	15 (1.0)
	H	4.6	3.8 (1.9)	0.08 (0.3)	25 (1.0)	11.5 (0.8)
5	V	1.8	1.5 (0.8)	0.33 (1.4)	25 (1.1)	10 (1.1)
	I	2.2	1.8 (1.0)	0.24 (1.0)	23 (1.0)	9 (1.0)
	H	2.9	2.4 (1.3)	0.18 (0.8)	18 (0.8)	7 (0.8)
6	V	2.1	1.7 (0.6)	0.25 (1.5)	28 (1.0)	12 (1.1)
	H	3.3	2.7 (1.0)	0.17 (1.0)	27 (1.0)	11 (1.0)
7	V	2.6	2.1 (0.8)	0.24 (1.5)	38 (1.2)	15 (1.2)
	I	3.1	2.5 (1.0)	0.16 (1.0)	32 (1.0)	12.5 (1.0)
	H	4.3	3.5 (1.4)	0.12 (0.8)	30 (0.9)	11 (0.9)

turbed residual schist saprolite differ appreciably for different inclinations of the planes of schistosity. It is therefore of great importance that the inclinations of the planes of schistosity are noted for every sample tested and should be borne in mind in extrapolating the results obtained.

REFERENCES

Arthur, J.F.R. & B.K. Menzies 1972. Inherent anisotropy in sand. Geotechnique 22 (4): 115 - 128.
B.S. 1377 1975. Methods of testing soils for civil engineering purposes. British

Standards Institution.

Casagrande, A. & N. Carillo 1944. Shear failure of anisotropic materials. Proc. Boston Soc. of Civ. Engrs. 31: 74 - 87.

De P.K. & B. Furdas 1973. Discussion on Wallace (1973). Geotechnique 23 (4): 601 - 603.

Lohnes R.A. & T. Demirel 1973. Strength and structure of the laterites and lateritic soils. Eng. Geol. 7: 13 - 33.

Maini K.S. 1979. Anisotropie in Böden, Festschrift zum 65 Geburtstag von o.Prof. Dr.-Ing. Richard Jelinek, T.Ü. München : 265 - 285.

Wallace K.B. 1973. Structural behaviour of residual soils in the continually wet highlands of Papua New Guinea. Geotechnique 23 (2): 203 - 218.

The creep model of the intercalated clay layers and the change of their microstructure during creep

Modèle de fluage de bancs intercalaires d'argiles et la modification de leur microstructure pendant le fluage

Xiao Shufang, Wang Xianfeng, Cheng Zufeng & Nie Lei, *Changchun College of Geology, China*

Abstract: In this paper, the behavior and the experimental equations of creep of intercalated clay layers distributed in the Permian sandstone-shale system have been studied. The natural microstructure and, in particular, the changes of microstructure of intercalated clay layers during different stages of creep have been examined by electron scanning microscope. The model which describes the whole process of creep is given. This model would make a contribution to a better understanding of the creep mechanism and predicting the long-term behavior of creep by using the short-term testing results.

RÉSUMÉ: Dans cet papier, la caractéristrique et les équations expérimentales du cheminement des tendeurs argiles intercalatés dans le systèm du strate grès-schistesont été etudié. La microstructure naturelle et, en particulier, le changement de la microstructure de tel intercalation durant cours diffirents du cheminement a été examiné par électron microscope. Le modéle difinant le procédé entier du cheminement est rendu. Cet modéle contribuera à meilleur comprendre du mécanisme du cheminement et predir la trait du cheminement de longue durée par employer resultates des expériments à court terme.

The intercalated clay layer is a weak intermediate clay layer with high water content, which was formed from the thin soft clayrock crushed by bedding slip and weathered by the physical-chemical process of groundwater. The existence of intercalated clay layers brings a serious threat to the stability of the slop or the foundation of dam.

The special genesis of intercalated clay layers makes their microstructure and mechanical properties differ from matural clay. Because of the difficulty in cutting the undisturbed samples, so far the study on this subject is insufficient.

We have cut down a number of undisturbed samples with success from the sandstone-shale system by means of the cutting machine which was designed by ourselves, conducted the drained shear creep test to study the creep behavior and equation of them and, in the meantime, examined the change of the microstructure during creep by the electron scanning microscope so as to make a series of typical pictures of microstructure which are corresponding to every stages of creep. According to the results of the test, we drew a complex model of creep.

1 The creep curves and experimental equations of intercalated clay layers

1.1 The creep curves

After analysing 10 sets of testing curves at different stress level, we can sum them up as three types (Fig. 1)

1. The stable creep curve: It is the typical curve while the stress level is low.

2. Type A curve: It consists of four sections, i.e. the instantaneous strain, the primary creep, the steady state creep and the accelerated creep. The strain rate at steady state creep is the minimum which is denoted by the signγ_{min}, and the time at the beginning and terminal of the steady state creep are expressed respectively as t_{min1} and t_{min2}. While the intercalated clay layer is quite thin, its creep curve apperars to be the type A.

3. Type B curve: There is no steady state creep in the whole process of creep,

the primary creep is followed directly by accelerated creep. The plot turns from concave-downward to concave-upward. The strain rate at the turning point is the minimum. It is the type of curve while the clay layer is comparatively thick.

1.2 The experimental equations

It is obvious that the total strain is the sum of the strain of each stage, i.e.

$$\gamma = \gamma_0 + \gamma_1 + \gamma_3 \qquad (\text{ for Type A })$$

$$\gamma = \gamma_0 + \gamma_1 + \gamma_2 + \gamma_3 \qquad (\text{ for Type B })$$

Where
$\gamma_0 =$ the instantaneous strain
$\gamma_1 =$ the primary creep
$\gamma_2 =$ the steady state creep
$\gamma_3 =$ the accelerated creep

The curve fitting expressions for each stage of creep can be obtained by regression analysis.

1. The primary creep γ_1

For Type A, the curve fitting expressions are power laws of the form

$$\gamma_1 = at^n$$

Where
$t =$ time
$n =$ constant, less than 1
$a =$ constant

For Type B, the strain-time behavior can be approximated by the following

$$\gamma_1 = [1 - \exp (-bt)]$$

2. The steady state creep

For Type A, the creep could be represented by the equation

$$\gamma_2 = \dot{\gamma}_{min} \cdot t$$

3. The accelerated creep

So far a number of empirical equations have been developed to express the primary and steady state creep, but no simple equation has been found for accelerated creep. By applying the linearization method suggested by D. Varnes, we examined the simple relations between γ, t, $\dot{\gamma}$, t, and their logarithms, exponential and powers to see if they were linear. We found that the reciprocal rate-time ($t \sim 1/\dot{\gamma}$) plot approximate to linear in accelerated creep stage. It means that the famous pure Saito relation is fit, hence we can express the result as follow

$$\dot{\gamma} = C / (t_f - t)$$

Where
$\dot{\gamma} =$ shear strain rate during accelerated creep
$C =$ constant
$t_f =$ the time to failure, it can be estimated by extending the line to the t-axis (Fig. 2)

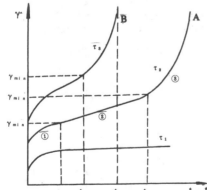

① Primary creep
② Steady state creep
③ Accelerated creep

Fig.I The creep curve of intercalated clay layer

472

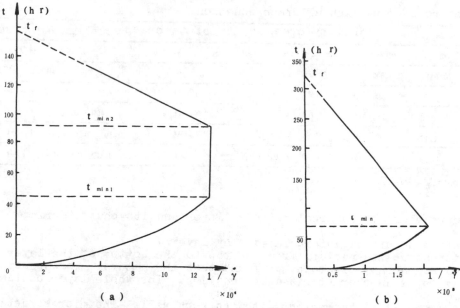

Fig. 2 The $t \sim 1/\dot{\gamma}$ curve of intercalated clay layer

Intergration of these equation yields

$$\gamma = \gamma_1 + C \ln [(t_f - t_1)/(t_f - t)]$$

Where the constant of intergration C could be determined from an observation that $\gamma = \gamma_1$ at $t = t_i$

Finally the strain-time relationship of whole process of creep was expressed in the form.

For Type A

$$\gamma = \begin{cases} \gamma_o + at^n & [0, t_{min1}) \\ \gamma_{min1} + \dot{\gamma}_{min}(t - t_{min1}) & [t_{min1}, t_{min2}) \\ \gamma_{min2} + C \ln[(t_f - t_{min2})/(t_f - t)] & [t_{min2}, t_f], \end{cases}$$

$$\dot{\gamma} = \begin{cases} an t^{n-1} & [0, t_{min1}) \\ b & [t_{min1}, t_{min2}) \\ C/(t_f - t) & [t_{min2}, t_f] \end{cases}$$

For Type B

$$\gamma = \begin{cases} \gamma_o + D[1 - exp(-bt)] & [0, t_{min}) \\ \gamma_{min} + C \ln[(t_f - t_{min})/(t_f - t)] & [t_{min}, t_f] \end{cases}$$

$$\dot{\gamma} = \begin{cases} bt[exp(-bt)] & [0, t_{min}) \\ C/(t_f - t) & [t_{min}, t_f] \end{cases}$$

Where the parameters a, b, c, D are constant when a constant shear stress is applied for any time. They are given in table 1

Depending on the data in Table 1, we can make a plot of shear strss τ against the lg t_f, and express the result as the form $\tau = 0.0382 - 0.047$ lg t_f

In order to explain these experimental creep equation, we observed the natural microstructure and its change at different creep stages.

Table 1. The parameters of experimental creep equation

Sample	τ (MPa)	t_{min} min	$\dot\gamma_{min}$ min	t_r min	primary creep $af+\gamma_0$	steady state creep bt	accelerated creep $C-D\ln(t_r-t)$
1	0.0241	1200	0.0131	3600	$1.2\times10^{-4}\ t^{0.215}+0.026$	$0.0131t$	$0.70\ 0.5\ln(t_r-t)$
2	0.0263	74	0.476	320	$2.77\times10^{-4}\ t^{0.383}+0.0397$	$0.476t$	$0.752-0.0703\ln(t_r-t)$
3	0.0283	96	0.76	158	$0.174\times10^{-4}\ t^{0.42}+0.105$	$0.760t$	$0.409-0.029\ln(t_r-t)$
4	0.03	30	1.17	59	$1.57\times10^{-4}\ t^{0.70}+0.12$	$1.170t$	$0.301-0.00204\ln(t_r-t)$
5	0.0310	16	1.79	36	$4.49\times^{-4}\ t^{0.80}+0.036$	$1.790t$	$0.312-0.0212\ln(t_r-t)$

The normal stress $\sigma = 0.08$ MPa

2 The natural microstructure of intercalated clay layer and its change during creep

2.1 The natural microstructure of intercalated clay layer

After taking the microstructure picture from the top to the bottom of the layer by means of electron scanning microscope, we found that the natural microstructure of intercalated clay layer is characterized by zonation. The whole layer could be divided into three zone

1 The top contact zone, Tt is in contact with the rock wall, The microstructure of this zone is unoriented and loose. The water content is high.

2 Oriented zone, this is a very thin zone (50-500 μm). The aggregates of clay particles are arranged as scaly and imbricate.

3 Unoriented zone, it is relatively thick and the arrangement of the aggregates is unoriented. The microstructure is honeycomb structure having higher strength bond. Moreover, we found there were some cracks and partial oriented arrangement in this zone.

Some microstructure character of each zone is shown in Table 2

Table 2. The zonation of microstructure of intercalated clay layer

Zone	Thickness ( )	The contact of aggregate	The degree of orientation	The degree of compaction	Character
Top contact Zone	800	Edge-edge (major) Edge-face (minor)	No	Very loose Porosity 50%	Weak bond No micro-crack
Oriented Zone	400	Face-face	Oriented aggre-gete 70%	Relatively compactive	Scaly and Imbricate
Unoriented Zone	5200	Edge-face	Very few	Compactive	Stronger Bond

The character of this zonation make it easy to shear along the top contact zone. But under the condition of the higher stress level or, the quite thick layer, The shear will extend from microcracks in the unoriented zone.

2.2 The microstructure change during creep

By comparing the microstructure pictures we found that

1 In the stage of primary creep, the shear strain rate decreased with time, it is because that the aggregates flex, close up, inlay, push and squeeze each other with the lapse of time. The aggregates close up each other can be verified by compacting phenomenon during shear creep. (Fig. 3)

The picture 1 shows the phenomenon of push and squeeze of aggregates. From this picture, we can see some aggregates are squeezed into the pores. The edge-edge and face-edge contacts start to turn into face-face contact gradually.

The picture 2 shows phenomenon of inlay and interlock of aggregates.

In picture 3, we can see the flex of domaid of clay particles.

All of these led up to strain-hardening. On the other hand, the increasing orientation of microstructure would make the thickness of the absorbed water film increasing and the strength of bond decrease, thus led up to strain-softening. The strain-hardening and the strain-softening exist simultaneously in the whole process of creep. But in the stage of primary creep, the strain-hardening is major.

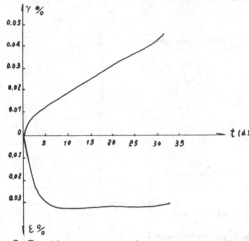

Fig. 3 The $\gamma \sim t$ and $\varepsilon \sim t$ curve of intercalated clay layer

2 In the stage of steady state creep, the strain rate is constant and the viscosity coefficient η is constant too. It means that the strain-hardening must reach equilibrium with strain softening. But it is not easy to reach such a balance unless the shearing creep progressed along the thin top contact zone, because the disturbed structure range is relatively thin, it is easier to keep the balance between strain-hardening and strain-softening for quite a long time. In such a case, the strain-time curve appears as Type A. On the other hand, in the case, of higher stress level or thicker layer, the shear progressed through whole clay layer, both strain-hardening and strain-softening were too complex to keep balance. Hence, the strain rate and the η would not be still constant. The strain-time curve appeared as Type B, i.e. no steady state creep stage.

3 The change of microstructure in accelerated creep stage was characteristic of increasing the degree of orientation. The strain-softening was major and, with the development of orientation, the strain rate increased rapidly, at the end, up to failure. The microstructure of the shearing surface arranged as scaly and imbricate.

4 The recovery of the microstructure
When we put the failure sample into the water for several days and observed the change of microstructure of the shearing surface. We found that after 10 days soaked in water, most of oriented microstructure in the shear zone was recovered, see picture 4.

3 The creep model of intercalated clay layer
In order to describe the creep behavior of intercalated clay layer, we tried to apply a complex creep model.
The construction of the model was based on the following recognization.
3.1 When the shear stress is less than the long term shearing strength   , the shear strain at t = 0 is instantaneous strain i.e. $(\gamma = \gamma_o)$ and when $t \to \infty$, the shear strain tends to a constant and the sample would not be failure. The creep behavior can be discribed by Kelvin model which give the creep in form

$$\gamma = \tau / G_1 + (\tau / G_2) \cdot [1 - \exp (-G_2 t / \eta_1)]$$

3.2 When shear stress is greater than τ_∞, the steady state creep occurs. By combining a Kelvin model and a Binghan model in series, we can obtain the complex model which can represent the primary and steady state creep stage. The equation is
$$\gamma = \tau / G_1 + (\tau / G_2) \cdot [1 - \exp (-G_2 t / \eta_1)] + (\tau - \tau_\infty) \ t / \eta_2$$

3.3 When strain reaches to the γ_{min} the steady-state creep will change into accelerated creep, the viscosity coefficient η will decrease with time.
A special dashpot having two swiveling holey piston is used to describe the change process of viscosity η . While the holes of two poston open onto each other, the η is minimum and the strain rate $\dot{\gamma}$ is corresponding to maximum. Conversly, swiving

475

one piston until the hole of two piston stagger each other, then, $\eta = \infty$ and $\dot{\gamma} = 0$
Combining the special dashpot and Kelvin-Binghan model in the way shown in Fig. 4

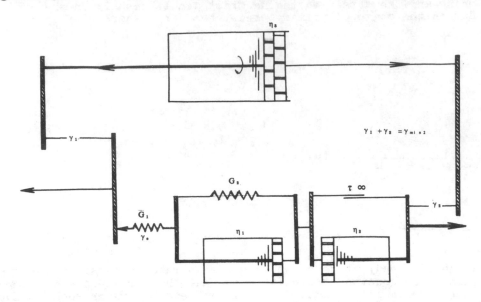

Fig. 4 The model of the whole process of creep of intercalated clay layer

From Fig. 4 we can see that when such a model is subjected to a constant stress which is less than τ_∞ , the strain will only cover the primary creep.
If $\tau > \tau_\infty$ and $\gamma < \gamma_{min_2}$ the strain produced will be the sum of the strain of Kelvin and Bingham model.
If $\gamma \geqslant \gamma_{min_2}$ strain produced will be given by

$$\gamma = \gamma_{min_2} + \tau (t_f - t) / \eta_3$$

Here the η_3 is not a constant. It is a function of the t. Having the aid of the experimental equation, we can express the η_3 in the form

$$\eta_3 = \tau t \quad [B - A \ln (t_f - t)]$$

The whole creep process of intercalated clay layer could be represented by the equation

$$\gamma = \tau / G_1 + (\tau / G_2) \cdot [1 - \exp (- G_2 t / \eta_1)] + (\tau - \tau_\infty) \quad t / \eta_2$$

$$\gamma = \gamma_{min_2} + B - A \ln (t_f - t)$$

Where γ_{min_2} is the shear strain at $t = t_{min_2}$, G_1 , G_2 , η_1 , η_2 , B,A are parameters being irrelevant to stress level and time. They can be calculated from several sets of testing curves by micro-computer. The flow diagram is given as following page The values of η_1 , η_2 , A, B are given in Table 13

Table 3. The values of parameters η_2 , η_f , A, B

	I Group				II Group		
σ (MPa)	0.057	0.070	0.092	0.120	0.063	0.085	0.104
η (P)	1.3×10^{3}	1.7×10^{13}	2.3×10^{13}	2.5×10^{13}	0.28×10^{13}	0.66×10^{13}	0.67×10^{13}
η_2 (P)	0.9×10^{13}	2.5×10^{13}	4.3×10^{13}	1.8×10^{13}	0.05×10^{13}	0.29×10^{13}	0.40×10^{13}
A (1/min)	0.13	0.0212			0.03		0.07
B	0.80	0.312			1.43		0.5

The flow diagram

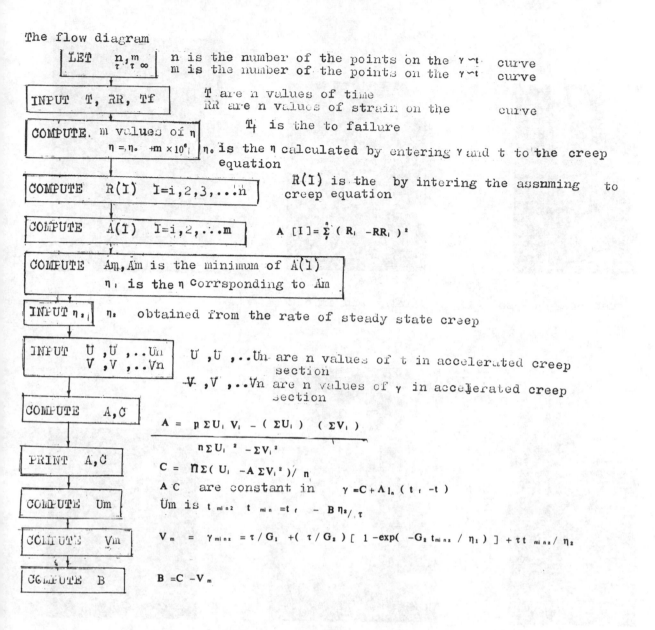

LET n, m τ, τ_∞ — n is the number of the points on the $\gamma \sim t$ curve
m is the number of the points on the $\gamma \sim t$ curve

INPUT T, RR, Tf — T are n values of time
RR are n values of strain on the curve
T_f is the to failure

COMPUTE m values of η
$\eta = \eta_0 + m \times 10^6$ — η_0 is the η calculated by entering γ and t to the creep equation

COMPUTE $R(I)$ $I = i, 2, 3, \ldots n$ — $R(I)$ is the by intering the assuming to creep equation

COMPUTE $A(I)$ $I = i, 2, \ldots m$ — $A[I] = \sum (R_i - RR_i)^2$

COMPUTE Am, Am is the minimum of $A(I)$
η_1 is the η corrsponding to Am

INPUT η_2 — η_2 obtained from the rate of steady state creep

INPUT $U, U, \ldots Un$
$V, V, \ldots Vn$ — $U, U, \ldots Un$ are n values of t in accelerated creep section
$V, V, \ldots Vn$ are n values of γ in accelerated creep section

COMPUTE A, C
$$A = \frac{n \Sigma U_i V_i - (\Sigma U_i)(\Sigma V_i)}{n \Sigma U_i^2 - \Sigma V_i^2}$$

PRINT A, C
$C = n \Sigma (U_i - A \Sigma V_i^2) / n$
A C are constant in $\gamma = C + A_{1_a}(t_f - t)$

COMPUTE Um — Um is t_{min2} $t_{min} = t_f - B \eta_2 / \tau$

COMPUTE Vm
$$V_m = \gamma_{min2} = \tau / G_1 + (\tau / G_2)[1 - exp(-G_2 t_{min2} / \eta_1)] + \tau t_{min2} / \eta_2$$

COMPUTE B — $B = C - V_m$

In conclusion, it seems to be that the model is conducive to analysising the mechanism of creep and predicting the long term behavior of creep, however, for practical purposes it is also neccessary to inquire further the relationship between the creep parameters and the index of microstructure change. This is a subject that we will deal with in the near future.

REFERENCES

Kubetsky, V.L., Certain laws of creep
Langer, M., 1983. Rheologic behavior of Rock and rock mass
Osepov, 1983. Practical application of the clay microfabric studies
Roy, 1980. Endochronic-critical state models for sand
Stepanion, G.T., 1981. Creep of a clay during shear and its rheological
Tan Jongkie, 1979. The discussion in 4th congress of ISRM
Varnes, D., 1983. Time deformation relations in creep to failure of earth materials
Wang Youlin, 1980. Character of microstructure of intercalated clay layer

1 push and squeeze   2 inlay and inlock   3   flex

4 the edge-edge contact   5 face-face contact in   6 unoriented zone
  in top contact zone   oriented arrangement

478

Engineering geological classification of fault rocks

Classification géotechnique de roches situées dans des zones faillées

Zhang Xian-Gong & Han Wen-Feng, *Department of Geology, Lanzhou University, China*
Nie De-Xin, *Chengdu College of Geology, China*

ABSTRACT: For the first time the authors put forward the idea that in engineering geology the term "fault rocks" should be used to describe the rocks directly related to rift movements instead of "structure rocks". And the authors also summarize the present situation of the classifications of fault rocks. Three basic principles are proposed for the engineering geological classification of fault rocks: 1) It should be based on the classification of fault rocks in petrology; 2) It should reflect the engineering geological features of fault rocks; and 3) It should be simple, clear and easy to be applied in practice.
The authors put forward a new engineering geological classification of fault rocks. In this classification, according to whether they are cemented, fault rocks are divided into two series: loose fault rock series and consolidated fault rock series. Each series can be further divided according to the grain size of fault rocks. Thus, fault rocks are divided into six basic types: fault cataclastic rocks, fault block rocks, unindurated fault breccias, fault gouge, fault breccias and mylonites. The basic geological characteristics and engineering geological properties of them are also introduced briefly in this article.

RESUME: L'article a premièrement proposé dans la géologie civile que le terme technique "brèche de faille" prennait la place de "tectonite" pour signifier les roches qui avaient une directe relation avec activité de faille. L'article a exposé l'état actuel de la classification de la brèche de faille et il a propose fooit principes essentiels de la classification de géologie civile de la brèche de faille: 1) Prendre la classification de la brèche de faille dans la Pétrologie comme base; 2) Prendre les caractères de géologie civile de la brèche de faille comme règle; 3) La classification doit etre simple, claire et, facile à utiliser.
L'auteur a proposé un noúveau projet de la classification de géologie civile de la brèche de faille, c'est-à-dire, selon que s'il y a une liaision dans la brèche de faille en la divisant deux systèmes: la brèche de faille dispersée et la brèche de faille compacte, de plus, selon que les grains de la brèche de faille sont gros ou menus, on ia divise six types essentiels: roches de rupture de faille, roches cataclastique de faille, non consolide brèche, argile de frottement, brèche de faille et mylonité de faille. L'article a sommairement présenté la caractéristique essentielle de la géologie et le caractère de la géologie civile de toutes sortes de brèches de faille divisées.

ORIGIN OF THE PROBLEM

Faults destroy intact rock masses and result in weak zones. The fault-formed weak zones always have very poor engineering properties. Usually, they can not satisfy the requirements of heavy buildings for foundations and need to be disposed for engineering purposes. We can say that fault zones are usually one of the fundamental factors controlling the economic respect and the safety of engineering constructions in bed rock areas. Therefore, both engineering geologists and designers should pay much attention to fault zones and it is of great importance for them to develop a unified proper classification of fault rocks for engineering geological purposes.

For a long time rocks related to faults are considered as a type of metamorphic rocks in petrology. And they are called "dynamic metamorphic rocks", or called "cataclastic rocks". In structural geology they are called "tectonites" or "fault tectonites". There are some other names but not a unified one exists. The engineering geologists of China have been using the term "structural rocks". But this term covers broad meanings and it is not limited to the rocks of fault zones. According to Sander's definition (1930), which is universally accepted in geology, it is "the irreversible constituents move in their fabrics to reflect the rocks and the rock masses" [1]. That means they are not limited to the rocks only related to rift movements. In a descriptive manner

Sander divided the structural rocks into S-structural rocks, B-structural rocks and R-structural rocks [17]. From all the above, we can see that the term "structural rocks" is not clear enough to stand for the rocks related to faults.

In 1977, Sibson R.H. put forward the term "fault rocks" [14] and he pointed out that "fault rocks" is a competitive term which describes those rocks in the shear processes of the superficial part or the deep part of the crust, the fabric of those rocks are generally considered to be formed in the shear processes. The meaning of the term is quite clear. It indicats the origin and the structural properties clearly. Some of the structural geologists in China have already accepted this term and explained it more clearly. We consider that the term "fault rocks" is suitable to use in engineering geology. It had a clear meaning and we can infer literally that the rocks are related to faults. Because the name matches the reality, confusion is avoided. So we suggest replacing "structural rocks" by "fault rocks" and hope that through discussion among engineering geologists, a unanimous opinion can be reached.

THE PRESENT SITUATION OF THE CLASSIFICATION OF FAULT ROCKS (CATACLASTIC ROCKS AND STRUCTURAL ROCKS)

Although the amount of fault rocks is not large, but they are distributed widely. Besides the petrologists' classifications, there are some other classifications made by structural geologists, seismic geologists and engineering geologists. The authors do not intend to comment on various classifications in detail, but consider it necessary to introduce several classifications which are used widely at home and abroad in order to lay a foundation for putting forward a new classification.

A) The classification based on degree of crushing and cataclastic structures

This classification belongs to descriptive classifications and it is used in geology comparatively widely. The typical classifications of this kind include Y.C. Cheng's (1963) and A. Spray's (1969).

Dr. Y.C. Cheng, the famous Chinese petrologist, divided this kind of rocks into five types according to their degree of crushing: [12] a) partly-crushed (or broken) rocks; b) cataclastic rocks; c) mylonite; d) phyllite; e) buchite.

A. Spry divided fault rocks according to cataclastic structures (table 1) [16]

Table 1

Nature of matrix		Proportion of matrix			
		0-10%	10-50%	50-90%	90-100%
Crushed	foliated massive	crush (tectonic) breccia or conglomerate	protomylonite protocataclasite	mylonite cataclasite	ultramylonite ultracataclasite
Recrystallized	minor major	hartschiefer blastomylonite			
Glassy		pseudotachylyte (hyalomylonite)			

From A. Spray, 1969, p.229.

B) Classification based on origin

It should be said that this kind of classification is the most reasonable classification. Due to different points of view, there exist many classifications. The typical one is made by Professor Wang Jiayin (1951, 1978) of Beijing University. This classification is based chiefly on stress features from the view point of geomechanics. H.P. Zeek's classification (1974) is based on the relation between crushing deformation and recrystallization. and the third one is R.H. Sibson's classification (1977), which is based on the deformation features.

Zeck divided cataclastic rocks into three classes: cataclastics, myloblastites and blastocataclastites. And each of the classes was divided into several types according to the degree of crushing. (Fig.1) [18]

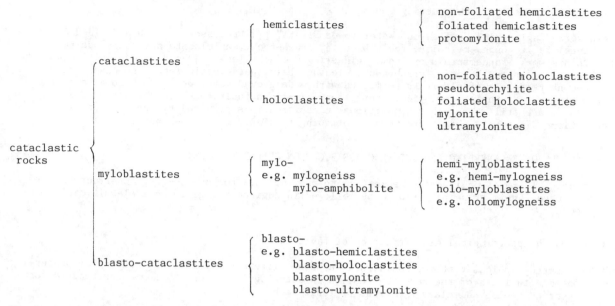

```
                                          ┌ non-foliated hemiclastites
                        ┌ hemiclastites  ┤ foliated hemiclastites
                        │                 └ protomylonite
         cataclastites ┤
                        │                 ┌ non-foliated holoclastites
                        │                 │ pseudotachylite
                        └ holoclastites   ┤ foliated holoclastites
                                          │ mylonite
                                          └ ultramylonites

cataclastic                ┌ mylo-                ┌ hemi-myloblastites
rocks      ┤ myloblastites ┤ e.g. mylogneiss     ┤ e.g. hemi-mylogneiss
                           └   mylo-amphibolite   │ holo-myloblastites
                                                   └ e.g. holomylogneiss

                                    ┌ blasto-
           └ blasto-cataclastites  ┤ e.g. blasto-hemiclastites
                                    │ blasto-holoclastites
                                    │ blastomylonite
                                    └ blasto-ultramylonite
```

Fig.1 A schematical diagram presenting the classification of cataclastic rocks (Form H.P.Zeck)

Based on fabric features (random fabric and foliationized fabric) and original cohesion, Sibson divided fault rocks into some classes. And according to the properties of the groundmass he divided the classes further into different types (approximately the same as Spray's classification standard). (Table 2) [15]

Table 2. Textural classification of fault rocks (R.H. Sibson, 1977)

		RANDOM - FABRIC		FOLIATED	
INCOHESIVE		FAULT BRECCIA (visible fragments >30% of rock mass)		?	
		FAULT GOUGE (visible fragments <30% of rock mass)		?	
COHESIVE	Glass/devitrified glass	PSEUDOTACHYLYTE		?	
	NATURE OF MATRIX — Tectonic reduction in grain size dominates grain growth by recrystallisation & neomineralisation	CRUSH BRECCIA / FINE CRUSH BRECCIA / CRUSH MICROBRECCIA	(fragments > 0.5 cm) / (0.1cm < frags. < 0.5cm) / (fragments < 0.1 cm)		0 - 10%
		PROTOCATACLASITE	PROTOMYLONITE		10 - 50%
		CATACLASITE	MYLONITE	Cataclasite Series / Phyllonite variety / Mylonite Series	50 - 90%
		ULTRACATACLASITE	ULTRAMYLONITE		90 - 100%
	Grain growth pronounced	?	BLASTOMYLONITE		

PROPORTION OF MATRIX

481

C) The classification in engineering geology

In his work "Engineering Geology for Water Power Plants" [7] Professor Gu Dezhen divided fault rocks into: fault gouge, mylonite, fault breccia, crush-rock, cataclasite and schistose rock. And later in his work "Fundamentals of Rock Engineering Geomechanics" [9] he divided fault rocks into: fault gouge, mylonite, breccia, crush-rock, cleavage belt, cataclastic rock, the cataclastic rock in fold-contorted belt, fault rocks permeated with underground water etc. He also gave a brief account of the engineering geological properties of various fault rocks in his work. Besides these, there are still some other transitional classifications proposed by Sun Yan, [6] Yang Zhuen*, Liang Jincheng**, etc. But we are not going to elaborate on them here.

THE PRINCIPLE OF ENGINEERING GEOLOGICAL CLASSIFICATION OF FAULT ROCKS

The fundamental aim of engineering geology is to provide data for engineering construction. We hold that three basic principles should be followed in engineering geological classification of fault rocks.

A) Based on the petrological classification of fault rocks

Different petrographical types of fault rocks possess different physical and mechanical properties. To make full use of the research results in petrology is essential for conducting reasonable engineering geological classifications of fault rocks.

B) Based on the engineering geological properties of fault rocks

Each engineering geological type of rock and soil should possess its peculiar engineering geological properties (mainly physical and mechanical properties). This should also be true for fault rocks. Therefore in engineering geology, the similarity of engineering geological properties must be taken as the criterion for the classification of fault rocks.

C) The classification should be simple, clear and convenient to use

It is well-known that all engineering geological data will be presented to designers and constructors at last. Generally speaking, they have little knowledge of geology, so it is not convenient for them to use a complex classification. Thus the classification should be as simple as possible. Moreover, engineering geologists do not mainly rely on microscopes to determine the types of rocks, the signs for classification should be obvious to see so that the types of rocks can be easily determined in the field, since a too complex classification will cause difficulties for determinations with naked eyes. Each type of the classified rocks should be indicated with quantitative indices which can describe the engineering geological properties of the rocks properly. Thereby, we can compare fault rocks with other types of rocks and soils and use the indices to evaluate rock masses by the method of engineering geological comparison. In this way we may attain the aim of classification.

THE PROPOSED SCHEME OF CLASSIFICATION FOR FAULT ROCKS IN ENGINEERING GEOLOGY

Cohesiveness is the key controlling engineering geological property of fault rocks. Therefore, we set the degree of consolidation as the first-order basis for our classification, and thus fault rocks can be divided into two series: loose fault rock series and consolidated fault rock series. The most important characteristic of the consolidated fault rocks is that they are connected crystallizedly and to some extent water-stable. The grain (minerals and debris) size and grading of fault rocks are related to their origin. From experimental data we can see that the engineering geological properties of fault rocks are controlled by their granularity, especially for the loose fault rocks. So referring to the classification of loose soils, we set granularity as the second-order basis for our classification. Based on experimental data and referring to the standard of geotechnology and petrology we have determined the grading composition of different fault rocks [11]. To be simple, we take diameter 0.5 mm, which is the limit for eyes to discern grains, as the grading limit between coarse grains and fine grains. According to the content of coarse and fine grains, different fault rocks are classified. When grading-analytical data are available, we can use the terms used in classifying fine grained soils as modifiers to modify fault rocks, for example, clayey fault gouge. According to the above principles we divide

---

* Yang Zhuen, 1981, Approach on Fault Rocks, paper not published yet.
** Liang Jincheng, 1984, On Fault Rocks of Xin Zhi Fault Zone and Its Evolution, paper not published yet.

fault rocks into six basic types (table 3). It is well-known that the mother rocks of fault rocks have some influence on their engineering geological properties, but because of their variety it is not convenient to take them as the standard of classification. However, we can use the names of the mother rocks to modify the basic types, especially for loose coarse fault rocks, for example, granite-fault-block-rock, granite-fault-breccia etc. Clay minerals play an important role in the engineering geological properties of fault gouge. When a certain clay mineral is dominant in fault gouge, the name of the clay mineral also can be taken as a modifier, for example, montmorillonite fault gouge, illite fault gouge etc.

Table 3. Engineering Geological classification for fault rocks

| Degree of consolidation | types of rocks | grading composition | | remarks |
		diameter (mm)	content (%)	
Binding force between cracks	fault cataclastic rock	>200		the rock within the fault influenced zone
	fault black rock	2—200		
Loose fault rocks	unindurated fault breccia	0.5—2	>50	
	fault gouge	<0.5	>50	
consolidated fault rocks	fault breccia	>0.5	>50	
	mylonite	<0.5	>50	

In fact there are many transitional types which are not listed here. In detailed study we can use modifying terms to make further classifications of fault rocks, based on the results of grading analysis and referring to the classification standards for soils in geotechnology. For example, if a gouge has a relatively large amount of grains over 0.5 mm in diameter, it can be called gravelly gouge.

THE FUNDAMENTAL PROPERTIES OF VARIOUS TYPES OF FAULT ROCKS

A) Fault cataclastic rock

Fault cataclastic rocks are the dynamic metamorphic rocks with the lowest degree of metamorphism. They are in the fault-influenced zone or in the crack-concentrated zone beside the fault zone. The characteristics of the mother rocks remain unchanged. The distance between cracks is generally greater than 20 cm and the occurence of cracks can be along some orintations. The density of cracks gradually becomes smaller as the distance to the chief fault plane getting greater. The engineering geological properties of fault cataclastic rocks are similar to that of the mother rocks except permeability, which is higher for the former than for the latter. The modulus of deformation is smaller than that of the mother rock. Rock masses composed of this type of rock are commonly called fault-influenced zone.

B) Fault block rock

This kind of rocks is composed of blocks of various sizes which are formed through the breaking of mother rocks. They generally apear near fault planes and sometimes in lenticular form. The diamenter of the rock blocks is generally 2—20 cm. There is little fine-grained filling or no filling exists at all; the blocks inlay each other. The shapes of the blocks vary with the nature and magnitude of the stress acted upon the rocks. They can be regular, irregular or lenticular. The cracks are generally along a fixed orientation. There are weak secondary minerals on the planes of cracks. In the blocks there exist microcracks and other microstructures. The engineering geological properties of this type of fault rocks are mainly controlled by their mother rocks. But there are notable differences between fault block rocks and their mother rocks. For example, fault block rocks have smaller specific gravity and bulk density but higher porosity and permeability compared with mother rocks. The ability of anti-wheathering is lower and the degree of wheathering is higher than that of the mother rocks. Both the modulus of deformation and the shear strength are smaller than those of the mother rocks. Generally some measures of treatment are needed in order to satisfy the requirements of construction. The engineering geological properties of fault block rocks are closely related to their mother rocks and stress features. For example, fault block rocks that formed under compressive stress generally possess better engineering geological properties than those formed under tensional stress.

C) Unindurated fault breccias

They are formed through intensive crushing of mother rocks and they are composed of angular fragments, rock debris, a little rock powder and clayey materials. Particles with the diameter 0.5—20 mm are dominant. The primary structures of the mother rocks have been destroyed completely, their textures and minerals remain only in the angular fragments. The shape and size of the particls vary greatly with the nature and magnitude of the stress acted on their mother rocks. For example, the fragments in the unindurated breccia formed in a tension fault is angular in shape and large in diameter. Silt and finer grains occur in this type of rock. But in unindurated breccias formed in a compression-torsion fault the fragments are lenticular in shape and generally parallel to the trend of fault. There are more silt and finer grains, which are arranged around the lenticular fragments.

The differences between the engineering geological properties of the unindurated fault breccias and their mother rocks are apparent. The physical and mechanical properties of unindurated fault breccias are similar to those of gravel soils: loose structure, no tension strength, small modulus of deformation and large coefficient of compression. The friction between the particles plays a leading role in the shear strength. Because the angular fragments is diversified in mineral composition, size, shape and content, the engineering geological properties of the unindurated fault breccias vary within a certain range.

D) Fault gouge

Fault gouge is formed through intensive grinding of mother rocks. The textures and structures of the mother rocks have been destroyed completely. The minerals have generally changed. Gouge is composed of rock powders, clayey materials and rock debris. The diameter of most particles is smaller than 0.5 mm. A remarkable characteristic is that gouge contains some clay minerals. Gouge is generally distributed along the chief fault planes and there can exist many gouge belts in a large fault. The thickness of the belts varies greatly from millimeters to tens of centimeters and is generally related to the scale of the fault. The composition and the content of the clay minerals are related to the properties of the mother rocks, the degree of dynamic metamorphism, the hydrogeochemical environments, as well as the degree of weathering etc. The minerals of fault gouge are generally arranged parallel to the trend of fault, forming linear textures and foliated structures. According to the nomenclature in petrology this type of rock is called mylonite.

Fault gouge is the chief weak belt in rock mass. Its engineering geological properties are the poorest of all fault rocks and vary within a certain range depending on the content of clay particles and the kind of clay minerals. The physical and mechanical properties of fault gouge should be studied through special experiments. The spatial variations of its thickness and occurrence should be investigated and sorted out by statistics. The composition and content of clay minerals in fault gouge should also be studied specially. It is necessary to point out that besides being related to the composition of clay minerals and granulometric composition, the shear strength of the fault gouge is controlled by its density and consistency. Like super-consolidated clay fault gouge has structural strength. Experiments have shown that the density and the consistency of fault gouge are closely related to its natural stress state and its weathering degree (Table 4). Therefore, in order to evaluate the engineering geological properties of fault gouge we should determine its physical indexes under natural confining pressure, especially under saturated confining pressure. To analyse the natural stress subjected by fault gouge in different areas is the key to evaluating fault gouge correctly.

E) Fault breccia

Fault breccia is a kind of consolidated fault rock. It is composed of rock fragments with fine grained matrix and cement materials. The fragments may be angular-shaped, but often lenticular to form distinct parallel structure. The composition of the cement varies in different rocks. For example, the cement materials of fault breccias in igneous rocks are usually siliceous or calcareous, which are formed through precipitation of hysterogenetic or hydrothermal fluid. In the matrix, chlorite and sericites sometimes appear. Fault breccias in sedimentary rocks are generally cemented by calcareous and ferriferous materials. The engineering geological properties of the fault breccia are rather good, often depend on kind and nature of the cement materials.

F) Mylonite

This is the equivalent of gouge, but consolidated and with more remarkable parallel structure called mylonitic structure. Mylonite may also show foliation, lineation and banded structure. In the international academic symposium on mylonite held in U.S.A. in 1981, it is thought that the deformation of mylonite is ductile and there is no or little microfracture in it. According to the

Table 4. Relationship between natural physical properties of gouge and their buried depth and weathering degree

Number of fault	Sampling spot	Natural state					Thickness		
		bulk density (g/cm³)	moisture content (%)	dry bulk density (g/cm³)	void ratio	saturation	liquid limit (%)	plastic limit (%)	plasticity index
Fault F₄ at a damsite on Chuan River	At an outcrop of 254 m level	1.53	80	0.85			88	61	27
	At the tunnel portal at 248 m level	1.72	46	1.18			65	43	22
	In the tunnel at 248 m level	1.90	30	1.46			57	38	19
	In the No.3 tunnel under the river bed	1.81	25	1.45			45	32	13
Fault F₁₈ at a damsite on Yellow River	At strong weathering zone near the river bed	1.76	7.89	1.63	0.667	32.2	31.1	114	17.1
	At the slight weathering zone of an excavated slope	2.20	11.1	1.98	0.34	91.2	26.1	14.1	12.0
	At weak weathering zone where just removed compressive force	2.29	7.6	2.13	0.277	74.6	28.9	16.2	12.7
Fault F₁₈ at Ankang damsite on Han-jiang River	At an excavated slop, upper part of weak weathering zone	1.99	24	1.61	0.747	88.6	38	20	18
	At an excavated slop, low part of weak weathering zone	2.27	18.6	1.91	0.463	111.8	29.8	17.2	12.6

opinions of many researchers in recent years, mylonite is a kind of fault rock formed through plastic deformation of the rock in crust. But in the concepts of traditional petrology and engineering geology mylonite was mixed up with fault gouge.

The engineering geological properties of mylonite are comparatively good, their specific gravity and bulk density are high. The porosity is small and this kind of rock is comparatively water-stable. Both the modulus of elasticity and the shearing strength are rather high. The engineering geological properties of mylonite from some damsites in China are listed in table 5.

CONCLUSION

1. In view of its clear and definite meaning and being easy to be understood directly from the literality, the authors think that rather than "structural rock" or "cataclastic rock", the term "fault rock" is better to be applied in engineering geology, which was proposed by R.H. Sibson (1977).

Table 5. Physical and mechanical properties of mylonites

No.	Physical properties (g/cm³)				Compressive strength (kg/cm²)		Coefficient of softening	Shearing strength		Modulus of elasticity	Poisson's value
	Specific gravity	Dry density	Saturated density	Ratio of water absorption	Dry	Saturated		tgφ	C		
1	2.8	2.59		0.38	652	467	0.72	0.650	0.89	52	0.21
2	2.94	2.60	2.41		311	78	0.25	0.660	0.22		
3	2.92	2.75	2.84		299	41.6	0.14				
4	2.91	2.70	2.82		456	30	0.07			2.7	
5	2.96	2.72	2.83	1.76	173	46	0.27	0.45	0.1	16.5	0.53

2. In petrology, structural geology and seismic geology there are many kinds of classifications of fault rocks from different points of view. In this paper only some of them related to ours have been listed in order to correlate each other.

3. The three pinciples of our classification imply clearly that both the existing classifications and the aim of engineering geology should be taken into account in classifying fault rocks. Besides, simplicity and convenince for survey purposes is especially important. A classification would hardly be accepted if it is too complex to use in practice.

4. The proposed classification divides fault rocks by two orders. First according to wheather the rock is consolidated, and second its granulometric composition. The grade of 0.5 mm is a dividing line between coarse and medium sized sand grains, and also the limit for identification with naked eyes. According to the content of grains larger or finer than 0.5 mm gouge (mylonite) and unindurated fault breccia (or indurated fault breccia) can be distinguished.

5. Brief descriptions have been made for the understanding of the general characteristics of each type of fault rocks we have divided.

REFERENCES

1. International tectonic dictionary, English terminology 1983, Geological Publishing House.
2. Wang Jiayin, 1949. Approach to classification of cataclastic metamorphic rocks. Geol. Rev. 1-4.
3. ————— 1974. Preliminary study on New Huaxia faulting belt at Beishicheng. J. Geol. Vol.1, 1974.
4. ————— 1978. An introduction to stress minerals. Geol. Pub. Hou.
5. Tullis, J. et al., 1982. The meaning and origin of mylonites. Geology Vol. 10, 227-30.
6. Sun Yan et al., 1979. On classification and nomenclature of tectonites. J. Nanjing University, No.2, 83-100.
7. Institute of Water Resources and Electric Power, Institute of Geology Academia Sinica (IGAS) 1974. Engineering Geology for water resources and hydro-electric power. Scientific Publishing House.
8. IGAS 1978. Engineering Geological Mechanics of Rockmass No.2. Scientific Publishing House.
9. Gu Dezheng, 1979. Fundamentals of rock engineering geomechanics. Scientific Publishing House.
10. Zhang Yetao, 1980. Summary of research on tectonites. J. Chinese Academy of Geology, Sian Division.
11. Zhang Xiangong, 1979. Engineering Geology Vol.1. Geol. Pub. Hou.
12. Zheng Yuchi et al., 1963. Some basic problems of metamorphic rocks and method of study. Geol. Pub. Hou.
13. Yuan Kue-rong, 1982. Preliminary study on fault structure in rock masses of granite. J. Guiling College of Metallurgical Geology No.2, 1-14.
14. Engelder, J.T., 1974. Bull. Geol. Soc. Am. 85.
15. Sibson R.H., 1977. Fault rocks and fault mechanisms. J. Geol. Soc. Vol.133, Part 3, 191-213.
16. Spray, A., 1969. Metamorphic Textures. Pergamon, London.
17. Turner, F.J. & Weiss, L.E., 1963. Structure analysis of metamorphic tectonites.
18. Zeck, H.P., 1974. Am. J. Scie. Vol.247.

2.2 Foundations, grouting and draining in weak rocks
Fondations, problèmes d'injection et de drainage
dans les roches tendres

Foundation treatment and grouting works for Alicurá Dam, Argentina

Traitement des fondations et travaux d'injection pour le barrage d'Alicurá, Argentine

Carl-Anders Andersson, *SWECO, Engineers, Architects & Economists, Stockholm, Sweden*
Oscar A. Vardé, *Vardé y Asociados, Buenos Aires, Argentina*

ABSTRACT: This article describes the core foundation treatment of Alicura earthfill dam on potentially weak layers of claystone (pelite) interbedded in layers of sandstone. Details of the placing of dental concrete in the foundation and of the grouting from the surface and from a special grouting gallery are given. The efficiency of the grouting during different phases is discussed based on observations of seepage and pore pressures in the foundation due to the reservoir. Grout mixes thicker than normal were used.

RESUME: Le rapport décrit le traitement de la fondation du noyau du barrage en terre d'Alicura, sur un massif rocheuse constitué des couches entrecoupées de grès et d'argilite (pelite). On donne des détailles de la mis en place du beton au contact, et du rideau d'injection fait d'une galerie especiale. L'eficacité de ce rideau a été analisée par les observations entre les pressions interstitielles et la percolation avec la retenue. On ajoute différentes considerations sur les coulis plus riches utilisés.

1. INTRODUCTION

The Alicura Hydroelectric Proyect owned by Hidronor S.A., has been constructed on Río Limay, 100 km NE of the town of San Carlos de Bariloche, Argentina. The construction started in the late part of 1979, and the proyect was commissioned in 1984.

An earth dam with a volume of 13×10^6 m3, a maximum height above the foundation level of 130 m and a crest length of 950 m is part of the proyect (Figure 1). The reservoir has a total volume of 3.15×10^9 m3. The dam crest is at El.710 m above M.S.L., the upper storage level at EL.705 m, the minimum operation level at EL.692 m and the average downstream water level at El.585 m.

Two 9.0 m diameter concrete-lined tunnels have been constructed in the right abutment for diversion during the construction period. One of those has been subsequently converted to a bottom outlet for a maximum capacity of 600 m3/s. The installation on the left bank include an open-air power station with an installed capacity of 4 x 250 MW at a gross head of 118 m and a surface spillway with a maximum discharge capacity of 3,000 m3/s. The power intake and the spillway gate structures are located about 300 m downstream of the axis of the dam on the rim of the slope. A concrete lined approach channel has been constructed upstream from these structures.

Partially mylonized thin claystone layers between more competent sandstone layers presented potentially unfavorable stability conditions. Extensive drainage and grouting works have been constructed to ascertain the integrity of the dam and the left bank with appurtenant structures. These include 5.500 m of drainage galleries with 35.000 m of drain holes to form 150.000 m2 of drainage curtains, a 1.360 m long grouting gallery, and 65.000 m2 of grout curtain.

This paper briefly describes the rock mass of the dam site and its main characteristics, the dam design and in more detail the foundation treatment by means of grouting. Relatively dense but fluid mixes were used for grouting the low permeability foundation. The washing effect of the reservoir in the rock mass is discussed on basis of results from regrouted areas.

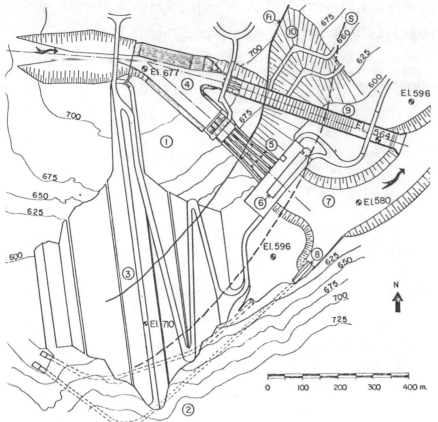

Figure 1. General layout

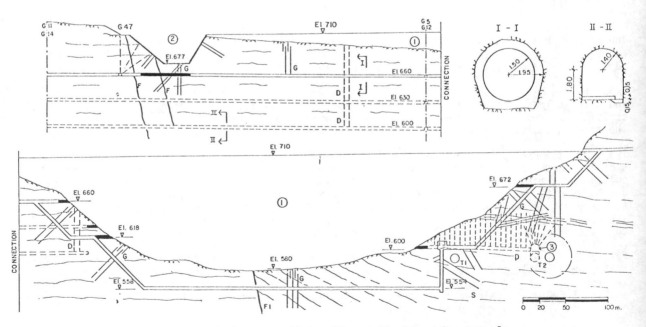

1. Dam   2. Approach channel   3. Bottom outlet   T1 and T2. Diversion tunnels.
G. Grouting gallery and direction of grout holes D. Drainage galleries with holes F. Fault
S. Syncline.

Figure 2. Profile of grouting gallery

2. GEOLOGICAL AND GEOTECHNICAL FEATURES

2.1 Description of the rock mass

The rock mass at the site consists of sandstone (psammite), interbedded with silstone and claystone (pelite) formed during Lower Jurassic. The sandstone is generally medium grained and competent and is encountered in beds of 0.5-1 m thickness, which may reach about 5 m in a few places. Orthogonal joint sets, perpendicular to the bedding, exist in the sandstone with spacing approximately equal to the thickness of the layer.

The pelite beds typically vary in thicknees from a few centimetres to 0,5 m. They are often continuous and can sometimes extend with minor ondulations for some hundred metres (Vardé, Andersson, Pi Botta, Paitovi 1986). Depending mainly upon grain size, the competence of the pelite varies considerably, the finer grained pelites showing weakness planes parallel to the bedding and being slickensided and mylonized. In contact with air and water the weak pelite disintegrates quickly and forms a sticky paste. However, the thickness as well as the texture of a specific pelite layer vary significantly from one place to another, and it proved impossible to make realiable stratigraphic correlations between investigation borings, even when only some ten metres apart.

The bedding of the psammitic-pelitic layers is generally nearly horizontal. However, a zone of inclined bedding exists and is delimited from the areas of horizontal bedding further to the left by the steeply dipping fault 1 (Figure 1 and 2) which crosses the foundation of the dam core in the middle of the river channel. The transition from inclined to horizontal bedding in the right abutment is formed by a syncline. The dipping of the inclined layers is approx. 40° from the horizontal near to the fault and decreasing towards the right. The fault 1 proved to consist of a zone of plastic mylonized rock, about 1 m thick in the dam area.

For the right abutment, a relatively deep core trench was excavated as to widen an existent gully while in the smooth left slope the depth of the excavation was 5 to 10 m. In the right core abutment a layer of very soft and friable sandstone-siltstone was encountered between E1.629 and 634. Below this layer the rock is very competent while above the soft layer it is rather fractured. A vertical fault, parallel to the dam axis and about 10 m downstream of the same, extends all the way from the weak layer to the level of the damcrest at E1.710. The thickness of this fault is generally less than 0.2 m. A second fault parallel to the other and located another 10 m downstream was encountered above E1.670. The left abutment presented quite normal and favorable rock conditions.

2.2 Permeability

The permeability of the rock mass is very difficult to assess. Without doubt, there exists an important anisotropy in the mass, due to more pervious psammitic layers and the impervious pelitic layers. The permeability of the sandstone is dependent upon the joint opening and joint fill. It may be reasonable to assume about the same permeability horizontal and vertically for the sandstone members. For the rock mass a relation of horizontal to vertical permeability of 10: 1 has been used in the seepage analyses.

In situ measurements of permeability have been made by water pressure tests in boreholes during the investigations and during construction. Most values are well below 5 Lugeons and seldom above 25. In general, it has been impossible to find a pattern between different borings with respect to water absorptions, and the general impression during the investigations was that the rock had a low permeability. The grouting and drainage drillings have confirmed this assumption. For the purpose of analysis the horizontal permeability adopted was 3.10^{-7} m/s.

However, the permeability of the rock mass have been afected, in some areas, by the presence of the reservoir. Regrouting performed after impounding showed in some cases, higher grout absorptions than the initial grouting without the reservoir as described in chapter 4.

2.3 Shear strength

It has been observed that exfiltration is often concentrated to the upper contact zone of a pelitic layer and to the lower points, i.e. the wave bottoms, along a layer. In such locations the pelite often appears soft and plastic. Therefore, it is not very common to encounter extended layers of soft and plastic pelite, but nevertheless it was necessary to take this possibility into account in the stability analyses.

A slide in the zone of inclined bedding occurred in the beginning of construction and led to a re-evaluation of the shear strength parameters based on laboratory test of material from the sliding surface. The slide had developed in a plastic pelite layer with typical plastic and liquid limits of 10 and 30% respectively. The shear strength assigned to continuous plastic pelite layers was $\emptyset = 17°$, $c' = 0$ for static loading based on residual values, and $\emptyset = 12°$ $c' = 5.6$ t/m2 for dynamic (seismic) loading.

491

3. DESIGN FEATURES OF THE DAM AND ITS FOUNDATION

In general, the foundation conditions for the dam could be considered favorable except for the potentially weak and continuous pelitic layers which influenced very much on the stability of the dam. Therefore, it was indispensable to assure effective drainage of the left slope and the dam foundation, as well as reasonably low seepage. The difficulty to determine the permeability of the rock resulted in a preliminary design of the drainage and grouting works that, in part, was modified during construction.

The Alicura Dam has a symmetrical central core (Figure 3) constructed with screened morainic material. The core was founded on rock in its full width of maximum 80 m. The overlaying coarse alluvium, with a thickness of up to 15 m in the river valley was removed in the area of the core foundation and the relief wells, but left in place under the remaining part of the shells (Vardé, Andersson 1983). It was foreseen to excavate the upper 3 to 5 meters of rock under the full width of the core in the river valley to obtain a contact of competent rock. For the abutments of the core the depth of the cut-off had to be determined in relation to the rock quality. The rock contact was to be sealed with dental concrete and shotcrete.

1	CORE	A	GROUTING GALLERY
2	CORE CONTACT	B	RELIEF WELLS
3	FILTER	C	DRAINAGE TRENCHES
4	DRAIN		
5	SHELL		
6	RANDOM FILL		
7	RIPRAP		
8	SLOPE PROTECTION		

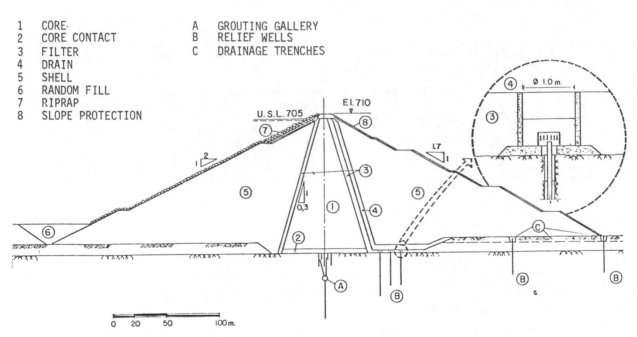

Figure 3. Dam section with detail of relief well top.

The design included a system of relief wells with three lines directly downstream of the core and two lines further downstream, the last one at the dam toe, having the possibility of future inspection. The design of the embedded relief wells, diam. 200 mm, required special attention in the design and construction as to assure their efficiency during the life time of the dam. In the left bank a deep drainage curtain was to be placed about 20 m downstream of the axis. This curtain comprised two galleries at El.600 and El.630, forming part of the large system of drainage galleries extending further downstream in the left bank in order to drain the critical left slope (Andersson, Jansson, Vardé 1985). Drainage holes, diam 85 mm, were to be drilled between the galleries and the surface every 6 m, and 3 m in the vicinity of the approach channel. A shorter drainage gallery 25 m downstream of the dam axis was provided in the right abutment at El.600 with upwards drainage holes to approx. El.640 Above this elevation no drainage is provided.

A grout curtain below the core was designed as to assure a reasonable homogeniety in the permeability of the foundation and a reasonably low seepage by sealing wider fissures. It was decided that a grouting gallery, about 25 m under the core contact (Figure 2), would be advantageous. This would allow grouting without interference with the core fill and, later, the reservoir. The grouting gallery and grout curtain extend into the abutments and into the left bank at a higher level, El.660. Consolidation grouting was planned to be performed under the central one third of the core width and to a depth of 6 m.

4. DESCRIPTION OF PERFORMED FOUNDATION TREATMENT

This description concentrates on the sealing of the contact surface of the core, and the grouting works during different construction periods. The execution of the drainage systems is not described here.

4.1 Sealing of the contact surface

Already in the first phase of the rock excavation for the contact surface in the river valley it turned out that the depth of the excavation could be greatly reduced. The quality of a deeper foundation surface was, in general, no better than a more superficial one. It was very important to use an excavation method that left the rock surface as little disturbed as possible.

In the river valley the rock excavation was done mainly with ripper and backhoe to remove loose rock. Blasting was used only as a complement to remove rock which could not be ripped and which would cause unacceptably high steps in the foundation unless using large volumes of concrete. In the area of inclined bedding, to the right of the fault 1, the excavation profile resulted very uneven with V-shaped depressions formed by the weaker pelite planes on one side and fracture planes in the sandstone layers on the other side. The pelitic layers fissured easily when exposed to dry air and softened when becoming moist. Therefore, all weathered pelite had to be carefully removed by using pneumatic hammers, spades and jets of air and water immediately before re covering the cleaned area with concrete. The concrete was filled into the depressions or at the sides of steps, giving a smooth surface without high and abrupt steps. Areas with more horizontal pelitic layers in the river bed were covered with concrete slabs; it was deemed necessary to use no less than 0.3 m thickness in the slabs to avoid excessive fracturing due to the overburden load. Consequently, competent sandstone layers were left without concrete and core material was filled directly on the cleaned surface. The plastic material in fault 1 was cleaned out to approx. 2 m depth and replaced by concrete.

In the core trenches of the abutments the contact surface was done by smooth blasting. Horizontal steps, in general 1 to 2 m wide, occurred every 8 m due to the necessary displacement between the contour holes from one bench to the other. These steps were smoothed with concrete. Most of the contact surface was covered by 50 to 100 mm of shotcrete, which was applied working from a platform hanging in a mobile crane parked on the core fill. The shotcrete was used in order to protect the pelitic layers and to smoothen the surface where neccessary. The quantity of shotcrete used on 22.500 m2 in the abutments was 1.200 m3. It should be mentioned that the layer of weak sandstone-siltstone at El.629 - El.633 in the right abutment was left without shotcrete and the core fill was placed in direct contact with the foundation. It was deemed that a layer of shotcrete on this soft foundation would have broken and caused undesirable potential seepage paths.

The volume of concrete used to prepare the 21.000 m2 of total foundation surface in the river valley was 5.300 m3 and another 2.700 m3 was used in the abutments. The concrete used had a maximum agregate size of 19 mm and a characteristic resistance of 170 kg/cm2. In order to place the concrete without formworks in slopes of about 1 (vert) : 2 (hor.) the concrete was delivered with a settlement of only 2-3 cm. Tubes were embedded in the concrete in areas with fissured rock to allow for contact grouting. In areas with finely fissured but competent rock, cement slurry was poured and worked into he fissures immediately before placing core fill.

4.2 Drilling and grout tests

Different drillings methods were tested and evaluated on basis of water pressure tests and grout absorptions. The results of the tests indicated that the drilling with preference could be done with the rotary-percussion method. This method did not show any appreciable difference in water absorption and groutability compared to the originally specified rotary drilling; conventional percussion drilling, however, proved very unsatisfactory as the pelite tended to clog the drill bits and caused lower grout absorptions. Moreover, rotary percussion drilling with 85 mm diameter showed a cost advantage of 20% in comparison with NX rotary drilling.

Laboratory tests on varios grout mixes were performed to study the most important characteristics, i.e. viscosity by means of the Marsh cone, density, sedimentation, setting time and unconfined compressive strength on cylinder samples. Also two different types of cement were tested, one was the normal portland cement with a Blaine fineness of 3200 and the other a finer cement with Blaine 4500.

The choice of most suitable grout mix was based on the, by 1980, relatively new criterion of using thick and stable mixes (Lombardi 1985). The mix that appeared to have the most desirable properties had a w/c - ratio of 1:1 by weight and 2% (by weight of cement) of bentonite, which gave typical values of 35 to 38 sec. for the Marsh cone, density 1.50 g/cm3 and 3-5% of sedimentation. This was used for the tests and initial grouting works. Even though the finer cement gave somewhat more favorable characteristics in the laboratory than the normal cement for the same type of mix, it was decided to use normal cement because of better regularity and security in supply.

Later during the construction the grout mix was changed to a denser one with a w/c - ratio of 0.67: 1 but with 1% Sika Intraplast instead of 2% bentonite. Though the density increased to 1.67 g/cm3 the Marsh value was lowered to 32-34 sec. thus facilitating the grout penetration. This latter mix also showed much higher strength with typical values of 180 kg/cm2 than the corresponding mix with bentonite. While the sedimentation values were even lower with this additive, the initial setting time increased from tipically 3 to 15 hours.

One of the advantages of using Intraplast was that it could be added directly to the mixer and did not have to be separately mixed, hydrated, stored and wasted, when overaged after 2 days, as in the case of using bentonite. The mixes with Intraplast also proved to be more homogeneous according to the test results.

4.3 Grouting before the impounding

Consolidation grouting to 6 m depth and grouting of the upper 12 m portion of the curtain were done from the surface. Later the grout curtain was extended upwards and downwards from the gallery; the upward holes were overlapping 6 m of the holes drilled from the surface.

The grouting from the surface was made in 3 primary lines, one in the dam axis and the others 6 m upstream and downstream of the axis, respectively, to 12 m depth. Initially, also primary lines to 6 m depth were made 12 m upstream and downstream of the center line, but the low grout absorptions indicated that it was sufficient to concentrate the grouting to the 12 m wide strip. The hole spacing was 6 m. Secondary lines were performed to 6 m depth at 3 m distance upstream and downstream of the center line with 6 m spacing but the holes were positioned inbetween those of the primary holes. Additional holes were added inbetween in a tertiary stage when the grout absorptions so indicated.

The pressures commonly used were: 0.2 to 3 m - 1 kg/cm2, 3 to 6 m - 2.5 kg/cm2 and 6 to 12 m - 5 kg/cm2. The grout absorptions were generally low with average takes of 25 kg/m in the first phase and 21.5 kg/m in the second, discounting a zone of higher takes on the right abutment.

The grout curtain extending upwards from the gallery consisted, in principle, of 3 lines, a vertical one and two inclined slightly upstreamwards and downstreamwards respectively. Later one of these lines was omitted as the grout absorptions were very low, in average 10 kg/m except for near the gallery, indicating a homogeneous rock with low permeability when undisturbed. The longitudinal spacing of the holes was 6 m; in the single line curtain extending 30 m downwards from the gallery the spacing was reduced to 12 m in part of the river valley. The grout pressures in the curtain were 10 kg/cm2. The grouting was made in steps of 6 m; in about 5% of the steps, in areas with normal grout takes, the absorptions were higher than double the average. In the right abutment, between the inclined gallery and the consolidation grouting, the takes were higher, and averaged 40 kg/m in the third stage.

In the area around the approach channel in the left bank the poor rock conditions were interfering with the execution of the grouting gallery at El.660. Due to a tight time schedule it was finally decided to plug the gallery with concrete over a 60 m long portion below the channel. The grouting in this area was made from the surface and included also consolidation grouting to 40 m depth in an area 50 x 30 m2 of the bottom of the channel. The average absorptions for the consolidation grouting was 50 kg/m and for the curtain 37 kg/m; in some levels the absorption was more than 200 kg/m.

4.4 Grouting during impounding

During the impounding, which started in Nov. 1983, the grouting works continued in the upper portions ahead of the rising reservoir. In the right abutment additional grouting was initiated when the seepage from some drains suddenly increased when the reservoir reached El.641. The quite normal takes before impounding,in the curtain below the inclined gallery, increased substancially, up to 500 kg/m in some portions of the boreholes, in a layer between El.630 and 640 about 70 m away from the surface of the core abutment. It was possible to abtain closure conditions with average takes of 40 kg/m in a fifth stage (Figure 4). The additional grouting which showed to be most effective was made with subhorizontal holes from the inclined grouting gallery. Also subvertical boreholes were drilled from the horizontal portion of the gallery at El.670. The hole spacing was reduced to 0.75 m. As a result of the additional grouting the flow of approx. 5 l/s in the 100 m long drainage gallery in the right bank did not increase in spite of the rise of the reservoir.

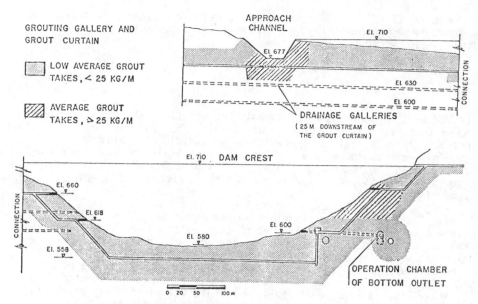

GROUTING GALLERY AND
GROUT CURTAIN

LOW AVERAGE GROUT
TAKES, < 25 KG/M

AVERAGE GROUT
TAKES, > 25 KG/M

Figure 4. Schematic
distribution of grout
takes.

4.5 Grouting with the reservoir at operating level

When the reservoir rose, in the end of 1984, from its initial level, El.695, to full reservoir level, El.705, the seepage increased in the drainage gallery at El.630 in the left bank and also in the right abutment. The increments in seepage were proportionally higher than the increase in hydrostatic pressure.

It was considered appropriate to check and, if possible, improve the impermeabilization, specially in the area around the aproach channel where the most significant seepages were observed.

During 1985 a campaign of additional grouting was performed from the gallery at El.660. In the portion between the approach channel and the left dam abutment partial regrouting was performed in up to four phases by split spacing, i.e. in some places the spacing was only 0.75 m. Some interesting observations were made on basis of the results obtained and are summarized as follows:

i)   The average grout absorptions were in general higher in the second campaign than in the first before impounding.

ii) The absorptions were higher around the gallery, average 46 kg/m, diminished with distance from the same and were always low in the contact with the core (20 kg/m).

iii) The average absorptions are practically the same in the first and second phase but with a tendency to lower values in the second phase if the absorptions in the first phase were higher than the average.

iv) In the third phase several boreholes showed to be in communication with each other though the holes were 6 m apart.

v)   The pressures used in the first and second compaign as well as the grout mixes were approximately the same.

vi) The seepage in the drainage curtain 25 m downstream of the grout curtain did not decrease.

Further comments on these findings are included in chapter 5.

It was also decided to check the grouting and extend the curtain somewhat further into the right abutment. This was done more than one year after finishing the extensive additional grouting in this abutment. The regrouting showed that the absorptions in the already treated area below the inclined and horizontal gallery at El.672 were in average 30 kg/m except for in the 6 m portion inmediately below the horizontal gallery. In this portion the absorptions sometimes exceeded 100 kg/m in the primary holes spaced 6 m. In the extended portion of the curtain the absorptions were low demonstrating that the seepage in the layer between El.630 and El.640 could not be reduced further by grouting in a narrow curtain.

The area of the grout curtain above E1.672 in the right abutment was not treated extensively due to low absorptions. Check grouting in the beginning of 1986 resulted in absorptions frequently above 100 kg/m and additional grouting had to be performed.

5. OBSERVATIONS

Though the foundation of the core can be characterized as only potentially weak, ample precautions had to be taken in order to safeguard the integrity of the dam. The drainage curtains in the abutments offer good possibilities to register seepage both in individual drains and the total flow from the galleries. This has been done in a regular way since beginning the impounding. Also a great number of piezometers in the foundation (Pujol, Andersson 1985) monitor the behavior. In this chapter the grouting results are analyzed in the light of monitored pore pressures and seepages in the foundation.

The piezometers installed about 10 m below the core contact in the river valley show that the difference in piezometric levels between those installed about 15 m upstream and 15 m downstream of the central grout curtain is in average only 2 m. However, the absolute average piezometric level of approx. E1.630 coincides well with the calculated one, demonstrating that the overall permeability of 3.10^{-7} m/s horizontally and 3.10^{-8} m/s vertically is reasonable. On the other hand it is evident that the effect of the grout curtain is very small as could be expected from the low absorptions. Still, after 2 years with nearly full reservoir, the piezometric levels in these piezometers are increasing very slightly, but more in those upstreamwards than downstreamwards. This is also a verification of the low overall permeability in the foundation. The seepage to the relief wells can not be determined in this portion of the foundation but piezometer readings prove that these wells are very efficient in reducing pore pressures.

In the left bank the downstream piezometric levels are moderate and below acceptable limits. It is, in general, difficult to differentiate the effects of the grout curtain from those of the drainage curtain. However, there are evidences that the grout curtain is very effective in areas of high absorptions, e.g. below the approach channel.

In the right abutment, above E1.640 where there is no drainage curtain, the piezometric levels upstream and downstream of the grout curtain are practically the same. This is somewhat surprising as in this area the grout absorptions were high and the grouting was continued until the absorptions became reasonably low. An explanation may be that the freatic level in the rock further behind the abutment is high.

Water pressure tests were performed with a relatively low frequency as, in general, low water absorptions corresponding to Lugeon values below 2 were registered. Some of the water pressure tests were made in the drain holes spaced 3 m but communications between the holes were rare. Notwithstanging, it was observed that sometimes the piezometric level lowered several tens of meters away from a drilled hole but increased to former value if grouted. This confirms the observation that the permeability is principally created from fine fissures and a few wider ones. Furthermore, the fissures have a limited extension vertically as they generally appear in the more brittle sandstone layers and are interrupted in the more flexible pelitic layers.

6. DISCUSSION

From the previous description it is evident that the rock foundation, in general, can be classified as relatively impermeable. The concept followed was to see the grouting works before impounding as the necessary final investigation to detect areas which might require more extense treatment. Two such areas were encountered one below the approach channel and one in the right abutment. The latter one was only partially detected before impounding. It was a sudden increase of the seepage caused by the rising reservoir which indicated that the area of high permeability was larger than first detected.

Of course, the fact that an area of high permeability like the one in the right abutment could pass undetected though penetrated by several boreholes is a matter of concern. At the same time it is satisfying to note that the extensive grouting had a very positive effect and detained the seepage. On the other hand the additional grouting in the left bank did not result in any visible reduction of the seepage.

The aforementioned viewpoints emphasize the difficulty to grout a rock of low permeability and also rises the question to what degree it is necessary to perform grouting. In this context, it should be noted that the excavation of the grouting gallery introduced a disturbance in the surrounding rock. Imperfections in the contact grouting of the gallery lining and the joints in the gallery, in some cases, produced drainage towards the lining with relatively steep gradients. The higher grout absorptions around the gallery, registered in the presence of the reservoir, probably were caused by a washing effect due the mentioned gradients. Also the development of the drainage system probably has caused a certain progressive cleaning of fissures which could account for the higher absorptions in the later grouting campaigns. The control of the water from the drain holes does not indicate transport of solids but a content of dissolved salts (mainly bicarbonate ions) about 400 mg/l. The salt content in the drainage water from different areas in the rock mass, with and without grout curtain, is about the same and much higher than in the reservoir. This fact may indicate that the surfaces of the fissures are successively cleaned from deposits of salts and oxides by the percolating water.

In summary, it can be said that only limited areas of the grout curtain can be considered as a positive cut-off, e.g. in the right abutment and around the approach channel. The rest of the grout curtain can be considered as a sequence of grouted holes without interconnections or with fissures too fine to grout with cement. The use of a standard dense grout mix has not proved to be a reason for the generally low grout takes in the first campaign. This was shown by the generally very low takes in water pressure tests.

The execution of a grouting gallery below the dam foundation is of utmost importance for the execution of complementary grouting after impounding as well as a good insurance against construction delays by reducing surface grouting. Finally, it should be kept in mind that the final evidence of a succesful grouting can be seen only in presence of the reservoir. Therefore, it seems logical that the final extension of the grouting works should be based on the observations of seepage and pore pressures due to the reservoir. In this way, the works can be optimized both technically and economically.

7. ACKNOWLEDGEMENTS

The authors wish to thank HIDRONOR S.A. for their permission to publish this paper. Thanks are also extended to Consorcio Consultores Alicura (Franklinconsult, Latinoconsult, Tecnoproyectos, Esin, Elektrowatt, SWECO) as responsible for the design and supervision, and HIDRONOR's Board of Consultants (Drs Deere, Hilf, Lombardi, Lyra and Seed) who reviewed the design, construction and performance.

Civil works were performed by E.C.A.S.A. Empresa Constructora Alicopa S.A. under the leadership of Impregilo S.A.

REFERENCES

Andersson, C.A., Jansson, S. & Vardé O.A. 1985. Grouting and drainage systems at the Alicura Hydroelectric Project, Argentina. Proc. 15th International Congress on Large Dams, Q.58, R. 38.
Lombardi, G. 1985. The role of cohesion in cement grouting of rock. Proc. 15th International Congress on Large Dams, Q.58, R 13. Lausanne.
Pujol, A. & Andersson, C.A. 1985. Instrumentation of the dam and left bank at the Alicura Project, Argentina. Proc. 15th International Congress on Large Dams, Q.56, R.20. Lausanne.
Vardé, O.A. & Andersson, C.A. 1983. Investigation of gravels for an earthfill dam. Proc. 7th Panamerican Conference on Soil Mechanics and Foundation Engineering, P.1-14. Vancover.
Vardé, O.A., Andersson, C.A., Pi Botta, L. & Paitovi, O. 1986. Foundation of the Alicura spillway on weak pelites. Proc. 5th International Congress IAEG. Buenos Aires.

Engineering geological characteristic of 'red beds' in the south of China

Caractéristiques géotechniques des 'couches rouges' dans le sud de la Chine

Hou Shitao, *Investigation & Designing Institute, Hubei, China*

ABSTRACT: There are kinds of "Red Beds" in the south of China, and the areas have become more important for Chinese construction day by day. The stratigraphy behaviour, mineral-chemistry composition, structural tectonic and physico-mechanical properties of "Red Beds" have been studied for projects in this paper. And the strata have been classified from the point of view of engineering geology.
  The treatment experience for housing soil foundation in "Red Beds" in the south of China has been summarized, and some soil foundation designing examples of high buildings on such strata are also presented.

Résumé: Il y a toutes sortes de couches rouges dans le sud de la Chine, et la région de ces couches est devenue plus importante pour la construction chinoise de jour en jour. Le comportement stratigraphique, la composition minérale et chimique, la tectonique structurele et l les propriétés physicomécanique de "Couches rouges" ont été étudiés pour les projets dans ce papier. Et les couches ont été classées au point de vue de géologie de l'ingénieur.
  L'expérience de traitement pour les sols de fondation de maisons dans les couches rouges dans le sud de la Chine a été résumée, et certains exmples de sols de fondation ce bâtiment haut sur telles couches sont encore présentés dans ce papier.

INTRODUCTION

A series of the red beds, which comprise Pre-quaternary red beds and Quaternary red soils, are widely spreaded in southern China. Many valuable data about the red beds which are fully of reference to the geological properties are now being accumulated as the urban areas, especially some specific economic zones are rapidly developing in construction. In this paper, from the view of the fundamental designing, the engineering geological characteristics of the red beds are summarized according to the data which have been obtained, which provides a useful reference for studying the red beds and a base for designing building foundations.

PRE-QUARTERNARY RED BEDS

1. Lithology

The Pre-Quarternary red beds mainly belong to the clastic formation sedimented in the interior river-lake in Cretaceous-Tertiary Period and their outcrops cover an area of 200,000km^2 with a max thickness of several thousand meters. These red beds, composed of sandy conglomerate, sandstone, siltstone and claystone, are distributed in a series of fault-block depressions varisized in shape due to tectonic movements. The red beds display a gradual change in particle sizes, usually from coarse at the margin of the basins to fine at the center because of basin sedimentation, but they are charaterized by poorly sorting, intensive change of facies and thickness, and the strata occur in lens shape. All these characteristics cause the red beds to display nonhomogeneous engineering properties. Main types of the clastic rocks occurring in the red beds are described as follows:
  (a). Conglomerate.
Thick-bedded mainly, composed of gravels, grain sizes and roundness are dependent on the source rocks' nature and transporting distance of clastic sediments. Gravels may form framework or be dispersed in cements. The fragments consist mainly of dolomite, quartz, feldspar and chert. The cements are mainly calcite and quartz. The chemical composition are: SiO_2-55%, CaO-16%,

Al_2O_3-6%, $K(Na)_2O$-5%, Fe_2O_3-2%.

(b). Sandstone.

Main types are quartzarent, arkose and argillaceous sandstone. The percentage of the particles with grain sizes of 0.05-2.0mm is over 5%. The cements are calcareous, siliceous, argillaceous as well as calcareous-argillaceous. The mineral compositions are mainly quartz, calcite, clay minerals, and chemical compositions are: SiO_2-55%, CaO-16%, Al_2O_3-6%, $K(Na)_2O$-5% and Fe_2O_3-2%.

(c). siltstone.

The grains of 0.05-0.005mm in size make up for over 50% and those of below 0.005mm still make up for over 15%, the rest are sand grains (calcareous or argillaceous-cemented). The main minerals are quartz, calcite and clay minerals. The chemical compositions are: SiO_2-52%, CaO-12%, Al_2O_3-11%, Fe_2O_3-3%.

(d). Claystone.

Clay grains make up for over 30% and even up to 80% if silt ones are included, some claystones display lamina apparently and usually be called argillaceous shale. The minerals mainly contain clay minerals, calcite, quartz and feldspar. The main chemical compositions are: SiO_2-58%, Al_2O_3-11%, CaO-9%, $K(Na)_2O$-6%, Fe_2O_3-3%.

In the red beds mentioned above exist a great amount of clay minerals, of which the content increases as the grain sizes decrease, say, can be up to 30%-50% in argillaceous siltstone and clayrocks. This increasing content would cause the hydrophilicity and the softening coefficient of rocks increasing while their compressive and shear strengths decreasing. In addition, the rocks with a higher content of argillaceous matter are liable to argillation during the tectonic process. Argillation is a unique engineering geological characteristics of the red beds.

From the view of structrue deformation, the red beds themselves are characterized by their gentle folding, tilted fault blocks, from aspect of sedimentary rhythms, a gradual transition of lithological characters from top to bottom is generally presented in this red beds, which is utterly different from the pre-cretaceous sediments that have been intensively folded and have a conspicuous angle discordance. These structure characters do control various set-up forms of the interstratal shear within the rock bodies of the red beds and therefore decide the forming of some special engineering geological properties in the beds.

2. Physio-mechanical Properties

As described above, owing to the intensive change of the facies and lithology in the red beds, a great difference in physio-mechanical properties is displayed in them. The physio-mechanical properties of several main rocks of the red beds are listed in following table:

The physio-mechanical properties of main rocks in the red beds

name of rock	dry unit weight g/cm^3	percentage of claygrains	soften. coefc.	saturated compress. strength	shear strength fric co-effic ience	kg/cm^2 cohes.	Young's modulus kg/cm^2
conglomerate	2.3-2.6	little	0.4-0.8	100-800 $\frac{kg}{cm^2}$	0.5-0.8	3.0-40.0	$n \times 10^{4-5}$
sandstone	2.0-2.7	5%	0.2-0.8	40-900	0.1-1.0	2.0-50.0	$n \times 10^{4-5}$
siltstone	2.0-2.5	10-30%	0.2-0.6	30-300	0.4-0.7	0.1-20.0	$n \times 10^4$
claystone	2.4-2.7	20-50%	0.1-0.7	30-350	0.3-0.5	0.1-15.0	$n \times 10^4$

These properties appear to display the following rules in the relation between grain size or cement composition and strength:

to grain size: the greater the amount of the coarse grain content, the higher the shear and compressive strength.

to clay fraction: the greater the amount of clay content, the lower the physio-mechanical indexes, but the higher the hydrophilicity.

to cement: siliceous cement has the greatest strength, ferroceous next, argillaceous lowest.

to stratification: lower strength when parralel to the bedding plane, higher when perpendicular to the bedding plane.

3. Major Engineering Geological Problems

The major engineering geological problems of the red beds are: nonhomogeneity, weak intercalation of argillation, weathering and solution of the soluble salts, etc..

The nonhomogeneity is due to facies change and lithological difference in the red beds in addition to all argillation and weathering of the weak intercalation under variant geological conditions which also strengthen the nonhomogeniety of the red beds.

The argillation of the weak intercalation of the red beds is a special problem which has been studied thoroughly in investigating the site of Gezhouba Dam acrossing the Changjiang River. Such a argillation procedure results from interbedded shearing which may be caused by the relative displacements between the beds when the horizontal strata were folding and tilting.

The original structure within the interbedded shearing zone of the red beds has been destroyed, the width of the fissure within the zone increased, the rock permeability rised and the shear strength reduced to the residual one, so that a sliding surface formed damaging the stability of building foundations.

The climate in south China is warm and wet, the weathering processes are therefore very intensive in this region. The sickness of this intensive weathering zone of claystone is about 5 meters in general, and the greatest one can be over 10 meters. The claystone usually produces volumetric swelling under weathering process due to its main compositions of montmorillonite or illite, which is reflected not only in natural weathering zone but also in the case of artificially changed natural conditions such as excavations of basements and underground openings. The contracting will follow once the swelling takes place. This alternation of swelling and contracting take places repeatedly as the change of climate and natural conditions, which will terribly be harmful to buildings.

It is well common to find rock salts, gyps and mirabilite, and some of them even form large industrial deposits. In engineering, solution of salt-bearing red beds as well as migration and redistribution of SiO_2, Al_2O_3 fraction, ect., in argillation process must not be negligible, this process is, in engineering construction called "Red beds' karst". It is also notable that karstenite can intensively swell when encountered with water, changing into gyps.

4. Assessment of the Pre-quaternary Red Beds as a Ground Base of Buildings

Although the red beds formation is a group of modern sediments, they had become rocks after lithification under diagenesis, and they have deformed slightly, so they can be used for building bases.

The red beds are characterized by low strength, high hydrophilicity, intensive facies change along with difference in lithological properties and liability to being weathered. These characters bring about the obvious nonhomogeneity of the red beds. And perhaps due to composition of gyp-salts of the red beds and their Karst phenomenon, the engineering conditions are very complicated. Fortunately, only if a better understanding of the characters mentioned above and a careful investigation aimed at engineering geological considerations are available, a right assessment of the red beds as a ground base can be obtained.

Taking the City of Guangzhou as an example, where the greatest burried depth of the red beds is about 30m, and the beds outcrop at somewhere. Many tall buildings use the beds as competent bearing strata of their footings, for instances, the China Restaurant is immediately lying on the beds' outcrops and the White Cloud Hotel, with a weight of about 100,000 tons, is supported on shallow-burried strata by 287 concrete piers of 1.0-2.0m in diameter. For the White Swan Hotel however on the Pearl River, since it is located in a region with recent thick sedimentary and a high underground water table, a group of large diameter piles were used to transfer the total loads of the building to the deeply burried red beds. All these multistorey buildings have been safe in service since completion and up to now no abnormal phenomena have been found yet, they are the typical successful instances of using the red beds as ground bases.

In recent years, large diameter drilling machinery has gradually been employed by civil engineering departments, and it has been achieved good results that by means of the red beds' nice drillability, piles are drilled to such a depth that the friction between pile bodies and the red bed rocks can be fully brought into play to raise the load-bearing capacity of individual piles when the upper parts of the red beds are bearing weights. In a certain project for example, a long pile of 1.8m in diameter drilled to the depth of 20m bears a weight of over 1,000 tons; and it has become turn to support each column only by a single pile.

QUARTERNARY RED BED

The quarternary red beds are most widely spreaded in south China and of significant value to engineering practice, which consist of alluvial-deluvial clay and resisual clay of granite and carbonate rocks. All these soils are looking reddish, which reflects their common "experiences" in some time intervals of geological history. Although the soils have some common characters because of the common experience mentioned above, their engineering properties are extremely different from each other owing to different geneses and compositions.

1. Early Quarternary Alluvial-deluvial Clay

The early-middle Pleistocene clay is mainly represented by the "Baishajing" network red soil-a brown-red to yellow-red stiff clay in Hunan province. The texture of the soil is very fine. The content of the clay grains with sizes of below 0.005mm is 30-60%. There is a tendency that the grain sizes are getting coarser downwards, and somewhere the soils have transformed into sandy gravels and formed a double-strata structure. The void ratio of these soils is 0.700-0.750, the internal friction angle 26°-30°, the cohesion 0.60-0.80 kg/cm², the compressibility coefficient 0.010-0.020, and in-situ critical plastic load (P_c) through plate test as high as

8-10 kg/cm². However, the strength of the deep soil strata displays a decreasing tendency. The network red soil almost consists of kaolinite, and a small amount of illite is found only in some individual samples. The chemical compositions of the soil are; $SiO_2 + Al_2O_3 + Fe_2O_3 > 90\%$ ($Fe_2O_3 > 90\%$), and the chemical properties are by far more stable.

The network red soil commonly outcrops at high terrace, in Hunan province the soils usually form the third step terraces. The terraces are often cut by later erosion, forming low hills. Besides in Hunan, these geomorphic landscapes are widely dispersed in the provinces of Zhejiang, Anhui, Jiangsu and so on. Because of the Neotextural motion, however, this kind of soil is somewhere burried beneath the younger strata. In this case, the water content increases while the strength decreases, but compared with other soil strata, the red soil is characterizes by regular laying down, simple mineral composition, high clay grain content, stable physiochemical and mechanical properties. The previous underestimation of the soil's strength in architectural engineering (load-bearing capacity in use is only 2-3 kg/cm²) has resulted in wasting a lot of money. It has proved to be suitable to make full use of the high strength of the red soils as natural bases for multistoreyed buildings and to introduce a compensating or strengthening base for 10-20-storeyed building. Even for much higher buildings, priority should be given to excavated piles not penetrating through the red soils or large diameter grouting piles in order to fully use the soils' strength, this is because it is convenient to construct buildings in this red soil basically with no underground water.

2. Granite Residual Soils

The granitic residual soils, another type of red soil, are dispersed in the Naning Mountains and along the coastal areas of the South China Sea. The original structure still remains in them, when these breccia-bearing clays are imperfictly weathered. It is weell-known that the weathered zone of granite is great in thickness and very nonhomogeneous in properties. Such phenomena are closely related to the texture of the rocks themselves and, to be more important, to the structure zone - a local deep weathering one along a tectonoclastic zone. In that weathering zone, the void ratio is as high as 2.000 or more, the liquidity index is also over 2.0, which in general will bring about a badly hidden danger to the buildings.

This granite weathering zone can generally be classified into three subzones, for example in Shenzhen city, the zone near surface, the middle zone and the zone near parent rocks, which are described as follows:

(a). The zone near surface.
It consists mainly of sand-gravel-bearing clay, most of which are brown-red, vermiform clay soils with white net-liked lines in them, having a void ratio of about 0.800 and a compressibility coefficient (a_{1-2}) of 0.025 and an in-situ plasto-critical load of 5 kg/cm².

(b). The middle zone.
This zone comprises clay or subclay with sands and gravels, the white networks are heavy and vague. Most soils look brown-reddish and there are many weathered residual blocks in this zone. In some local areas, the zone enriched with finely granular granite residual soils and kaolinite soils weathered from feldspars is quite different from the surrounding rocks either in forms or in physio-chemical and mechanical properties. This zone as a whole, has a void ratio of over 0.009 and a compressibility coefficient (a_{1-2}) of up to 0.040 and an in-situ plasto-critical load of 3.0 kg/cm² or so.

(c). The zone near parent rocks.
The subclay dominates in this zone, containing more gravels and fragments weathered from rocks which are briefly brown-coloured with a void ratio of 0.900, a compressibility coefficient of about 0.030 and an in-situ plasto-critical load of 8.0 kg/cm² or more. This zone is gradually transforming into the parent rocks.

The existence of such three zones reflects, on one hand, the weathering tendency from strong extent to gentle as the increasement of the depths, and on the other hand, the tendency of lateritization to becoming gentle from surfaces to deep parts in a damp and hot climate.

From aspect of building foundation, the main engineering geological problems are nonhomogeneity and high compressibility of the granite residual soils. While using the high strength of the zone near surface, the high compressibility of itself and the relative weakness of the middle zone must be taken into consideration, both of them are unfavourable factors.

3. Residual Red Soils of Carbonate Rocks

There exists another type of "red soil", i.e., the one which is formly named as "The Red Soil" in "Chinese Designing Standard Concerning Nature Bases". It is now considered that the "true" red soil should be residual materials of the carbonate rocks, which are gradually becoming "secondary red clay" through retransporting and re-sedimenting. The typical engineering geological properties should be represented by the "true" red soils.

The main mineral compositions in this typical red soils are kaolinite and a small amount of illites. The colloidal clay (grain sizes less than 0.002mm) makes up for 60% or more of the total composition and its high dispersibility decides a high hydrophilicity of itself and consequently, its liquidity index is generally over 80% and void ratio is over 1.000. Althouth the physical properties of the clay are as such mentioned above, its bearing capacity near the surfaces may be up to 4-5 kg/cm² or higher. It should be noted that most red clay has a swelling-contracting property to a certain degree, its swelling force is about 0.4 kg/cm² and swelling capacity about 0.8% or so and contraction ratio about 3.5%.

This red soil displays another conspicuous character, i.e., its considerably changeable physical and chemical properties along vertical direction. It is within the strata with a total thickness of 10m that the soil's state changes from solid one at the surface to plasto-flow one near the bedrocks and the strength reduces to 1.0 kg/cm² or less. Because the underlying bedrocks are carbonate rocks. Karst phenomenon is another major troublesome engineering geological problem in this red soil region. The earth caves or so-called "red caverns" are most harmful to buildings.

The high strength of the surface soils is sometimes fully used in designing foundations on red soils; however, this is just suitable for not very high building or not very heavy structures. In this case, it is important to compromise between the shallow-burried base and the zone with a violent swelling-contracting property; but for multistorey buildings or fairly heavy structures with deeply-burried bases, what is essential is to assure the engineering geological problems associated with Karst topography.

4. Assessment of the Engineering Geology of the Quarternary Red Bed

From the statements hereinabove, it can be seen that the Quarternary red beds have a common character of surface layers' having high strength, even for higher void ratio and smaller dry specific gravity, the strength of the red soils is still high compared with other kinds of soils, this results from lateritization.

The chemical analyses show that the ferric oxidation abuntantly concentrated during the lateritization process. In these ferric oxidations, the free ferric iron takes over two thirds of the total iron content. Most of ferric oxidations precipitated around the surfaces of other clay grains in the form of "enclosure membranes" or filled into the grain voids, cementing the clay grains together. In addition, this red soil contains great amount of clay grains and the static electric attraction exsists between the grains due to plat-like form of the clay grains having different charges at their edges. On the other hand, the kaolinite is a dominating clay mineral in the red soils, and the oxygen of silica tetrahedrons and hydroxyl of alumina octahedrons are exposed outside in crystal structures, so the oxygen ions can be closely bonded with hydroxyl ions by hydrogen chains, forming a micro poly-aggregations with a face-to-face orientation. In turn, these elementary aggregations combine together to form large assemblages with edge-face, face-face and edge-edge orientations. These large assemblages are the best form in texture which bring the cementation of the free ferric oxidates into play. All the facts stated above are useful to raise the strength of the red soils.

As concerns the red bed of early-middle Pleistocene, represented by "Baishajing" red strata of Hunan province, its strength changes slightly only when the extents of lateritization are different (strong or weak), nevertheless from the point of view of the whole strata, its physical-mechanical properties are uniform and stable enough, in other words, its high strength should be fully used when designing a foundation.

The layers near the surfaces of the granite residual red soils were strongly weathered though, their strength are high due to strong lateritization. For the layers near parent rocks, most of their structure strength still remains, which makes their strength high; however, for the middle zone, because their weathering is stronger than that of the zone near parent rocks and the lateritization is weaker than that of surface layers, its strength is relatively low. In addition, the lithological variation of granite mass itself results in quite different physiochenical properties of the weathered strata, usually forming a underlying weak bed beneath the very strong surface layers. Consequently, this fact must fully be considered when the granite residual soils are used as foundations.

The surface layers of the residual red soils of carbonate rocks have a high strength, but the soil grains in the layers are so highly dispersed that the hydrophilicity is also very high, which makes the deep red soils soften rapidly, creating an underlying weak bed unfavourable to engineerings. Besides, because of illite exsisting in the soils, they have a certain swelling-contracting property to which full attention should be paid if this soil is considered as ground foundations.

REFERENCES

Hou Shitao., (1964): Quaternary geology and new structural movements in the middle-lower reaches of the Xiang Jiang River. Proceedings of the 2nd congress of the Quaternary period research commission Academia Sinaca.
Hou Shitao., (1980): Origin types of the soil with swelling and shrinkage characters and its Engineering geological properties.

Ju Huiyang., Gu Renjie., (1984): Geological properties of the red beds and their engineering geology consideration in the south of China. Proceedings of the 2nd congress of engineering geology made by Geological society of China.

Sun Weidong., (1984): Comprehensive testing research on physico-mechanics property of granite eluvial soil in city Shen Zhen. Proceedings of the 2nd congress of engineering geology made by Geological society of China.

Wang Youlin., (1979): Composition and structural behavior of the Quaternary red clay in city Pugi related to engineering properties.

Wang Youlin., Xiao zhenshun., (1982): The microstructure and the behavior of the mudded faulted zone in weak intercalation. Chinese Journal of rock mechanics and engineering. I.S.R.M National group of China.

Xiang Chunyao., (1984): Research roport on engineering geological properties of mottled pattern red soil in province Hunan and evaluation of foundation on it.

Geotechnical investigation for 320 km transmission facilities
Investigations géotechniques pour facilités de transmission

Kin Y.C.Chung, *Gilbert/Commonwealth, Jackson, Michigan, USA*

ABSTRACT: In order to adequately design and construct more than 1500 transmission towers, several large substations and a 1220m long submarine cable crossing under the Hudson River, a complete geotechnical investigation was conducted in east central New York, USA. This paper describes the application of engineering geology to the practical foundation design of transmission facilities within a geologically variable region.

RÉSUMÉ: Pour projeter et construire plus de 1500 tours de transmission, quelques grandes sous-stations et une traverse de cable de longueur de 1220m au dessous de la riviére Hudson, une investigation complete était conduite a est-centrale New York, USA. Cet exposé va décrire les applications de géologie-ingenieur en desseinant fondations practiques de facilités de transmission dans une région complexe géologique.

1.0 INTRODUCTION

Geotechnical investigations were conducted to determine the surface and subsurface conditions that would affect the design and construction of the foundations of 345 kV transmission facilities. The proposed facilities, comprising Facilities A, B, C, D and a submarine cable crossing (Figure 1), will interconnect the Marcy/Edic Substations near Utica, New York, to a new substation at East Fishkill, New York.

The transmission line traverses several of the formational and structural features that characterize the geology of Eastern North America. The rock succession crossed by the transmission line includes all of the common types of sedimentary rocks. Glacial deposits of Pleistocene age sand, gravel, clay and till mantle much of the route.

Field investigations included a seismic survey, an earth resistivity survey and test borings. Test borings were made both on land and offshore in the Hudson River and included sampling and in-situ testing. In-situ tests included vane shear tests in the offshore borings and Susquehanna River Valley, and water pressure and pressuremeter tests at problem locations in soft rock formations along the route. Laboratory testing consisted of soil and rock classification and strength tests.

As a result of the investigation, general and specific soil, rock and groundwater conditions were identified. Soil and rock design parameters, foundation design criteria and construction methodologies were established. Typical foundation types for the transmission structures such as drilled piers, steel grillages and rock anchors were designed. Trenching and backfilling were selected for installation of the cables at the submarine cable crossing. To increase the bearing capacity of the extremely soft plastic clay of the river bottom, filter fabric will be placed on the trench bottom.

2.0 FIELD AND LABORATORY INVESTIGATIONS

2.1 Borings

More than 400 borings were drilled and sampled along the original and alternate overhead transmission routes, and in the Hudson River at the submarine cable crossing. The borings varied from 1.8 to 37m deep and were drilled using either track-mounted or truck-mounted air/mud rotary equipment. Drilling in the river was conducted from an 18 by 7.6m barge. Shallow soil drilling was accomplished using hollowstem augers while deep drilling utilized casing and rotary wash methods with water or bentonite drilling fluid. Rock was drilled using NX diamond bits and core barrels. Soils were sampled at 1.5m intervals using the Standard Penetration Test (SPT). The SPT value is defined as number of blows of a 63.6 kg hammer falling 76.2cm required to drive a 5cm outside diameter split-spoon sampler 30.5cm. Relatively undisturbed samples using Shelby tube, Osterberg and Denison samplers were obtained at selected elevations in some borings at the discretion of the Engineer/Geologist during inspection.

Rock cores using NWX core barrels were taken continuously at the Engineer's direction. SPT, undisturbed sampling and rock coring were performed in accordance with ASTM methods D1586, D1587, and D2113, respectively. Soils encountered during drilling were described in accordance with the Unified Soil Classification

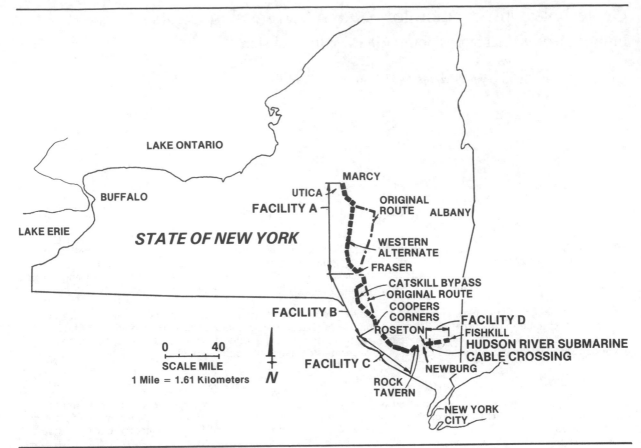

Figure 1. Project location.

System (ASTM D2487). Rock was classified by its type and its Rock Quality Designation (RQD). The RQD is based on a modified core recovery procedure which, in turn, is based indirectly on the number of fractures and the amount of softening or alteration in the rock mass as observed in the rock cores from a drill hole. Instead of counting the fractures, an indirect measure is obtained by dividing the total length of hard and sound core pieces which are 10.2cm or greater in length by the total length of the core run (Stagg & Zienkiewicz, 1968).

2.2   Seismic refraction survey

A seismic refraction survey was performed to determine the depth to rock at some tower locations between borings along the transmission route. The seismic crew utilized a Geometrics Seismograph, Model ES-1210F, 12-channel signal enhancement seismograph with built-in hard copy recorder.

2.3   Earth resistivity survey

To measure the actual resistivity of the soils for design of grounding systems, an earth resistivity survey was performed at each boring location. Survey crews utilized a Bison Offset Sounding System (BOSS) Model 2365 and Earth Resistivity Meter, Model 2350. The survey applied Wenner configuration using spacings of 1/2, 1, 2, 4, 8 and 16m.

2.4   In-situ vane shear tests

In-situ vane shear tests were performed at several soft ground locations along the overhead transmission line and in the Hudson River along the submarine cable crossing. Drilling in the river was conducted from a barge. A special submersible platform which supported the drilling casing and the vane shear device was provided to isolate the testing from the operations of the barge. This submersible platform has a base 1.7m square, made of .64cm steel plate. Two types of vane shear devices (Geonor and Roctest M-1000) were used for determining the undrained shear strength of the river sediments.

2.5   Pressuremeter tests

More than 15 boring locations were selected for pressuremeter tests in soft rocks using a Menard Pressuremeter Model G-Am from Roctest. Working pressures up to 100 bars were utilized for the tests.

2.6   Laboratory testing

Soil laboratory testing programs included classification, unconfined and triaxial compression strengths, moisture-density, consolidation, pH value, organic content, sedimentation, conductivity, salinity, and thermal resistivity of the soils. Rock laboratory testing programs included unconfined compression and pulse velocity.

3.0   GEOLOGY

The area of southeastern New York State crossed by the proposed overhead lines shares several of the formational and structural features that characterize the geology of eastern North America. The rock succession includes all of the common kinds and types of sedimentation. Plutonic and dike rocks are also present in the eastern region. Glacial deposits of Pleistocene-age sand, gravel, clay and till mantle much of the proposed route (Newland, 1933, and SUNY, 1966).

The rocks exposed in the areas of Facility A are of the Ordovician age in the north, and Silurian and Devonian in the south end. Most of the rock types are shale, siltstone, sandstone, limestone and dolostone. The strata are essentially flat-lying in Facility A.

The rocks exposed in the areas of Facility B and the northern part of Facility C are of early Upper Devonian age, and consist of shale, siltstone and sandstone. These rocks make up most of the Catskill Mountains. As described by Rich (1934), the Catskill Mountains stand at the northeast end of the great Allegheny Plateau, which extends as a definite physiographic unit all the way from Tennessee along the western border of the folded Applachians mountains. The present mountainous form of the Catskills is due entirely to the action of streams in carving deep valleys in the flat-lying rocks of the uplifted plateau. The fact that the mountains rise about 610m above the adjacent parts of the Allegheny Plateau seems to be due to the superior resistance of the rocks of which they are composed.

The larger part of the area of the Catskills is covered by till or, as it is sometimes called, ground moraine. The till is an unsorted mixture of clay, sand, cobbles and boulders of all shapes and sizes. The relative proportions of the constituents depend upon the local conditions and upon the mode of deposition. The glacial till generally shows a more compact portion, "the hardpan," which is overlain by a less compact portion that grades imperceptibly upward into the surface soil (Rich, 1934).

Rocks exposed in the southern part of Facility C are predominantly of Ordovician, Silurian, and lower Devonian ages, which consist of shale, siltstone, sandstone and limestone. The rocks at the south end of Facility C are of Cambrian-Ordovician age and consist of intensely folded and faulted shale, limestone and graywacke.

The proposed transmission line in Facility D is located at the northern edge of the Hudson Highlands. The rocks, which are slightly metamorphosed and recrystallized, are of Cambrian-Ordovician age and consist of limestone, folded and faulted shales, slate, sandstone and mudstone.

4.0   SEISMICITY AND WIND

Seismic risk evaluation for the proposed transmission route based on the seismic risk map (Algermissen and Perkins, 1973), indicates the region as having earthquakes of low to moderate magnitude and intensity that correspond to Zone 1 and Zone 2.

The largest historically recorded seismic event occurring nearest the route is the Lake George earthquake of April 10, 1931 (Eppley, 1965), with an epicentral intensity VII on the Modified Mercali Scale (MM). The epicenter was about 137km east of Utica and about 258km north of Newburgh.

Taking into account the frequency of earthquake occurrence (Applied Technology Council, 1978), an average earthquake peak acceleration value of 0.1g was obtained. All structures were designed to withstand Zone 2 seismic load.

A mean recurrence interval of 100 years has been selected due to high sensitivity of transmission structures to wind loading and the reliability required by this project. The basic wind speed selected was 136.8km per hour at 9m above ground. Velocities were corrected to the average wire and structure heights using one-seventh power law (Thom, 1968).

The foundation design was governed by the wind loading in this project.

5.0   TOPOGRAPHY AND SURFACE CONDITIONS

The topography along the proposed transmission route varies from steeply to gently sloping. Topographic profiles (G/C reports, 1984a, b and c, 1986a and b) indicate a maximum elevation of 740m NGVD (National Geodetic Vertical Datum) at the Catskill Mountains in Facility B and a lowest elevation of -18m NGVD at the Hudson River. The surface conditions at boring and seismic survey locations along the route have been identified by rock outcrops, boulders and seasonally wet areas.

6.0   SUBSURFACE CONDITIONS

6.1   Facility A

Overburden soils vary from zero to more than 37m thick and consist of glacial deposits such as sand, gravel, clay and till. A major portion of the overburden soil is composed of glacial till which is a term applied to the heterogeneous mixtures of particles ranging from boulders to clay. Except for soft soil locations in the valleys of the Mohawk River, Cherry Valley and the Susquehanna River, the overburden soils have standard penetration blow counts ranging from 15 to over 100 blows per 30.5cm.

Glacial soils are classified as stiff to hard consistency and dense to very dense compactness for cohesive and cohesionless soils, respectively. The depth to rock ranges from zero at surface outcrops to greater than 37m in the Mohawk River Valley. Bedrock consists of black shale, gray limestone, and gray or reddish brown siltstone, sandstone and mudstone. RQD varies from 0 to 100 percent. In most of the boring locations, black broken shale with 0 percent RQD was found during coring. Most of the unbroken rock cores recovered from drilling such as sandstone, limestone, siltstone and mudstone were relatively hard with pulse velocities higher than 3050m/sec. Low RQD was caused mostly by the often interbedded shale layers.

6.2   Facility B

As stated in the geology section, the overhead transmission route in Facility B crosses the Catskill Mountains. Surface rock outcrops and boulders were observed throughout the route. Virtually all of the ridge tops and steep slopes were underlain at shallow depths by bedrock. Low areas between ridges usually contained more than 13m of glacial till. Most of the borings in the till encountered numerous boulders and cobbles. SPT values in the till range from 7 to over 100 blows per 30.5cm. In most areas, the till is classified stiff to hard in consistency and dense to very dense in compactness for cohesive and cohesionless soils, respectively.

The rock in Facility B was found relatively shallow at most of the locations along the route. Data obtained from the seismic survey indicate that rock depths range from the surface to 21.6m. The rock consists of red shale, gray to reddish brown sandstone, siltstone, claystone and mudstone. Sandstone is the predominant rock type in Facility B. The RQD varies from 0 to 100 percent. The low RQD was mainly caused by the interbedded soft shale within rock cores.

The boring program revealed that there is at least one loose soil location along the route in Facility B which requires particular attention. The soil at Mongaup River consists mainly of loose to medium dense silty sand with fine gravels overlying the sandstone bedrock at 16.2m below ground surface.

6.3   Facility C

The northern segment of the transmission route (north of Neversink River) in Facility C is still in the Catskill Mountains and, therefore, has geology similar to that in Facility B. Surface outcrops and boulders are found along the major portion of the route. The overburden soil is glacial till having a consistency or compactness of stiff to hard or dense to very dense for cohesive or cohesionless soil, respectively. Rock depths are relatively shallow at most of the locations along the route and range from the surface to over 13.7m. The rock consists mainly of gray conglomerate, sandstone and red shale with RQD varying from 0 to 95 percent. In most cases, the conglomerate and sandstone were interbedded with soft shale.

The southern segment of the transmission route in Facility C is in a different geological region where the Great Valley of the Appalachian Ridge and Valley Province are located. The rock in this area consists mainly of dark gray mudstone, slate, siltstone, red shale and gray sandstone. The unbroken pieces of rock cores are hard with high pulse velocities. The overburden soil is glacial till with similar characteristics to the northern portion of the route. Artesian conditions existed at several locations during drilling operations.

6.4   Facility D

The overburden soil consists of dense or hard glacial till along the transmission route. Similar to that found in other facilities, this till is comprised of a heterogeneous mixture of soil particle sizes; however, fewer boulders were encountered during drilling.

The rocks encountered were brown hard siltstone on the west side of the Hudson River and gray broken slate and shale, siltstone, sandstone, and mudstone on the east side of the river. The RQD is very low due to extensive fracturing, ranging from 0 to 40 percent.

During the field investigation, one boring was found to be situated within a large man-made debris dump area. The results obtained from the earth resistivity survey indicate an extremely low apparent resistivity (2 to 3 ohm-meters) within the debris dump material. This reveals that the debris will be highly corrosive to tower foundation materials.

6.5   Submarine cable crossing

The Hudson River sediments consist of gray soft plastic clay occasionally interbedded with silt layers. Shell fragments were also noted. The organic content of the sediments decreases as depth increases. The thickness of this clay varies from zero to 6.1m at the shoreline and 18.3 to 29m in the middle of the channel. Underlying the clay layer is a layer of very dense silty sand (Figure 2).

Field and laboratory test results indicate that the river clays are extremely soft. The undrained shear strength (Su) of these deposits varies from about 2.4 kn/m^2 near the river bottom to 38 kn/m^2 at a depth of 19.8 to 21.3m. The top 3.05m of sediment is gray organic plastic clay and the natural water content ranges from 70 to 100 percent, which is higher than the liquid limit of the soil. The average liquid limit is about 60 percent. The natural water content of the clay decreases slightly with depth. With natural moisture contents near and higher than the liquid limit, these clays are sensitive with a sensitivity value of 2 for the top 1.83m to 6 for a greater depth. Sensitivity is defined as the ratio of the undisturbed and the remolded shear strengths.

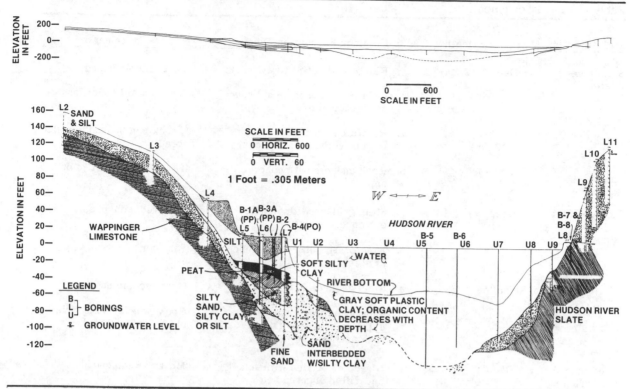

Figure 2. Hudson river crossing geologic cross-section.

Remolded undrained shear strengths vary from 1.3 kn/m^2 at top to 5.75 kn/m^2 at a depth of 21.3m. The clays are slightly overconsilidated with the overconsolidation ratio (OCR) of 1.5 to 2 in the upper 9.1m and of 1.1 to 1.3 below 9.1m. Test data indicate that the submerged unit weight of the river sediment ranges from .45 to .51 g/cm^3.

In order to select a shear strength value for design of the foundation system, the field and laboratory strength data were evaluated and compared with typical values from available literature (Bjerrum, 1972, Duncan & Buchignani, 1973). The top 3.05m soils were found constantly disturbed by the tidal action and currents. It is worthwhile to note that the results obtained from field vane shear tests were under conditions of large failure strains (more than 20 percent) and fast rate of loading. A design shear stress ratio (S_u/σ_{vo}) of 0.32 for the top 3.05m has been selected based on field vane shear results. σ_{vo} here is the effective overburden pressure of the soils.

Results obtained from laboratory thermal resistivity tests on grab samples indicate that the top river sediment has thermal resistivity values ranging from 70 to 92°C-cm/watt (Thermal ohms or Rho) and from 151 to 188° C-cm/watt for needle input power of 0.2 watt/cm and 1.0 watt/cm, respectively (G/C Report, 1984a). These tests were employed on a 100mm long, 1.6mm diameter stainless steel thermal needle. The test procedure was conducted in accordance with IEEE Standard (1981). For lower power case, the temperature applied at the needle was 19 to 20°C at the start of test and rose to 27 to 30°C at the end of test. For high power case, the temperature was 28 to 30°C at start of test and rose to 53 to 60°C at the end of test. The increase of the thermal resistivity values with respect to the increase of the input power indicates the material is thermally unstable.

7.0 FOUNDATION CONSIDERATIONS

7.1 Transmission towers

7.1.1 Foundation types

Foundation types for the proposed transmission towers are mainly dependent upon the structure type, governing loads, the soil and rock conditions, site accessibility, constructibility and economics. In this project there will be more than 40 different types of transmission structures along the route. However, for the purpose of foundation design, structures can be grouped into three categories: Steel pole(s) or steel H-frame; Lattice steel; and Wood pole(s) or H-frame (permanent and temporary).

A study of the recommended foundation types based on proposed structure types, and available soil and rock conditions for this project is summarized in Table 1.

Table 1. Foundation types.

Structure Type	Rock Depth	Common Foundation Type	Remarks
Steel pole(s) or steel H-frame	Deep (>12.2m)	a. Reinforced concrete cylinder	Most recommended
		b. Reinforced concrete cylinder or steel H-pile	For poor soil conditions
	Shallow (<12.2m)	a. Reinforced concrete cylinder with or without rock socket	If rock can be easily drilled or cored.
		b. Rock anchors with concrete cap	If rock is difficult to drill or core through
Lattice steel	Deep	a. Steel grillage with or without concrete slab	Most recommended
		b. Reinforced concrete cylinder or steel H-pile	For poor soil
	Shallow	a. Steel grillage	Most recommended
		b. Short reinforced concrete cylinder with or without rock anchors	Rock excavation by drilling or blasting is required
Wood pole(s) of H-Frame	Deep	a. Direct-embedded with select backfill in augered hole	Most recommended
		b. Above plus uplift soil anchors and/or bearing pad	In very soft soil
	Shallow	a. Direct-embedded with select backfill with uplift anchors and bearing pads as required	Most recommended
		b. Rock anchor	In hard rock

7.1.2 Design methodology

The unique characteristics of transmission structure foundations, in contrast with foundations for other structures, is that they are usually subjected to proportionally large moments, lateral forces, and/or uplift forces. For examples, the ground line moment may be as high as 40,700 kn-m for one of the heavy dead end steel pole structures in this project. Reinforced concrete cylinders, steel grillages and rock anchors are the most suitable foundation types for these loadings.

For design of 1.8 to 3.4m diameter reinforced concrete cylinders in soils, methods developed by Broms (1964a, b, and 1965) were utilized. The Broms methods were modified to take into account rigid drilled piers embedded in soft rock. Rock engineering properties were obtained and evaluated from results of pressuremeter tests and RQD value. For design of steel grillage, the uplift capacity was established using the cone-of-earth method and the bearing capacity was evaluated using the Modified Terzaghi Equation (Lambe & Whitman, 1969). For design of rock anchors, the grout length was established based on one of the following criteria: a) bond strength between grout and the rock, b) bond strength between grout and the rebar, and c) tensile strength of the rebar. For rock anchors in highly weathered and fractured rock such as slate and shale in Facilities C and D with RQD ranging from 0 to 40 percent, the condition (a) generally governs. The bond strength between the anchor and the rock used in design was verified by several in-situ pull-out tests in soft rock at each Facility.

7.1.3 Soil corrosion potential

A number of factors affect the corrosive potential of soils. These factors are: porosity (aeration), soil resistivity, dissolved salts, moisture, acidity or alkalinity, amount of sulfates, and amount of cinders. There is no exact relationship or formula to evaluate soil corrosion potential. However, the common practice based on experiments indicates that highly corrosive soils usually have poor aeration, low resistivity, low or high pH, and high contents of dissolved salts, sulfates, and cinders.

The following criteria have been applied for this project for evaluation of soil corrosion potential along the transmission line route:
*Based on Earth Resistivity (NACE, 1980)

Resistivity, Ohm-m	Category
Below 5	Very corrosive
5-10	Corrosive
10-20	Moderately corrosive
20-100	Mildly corrosive
Above 100	Progressively less corrosive

*Based on pH Values

The soil can be classified as having relatively high corrosion potential, if pH value is less than 5 or higher than 9. A pH value less than 5 indicates that the soil, when saturated, contains relatively high concentration of hydrogen ions. Very high pH values indicate that the soil, when saturated, usually contains sodium-carbonate-bicarbonate waters.

7.2 Submarine cable crossing

7.2.1 Trench stability

Results of the investigations confirm that the original design concept to install six pipe-type submarine cables across the Hudson River in an excavated trench is feasible providing that certain conditions are met. Trench stability is governed by the following factors: thermal backfill, trench depth, trench side slopes and settlement.

Results obtained from laboratory thermal resistivity tests indicate that the top river sediments are thermally unstable. Since these soils are organic and plastic clay in nature, moisture content and permeability will change under the long-term action of any heat source. Special backfill material consisting of sand with a thermal resistivity under saturated conditions less than 70 Rho has been identified in the construction specification.

Based on the bearing capacity analysis, the safety factor against bearing capacity failure of the trench bottom using the relatively undisturbed undrained shear strength of the river sediment is 1.3 for a 1.83m deep trench with 1.83m backfill. Due to the uncertainties associated with the construction operation (nonuniformity of the backfill thickness), the effect of currents and wave action, and the unknown soil strength of the top 0.61m of sediment, this safety factor (1.3) is considered to be too low. At low factor of safety, creep to failure has been observed. If the remolded strength is used, the factor of safety reduces to 0.87. The most likely bearing capacity failure mode for soft soil at top and stiffer soil at bottom would be lateral plastic flow. Therefore, should the trench slope and bottom fail, some granular cover will be lost laterally. In order to satisfy environmental requirements and to increase stability during construction, three alternatives were considered during design:

1. Maintain the 1.83m trench depth and cover the trench with the filter cloth prior to placement of the sand backfill and the pipes

2. Maintain the 1.83m trench depth and use light-weight aggregate to replace the granular backfill originally proposed. The thermal characteristics of the light-weight aggregate should, therefore, be checked against the design criteria

3. Increase the trench depth to 2.44m and maintain the thickness of granular backfill at 1.83m. Based on the cost comparison and the limit size of the upland disposal area for excavated river sediments, alternate (a) was selected for construction.

Settlement analysis for the combined pipe-backfill system was performed. The maximum total settlement consists of 47cm of consolidation settlement and 25.4cm of secondary settlement (creep). The differential settlement between two points 30.5m apart along the cable alignment can then be estimated to be less than 30.5cm. Based on Terzaghi's one-dimensional consolidation theory, an estimated three years is required to obtain 90 percent of total consolidation settlement for an average clay thickness of 18.3m. Due to the presence of silt layers in the sediment and three-dimensional effects, the actual settlement rate will be faster than the predicted value. The maximum amount of settlement is not expected to impact the pipe cables. The stress in the pipe induced by these total and differential settlments has been verified by conventional analysis uisng beam om elastic foundation approach and is negligible.

The final side slopes of the trench are governed by the combination of river currents, tidal variations and strength of the fine-grained soft river sediments. Using the undrained shear strength of the soils, the slope stability analysis indicated that the submerged slope on 1 vertical and 9 horizontal will have a safety factor of 3.6 assuming no current or tidal action is involved. Therefore, with current and tidal action taken into account, a 1 on 9 slope will provide a reasonable estimate of trench excavation and backfill quantities.

7.2.2 Cable exit system

Six 30.5cm diameter pipe cables at the east shore of the Hudson River will have to cross underneath two railroad tracks and rise to the top of the hill and be connected to an East Transition Station. In order to protect the railroad embankment and the hill-side slope, a cofferdam and a bore pit with three layers of tie-back anchors were designed and constructed. Analysis indicated that the present hill-side slope should be excavated to 1.5 horizontal on 1 vertical.

Results obtained from the geotechnical investigation at the west shore indicate that the site is covered by a random brick fill with a thickness varied from 4.6 to 9.1m. Underneath the brick fill, there is a layer of peat with a thickness varying from 3.6 to 7.9m. Soft silty clay and loose to dense sand were found between the peat and the bedrock which is sloping down to the river at an angle of 10 degrees and at depths varied from 29 to 35m. Ground water exists under artesian conditions. To avoid future flooding, 2m average additional fill must be placed on the existing brick fill in order to raise the final grade of the site. Pile foundation was designed and constructed to support the cable exit system and all the structures in the West Transition Station.

8.0 CONCLUSIONS

8.1 Transmission tower foundations

Based on the results from this geotechnical investigation, it is concluded that:

1. The overburden soils along the overhead lines are mainly dense or hard glacial till which is a heterogeneous mixture with sizes ranging from boulder to clay. Poor soil locations and different geologic setting and characteristics along the lines have also been identified.

2. The line traverses a variable geologic setting and difficult surface conditions such as boulders, swamps, steep slopes, etc.; therefore, the depths to rock were precisely located only at test boring. The seismic survey provided depth to rock information which will be used as a supplement to the boring data.

3. Typical soil and rock parameters were developed only as a guide to the foundation design. The actual foundation design was based on the local surface and subsurface conditions as identified on subsurface profiles (G/C Reports, 1984a,b and c and 1986a and b). For poor soil locations, design parameters were developed according to the individual field and laboratory test results.

4. Foundation types and constructibility for each structure location are greatly dependent upon the soil and rock conditions, access availability and site conditions such as steep slope, wet surface areas, artesian conditions and high water table. Therefore, the foundation design is flexible. Alternative schemes were developed for each foundation according to the information as presented in Table 1.

5. Based on results from the earth resistivity survey and pH values at each boring location, the design criteria to identify the corrosion potential of site soils were developed. Protection methods such as special coating, backfilling with offsite materials, or cathodic protection for foundations were considered in design.

8.2 Submarine cable crossing

1. Due to the soft soil condition of the top 3.05m of river sediment, the factor of safety against bearing capacity failure of the trench bottom after placement of the granular backfill was found to be lower than the normally acceptable level. Filter fabric was added in a 1.83m deep trench in the river bed approximately 47.3m wide at the top and 13.7m wide at the bottom to raise the factor of safety and provide a stable base for the backfill material.

2. Stringent control of excavation and backfill trench operations is required in order to minimize large reduction of soil strength and overdredging of the trench bottom. All geotechnical concerns as mentioned previously have already been incorporated into the construction specification.

9.0 REFERENCES

Algermissen, S. T., and D. M. Perkins. 1973. A probabilistic estimate of maximum acceleration in rock in the contiguous United States. U.S. Geological Survey Open File Report No. 76-H16.

Applied Technology Council. 1978. Tentative provisions for the development of seismic regulations for buildings. NIS special publication 510, Washington, D.C.

Bjerrum, L. 1972. Embankments on soft ground. ASCE Spec. Conf. on performance of earth and earth-supported structures, Purdue University. 2:1-54.

Broms, B. B. 1964a. Lateral resistance of piles in cohesive soils. ASCE J. of SMFD. SM-2:27-63.

Broms, B. B.. 1964b. Lateral resistance of piles in cohesionless soils. ASCE J. of SMFD. SM-5:123-157.

Broms, B. B. 1965. Design of laterally loaded piles. ASCE J. of SMFD. SM-3:79-99.

Duncan, J. M., and A. L. Buchignani. 1973. Failure of underwater slope in San Francisco bay. ASCE J. of SMFD. SM-9:687-703.

Eppley, R. A. 1965. Stronger earthquakes of the United States. Washington, D.C.: U.S. Government Printing Office.

Gilbert/Commonwealth. 1984a. Design study of the Hudson river submarine cable crossing-offshore geotechnical investigation. Study No. DS-19. Jackson, Michigan.

Gilbert/Commonwealth. 1984b. Design study of the Hudson river submarine cable crossing-onshore geotechnical investigation. Study No. DS-19. Jackson, Michigan.

Gilbert/Commonwealth. 1984c. Geotechnical investigation of Marcy-South 345 kV transmission facilities, overhead transmission lines. Report No. 2680. Jackson, Michigan.

Gilbert/Commonwealth. 1986a. Geotechnical investigation report for Marcy-South 345 kV transmission lines - western alternates. Report No. 2680 A 2. Jackson, Michigan.

Gilbert/Commonwealth. 1986b. Geotechnical investigation report for Marcy-South 345 kV transmission facilities, overhead transmission lines - Catskill bypass. Report No. 2680B. Jackson, Michigan.

Institute of Electrical and Electronic Engineers. 1981. IEEE guide for soil thermal resistivity measurements.

IEEE Standard 442-1981.

Lambe, T. W., and R. V. Whitman. 1969. Soil mechanics. New York: John Wiley & Sons, Inc.

National Association of Corrosion Engineers. 1980. Basic corrosion course.

Newland, D. H. 1933. The Paleozoic stratigraphy of New York. International Geological Congress, XVI Session, USA.

Rich, J. L. 1934. Glacial geology of the Catskills. Albany, New York: The University of the State of New York.

Stagg, K. G., and O. C. Zienkiewicz (ed.). 1968. Rock mechanics in engineering practice. New York: J. Wiley & Sons.

Thom, H. C. S. 1968. New Distributions of extreme winds in the United States. J. structural division, ASCE. ST7:1787-1801.

The University of the State of New York (SUNY). 1966. Geology of New York - a short account. Educational Leaflet No. 20.

10.0 ACKNOWLEDGMENTS

The author wishes to thank Gilbert/Commonwealth Senior Geologist, A. Brewster; Engineers, M. Razi and P. Mundy; Technician, B. Bearinger; and former Gilbert/Commonwealth Engineer, R. Wagner for their contributions in boring and laboratory test inspections, in performing field and laboratory tests, and in preparing geotechnical reports. The author particularly thanks R. Broad, Project Manager of the entire transmission project for his valuable review and suggestions in this geotechnical investigation program.

La réparation d'un barrage fondé sur rocher tendre: Le barrage Itiyuro
The reinforcement of a dam based on weak rocks: Itiyuro dam

Luis R.Siegrist, *Administración General de Aguas de Salta Argentina*
Luis Garcia Tobio, *Cimentaciones Argentinas-Soletanche, Argentina*

RESUME: En Décembre 1981 le barrage de grande hauteur d'Itiyuro (Argentine) subit un accident de caractéristiques catastrofiques.Le barrage a pu cependant etre sauvé,par la mise en oeuvre de travaux de confortement,particulièrement des injections du terrain de fondation.
ABSTRACTS: On December 1981 the Itiyuro dam suffered a catastrophic accident.The dam could be recovered by works of consolidation,in particular grouting works in the foundation.

1) CARACTERISTIQUES DU BARRAGE ITIYURO

Le Barrage Itiyuro est situé à l'endroit appelé Itaque,dans le nord-ouest de la Province de Salta (République Argentine),à environ 12 km de la frontière avec la Bolivie
Son objectif est de regulariser la rivière Itiyuro,de régime torrentiel,pour irrigation,eau notable et génération d'énergie.
Le Barrage est fait en enrochement,avec un écran en béton sur le talus amont.Il se situe dans une gorge étroite,le talus rive droite étant de 45°et celui de rive gauche presque vertical.
Ses caractéristiques principales sont (fig.1 et 2):
- Hauter maximale depuis le lit de la rivière
  (cote 592,30)   : 57 m
- Longueur du couronnement   :270 m
- Largeur du couronnement   : 6 m
- Talus Amont   :1:1,25
-Talus Aval   : 1: 1,3
-Volumen d'enrochement:
a) Jeté   : 535.000 m3
b) Placé   : 95.000 m3

L' écran d'imperméabilisation à l'amont est constitué d'un parafouille en Béton de profondeur variable (8 a 15m) qui se continue vers le haut par un écran aussi en béton armé appuyé sur des poutres longitudinales et verticales,constituant des panneaux,l.étanchéité desquels est assurée par des joints en cuivre.
En dessous du parafouille un écran d'injections de ciment avait été exécuté,de profondeur variable.
Dans la partie supérieure du parafouille une galerie d'inspection a été construite sur une partie de barrage,pour contrôl de filtrations et éventuelles re-injections du terrain de fondation.
La tour de prise se situe à proximité de la rive droite,sa forme étant circulaire de 3,20m de diamètre,avec trois prises à des niveaux differents.
Un tunnel en béton armé de petite longueur et d'1,80m de diamètre relie la partie inférieure de la tour de prise avec le tunnel de déviation,dans lequel se continue avec une conduite d'acier d'1,50m de diamètre.Cette conduite sert pour alimenter la central de potabilisation.
Sur la rive droite,un tunnel de 4,50m de diamètre a été construit,en béton armé,et d'une longueur de 302m.Pendant la période de construction ce tunnel a été utilisé pour la déviation de la rivière et plus tard pour loger les conduites de la vidange de fond et de l'utilisation.
Le déversoir,situé sur la rive gauche du Barrage,est une structure du type écoulement libre dont le couronnement se situe à la cote 586.
C'est une sutructure circulaire en plan,en Béton armé,avec un développement de 140m et un canal d évacuation ensuite,avec retrecissement graduel,pour terminer dans une rapide qui décharge dans la rivière.
La capacité d'évacuation de projet du déversoir est de 2.300 m3/s.
La construction du Barrage s'est terminée en 1972

2) CARACTERISTIQUES GEOLOGIQUES DES FONDATIONS

- Lit de la rivière
Dans le lit de la rivière on trouve fondamentalement des grès très peu consolidés,avec grains--- siliceux et ciment calcaire très faible.

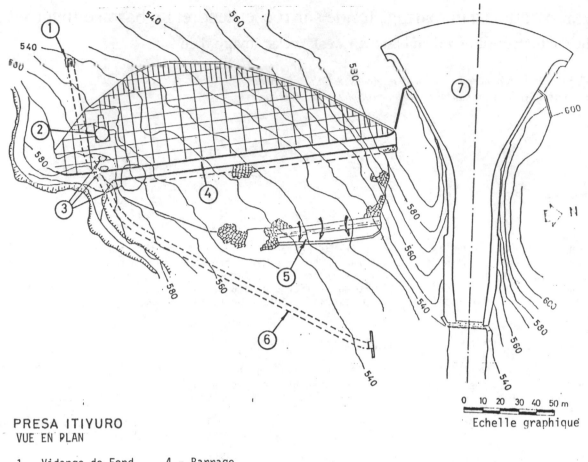

PRESA ITIYURO
VUE EN PLAN

1 - Vidange de Fond   4 - Barrage
2 - Tour de Prise   5 - Fuites   7 - Déversoir
3 - Effondrements   6 - Galerie

FIG.1

Une coupe transversale à l'axe du barrage presente la succésion de couches suivantes, de l'aval a
l'amont:
. Grès et argiles rouges inférieures
. Conglomérat manganésifère
. Conglomérat argileux
. Grès blanchâtres et rosés
. Conglomérat de "Galarza"
. Grès verdâtres
Ces couches presentent une inclinaison vers l'amont de 30°a 35°.
La couche la plus puissante est celle des grès blanchâtres et rosés qui constitue la plupart de la
fondation du barrage.
- Rive Droite:
Dans cette rive le barrage s'appuie sur les mêmes formations de grès blanchâtres et rosés,friables
et perméables,surmontés à l'amont par le conglomérat de "Galarza" (conglomérat dur de graves de
silice).
- Rive Gauche:
Dans la zone d'appui du barrage dans cette rive,les grès sont recouverts d'une couche de calcaire,
disparaissant le conglomérat.
Le Barrage a été situé de façon que le parafouille amont soit fondé dans la couche de conglomérat,
s'agissant d'une couche resistante à l'érosion,bien que de faible épaisseur.
A l'aval de l'axe du barrage et a proximité de son pied,on observe l'affleurement du conglomérat manga
nésifère,très dur qui enferme avec le conglomérat de "galarza"à l'amont les couches de grès erodables.
Au début de la construction du barrage,devant la mauvaise qualité des grès en presence,differentes
solutions on été etudiées pour protéger les fondations de l'érosion. La disposition finalement
retenue a été la suivante:

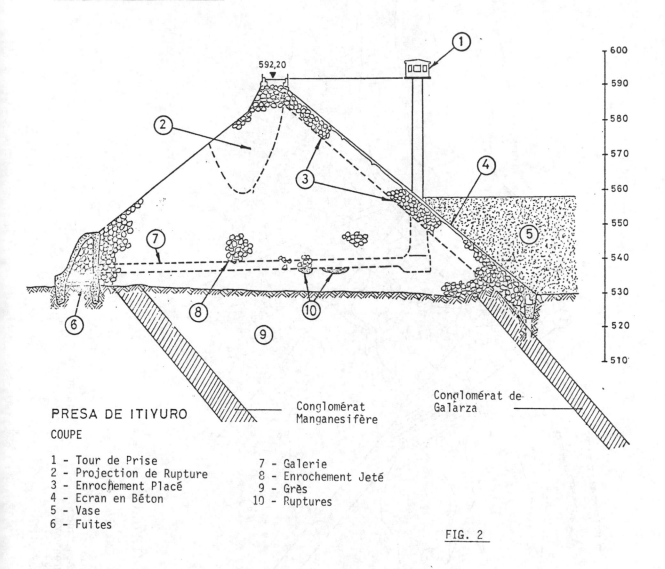

PRESA DE ITIYURO

COUPE

1 - Tour de Prise
2 - Projection de Rupture
3 - Enrochement Placé
4 - Ecran en Béton
5 - Vase
6 - Fuites

7 - Galerie
8 - Enrochement Jeté
9 - Grès
10 - Ruptures

FIG. 2

- Construction du parafouille amont encastré dans le conglomérat
- Recouvrement de la surface du terrain entre le parafouille à l'amont et le conglomerat mangané-sifère à l'aval, par une couche de béton poreux de 50cm d'épaisseur.
- Construction d'un réseau de drains superficiels conduisant l'eau drainée a un collecteur qui dé charge au pied du barrage.
Le débit drainé n'a pas été controlé jusqu'en 1981,mais peût être estimé en 100 1/s.
On espérait de cette façon éviter l'entraînement des grains du grès,en l'enfermant avec des couches resistantes à l'érosion,jouwant le rôle de filtre extérieure.
A la mise en eau,le remplissage du lac se produit très violemment par une forte crue qui remplit le lac en 24 heures.
Comme conséquence, des fissures apparurent sur le talus aval,ainsi que des résurgences d'eau et-sable par zones,pendant quelque temps.Puis les résurgences disparurent et les fissures se sont---stabilisées.
Probablement, en ce moment s'est produit la rupture de la couche de béton poreux.

3) DESCRIPTION DE L'ACCIDENT

Etant à peu près 20.00 heures du 14 Janvier 1981,se produit un effondrement d'un secteur du barra ge sur l'appui rive droite, sur une longueur d'environ 20 m, et sur toute la largeur de la ------chaussée ducouronnement,empiétant aussi sur le talus aval.
La profondeur du fontis était de quelques 20m.

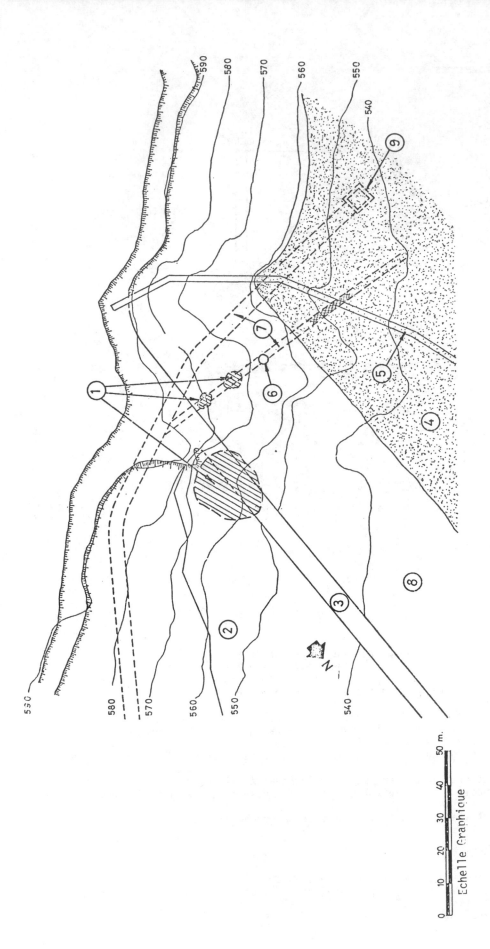

PRESA ITIYURO

APPUI RIVE DROITE-VUE EN PLAN

1 – Ruptures visibles   6 – Prise
2 – Enrochement   7 – Galéries
3 – Couronnement   8 – Ecran en Béton
4 – Vase   9 – Vidange
5 – Parafouille

Echelle Graphique

0   10   20   30   40   50 m.

FIG. 3

Le fontis presente les caractéristiques d'un effondrement vertical,et non pas de glissement du --
talus.Le matériel tombé sur le talus est minimum.
Le volume estimé d'enrochement effondré est de 3000 m3.
Une inspection détaillée a permit d'observer:
a) Près du fontis,des fissures sur le couronnement s'aprochant du milieu du barrage,montrant la-
rupture de l'enrochement par l'effet de talus vers le creux.Ces fissures se prolongeaient jusqu'à
environ 20m en direction parallèle à l'axe du barrage.
b) Augmentation très importante du débit des drains existants avec entraînement de matières solides
c) Talus amont totalement déchaussé dans sa partie supérieure.
Ceci produisait l'affaiblissement de l'appui de l'écran en béton armé existant,recouvrant le talus
amont et qui constitue l'organe d'étanchéité du barrage.
Le barrage était donc en risque de rupture totale imminente en cas de remontée de la retenue,ce --
qui pouvait arriver dans un court délai,étant donné le volume relativement petit de la retenue---
et la grandeur des crues normales dans cette période de l'année.
En effet, la pression de l'eau sur la dalle déchaussée aurait entraîné sa rupture et en conséquen
ce le déversement au-dessus du corps du barrage.

4- MESURES D'EMERGENCE

Devant le danger immédiat de rupture du barrage,les suivantes mesures d'émergence ont été prises:
a) Reconstituer dans le plus bref délai le talus aval utilisant pour remplir le fontis des grès--
friables excavés sur la rive droite à proximité du barrage,ainsi que de gros blocs de calcaire d'
une carrière proche.Ces blocs ont été placés sur le grès,facilitant leur mise en place avec des -
jets d'eau à pression.
Cet enrochement devait garantir en principe l'appui de l'écran en béton,devant le risque des crues.
b) La descente du niveau de la retenue.Cette opération a été retardée par le fait de se trouver--
la vidange de fond bloquée par la vase,l'évacuation d'eau se faisant donc seulement à travers la
conduite de prise.

5- TRAVAUX DE RECUPERATION

Une fois realisés les travaux d'émergence décrits,un plan de travaux a été lancé pour la recupéra
tion definitive du barrage,avec principalement les phases suivantes:
- Réparation du tunnel de prise
- Recherche des causes de l'accident.
- Travaux de confortement des fondations,suivant les résultats des recherches.
Etant donné l'urgence des travaux,les activités indiquées se sont réalisées en partie simultané-
ment.

5.1- Reparation du tunnel de prise
     L' inspection du tunnel de prise montra la présence de ruptures d'importance (fig.3) dans la
zone entre la tour de prise et le début de la conduite en acier,dues à la présence de grandes---
vides a l'extérieur du tunnel.
La reparation du tunnel a nécessité la mise en oeuvre des travaux suivants:
- Remplissage au béton des zones de rupture pour récuperer le profil original du tunnel.
- Injections de consolidation autour du tunnel.Ces injections ont été réalisées avec coulis de --
ciment pur ou de bentonite-cimet,en trois phases succesives:
1- Injections de contact:à partir de perforations de 0,50m de profondeur,destinées a remplir les
vides existants derrière le revêtement en béton.
Dans cette phase des absorptions moyennes importantes ont été obtenues,surtout dans la zone proche
à la tour de prise,où se trouvaient les ruptures.
La valeur moyenne des absorptions a été d'environ 0,5 tonnes de ciment par mètre linéaire de ----
tunnel,avec des pointes dans les zones de rupture allant jusqu'à 1,03 t/m.l.
Ces valeurs sont a rapprocher des valeurs habituelles de 50 kg/m2 de tunnel (soit dans notre cas
0,3 t/ml) pour tunnels injectés par première fois,après leur construction.
La pression de refus utilisée a été de 15 kg/cm2.
Les perforations ont été disposées en auréoles de 3 perforations tous les 2m de tunnel.
2- Injections de remplissage: à partir de perforations de 2m de longueur situées en auréoles de 3
perforations tous les 3m de tunnel.
Ces injections étaient destinées à completer la consolidation du terrain sur une bande de 2m d'
épaisseur avec des coulis de ciment pur.
La pression de refus a été fixée à 12 kg/cm2.
Les absorptions ont été aussi importantes avec une moyenne de l'ordre de 1t/m.l. de tunnel,avec
un maximum de 2,54 t/m.l.
3- Injections de consolidation: à partir de perforations radiales de 15m de longueur,destinées à-
consolider les zones de terrain altérées au sein du massif,en complément des injections de consoli
dation réalisées depuis le couronnement.

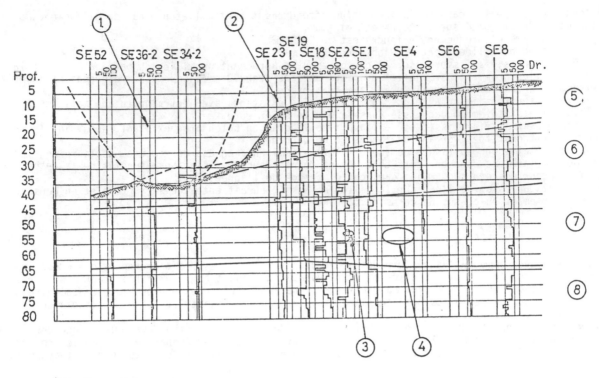

PRESA ITIYURO

ENREGISTREMENTS ENPASOL

NOTE:

Dr= Dureté relative = $\frac{T \times E}{V}$

T = Couple en Kg/cm2

E = Poussée Verticale en kg/cm2

V = Vitesse d'avancement

1 - Remplissage
2 - Enrochement
3 - Galérie tour de Prise
4 - Vidange de Fond

5 - Grès Rosés
6 - Grès Blanchâtres
7 - Grès à Bandes Rosées
8 - Grès Rouges

FIG.4

Ces injections ont été faites suivant la même métodologie qui est indiquée plus bas pour les in---
jections de consolidation avec des coulis thixotropiques de bentonite-ciment.
L'absorption moyenne obtenue dans cette phase a été de 0,22 m3/m de perforation.
c4-Blindage du tunnel par mise en place d'une conduite en acier d'1,30m de diamètre. En même temps
la totalité de la conduite d'acier dans le tunnel de déviation a dû être remplacée, à cause du --
mauvais état de la conduite originale.
d5-Injections d'étanchéité autour du tunnel pour imperméabiliser les zones proches au même,évitant
les zones de circulation préférentielle de l'eau,et donnant un peu de cohésion aux zones sableuses
provenant de l'altération du grès qui sont resté englobées dans les injections de ciment effectuées.

5.2- RECHERCHE DES CAUSES DE L'ACCIDENT
    Dans le but de rechercher les causes de l'accident,une série de travaux de reconnaissance ont
été réalisés,consistant fondamentalement en:
- Puits de reconnaissance
- Sondages avec prise d'échantillons
- Sondages avec enregistrement continu ENPASOL.Cette méthode consiste à enregistrer éléctronique-
ment et de façon continue les paramètres de perforation,ce qui permet d'obtenir,quand l'on dispo-
se d'une perforatrice de haut rendement,une radiographie continue de la qualité et consistance
des terrains traversés,avec une grande précision et en bref délai.
L'interprétation de ces mesures a permis de délimiter clairement,tant en plan qu'en profondeur,
l'étendue des zones altérées (fig. 4).
- Installation d'un réseau de phréatimètres.
- Installation d'un ensemble de témoins de mouvement superficiels et en profondeur.

- Exécution de nombreux analyses ioniques de l'eau,mesures de résistivité et temperatures,essais avec traceurs,contrôl du débit des drains,etc.
Ces études ont montré l'existance d'une zone alterée de grandes proportions dans le sein du massif de grès de la fondation de l'appui rive droite, à proximité de la zone de l'effondrement.
En plus de zones sans cohésion à partir d'environ 20m de profondeur,avec grandes fissures --- ouvertes, de nombreuses cavités ont été trouvées entre 40 et 65m de profondeur,dans la zone des grès blanchâtres et rosés.
Quelques unes de ces cavités étaient de grandes dimensions (jusqu'à 9m de hauteur vide),menaçant de nouveaux effondrements dans le corps du barrage (figs. 4 et 6) L'étendue en plan de la zone alterée dans les fondations était d'environ 70m x 50m (fig. 7).
Par ailleurs,les études réalisées sur les écoulements souterrains ont montré l'existence de deux types d'écoulement: un latéral,provenant d'une nappe naturelle en rive droite, un autre général de filtration en provenance de la retenue. Ce dernier écoulement montre une longueur de percolation importante autour de l'écran d'étanchéité en rive gauche;dans la zone centrale,l'existence d'une couche de vase a produit l'imperméabilisation du fond du lac,allongeant les lignes de --- courant dans cette zone,diminuant en conséquence le gradient de l'écoulement,améliorant ainsi les conditions de stabilité.
Par contre, en rive droite on observe un contournement latéral immédiat à la digue,ainsi qu'une filtration directe sous le parafouille,comme conséquence de l'inéfficacité de l'écran d'injection réalisé pendant la construction du barrage et de l'inexistance de voile au large.(fig.5).
Ces faits, joints a:
- l'existence de chemins préférentiels de circulation de l'eau à l'extérieur des galéries.
- l'existence de fissures ouvertes dans le rocher de fondation.
- La grande érodabilité des grès par l'eau de filtration.
- La rupture de la couche superficielle de béton poreux de protection,probablement à la mise en eau du barrage.
Ont conduit au déclenchement d'un processus d'érosion regressive avec entraînement de matériaux vers le corps de la digue.
Les circulations préférentielles le long des galéries ont fait apparaître des vides d'importance derrier leur recouvrement,provoquant leur rupture et l'accélération du processus d'érosion par l'eau a pression venant des galéries.
Les énormes vides ainsi creés,spécialement dans la zone d'affleurement de l'eau d'infiltration sous le barrage,où le processus d'érosion regressive a demarré,ont finalement provoqué l'effon drement de proportions catastrophiques enregistré,par la rupture du toit d'une des cavernes --- formées.

5.3- Travaux de réparation des fondations.
A la vue des resultats des études un projet de réparation des fondations du barrage a été établi comprenant fondamentalement les phases suivantes:
- Une première phase très critique dans laquelle on devrait rendre au massif de fondation érodé sa capacité mécanique pour encaisser le poids du barrage (phase de consolidation).
- Une deuxième phase de travaux conduisant à éviter la reproduction du phénomène d'érosion dans d'autres zones (phase d'étanchéité et drainage).

5.3.1- Phase de consolidation (fig. 6)

5.3.1.1- Traitement de la zone entre galéries et effondrement.
Dans cette phase,la zone comprise entre les tunnels et l'endroit de l'effondrement a été considerée comme la zone plus critique,étant donné les dimensions et le nombre des vides - rencontrés,et elle a été donc traitée en priorité.
Le traitement a été fait par des injections,à partir d'auréoles de perforations,réalisées depuis le couronnement du barrage jusqu'à une profondeur de 80m.
Les perforations ont été exécutées par primaires et secondaires,avec une maille de 5m pour les primaires et de 2,50 m pour les secondaires.
Dans cette zone (zones 1 et2) le processus d'injection suivi pour atteindre la consolidation fina le a été le suivant:
1- Remplissage des cavités par gravité à mesure de l'avancement de la perforation au moyen de coulis de bentonite-ciment-sable additionnés de silicate de soude,dosés de façon à obtenir un -- produit suffisamment rigide en état frais pour ne pas être entraîné par l'eau du terrain,d'une résistance à compression simple à 28 jours supérieur à 10 kg/cm2,avec peu de décantation et --- économique,permettant de remplir de grands volumes vides avec le minimun de matière sèche.
Après de nombreux essais avec les matériaux de base disponibles, le dosage le plus utilisé de ce coulis de remplissage a été le suivant:
Par m3 de coulis: ciment: 350 kg
                   bentonite: 30 a 40 kg
                   silicate de soude: 6 a 10 l
                   sable :   500 l
2- Injections de blocage à pression en phases successives,limitant les quantités injectées dans chaque phase pour éviter les déperditions.

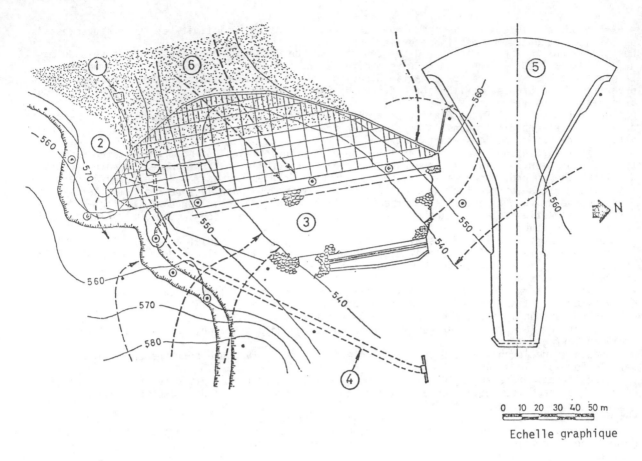

PRESA DE ITIYURO

LIGNES DE COURANT ET PIEZOMÉTRIQUES

1 - Vidange de Fond
2 - Tour de Prise
3 - Barrage
4 - Galérie
5 - Déversoir
6 - Vase

FIG. 5

Ces injections ont été faites à partir des mêmes forages que les injections de remplissage, en reperforant le coulis déjà injecté.Elles ont été conduites aussi par primaires secondaires,par passes remontantes de 5m à partir du fond du forage.
Les coulis utilisés pour le blocage ont été des coulis de bentonite-cimet.Le dosage le plus ---- souvent utilisé a été,par m3:

          Ciment: 300 kg
          Bentonite: 30 kg

Les pressions de refus fixées pour l'injection augmentent avec la profondeur jusqu'à une valeur maximale de 30kg/km2.Les absorptions totales obtenues par forage pour atteindre la pression de- refus dans cette zone ont varié entre 2 et 4 M3 de coulis par m.l. de perforation.
Le total de coulis injecté dans cette zone est d'environ 10.000 m3,ce qui représente une propor tion de 22% du volume du terrain traité.
Ce dernier chiffre est à rapprocher des absorptions d'un rocher normalement fissuré (1 a 2%), -- atteignant en cas très exceptionnels de roches très altérées valeurs de l'ordre de 10% (cas du-- barrage el Bosque,México).

5.3.1.2- Traitement de la zone de l'effondrement
        Dans cette zone des injections de consolidation ont aussi été réalisées à partir d'auréo les de perforations exécutées depuis le couronnement avec un double objetif:
1- Augmenter le coefficient d'élasticité global du remblai en enrochement de la zone effondrée,en lui donnant un certain dégré de cohésion,au moyen d'injections limitées,réalisées par ciment et

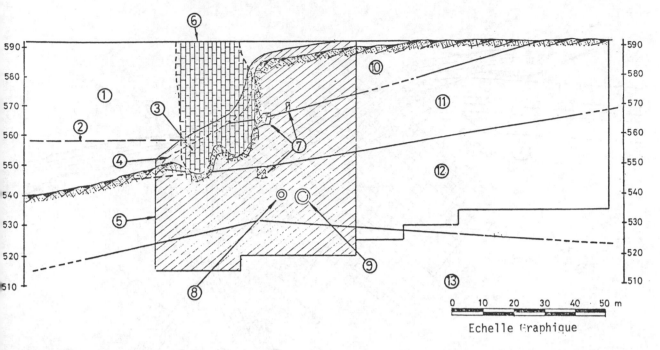

PRESA ITIYURO

TRAITEMENT APPUI RIVE DROITE-COUPE

1 - Enrochement   7 - Cavernes   12 - Grès à Bandes Rosées
2 - Vase   8 - Tunnel de Prise   13 - Grès Rouges
3 - Enrochement Tombé 9 - Tunnel Vidange
4 - Terrain Original 10 - Grès Rosés
5 - Zone Traitée   11 - Grès Blanchâtres
6 - Effondrement

FIG. 6

silicate de soude,dosés de façon à avoir une rigidité telle que le volume injecté soit d'environ
2 m3/m.l. de perforation.
2- Consolidation du terrain sous l'enrochement avec la même méthode décrite plus haut pour le --
traitement de la zone entre galéries et effondrement.
Dans cette zone les absorptions dans le rocher de fondation ont été beaucoup moindres que dans
la zone antérieur,se situant les moyennes entre 0,5 et 2m3/m.l. de perforation.
La proportion du volume total de coulis injecté par rapport au volume de terrain traité est ici
11%.
Le volume total injecté dans cette zone a été d'environ 6.000 m3.

5.3.1.3- Exécution des travaux d'injection
        1- Matériel de perforation.
Pour l'exécution des perforations nécessaires pour les injections,d'un diamètre compris entre 5
et 15 cm,suivant les cas,des outillages de perforation très variés ont dû être utilisés,à cause
de la grande variété de cas de perforation qui se présentaient (en galerie,sur talus,avec échan
tillons,destructifs,enrochement,terrain altéré,etc.)
En particulier, on a dû repondre au défi technique de perforer à travers de l'enrochement du ---
barrage,jusqu'à 50m de profondeur,en zones totalement ouvertes,avec alternances de blocs durs
de quartzite et vides,ce qui pose de problèmes techniques importants qui ont dû etre résolus avec
la mise au point de méthodes de perforation adaptées et des outillages puissants.
La récupération de témoins dans les grès alterés a aussi posé des problèmes délicats à cause de la
présence de petites couches résistantes intercalées dans le grès rendu à l'état de sable. Des
carottiers spéciaux a trousse coupante dépassant ont dû être utilisés.
Les types d'outillages de perforation utilisés ont été les suivants.
- Rotopercusion sur camion,avec marteau fond de trou à air comprimé,pour grandes profondeurs d'
enrochement.
- Rotopercusion avec marteau à air comprimé en tête pour petites perforations (Jusqu'à 15m) des
tructives en galérie,et sur talus.
- Perforatice hydraulique rotative sur chenilles,de haut rendement,pour perforation destructive
au tricône à l'air libre dans le rocher in-situ et profondeurs moyennes d'enrochement.

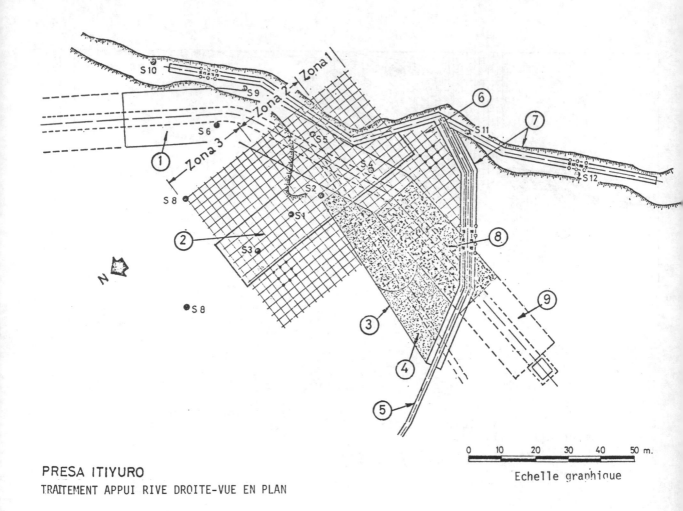

PRESA ITIYURO

TRAITEMENT APPUI RIVE DROITE-VUE EN PLAN

1 - Contact et consolidation dans galérie
2 - Zone Consolidée
4 - Traitement tunnel prise
5 - Parafouille
7 - Ecrans d'étanchéité
8 - Bouchon étanche dans galérie

FIG.7

Cet outillage a aussi servi à réaliser les enregistrements ENPASOL.
- Perforatrice hydraulique sur bâti pour perforations destructives profondes ou sondages en talus
et galéries,réalisées avec couronnes à diamants pleines ou circulaires.
-Sondes pour l'exécution de forages carottés à l'extérieur.
2- Matériel d'injection
Pour l'exécution des injections une centrale d'injection fixe a été installée comprenant:
- 2 digesteurs pour préparation des coulis
- 4 injecteurs à air comprimé à piston
- 2 injecteurs à vise
- Agitateurs,compteurs,doseurs,cuves de stockage,etc.
En plus on dispose de deux unités d'injection mobiles formées chacune de:
- 1 melangeur à haute turbulence
- 1 injecteur hydraulique à piston
Cet ensemble proportionnait une capacité d'injection d'environ 40 m3/h, comptant avec differents
types d'injecteurs adaptes aux exigences tres variees des injections a effectuer telles que:
. Possibilité d'injecter des coulis même très rigides.
. Possibilité d'atteindre des pressions d'injection très élevées (jusqu'à 100 kg/cm2.) et avec
  réglage fin aux petites pressions.
. Possibilité de reglage des cadences d'injection ainsi que d'atteindre des vitesses élevées.
3- Méthodologie générale de l'injection

524

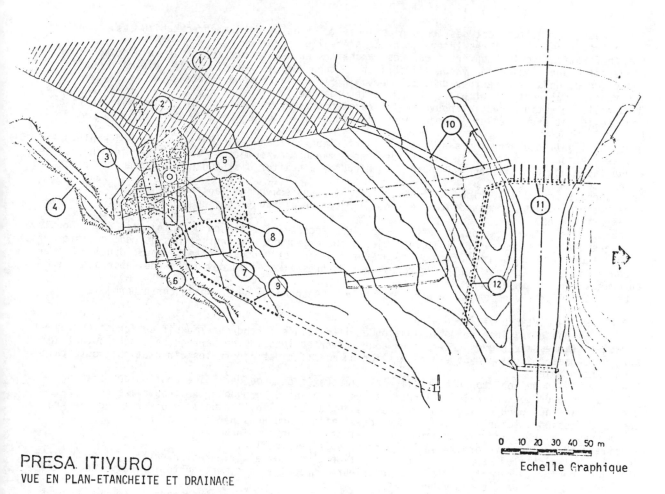

PRESA ITIYURO
VUE EN PLAN-ETANCHEITE ET DRAINAGE

Echelle Graphique

0 10 20 30 40 50 m

1 - Vase
2 - Bouchon étanche en galérie
3 - Voile d'étanchéité
    sous parafouille
4 - Voile au large
6 - Consolidation zones 1-2-3

8-9 - Ecran drainage rive droite
11-12 Ecran drainage rive gauche

FIG.8

Des differentes méthodes d'injection ont été utilisées,suivant la zone:
a)-injections par gravité:
  A travers de tubes ouverts placés dans les perforations,pour le remplissage sans pression des grandes cavités.
b) Injections a pression à trou nu:
Pour le coulis de ciment;ces coulis,à differentes compositions et caractéristiques,ont été injectés par la méthode de phases ascendentes,c'est à dire:
- Exécution de la perforation jusqu'au fond
- Injection des 5 derniers mètres par mise en place d'un obturateur gonflable en haut de la --- tranche.
- Remontée de l'obturateur de 5m et injection de la nouvelle tranche et ainsi de suite.
- En chaque phase l'injection est poursuivie jusqu'à atteindre la pression de refus fixée d'avance,ou le volume maximal donné pour la tranche.
Etant donne la grande altération du terrain dans quelques zones,l'injection en pression,dans les premières phases du blocage,conduisait souvent au claquage du terrain au dessus de l'obturateur et le contournement du même par le coulis.
Dans ces cas l'injection était arretée pour permettre la prise du ciment,se poursuivant postérieurement dans une opération d'injection ultérieure après reperforation du coulis remplissant le -- forage.
Répétant le processus les fois nécessaires une complète consolidation du massif a été obténue,- atteignant partout les pressions de refus fixées pour l'injection.
c)- Injections à travers de tubes à manchettes.
Cette méthode est utilisée pour injecter des produits chimiques alternés avec coulis de ciment.

525

Les tubes à manchettes sont des tubes plastiques ou en acier de diamètres habituelès compris
entre 1 et 2 pouces,perforés lateralement à intervalles réguliers.
Ces perforations sont recouvertes par des manchettes en caoutchouc,jouant le role de soupapes --
s'ouvrant quand l'injection va de l'intérieur du tube vers le terrain,et se refermant quand l'
injection a tendance à rentrer dans le tube,à la fin de la phase d'injection.
Ces tubes à manchettes sont placés à l'intérieur des perforations et scellés au terrain au moyen
d'un coulis plastique spécial pouvant être cassé pour laisser passer l'injection,mais suffisamment resistant pour sceller l'espace annulaire tube-forage,empêchant,le coulis de circuler le
long du forage
Avec lútilisation d'obturateurs spéciaux à l'intérieur du tube à manchettes,il est possible d'--
isoler des zones du même, à la hauteur de chaque manchette,permettant de réaliser les injections
d'une façon homogène le long de toute la profondeur à traiter,distribuant correctement les coulis
à injecter.
Ce système permet l'injection de coulis differents successivement sans avoir besoin d'autres ---
perforations,pouvant ainsi adapter le type d'injection à chaque zone de terrain,suivant ses ----
caractéristiques.
En particulier,l'utilisation de cette méthode est nécessaire quand on doit injecter des produits
chimiques,car dans ces cas on injecte toujours successivement plusieurs types de coulis differents.
L'injection par tube à manchettes a été prévue pour injecter du silicate de soude dans le but de:
- Donner de la cohésion à certaines zones de grès très alterés où des poches de sable sont restées
enfermées par les injections de ciment.
- Etancher le grès dans sa masse dans les voiles d'étanchéité de la phase correspondante

5.3.2.- Phase d'étanchéité et drainage
          La phase initiale de consolidation était destinée à rendre au massif de fondation les -
caractéristiques mécaniques nécessaires pour supporter le poids du barrage,sans pour autant agir
sur les causes qui ont provoqué l'altération du massif,c'est à dire les circulations souterraines
d'eau incontrolées
Une fois les fondations consolidées,il était impératif de procéder à l'exécution des travaux --
nécessaires pour contrôler les filtrations les empêchant de provoquer des nouvelles érosions,--
d'autant plus que,pendant l'exécution des travaux de consolidation,on a pu observer que le processus d'érosion interne expérimente des accélérations brusques pendant les crues.
Cette phase comprend fondamentalement les éléments suivants (fig. 8):
- Voile d'injection rive droite au large et sous le parafouille.
Ce voile est constitué par 3 files de perforations,les deux extérieures injectées à trou nu avec
coulis de bentonite-ciment, et la file intérieure au coulis de ciment-bentonite,suivi de silicate
de soude à travers de tubes à manchettes.
Ces perforations sont effectuées à partir de la surface jusqu'à pénétrer dans la couche de grès
rouges résistants.
- Anneaux injectés autour de la galérie de vidange de fond et de la tour de prise afin de couper
les circulations le long de ces éléments.
- Ecran de drainage en rive droite et en rive gauche permettant de capter l'eau de filtration a
l'intérieur du massif,empêchant l'entraînement de matières solides à la sortie.
Ces travaux sont actuellement en cours d'exécution,le projet de détail étant en revision a cause
de la forte remontée du niveau de la vase les derniers mois,qui conseille diminuer l'ampleur des
travaux prévus.
A la fin des travaux de consolidation et d'étanchéité et drainage,il est prévu de mettre en place
un système définitif de contrôl du barrage à base de piézomètres à corde vibrante et de déforme-
mètres,permettant de suivre dans le temps le comportement du barrage.

Foundation of the Alicurá spillway on weak pelites

Fondation du canal d'Alicurá sur des pelites à faibles caractéristiques

Oscar A. Vardé, *Vardé y Asociados, Buenos Aires, Argentina*
Carl-Anders Andersson, *SWECO, Engineers, Architects & Economists, Stockholm, Sweden*
Luis Pi Botta, *Franklin Consultora SA, Buenos Aires, Argentina*
Oscar Paitovi, *Consorcio Consultores Alicurá, Buenos Aires, Argentina*

ABSTRACT: A surface spillway has been constructed on the left slope at Alicura dam site, Argentina. The rock consists of a sequence of sandstone and pelite. The locally mylonized pelite layers dipping parallel to the slope presented unfavorable stability conditions. In the original spillway design the energy dissipation of up to a 3000 m3/s maximum discharge was to be attained by a skijump which, however, had to be replaced by a concrete stilling basin. This article presents the geology and the geotechnical properties of the foundation and describes the performed stability analyses. These analyses showed that a large volume of excavation was necessary to assure the stability of the lateral areas. Also post-tensioned anchors had to be installed.

RESUME: Sur la rive gauche de l'aménagement hydroélectrique d'Alicura, Argentine, on a construit un evacuateur de crue. La fondation est formée par couches de pelite qui entrecoupent couches de grès. Les couches de pelite mylonisées localment avec une litage d'inclination parallele a la pente, ont presente des conditions de stabilité desfavorables. Dans le dessin original de l'evacuateur la dissipation d'energie (jusqu'a 3.000 m3/s) était obtenue par un saut de ski qui, cependant, fut changé par une bassin de tranquillisation fait en beton. Cet article donne les characterisques geologiques et geotechniques de la fondation et décrit les analyses de stabilité effectués. Ces analyses ont montré qu'il fut necessaire des grandes excavations pour assurer la stabilité des zones prochaines. Aussi, il fut necessaire installer des ancrages post-tendus.

1. INTRODUCTION

The Alicura Hydroelectric Project owned by HIDRONOR S.A. has been constructed on Río Limay, 100 km NE of the town of San Carlos de Bariloche, Argentina. The project includes a 130 m high earthfill dam with the principal appurtenant structures located on the left bank. A general plan of the site is included in another article to this congress (Andersson, Vardé 1986). The average annual flow in the river is 264 m3/s and the maximum probable flow 3.000 m3/s. The maximum storage level of the reservoir is at El 705 m above M.S.L. and the average downstream water level at El 585 m.

This paper deals, principally, with the spillway foundation and the stabilizing measures in the left slope. This surface spillway on the left bank was designed for the maximum probable flow and comprised an intake structure, a chute and a skijump in the original layout (Figure 1). Due to more unfavorable geological conditions than anticipated originally the stability of the slope turned out to be critical as the plunge pool of the skijump could undercut the slope. Hydraulic model tests and stability analyses of this solution and of an alternative one with a stilling basin in concrete were performed so as to determine a safe solution. The decision was taken to replace the skijump by a stilling basin.

Chapter 2 summarizes the geology of the left bank and chapter 3 the material properties of the foundation. Chapter 4 gives a description of the stability analyses performed in different stages of the design with details of the models and the dimensions of the prototype. The results are analyzed in chapter 5. Finally, chapter 6 contains some details from the execution of the excavation for the spillway and the installation of post-tensioned rock anchors.

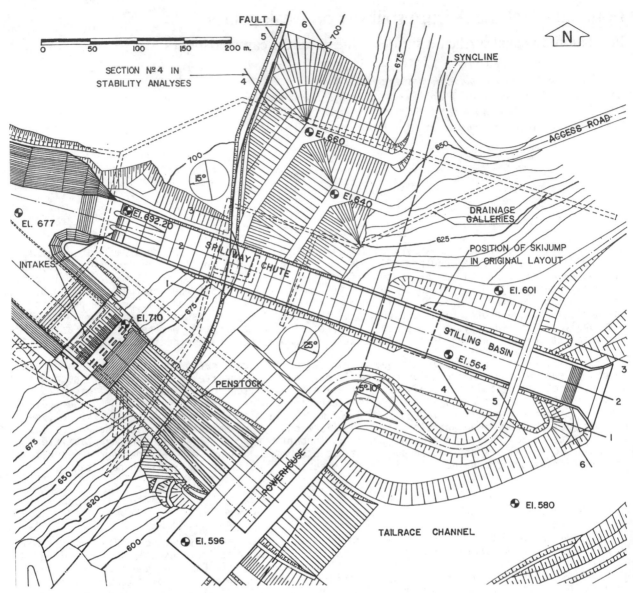

Figure 1. Layout of the left slope in Alicura.

2. GEOLOGICAL CONDITIONS

2.1 Lithology

The bedrock of the project area consists of a succession of psammitic (sandstone, arkose, conglomerate) and pelitic (mostly mudstone and siltstone, some claystone) rocks of Liassic (Lower Jurassic) age. The depositional environment of this sedimentary series is interpreted as having been a flat swampland subjected to frequent flash-floods which carried with them the coarser sediments. The deposits created by these repeated events are a succession of sand banks of widely varying extensions, but virtually always in elongated lenticular shape. The periods inbetween floods produced the deposition of the finer particles (silt, clay) and also of organic material (preserved in the form of a black pigment and as thin layers of coal). These interlayers vary considerably in their apparent composition and consistency. Planar sedimentary structures do not persist over any considerable distance and individual sandstone, arkose, conglomerate or pelitic layers cannot be correlated with certainty between drillholes and outcrops.

2.2 Structural geology

The bedding interfaces are the dominant structural element in the bedrock, being in general horizontal with some very gentle folding. A major fault, denominated fault 1, running roughly

NNE to SSW and dipping steeply SE intersects the penstock trench and spillway chute downslope of the corresponding intake structures. At the downhill side of this fault, in the penstock as well as in the spillway area, the bedding of the rock is abruptly changed to a dip of 20° to 38° SE to E, i.e. parallel to the slope.

The existence of the zone of inclined bedding on the left bank was not known at the initial design stage, when the investigaions were based mainly on geological mapping of outcrops and borings. The inclined bedding was subsequently encountered in investigation trench excavations in the area of the penstocks, and considerable efforts were directed towards exploring this feature in detail. An exploratory gallery at approx. El 590 m and a vertical shaft were excavated. The gallery crossed fault 1, which proved to consist of a zone of plastic mylonized rock, 2 m thick. Fault 1 appeared quite impervious, seemingly forming a barrier that cut off seepage from the uphill side. The trenches, though being superficial, in the area of the penstocks and along the spillway provided useful information together with investigation borings and laboratory testing of samples. Drillholes, shafts and trenches in the valley floor indicated a return to flat dips probably due to the presence of additional faults.

In an initial stage of the excavations on the left slope, a slide of about 120,000 m3 occurred in Aug. 1980 in the lower spillway area. The sliding plane was visible in an upper portion of the inclined bedding and it could be concluded that the active part of the slide had a lower boundary consisting in a soft pelitic layer.

Undisturbed samples of this material were tested in the laboratory and gave shear resistance values lower than earlier tests as described in chapter 3.

Also, horizontal layers of weak pelite were observed upstream of fault 1 in some investigation drillings. The most important one at El 655 to El 660 below the penstock and spillway intakes became visible during the penstock trench excavation. It was deemed necessary to add extra safety by constructing shear keys under these structures. The shear keys were formed by excavating galleries and filling them with concrete. During the construction it was found that this bedding was slightly subhorizontal with a favorable dip of up to 15°W near to fault 1.

The accurate configuration of the tectonically disturbed section coinciding with the lower part of the slide, could not be fully determined until the deeper excavations for the powerhouse and spillway were under way. It was found that the inclined bedding in the slope changed to a bedding with a dip of between 5° and 10° to the NW. The transition zone can be described as a fold (syncline) approximately parallel to fault 1 and located 200 m from the latter. In the spillway area it was noted that the fold was more abrupt and accompanied by a small inverse fault as a sign of large compressive forces. The normal dip of 25° increases to a value of about 45° near the fault.

Steeply dipping fractures (joints) intersect the sandstone beds, mostly in clearly defined sets and predominantly normal to the bedding planes and at right angles to each other. Their frequency of occurrence varies but is mostly between 2 and 4 fractures per linear meter of exposed rock face. In most cases these joints are planar to ondulated with rough surfaces, in the weathered zone invariably stained or coated with iron hydroxide ("limonite").

A large number of the pelitic interlayers are more or less intensely sheared, predominantly along the upper contact with the sandstones. This shearing produces slickensides parallel to the bedding planes but in places a layer of up to 5 cm thickness is totally crushed. This clay/silt gauge, mylonite, is quite frequently squeezed out, thus accumulating in some and being absent in other places along the same shearzone. Due to the lenticular sedimentary structure, the pelite layers have a pronounced wavy surface, documented by the scatter of dip and strike values of bedding structures on the stereographic projections. In the lower parts of the pelite beds, steeply dipping, randomly oriented "mirror" faces can be observed, which can be attributed to movements associated with the compression and the consolidation of the pelite deposits after the sedimenta tion.

2.3   Hydrogeology

Groundwater levels were observed in drillholes during the investigation stage. Water levels in five drillholes at the left bank during the period July 1978 - September 1979 showed variations of 6 to 11 m. The mean groundwater level in the area of the intake was at 675 m, i.e. about 30 m below natural surface. Along some cut rockfaces water seepage could be observed in an early stage in connection with pelitic layers which act as impermeable membranes (aquacludes) between the sandstone layers which are moderately permeable. The permeability parallel to the bedding was considered at least 10 times higher than normal to the bedding. The water pressure tests in drillholes showed generally low to moderate permeability with most Lugeon-values below 5.

Casagrande type piezometers were installed in the left bank at various depths in the first phase of construction. The water columns in these pipes rised only a few meters above the bottom of the hole independent of the level of the installed depth. It became evident that there existed several different ground water tables reinforcing the impression of the pelitic layers acting as impermeable membranes. These observations were confirmed during the excavation of more than 5.000 m of drainage galleries in the left bank and the drilling of 35.000 m of drain holes (Andersson,Jansson, Vardé 1985).

In general, the water levels in the inclined beddings were even lower than upstream of the fault 1 which also acted as an impermeable membrane. A drainage curtain with a gallery at the lower end was constructed inmediately upstream of the fault to assure low water pressures in this area.

However, as an exception, piezometer readings indicated very high water levels along the syncline to the left of the spillway axis. It was probable that the aforementioned slide occurred due to high water pressures. The syncline which extends as a gully for several hundred meters to the north is catching the water from the higher terrain.

3.   GEOTECHNICAL PROPERTIES

The characteristics of the sandstones vary greatly; the results of testing on NX-core drilled samples, obtained during the early design stage, gave the following typical values for sandstone:

density	2,40 g/cm3
unconfined compressive strength	300-400 kg/cm2
compression wave velocity	2.800 m/s
Young's modulus	180.000 kg/cm2
Poissons's ratio	0.05
Shear strength:$\emptyset$	35°-55°
C	2-25 kg/cm2

The competent pelites gave test values only somewhat lower than those given above. Later, when the geological features became better known, most of the efforts of assessing the rock properties were concentrated on the weaker pelites due to their crucial importance in the stability. According to the observations made at the site and in the laboratory, the pelite samples were classified in 4 different categories:

Type A: Very fine grained pelites, high degreee of diagenesis and very well lithified.

Type A': Same as before but coarser grained and presenting micaceous materials.

Type B: Intermediate state between lithified rock, type A or A' and type C. These pelites do not show preferential failure planes.

Type C: Rock with high degree of mechanic alteration. It does not keep the original structure and sometimes contains hard lumps.

Type C is found, in almost every case, in the contacts between pelite and sandstone layers, whereas A or A' behave as the main core of pelitic layers. Therefore, type B has a transitional behaviour.

It was necessary to assume conservatively in the stability calculations that weak pelite existed all along the sliding planes. The occurence of the slide mentioned in 2.2 led to a reassessment of the shear resistance to be used in the analyses. Test results from samples taken in the sliding surface and from the horizontal seam at El.655 behind fault 1 were used to establish the values for the final design verification. These results are discussed below.

The weak pelite can be classified as a clay of low to medium plasticity with plasticity indices of 5 to 20 and liquid limits between 20 and 40 (Figure 2a). The most plastic samples contained about 40% of clay ($<2\mu$) and 25% was finer than the sieve № 200 (0.074 mm).

The drained strength parameters were determined by direct shear tests, simple shear tests and triaxial tests on, mainly, remolded samples. It was very difficult to extract good undisturbed samples from the generally thin seams and these samples sometimes contained small lumps. The remolded samples were reconsolidated to a density similar to the undisturbed ones. The samples taken in the field had a natural water content at or below the plastic limit and were practically fully saturated. The dry in situ density of the weak pelite was between 1.80 and 1.90 g/cm3 while the specific density was varying from 2.44 to 2.68 g/cm3.

The testing was performed at different laboratories on samples that for practical reasons not always were of the same quality. In spite of this, the variations in shear resistance of pelite type C were reasonable as can be seen from the results compiled below:

Type of test	Peak shear strength		Residual shear strength	
	C (kg/cm2)	∅°	C (kg/cm2)	∅°
Direct shear tests	0,0 - 0,3	22-31	0,0 - 0,1	17-21
Triaxial tests	0,0 - 0,3	22-30	0,0 - 0,1	21-24

The residual friction angles are plotted against the corresponding range of plasticity indices (Figure 2b). On the same chart the empirical boundary curves proposed by Deere and Seycek, respectively, are indicated. It can be observed that the values obtained by direct shear tests are lower than those by triaxial tests and are rather close to the lower boundary. Routine testing in the field laboratory of the Atterberg limits was then used so as to check that the plasticity indices fell in the known range.

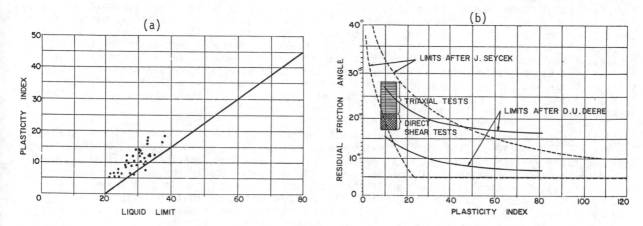

Figure 2. a) Plasticity diagram for weak pelites.
         b) Residual friction angle versus plasticity index.

The direct shear tests were performed with controlled strain at a velocity of 3.3 μ/min. for vertical stresses between 1 and 5 kg/cm2. The difference between peak and residual strength values varied between the samples. Most samples gave a clear peak value for about 4% of strain but a few ones did not show any peak. Some samples were subjected to several shear cycles in order to simulate the disturbance due to large displacements. The samples with higher plasticity had a more pronounced reduction of the residual shear strength.

Consolidated undrained triaxial tests were carried out in order to assess the shear strength parameters for the seismic verification of the slope stability using the methods of Newmark and Makdisi-Seed. In this approach, the verification is made using the cohesion and angle of friction as obtained in aforementioned tests but applying a reduction factor of 0.8. The parameters finally obtained were C = 0.56 kg/cm2 and ∅ = 12°. These results were in agreement with those obtained in comparative special direct shear tests. In these tests the samples were first deformed to a significant degree of strain at a slow rate and then an instantaneous load was applied. The corresponding increases in shear strength for different vertical loads gave the cohesion and friction angle values looked for.

Cyclic tests were performed on remolded pelite samples at the Cyclic Testing Laboratory, Rensselaer Polytechnique Institute, Troy, N.Y. For the tests a modified NGI direct simple shear device was used. The analysis of the results was made by Dr. H.B.Seed who concluded that the material was not susceptible to strength degradation under cyclic loading.

Finally, the susceptibility to creep was controlled in direct shear tests. The samples were subjected to a constant shear stress equal to 80% of the residual value during several weeks. The results indicated a negligible creep.

4.   STABILITY ANALYSES

4.1   General

The geological and geotechnical caracteristics of the left bank required a thorough design work in order to guarantee the stability of the slope and its structures. Descriptions of general design considerations (Andersson, Jansson, Vardé 1985) and of the stability analysis of the penstock area (Vardé, Pi Botta 1982) have been published. This chapter deals with the stability calculations for the lower part of the spillway chute and energy dissipator. However, it should be mentioned that also the intake structure for the spillway, as well as the penstock intake, required extensive stability analyses to guarantee that sliding would not occur on the weak horizontal pelite layers.

The main reason for discussing the stability of the lower part of the spillway only in this paper is that this part proved to be the most complex and critical from several points of view. Besides geological, geotechnical, and hydraulic problems, economical and programming reasons weighed heavily.

4.2 Choice of type of spillway energy dissipator

Hydraulic model tests were first performed with the skijump alternative. These tests showed that the plunge pool downstream of the skijump, at El. 594, was likely to be eroded to about 30 m depth below the bottom of the tailrace channel at El. 580. The plunge pool would reach it maximum depth about 170 m downstream from the skijump in the area of horizontal bedding. The pool would create unacceptable low safety factors unless exceptional stability measures were taken in form of excavation and rock strengthening. Even so it was felt that the erosion was a factor of uncertainty as well as the rock conditions at that time. Consequently, an alternative with a concrete stilling basin was introduced and tested in the hydraulic laboratory in order to verify its dimensions and function. This latter alternative also required stability measures but to a lesser degree than in the other alternative.

Stability calculations and layouts of the stabilizing measures as well as cost estimates were made for the two alternatives. The cost estimates indicated that for the same cost as for the concrete stilling basin an important amount of strengthening could have been done in the skijump alternative, mainly by use of additional rock anchors. Even though theoretically the safety factor could have been brought to similar levels as with the stilling basin, it was felt that the primary dependence on anchors involved a higher risk. It should be mentioned that the stability during construction was more critical for the stilling basin than for the skijump but that in the operational stage the situation was reversed. The dynamic stability was also checked.

The main characteristics of the spillway with the stilling basin are given below:

Level of tail water	El 582 - El 594
Level of spillway ogee	El 692.2
Number of gates	3 tainter gates
Specific discharge	77 m3/s
Spillway chute: length	300 m
width	39 m
inclination	19°
aireators	4 Nos
Stilling basin: length	223 m
width	39 m
wall height	35 m
max. wall and floor thickness	6 m
finished floor level	El 564
Excavation: in soft (for structures and stabilization of left slope)	900.000 m3
in rock (in total)	480.000 m3
" (for stilling basin)	220.000 m3
Post-tensioned anchors	500 Nos or 2.10^6 ton.meters
Passive rock anchors	4.700 Nos.
Concrete: intake structure	24.000 m3
spillway chute	34.000 m3
stilling basin	90.000 m3

4.3 Mathematical model and analyzed sections

The stability analyses on computer were performed using a two-dimensional model composed of an active block and a passive one. The active block was delimited by the subvertical plane of fault 1 and by an inclined plane corresponding to a possible weak layer of pelite (Figure 3). The contact between the active and passive blocks corresponded to the syncline observed in the field. The lower boundary of the passive block was first horizontal and corresponded to the weak pelite in the bedding of the river valley as known at that time. In the alternative of the concrete stilling basin the resistance of the concrete basin was included in the passive block. In the final stage of the design verification the lower boundary of the passive block was taken to be a 7° ne

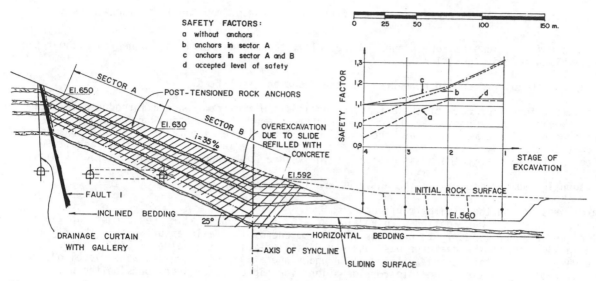

Figure 3. Profile along spillway axis with safety factors in the excavation stage.

gative slope as revealed when the excavations advanced.

In total six longitudinal sections were analyzed in both alternatives, three parallel to the spill way axis and three oblique to the same (Figure 1). Of the former sections one was located in the spillway axis and the two others at each side so as to estimate the three-dimensional effect from the more stable lateral sections; the angle of the sliding plane was 25°with respect to the horizontal which was equal to the maximum dip. The three oblique sections were analyzed in order to determine the necessary lateral excavation to the left of the spillway in order to unload the slope and obtain an acceptable factor of safety.

4.4   Material properties, failure mechanisms and external loads

The parameters used in the calculations were chosen from the test results described in chapter 3 after a thorough analysis. The following shear strength values were adopted:

Type of sliding surface	Shear strength	
	C/(kg/cm2)	$\emptyset°$
Shear across sandstone-pelite: - static and dynamic cases	2.0	30
Shear along pelite: - static case - dynamic case	0 0.56	17 12
Shear across concrete of stilling basin	Variable, see below	

These parameters were used for all the length of the relatively long sliding surfaces which added to the conservatism as it was very unlikely to encounter weak planes of that extension and without macro-and microondulations. However it must also be pointed out that, in view of this, the minimum overall safety factor for the operation stage was accepted to be only 1,30. The safety factor along the spillway chute was accepted to be even lower relying on a certain three dimensional effect.

The failure mechanism was defined by the aforementioned sliding surfaces along weak pelite layers and, for the final stage, by the shear surfaces across the concrete of the stilling basin. The following four failure mechanisms of the basin were considered:

a) The basin fails along its block joints through the rupture of the shear keys in the vertical block joints; the blocks remain disconected without the right block changing position. (compare Figure 5).

b) The shear keys in the block joint do not fail and the force from the left block tends to lift the right one.

c) The basin fails along oblique planes in the walls; the blocks remain disconnected without the right block changing position.

d) As c/ but the blocks remain connected and the right one is lifted.

The resisting force for each of these failure mechanisms was determined and introduced into the model as a ficticious cohesion acting on the fracture plane. This cohesion varied in each case due to type of failure mechanism and the location of the failure plane. The basin was considered as a unit in the analysis and replaced in the model by a material with a ficticious density determined as the weight of the basin divided by the volume contained within the exterior limits.

The external loads considered in the analyses are listed below:

i) Water pressure on the sliding surface:
   The water pressure was applied with a value of 1/6 of the hydrostatic pressure on the fault plane and reduced linearly to 0 at the downstream end of the failure surface. To the right of the spillway where the drainage was considered less efficient the value of 1/3 was used. In the case of tailwater in the stilling basin under operation, the depressed water level due to the hydraulic jump was taken into consideration; the higher water level outside the basin would produce an uplift force.

ii) Anchor forces:
   The forces from the prestressed anchors were applied as a uniformly distributed load on the surface of the active block. The direction of the forces was determined so as to give optimum use of the anchors. The angle was chosen to be 30° from the horizontal.

iii) Seismic forces:
   The forces produced by seismic activity were treated as pseudostatic horizontal forces acting on the sliding mass, see paragraph 5.3.

5.   RESULTS

5.1   Comparison between the skijump and stilling basin alternatives

The stability analyses made to compare the skijump and stilling basin alternatives were made assuming horizontal bedding in the passive block as the exact geological setting was unknown at this stage.

For the skijump alternative the factor of safety was calculated for three different depths of the plunge pool. With the most severe assumption, i.e. the bottom of the plunge pool at El 545, the minimum factor of safety in the spillway axis section was 1.02 and in the sections located 40 m from the axis at each side only slightly higher, i.e. 1.06 and 1.08 respectively. Evidently, the lateral strips did not give any important stabilizing effect on the central one. All three strips were assumed to be provided with tensioned rock anchors. In the active blocks the applied anchor load varied between 3 and 6 t/m2 at an angle of 30° from the horizontal; the effective anchor length was 30 to 40 m. The passive block, on which the skijump was seated, was assumed to receive a uniform load of 10 t/m2 at an angle of 45°, by 40 m long anchors. Furthermore, 10 to 20 m high berms were placed at the foot of the slope. In conclusion, the strengthening and stabilizing works assumed in the stability analysis were not sufficient to obtain acceptable stability factors.

In the case of the stilling basin alternative, the three sections located as those in the skijump alternative clearly showed that the overall stability in the final stage was much better than for the skijump. This fact was due to the higher safety factors of about 1.60 obtained in the lateral sections; it was assumed that both lateral strips of 40 m width were provided with an anchor load corresponding to 2.4 t/m2 along the active part. These safety factors would be only somewhat lower during construction due to lack of backfill around the basin. A safety factor of 1.40 was calculated for the central section with the stilling basin in place assuming that the central strip corresponding to the chute had a load of 4 t/m2 from post-tensioned anchors. It was also shown that the efficiency of the drainage significantly influenced on the factor of safety, e.g. applying a water pressure of 1/3 instead of 1/6 reduced the safety factor from 1.40 to 1.15 in the central strip.

From this comparison it is evident that the skijump alternative gave unacceptable low stability even with very extensive strengthening measures. The skijump alternative would also requiere a larger lateral excavation to the left of the spillway so as to maintain a minimum safety factor

534

of 1.30 in the analyzed oblique sections.

The decision was made to opt for the stilling basin alternative which is further discussed in the following paragraphs.

5.2   Stability during construction for the stilling basin alternative

From stability point of view the easiest solution would have been to proceed with the excavation in a downward direction. However, the programming of the works required the excavation to be executed so as to meet certain key dates. As a consequence, it was necessary to start excavating for the basin before   being able to fully complete the excavation of the slope and the installation of anchors. Therefore, stability calculations were made in order to determine a sequence of works so that the factor of safety in no occasion would be smaller than 1.10 in the central section which was the most critical one.

Basically, the more than 200 m long excavation for the dissipator was subdivided into four stages, assuming that the furthest downstream portion would be excavated in the first stage. For each stage of excavation in the basin, the corresponding volume of excavation and reinforcement in the slope was determined. The minimum factors of safety in each stage with and without anchors were calculated (Figure 3). In these calculations made before the excavation it was assumed that the sliding surfaces in the passive block were horizontal. The minimum acceptable safety factor during construction was 1.1. The full line, d, in Figure 3 shows how this factor was maintained by requiring the anchors in sector A to be installed before excavating stage 3 and in sector B before stage 4. The length of the anchors was increased from 30 m in the upper portion to 50 m in the lower portion. This was done in order to include all the length of the sliding surface that has its horizontal part at El 560 and because the inclination of the chute is less steep (19°) then the bedding (25°). The steeper angle (45°) of the lower anchors was chosen so as to reduce the length of the anchors and the problem of maintaining the drill holes open in this zone of very fractured rock.

The lateral excavation was designed so as to maintain a minimum safety factor of 1.30 during the excavation stages. It was assumed in the calculations that the stilling basin did not contribute to the stability in the final stage.

5.3   Dynamic analysis due to seismic loading

The dynamic analysis was done for both alternatives but was finally more elaborated for the stilling basin alternative. To begin with, the seismic acceleration, k_y , which would cause limit equilibrium was determined for the different sections. Dynamic parameters were used for the pelite and the static parameters for the other materials. A certain interaction between the analyzed central spillway section and the lateral ones was considered through the introduction of a cohesion of 1 kg/cm2 for the failure surfaces deeper than 40 m.

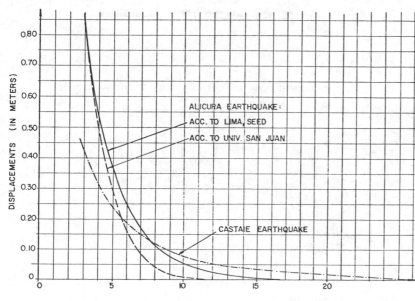

Figure 4. Displacements versus the limit equilibrium value, k_y, for three different earthquakes.

The displacements caused by seismic acceleration peaks greater than the limit value k_y were calculated with the Newmark's method. This method is considered to be very appropiate for rocks as, in general, the acceleration is acting uniformly in the mass without pronounced amplifications. The design earthquake   established for Alicura formed the basis for these calculations but comparisons were made with two other earthquakes (Figure 4). The k_y-value for the spillway section is 0,12 and would correspond to a displacement of 3 cm. This value can be considered very acceptable.

5.4.   Verification of the design as built

A final control was made taking into account the effect of subhorizontal layering in the stilling basin area (Figure 5). It was also assumed that the high density of passive anchors in the basin bottom reinforced the rock so that the shear strength in that portion of the rock increased to the values used for sandstone. In these conditions the safety factor for the central spillway section increased to 1.30 for the operational stage.

6.   CONSTRUCTION DETAILS

When the excavation for the stilling basin started in Sept. 1982 by removing the alluvial deposits in this area a superficial movement was observed in the lower part of the slope. It was the deeper portion of the old slide from Aug. 1980 that was reactivated. However, the rock excavation for the basin could be executed in accordance with the aforementioned stages. Only some minor modifications were necessary, mainly in order to improve the access as the blasting was made in three benches. Therefore, the central one-third of the width was advanced as a narrow trench somewhat ahead of the stipulated stages. This could be allowed due to the reduced width and by leaving a compensating load in the previons stage. The access to the basin was maintained by a ramp at the downstream end of the excavation. In a final construction stage this ramp was filled with roller compacted concrete.

The slide material was located in the area that corresponded to the last (4th) stage and an overexcavation had to be accepted so as to clean out all material that belonged to the slide. The overexcavated area was backfilled with approx. 8.000 m3 of lean concrete; the maximum depth of backfill was 4 m.

The water flow through the syncline on the left side of the spillway was intercepted by drainage curtains between the surface and galleries constructed at El. 594. It was required to have this drainage working before starting the deep spillway excavation.

The stabilizing excavation to the left of the spillway was initiated downstreaw of fault 1 in an early stage before advancing the rest of the excavation. An about 25 m high vertical wall was created by smooth-blasting technique immediately behind fault 1 (Figure 1) and the excavation advanced in a downward direction. It turned out that a great part of the excavation could be executed by ripping with dozers. Stability problems occurred only in a small portion of this

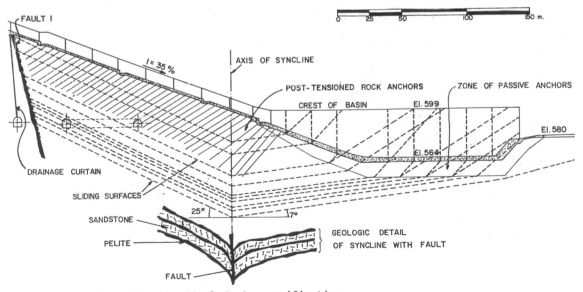

Figure 5. Model used in the final design verification

excavation at the toe of the slope to the left of the stilling basin. Minor volumes of material moved without causing any inconveniences. Notwithstanding, such movements were indicators of the precautions that had to be taken so as to avoid major problems.

The drilling of holes for anchors, casting of anchor heads and installation of cables were done as soon as possible after the cleaning of the excavated rock surface was complete. The spacing of the 100 tons anchors used in the spillway area was tipically 5 m x 5 m measured in a plane perpendicular to the anchors.

The Tensacciai system, similar to the Freyssinet system, which works with wedges to lock each cable at the anchor head, was used. Each 100 tons anchor consisted of 7 cables; each cable in its turn consisted of twisted strands. In the active part, with a length of 6 m, the cable configuration   was ondulated so as to improve the adhesion. The active portion was delimited from the free one by an expandable packer and protected by a corrugated PVC-tubing. The free portion was also protected by a PVC-tubing which in this case was smooth. The PVC-tubing served as a protection against corrosion by water and/or stray currents. The corrugated part would guarantee the transfer of shear forces between the grout fill inside and outside the tubing when installed. The hole diameter used for the anchors was 135 mm. The bunch of cables was prepared outside and then inserted in the hole. The packer was expanded by introducing grout through a plastic hose. Another hose was used to fill the active portion with cement grout.

The tensioning of the cables was made in two stages to 70 and 100 tons respectively with an interval of some days. Those anchors which showed anormal elongation after being tensioned to 100 tons were later retensioned. About 10% of the anchors were then retensioned as a control. In order to finally protect the cables from corrosion the holes were completely filled with a cement slurry.

About 1% of the anchors were provided with load cells in order to monitor the anchor forces. Also subhorizontal extensometers were installed so as to detect any movement in the slope. (Pujol, Andersson 1985). The results from the first half year of observations have indicated a loss of up to 10% in the 100 ton anchors combined with compression movements of up to 2 mm in the direction of the extensometers. In anchors with 50 ton load the loss was less than 5% and the related movements less than 0.5 mm. It is believed that these phenomena partly might be caused by a certain accomodation of the rock and to relaxation in the steel of the cables.

Passive anchors were installed so as to bind the spillway chute slab to the rock foundation. These anchors, $\emptyset$ 25 mm and 6 m long, were intercalated between the post-tensioned anchors in a 2.5 m x 2.5 m pattern. For the stilling basin, 12 m long reinforcement bars, $\emptyset$ 32 mm, were installed in the bottom of the excavation in a dense pattern of up to 1.7 m x 1.7 m. These anchors serves to add weight to the basin in order to improve the safety against flotation. The passive anchors sew together the upper portion of the rock which thereby recieves the benefit of the stabilizing effect of the more sparsely spaced active anchors

Drainage was provided in the contact surface between the rock and concrete structure by placing lines of half concrete tubes ($\emptyset$ 200 mm) on a thin bed of porous concrete. Along these lines drainage holes, $\emptyset$ 50 mm, were drilled, subhorizontally below the spillway chute and vertically in the stilling basin. The drainage holes were provided with slotedd PVC-pipes wrapped in geotextile. The reason for the drainage below the chute was to avoid water pressures in the rock under the slab and the drainage was produced by gravity. The drainage system below the basin is connected to deep pumping wells which served during construction and will be used in the future in case the basin should have to be emptied.

For practical reasons the length of the blocks in the spillway chute floor was determined to be as long as 63 m, i.e. the distance between the aireation steps. It was considered that transversal fissuring of the chute floor could not be avoided even with very short blocks due to irregularities in the excavated surface and to variations in rock properties. The quantity of reinforcement was determined so as to distribute the fissures evenly along the block. The actually observed fissures in the floor of the finished chute occur in a relatively regular pattern approximately every 5 m. It should be mentioned that the walls of the chute have joints every 10.5 m. The length of the blocks in the stilling basin are 20 m both in the floor and walls.

In summary, the excavation and strengthening works in the spillway area were executed without any special difficulties or problems. The quality of the rock in the stilling basin was in general better than expected. The water seepage towards the excavation was very limited taking into account the depth and extension of the same.

7. ACKNOWLEDGEMENTS

The authors wish to thank HIDRONOR S.A., for their permission to publish this paper. Thanks are also extended to Consorcio Consultores Alicura (Franklin Consultora S.A., Latinoconsult S.A., Tecnoproyectos, Esín, Electrowatt, SWECO) as responsible for the design and supervision, and HIDRONOR's Board of Consultants (Drs Deere, Hilf, Lombardi, Lyra and Seed) who reviewed the design, construction and performance. The hydraulic model tests were performed by INCYTH - Instituto Nacional de Ciencia y Técnica Hídricas, Laboratorio de Hidráulica Aplicada-and thanks are due to its personal and to Dr. Angelin for his assistance.The valuable contribution to the geological survey by Dr. Baumer is also recognized.

Civil works were performed by E.C.A.S.A. - Empresa Constructora Alicopa S.A. under the leadership of Impregilo S.A.

REFERENCES

Andersson, C.A. & Vardé O.A. 1986. Foundation treatment and grouting works for Alicura Dam, Argentina. Proc. 5th International Congress IAEG. Buenos Aires.
Andersson, C.A., Jansson, S. & Vardé, O.A. 1985. Grouting and drainage systems at the Alicura Hydroelectric Project, Argentina. Proc. 15th International Congress on Large dams, Q.58, R. 38. Lausanne.
Pujol, A. & Andersson C.A. 1985. Instrumentation of the left bank at the Alicura Project,Argentina. Proc. 15th International Congress on large Dams, Q.56, R.20. Lausanne.
Vardé, O.A. & Pi Botta, L. 1982. Estabilidad de la margen izquierda de la presa de Alicura. Proc. 1er. Congreso Suramericano de Mecánica de Rocas. Bogota.

2.3 Underground excavations
Fouilles souterraines

Construction of a large section tunnel in weak rock

Construction d'un tunnel à grande section dans une roche tendre

Juan C.del Río, Raúl E.Sarra Pistone & Roberto I.Cravero, *Agua y Energía Eléctrica, Inspección Obras Complejo Hidroeléctrica Río Grande I, Córdoba, Argentina*

ABSTRACT: The solutions applied to the construction of a large cross-sectional area tunnel for hydraulical purposes through zones of weak rock, are evaluated. Starting hypotheses, their implementation and obtained results are discussed. Finally, a construction methodology is proposed, based on a strict control of the construction through measurements of particle velocity during blasting operation, the insertion of a support installation task in the excavation cycle and the rock mass and applied monitoring.

RESUME: Dans ce travail on evalue les solutions apliquées dans la construction d'un tunnel hydraulique de grand section, excavé avec explosives particuliermment les adoptées en zones des failles et des roches faibles. On expose ici des hypoteses de déport son implementation et les resultats obtenus. Finalement on propose une méthodologie de travail bassée sur un strict contrôl de la construction à travers de la medition de vitesses des particules pendont les soutages; de l'inclusement de tachês de colocation de support dans le cicle de construction et d'auscultation du comportement de la section et du support installé.

1 INTRODUCTION

The Tailrace Tunnel of the Río Grande T Hydroelectric Complex, a pumped-storage scheme with an underground powerstation, has a cross-sectional area of 206 m2 and a total length of 5,5 km.

Its excavation was planned in two stages: vault, of 86 m2 and bench of 120 m2 of section. The tunnel was concluded in 1985 and at present is in operation.

The Tunnel's function is to conduct water to the lower reservoir after being turbined through the four machines installed in the cavern and from there to the upper one, during the pumping stage (Fig. 1).

Concrete and/or gunite lining was constructed in the 16% of its length, only, at fair to bad quality rock stretches, mainly constituted by foliated gneiss and fault zones.

The remaining was excavated in good rock, massive gneiss, where structural and spot bolting was required as main support (del Río, S. Pistone, Cravero, 1983), and 5 to 10 mm - thick gunite was spread at isolated fault belts.

Is this paper, the construction methodology implemented to traverse a regional fault, named Transversal Sur Fault, is discussed. This fault affected the tunnel for space of 190 m. Nevertheless, emphasis is made in the 40 m-weakest rock stretch, where major difficulties were arosen.

2 GEOLOGY

2.1 Lithology

The complex is located on the eastern slope of the Comechingones Range which belongs to the Sierras Pampeanas system constituted by Precambriam rocks, specifically at site by massive tonalitic-cordieritic gneiss.

2.2 Structure

The Transversal Sur faults has a straight trace 7 km long, with a strike of 110°. As the tunnel has a general direction of 95° of azimuth, the fault goes almost parallel through it. This unfavourable

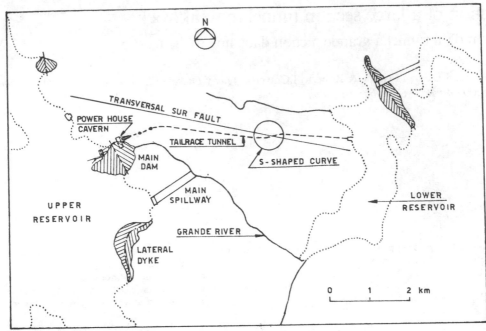

FIG.1 GENERAL PLAN VIEW

situation gave rise to a modification of the layout of the tunnel with the view to traversing it as perpendicularly as possible. For this purpose a S-shaped curve was designed (Fig. 1).

The principal faults of the TSF belt, which contains the weakest rock-zone dealt with in this pa_per, was named FN 2937 and has a strike of 150° and an inclination of 35° to the Southwest.

The rock of the fault is: heavily jointed cataclastic gneiss, decomposed and friable; filonites and sandy clay breccia. The zone is traversed and delimited by shear features containing montmorri_llonitic red clay (Fig. 2).

2.3 Underground water

Fortunately, the fault yielded small quantities of water (10 to 20 lts/min). However, the rock mass was saturated and problems due to water thrust, internal erosion and shrinkage of the clay ex_posed to the air were arosen, given rise to progressive instabilization of the excavation.

3 EXCAVATION. STAGES, PROBLEMS AND SOLUTIONS

A detailed description of the 1st. stage excavation methodology and support utilized, is published elsewhere (Sarra Pistone, del Río, 1982). The following is a summary introduction to that, to deal, afterwards, with the 2nd stage in particular, and the total section in general.

3.1 First stage

The uncertainty about the precise location and characteristics of this regional fault was one of the main problems arosen at this stage.

When rock quality deterioration signs(increasing weathering and joint frequency) appeared, coin_cident with the S-shaped curve zone as foreseen, the excavation was continued by means of a pilot tunnel, half the design width (Fig. 3-I).

Despite the change in the excavation section, a 4 m-high-bell-shaped cave in developed in the crown. Because of the fault inclination in the advance direction, it was not possible to foresee the exact location of the decomposed rock zone.

Inmediately after the cave in had taken place and due to the impossibility of installing perma_nent support in the provisional excavation, the pilot tunnel began to be enlarged in the upper mid-section (Fig. 3-II), in order to set up a group of steel arches fixed in the walls. Concrete mor_tar was pumped into the extrados for assuring a full contact with the rock mass.

Besides, three drillings were bored at the face that allowed foresee location and dimensions of the fault and appreciate the rock characteristics and underground water quantities to be met in

542

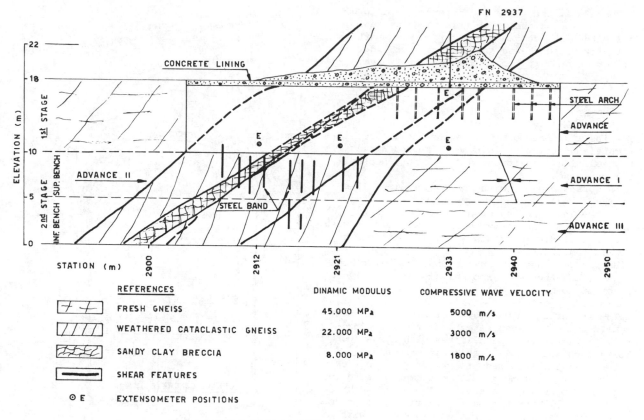

FIG. 2 LATERAL VIEW OF TUNNEL NORTHERN WALL

subsequent advances.

First stage completion demanded remotion of the lower parts of the walls (Fig. 3-III) in alternative advances one side each time, installation of rock bolts and eventually construction of a 0,5m thick concrete lining.

3.2 Second Stage

Before starting excavation of second stage consisting in removing a 10 m-high and 11 m-wide bench (Fig. 3), consolidation groutings were performed at a maximum pressure of 0,3 MPa in drillings up to 10 m long and 5 cm in diameter. No absorptions were observed, therefore it was interpreted as an indirect evidence of absence of open fissures produced by weak rock mass colapse.

Nevertheless, excesive strains could even occur as a consequence of the abrupt change in the section geometry that bench excavation would eventually give rise. In order to limit possible deformations and make the vertical arch sides of concrete lining to work together with the rock mass, a double line of prestressed rock bolts was installed (T = 140 kN, L = 5 to 6 m) 1 to 3 m high from the invert, so it would become to be 11 to 13 m from the final invert, once the second stage was completed (Fig. 3).

The second stage was designed to be excavated in two benches 5 m high and 11 m wide each one. The basic hypothesis used to discard any provisional excavation was that the upper concrete lining should inteact with rock mass responding solidarily with it.

The smooth-blasting line was drilled vertically following the concrete wall, resulting in a composited section (Fig. 3). On this way the originally schedulled 2nd stage concrete lining was eliminated because it would not contribute in anything to rock mass support.

So the excavation of a central trench and subsequent lateral enlargements proposed to support the upper concrete lining by underpining it with concrete walls, was discarded too.

This method of excavation did not offer reliable advantages at all and, at the same time, implied serious difficulties: provisional excavation surfaces would not admit permanent support; usually such surfaces are unstables due to negative incidences the form introduces in stress distribution; delays might be provoked by task diversities and contractual problems arose eventually.

To start second stage a full-width section excavation and an essentially active support was decided to be used.

Therefore, the following resources had to be employed, on account with this singular geological situation: <u>controlled blasting</u> with particle velocity measurements and selection of a damage criterion and, <u>rock mass displacement survey</u> using rod extensometers.

4 SUPPORT

The installed support was in the whole cases intimately associated, in type, time and amount, to geological conditions and rock mass mechanical response.

In correspondence with the support installed at the 1st-stage concrete walls, it was assumed that the 2nd-stage upper bench support (Fig. 2,3 El. 5 to 10 m) should be stiff enough to cope with the lateral displacements which was expected to be maximum there.

Besides, to advance in the opposite direction to that of a possible rock sliding was resolved, in order to mantain wall axial confinement by means of support.

The installed support consisted in rock bolts, gunite and steel mesh. At the most critical zones, vertical steel bands fixed with three rock bolts each and double steel mesh were added.

Mechanically-anchored and cement-grouted rock bolts, 4 to 8 m long and 25 mm diameter, were installed. They were tensioned up to 90 or 140 kN depending on jointing degree and bearing capacity of the rock. When the rock was strong enough with low frequency of joints, charge could be increased up to 140 kN (del Río, Sarra Pistone, Cravero, 1983).

Fast installation methods using polyester resin were rejected because it could not be possible to confirm the long term performance of such resins in weak rock.

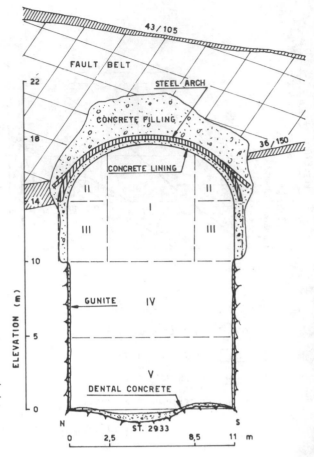

FIG. 3 - TUNNEL SECTION - EXCAVATION STAGES

Gunite was spread on the wall inmediately after blasting to prevent sudden deterioration of the saturated and decomposed rock mass. Variable thicks were used from 50 to 100 mm.

The lower bench support differred substancially from that of the upper one. It turned out being less stiff as a consequence of smaller strains registered and, moreover, of the trend of such strains whose variations decreased as the excavation went ahead (see paragraph 6). Hence, the support consisted in gunite without steel mesh and shorter rock bolts in less quantities. One-hundred mm-thick reinforced gunite was used eventually, as definitive lining of the walls.(Fig. 5).

5 CAREFUL BLASTING

In order to preserve rock mass integrity and to reduce damages caused by explosions, the careful blasting technique was implemented.

The velocity that a rock particle acquire during a near blasting,was used as main reference parameter. It was assumed a velocity up to 20 cm/seg for the most critical zones, beyond that, important damages could occur (del Río, Cravero, 1986).

Such a criterion allowed to improve blasting patterns and ignition sequence, limiting the charge per delay, chosing the adequate explosive and minimizing the specific charge (170 gr/m3) according to geological conditions.

The use of the smooth blasting technique instead of prespliting, helped to lessen damages notoriously in the decomposed and friable rock and almost eliminate wall overbreaks.

Besides, the use of explosives with detonation velocity similar to that of the compressive wave in weak rock mass (average 3000 m/seg), apparently contributed to same effects.

As an extension of that study and with the purpose of improving the rock mass elastic response interpretation, a microseismic survey was carried out at fault belt. Different compressive wave velocities and dynamic module are plotted in fig. 2.

Good correlation with different qualities of rock was obtained which allowed to deepen the understanding of rock mass behaviour and to adjust evenmore the excavation methodology.

Measurements of superficial and internal displacements referred to a deep point assumed to be "fix" (at a depth of 16 to 18 m) were carried out utilizing tree position-mechanical rod extensometers, cement grouted and read with a deflectometer of 0,01 mm.

Two station were instrumented with three extensometers each one. They were installed at the 1st. stage to check the behaviour of:
- concrete lining of the 1st. stage,
- faulted rock mass at the point where greatest deformations were expected because of the second stage excavation,
- slightly weathered and jointed rock mass adjacent to the fault.

Figures 2 and 4 show extensometer locations and some outstanding characteristics of the monitored rock mass.

This two stations integrated a group of seven extensometric section distributed along the tunnel in other fault zones and undisturbed rock mass used as reference patterns for the strain interpretation. At the same time a mathematical model of boundary elements (B.E.M.) (Hoek and Brown, 1980) was run, approaching to the problem within the elastic field, simulating a continous, homogenous and isotropic rock mass.

The extensometers indicated appreciable displacements from the beginning of the upper bench exca vation (Fig. 3-IV), though the front-heading was 26 m away. Those displacements matched with first advances, carried out without using the careful blasting technique.

After being fully implemented and having installed active support just after each blasting, move ments occurred associated almost exclusively to geometric changes of the section. The 70% of total displacements happenned when the top-heading was within 5 m before and after each instrument station, and the general trend of deformations showed a progressive increment inward the rock mass as the excavation went ahead.(Fig. 4 a).

Eventhough the two extensometric station monitored different quality rock masses, their results were very similar. Station I, at the fault zone, indicated displacements slightly greater than tho se of the Station II, located in an almost unaltered rock mass. Such a behaviour was interpreted as

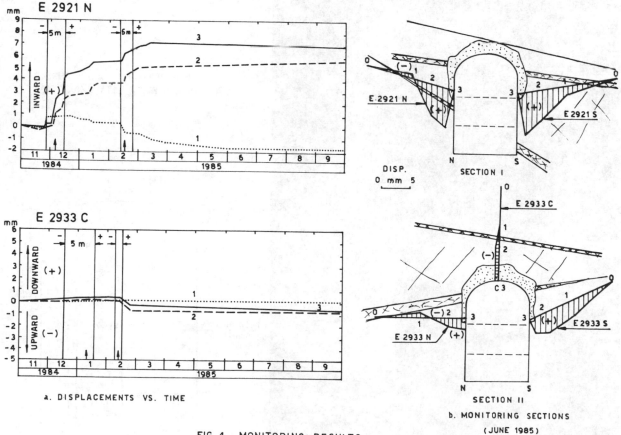

a. DISPLACEMENTS VS. TIME

b. MONITORING SECTIONS
(JUNE 1985)

FIG. 4   MONITORING RESULTS

being a consequence of the stabilizing effect of the installed support.

Displacements measured at the vault indicated a gradual descent of the crown during the excavation of the 2nd stage upper bench.

Once the full section was achieved an absolute ascent was registered at the same point (Fig. 4). The velocity of deformation remained practically the same. The magnitude of these displacements was smaller than that of the rest of extensometers.

The location of this instrument was precisely chosen for checking the 1st. stage cave in zone, where a 4 m thick concrete lining was built to restore the originally design tunnel section.

Rock mass behaviour turned out to be better than expected as long as aforesaid crown ascent had the desired effect of compressing the rock above the concrete up to a height of 10 m.

7 DISCUSSION

- A substancial change of ideas about the way the excavation control had to be held, was operated between the first and the second excavation stage (Fig. 5). The former was constructed within the years 79-80, the latter in 84-85. During all this time, experience was carried out dealing dayly with the whole excavations of the complex. It was concluded that only providing suitable assistance, just in time, to every particular situation, it could be possible to prevent potential instability problems.

In the first stage it was used to excavate without a sistematic control. Such a methodology implied high risk assumptions and eventually demonstrated to be ineffective to provide a sustained production rate, as long as instability situations were encountered, problems grew quickly out of control and execution time turned to be impredictable.

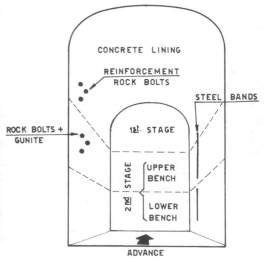

FIG. 5 PHOTOGRAPH TAKEN DURING FTS FAULT BELT EXCAVATION

On the contrary, in the second stage a systematic control was implemented as a part of the production cycle. So, preventive support was installed with the aim of helping the rock mass to develop the greatest available strength capacity (self-support).

- The first-stage coaxial pilot tunnel resulted to be less effective than that excavated in other weak-rock tunnel site, made by blasting a lateral half section instead of a central one (Sarra Pistone, del Río, 1982). In this case minor definitive surface was exposed which hindered the inmediate installation of necessary support, originating delays in the excavation rate.

From this point of view, the trench designed to be excavated at second stage, posed similar disadvantages. As an alternative, full-width-bench excavation, horizontally divided in two stages, turned out to be more advantageous as long as it permitted to cope with a smaller section as well as to use the adequate support on the just time.

- Careful blasting designed according to damage criteria and a subsequent strict control of every step involved in the process, allowed to assure a successful excavation with a minimum overbreak.

- The monitoring results demonstrated that the tunnel was under control during the whole excavation. A realistic interpretation of a huge quantity of sistematically collected data yielded a good guide to chose suitable support particularly for the second stage.

It also allowed to confirm some hypotheses and to appreciate general behaviour step by step. When the upper bench was removed instruments registered inward displacements (Fig. 4a). At this point the excavation had a width/height rate of 0,85 and total displacements at walls were almost ten times larger than those measured at crown (Fig. 4b) due to greater vertical stiffness provided by the form-effect and by 0,5 m-thick concrete lining of the first stage.

When the tunnel second stage was completed the ratio width/height lowered to 0,61. Therefore the tunnel became horizontally less stiff and the displacements at the walls surpassed eighteen times to those at the crown which, on the other hand, reversed their trend indicating the eventual crown uplift.

8 REMARKS

- Suitable geologic-geotechnical rock mass knowledge is the indispensable frame to accomplish an economic and safety excavation.

- Usually a large tunnel must be excavated in different stages. When such a tunnel is constructed in a weak rock mass, sub-stages should be designed, too, trying to expose as much definitive surface as possible in each sub-stages in order to install permanent support at once.

- Systematic excavation control and support installation tasks must be included in the production cycle.

- Support must be aimed to contribute to rock mass self-support, instead of stabilizing it after failure occurrence using passive and heavy support.

- Gunite as inmediate lining is advisable to avoid the humidity loss and progressive block loosenning of weak and decomposed rock and clayley breccia.

- Reinforced gunite with steel mesh anchored to the rock, is a suitable solution to obtain an adequate surface for structural as much as hydraulical performance.

- It is highly convenient that support design (Active Design, Morfeldt, C.O. 1983) be permanently confronted with monitoring results all along the construction period.

- In weak rock tunnelling it is strongly recommended that blasting be controlled using particle velocity and wave frequency as reference parameters in order to prevent charge excess and consequent rock mass damage.

- Excavation monitoring is particularly necessary in weak rock mass. Displacement measurements made with multiple-position rod extensometers are difficult but very illustrative because they allow to draw a general view of average deformations suffered by the tunnel surrounding rock mass.

Though the authors understand that the proposed recommendations do not introduce outstanding news in modern tunnelling, they believe their utility is the confirmation of such concepts for large tunnels endorsed by an accomplished underground scheme and insistence in their application will allow to bring up to day contractual specifications for this kind of project particularly in countries, as Argentina, with less development in tunnelling.

9 ACKNOWLEDGMENTS

The authors wish to thank the authorities of AGUA y ENERGIA ELECTRICA for having permitted the publication of this paper.

10 BIBLIOGRAPHY

del Río, J.C., Cravero, R.I. 1986. Monitoring of a large portal excavated in rock. Vth International Congress of IAEG. Buenos Aires, Argentina.

del Río, J.C., Sarra Pistone, R.E. Cravero, R.I. 1982. Estudio de la estabilidad del túnel de restitución del Complejo Hidroeléctrico Río Grande I. Actas ASAGAI, Vol. II, 49-67. Buenos Aires, Argentina.

del Río, J.C., Sarra Pistone, R.E., Cravero, R.I. 1983. Rock bolting used in the Underground Excavations of the Río Grande I. International Symposium on Engineering Geology and Underground Construction. IAEG. Vol. I. Lisbon, Portugal.

Dorso, R., del Río, J.C., de la Torre, D.G., Sarra Pistone, R.E. 1982. Powerhouse cavern of the Hydroelectric Complex Río Grande I. Symp. Rock Mech. Caverns and pressure shafts. Aachen, W. Germany.

Hoek, E., Brown, E.T. 1980. Underground excavations in rock. London. Institution of Mining and Metallurgy.

Morfeldt, C.O. 1983. The influence of engineering geological data on the design of underground structures and on the selection of construction methods. International Symposium on Engineering Geology and Underground Constructions. IAEG. Panel Report, theme II. Lisbon, Portugal.

Sarra Pistone, R.E., del Río, J.C. 1982. Excavation and treatment of the principal faults in the Tailrace Tunnel of the Río Grande I Hydroelectric Complex. IVth International Congress of IAEG. Vol. IV, th. 2. New Delhi, India.

Problèmes posés par un calcaire très poreux dans l'étude de la stabilité d'une carrière souterraine

Influence of the high porosity of a limestone on the stability of an underground quarry

André Denis, Jean-Louis Durville, Etienne Massieu & Robert Thorin, *Laboratoires des Ponts & Chaussées, France*

RESUME : La mise en place d'un remblai au-dessus d'une carrière souterraine abandonnée a nécessité une analyse géomécanique de stabilité des piliers et du toit. Le matériau, un calcaire d'âge éocène, pose des problèmes particuliers liés essentiellement à sa porosité, qui varie de 28 % à 42 %. La variabilité de cette porosité, sur des distance de l'ordre du mètre, entraîne une variabilité des propriétés mécaniques (résistance à la rupture, déformabilité); se posent alors les problèmes de la représentativité des mesures et de l'introduction de l'hétérogénéité dans la modélisation géomécanique. D'autre part, la forte sensibilité des propriétés du calcaire à l'état de saturation conduit à prendre en compte les valeurs minimales obtenues à saturation. En bref, la conduite de l'étude géomécanique du matériau a été fortement conditionnée par ces particularités, liées au faciès détritique grossier du matériau.

ABSTRACT : Before laying an embankment above an underground quarry, a stability analysis of roof and pillars was performed. The rock, a limestone of the eocene period, requires to face with some particular problems related with its porosity, wich lays between 28 and 42 %. The varying pattern of the porosity at distances of some meters causes related variations of the mechanical properties (strength and deformability). Performing representative in situ or laboratory tests implies to increase the testing points, and the heterogeneity has to be introduced in the geomechanical model. The mechanical properties of the limestone are also influenced by the water content. Such particular properties of the limestone, which can be related to its coarse detrital facies, strongly conditioned the geomechanical survey.

1 INTRODUCTION

La carrière souterraine de Villiers-Adam, située en région parisienne, a fait l'objet d'une exploitation pour pierre à bâtir, suivant la méthode dite par chambres et piliers, depuis le siècle dernier jusqu'à l'arrêt de l'exploitation en 1962. La mise en place, à l'occasion de travaux routiers, d'un important remblai en surface, au-dessus de cette carrière, apporte à son niveau une surcharge de 50 % environ. Aussi, une analyse géomécanique de la stabilité à court ou long terme a-t-elle été entreprise avec instrumentation dans la carrière et suivi des contraintes et des déformations. Les paragraphes qui suivent présentent l'étude mécanique en laboratoire du matériau, un calcaire poreux hétérogène, avec les particularités qu'il possède et leurs conséquences pour l'analyse géomécanique du site.

Le "calcaire grossier", d'âge Lutétien (Eocène) est un dépôt biodétrique, formé par une accumulation en mer peu profonde de débris coquilliers plus ou moins cimentés. L'observation au microscope montre des débris organogènes ($\emptyset$ 0,1-0,5mm) difficilement identifiables, quelques grains de quartz ($\emptyset$ 0,1mm en moyenne) et une porosité importante laissée libre par le ciment calcitique peu abondant. Les couches calcaires, épaisses d'une vingtaine de mètres au total, se présentent en bancs quasi-horizontaux; la partie exploitée correspond à un niveau tendre ($R_c < 15$ MPa), en strates d'épaisseur métrique. Les couches sont perturbées par de rares accidents karstiques et par quelques fractures quasi-verticales. L'exploitation souterraine se fait sur une hauteur de 2,5 à 3 m, avec des piliers de 10 à 12 m^2 de section.

2 ETUDE DE LA POROSITE

Comme cela est classique, la porosité est le facteur explicatif prépondérant des propriétés du calcaire de Villiers-Adam; sa forte variabilité a imposé une étude approfondie de ce paramètre. Plus de quatre cents mesures de porosité ont été effectuées sur carottes $\emptyset$ 40 mm prélevées en différents points de la carrière. L'histogramme correspondant est présenté sur la figure 1 : les valeurs s'étalent entre 28 et 46 %.

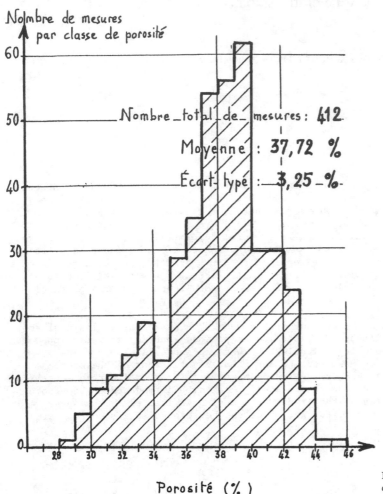

Figure 1. Histogramme des porosités du calcaire de Villiers-Adam.

Autant que la distribution statistique des mesures de porosité, c'est la répartition spatiale de celle-ci qui est importante. La figure 2 montre l'évolution de la porosité mesurée le long d'un sondage horizontal au travers d'un pilier de 4 m de large. L'hétérogénéité se manifeste à l'échelle du décimètre (dispersion des mesures sur les trois carottes Ø 40 mm issues d'une même carotte Ø 100 mm) comme du mètre (variation de près de huit points de porosité sur une distance d'un mètre). Le redécoupage d'une carotte Ø 40 mm, L = 100 mm révèle une variabilité encore sensible à l'échelle de 20 à 30 mm.

Un variogramme expérimental (figure 3), obtenu à partir des données du sondage mentionné ci-dessus et du même type, semble confirmer l'existence d'au moins deux échelles d'hétérogénéité emboîtées, la "portée" la plus longue dépassant trois mètres (autrement dit, il subsisterait encore une certaine corrélation entre deux prélèvements distants de trois mètres).

Ces variations spatiales peuvent s'expliquer par le type de faciès du calcaire grossier : accumulations plus ou moins compactes de grains plus ou moins fins, avec cimentation ultérieure plus ou moins poussée. Des stratifications obliques, d'amplitude plurimétrique, sont visibles en certains endroits.

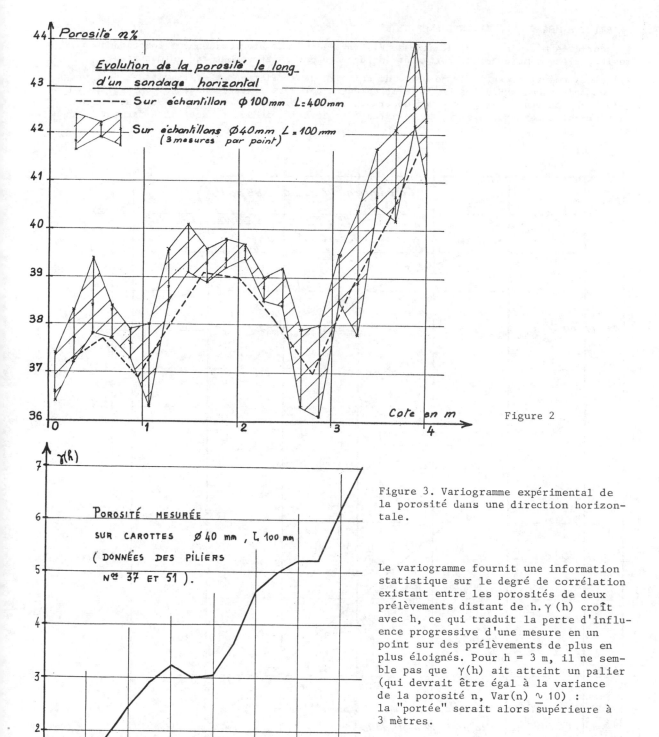

Figure 2

Figure 3. Variogramme expérimental de la porosité dans une direction horizontale.

Le variogramme fournit une information statistique sur le degré de corrélation existant entre les porosités de deux prélèvements distant de h. γ (h) croît avec h, ce qui traduit la perte d'influence progressive d'une mesure en un point sur des prélèvements de plus en plus éloignés. Pour h = 3 m, il ne semble pas que γ(h) ait atteint un palier (qui devrait être égal à la variance de la porosité n, Var(n) $\simeq$ 10) : la "portée" serait alors supérieure à 3 mètres.

Les propriétés mécaniques du calcaire de Villiers-Adam sont bien évidemment fonction de sa porosité; elles sont de plus, fortement influencées par la teneur en eau du matériau.

3.1. Les corrélations entre la porosité du calcaire de Villiers-Adam et la vitesse de propagation des ondes longitudinales Vℓ, de même qu'entre porosité et résistance en compression uniaxiale, sont présentées figure 4. On observera la sensibilité des valeurs de résistance aux variations de porosité. De même, la résistance en traction indirecte décroît nettement lorsque

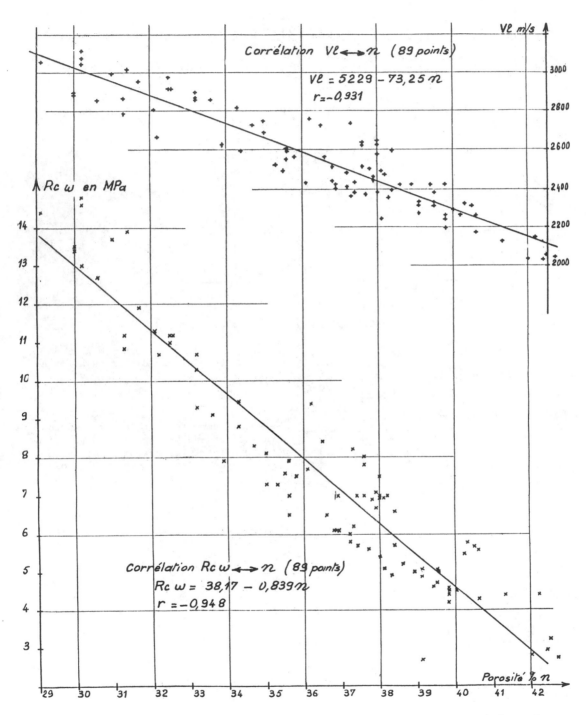

Figure 4. Corrélations entre la porosité n et la vitesse des ultrasons Vℓ (mesurée sur roche sèche), et entre la porosité et la résistance en compression simple $R_{c\omega}$ (mesurée sur roche saturée).

la porosité augmente (figure 5). En ce qui concerne les coefficients d'élasticité, le module d'Young chute lui aussi lorsque n croît, mais avec une dispersion importante; les mesures du coefficient de Poisson, situées autour de 0,2, sont trop peu nombreuses pour être reliées à la porosité.

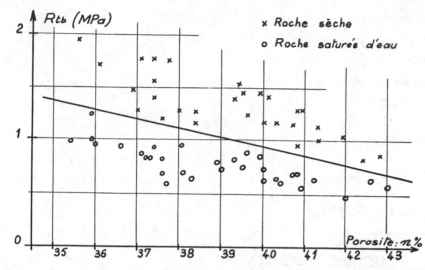

Figure 5. Résultats de traction indirecte (essai brésilien).

3.2. L'influence de l'eau sur les propriétés mécaniques des roches a été observée maintes fois, mais les interprétations sont loin d'être entièrement satisfaisantes. Elles seront discutées ci-dessous à propos du calcaire de Villiers-Adam : propriétés élastiques (statiques ou dynamiques), limites de rupture.

La vitesse de propagation des ondes $V\ell$ est relativement peu affectée par la présence d'eau (figure 6); on observe en général une légère diminution : $V\ell_w \sim V\ell - 150$ (m/s). Ce comportement du calcaire de Villiers-Adam est typique des roches à forte porosité, et s'oppose à celui des roches microfissurées (Tourenq et al., 1971). Il peut s'expliquer dans le cadre d'un modèle de diffraction d'ondes par des pores sphériques petits devant la longueur d'onde (ici le rapport est de l'ordre de 1/100 à 1/10) : dans le milieu composite, matrice solide + pores remplis de fluide, le remplacement de l'air par l'eau entraîne une augmentation de l'inertie du milieu, qui ralentit la propagation, et cet effet l'emporte sur la diminution de compressibilité, qui a tendance à accroître la vitesse des ondes (Kuster et Toksöz, 1974).

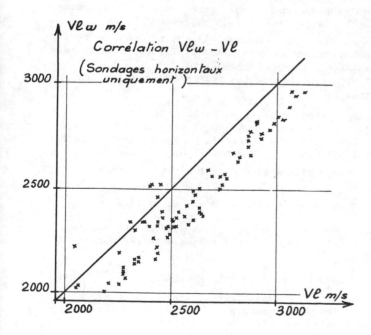

Figure 6. Influence de l'eau sur la vitesse de propagation des ondes ultra-sonores (ondes longitudinales) dans les calcaires de Villiers-Adam.

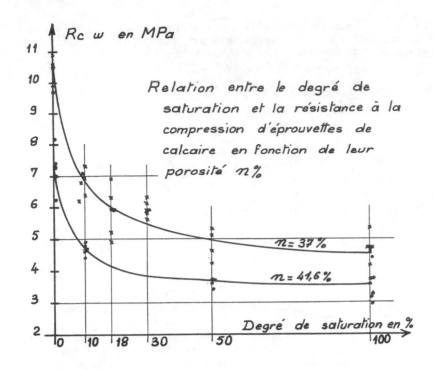

Figure 7

La figure 7 montre la variation de la résistance en compression uniaxiale en fonction du degré de saturation, les essais étant effectués sur éprouvettes séchées très progressivement à partir de l'état saturé : la valeur-plancher de la résistance se maintient pour un degré de saturation de 100 % jusqu'à 50 % environ; le rapport R_ω/R_c dépasse 2. De même, en ce qui concerne la résistance en traction indirecte (essai "brésilien"), la différence entre valeurs "sèches" et "saturées" est nette malgré la dispersion (cf. figure 5). Les mesures de module d'Young révèlent également un amollissement du matériau en présence d'eau.

Deux interprétations sont a priori disponibles pour expliquer l'affaiblissement du matériau mis en présence d'eau :
- diminution de l'énergie de surface des minéraux (ici la calcite) lors d'une mise en contact avec l'eau. Par exemple, certains auteurs ont mesuré une baisse de 30 % du coefficient de tension superficielle (de la phase solide) dans des grès quartzeux lorsque l'on passe du milieu "air" au milieu "eau" (Dunning et Huf, 1983). Ce phénomène a pour conséquence la propagation plus facile des fissures, donc une diminution de la résistance à la rupture.

- action de minéraux argileux : intercalés dans les microfissures, ils ont tendance à ouvrir celles-ci lorsqu'ils adsorbent de l'eau, à leur surface ou dans leurs feuillets, et ceci même lorsque les quantités d'argile sont faibles (Denis, Tourenq, Tran, 1980).
Concernant le calcaire de Villiers-Adam, il est probable que les deux phénomènes ajoutent leurs effets, compte-tenu des observations suivantes :

. pureté imparfaite du calcaire : la teneur en carbonate se situe entre 88 et 92 % (mais le quartz visible en lame mince, qui est inerte, constitue une bonne part des non carbonates); l'essai d'adsorption du bleu de méthylène sur roche broyée dénote la présence, en très faible quantité probablement, de minéraux argileux (VB = 0,07 à 0,09 g pour 100 g de matériau),

. gonflement non négligeable d'éprouvettes lors de l'immersion dans l'eau ($0,9.10^{-4}$ à $1,6.10^{-4}$), significatif de la présence de minéraux argileux,

. influence quasi-nulle, sur la résistance en compression simple, d'une saturation avec un liquide tel que l'heptane.
Enfin, des études sur différents calcaires ont montré des corrélations, de qualité moyenne il est vrai, entre gonflement, valeur au bleu de méthylène, et chute relative de résistance $(R_c-R_{c\omega})/R_c$ (Denis, Tourenq, Tran, 1980).

4 INCIDENCE DES PROPRIETES DU CALCAIRE SUR LE CONTENU DE L'ETUDE GEOMECANIQUE

Les principales caractéristiques du calcaire de Villiers-Adam (sensibilité à l'eau, forte porosité, hétérogénéité) ont des répercussions sur les différentes parties de l'étude géomécanique : essais de laboratoire, mesures in situ, modélisation numérique.

4.1. La sensibilité des propriétés mécaniques à la teneur en eau a conduit, après mesure du degré de saturation in situ (S_r varie entre 50 et 100 %), à prendre en compte dans l'évaluation de la stabilité uniquement les valeurs prises sur roche saturée.

4.2. Le caractère très poreux du calcaire de Villiers-Adam entraîne un certain nombre de particularités ou de difficultés concernant les essais de laboratoire, telles que :

- l'utilisation, sur la roche saturée, de jauges à fil résistant pour la mesure des déformations est très délicate, ce qui rend difficile la mesure du coefficient de Poisson par exemple.Même dans le cas du matériau sec, le collage des jauges, en raison de la pénétration de la résine dans les pores, risque de fausser les mesures : on a vérifié que cette erreur était négligeable, en comparant avec des mesures avec extensomètres, à condition d'utiliser le minimum de colle,

- la dimension de l'échantillon a une influence sur les valeurs mesurées de la porosité : la figure 2 montre,en dépit de la variabilité de la porosité, qu'une différence systématique existe entre les carottes Ø 40 et Ø 100 mm. Cet accroissement des valeurs mesurées, lorsque l'on passe d'une grande éprouvettes à une petite, est probablement lié à l'augmentation concomitante du rapport : surface extérieure/volume : le carottage en présence d'eau de roches très poreuses produit un arrachement de petites particules en surface, ignoré par les mesures au pied à coulisse,

- l'interprétation de l'enveloppe de Mohr déterminée par essais triaxiaux est particulière : une compression isotrope drainée du matériau produit, à partir d'un certain seuil, un effondrement du squelette minéral, avec microfissuration généralisée (figure 8). Au delà le matériau rejoindrait progressivement un comportement de type "sol" (Hazebrouck et Duthoit, 1979).

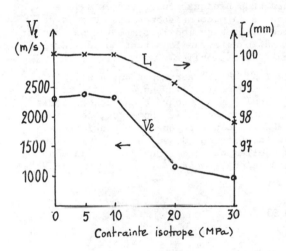

Figure 8. Chargement sous pression isotrope d'un échantillon de calcaire de Villiers-Adam (porosité initiale : 39,2 %) : à partir d'un seuil situé entre 10 et 15 MPa, on observe une chute considérable de $V\ell$ (mesurée après déchargement) ct une déformation permanente notable. La résistance en compression simple de cette éprouvette est estimée à 5,5 MPa (cf. figure 4).

4.3. L'hétérogénéité du calcaire de Villiers-Adam, qui se manifeste à différentes échelles, pose un certain nombre de problèmes :

- la détermination en laboratoire des paramètres d'identification et des propriétés mécaniques du matériau est conditionnée par la représentativité très réduite d'un prélèvement ponctuel : la variabilité de la porosité, et donc des autres propriétés, oblige à multiplier les points de mesure, de façon à obtenir des résultats statistiquement significatifs. Le cas des essais de fluage et de rupture différée sous charge constante est particulier : si l'on souhaite charger à un pourcentage donné de R_c, seule une estimation de R_c à partir de la porosité est possible, dont la précision est à peine suffisante pour différencier les comportements observés,

- l'interprétation des mesures in situ nécessiterait, pour être vraiment fiable, la détermination au même endroit de certains paramètres; la mesure de la seule porosité, qui commande l'ensemble des propriétés mécaniques, peut être un palliatif pour minimiser le nombre des prélèvements. Par exemple, la mesure de la vitesse du son en paroi peut fournir des indications sur la fissuration, à condition de connaître la porosité; la mesure de contrainte au vérin plat doit pouvoir être rapportée à la résistance R_c du matériau; l'extensométrie sur les piliers permet, avec l'aide des modules, d'évaluer les contraintes, etc...

- les méthodes d'estimation du risque d'effondrement de la carrière souterraine sont compliquées par l'hétérogénéité dans la répartition des contraintes et des déformations : les modèles utilisés, qu'ils soient simplifiés ou très complets, accordent en général peu ou pas de place à l'hétérogénéité du matériau. Le paragraphe qui suit s'attache à évaluer les conséquences possibles de l'hétérogénéité à l'échelle d'un pilier.

4.4. L'étude de la répartition des contraintes dans un pilier a été menée sur un modèle à deux dimensions: pilier rectangulaire, de 3,00 m de large et 2,50 m de haut, soumis à une contrainte verticale uniforme de 1 MPa sur sa face supérieure, les déplacements étant bloqués à la base.

La simulation numérique a été obtenue en tirant au sort une distribution de modules d'Young dans un quadrillage 25 cm x 25 cm de la façon suivante :

- affectation aléatoire d'un module à chaque carré, avec une distribution gaussienne de moyenne 7200 MPa et d'écart-type 4925 MPa,

- lissage par moyenne glissante adéquate, de façon à obtenir les caractéristiques statistiques suivantes (les modules de chaque carré suivent encore une loi gaussienne) :

  moyenne de chaque carré $\bar{E}$ = 7200 MPa
  variance de chaque carré $Var(E) = 3,35.10^6$ $(MPa)^2$ (écart-type : 1830 MPa)
  covariance à 25 cm de distance : $Cov(E_i, E_j) = 2,2.10^6$ $(MPa)^2$
  covariance à 35 cm de distance : $Cov(E_i, E_k) = 1,8.10^6$ $(MPa)^2$
  covariance à 50 cm de distance : $Cov(E_i, E_\ell) = 0,6.10^6$ $(MPa)^2$

On obtient ainsi une distribution de modules, sans sauts trop brusques entre carrés adjacents, se rapprochant ainsi de la réalité. Toutefois, en raison de la méconnaissance des données statistiques conplètes concernant le module, et à cause des limites du modèle simple de génération utilisé, les valeurs numériques ci-dessus ne reflètent pas exactement la situation réelle (par exemple, on a supposé les covariances indépendantes de la direction).

Mille calculs ont été effectués (programme aux éléments finis ELIP) correspondant à mille affectations de modules aux 120 carrés du modèle. A titre d'exemple, la figure 9 présente l'histogramme des contraintes verticales dans un carré situé à mi-hauteur et en façade, c'est-à-dire en un point où les mesures au vérin plat sont effectués; un pilier homogène, de module 7200 MPa constant, donnerait à cet endroit σ_z = 0,958 MPa. On constate que la variabilité au niveau de la contrainte verticale n'est pas négligeable (coefficient de variation : 0,124); ceci pourrait donc contribuer aux différences mesurées réellement entre deux faces d'un même pilier, indépendamment du rôle probable des dissymétries de chargement.

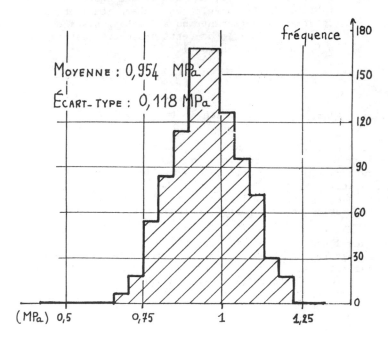

Figure 9. Histogramme des contraintes verticales mesurées sur une face de pilier

5. CONCLUSIONS

Bien qu'apparaissant homogène, lors d'un examen rapide à l'oeil nu, le calcaire de Villiers-Adam, roche de faible résistance et sensible à l'eau ($R_{c\omega} \simeq 4$ MPa), s'est de plus révélé très hétérogène; la composition minéralogique étant constante, c'est la porosité comprise entre 28 et 42% qui constitue le facteur variable dans ce matériau et qui commande ses propriétés mécaniques.

Les variations spatiales se manifestent à l'échelle de quelques décimètres, donc des prélèvements pour essais de laboratoire. Un échantillonnage classique, avec réalisation d'un petit nombre d'essais, est donc inadapté, et il est nécessaire d'augmenter le nombre de points de mesures, en favorisant les essais rapides tels que la mesure de la vitesse du son $V\ell$ sur carottes : celle-ci fournit une indication sur l'homogénéité d'un ensemble de prélèvements, et permet d'estimer les autres propriétés du matériau par corrélation.

L'hétérogénéité est également présente à une échelle supérieure au mètre, en rapport avec les dimensions d'un pilier : celles-ci ne sont donc pas suffisantes pour que la variabilité soit annulée statistiquement. Une simulation numérique par méthode de Monte-Carlo a démontré l'influence non négligeable de cette hétérogénéité sur la répartition des contraintes dans un pilier : l'exploitation des mesures in situ et l'estimation de la stabilité des piliers devront en tenir compte.

REFERENCES BIBLIOGRAPHIQUES

Beaufrère C.(1985). Etude du rôle des facteurs géologiques dans la stabilité des carrières souterraines. Bull. AIGI n° 32, p. 11-24

Denis, Tourenq, Tran (1980). Capacité d'adsorption d'eau des sols et des roches. Bull. AIGI n° 22, p.201-205

Dunning et Huf (1983). Effects of aqueous chemical environments on crack propagation. J.Geophys. Res., B8, Vol.88, p. 6491

Hazebrouck et Duthoit (1979). Particularités du comportement mécanique des craies : rôle de l'eau. Rev. Fr. Géot., n° 8, p. 45-50

Kuster G., et Toksöz N. (1974). Velocity and attenuation of seismic waves in two-phase media : Theoretical formulations. Geophysics, vol. 39, n° 5, p. 587-606

Tourenq et al. (1971). Propagation des ondes et discontinuité des roches. Symposium S.I.M.R., Nancy

Foundation treatment of the earth dam of the Balbina hydroelectric power plant – Grouting with hydraulic fracturing in residual soil

Traitement des fondations du barrage en terre de la centrale hydroélectrique de Balbina Injections après fracturation hydrauliques dans des sols résiduels

Gilson H.Siqueira, Lourenço J.N.Babá, João Marcos de Siqueira, *ENGE-RIO, Engenharia e Consultoria SA, Rio de Janeiro, Brazil*

ABSTRACT: The Balbina hydroelectric power plant is located on the Uatumã river, left tributary of the Amazon river, approximately 170 km North of Manaus in Brazil. The foundations of the earth dam are on clay-silty residual soil originating from vulcanic rocks. Tubular cavities, denominated "canaliculos" were encountered in these materials. The permeability of the soil matrix is in the range of 10^{-4} to 10^{-5} cm/s, reaching values greater than 10^{-3} cm/s in the areas with "canaliculos". In order to reduce the permeability of these critical zones, a foundation treatment using soil cement grouts for the injection with hydraulic fracturing was used. This paper presents the most relevant aspects of the foundation treatment design on the "canaliculos" zones as well as the tests performed on an experimental grout curtain and the results obtained to date on the grout curtain itself.

RÉSUMÉ: La fondation du barrage de terre de l'usine hydroéléctrique de Balbina, située dans le fleuve Uatumã, affluent de la rive gauche de l'Amazonas, et eloignée de 170 km de Manaus, Brésil, a eté implantée principallement sur une couche de sol silt-argileux, résiduel de roches volcaniques. Dans cette couche il a eté rencontrée des cavités de forme tubulaire, dénominées "canaliculos". Dans les zones d'occurrence des "canaliculos" la permeabilité atteint des valeurs supérieures à 10^{-3} cm/s. Avec le but d'éliminer les zones critiques de haute permeabilité, il a eté défini, basé dans des études techniques et économiques, l'exécution du traitement de la fondation à travers en écran d'injections par claquement, en utilisant un coulis de sol et ciment, dans la couche résiduelle. Ce travail présente les concepts considerés au projet de traitement de la fondation, les tests éffectués pendant la phase expérimentale et les résultats atteints dans la phase d'exécution.

1 INTRODUCTION

The Balbina hydroelectric power plant is under construction on the Uatumã river, left-bank tributary of the Amazon. It has been designed to supply the electric power requirement of Manaus, capital of the Amazonas state. Its site is about 170 km to the North of Manaus (figure 1).

The main features concerning the plant are: installed capacity 250 MW; concrete volume 350,000m³; volume of the compacted soil 5,000,000m³; total length of dam 3,300 m, 330 m of which correspond to the concrete structure; maximum height of the structures 34 m and 33 m for the earthfill dam.

The location of the dam, far from the main industrial centres of the South and Southeast regions, together with the scarcity of available data concerning topographic conditions as well as geological and geotechnical behaviour of materials are specific difficulties for the dam construction due to particular events which occurred during the project development.

Among the conditioning factors involved, this paper reports one of the aspects deriving from the last of the abovementioned difficulties: the existence of points of high permeability coefficient (above 10^{-3} cm/s) found in the earthfill dam foundation soil, where the expected values were 10^{-4} cm/s to 10^{-5} cm/s considering the clay-rich soils. Investigations carried out during basic and executive design stages led to identification of the causes for such high permeabilities related to the presence of tubular cavities, from millimetric to centimetric, erratically distributed and often inter-communicating, called "canalículos", which would be responsible for localized concentrations of water flow, leading to high permeability values.

The presence of such features in the earth dam foundation soil determined an accurate analysis of the foundation treatment procedures considered technically and economically more adequate to ensure the necessary imperviousness to this soil.

Studies were made, the adopted criteria as well as the physical models were theoretically developed, besides the costs involved; and all indicated as more suitable a treatment by hydraulic fracturing grouting, which would improve the soil conditions, leading to the reduction of those

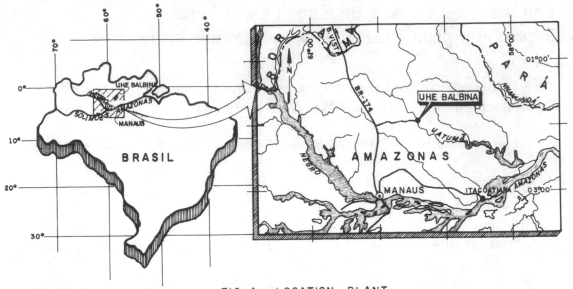

FIG 1 - LOCATION PLANT

LONGITUDINAL SECTION ALONG CENTER LINE OF CUT-OFF IN THE LEFT ABUTMENT

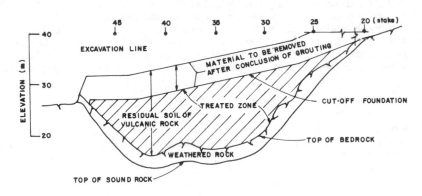

TYPICAL CROSS SECTION

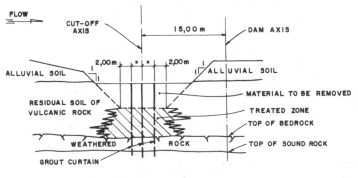

— FIG 2 —

high permeability zones.

Therefore, this paper presents studies, tests and services carried out during all stages of research and implementation of design criteria, as well as evolutions and modifications gradually introduced according to observation, analyses and evaluation which accompanied all these stages. This paper specially details the analysis of the results obtained with the left-bank earth dam foundation treatment, which has been completed.

2 GEOLOGICAL SYNTHESIS OF THE DEVELOPMENT SITE/CHARACTERISTICS AND GEOTECHNICAL FACTORS

Within the geotectonic model that is now accepted, the plant site is located in the northern part of the central Amazonian province. The dam building area includes two structural macro-features respectively represented by the Guyanas shield, located in the northern part, and the Amazonian synclise, which forms the southern area.

In order to provide a general view of the involved and occurring formations at the site, there follows a simplified synthesis of the local lithological and chronostratigraphic column.

Period	Litho-stratigraphic units	Lithology
Quaternary	Alluvia and colluvia	Gravel, sand, silt and clay
Tertiary	Lateritic covers	Concretions and lateritic levels
Tertiary Cretaceous	Barreiras formation	Sandstones, siltstones and claystones
Ordovician Silurian Devonian	Urupadi group	Arenites, siltites and shales
Upper Pre-Cambrian *Uatumã group*	Iricoumé formation	Rhyolites, rhyodacites, dacites, andesites, trachytes and pyroclasts

The Uatumã group is represented by an extensive and intensive acid to intermediate vulcano-plutonism, with some basic components which occurred on a large area of the Amazonian platform from the end of the lower Pterozoic to the beginning of the middle. By its extension, it was the largest acid to intermediate igneous event ever occurred on the crust.

The Iricoumé formation (represented by vulcanic rocks) is the dominant phase in the Balbina area. It is petrographically represented by several lithological types, herein generically called vulcanite to keep the same simplified terminology used during the works.

The Paleozoic formation which was formed by the sedimentation of the Amazon basin is predominantly represented by alternate sandstones/siltstones and subordinately by shales which form the Urupadi group.

Above these sediments, in some areas there are occurrences of extensive outcrops of a sandy-clay layer capping the older formation and related to the Barreiras formation, from the Tertiary period.

The upper layers, which are formed by unconsolidated sediment, cover the lower areas now flooded or susceptible to flooding during the rainy season. They are the colluvio-alluvial deposits from the Quaternary period, represented by alluvia and colluvia of sandy-clay and/or silty composition, including gravel levels.

Due to their topographic position which certainly reflects the historical conditions of these sediments, they are divided in recent alluvia and alluvial terraces. The former usually present a plastic aspect, being mainly clayey-silt with sand, and the latter are basically formed by sand and/or gravel.

The physiographic and climatic conditions of the region, acting on these materials, led to differently altered horizons, originating products which differ in genesis, nature and characteristics.

Vulcanic rocks basically show all their alteration zones spectrum. The main characteristic of the materials is the relatively low thickness of the altered rock (usually fractured), about 2 metres thick. Besides, the saprolite and residual soil capping (figure 2) shows considerable thickness, often above 10 metres. In this stratum saprolite and saprolitic soil predominate and the upper portion is 80% of the layer. The upper layer (mature residual) is often eroded, remaining the saprolitic soil only. Another important aspect is the extremely irregular topography presented by the structural contour of the rock top, of the vulcanic rock, due to an erosive process only or to the tectonic features which condition higher or lower rock alteration depth levels. Another characteristic of this rock which was favourable to investigation and/or soil grouting is the almost total absence of boulders or blocks in the soil mass. This material forms the main layer where the occurrence of "canalículos" was noted, and all the dam foundation treatment carried out through grouting by soil hydraulic fracturing was aimed at reducing its permeability level. Considering this clayey-silt soil, acceptable natural permeability values would range from 10^{-4} to 10^{-5} cm/s;

the occurrence of these features lead to secondary percolation with permeability values above 10⁻³ cm/s, showing the need for a treatment to fill these cavities erratically distributed in the residual soil.

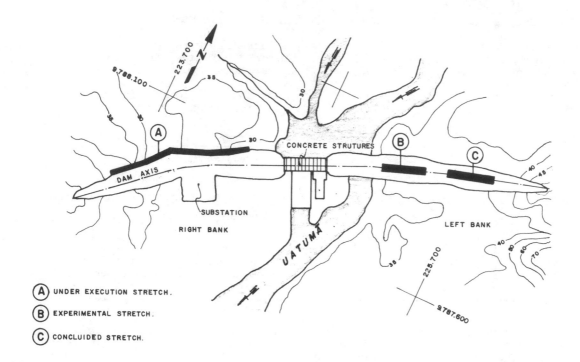

(A) UNDER EXECUTION STRETCH.

(B) EXPERIMENTAL STRETCH.

(C) CONCLUIDED STRETCH.

FIG 3 – ZONES OF FOUNDATION TREATMENT BY GROUTING

3 WORK CARRIED OUT FOR THE DEFINITION OF GROUTING CRITERIA

3.1 Experimental stretch

The unusual characteristics of the proposed kind of foundation treatment in residual soil led to scheduling and performing an experimental grouting stregch about 180 metres long to permit selection of the main criteria and methods for the abovementioned treatment. This stretch was carried out in the dam foundation area so as to be a part of the whole treatment of the area. The main aspects considered on the experimental stretch are summarized in the following table, as well as the variables which were analysed on each of the four sub-segments executed as presented by Santos et al (1985).

| Characteristics | Parts of the experimental grouting line | | | |
	I	II	III	IV
Length (m)	40	24	50	68
Borehole disposition (m)	2 x 4	2 x 2	2 x 4	2 x 2
Phases of grouting (un.)	3	2	1	1
Volume of grout per phase ( )	333	125	1000	300
Total grout volume ( )	1000	250	1000	300
Order of execution	soil rock	rock soil	soil rock	soil rock

The abovementioned authors described the work performed, the results obtained and the main definition chosen after the experimental stretch.

3.2 Complementary tests

During the experimental stretch work and even after its completion, several complementary tests were made, in order to improve the procedures adopted and to try to clarify some mechanisms which would condition soil grouting by hydraulic fracturing.

The soil-cement grout for injection and the sealing of the PVC pipe hole deserved an extensive and detailed study schedule developed simultaneously to the implementation of the experimental stretch, and that led to the definition of mixes with low cement content and with geomechanical characteristics compatible with the soil to be treated.

Siqueira, G.H. et al (1985) present in detail the above mentioned study on soil-cement grout.

Among the tests made the following proved to be most important:

3.2.1 Hole seal model tests

The physical characteristics of the hole seal, which correspond to the annular space filled with soil-cement grout formed between the wall of the grout hole and that of the PVC pipe with manchettes, partly condition the pressure values required for its rupture.

In order to know better these conditioning factors for grouting, several hole seals were made in PVC pipe moulds, with variable thickness and grout resistance. The tests simulated rupture processes of the hole seals and of grouting. They were placed in a close chamber which made it possible to test a third variable corresponding to the hole seal confinement pressure due to the surrounding soil.

The results of these tests, presented in figure 3 graphs, showed that the increments of the hole seal rupture pressure as a function of the three variables tested were higher with the confinement pressure and with the hole seal thickness, for rupture pressure values within the same order of magnitude.

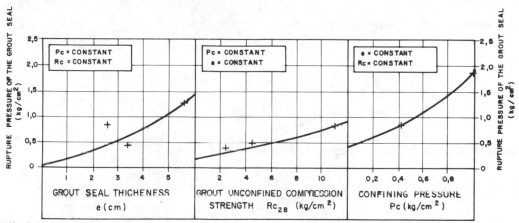

Figure 4. Results of tests with grout seal models

3.2.2 Grouting reach test

Trying to optimize and establish the criteria for grouting, e.g. bore-hole disposition and grout volume per manchette, a test was made in order to determine the reach of this grouting.

A hole was drilled with only one manchette installed at a depth of 7 metres surrounded by 24 observation holes 9 metres deep, the 6 upper metres of which protected with PVC pipes. The observation holes where it was possible to install manometers, were distributed in three circles concentric to the grouting hole. The outer circle had a radius of 3 metres. The central hole was grouted until grout showed in all of the observation holes.

The main aim of the test, however, is considered not achieved, since the appearance of grout in the observation holes was quite erratic, with no definite direction or logical sequence. This can be possibly due to the fact that when the grout reached an observation hole this would become a new grouting hole, more efficient than the first because of the absence of a hole seal, thus generating new hydraulic fracturing plans.

Among the aspects observed during the test there are the following:
- average grouting manometric pressure was 5.5 kg/cm² and the manometric pressure measured at the observation holes was seldom higher than 3.0 kg/m².
- the loss of head which was measured appeared mainly at the manchette-hole seal set, and there was no noticeable loss between observation holes.
- at each pause during tests, the pressure dissipation was almost immediate in most of the observation holes. This confirms the assumption that those holes had become grouting holes.

GEOMETRIC HOLE DISPOSITION IN THE CURTAIN

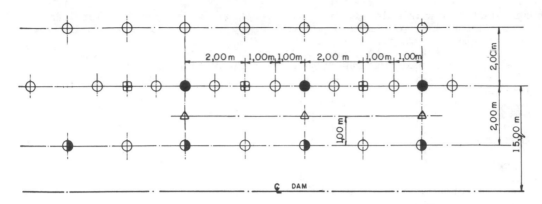

GROUTING HOLE.

HOLE FOR PREVIOUS PERMEABILITY TEST AND SUBSEQUENT GROUTING.

HOLE FOR SECONDARY CONTROL PERMEABILITY TEST.

HOLE FOR COMPLEMENTARY CONTROL PERMEABILITY TEST.

COMPLEMENTARY GROUTING HOLE.

TYPICAL MAIN EARTH DAM CROSS SECTION — LEFT BANK

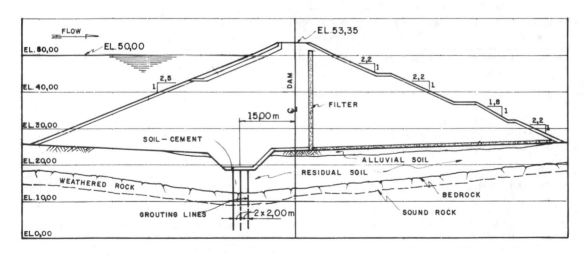

— FIG 5 —

3.2.3 Test to determine the rupture pressure of the 3 inch diameter PVC pipe

Some manchettes had very high opening pressures, up to 45 kg/cm^2, and this was considered to be a consequence of the wrong positioning of the packer outside the manchette. In order to check this assumption a test was made, causing the rupture of three PVC pipes with 3 inches in diameter by internal pressure with grout. Two of these pipes were installed in grouting holes and protected with hole seal and the third one was tested with no lateral confinement.

The rupture pressures of the PVC pipe ranged from 70 to 76 kg/cm^2 and the hypothesis was not confirmed.

3.2.4 Residual soil grouting test under a compacted earthfill

The grouting test under a compacted earthfill was planned when the left-bank earth dam foundation treatment was underway and its main grouting criteria had been defined considering the experimental stretch, as it was settled that the first five metres of depth from the natural ground surface would not be grouted due to the high incidence of grout surgences on the surface. This part would be excavated and filled with compacted soil after treatment of the underlying horizon.

The test had the purpose of creating an alternative to grouting, especially where the vulcanite residual soil horizon was more superficial. A 3 metre high fill was then made on the residual soil foundation, which was compacted under the same conditions as the dam. A curtain was then made, a grid of three 2 x 2 metre lines, totalling 14 grout holes. The residual soil grouting with soil-cement was made to the upper manchette, which was installed 30 cm below the fill foundation.

Evaluation of the grouting influence within the compacted earthfill was made by mass permeability tests before and after grouting and also by careful visual inspection in trenches. The results showed no variation in the earthfill imperviousness due to residual soil grouting, in spite of some grouting plans in the earthfill which were arranged, in most of the cases, in sub-horizontal planes going up to about 30 cm above the foundation.

Data obtained with this test permitted the grout curtain of the right bank earth dam to be made upstream from the dam body, to have its execution largely optimized in terms of planning and logistics of the plant.

4 GROUTING CRITERIA DEFINITION

Grouting criteria were settled from data obtained mainly with the experimental stretch and tests made. However, minor modifications and optimizations were still introduced during the works, due to partial data obtained and to observations in inspection trenches. Evolution and optimization of several topics and variables analysed until definition of final criteria are described below.

4.1 Curtain geometry

It was decided that the curtain to be made would inlcude three grouting lines with a 2 m interval between them, so that grout holes could be also used for altered rock treatment, a case when this curtain geometry is usually adopted. So, the holes were located obliquely in relation to the grouting axis (diagonal grid curtain) with boreholes at every 2 metres was defined (figure 5).

This disposition has the advantage, over a regular grid of equivalently spaced holes, of requiring, theoretically, a lower grout volume for the curtain, besides permitting better distribution of control holes for permeability tests.

4.2 Grout volume per manchette

The 1,000 litre grout volume per manchette was estimated by analogy with consolidation grouting in sand, that is, as a function of the soil porosity. Tests in the diagonal grid curtain with 4 metre spaced holes proved that volume to be excessive, due to the long reach of the hydraulic fracturing planes which were detected in inspection trenches, often at points over 10 metres away from the curtain.

Grout volume per manchette was now settled at 300 litres, now associated to a curtain made by a diagonal grid curtain with holes at every 2 metres (item 4.1). This represented an actual reduction of 40% in grout volume per linear metre of curtain, comparing with the formerly estimated volume.

Theoretical calculation procedures related to expressions of fracture initiation and propagation induced by hydraulic fracturing were presented by Barradas, S. (1985) and were proved more suitable for a preliminary grout volume evaluation than analogies made with grouting to fill the soil pores.

SOIL PERMEABILITY VALUES ALONG THE FOUNDATION OF THE DAM – LEFT BANK

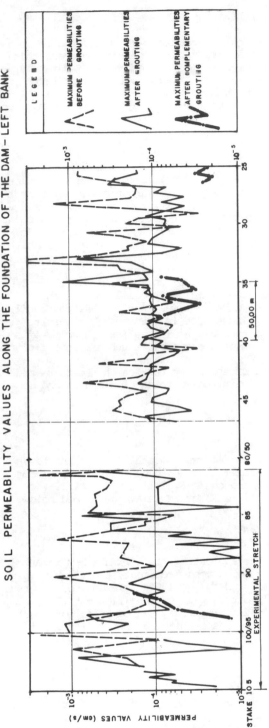

HISTOGRAMS OF PERMEABILITY BEFORE AND AFTER TREATMENT

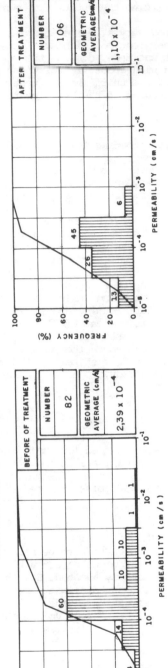

– FIG 6 –

4.3 Number of grouting stages

Grouting, in stages, until the calculated volume per manchette is completed, was intended to create hydraulic fracturing planes in several directions, corresponding to each stage.

Trenches were open in segments which were grouted in not more than three stages. Visual inspection in these trenches showed that grouting planes of different stages generally formed parallel planes, often overlapping.

These observations showed that grouting should be made in only one stage per manchette.

4.4 Grouting sequence

Once defined the altered rock horizon treatment by the conventional process of grouting, soil-rock and rock-soil grouting sequences were tested to find which one would be more efficient, especially at the interface zone of these horizons.

Water loss tests were made in the contact zone between horizons, in segments treated according to both sequences. A comparison of results showed that grouting was more efficient when the altered rock was grouted first, thus settling the rock-soil grouting sequence.

The explanation for the lower effectiveness of grouting close to the horizons interface can rest on the probable soil-cement grout migration to the untreated altered rock fractures. As residual soil grouting is made with constant grout volumes, such migration would reduce grout quantity at the base of the residual soil, consequently reducing the effectiveness of this treatment.

4.5 Spacing of manchettes

Spacing between manchettes was settled on 0.5 m, according to a similar granular soil grouting, with spaces from 0.33 to 1 metre.

4.6 Grouting manometric pressure

Grout suffered loss of head during grouting, mainly at the manchette hole seal set and also at the hydraulic fracturing planes. This loss depends on the physical characteristics of the hole seals and the grout rheological properties and on the soil geomechanical characteristics. Such loss does not permit an effective control of the grouting process.

Therefore, grout flow was under control during the works, just by recording the manometric pressure resulting from the process. During the grouting process the only condition concerning the resulting manometric pressure was that it did not exceed in more than 50% the pressure recorded immediately after hole seal rupture.

Such condition derived from results of the relations between initial and final manometric pressure recorded during the execution of the experimental stretch. It was then statistically observed that the occurrence of cracks at the surface followed by surgence of grout, generally happened when that relation went above 1.5.

Grout flow was generally 60 l/min, and this value was sporadically reduced to control and reduce occasional grout surgence at the surface.

4.7 Grouting sequence and process

The upward grouting process was established.

The downward process was discarded for executional reasons, because the grouting equipment would have to be taken out after each grouted manchette to check the occurrence of grout leakage through the lower part of the double packer.

The grouting executional sequence was followed by the initial grouting of the outer lines regardless of their sequence. The central line grouting kept a minimum distance of 12 metres to the closest outer line grouted hole.

4.8 Embankment above grouting

With the purpose of grout surgences at the surface residual soil grouting was carried out starting from a minimum depth of 5 metres below the ground surface, in order to apply, with this soil covering, an overload on the hydraulic fracturing plans, specially when grouting the shallower manchettes.

As mentioned in 3.2.4, with the purpose of obtaining an optimization and compatibilization with regular fill placing activities, grouting can be alternatively made under a compacted earthfill 3 metres high from its foundation, in residual soil.

4.9 Water loss tests

4.9.1 Preliminary tests

Preliminary water loss tests were conceived to determine the virgin residual soil horizon permeability.

At the initial phase of the work tests were also useful for the evaluation of the order of magnitude of the soil hydraulic fracturing pressure. In this case the test was carried out in increasing pressure stages of 0.3 kg/cm², starting with 0.1 kg/cm² until the soil hydraulic fracturing was reached, which happened from 0.7 to 0.9 kg/cm² effective pressure.

With the evolution of the work testing procedures were being improved, until the definition of test holes at every 4 metres of the curtain with tests being carried out at each one in consecutive 4 metre stages up to the rock top.

4.9.2 Control tests

Control water loss test results permit to check the treatment efficiency when compared to permeability values obtained in previous tests.

The development of testing procedures went from tests carried out by increasing pressure stages until soil hydraulic fracturing to tests limited to 1 kg/cm² pressure avoiding hydraulic fracturing of the treated soil.

With the purpose of standardization, the frequency and procedure criteria chosen for control tests were the same as those previously applied.

4.10 Treatment acceptance criteria

Treatment is considered completed when control test results reach permeability values that do not exceed 10^{-4} cm/s. It must be noted, however, that the need for supplementary grouting at the curtain is defined considering the existence of zones with a concentration of permeability coefficients above 10^{-4} cm/s not corresponding to localized points with a permeability coefficient closely above the settled limit.

Curtain reinforcement is made by grouting control holes as well as new grout holes, closing the grid, as shown in figure 4. After the curtain is reinforced new control holes called second stage must be made.

4.11 Grout

Soil-cement grout used for residual soil treatment underwent a long development process to be considered adequate for the treatment.

The optimization of the soil-cement grout mix brought cement ratio from 375 kg at the beginning of the studies to 73 kg per grout cubic meter. Some important consequences of this reduction were the increase in grout strength through consolidation and water absorption by the surrounding soil, a phenomenon called "presso-filtration". Thus studied, grout improved economically and had its geomechanical characteristics suited to the soil to be treated.

Grout mix used in the left bank dam foundation treatment for hole seals and for grouting had the following ratios per grout cubic meter:
- for hole seals: cement : 285 kg;   soil : 290 kg;   water : 805 kg
- for injection : cement : 73 kg;   soil : 310 kg;   water : 860 kg

5 RESULTS OBTAINED AND RESPECTIVE ANALYSIS

With the completion of the experimental stretch and definition of the main criteria as well as those applied to residual soil grouting, foundation treatment of the left bank earth dam was actually started on a segment about 230 metres long. At this segment, a total of 5300 metres was drilled with maximum treated soil thickness reaching 15 metres. Soil-cement grout total volume was about 2000 m³.

Results obtained from the treatment are presented in figure 6.

Control tests indicate that after grouting there was considerable reduction in permeability coefficients, especially with the elimination of coefficients above 5.10^{-3} cm/s. However, two zones remained above 10^{-4} cm/s. After curtain reinforcement in both zones no permeability coefficients above 1.10^{-3} cm/s could be found. The average foundation permeability was reduced from 2.39×10^{-4} to 1.10×10^{-4} cm/s.

An analysis of these data demonstrates the success of the treatment applied considering the elimination of high permeability coefficients and the consequent foundation homogenization.

Manometric pressure records, with merely informative purposes, presented for hole seal rupture pressure coefficients about 10 kg/cm², reaching in some cases to coefficients above 40 kg/cm². Grouting pressures were below 10 kg/cm² in most of the cases.

6 NOTES ON FOUNDATION TREATMENT BY HYDRAULIC FRACTURING GROUTING

Observation data and information obtained while the work was being carried out, inspection in several trenches excavated in the treated foundation and the results obtained with the treatment permit us to comment on some important aspects to be considered in the elaboration of final specifications for foundation treatment by hydraulic fracturing grouting.

6.1 Curtain geometry

Curtain geometry definition is conditioned to other interdependent aspects such as: grout volume per manchette, distance between manchettes and treatment of the underlying rock. The loading conditions of the grout curtain foreseen in the design will have an influence on the choice of the number of grout lines and their spacing.

Due to the influence of the curtain geometry on the treatment execution delays, some aspects must be taken into consideration for its definition, as foundation treatment within the general planning of the job as well as the contractor's equipment for this kind of work.

6.2 Distance between manchettes

As mentioned above, distance between manchettes must be defined considering other variables as grout volume per manchette as well as the geomechanical characteristics of the soil to be treated. According to trench inpsection in low-strength soils manchettes too closely installed will probably generate overlapping grouting planes.

6.3 Grout volume per manchette

Together with the curtain geometry grout volume per manchette can be determined by theoretical processes at a preliminary estimate.

The suitability of this volume can be estimated through observation of grout surgences at the surface as well as through grout verifications at inspection trenches.

During the grouting of the more superficial manchettes there is generally an increase in the surgence of grout at the ground surface. This occurrence suggests the need for specifying lower grout volumes for the upper manchettes.

6.4 Grouting stages

Grouting of a manchette in more than one stage has shown low efficiency for this type of soil, since the grouting planes of each stage in most of the times tend to become overlapping and parallel.

6.5 Embankment as an overload

Grouting effects on the covering layer which acts as an overload and confinement to the curtain can be empirically evaluated through the occurrence of grout and cracks at the surface. They can also be determined by vertical strain measuring.

In the case of grouting under the embankment, where vertical strain is damaging, strain can be correlated to grouted volumes and to the corresponding grouting depth, so as to establish grouting limits in relation to these movements.

In the natural soil covers, where grouting is carried out 5 metres below ground surface, strain and cracks are less important. However, grouting effects on this layer must be researched, so as to optimize its excavation keeping part of this layer, probably treated by hydraulic fracturing planes generated by underlying grouting.

6.6 Grouting manometric pressures

Grouting manometric pressures are associated to grout rheological characteristics, to grouting flow, to physical characteristics of the hole seals as well as to the soil geomechanical characteristics.

As the grouting model is by hydraulic fracturing, measured pressures have an informative character, unless vertical strain implies some kind of risk to structural elements near the treatment area. In this case some kind of pressure control must be studied, either through grouting flow or through grouted volume.

6.7 Acceptance criteria

Treatment acceptance criteria can be better evaluated through a broad range of tests (pumping test type) due to the punctual characteristics of water loss tests carried out in drill holes.

7 FINAL CONSIDERATIONS

This kind of work requires permanent evolution in its executive process in relation to the treatment dimensions and characteristics, as well as in its specifications, considering the type of soil to be treated, the grout materials and mainly the observation and analyses of preliminary results and trench inspection. As a matter of fact, after establishing design criteria, persistent accompanying of all work stages, together with a complete re-evaluation schedule, adaptation of the conceived models and through instrumentation to face new situations, important technical and economical evolution can be reached.

8 SUMMARY

The Balbina hydroelectric plant earth dam foundation, on the Uatumã river, left-bank tributary of the Amazon and located about 170 km from Manaus, in Brazilian territory, is predominantly made up of a clayey-silt soil layer, vulcanic rock residual. In this layer were found tubular shaped cavities called "canalículos", often inter-communicating and erratically distributed, with diameters varying from millimetric to decimetric values.

At the occurrence zone of these "canalículos", the foundation layer permeability, with a matrix presenting coefficients between 10^{-4} and 10^{-5} cm/s, reached values above 10^{-3} cm/s. Technical and economical studies were made with the purpose of eliminating such high permeability localized zones. A decision was made for a foundation treatment by means of a grout curtain by hydraulic fracturing, using soil-cement grout in the residual soil layer.

This paper presents an analysis of the results obtained with the left-bank earth dam foundation treatment.

9 REFERENCES

Balbina hydroelectric plant. Quality control division. Internal reports 02-70-001, 02-70-0012, 02-70-0043 and 02-70-0122.
Remy, J.P.P. et al 1985. Choice of foundation treatment of Balbina earth dam. 15th International congress on large dams. Lausanne, Switzerland.
Sathler, G. & F.P.Camargo 1985. Tubular cavities 'canalicules' in the residual soil of the Balbina earth dam foundation. 15th International congress on large dams. Lausanne, Switzerland.
Santos, O.G. et al 1985. Experimental grouting of residual soil of Balbina earth dam foundation, Amazon, Brazil. 1st International conference on geomechanics in tropical lateritic and saprolitic soils. Brasilia, Brazil.
Barradas, S., 1985. Iniciação e propagação de fraturas induzidas por injeções em solos argilosos com canalículos. M.Sc. degree thesis presented at COPPE, Rio de Janeiro Federal University.

2.4 Behaviour and treatment of weak structural zones
Comportement des zones structurales à faible résistance
et leur traitement

Clay barrier behaviour on heavy metal diffusion

Comportement d'une barrière argileuse vis à vis de la diffusion des métaux lourds

Kurt A.Czurda & Jean-Frank Wagner, *Department of Applied Geology, Karlsruhe University, FR Germany*

ABSTRACT: Clay and clay rocks show the properties of semipermeable membranes and are therefore suitable as geological barriere against problematic waste sewage, sewage mud and solutions of waste disposal sites. Geological membranes delay or prevent the solution discharge in a very different way whereby the solution parameters, the rock parameters and the parameters of the geological environment determine the degree of the suitability of the barrier. By means of small scale column experiments the ad- and desorption behaviour of various clays in dependence on the diffusion potential and other rock parameters as mineralogical composition, rock acidity, microtexture etc. was qualitatively and quantitatively determined. The percolate consisted of weakly acid, heavy metal loaded solutions. According to the results, model calculations for the security of a waste disposal site can be drawn.

RESUME: Argiles et roches argileuses ont les qualités de membranes semiperméables et sont pour cette raison des barrières géologiques excellentes vis-à-vis des eaux d'égout problématiques, des boues de décantation et des eaux de décharge. Les membranes géologiques retardent et empêchent le passage d'une solution de manière très différente; le degré de qualification de barrière est défini par les paramètres de la solution, de la roche et du milieu géologique. Le comportement d'adsorption et de désorption de differentes argiles en dépendance du potentiel de diffusion et d'autres paramètres de la roche comme la composition minéralogique, l'acidité de la roche, la microtexture etc. a été déterminé au moyen d'essais de colonne à petite échelle. Le percolat était composé de solutions chargées de métaux lourds faiblement acides. A partir des résultats on peut dresser des modèles de calcul pour la sureté d'une décharge.

1 INTRODUCTION

Besides technical barriers like backfill material, container, conditioning mass etc. a geologic host rock has to be available in order to dispose hazardous waste. Plastic clay and the wide range of clay rocks are beside rock salt, anhydrite, granite etc. highly appreciated barrier rocks.

Clay formations can generally be characterized by the following properties:

1. Low permeability; a solution with a dangerous concentration of toxic chemicals, which enters into the host rock through a leak, will circulate in a clayey subsoil extremely slowly.
2. Excellent sorption properties; clays show the favourable property of sorbing different ions in different ways and they exchange hazardous cations for harmless ones.
3. Sealing ability; eventual occuring joints and cracks (e.g. at different amounts of sediment consolidation) will close if solutions circulate. This is due to the swelling properties of most of the clay minerals.

The seepage velocity of waste loaded solutions and the sorption ability of the clay barrier rock depend on different parameters:

a) Solution parameters, concentration, pH, electrolyte content, kind of solved hazardous material.
b) Rock parameter: grain size distribution, permeability, porosity, exchange capacity, original ion fixation, rock pH, organic compounds, stratification, schistosity, microtexture and cleavage.
c) Geological parameters: degree of diagenesis, lithostatic pressure, depth and fluctuation of the ground water level and the geothermal gradient.

The sorption properties of different soils and clays as a function of toxic ion concentration and the solution pH was studied by means of so called batch tests. The results are plotted as adsorption isotherms (e.g.Hahne & Kroontje 1973; Herms & Brümmer 1978; Gerth & Brümmer 1979;

Laske 1979; Gerth et al. 1981; Meier et al. 1984, etc.). During batch experiments different rock parameters like permeability and the solution contact time resulting from it as well as stratification and microtexture effects will not be taken into consideration. Those relations can only be obtained on undisturbed clays or soil samples. The most reliable results can certainly be obtained from in-situ tests. In this study exchange column tests with undisturbed clay samples will give basic evidence on the dependences mentioned above. Heavy metal loaded solutions are the type of hazardous waste which is deposited on special toxic waste sites. The following chapters deal with diffusion-, flow- and adsorption processes of low acid, aqueous Cu- and Zn solutions in two different limnic clays.

2 DESCRIPTION OF THE CLAY BARRIER ROCK

As mentioned above, the adsorption behaviour depends on several rock parameters. Therefore, a detailed analysis of the clay rocks in question is necessary.

One type of rocks investigated belongs to the eastalpine molasse region and may be characterized as freshwater clay of the Upper Tertiary. This clay appears as plastic pelite without diagenetic lithification and forms the 2 m thick interlayer between two coal seams. In the following, the rock is simply called "organic clay".

The second rock type is a varved clay. This varved (lacustrine) clay is a sediment of one of the alpine interglacial periods of the Quaternary. It originates from the Austrian Inn Valley near Innsbruck.

According to its grain size distribution, the organic clay has to be designated as fine sandy silt (fig. 1) which occasionally contains small coal particles. The varved clay is built up of clayey and fine sandy silt layers (fig. 2).

The permeability coefficients (k_f-values) are for both clay types $5 \cdot 10^{-9}$ m/s on average. For the varved clay the k_f may rise up to $1 \cdot 10^{-7}$ m/s. This is the case if the fine sand strata portion rises significantly.

In terms of mineralogical composition, the two clay-types are somehow different. The organic clay consists mainly of illite (ca. 50 %) and chlorite (ca. 20 %). Further clay mineral phases are some kaolinite and up to 9 % montmorillonite . The non-clay phases consist of quartz and some feldspar. The varved clay consists of mainly illite respectively muscovite (45 %), chlorite, quartz and feldspar. Moreover it contains 15 % dolomite and 6 % calcite on average.

Of special importance for the adsorption ability of the organic clay are the swelling clay phases, and in the case of varved clay presumably the carbonate content.

The cation exchange capacity (CEC) for the organic clay was determined between 15 - 24 meq/100g and for the varved clay between 3 - 15 meq/100g. The CEC values are rather low and therefore both clay types show relatively poor exchange capacities compared with pure plastic clays.

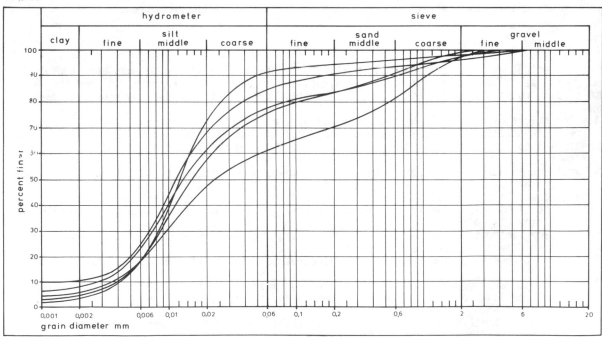

Figure 1.Grain size distribution of the Tertiary organic clay (Upper Freshwater Molasse) from the Hinterschlagen coal mining area (province of Upper Austria). The occurence of mm-sized clay pebbles simulates enhanced portions of sand fraction.

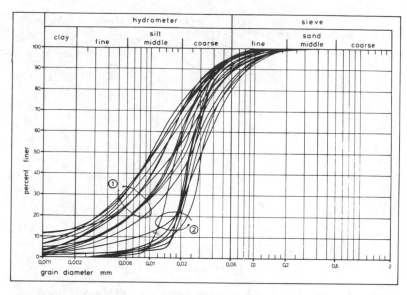

Figure 2. Grain size distribution of the Quaternary varved clay from the Arzl clay pit near Innsbruck. It has to be distinguished between sections with predominantly fine - middle silt (1) and those with middle - coarse silt (2).

3 DIFFUSION AND ADSORPTION EXPERIMENTS

All of the experiments were performed on undisturbed clay rocks. The clay column diameter always was 50 mm, the sample length changed within the range of 24 and 56 mm. The clay was cut out by steel cylinders and placed in plexiglass tubes (exchange columns) for the experiment. All samples were water saturated first and finally the diffusion experiments with heavy metal solutions were performed. A slight overpressure of 0.2 bar was applied. This accelerates the experiment length but shortens the contact time - one of the parameters which are of importance.
   The filtrate was collected and the heavy metal concentration determined by AAS. After completing the diffusion respectively the perfusion experiment, the sample was removed from the exchange column, cut in mm-slices, oven dried and ground. The adsorbed heavy metal ions were washed out of the rock powder in order to gain a vertical distribution pattern within the sample column.
   By this, two different distribution curves showed the quite different cation retaining behaviour (fig. 3). The varved clay, with pH-values between 8 and 9,5 has proved to precipitate heavy metals out of low acid solutions very rapidly. This because of the high carbonate content of the clays. This leads to a maximum in heavy metals concentration within the upper portion of the sample column and decreases with column length.

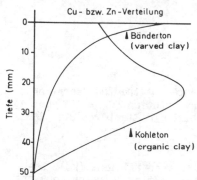

Figure 3. Vertical heavy metal distribution within a 50 mm thick clay column after the percolation of about 100 ml of a low acid heavy metal solution.

   In the case of the acid organic clay (pH 4-5), the maximum of heavy metal concentration is to be detected about in the middle of the sample. That means that a good deal of the heavy metal ions migrate through the clay until it will be adsorbed. These different heavy metal distributions are a function of solution migration distances, which means different contact times, and are valid for Cu and Zn as well. The dependence of heavy metal adsorption from the rocks pH became obvious.

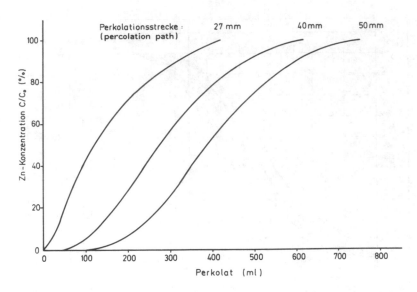

Figure 4: Organic clay migration curves at different lengths of percolation path (C_0 = 750 ppm Zn, pH 4).

By means of in portions collected filtrate quantities and the adequate changes in concentration, passage lines can be drawn. In fig. 4 Zn-passage curves for different lengths of the percolation path are shown.

As it can be seen distinctly, after the migration of about 100 ml of a Zn-solution with initial Zn-concentrations of 750 ppm, first Zn-concentrations occur within the filtrate of the 50 mm sample column. After about 750 ml solution passage, the filtrates concentration equals the initial solution concentration. In this state no further Zn adsorption takes place, the clay appears Zn-saturated. The dependence of the Zn-adsorption from the pH of the Zn-solution is to be seen in fig. 5. A Zn-solution with pH 2 migrates faster respectively shows lower adsorption quantities than a pH 4-solution.

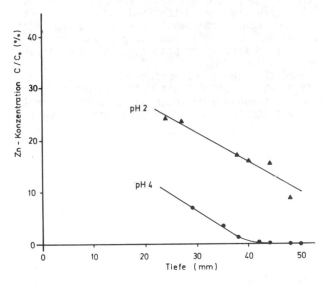

Figure 5. Dependence of the Zn-adsorption on pH of the primary solution (C_0 = 750 ppm Zn; C/C_0 = Zn-concentration of the percolate; test rock = organic clay).

Furthermore, the adsorption ratio depends strongly on the permeability coefficient (k_f). See fig.6. The more impermeable the clay - expressed by periods of longer lasting solution contacts with the rock - the more effective the adsorption.

4 DISCUSSION OF RESULTS

The sorption properties of the two different natural clay-silt-sediments against heavy metals shown in this study, are to be understood as one small step towards the understanding of the

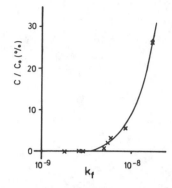

Figure 6. Example of Zn-adsorption dependences from varved clay samples of different permeability (k_f). C_0 = 5700 ppm Zn (pH 4); C/C_0 = Zn-concentration of percolate.

very complex processes of ad- and desorption of toxic material by barrier rocks. Only a few of the great variety of parameters have been taken into account: solution pH, rock pH and permeability. Besides, only two heavy metals, Cu and Zn, were used for the experiments. For many other cations the adsorption behaviour will be different and for different rock types as well. The tests reported were done with aqueous solutions. However Anderson 1982 shows that for organic solvents e.g. acetone, methanol, xylene etc., as they are to be detected frequently in hazardous waste sites, the permeability may be lowered respectively improved. Other regularities become involved.

Furthermore, the mechanism of the ion retention is not fully understood yet. It has to be distinguished between the adsorption at the particle surface, within the interlayer space and the installation within the lattice of sheet silicates. Exchange processes, precipitation and complex formations have to be taken into account. Of similar complexity appears the process of desorption.

According to the grain size distribution, the sediments under investigation are silts and not clays. Nevertheless, this type of sediment is appropriate for hazardous waste deposits and is used for this already. Pure clays are comparatively rare.

Concerning the column experiments it has to be noted that adsorption values gained by this technique are lower than those from batch experiments and batch sorption coefficients may be 10 times higher than those of in-situ tests (Magrita 1985). It has to be recommended vigorously, based on batch- and column experiments, to diffuse solutions in-situ respectively to detect the path of diffusion of solutions out of hazardous deposits. The columns used for experiments should be as big as possible in order to reduce margin effects.

REFERENCES

Anderson, D. September 1982. Does landfill leachate make clay liners more permeable? Civil Engineering, ASCE, p.66 - 69.
Gerth,J.& G.Brümmer 1981. Quantitäts - Intensitäts - Beziehungen von Cadmium, Zink und Nickel in Böden unterschiedlichen Stoffbestandes. Mitt.Dtsch.Bodenkundl.Ges.,30,p.19 - 30.
Hahne,H.C.H.& W.Kroontje 1973. Significance of pH and chloride concentration on behaviour of heavy metal pollutants: mercury (II), cadmium (II), zink (II) and lead (II). J.Environ.Quality,2,p.444 -450.
Herms,U.& G.Brümmer 1978. Löslichkeit von Schwermetallen in Siedlungsabfällen und Böden in Abhängigkeit von pH-Wert, Redoxbedingungen und Stoffbestand. Mitt.Dtsch.Bodenkundl.Ges., 27,p.23 - 34.
Laske,D.1979. Bestimmungen von Verteilungskoeffizienten (k_d-Werten) bei der Adsorption von Cs und Sr an möglichen Zusätzen zu mineralischen Verfestigungen. NAGRA Techn. Ber.,28, 110 p..
Magrita,R.1985. Etude méthodologique de la migration des radioéléments dans les barrières géologiques. OECD Nuclear Energy Agency, Newsletter 11,p.13 - 14.
Meier,H.,E.Zimmerhackl,G.Zeitler,P.Menge & W.Hecker 1984. The dependence of distribution coefficients on solution-to-solid ratios. Int. Conf. on Nuclear and Radiochemistry, Lindau, FRG, 1984.

Study of the law of distribution of clay intercalations due to interlayer shearing at Xiaolangdi Project on the Yellow River

Etude de la régularité de répartition de l'intercalation argileuse cisaillée entre les couches de la centrale hydraulique de Xiaolangdi du Fleuve Jaune

Ma Guoyan, *Design Institute, Yellow River Conservancy Commission, China*

ABSTRACT: The law of distribution of thin clay intercalations (henceforth CI, for brevity) can hardly be cognized in an overall way by using conventional methods of geologic exploration. The paper deals with findings on the types, characteristics, continuity and law of distribution of CI due to interlayer shearing, among others, having employed effective methods such as shaft and adit exploration and "sleeve drilling" for 100% recovery. The coefficient of friction along potential rupture surface is determined in an overall way, taking into account the area of CI and non-siltized intercalations. This is necessitated by the fact that through comprehensive analysis of the degree of continuity, classification and distribution of CI with respect to the rupture surface as well as the stress condition, the coefficient of friction determined comprehensively may be more or less appropriate to actual condition.

Résumé: Il est difficile, avec les méthodes de prospection normales de connaitre complétement l'état de répartition d'intercalation argileuse. Cet article présente le résultat d'une recherche de la espèce, de la caractéristique, de la continuité et de la répartition de l'intercalation argileuse cisaiIée, par les méthodes suivantes: grand puits de recherche et chennal de recherche, "sondeuse par tubage" qui permet le carottage de 100%. L'auteur propose qu'il faut estimer le coefficient de friction de la section de rupture potentiel, en combinant les surfaces de l'intercalation argileuse et de l'intercalation non argileuse. C'est à dire, après avoir étudié, dans tous les domaines, la continuité, l'espèce, la répartition dans la section du rupture et l'état de contrainte de l'intercalation argileuse, on peut choisir correctement le coefficient de friction générale.

1 INTRODUCTION

Red bed of Triassic system (T) at the Xiaolangdi Project, the last cascade on the middle reaches of the Yellow River, comprises of fragmental deposits of continental facies of extremely complicated lithological characters and petrographical facies. Comparatively hard medium and fine sandstone and calcareous siltstone (henceforth hard rocks, for brevity) are intercalated with softer rocks such as clay stone, silty clay stone and pelitic siltstone (henceforht soft rocks, for brevity), quite characteristic of the area. The CI formed mainly by interlayer shearing are of much significance, affecting slope stability, safety of project, choice of alternatives, cost, as well as construction period.

During the past years, particular means such as large-diameter boring, capable of taking all the cores from holes 1 m in diameter, and "sleeve drilling" of 46/110 mm have been used together with conventional methods of exploration in trenches, adits and shafts. Detailed studies have been made on the law of distribution of CI, dealt with briefly as follows:

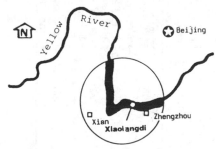

Fig.1 Sketch map showing location of Xiaolangdi Project on the Yellow River

2 LITHOLOGICAL CHARACTERS OF MOTHER ROCKS OF CI

In accordance with lithological characters of the strata (see Table 1) and analysis of data of the mineralogical identification of the rocks, Triassic formations of Liujiagou group exhibits rhythmic

Table 1.   Lithologic Characteristics of Rock Formations

System	Formation	Litho-group Symbol	Mean thickness (m)	Main lithologic characteristics	Contents (%) Medium and fine sandstone, conglomerate	Siltstone	Clay stone	Distribution of hard and soft inter-bedding	Remarks
Triassic	Heishang-gou group	T_1^{6-3}	37.50	pelitic siltstones somewhat intercalated with a few thin layers of clay stone	13.0	75.5	11.5	only a few	Clay rocks comprise clay stone & silty clay stone.
		T_1^{6-2}	10.00	calcareous sandstone with slight intercalations of thin clay stone layers	61.4	38.3	0.3	ditto	Siltstone with calc. & arg. cementation agents.
		T_1^{6-1}	49.50	Pelitic siltstone intercalated with fine calcareous sandstone	21.8	66.1	12.1	ditto	
	Liujiagou group	T_1^{5-3}	33.65	calcareous & siliceous silty & fine sandstone somewhat intercalated with clay stone	47.0	43.0	10.0	spread in upper & middle part	Soft rock layers are intercalations under 30 cm thick.
		T_1^{5-2}	15.50	medium siliceous sandstone of medium or large thickness somewhat intercalated with clay stone	75.0	12.0	13.0	spread in entire formation	
		T_1^{5-1}	12.50	intercalations of thin layers of fine calcareous sandstone, calci-pelitic siltstone and clay stone	46.0	27.0	27.0	ditto	
		T_1^{4}		extremely thick and thick medium & fine siliceous sandstone with a few thin layers of clay stone	93.6	2.5	3.9	within lower 10 m or more	
		T_1^{3-2}	26.10	fine calcareous sandstone & siltstone somewhat intercalated with pelitic siltstone & clay stone in thin layers	62.0	27.0	11.0	in upper part	
		T_1^{3-1}	30.81	fine siliceous sandstone with calcareous silty & fine sandstone with a few intercalations of clay stone	82.6	10.0	7.4	at the bottom	
		T_1^{2}	30.90	calcareous and calci-siliceous siltstone & fine sandstone somewhat intercalated with pelitic siltstone and clay stone	70.0	20.0	10.,0	in upper & lower part	
		T_1^{1}	30.22	calcareous siltstone & fine sandstone & pelitic siltstone with a few thin layers of clay stone	27.5	65.2	7.3	ditto	
Permian	Shiqian-feng gr.	P_2^{4}	61.50	mainly silty clay stone intercalated with thin calcareous sandstone at the top				at the top	

variation as a result of different circumstances of deposition and depths of streamflow in the process of depositing.

2.1 Interbedding of thin layers of hard and soft rocks

Sandwich structure of alternate hard and soft rock appears repeatedly in the profile, most of the cementation agent being argillaceous, calcareous or siliceous, alternating in the profile. With respect to microstructure, the soft rocks are intercalations of very thin formations of siltstone with argillaceous, calcareous or siliceous cementation and clay stone with silty or pelitic structure.

2.2 Complicated bedding structure of both hard and soft rocks

On the upper surface of bedding of fine silty sandstone and siltstone, there are often current ripple marks with undulations of 1 - 7 cm. Warty masses of overlying sandstone may sometimes be included in the upper part of clay stone beddings. Shrinkage cracks may also be observed. The bedding structure reflects variation of hydrodynamic conditions at different times in the longitudinal direction.

2.3 Petrography of soft rocks varies widely

Thin layers of soft rock scattered in the bed are generally several tens of meters in length, seldom to exceed 100 m. The thickness is often 1 - 3 cm. The thicker the stratum, the longer it extends. In the direction of extension of the bedding plane of the soft rock, the grains often become first coarser and then finer. Such reflects variation of hydrodynamic conditions in the transverse direction within a certain period of time.
   The mother rock of CI is the soft rock adjacent to the hard.

3 EXTRINSIC FACTORS FOR FORMATION OF CI

The general feature of tectonics in the region is characterized by slight degree of folding and developed rifting at steep angles of inclination.

3.1 Kuangkou anticline

Within the scope of the dam site, the anticlinal axis at gentle slope extends and dips toward east along the right bank of the Yellow River. The northern limb dips toward NNE at angles of 10 - 15°, and the southern towards SE at 8 - 10°. The anticline was formed by upheaval due to principal stresses approximately in the direction of north to south. All envisaged structures are to be located on the northern limb of the anticline.

3.2 Faults

Faults within the region may be shear faults or bedding plane faults (at the CI) in accordance with their relation with the bedding planes, and are classified as major ones (throw exceeding 100 m), medium (nearly 100 m or less) abd minor ones (several meters or still less).
   F_1 and F_{28} (major faults), F_{236} F_{238} etc (medium ones) are more or less closely related to the engineering structures. On both sides of the aforesaid faults, there are still smaller ones. The faults are mostly steeply inclined at angles of 75 - 85°. Fault F_{28} was subjected to tension and afterwards to limited compressive and shearing stresses, whereas the others are mainly compression and shear faults. There are fractured zones of different widths on both sides of the faults, the harder the rocks the wider they are.
   As the faults were formed when anticline in the north-south direction heaved up under the action of principal stresses, shearing stresses, large or small, were induced in the bedding plane. These are the extrinsic factors inducing interlayer shearing to form the CI.

4 DIFFERENT TYPES OF CI

It may be of benefit to have a brief account of the classification of CI as an illustration. From the viewpoint of studying the law of distribution of CI, they may be divided into five types in accordance with grain size, minerological composition and thickness, as shown in Table 2.

Table 2. Classification of CI

Type of CI Symbol Nomenclature	Thickness, grain size and composition	Lithology of mother rock	Coefficient of friction
I mud	over 0.3 cm thick, cohesive soil with breccia under 3 %	mainly clay stone 0.3 - 1 cm thick	0.20 - 0.235
II mud with debris	cohesive soil with debris content of 7 - 10 %, debris of size mainly 0.2 - 1.0 cm, with angularity, more or less, mostly packed in soil, some oriented	mainly clay stone over 3 cm thick	0.205- 0.240
III mud with silt & silt with mud	sandy loam with debris content under 10 %	siltstone with argillaceous or calcareous cementation	0.280
IV mud membrane	clayey mud membrane of thickness under 0.3 cm	clay stone of thickness under 0.3 cm	0.290
V debris with mud	debris of sub-angularity, mainly in plate form, at contents of 10 - 50 %	interbedding of siltstone and very thin clay stone	0.315

It can be seen from the table that CI are distributed in certain soft rocks (mother rocks), and their coefficients of friction vary within certain ranges.

From the viewpoint of mechanical properties of CI, type I and II with roughly the same coefficient of friction may be merged into one, as also type III and IV. In doing so, the five types shown in Table 2 may be reduced to three.

5 GRAIN SIZE AND CHARACTERS OF CI

Grains of CI mainly comprise of sand, silt, clay and colloidal particles (fines, for brevity) and rock fragments. The content of fines is dependent on the composition of the mother rock, prior to the formation of which there existed single particles and clusters. In order to know whether there was crushing of such due to interlayer shearing, samples were collected from the same formation, both from the soft and weak interbeds and from the CI, for comparative tests of grain sizes. It has been shown that as the weak and soft strata were turned into CI, the sand grains were subjected to some degree of crushing, the min 0.002 mm particles (colloidal) increasing 3.89 % on the average, min 0.05 mm grains increased by 4.5 % and sand grains diminished by 6.71 %, averagely speaking.

Very thin siltstone in the weak and soft interbeds was fragmented during tectonic movements. The amount and size of fragments in the CI are therefore closely related to the presence of siltstone in the interbed, its thickness and composition of cementing agents, see Table 3. There were primary rock fragments and pieces of single particles prior to the forming of the rock in the interbed. There are warty sandstone inclusions in the upper portion of soft intercalation. Sandstone near the top of the wave marks above the lower interface may be crushed and take the form of flaky with angularity and subangularity. Rod-formed fragments are also included in the CI.

Table 3. Relation between composition of cementing substance and debris mixed with mud

Item Symbol	Main cementing agent in thin siltstone and adjacent sandstone strata	Fragments mixed with mud (% of all CI)
T_1^{5-3}	calcareous, argillaceous	19
T_1^{5-2}	calcareous, siliceous	23.9
T_1^{5-1}	calcareous, argillaceous	18.5
T_1^4 to south of F_{236}	siliceous	63

The degree of crushing of thin siltstone in the interbed under action of the shearing force mainly depends on the composition of the cementation agents, as the strength of the rock is governed by the strength of the latter as well as the bond strength to hold the fragments together. For instance, fine sandstone with siliceous cementation has a strength of 1150 kg/cm^2 when under compression in dry, whereas that cemented by argillaceous substance only reaches 630 kg/cm^2. When saturated, the latter drops abruptly to approach zero. Siltstones cemented by siliceous, calcareous and argillaceous matters respectively have compressive strengths of 1200, 900 and 440 kg/cm^2 when in dry. The former two do not change appreciably when saturated, while the last-mentioned dropped to 0 - 35 kg/cm^2 subsequent to saturation. Obviously, under similar interlayer shearing, siltstone cemented by argillaceous matter may partly become mud, whereas that cemented by siliceous matter is only crushed into fragments.

The characters of CI are such that the mean values of specific gravity, void ratio and dry density of soft rock intercalations are respectively 2.84, 0.18 and 2.42 g/cm^3. This shows that the soft rock intercalations are structurally well consolidated and maintain the dense condition. It can be seen from the Table 4 that CI are sound and dense. The swelling and expansive force are respectively in the ranges of 3.1 - 5.1 % and 0.069 - 0.19 kg/cm^2. As non-expansive soil, the CI have low rate of slaking, 44 % of which being under 5 %. The water-resisting capacity of the clay intercalations in the region is high The modulus of deformation of silty clay, silty loam with small amounts of debris and mud membrane are respectively 80 - 160, 170 - 500 and 500 kg/cm^2. The CI are therefore of medium compressibility.

Table 4. Physical properties of CI

Natural moisture content (%)	Bulk density (g/cm^3)	Dry density (g/cm^3)	Void ratio	Degree of saturation (%)	Consistency	Sp.gr.	Plasticity index (%)
14.16	2.23	1.97	0.355	81.5	- 0.28	2.83	11.9

Mineralogical composition of clay in soft rock intercalation is mainly illite, with small amounts of kaolinite and ferrohydride, among others. After it is changed into CI, the composition remains unaffected.

6 LAW OF DISTRIBUTION OF CI

Where are layers of weak rock intercalation susceptible to siltization (henceforth ISS, for brevity) to be found, where are they most likely to occur, and how about the continuity? The answers are dependent on the regularity of the distribution of ISS, which has already been studied.

6.1 CI may be found in rock formation of intercalations susceptible to siltization

Intercalations of relatively thick sandstone and soft rock of limited thickness (ISS) hint the possible occurrence of CI, which are seldom found in thin interbeddin strata. This is substantiated by huge quantities of field data, showing that CI generally occur at interbedding strata of thick sandstone and thin layers of soft rock, see Fig.2.

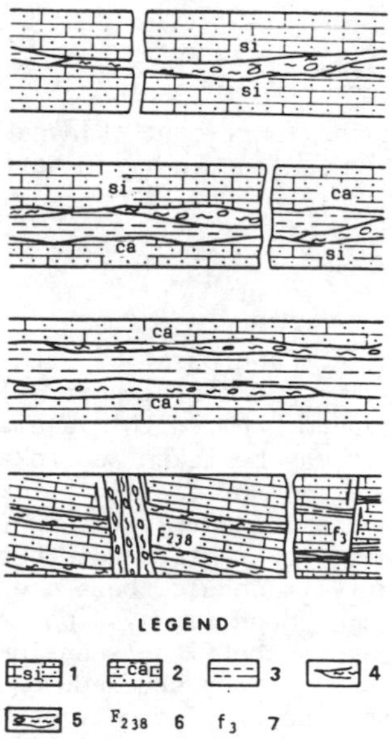

1 --- fine siliceous sandstone, silty and fine sandstone, and siltstone;
2 --- fine calcareous sandstone, silty and fine sandstone, and siltstone;
3 --- clay stone and silty clay stone;
4 --- all-mud CI with thicknesses of 0.3 - 1.0 cm;
5 --- CI with debris, 1 - 3 cm thick;
6 --- fault with throw exceeding 100 m, with numbering;
7 --- ditto with throw of several meters.

Formations T_1^{6-3} and T_1^{6-1} do not contain intercalations of soft rock under 10 cm thick, so that there will not be ISS. T_1^{6-2} containing sandstone of considerable thickness though, has only 0.3 % of siltstone, hence seldom ISS. ISS often occur in T_1^{5-3}, T_1^{5-2} and T_1^{5-1}. Although thick siliceous sandstone strata do appear in the upper and middle part of T_1^4, there are few ISS because of absence of soft rock intercalations. This is not so within 10 m from the bottom of T_1^4, where quite a number of ISS were found. T_1^3 and T_1^1 contain some ISS, more so in T_1^{3-1} and T_1^2. In short, CI comes hand in hand with ISS, so that the latter may be taken as "key horizon" for locating CI.

LEGEND

si 1   ca 2   3   4

5   F$_{238}$ 6   f$_3$ 7

Fig.2 Typical position of CI in ISS

6.2 Conditions for occurrence of Ci in ISS

1. ISS near major faults. Statistics of density of CI in the dam area, at major and medium-sized faults where the hanging wall and foot wall were formed at the same time, has shown that the width of hanging wall is larger than that of the foot wall in sections where CI often occur. The average width of hanging wall is 26.33 m, while that of foot wall is only 10 m or so. Minor faults with throw of only a few meters do not affect formation of CI in a significant manner. There is also no direct relation between distribution of CI in foot wall of fault F_{28}, markedly subjected to tension, and the distance from the fault zone. Quite a number of CI exist at places near major faults where the dip exceeds 15°.

2, Low strength of soft rock in ISS. It can be seen from Table 5 that the probability of siltization of soft rock intercalation is directly related to grain size. The higher the clay content, the more readily is the stratum to siltize.

Table 5. Statistics of number of CI in different soft rock interbeds

Lithology of soft rock intercalation	Number of CI discovered	% of total number of CI
Clay stone, shale	128	56.4
Silty clay stone,pelitic siltstone	78	34.4
Calcareous siltstone	21	9.2

3. Relatively high strength of sandstone on one side of ISS. There are often current ripple marks on the lower interface of interbedding soft rock, and warty sandstone inclusions on the upper, forming undulation of the interfaces. With cementation agents of different composition in formations over-+ and underlying the soft rock intercalation, CI is generally to appear on that side of the sandstone where the siliceous or silico-calcareous cementing matters possess higher strength. Mutual action of difference in horizon and strength causes the CI to appear on the side of sandstone where the comprehensive strength parameters are higher.

4. Thin soft intercalation in ISS. It has been demonstrated that the thinner the intercalation, the more liable is it to siltize, see Table 6. It is seen that the majority of CI appeared in the

Table 6. Statistics of intercalation thickness vs location of CI

Thickness of soft rock intercalation (cm)	Number of CI discovered	% of total number of CI	% of a specific type of CI			
			A	B	C	D
<10	144	63.4	8.6	65.5	19.7	6.25
10 - 30	53	23.4	0	63.1	34.0	1.89
>30	30	13.2	0	70.0	20.0	9.99

upper part (Class B) of soft rock intercalations, and the rest in the lower part (Class C), very few exist in the middle part. CI that are almost all mud (Class A) are found only in soft rock intercalation under 10 cm thick.

6.3 Continuity of CI

The uninterrupted length of a CI is closely related to its thickness. Lengths of CI with thickness under 0.3 cm never exceed 50 m, whereas those thicker than 2 m may be 180 m or longer. There may be 5 types of CI in a single layer, differing in thickness from 0.3 to 3 cm. Within a small range, the CI may be integral and continuous, whereas in a wider range, it may be taken as broken pieces or scattered. We are of the opinion that within a strip-formed weak zone of certain thickness containing CI, shearing failure of rock mass may occur along surface passing partly through the CI and partly through the soft and hard rock, or through the cracks in hard rock interbedded between CI and subsequently along the surface of another layer of CI. Any analysis should therefore take into account the strength parameters of both CI and rock strata in a comprehensive way. The relative importance of each is governed by the proportion of respective areas in the final shearing surface, for which purpose the degree of continuity of strip-formed weak zone must be studied, see Table 7.

It can be seen from the table that under the prerequisite of ensuring precision of sampling, the precision of measuring continuity is dependent on the arrangement of position and spacing of outcrops. The types of IC and their continuity in the weak zones, if in all, are also shown in the table.

Table 7.   Continuity of main strip-formed weak zones in building area

Position

Group	Distance from bottom (m)	Number of outcrops,M	Ditto,where CI are found,N	\% of particular type of CI — I	II	III	IV	V	Continuity (N/M) in \%
T_1^{5-3}	29.9 - 31.50	10	3		50	37.5	12.5		30.0
	27.06 - 28.70	11	3	25.0	8.3	16.7		50.0	27.0
	22.35 - 24.59	14	3	66.6				33.4	21.4
	18.64 - 20.75	16	6		42.8			57.2	37.5
	0 - 1.17	24	8	16.7	16.7	33.3	16.6	16.7	33.3
T_1^{5-2}	17.46 - 19.58	27	13	15.4	38.5	4	7.6	34.6	48.1
	13.15 - 14.60	29	10	14.3	57.1	14.3		14.3	34.5
	8.88 - 11.80	26	13	54.5	18.2		9.1	18.2	50.0
	2.27 - 3.55	28	10	25		25	25	25	35.7
T_1^{5-1}	8.25 - 10.84	24	13	72				28.0	54.2
	2.05 - 3.24	21	8	25	33.3		33.3	8.4	38.1

7 MECHANISM OF DISTRIBUTION OF CI DUE TO INTERLAYER SHEARING

In the process of upheaval of Kuangkou anticline, interlayer shearing stresses were induced, the magnitude of which being related to the intensity of folding. The steeper the inclination of the two limbs, the higher the stresses. Shearing stresses were also induced when faults were formed, their magnitudes being dependent on the nature and scale of the faults. The smaller the angle included between the fault plane and the rock interfaces, the higher the stresses. Reverse faults bring about higher interlayer shearing. During tectonic movements, in sections of sandwiching of sandstone and soft rocks, there prevailed interlayer shearing in every stratum, more or less. As there are so many interlayers, the slip surfaces were spread over different layers, so that interlayer shearing was not concentrated. In areas of thick sandstone without frequent occurrence of soft rock strata, the sandstone of high strength does not deform readily, so that the stresses may be transmitted. For soft rocks, the contrary holds true, and slipping occurs where adjacent rock formations differ widely (at the interfaces). Striae in the direction of NE 20° and parallel to the dipped strata are often seen on the top and bottom of a CI. Using microscope, one may observe the torsional deformation and damage to the primary structure of the parent rocks of the CI.

Under the action of similar interlayer shearing, the shearing strength is low and throw due to shear is liable to occur in case of high colloid and clay content. Therefore, siltization is more likely to take place in the ISS. With substantial undulation of interfaces and marked difference in strength of adjacent strata, siltization is more likely to occur on that side where the undulation is more pronounced(where grains of sandstone are coarser). Here, the modulus of deformation and strength parameters are both comparatively large and interlayer shearing is more liable to be concentrated.

In order to verify the correctness of the analysis of mechanism of distribution presented above, two other findings may be mentioned:

Results of computer-aided investigations of bedded geologic model: Based on Mohr-Coulomb law, with no tensile stress along bedding plane and in the normal direction (i.e. $|\tau_{xy}| \leqslant |f\sigma_y|$) the initial stresses being computed on the principle of gravitational field, as also the distribution of stresses at tunnel walls brought about by excavation in stratified formations of alternate hard and soft rocks, it was found that at the interface between layer of soft rock in T_1^{5-1} and relatively hard rock in T_1^{5-2}, the horizontal stress σ_x in the two formations differed by 200 %, and between T_1^{5-1} and hard stratum T_1^4 by 700 %. After completion of full-face tunnel driving, the horizontal stresses between T_1^{5-1} and T_1^{5-2} were released, with some residue in the vicinity of the interface, so that there still remained difference of horizontal stresses.

Findings of model tests: Model of cavern for main power plant made of a mixture of gypsum and sand was subjected to deformation under different loading conditions and tested until rupture occurred. It was found that in the relatively soft T_1^{5-2}, with modulus of deformation only half as much as that of T_1^4, all radial deformations, of considerable magnitude, were tensile. This shows that there occurred deformation toward tunnel axis affecting rather deeply into the model under compression. Values of tensile strains at bedding planes of T_1^4 and T_1^{3L2} were both rather high at the side walls, a maximum of 145 $\mu\varepsilon_v$ being manifest (ε_v is the volumetric strain, and discontinuities of observed values were quite obvious for upper and lower bedding plane, a largest deviation of 300 % having been recorded. To a degree, this reflects the particular features at the bedding planes and serves as symptom of possible dislocation.

The aforesaid show that, under the action of own weight and horizontal stresses, concentration of interlayer shearing is not to take place on interface between hard and soft rock. This essentially agrees with the result of a large amount of data on the distribution of CI acquired through field observations.

CONCLUSIVE REMARKS

1. CI in this region are mainly thin layers of soft rocks in the ISS, formed during folding and shearing of rock strata. Besides, creeps at slopes, rebound due to unloading, groundwater erosion as well as the earthquake forces, among others, are all of influence. The distribution of CI is mainly governed by tectonic stresses and distribution of ISS.

2. CI are mostly to be found near major faults, in horizons where the ISS occur less frequently, and where rock strengths differ most in the ISS.

3. The type of CI is governed by size of grains in thin soft rock in the ISS and the strength and minerological composition of the cementation agents.

4. Thin layers of soft rocks as mother rocks of CI are distributed in slate form, the extent of CI being limited by that of the ISS.

5. Based on detailed investigation of lithology and tectonic features, it is possible to cognize the law of distribution and continuity of CI. Furthermore, through comprehensive analysis, the areal continuity of CI, their types, their location on the potential rupture surface under shearing force may be determined , so as to choose a global value of the coefficient of friction embracing all the factors and being appropriate to actual condition.

REFERENCES

Geologic Section, Design Institute, Yellow River Conservancy Commission, 1983. Report of engineering geologic exploration for preliminary design of Xiaolangdi Reservoir on the Yellow River (restricted).

Dept of Hydrogeology and Engineering Geology, Changchun Geologic Institute, 1982. Report of practising fieldwork at Xiaolangdi Reservoir site.

Planning and Exploration Brigade, Y.R.C.C.,1975. Preliminary study of weak interbeds at Xiaolangdi dam site on the main Yellow River (restricted).

Ma,G.Y. 1984. Study on clay intercalation at left abutment of dam site No.3 of Xiaolangdi project on the Yellow River, Selected Articles on Hydrogeology and Engineering Geology.

Dept of Water Conservancy, Qinghua Univ. 1980. Research on rational spacing of tunnels forr Xiaolangdi Hydropower Station (restricted).

Don,U.D.,1978. Applied rock mechanics --- the importance of weak geological features. Proc. of the 4th Congress of the International Society of Rock Mechanics, Vol.3,pp. 22-25.

2.5 Dynamics and tectonics of weak rocks
Phénomènes tectoniques dans les roches tendres

Buckling and shearing of basalt flows beneath deep valleys
Flambage et cisaillement de couches basaltiques sous les vallées profondes

John G.Cabrera, *Las Cruces, New Mexico, USA*

ABSTRACT: It has been generally thought that the type of deformation due to stress relief by unloading, described herein, occurs primarily in weak, horizontally-bedded sedimentary rocks. However, at some major dam sites on the basalt flows of the Upper Paraná River Basin in southern Brazil and southeastern Paraguay, the presence of intersecting, inclined shear zones has affected the foundations for the central sections of the dams. These thrust faults, containing gouge and mylonite, appear to have originated as the result of rapid downcutting of the valley bottom by the rivers and consequent release of stored strain energy acting in a horizontal direction. In situ stress measurements made in the valley flanks prior to foundation excavations at some of the sites investigated, revealed a decrease of horizontal stresses, most likely due to the buckling, fracturing and overthrusting that took place.

RÉSUMÉ: On pense généralement que le type de déformation du au soulagement des contraintes par déchargement ici décrit, a lieu d'abord en roches sédimentaires faibles, en lits horizontaux. Néanmoins, à quelques sites de barrages majeurs en couches basaltiques dans le haut bassin du fleuve Paraná au sud du Brésil et au sud-est du Paraguay, la présence de zones de cissaillement inclinées qui s'entrecoupent a affecté les fondations des sections centrales des barrages. Ces failles inverses, avec remplissage et mylonite, semblent être originées par une érosion rapide des talwegs des fleuves et la conséquente libération d'énergie de décompression en sens horizontal. Des mesures des contraintes in situ faites sous les flancs de la vallée avant les fouilles de la fondation en quelques sites recherchés, ont revelé une décroissance des contraintes horizontales, probablement due a l'arcboutement, la fracture et le chevauchement qui ont eu lieu.

1 INTRODUCTION

Investigations made at several of the sites proposed for dams on the basalt flows of the Upper Paraná Basin have revealed the occurrence of shear zones in the valley bottoms.

Among the most important dam sites where thrust faulting has been encountered beneath the river bed are the Itaipu and Salto Santiago sites, respectively on the Paraná and Iguaçu Rivers (Figure 1).

Determination of the in situ stresses existing at some of these sites has led to a correlation between singularly low horizontal stresses and the occurrence of inclined shear zones in the valley bottoms, 10 to 20 meters below the top of rock at the bottom of the river channel.

These structural features have generally been related in the literature to weak, incompetent sedimentary rocks (Ferguson 1970, Ferguson & Hamel 1981), but foundation studies in the Upper Paraná Basin indicate that rebound stress relief structures can occur in subhorizontal flood basalts, provided the unloading conditions are similar.

2 THE PARANÁ BASIN BASALT FLOWS

2.1 Lithology

These lava flows are a succession of extensive tholeitic basalt fissure extrusions that occurred from late Jurassic to Early Cretaceous time within the broad, synclinal intracratonic Paraná Basin. They generally are separated by interflow deposits of eolian sand, which when mixed with subrounded, scoriaceous fragments from the wind-eroded surface of an underlying flow, became a sedimentary breccia resembling a "pudding stone" or conglomerate.

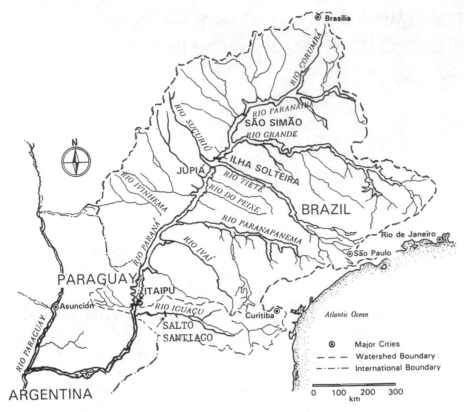

Figure 1. Upper Paraná Basin site location map.

2.2 Structural Features

In addition to structural features related to cooling (e.g. prismatic jointing), to mineral differentiation or lava flow characteristics, gouge-filled shear zones have been found within the valley bottoms at several sites. In some cases the valley has undergone considerable widening since the occurrence of shearing and the fault zones are vestiges of what took place when the river had much higher and steeper valley slopes (Figure 2).

There is some evidence (De Loczy 1970) indicating that the Paraná Basin is slowly being constricted by compressional tectonic forces acting essentially in an E-W direction. These regional pressures may be partially responsible for the existing distribution of initial stresses, however it appears that the in situ stress fields measured at dam sites in the basin are principally due to lithostatic pressure. Hontoon and Elston 1980 believe that anticlines occurring in a reach of the Colorado River 97 km in extent resulted from unloading, the driving mechanism being the stress gradient arising from the difference in lithostatic load between points deep under the canyon slopes, overlain by nearly 200 m of strata, and the unloaded valley bottom. Low angle thrust faults have formed in the limbs of the anticlines in response to the folding of the relatively brittle limestone.

A considerable amount of strain energy can also be stored in igneous rock such as granite (Swolfs, Handing & Pratt 1974), a more competent rock than basalt, which can lead to buckling and shear fracturing. Residual strain energy may also be due to non-uniform cooling of lava flows or to deuteric alteration of minerals accompanied by volume changes. Therefore, the state of stress at any point below valley slopes depends on the loading history of the rock (Harper & Szymanski 1983) and the nature of the intergranular bonds. The "locked in" strain energy can be released either slowly or rapidly (Varnes & Lee 1972) depending on whether the stored energy is elastic, plastic or partially both (Nichols 1980, Nichols & Abel 1975).

3 INVESTIGATIONS

During the investigation of the São Simão Dam site on the Paranaíba River, which is joined by the Rio Grande from the east to form the main stem of the Paraná (Figure 1), in situ stresses were measured using the stress tensor gage developed by the Lisbon National Laboratory of Civil Engineering. High horizontal stresses were determined, in many cases as much as 17 MPa.

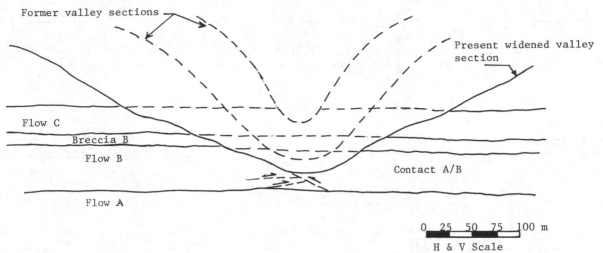

Figure 2. Itaipu Dam Site. Stages in the erosional unloading of the valley bottom and resulting rebound, buckling and shearing.

When the same tests were made in the abutments of the Itaipu site, it was found that the horizontal stresses were surprisingly low (0.5 MPa) or were almost totally relieved near the river banks, therefore it was concluded that they had been dissipated in some manner. Borings penetrating the flow (flow B) just below the river bed encountered a zone of broken, blocky basalt near the base of the flow (Figure 3). Direct observation of that zone in a tunnel radiating from a 110 m deep shaft in the right bank confirmed that the zone had undergone shear displacement and contained tabular fragments of basalt in a matrix of mylonite (Cabrera & Barbi 1981).

An intensive program of core borings (some using the integral core recovery method) and direct observations from four shafts and tunnels was carried out after unwatering and removing the alluvium within the foundation area of the high blocks of the main Itaipu Dam.

It soon became apparent that the shear zones were situated in a pattern resembling the intersecting shear fractures resulting from a triaxial test, with the principal stress acting in a horizontal direction (Figure 4). These thrust faults, which dip between 25° to 35°, were followed longitudinally and also crossed by several galleries that were later incorporated into a warped grid of tunnels which were backfilled with concrete and grouted tight. By this expensive foundation treatment, approximately 25% of the foundation area of the high blocks, containing shear zones and highly-fractured rock, was excavated and replaced by concrete, thus converting the tunnels into "shear keys" which prevented renewed deformation and displacement. At Itaipu it was observed that the contact between the approximately 20 m thick remnant of flow B and the much thicker underlying flow A was opened wider under the right bank and the river thalweg, allowing weathering to proceed at a faster pace. This appears to indicate a very slight uparching of flow B. Indeed, the contact was found to be tightly closed beneath the left river bank.

4 ORIGIN OF THE SHEAR ZONES

Igneous rocks, including extrusives, have traditionally been regarded as massive and very resistant, although they generally have cooling tension joints. In contrast, sedimentary rocks with a distinctive stratification have been thought to be less resistant to high horizontal compressive stresses (Patton & Hendron 1974). However, in a basalt flow containing bands of microscopic anisotropy, caused by concentrations of clay minerals derived from deuteric alteration and/or by mineral differentiation, inherent planes of weakness form, which act as contacts between strata.

Another possible cause of subhorizontal planes of weakness would be the development of flow shear zones due to variations in the temperature and viscosity of the lava. When these structural conditions are allied to a decrease in the thickness of a basalt flow, resulting from rapid down-cutting by a river, the horizontal initial stresses may in some cases be sufficiently high to cause buckling and overthrusting of the flow remnant. Even deformation of massive intrusive igneous rock due to unloading and valley bottom stress relief has been reported by some observers (Watters & Inghram 1983).

A careful examination of the surfaces of shear zones occurring within the remaining thickness of the flow immediately beneath the river channel at Itaipu showed slickensides and evidence of relative movements indicating several episodes of shear displacements. Since more detailed investigations of the shear zones were made at this site, a brief description of the conditions encountered will serve to illustrate the complexity of the foundation problems that can result

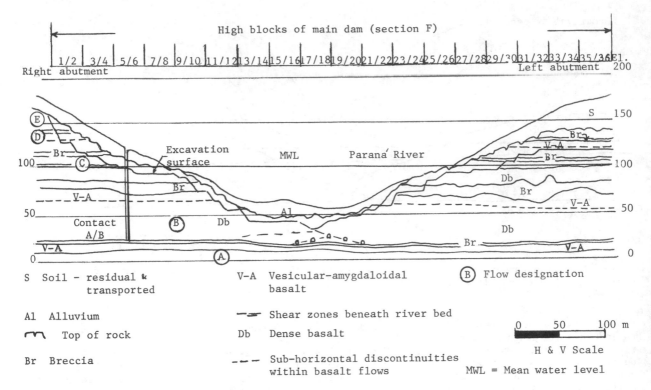

High blocks of main dam (section F)

1/2 3/4 5/6 7/8 9/10 11/12 13/14 15/16 17/18 19/20 21/22 23/24 25/26 27/28 29/30 31/32 33/34 35/36 El.
Right abutment   Left abutment 200

S   Soil - residual &
    transported

Al Alluvium

⌒ Top of rock

Br Breccia

V-A Vesicular-amygdaloidal
     basalt

━ ⚬ Shear zones beneath river bed

Db Dense basalt

━ ━ ━ Sub-horizontal discontinuities
       within basalt flows

Ⓑ Flow designation

0   50   100 m

H & V Scale

MWL = Mean water level

Figure 3. Itaipu Dam site. Geologic section across the river valley.

from stress relief.
   Five shear zones were defined by the investigations and by the systematic three level grid of
tunnels parallel to and at right angles to the dam axis. Their spatial relationship is shown in
Figure 5.
   Slickensides and relative displacements observed in these shear zones indicate that the rock
mass above the upper shear zone (No. 1 in Figure 5), which dips eastward, was thrust towards the
west (Paraguay).
   The shear zone that is tangential to the contact between flows A and B below the right (west)
side of the river channel (No. 2 in Figure 5) shows evidence of movement of the rock mass above
it towards the NE.
   There are two imbricated shear zones (Nos. 3 and 4 in Figure 5) that dip downstream to the S-SW,
which show evidence of the displacement of the rock slices above them in an easterly direction,
that is, the direction of movement was generally parallel to the strike of these zones. Both
these zones are cut off by the eastward-dipping upper zone (No. 1).
   A fifth, very localized shear zone, was found in the downstream portion of the main dam foundation area. The rock mass above this fracture appears to have been displaced in a westerly
direction.

5 ENGINEERING SIGNIFICANCE

The detailed investigation of the foundation rock directly beneath the river channel at Itaipu and
other sites within the Upper Parana Basin indicate that if a river accomplishes rapid downcutting,
the remaining thickness of a basalt flow may undergo buckling due to unloading and even shearing,
if the in situ horizontal stresses are sufficiently high.
   The relationship between the decreased thickness of a basalt flow and cambering or buckling and
even failure of the flow remnant due to decreased normal load and high horizontal stress should be
borne in mind by the engineering geologist. In the case of the Itaipu site, the existence of a
highly fractured zone just above the contact between flow B in the river bottom and the thick
underlying flow A had been determined as far back as the feasibility stage borings. What was not
realized then was the extent of this fractured rock and its cause. If it had been known, an
alternative axis nearly 3 km farther downstream could have been selected, in spite of there being
some drawbacks with respect to spillway location and future navigation facilities. The borings
made in the river channel at that site indicated that the dam would be founded at the top of
flow A, a flow that is very thick and singularly free of discontinuities.
   Determination of the orientation of the virgin or "locked-in" stresses in the rock mass prior to
deciding the location of a dam axis can aid in evaluating the possibilities of the existence of an

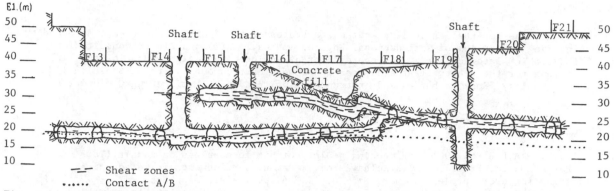

```
El.(m)
50 ___                                                                          ___ 50
45 ___                                                                     F21
40 ___         F13      F14↓     F15↓    F16     F17     F18     F19  F20        ___ 45
35 ___                          Shaft   Shaft                        Shaft       ___ 40
                                                    Concrete                      ___ 35
30 ___                                              fill                          ___ 30
25 ___                                                                            ___ 25
20 ___                                                                            ___ 20
15 ___                                                                            ___ 15
10 ___                                                                            ___ 10
```

~~ Shear zones
..... Contact A/B

Figure 4. Section along axis of main Itaipu Dam showing principal shear zones below foundation level.

imbalanced stress field that may have caused deformation of the rock in the valley bottom.

The following are some features related to valley bottom stress relief observed at several dam sites on basalt flows that are presently steep, narrow valley sections, or were formerly deep and narrow but have become widened in recent geologic time.

1. Geomorphic evidence of rapid stream downcutting, reducing the superincumbent lithostatic load.

2. Warping and buckling of a remaining thickness of basalt in the river bed, in some cases reaching rupture and the development of conjugate shear zones. These sets of fractures have features indicating that they are derived from compressional stresses, the direction of the principal stress bisecting the acute angle of the intersecting zones.

3. Abnormally low in situ horizontal stresses measured within the site abutments, indicating the relief of stored strain energy.

4. Occasionally, there are raised valley rims, often with small stream valleys paralleling the main river channel for some distance to a point where they make a right angle junction with the river (Matheson & Thompson 1973).

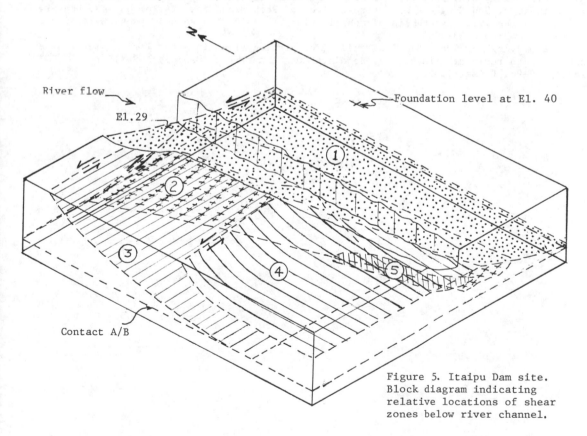

Figure 5. Itaipu Dam site. Block diagram indicating relative locations of shear zones below river channel.

ACKNOWLEDGMENTS

The investigations for the feasibility and design stages of the Itaipu project were directed by
the engineering consortium of International Engineering Co. (IECO) and Electroconsult of Milan,
Italy (ELC), which later on became the coordinator of engineering design. Most of the final
investigations for the treatment of the shear zones were carried out by Itaipu Binacional's
Division of Geology, Rock Mechanics and Instrumentation, headed by Engineering Geologist Adilson
Barbi.

REFERENCES

Cabrera, J.G. & A.L.Barbi 1981. Engineering geology of dam foundations on basalt flows of the
    Upper Paraná Basin, Brazil. In F.H.Kulhawi (ed.), Recent developments in geotechnical
    engineering for hydro projects: Amer. Soc. Civil Eng. p.177-191. New York.
De Loczy, L. 1970. Role of transcurrent faulting in South American tectonic framework. Bull.
    American Assoc. of Petroleum Geologists. 54, No. 11: p.2111-2119.
Ferguson, H.F. 1974. Geologic observations and geotechnical effects of valley stress relief in
    the Allegheny Plateau. Preprint, ASCE National Meeting on Water Resources Engineering, 31p.
    Los Angeles, Ca.
Ferguson, H.F. & J.V.Hamel 1981. Valley stress relief in flat-lying sedimentary rocks.
    Proceedings, Internat. Symposium on weak rocks, p.1235-1240. Tokyo.
Harper, T.R. & Szymanski 1983. Geological processes and the mechanical aspects of rock squeeze.
    Tectonophysics. 91: p.119-135 Elsevier.
Hontoon, P.W. & D.P.Elston 1980. Origin of the river anticlines, central Grand Canyon, Arizona.
    U.S. Geol. Survey Prof. Paper 1126A, 9p.
Matheson, D.S.& S.Thomson 1973. Geological implications of valley rebound. Canadian Journ. of
    Earth Sciences. 10, No. 6: p.961-978.
Nichols, T.C., Jr. 1980. Rebound, its nature and its effects on engineering works. Quarterly
    Journal Eng. Geology. 13: p.133-152.
Nichols, T.C., Jr. & J.F.Abel, Jr. 1975. Mobilized residual energy; a factor in rock deformation.
    Bull. Assoc. Eng. Geol. 12, No. 3: p.213-225.
Patton, F.D. & A.J.Hendron, Jr. 1974. General report on mass movements. Proc. 2nd Internat.
    Congress of Internat. Assoc. Eng. Geol. 2, Theme V: p.VGRI-VGR57. São Paulo, Brazil.
Swolfs, H.S., J.Handing, H.R.Pratt 1974. Field measurements of residual strain in granitic rock
    masses. Proceedings, 3rd Congress of Internat. Soc. of Rock Mechanics. 2A: p.563-568. Denver, Co.
Varnes, D.J. & F.T.Lee 1972. Hypothesis of mobilization of residual stress in rock. Bull. Geol.
    Soc. of America. 83: p.2863-2866.
Watters, R.J. & B.J.Inghram 1983. Buckling failure of granite slabs in natural rock slopes as an
    indication of high residual stresses. Proceedings 20th Annual Eng. Geology & Soils Eng.
    Symposium: p.83-96. Boise, Idaho.

Applications of microscopic analytical methods in studying coal bursting danger in No.8 coal seam of Beijing Mentougou mine

Application des méthodes d'analyse microscopique dans l'étude des dangers de rupture des niveaux de charbon dans la mine de Mentougou Pékin

Xin Yumei & Niu Xizhuo, *Central Coal Mining Research Institute, Beijing, China*

ABSTRACT: This paper deals with the important role of coal macerals, fabrics and microscopic features in the occurrence of coal burst. The paper presents the results of measurements of the structural planes of weakness, macerals, microfabrics and microhardness of individual coal slices in No.8 coal seam in the Mentougou mine of the Administration of Beijing Coal Mines as well as the results of comparison with the coals that have bursted.

This paper gives the results of electron-microscope and micropore analysis of the changes in the microstructures of the coal seam before and after getting wetted out. The paper suggests a reasonable range of water pressure when injecting water into the coal seam to eliminate the danger of burst.

RÉSUMÉ: Le présent article montre le rôle important que jouent le macéral, la petrofabrique et les caractéristiques micro-structurales des roches houilleuses dans la constitution du danger de coup de terrain des couches.

L'article donne des résultats de mesure des surfaces de structure à faible résistance de différents lits de la veine NO.8, des résultats d'analyse des macéraux, des microstructures et des microduretés des différentes couches de cette veine et des résultas d'analyse obtenus par comparaison avec des veines ayant subi des coup de terrain.

Pour connaître le changement de microstructure que la couche connaît après avoir été mouillée, nous avons fait l'analyse par microscope électronique et la mesure de micropore. Les résultats sont donnés dans l'article. Nous avons proposé une gamme de valeur de pression d'injection raisonnable pour faire disparaître le danger d'explosion géostatique.

INTRODUCTION

As mining depth increases, coal burst takes place in more and more coal mines. This serious problem causes great attention from major coal producing coutries worldwide. In China, serious impact ground pressure has been observed in a number of coal mines such as shengli mine in Fushueng Mining Area, Tao Zhuang mine in Zaozhuang Mining Area, Tangshan mine in Kailan Mining Area, Tianchi mine in Mianyang Mining Area and Fangshan, Mentougou and Chengzi mines in Beijing Mining Area and etc.. Obviously, study of characteristics of coal seams dangerous of coal burst, internal factors affecting the burst occurrence and analysis of coal burst danger have important and practical significance.

It is known that the physical Mechanical properties of coal seam as well as its roof and floor are important factors affecting coal burst. Mechanical property is one of the physical properties of coal and rock and depends on petrographic components, structures and the degree of decomposition by geological tectonic movements.

This paper presents results of the study of coal burst in NO.8 coal seam of Mentougou mine, where coal burst is a serious problem. Factors affecting coal burst danger have been analyzed and a rational range of water injecting pressure has also been recommended.

1 GEOLOGICAL CONDITIONS OF NO.8 COAL SEAM

Mentougou coal mine is located in the middle of Beijing western suburb coal field, where coal deposited during Jurassic Period. Geological structures here are very complicated. Folds and faults are well developed. There are five minable coal seams in the mine and three of them are being extracted now. The spacing between coal seams ranges from 60 to 120m, while thickness of coal seams from 0.7 to 3.5m. The roof consists of sandstone with thickness of 10–30m. Hard sandstone accounts for 80–85% of the total coal-bearing strata. The roof of No.8 coal seam with coal thickness of 2.85m is composed of fine-grained sandstone and powder sandstone with thickness of 6m, above which is lying medium and coarse-grained sandstone of 13–30m in thickness. The floor of this seam is powder sandstone of 2m in thickness. Coal of No.8 seam is highly metamorphic anthracite.

2 CHARACTERISTICS OF NO.8 COAL SEAM AND ANALYSIS OF ITS BURST DANGER

2.1 Structure features of the seam

The mechanical properties of coal are not only dependent on the strength of compactness of coal, but also closely related to the strength and direction of its weak planes. Geological folds, faults joints and crevices in No.8 coal seam are well developed. According to its ruptured degree, No.8 coal seam is divided into three natural slices: upper, middle and lower subseams.

With its specular planes well developed, coal in the upper slice is mostly decomposed into small fragments in 1 cm size by joints and crevices. Coal of the middle slice is in better conditions and harder in terms of structure and cut into large fragments of rhombuhedrons in 0.7m size by three groups of joints. Strip structures of this slice are clearly seen and there are very few fissures. Coal powder is often observed in between coal fragments in rhombuhedrons.

Seriously crumbled, the lower slice is mainly comprised of mylonitic planes. The thin coal slices are formed by extruding powder coal and fine-grained coal into coal fragments. No apparent crevices were found in this slice.

Of the three slices, the middle one is the hardest with compressive strength of 27.3–49.3 MPa while the upper one is the second in terms of hardness and the lower slice has the least structure strength because of the serious geological decomposition, crumbling and twisting. Structure failure reduces the built-up of elastic energy of coal seams and consequently cuts down the strength resistence to structure rupture. Therefore, during mining process, coal burst is most likely to take place in the lower slice of No.8 coal seam.

2.2 Microstructural features of No.8 coal seam

From the macroscopic point of view, the coal of this seam is mainly bright coal and semi-dull coal which accounts for 70–80% of the total. The coal is hard or medium hard and of steel-grey colour and strip structures. The lower part of the seam is soft and of granular structure.

1. Features of macerals and microcomponents

As observed by microscope, the micro-organic components of No.8 seam are mainly vitrain and less fusain. Microcracks in the vitrain are very well developed and there are lots of microholes in the coal. Part of the holes and cracks are filled with calcite and alittle amount of gypsum.

Coal of upper slice is mianly vitrain of strip structure and lens like fusain. The vitrain amounts to 63% of microorganic components while fusain about 37%. Cracks and microholes in the coal make up about 5% by volume, see Fig.1.

Fig.1 50ˣ transmission & reflectory light of upper slice

a. Structureless vitrain
b. Indistinct structure vitrain
c. Fusain

Micro-components of the middle slice are similar to those of the upper one, with vitrain making up approximately 74% while fusain about 26%. Vitrain and fusain are formed in combined narrow strip structures. Microcracks and microholes in the coal make up 5% by volume. Geologically, this slice is not seriously dislocated after coal structure was ruptured and only partially folded, see Fig.2.

Fig.2 50x transmission & reflectory light of middle slice narrow coal strips of vitrain and fusain

Vitrain in the lower slice makes up about 78% by volume, distributed in the form of broad strips. Compared with the upper and middle slices, the portion of structureless vitrain appears clearly bigger, accounting for half of total volume of vitrain in the slice. Microcracks and microholes make up about 11% by volume. The lower slice was seriously damaged by tectonic movements in later period. Dislocation and foldings are evidently found in it. The coal of this slice was compressed into fine grains with edges mostly worn out. Crack distributions in the coal are in great disorders. Complicated and crossing cracks in net form are also widely found, see Fig.3

Fig.3 50x transmission & reflectory light of lower slice

a. Structureless vitrain
b. Indistinct structural vitrain
c. Fusain
d. Narrow veins of calcite

Coal burst took place in part of NO.8 coal seam before, and the organic materals of bursted coal in appearance of broad strips in mainly vitrain that accounts for 78% of the total. Structureless vitrain is the major part of vitrain in the coal. Fusain makes up about22%. Microcracks in the bursted coal are very well developed. Two inclined groups of joints in net form are found. Small crush zones were seen and coal there was crushed into small fragments. Some of thim have evident edges. The seam is dislocated and partially folded. Cavities in fusain was deformed due to geological extruding action. The cracks and microholes make up 11% by volume and most of them were filled with calcite and some gypsum, see Fig.4.

The above analysis shows that the lower slice is quite similar to the part of the seam that has undergone burst in terms of coal macerals, microcomponents and microscopic features.

Fig.4 50x transmission & reflectory light of burst coal

a. Structureless vitrain
b. Indistinct structural vitrain
c. Calcite-filling materials in cracks

2. Microhardness of coal

As it is very difficult to analyize the mechanical properties of the upper and lower slices because of the decomposed structure, Wei's square cone diamond pressure trace method was adopted to measure the microhardness of each slice. It was discovered during the test that the mechanical properties of coal along the bedding were stable and its hardness was lower, while the hardness of coal perpendicular to the bedding was higher and varies a great deal. Under a given load, no pressure trace was observed on the specular planes of coal when measuring coal hardness within a short time and coal was fractured within a prolonged period. This indicates that the coal under a given load is first compacted, and then fractured. Coal crushing is a microscopicfeature of brittleness of the coal. It is also discovered that the microhardness of the filling mineral materials in microcracks of the coal varies greatly.

The results of measurements of microhardness of coal macerals from the three slices and the brusted coal (see Table 1) show that microhardness of the specular structureless vitrain in the same slice is the higherst. Joints and cracks are well developed and the coal has experienced serious structure decomposition. Microhardness of the hidden structural or structructural vitrain with some fusain in it is relatively lower. The coal in narrow and random strips which is made up of vitrain and fusain has the lowerest microhardness and was least affected by geological tectonic movements, with joints and cracks not developed at all.

Table 1. Microhardness of coal macerals (MPa)

Coal samples \ Coal macerals	Structureless vitrain *	Structural vitrain *	Vitrain and fusain in narrow and random strips	Mineral materials in crevices
Upper slice		1700–3330 / 2480	1770–2680 / 2380	1440–2090 / 1760
Middle slice			1270–2090	2090,3070
Lower slice	2210–2520 / 2360		970–1870 / 1260	230–720 / 440
Bursted coal	1670–3450 / 2500	1670–2440 / 1700–2400		
Bursted coal	640–1170 / 940		1120–1140 / 1260	1050–1440 / 1260

*Coal mixed with little other microcomponents.

Table 1 shows that macerals and microhardness of the lower slice is much similar to those of bursted coal.

3. Coal quality features

Moisture, ash, ash composition and volatile matter in coal are directly controled by coal macerals and these factors strongly affect the danger of coal burst.

Proximate analysis of the raw coal currently being mined from No.3, No.5 and No.8 coal seams and the analysis of coal ash composition indicate the following four points.

(1) Moisture content in No.8 coal seam is low (generally about 1.7%), ranging from 1.58% to 3.35% The moisture content of coal in No.8 seam in Mengtougou mine is dry and lower as compared with that in No.3, No.5 and No.8 coal seams of Qian Juintai mine.

(2) Coal ash content of No.3, No.5 and No.8 coal seams ranges from low to middle.

(3) The CaO and SO_3 contents (CO_2 not analyzed) of coal ash in No.8 seam is 4–5 times higher than that in No.3 and No.5 seams and also much higher than that in coals of general type, but the SiO_2 and Al_2O_3 content in coal ash is 100% lower, which indicates more calcite and gypsum but less clay minerals and other impurities are in No.8 seam.

(4) Volatile flux in No.8 seam ranging from 4% to 7.5% is 100% higher than that in No.3 and No.5 seams ranging from 2% to 4.5% and similar to the bursted coal.

The above analysis of coal quality features shows that coal in No.8 seam is dry and contains more hard and brittle calcite mineral impurities, less plastic clay minerals impurities and that the volatile flux in the coal is also similar to that of coal which is subject to coal burst. All these above-mentioned quality features are important factors which make No.8 coal seam more dangerous of coal burst than No.3 and No.5 coal seams.

3 WATER EFFECT ON COAL BRUST LIABILITY

3.1 Macroscopic and microscopic features of coal before and after water immersion

Before immersed in water, coal has good lustre in blackbrown colour. Microcracks are clearly seen and the thin coal slices of the edges of the cracks are easy to peel off. After immersed in water, coal becomes dull and in black collour. The microcrevices appear even smaller and hard to see. The thin coal slices of edges of crevices becomes dull, but no peeling off is found. Immersed coal in water looks to be dilated.

The following changes on coal macerals were noticed by electron-microscopic observation before and after water immersion. The coal between crevices is in good compactness and with coarse coal grains adhered to it. After coal is immersed in water, crevices become smaller and numerous microfissures are seen in the compact coal. The coal grains adhered become smaller while their amount increases substantially, see Fig.5 and 6 in electronmicroscope photoes.

After the wet coal is cut off, broad microcracks perpendicular to cutting direction could be seen, see Fig.7 in the electronmicroscope photoes. All this shows that the coal becomes loose after immersed in water.

Fig.5 Coal is more compact and there are well developed directional cracks and joints in coal before immersed in water.

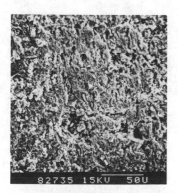

Fig.6 Coal becomes loose and has more microfissures after immersed in water

Fig.7 Fractured coal has more micro cracks perpendicular to cutting direction after immersed in water.

Visual and microscope observations of coal wetting best indicate that after coal is immersed in water, water gives a significant splitting effect on coal. While coal is undergoing structural changes, new cracks of wedge shape appear because of the water splitting effect and these fissures go down deep inside the coal. The more microfissures in coal, the more water comes in, which reduces the internal binding strength inside coal, thus makes coal loose, weakens coal strength, and as a result, alleviates the burst danger of the coal seams.

3.2 Diostributions of microholes in coal and changes of volume of the porocities before and after water immersion

Mercury pressure test was carried out on rom coal samples and wet coal samples with maximum mercury pressure up to 100 MPa 10^5Pa. The measurements of volumes and distributions of the microholes with diameters ranging from below 100 A to above 10,000 A are shown in Table 2. The measurements in dicate the following three points.

(1) Volume of permeable porocities whose diameters range from 100 A to 4,000 A enjoyed little changes aftewr water immersion (in terms of absolute value).

(2) Volume of permeable porocities whose diameters range from below 100 A to 4,000 A was greatly increased compared with that before water immiersion.

(3) After coal is immersed in water, volume of permeable porocities in the coal was increased by 0.0011 ml/g, i.e.by 10.3% against that before water immersion.

Comparison of distributions and volumes of permeable porocities of coal before and after water immerasion

Table 2. Comparison

Type of coal sample	rom coal		coal immersed in water	
No.	VIII-1 $_{(9)}$A		VIII-1 $_{(9)}$A'	
Total volume of permeable porocities (ml/g)	0.0107		0.0118	
diameters of procities A	volume distribution of permeable porocities			
	ml/g	%	ml/g	%
below 100	0.0013	12.18	0.0018	15.61
100–400	0.0049	46.02	0.0048	40.68
400–4,000	0.0025	23.24	0.0024	20.08
4,000–10,000	0.0003	2.53	0.0006	5.06
above 10,000	0.0017	16.03	0.0022	18.57

3.3 Volume changes of microholes in diameters of 2–126 microms in coal before and after water injection into coal seams

Structure analysis was made on over 30 rom coal samples and as well as water-injected coal samples by means of automatic microstructural analyzer. The results are given below:

(1) Before water injection, the volume of microholes in the upper slice accounted for 14.6%, while after water injection it was up to 18.3% ;

(2) Before water injection, the volume of microholes in the middle slice accounted for 14.75%, and after water injection it was increased to 14.85%;

(3) Before water injection, the volume of microholes in the lower slice accounted for 10.90%, and after water injection it was measured at 21.20%;

(4) The volume of microholes of bursted coal in the seam was measured at 20.10%.

Anaysis indicated that volume of microholes in the coal seam was more or less increased after water injection and the volume of microholes of the lower slice was most increased. On the other hand, the structure of microholes in the lower slice after water injection was quite similar to that of bursted coal.

3.4 Relationship between water injection pressure and ovlume of permeable porocities

Mercury volume in coal was measured under different pressures during mercury pressure test. Results of measurements are shown in Table 3. The following three points are derived from Fig.8

Table 3   Meeasurements of permeable porocities in rom coal

mercury pressure (MPa)		Volume of permeable porocities (ml/kg)		Permeability dfficiency
P	ΔP	V	ΔV	ΔV/ΔP
0.4	0.4	1.6696	1.6696	0.4174
4.4	4.0	2.9854	1.3158	0.0329
11.1	6.7	4.0350	1.0505	0.0157
19.6	8.5	4.9060	0.8737	0.0103
27.4	7.8	5.8521	0.9425	0.0121
35.0	7.6	6.8142	0.9621	0.0127
45.5	10.5	7.4913	0.6771	0.0064
55.5	10.0	8.2667	0.7754	0.0078
65.5	10.0	8.9439	0.6672	0.0067
75.0	9.5	9.7193	0.7754	0.0082
85.5	10.5	10.3964	0.6771	0.0064
100.1	14.6	10.6706	0.2742	0.0019

volume of permeable porocities 10.7ml/kg

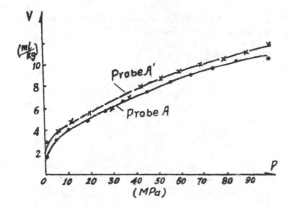

Fig.8 Relationship between pressure and volume of permeable porocities

As pressure increases, mercury volume in coal gradually expands.
Under the same pressure, mercury volume in wet coal is bigger than that in the rom coal.
Under different pressures, mercury volume, i.e. volume of permeable porocities is also diffenent. As pressure increases, volume of permeable porocities increase less and less. Table three shows that the permeability efficiency becomes lower and lower and drops down dramatically.

Permeability efficiency becomes high as pressure is below 10 MPa, therefore, the rational pressure value for injection should be lower than 10 MPa. Considering the fact that water penetration is stronger than that of mercury, injection pressure can be set at around 7.5 MPa or even lower than this figure.

Test results have proved that water runs into coal body along its crevices, and as a result, new microcrevices continuously appear inside the coal body because of water splitting effect, volume of microholes in coal is expanded, thus the coal becomes loose and its strength decreases. The maximum strength of coal before water injection was measured as 33.5 MPa and it was reduced to 22.4 MPa after water injection. This indicates that water injection decomposed its internal structure, reduces its internal binding strength and decreases the maximum failure strength by 33.1%. At the same time, the increase of number of microholes in coal makes the coal loose and reduces its elasticity. It is known by test that the coal elastic energy index (WET) was 9.65 before water injection, and it was reduced to 3.2 when coal was immersed in water for 4 weeks, and it was further reduced to 1.7 when coal was in water for 8 weeks.

As water runs into coal and the volume of permeable porocities in coal expands, coal becomes loose, its strength and elasticity gradually decrease. All this changes the coal burst tendency from strong (WET above 5) to weak (WET from 2 to 5) and finally eliminates coal burst tendency (WET below 2). The structural features of microholes in coal not only effect water injection efficiency, but also are decisive factors governing the intensity of coal burst tendency.

CONCLUSIONS

1. No.8 coal seam in mentougou mine is divided into three slices according to the degree of strc-
ture decomposition. Coal in the middle slice is hard and more compact, coal in the upper slice is
less hard and less compact, while the coal in the lower slice is seriously disturbed by tectonic
action and it is mylonitic type of coal.
2. It is mainly vitrain in the macerals of No.8 coal seam. Vitrain is brittle and has higher
microhardness. The lower slice contains more vitrain and microcrevices in this slice are mostly
developed and quite similar to those in the bursted coal in No.8 seam.
3. As compared with No.3 and No.5 coal seams in this mine, the coal in No.8 seam is drier and contains more carbonate minerals.
4. By analyzing coal structures, macerals and coal quality features of No.3, No.5 and No.8 seams
it is clear that coal burst tendency in No.8 seam appears most serious by comparison with other
two seams. In No.8 coal seam, the lower slice is the most dangerous of coal brust occurrence as
compared with other two slices. Therefore, water injection in the lower slice of No.8 coal seam
would be the most effective way against coal burst occurrence.
5. Analysis of microstructure of No.8 coal seam carried out after water injection indicated that
as the decomposition effect of water injection on the internal microstructure of coal is similar
to that of coal furst on its internal microstructure, the effects in both cases to alleviate or
eliminate coal burst tendency remain quite the same. Therefore, the analysis of internal microstructure of coal is recommended to be one of indexes to check water injection effect on coal seams and
to analyze coal burst tendency.
6. The rational range of water injection pressure for coal seam No.8 in Mentougou Mine should be
set below 7.5 MPa or close to it.

REFERENCES

Lu Zigan, Chang Huongsheng (1981) Brief Analysis of the Formation of Impact Ground Pressure in
    Mentougou Mine and its Control. Coal Science & Technology, in Chinese, Oct., 1981: 2-3.
Huong Gengwu (1980) Effect of the Direction of Structural Planes on the Strength of Rock. Technical
    Journal of Shandong Mining Institute, in Chinese, Feb., 1980:73-83.

2.6 Karst phenomena
Problèmes liés aux karsts

Models for base level of karstification · A basic problem in engineering geology

Modèles pour déterminer le niveau de base de la karstification
Problème de base en géologie de l'ingénieur

Chen Guoliang, *2nd Railway Surveying & Designing Institute, Ministry of Railways, Chengdu, Sichuan, China*

ABSTRACT: The lower limit of karst process is generally termed as karst base level(Ren & Lin1983: 161-163). It imposes much restraint on the depth of karst evolution and corresponding vertical zonation, and possesses a vital importance in evaluating the geologic settings of a construction site and especially of underground structures. Searching of karst base level is therefore an essential one of the problems in karst study.

There have been different versions about karst base level. Theoretically true as they are, however, they seem to be of less significance as a field guide to engineering geology. This is because what we want to know in civil engineering construction activities are the form of existence, the location and the elevation of a recent base level in the vicinity of the site for solving practical problems in engineering practice. Meanwhile, the concept of karst base level which means the lower limit of karst process seems apt to be misunderstood as a base level under which no karst would exist there. Hence, the present paper suggests the concept of base level of karstification, defining it as the lower bound of horizontal-flow zone in karstification, and summarises five models for it as: that with lateral limitation, with bottom limitation, in a bare valley, in a covered valley and that of deep karst below a valley, according to the features of occurrence and combination of soluble rocks with nonsoluble rock formations, to enable the concerned theoretical attacks to serve engineering practice in karst area.

RÉSUMÉ: Ce que nous entendons par la limite inférieure de la karstification, c'est le plan de référence(Ren & Lin 1983: 161-163) du karst que nous l'appolons. Il s'impose des restrictions sur la profondeur de l'évolution karstique ainsi que la zonation verticale, ces faits sont très important pour donner une appréciation sur les travaux de terrassement notamment sur les conditions géologiques des travaux souterrains. Par conséquent, le travail pour chercher le plan de référence du karst constitue donc une question fondamentale.

En ce qui concerne le plan de référence du karst, beaucoup de thèses ont été publiées et elles peuvent être bien fondées théoriquement, mais ne peuvent pas jouer un grand rôle indicatif pour les travaux en plein champ, car, pour résoudre les problèmes qui surgiront au cours des travaux de terrassement, on a besoin de connaître la forme d'existence, le site et l'altitude de plan de référence du karst de la région à l'époque contemporaine pour résoudre les problèmes pratiques des travaux. En même temps, la conception de la limite inférieure de karstification qui s'appelle le plan de référence du karst est exliqué aisément avec erreur comme s'il n'y ait plus d'existence du karst sous le plan de référence. Nous avançons dans cette thèse une conception sur le plan de référence de la karstification que est defini par la limite inférieure de la zone d'écolement horizontal, et en suivant la composition de la roche soluble et insoluble ainsi que la forme d'exposition à l'air du karst, on divise le plan de référence du karst en 5 modèles suivants: plan de limitation latérale; plan de limitation au fond; plan à découvert dans une vallée; plan couvert dans une vallée et plan profond du karst dans une vallée, ceci a pour but de mettre la théorie karstique au service des travaux pratiques de terrassement.

The lower limit of karst process, generally termed as karst base level, restrains the depth of karst evolution which underlies the hydrodynamic profile of karstwater movement together with karst vertical zonation. Different vertical zonation implies different size(quantity), scale and characteristics of cavern and karstwater: large cavern may often be found in the vertical vadose zone (equivalent to the aeration zone here) to cause the foundation of a structure suspended, mud and sand in the vertical karst conduit would bob up and karstwater would gush to menace the securities of construction; in the horizontal-flow zone(equivalent to the zone of saturation, with which the seasonal fluctuation zone has been classified by auther) large and steady discharge of karstwater and the large cavern transversely extended would cause serious damage to the structure; and in the deep-oozing zone the percolation of the karstwater remains the trouble though none of large caverns would exist there. The vertical zonation plays an important role in siting a civil engineering project, particularly an underground structure, and in evaluating their engineering geologic and hydrogeologic settings.

There have been different versions about karst base level at home and abroad, such as the lower bound of soluble rock formation, the lowest perched water table, the sea level or the river bottom in the concerned locality. They are all true, from a point of view concerning karstological and karst-hydrogeological principles and considering the ultimate limit of solution, but some problems to be solved in regard to the enginnering geology and hydrogeology in human's construction activities are:

—— The actual position of the base level in the area of carbonat rocks of exceptional thickness or the carbonat rocks stretching to unusual depth due to tectonic change while the base level does not yet reach its ultimate limit;

—— The actual position of the base level in an uplifting karst area, such as in southern China where karstwater is mainly of conduit flow and hasn't yet formed into a unified water table;

—— The actual position of the base level in the karst region which is far apart from coast and separated or isolated by nonsoluble rock formations, where direct control from sea level is obviously out of action though the argument of sea level restraint may be true in a global sense or in a coastal region;

—— The apprehension of karst base level on which some of the outlets of karstwater situate above the river bottom while others in the neighbourhood lie below.

The concept referring to the lower limit of karst process as karst base level is apt to be misunderstood as an implication that no karst would exist under the level.

As stated above, it is desirable to search the existing form of recent base level directly related to a project and investigate its position and elevation for evaluating the engineering geology and hydrology of the project under consideration. Therefore, in this paper, the term base level of karstification is regarded as the recent base level of karst that concerns engineering structures.

The termed base level of karstification, or karst base level, is referred to as the lower bound of horizontal-flow zone in karstification. Based on years-long practice in karst area, the models of base level of karstification have been summarized as follows:

1. Base level with lateral limitation

It is rather common that the mounts in mountainous area, particularly in divide belt, are usually consisted of both soluble and nonsoluble rock formation. In the case of a perennial stream running across the soluble rock formation which is confined with nonsoluble rocks on either side or both sides, the current through the nonsoluble part will lose its hydraulic connection with groundwater in the soluble. The groundwater in the soluble rock will percolate downward to a certain depth by gravity through joints, cracks and existing solution pipes and then move along the stretches of soluble rock until outflowing on the ground in forms of karst spring or outlet of subterranean stream. Obviously, it is one of the outflows as prescribed that constitutes a temporary karst base level in the locality rather than the main river valley itself. The base level may be a karst spring or an outlet of a subterranean stream emerged on the slope or river bottom of another stream valley which is much lower than the main river valley and far apart from it, but has a direct connection with the soluble rock formation in hydraulics.

As illustrated in Fig. 1, the Pin-guan Tunnel, 5.1km long, goes through the divide on the border between Yunnan and Guizhou Province, which is composed of an overturned anticline structure longitudinally. On either limb of the fold there exist nonsoluble rocks and a main stream. Soluble rocks outcrop in the kernel part of the fold where karst features such as karst mound-depressions, karst pits, sinkholes and dolines can often be observed. A saddle-shaped fold of nonsoluble rocks in the middle part of the tunnel has divided the tunnel into two hydrogeologically different sections——the east and the west. In east section, the tunnel has an elevation of 1826m. The bottom elevation of the Huo-pu Creek is 1820m. If the river bottom is taken as the karst base level, the depth of karstification in the east section should be about 1820m, and the tunnel must be situated in the horizontal-flow zone. But what have been revealed in tunnel excavation are featured of vertical-vadose zone with small amount of karstwater percolating downward through conduit. A later karst-cave connecting test with lycopod spore cast in the pilot tunnel was conducted and proved that the karst conduit in the east section are connected with the outlet of Long-tan subterranean stream, 16km apart from the southern of the tunnel. The subterranean stream, being the source of Chen-jia ba Creek, is a perennial one with average flow of 3-4m³/s and outlet elevation 1580m. Core tests adjacent to the tunnel have uncovered a subterranean stream with the bottom elevation about 1607m, LWL=1666m, rising of water level 20-40m in rainy season. From this results, it has been inferred that the tunnel goes through the vertical-vadose zone, with the horizontal-flow zone underneath, where should be over 1706m near the tunnel. Elevation of respective zones will gradually decrease toward south due to hydraulic gradient. Thus, the recent base level of karstification on south side of the tunnel should be 1580m. According to the vertical zonation determined by this karst base level, it is likely to interpret correctly the engineering geological conditions of the Pin-guan Tunnel. Similarly, the karst base level in north side of the tunnel is not at Sha-tuo Creek, but the karst spring named "Jiao-tian-long", 7km to the north of the tunnel. There seven groups of karst springs outflow from the contact of limestone and basalt, with a flow of 0.4-5m³/s and elevation of 1732 m, A comprehensive inference has been made on

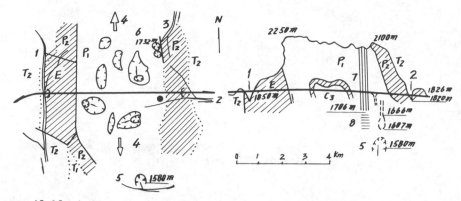

1--Yu-nan Creek, 2--Huo-pu Creek, 3--Sha-tuo Creek, 4--flow direction of groundwater, 5--Long-tan(outlet of subterranean stream), 6--Jiao-tian-long= (karst springs), 7-vertical vadose zone, 8--horizontal flow zone.

Fig.1 Sketch of Pin-guan Tunnel(profile included)

available informations that the subterranean divide will lie between the tunnel and the karst spring "Jiao-tian-long" in the north of the tunnel.

Long tunnels of the like, in addition, include the Zhong-liang-shan Tunnel, the Yan-zi-ya Tunnel, etc., and their common feature is that the surface streams in their vicinity do not play as a karst base level.

Another case of lateral limitation is shown in Fig. 2(a). The stream A on the nonsoluble rock side can't be a base level as mentioned above while stream B through the soluble rocks may gather much of the groundwater from the amount and forms a karst base level, but another subterranean stream or karst spring C elsewhere may also be the level, The base level of karstification on the east portal of the Sheng-jing-guan Tunnel is at the subterranean stream C, 10m lower than the surface river A, while on the west portal, the base level of karstification lies in the 6 karst springs group controlled by the river B (Fig.2(b)). Such are the cases with Mei-hua-shan Tunnel, Ka-la-zhai Tunnel, etc.

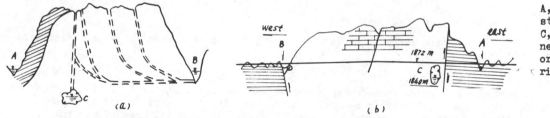

A, B, surface streams
C, subterranean stream or karst spring

Fig.2 Model of single lateral limitation

2. Karst base level with bottom limitation

The process of karstification in soluble-rocked massif can theoretically reach the top of underlain nonsoluble-strata as a limit, whether the nonsoluble is buried deep or outcropped on slope of the local valley. However, in case the nonsoluble lies very deep and much of groundwater in the soluble rock gathered by a surface stream, the vadose amount of karstwater will decrease; the hydrodynamic power for deeper circulation of groundwater and the content of free carbon dioxide will decrease gradually with its depth. Therefore, the intensity and depth of karstification are limited. No exploratory evidence has been available nowadays for confirming whether the top of a deep buried nonsoluble rock formation has become an actual lower bound of karstification. But it's very often that the top of nonsoluble strata outcropping in the valley has become a base level karstification.

As shown in Fig. 3, in the section from San-jia-zhai to Ai-jia-ping on the Quiyang-Kunming Railway, 4 subterranean streams outflow on top of nonsoluble rocks, 60-150m above the stream level of Ma-guo Creek---one of the tributaries of Nan-pan-jiang River. Geological survey and core test have revealed an intensive corrasion and transpotation together with a continual and sideward enlargement in the passage of subterranean streams. Deposits of block stones laid at most of the outlets. Karst features exposed after railway construction have convincingly shown that the bottom of the soluble strata is the very base level of karstification.

A further illustrative profile from one of the tributaries of Fu-rong River, Guizhou province, is shown in Fig. 4. The moving groundwater in soluble rocks is checked and discharged out on top of the underlain nonsoluble rocks. Outlets of subterranean streams are: Cave A with a perennial discharge of 0.2-2 m^3/s; Cave B with an accessible passage more than 1000 m long and a few blue holes with intrusion of karstwater in flood period; Cave C with groundwater never out-spilling and a lot of pearl-stones; Karst spring D with a perennial current discharged at the level of surface current.

As evidenced by preceeding descriptions and the like instances, it is very interesting that a unified free water table is unlikely to be formed on the ceiling of nonsoluble rocks to show a phteatic water table as a base level due to heterogeneity of karstification. However, by lithology,

607

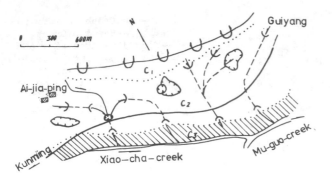

Fig.3 Part of drainage system of subterranean
streams from San-jia-zhai to Ai-jia-ping

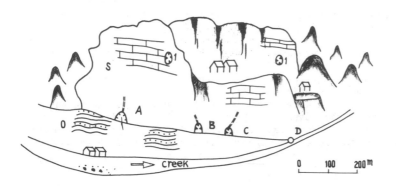

1--cavern,
A--cave Ma-wan-dong,
B--cave Ba-jiao-dong,
C--cave Shi-wan-dong,
D--karst spring Yu-quan

Fig.4 A section of profile from
tributary of Fu-rong River

the ceiling of nonsoluble strata wherever it is outcropping or thinly covered remains indisputable
base level of karstification.

3. Karst base level in a bare valley

In case of a soluble-rocked area experiencing a prolonged stability, the river bottom of the
main stream through the area will play in general a karst base level. But it is not the case
where the rate of river incision is obviously greater than that of downward karstification. In
such instances, the high-hanging subterranean streams or karst springs will, in most cases, cascade on the valley wall. Subterranean streams or karst springs in such cases may always be incapable of forming a unified free water table due to the heterogeneity of karstification. They are
possessed of a definite system of groundwater each, governing their respective and independent
domain of solution and robbing each other, and form a local base level of karstification(also
named temporary base level or local base level) respectively. The river bottom referred to as a
regional base level is merely the one of its kind. It is just the recent and local base level of
karstification, in different forms and at different heights, that should be considered in human's
engineering activities.

Fig. 5 shows a profile of Ma-bie river valley, one of the tributaries of Nan-pan-jiang River
in Xin-yi County of Guizhou Province. The basement of the cone-depression karst massif, formed in
Shiling Stage, 320 km^2 in area, has been incised by Nan-pan-jiang River and formed a gorge during
an intensive uplift in the subsequent Nan-pan-jiang Stage. There are 12 outlets of karstwater
(with 5 subterranean stream included) on the east wall of the valley in a distance of 13 km. The
three outlets K-7, K-9 and K-12 are 20-25 m above the stream level with the largest discharge in
K-7. The subterranean stream K-7 outflows in three cascades of separate height into the river:
the upmost with a flow of 150 L/s and 25 m above the stream level; the middle with a flow of
10 L/s and 10 m above the stream and the lowest with a flow of 4.5 L/s and 1 m above the stream.
They are all perennial. This fact shows that the three discharging points of karstwater, with the
strongly solutionized sitting anomalously above the weak one though the same cross-section as they
are in, are all hanging above the stream level but incapable of conforming with the main valley
as a base level of solution. The shortest distance between the two of the 5 outlets of karstwater
is merely 250 m, but corresponding difference of elevation is up to 30 m. All these phenomena
demonstrate the heterogeneity of karst evolution, the absence of a unified free water table and
the independent formation of respective local base level governing a certain extent of solution.
Although Ma-bie River is the major one of the area, yet it gives no restraint to the karst evolution in the area as a whole. Localities, both in plan and profile, for the three railway tunnels
(the Hong-bu-jin Tunnel, the Hua-di-wan Tunnel and the Wang-jia-dang Tunnel) to run through in
the vertical vadose zone were the favorable choice after a detailed analysis of karst base level
of different sections of the river valley.

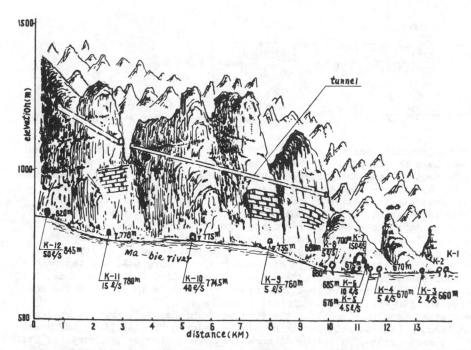

Fig. 5 The profile from
Ler to Ferry Zhao-jia-du

4. Base level of a covered karst valley

Karst hollows often filled up with loose materials can still be found within certain depth under the river bottom of a wide, flat and alluvial-deposited karst valley. In case of such instance, it is apt to take simplifiedly the stream level as a base level of karstification. As a matter of fact, it is in such cases not only the portion under the stream level that has been scarcely affected by solution on account of the feebleness of groundwater runoff due to the filling of hollows and the confining isolation effect of overlying soil, but also the soluble rocks above the stream level on both banks that have been subjected to a faint or null solution due to sedimental cover on the ground. In this way, where should be the base level then?

Analyses of core test information from many bridge sites have revealed that the karst under stream level, as a rule, belongs to the early karst or the palaeokarst before Cenozoic Era. Palaeokarst has been, in general, filled with Quaternary deposits and come in senitity to ceasedown, in the course of a regional Earth crust subsidence and consequent up-lift of the karst base levels. The present stream level may be an early base level after the event, but now has not a bit of significance as a base level for the underlain karst.

The base level of karstification for rock mass above stream level may possibly be a branch valley with discharging ability, a subterranean stream or a karst spring.

By the preceeding analysis, it is not difficult to understand that the mistaking of present stream level as a karst base level will inevitably lead to a mistaking of the underwater karst as dissolved fissures in deep-oozing zone and a consequent negligence to the severity of solution, as a result, some incorrect projects can be arised. Furthermore, by the recognition of the stream level as a karst base level of karstification, a horizontal-flow zone in the profile for karst above stream level can be set too low and causing possibly a damage for the project from an unexpected attack of groundwater.

Fig. 6 shows a major bridge across Xia-dong River in Guangxi Autonomous Region. The river bed is 500-800 m in breadth and overlaid by sand and gravel 5-10 m in thickness, with a perennial water level +15.9 m. In case the present water table of Xia-dong River is supposed to be the karst base level and still dominates the karstification underneath, there can only the karst fissures exist below the river bed. But data after five runs of exploratary boring have convinced that caverns were considerably developed in the underlain limestones. None of the boreholes hasn't encountered a chain of caves. The largest of the detected cavern is 12 m in height, extending to 21 m below the sea level and completely filled with gravels, sand, clay and none of block stones or detritus from collapse. The bridge site is 40 km to the shore of gulf Bai-bu-wan with nonsoluble rock formation lying between. There covered a thicker alluvium on the plain where the bridge sites. Evidently, no possibility of direct control on karstification therein from sea level can be expected. Hence this type of karst base level in a covered valley can only be the corresponding and earlier stream level which has been changed and transformed into the present stream level. However, it can give no restraint to the scale and depth of karstification below the level, hence plays not as a karst base level.

5. Base level for deep karst below valley

The karst below a recent base of surface water drainage is usually called deep karst. A deep karst may be formed in ancient time or developed in recent time, and ancient karst (or paleo karst)

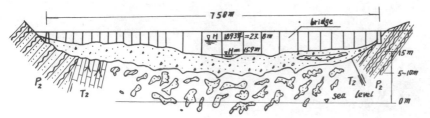

Fig. 6 Schematic geologic profile of Xia-dong-he bridge

may also be either deep-buried or outcropped and subjected to a series of sequential reformations.

No discussion will be given here for deep karsts whether developed along an abyssal fault or caused by deep circulation between hot and cold water, by hydrothermal process or influences from sulfide body; nor is given to the base level or a buried paleo-karst. The discussion will focus merely on the base level for deep karst relative to the overlying valley in hydraulic.

In the region in which deep karst exist, soluble rocks are mostly naked no matter the breadth of the valley. Surface runoff in this instance will percolate downward through steep bedding planes and faults to cause karstification within a certain depth. A part of conduits, unchoked during karstification, permit a depthwise solution to cause an invert-siphoned circulation, of which the final depth of solution is determined by the difference of pressure head and tectonic conditions as well as the floor position of the soluble rock. Theoretically, the ultimate depth of karstification may be up to the floor of soluble rock formation; but we need to identify the recent base level of karstification for solution of engineering problems.

Fig. 7 shows the double-tracked Nan-ling Tunnel on the Beijing-Guangzhou Railway, 6.1 km long, passing through the Mt. Nan-ling, a latitudinally striking divide between Yangtze River and Pearl River. The overburden of the tunnel is 35-60 m but up to more than 100 m near by the portals. The tunnel passes through an anticlinorium with $C d^2$ sandstone and shale by both portals and C_1d^1, C_1y^1 limestone throughout the middle. Runoff is considerable; shallow-sited caverns and subterranean streams exist in everywhere. Lian-qi Creek is perennial. Depth of karstification is within 40 m below the river bottom. Many surveys and researches have been made to find out the karst base level of the region in order to determine the locality of tunnel with reference to the vertical zonation and choose the method for gushing volume computation. Methods for tunnel gushing volume computation have been proposed such as the depressions-vadose-amount method for tunnels in vertical vadose zone while the water-amount-balance method for tunnels thereunder. A through study of background informations about tectonics and drainage has led to the conclusion that none of the aforementioned types of karst base level could exist in this region. Several caverns have been found after tunnelling, and causing large amount of gushing with mud and sand, of which artesian condition exists individually. There have occurred 30 odd sinks on the ground and the stream current has broken for several times with gushing of karstwater into the tunnel. The recovering of groundwater table to a considerable height after tunnel back-filling and grouting has testified the hydraulic connection between surface stream and covered karst thereunder. It seems thereby to suggest that river bottom in a karst region may be regarded as recent base level of karstification in case the surface stream has a hydraulic connection with karst thereunder, and the occurrence of Lian-qi Creek is one of such instances. The karst under the valley is therefore belonged to deep-oozing zone or invert-siphoned circulating zone under very base level.

With reference to the five models mentioned above, no significant trouble would be encountered in the searching of the form of existence and elevation of a karst base level in advance of the analysis for depth of karst evolution and vertical zonation. Based on such an analysis, the suitable

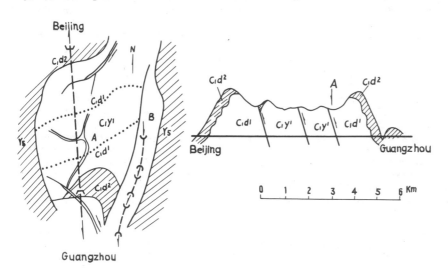

A—Lian-qi River
B—subterranean system of stream

Fig.7 Sketch of Nan-ling Tunnel on Beijing-Guangzhou Railway (plan and profile)

plan and elevation of a structure can be determined and the karst engineering geologic and hydrogeologic settings can be reliably evaluated, and hence some proper countermeasures can be taken. This is of significant importance in karst engineering geology studies, especially in the karst area in southern China where karstwater of conduit flow predominates. As for sea level and phreatic surface taken as the karst base level in coastal region and part of Northern China respectively, though not discussed here, are also the recent base level of karstification.

REFERENCES

Davis, W.M. 1930. Genesis of limestone caverns(Chinese). P.10-17. Nanjing University.
Karst research group, Institute of gedogy, Academia Sinica. 1979. Chinese karst research(Chinese). P. 132-137. Beijing: Science Press.
Ren Meie Lin Zhenzhong. 1983. The concept of karstology(Chinese). P. 161-163. Beijing: Commercial Press.
Rizhikov, D.V. 1956. Characteristics of karst and basic rules of karstification(Chinese version). P. 101-106. Beijing: Geology Press.
The 2nd railway surveying Designing institute. 1984. Karst engineering geology(Chinese). P. 44-62. Beijing: Railway Publishing House.

Geochemical investigation – Casa de Piedra Dam, Colorado River, Argentina

Etudes géochimiques sur le barrage de Casa de Piedra – Rivière Colorado, Argentine

Juan Carlos Ferrero, *Universidad de Córdoba, Argentina*
Alberto Carrillo, *Geotechnical Department, Ente Ejecutivo Presa de Embalse Casa de Piedra, Argentina*
Oscar Vardé, *Vardé y Asociados, Buenos Aires, Argentina*
Eduardo Capdevila, *Geotechnical Department, IATASA, Buenos Aires, Argentina*

ABSTRACT: With the purpose of getting to know with a sufficient approximation quantities of present salts within foundation materials for the dam a geochemical investigation program was elaborated and performed. In this case, special attention was given to gypsum presence, which had been detected on test pits and core samples from drillings in the terraces of both banks, and to the eventual existence of anhydrite. It was intended too to dispose information regarding its solubility and to obtain preciseness in respect of its mineralogical composition.

The investigation program included the following tasks:
Samples of crystals were taken, of materials close to the crystals, and of mixed foundation material. Samples were sent to a chemical laboratory, to a soil laboratory and to a petrographical laboratory. Water from Colorado River and distilled water were utilized on tests and determinations. In some tests water temperature was made become variable. Chemical compositions were evaluated in relation to soil dispersiveness.

Based on results obtained a conclusion was reached that, if computed infiltration flows of Casa de Piedra Project (less than 5 liters/min per meter of length of dam) is verifyed, a salt solubilization that can endanger the works within its lifetime is not to be expected.

NOTE

Present investigation exposed as following was authorized and financed by Ente Ejecutivo Casa de Piedra as owner of the Works. Authors of this report belong to different sectors acting on the Project and this presentation is made on behalf of all those people, organismis or companies related to this research, with the only purpose of diffusing the technical conclutions obtained.

1 INTRODUCTION

The purpose of this paper is to report the Geochemical Investigation Program that has been perfomed to determine the characteristics, under the point of view of its chemical behaviour, of materials constituting foundation of Casa de Piedra Dam.

The dam, integrating part of works that constitute Casa de Piedra Project over the Colorado River on Argentine Republic, has a length on its crest of 10.600 m, a maximum height from its foundation of 54 m in the river gorge and a volume of material fill about almost 11.000.000 m^3, being under construction at present.

During previons stages of the project, in the different geological-geotechnical explorations campaigns was observed evident presence of gypsum crystals and a high content of potentially soluble material in different degrees, in addition to suspecting eventual existence of anhydrite.

With the purpose of determining and evaluating these aspects and its incidence in time over works, this investigation was performed, which consisted in material and water sampling, soil laboratory testing, chemical laboratory testing, petrographical-mineralogical laboratory analyses.

Because of the length of the dam and variable geological characteristics of subsoil where dam is founded, to obtain representativity on the set of samples taken from borrow pits and drillings distributed on dam axis or its boundaries, on La Pampa and Río Negro terraces, was a duty of special significance.

To ensure representativity over the set of samples and the real behaviour of soils and waters, samples were taken at different levels of crystals, surrounding materials of crystals, mixed foundation materials (integral samples) y River Colorado water; at different periods.

These samples were sent, for its testing, to specialited laboratories.

Analysis and results evalluation of test performed in these laboratories, of known reputation and trajectories, and conclusions are described hence forth.

2 INVESTIGATION PROGRAM

2.1 Tests and of techniques types employed

2.1.1 Soil Laboratory.
Materials were defined by means of Atterberg limits determination, identification of soils tests and their classification under Unified System. On different samples, Pinhole tests using distilled water at 5-15 and 30ºC (with previous determination of pH) and Rio Colorado water at 15ºC were performed.

2.1.2 Chemical laboratory.
Chemical analyses and determinations were made, on materials and waters, aplying the following techniques.

2.1.2.1 Materials

a) They were described macroscopically, bringing up a rapid idea of its outstanding characteristics.
b) Material was dried at ambient air, at fairly normal ventilation, pollution, etc condition being the temperature in no case above 40ºC, with periodical removal of material. Then, material was ground as a previous step to appliance of technique described furthermore. Procedure followed up to here was with the purpose of not altering its mineral composition, provided that gypsum presence may produce erratic values.
c) Saturation extracts were prepared, obtained from material with distilled water at 15ºC (with similar moisture than liquid limid) paste, intimately mixed and left in contact during 48 hours with periodical removal. Once this time was over, liquid was extracted from the paste (called from here on saturation extract). Upon it were performed, previous a quantified dilution, chemical analyses of main ions present, such as calcium, magnesium, sodium, potassium, chlorides, sulphates, bicarbonates and nitrates. pH were as well determined and electrical conductivity expressed in micromhos/cm corrected to 25ºC.
From these results, removal of different ions was determined and its geochemical behaviour, so enabling to perform as well the principal hypothetical combinations.
d) Samples having a material-water ratio (1:50) were prepared on adequate recipients, over which strong agitations were performed during 48 hours with water at 15ºC temperature. Over the filtered floating liquid similar chemical determinations were perfomed than pointed in c). This was done with the purpose of observing variations in function of dissolutions with different solid-liquid ratio.
e) Soluble gypsum in percent (in crystals and surrounding material) was determined applying the Acetone Technique, with distilled water at 15ºC in sufficient and known quantity to dissolve present gypsum. This technique, same as in point c), corresponds to the Sodic Saline Soils, Manual Nº 60 of U.S.A. Salinity Laboratory. Result thus obtained stands for rigorous dissolution conditions for material is treated with enough distilled water and a vigorous agitation, aiming to mantain in contact the specific surface of gypsum crystals with dissolution water.
f) On a portion of material for analyses, moisture content was determined, to enable correction for material dried at 105ºC.

2.1.2.2 Liquids

a) Samples of Colorado River water taken in the Works area were analysed, upon which analyses of calcium, magnesium, potassium, sodium, chlorydes, sulphates, bicarbonates and iron ions were performed. pH and electrical conductivity, expressed in micromhos/cm, corrected, were determined, in order to evaluate superficial hydrogeochemical conditions and balance related to site materials, as well as obtaining values of liquid to be used on Pinhole test for enablying evaluations related to its flow, and its activity and equilibrium behaviour.
b) pH of distilled water used on Pinhole test was determined with the purpose of evaluating its flow.
c) Analyses of principal ions as well as solubilized gypsum on Pinhole flows and also its pH and electrical conductivity were performed, for evaluating equilibrium.

2.1.3 Petrographical-Mineralogical Laboratory

Petrographical descriptions of samples were carried out, with a microscopic description of material surrounding crystals. In the strictly mineralogical part description of gypsum crystals, anhydrate, glauconite and any other mineral having dissolution possibilities were required, at the same time physical properties (clivage, etc.) and fissurations or microfissures in general (corrosion) were inquired.

2.2 Gypsum - Generalities

Gypsum is an dihydrated calcium sulphate with diverse forms of appeerence, ranging from perfect crystals situated on large dimension monoclinal system to very fine dusty state can show up with flails, with diverse clivages usually with predominance of one of them (010) very perfect; vitreous or mother-of-pearl like shine, transparent or opaque, uncoloured or white and in some coses masked by other materials (pelites, iron, etc.) colours. Can also show up in lenticular, fibrous, compact, lumpy, scaled, rosetted, anhydrate amorphus forms or mixed with diverse chemical salts, generally soluble in water (i .e. halyte, glauconite, etc.). At 60ºC temperature begins loosing its composition water and with increase of temperature turns to intermediate states (very solubles) for finally transforming into anhydrate.

Solubility of selenitous variety of gypsum is in the average range of 0,22%, in saturated water at 33ºC, varying this value in function of form of appereance of mineral, activity and density of dissolution water (i.e. a sodium chloryded water with a 130 gr/l NaC1 content can solubilize up to 0,73%) (See graphics 1 and 2).

Gypsum has, in front of pelitic (fine) materials, a coagulant ability (due to calcium cation as such on one hand and to interchange of calcium ions to sodium in the other), but it is also characteristic that if water is renewed in a fairly rapid way as an effect of fissurations, etc, possibility exists that may develop an effect known as gypsum Karstification, which represent rapid disolutions, corrosion and fine particles dragging. (This would aggravate if permanent flow of renewed water into gypsiferous materials with important volumes of water exists).

2.3 Colorado River Water

From three samples analysed, even though saline content of one of then is remarkably lower, equilibrium of dissolved salts within water among then and in respect to natural ambitus of the site is fairly regulated, and one could state that, by means of geochemical computations, such waters posses at its maximum possibility a dissolving capacity lesser than distilled water, and that such capacity oscillates in the range of 30% with respect to distilled water, since predominating ions are calcium, chlorydes, sulphates and sodium, considering the samples as being within calcium sulphated chloryded to slightly sodio groups. (See graphic Nº 3, Salt contribution and modifying phenomenas by Davis and De wiest - 1966 and graphics Nº 4 and 5). Presence of solubilized gypsum on Colorado River waters was also detected which enhance thesis of 70% less agressiveness to mediums than distilled water.

If were add to this ionic interchange effect where water has to transfer calcium and captivate sodium, fundamentally on pelitic materials (fines). floculation capacity of materials is increased. This floculation capacity is of very slow development in function of perviousness of material and solid liquid contact.

2.4 Gypsum and anhydrite on materials of present study

Presence of anhydrite is scarce and was only detected on three samples having non significant values.

About gypsum one can say that three more or less defined varieties were detected.

2.4.1 Selenitous type, well crystallized with predominant clivage (010), closed almost in its totality, not showing signs of corrosion except in Station 10.200, between 2,70 to 3,70 m, where a sandy gypsolite shows cavities suggesting corrosion processes, but in this same material where are observed 0,42% of total soluble salts, appears a soluble gypsum value of 1,12%, which would indicate an erratic value in comparison to rest of analyses. Probably this is due to heterogeneity of total material on one hand, and a localized increase of solubility on tested sample thenomenum which induces the thought that a value of 0,30% to the whole for soluble gypsum on this sample must be taken and in general in the order of 0,18%.

2.4.2 Rosetted gypsum which is the one presenting a greater contact surface and better corrosion conditions, and hence a small dissollution increase, but this material is scarce in function of processed samples (average solubility 0,32%).

2.4.3 Fine dispersed gypsum in the pelitic (fine) was on cementing granular materials, this type of gypsum is the one presenting greatest solubility because of its ample contact surface and probably precarious crystallization, being very conmon as granular cement, representing karstification capability which is not observed on former varieties, specially of bearing material presents permeability higher than 7 liters/met/min (Average solubility 0,50%).

615

In all cases values represents extreme solubility tests, for river water, which is the most likely one to be in contact with materials, enters with 30% of mentioned values, and we have:

Selenitous gypsum: dissolves at a rate of 0,5 gr/l maximum in optimum solubility conditions at 30ºC.

Rosetted gypsum: dissolves at a rate of 0,96 gr/l in optimum solubility conditions at 30ºC.

Disperssed gypsum: dissolves at a rate of 1,5 gr/l in optimum solubility conditions at 30ºC.

It have to be taken into account that originally tested material was ground (passing sieve Nº 10º, treated with distilled water at 30ºc temperature with strong agitation (which enables maximum possible solubility). These are initial values and they diminish until, water meets ballance that would occur in the short term, in the estimated order of three months, if we take into account permeability in the order of 5 litter/meter/minute and average percentages of present gypsum, at a 15ºC temperature.

2.5 Lists with test results

These lists are included, as an annex, at the end of the text, in correspondence with stations.

3 CONCLUDING REMARKS

In function of evalluated results on the processed samples that can be seen on the annexed drawings the following conclution can be drawn:

3.1 Total soluble salt content on distilled water obtained from saturation extract is, in average, 0,65%.

3.2 Foundation materials under study do not show chemical signs of disperssiveness

3.3 Geochemical ballance observable in respect to tested samples is fairly regulated, except in some cases as following:

a) Granular materials in general.
b) Gypsolites (or gypsite) St. 1400 (4.20 - 5.05 m); St. 7650.
c) Dolomites. St. 6100 (5.50 - 6.50); St. 8300 (3.00 - 3.90).
d) Gypsiferous hematitic limestone - St. 2900 (5.20 - 5.40 m).
e) Marls and silttones. St. 3825 (6.40 - 7.20); St. 4000 (5.10 - 6.10 m).

3.4 Soluble material (gypsum, etc.), having an addecuate control of permeability and fissuration, will not present significant dissolution, corrosion and karstification problems, except on granular materials, down to an average depth in the order of 5.00 m, where gypsum cementation appears and at Station 10.200 (2.70 - 3.70 m), sandy gypsite, where infiltration flows will be controlled by means of proper engineering design.

3.5 Even though in the beggining there are certain possibililies of solubility, those are regulated by Colorado River waters characteristics, diminishing progressively in time until reaching solid-liquid equilibrium is met, which is estimated to become within three months of solid-liquid contact.

3.6 Iron oxide and its colloidal migration was analyzed, not showing values that may presume inconveniencies.

3.7 Regarding soluble salts and possibility of filter zones silting up it is estimated that no difficulties will become due to eventual precipitations on filter zones by water evaporation, a practically imposible case because a fairly permanent flow is considered and if some salt should precipitate, a greater water flow would again dissolve it because of its own disposition.

3.8 Saline content is heterogeneous, horizontally as much as vertically and same is gypsum content.

3.9 Gypsum in its diverse states, once solid-liquid equilibrium is reached, will remain almost unaltered considering lifetime of the project.

3.10 Finally, based on results of complete or thorough geochemical research, one can express that if design infiltrations flows are met, which for the case of Casa de Piedra Project are in the order of less than 5.00 liters/meter/minute of dam, it is not likely to forecast a salt solubilization that may indanger the dam within its lifetime.

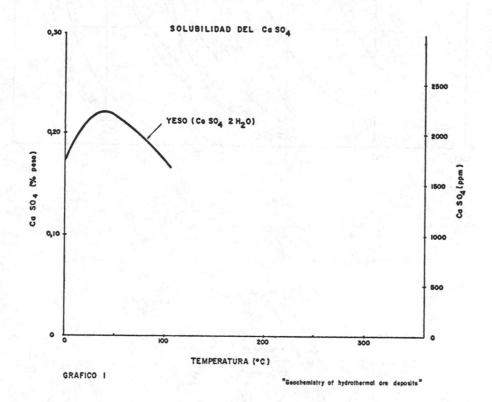

SOLUBILIDAD DEL $CaSO_4$

GRAFICO I

"Geochemistry of hydrothermal ore deposits"

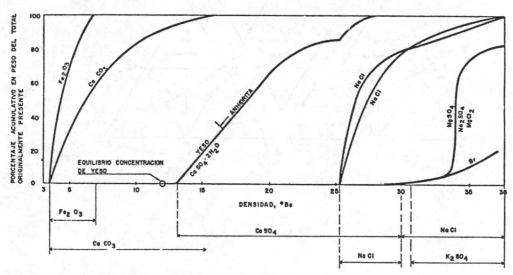

Depósito de sales durante la concentración del agua de mar

M.G. VENKATESH MANNAR 1982 _ (H.L. BRADLEY 1983)
GRAFICO 2

APORTE DE SALES Y FENOMENOS MODIFICADORES

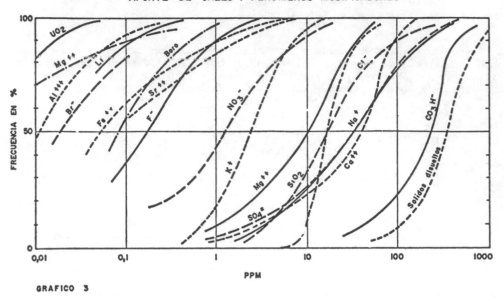

GRAFICO 3

TRIANGULO DE PIPER

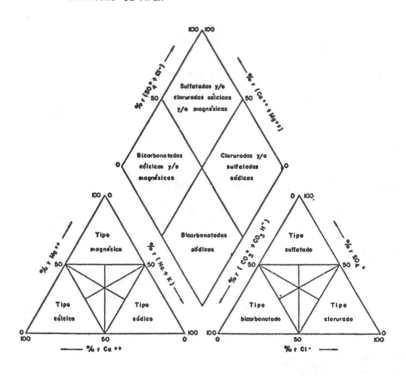

GRAFICO 4

618

CLASIFICACION GEOQUIMICA
AGUAS RIO COLORADO
(TRIANGULO DE PIPER)

AGUA DEL RIO COLORADO

■ MES ABRIL
● MES MAYO
+ MES JUNIO

ABRIL : CLORURADA
SULFATADA
CALCIFICADA

MAYO : CLORURADA
SULFATADA
LEVEMENTE SODICA
CALCICA

JUNIO : SULFATADA
CLORURADA
CALCICA

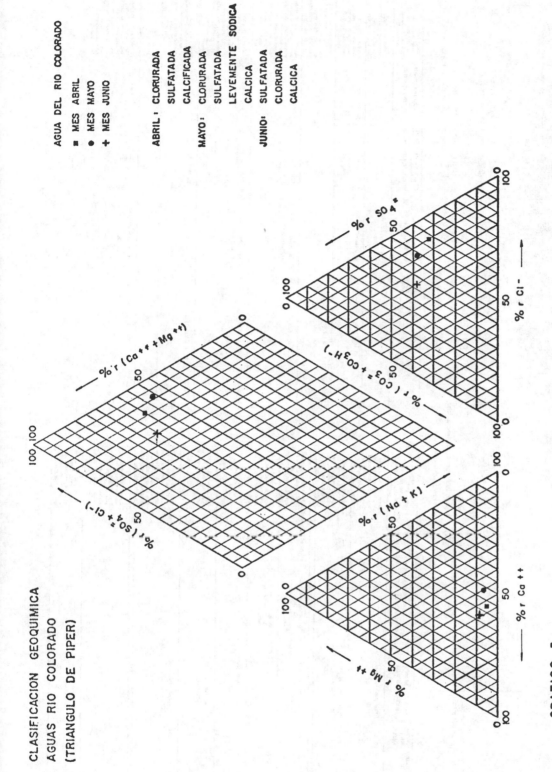

GRAFICO 5

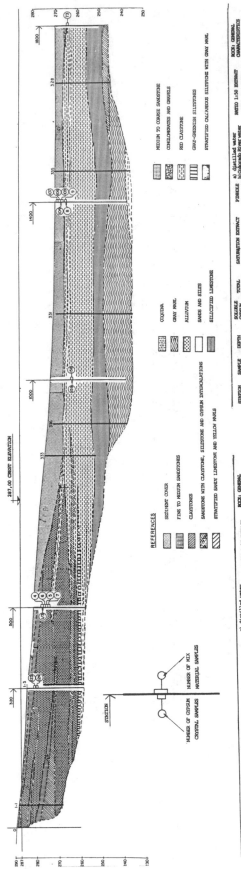

207,00 CREST ELEVATION

REFERENCES

- SEDIMENT COVER
- FINE TO MEDIUM SANDSTONES
- CLAYSTONES
- SANDSTONE WITH CLAYSTONE, SILTSTONE AND GYPSUM INTERCALATIONS
- STRATIFIED SANDY LIMESTONE AND YELLOW MARLS
- COQUINA
- GRAY MARL
- ALLUVIUM
- SANDS AND SILTS
- SILICIFIED LIMESTONE
- MEDIUM TO COARSE SANDSTONE AND GRAVELS
- CONGLOMERATES AND GRAVELS
- RED CLAYSTONE
- GRAY-GREENISH SILTSTONES
- STRATIFIED CALCAREOUS SILTSTONE WITH GRAY MARL

STATION

NUMBER OF MIX MATERIAL SAMPLES

NUMBER OF GYPSUM CRYSTAL SAMPLES

STATION N°	SAMPLE N°	DEPTH m	SOLUBLE GYPSUM %	TOTAL SOLUBLE SALTS %	SATURATION EXTRACT PREDOMINATING SALTS	PINHOLE a) distilled water b) Colorado River water PREDOMINATING SALTS ON FLOW	RATIO 1:50 EXTRACT PREDOMINATING SALTS	ROCK GENERAL CHARACTERISTICS
1400	107	0,40 a 1,40	-	0,03	Sulphated-Chloryded Sodic-Calcic	-	-	Gravels with sandy calcareous base, grayey brownish coloured. Source gypsum content.
1400	108	1,40 a 2,40	-	0,12	Chloryded-Sulphated Sodic-Calcic	-	-	Gravels with gray brown coloured sandy base. Gypsum content oscillates about 1%, in crystals.
1400	104	2,40 a 2,50	0,37	-	-	-	-	Gypsum crystals in compact plastic. Gypsum is found as crystals and as a very fine grain content in approximately total of 10%.
1400	109	2,40 a 3,55	-	0,19	Chloryded-Sulphated Sodic-Calcic	-	-	Gravels from vulcanites having scarce matrix, grayey brown coloured. Some source rolling thin semirows. Gypsum total content oscillates about 3%, with some crystals.
1400	9	4,20 a 5,05	0,18	1,01	Chloryded-Sulphated Sodic-Calcic	a)15%ClNa, Chloryded-Bicarbonated-Sodic-Calcic b)15%ClNa, Chloryded-Calcic-Sodic	Sulphated-Chloryded Sodic-Calcic	Clayey gypsolite, brownish red coloured. With a total gypsum content in the order of 50%, whereas predominate on it crystalline varieties of gypsum, there is dispersed gypsum.
1400	8	4,80 a 5,05	0,15	-	-	a) 15%Cl NO₃	-	Gypsum crystals surrounding gypsiferous siltstone, generally reddish coloured. Gypsum is mostly well crystallized.
1800	73	4,90 a 5,90	-	0,34	Sulphated-Chloryded Sodic-Calcic	-	-	Gravels with sandy base, fine sized gravels reddish brown coloured. It microscopic gypsum presence and anhydrite was observed in percentages of it well crystallized without corrosion.

STATION N°	SAMPLE N°	DEPTH m	SOLUBLE GYPSUM %	TOTAL SOLUBLE SALTS %	SATURATION EXTRACT PREDOMINATING SALTS	PINHOLE a) distilled water b) Colorado River water PREDOMINATING SALTS ON FLOW	RATIO 1:50 EXTRACT PREDOMINATING SALTS	ROCK GENERAL CHARACTERISTICS
320	123	3,00 a 3,90	-	0,36	Chloryded-Sulphated Sodic-Calcic	-	-	Limontic sandstone, little cohemence and scarce stratification. Gypsum not detected. Yellow gold planed colour.
320	124	4,00 a 5,00	-	0,13	Sulphated-Chloryded Sodic-Calcic	-	-	Compact sandstone, hardly cohesive light brownish yellow, few gustue of calcite. Having gypsum (R/3%) in crystals.
500	4	3,20 a 4,20	0,08	0,74	Sulphated-Chloryded Sodic-Calcic	a)15%ClNa, NO₃, Bicarbonated-Chloryded-Sodic-Calcic b)15%ClNa, Sulphated-Chloryded-Calcic-Sodic	Sulphated-Chloryded Sodic-Calcic	Calcareous siltstone bright greenish brown coloured, 10% of well crystallised rosetted gypsum.
500	1/3	3,15 a 4,10	0,08	-	-	a)15%ClNa, Bicarbonated-Chloryded-Calcic-Sodic	-	Gypsum crystals in calcareous siltstone, dark greenish gray coloured. Holds Gypsum (1 to 2%).
500	6	4,20 a 5,20	-	0,77	Sulphated-Chloryded Sodic-Calcic	-	-	Grey greenish coloured claystone. Gypsum is not observed at microscope. Very fine grained rock.
500	5	5,20 a 6,20	-	0,57	Sulphated-Chloryded Sodic-Calcic	-	-	Gypsiferous limonitic sandstone, bright to dark yellowish greenish coloured. Well crystallised gypsum at 35% contents.
500	7	6,20 a 7,20	-	0,94	Chloryded-Sulphated Sodic-Calcic	-	-	Marly dolomite; yellowish brown to reddish brown coloured, gypsum is not observed at microscope almost fine material but in certainly it enters in low proportions.
1000	159	9,75 a 9,90	-	-	-	-	-	Crystal surrounding material: Hematitic silt/Sandstone. Gypsum is found it well crystallised at it proportions. Deep red as a whole.
1000	158	9,00 a 10,00	-	0,75	Chloryded-Sulphated Sodic-Calcic	-	-	Clayey micritic limestone, reddish to greenish coloured with alternations. Source gypsum (less than 0,5%).

PLATE 1

620

287.00 CREST ELEVATION

REFERENCES: SEE PLATE 1

STATION N°	SAMPLE N°	DEPTH m	SOLUBLE GYPSUM %	TOTAL SOLUBLE SALTS %	SATURATION EXTRACT PREDOMINATING SALTS	PINHOLE a) distilled water b) Colorado River water PREDOMINATING SALTS ON FLOW	RATIO 1:50 EXTRACT PREDOMINATING SALTS	ROCK GENERAL CHARACTERISTICS (in the case of crystals, surrounding material)
2900	110	1,00 a 2,00	—	0,07	Sulphated-Chloryded Calcic-Sodic	—	—	Gravels with sandy calcareous base. Gypsum oscillates in the order of 3% being very well crystallised (spear point).
2900	111	2,00 a 3,00	—	0,10	Sulphated-Chloryded Calcic-Sodic	—	—	Gravels with gypsiferous sandy base. Gypsum oscillates in the order of 5% with source presence of gypsum adhered to coarser materials.
2900	103	3,00 a 3,15	0,61	—	—	—	—	Gypsum crystals on gypsiferous sandy conglomerate. Gypsum is well crystallised in general, but in a lesser proportion it exists as crystallised cement development and as matrix too.
2900	112	3,00 a 4,00	—	0,10	Sulphated-Chloryded Calcic-Sodic	—	—	Gravels with sandy base. Gypsum is well crystallised in general with source adherence of it to rolling stones. In total it does not exceed 1%.
2900	10/11	5,20 a 5,40	0,54	—	—	a)- 5%ClNa₂-Sulphated-Bicarbonated-Calcic-Sodic a)-15%ClNa₂-Sulphated-Chloryded-Calcic-Sodic a)-30%ClNa₂-Sulphated-Chloryded-Sodic	—	Gypsum crystals on medium brown coloured gypsiferous hematitic limestone. Gypsum content is in the order of 30% with massive gypsum presence, but in its major proportion it is well crystallised.
2900	12	5,30 a 6,30	0,20	0,25	Sulphated-Chloryded Calcic-Sodic	a)15%ClNa₂-Bicarbonated-Chloryded-Calcic-Sodic b)15%ClNa₂-Bicarbonated-Calcic-Sodic	Sulphated-Bicarbonated-Calcic-Sodic	Hematitic marl, deep red coloured with a gypsum content well crystallised of 20%.

STATION N°	SAMPLE N°	DEPTH m	SOLUBLE GYPSUM %	TOTAL SOLUBLE SALTS %	SATURATION EXTRACT PREDOMINATING SALTS	PINHOLE a) distilled water b) Colorado River water PREDOMINATING SALTS ON FLOW	RATIO 1:50 EXTRACT PREDOMINATING SALTS	ROCK GENERAL CHARACTERISTICS (in the case of crystals, surrounding material)
2000	15	5,70 a 6,80	0,08	—	—	—	—	Gypsum crystals surrounding the crystals we have siltstone and clayey siltstone, coloured ranging greyey green to greyey bright brown. Gypsum upon totality represents 5% and is well crystallised.
2000	14	8,30 a 8,55	0,15	—	—	—	—	Gypsum crystals on micritic limestone generally bright red coloured. Gypsum in general is well crystallised and showing no signs of alteration.
2000	13	9,50 a 10,35	—	0,36	Sulphated-Chloryded Sodic-Calcic	—	—	Micritic marly limestone, deep reddish brown coloured. Gypsum is not conserved at microscope.
2000	16	10,50 a 11,50	—	0,35	Sulphated-Chloryded Sodic-Calcic	—	—	Ferruginous marl, bright reddish brown coloured. Gypsum presence not noticeable at microscope.
2600	74	7,00 a 8,00	—	0,61	Sulphated-Chloryded Sodic-Calcic	—	—	Hematitic mudstone clayey siltstone, not very dense intense red coloured.

PLATE 2

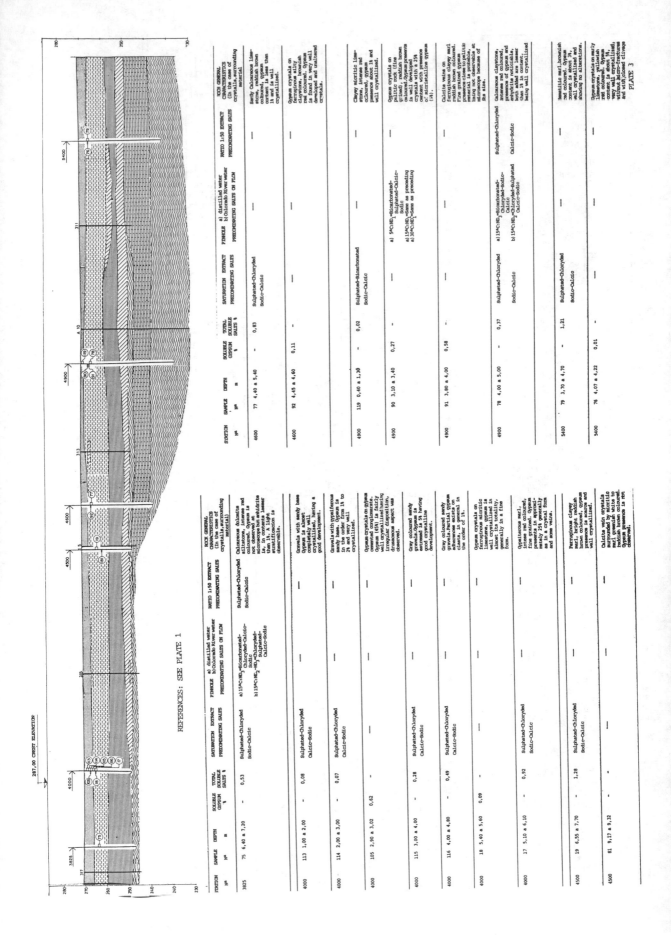

REFERENCES: SEE PLATE 1

PLATE 3

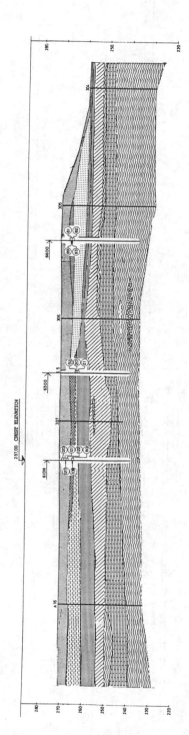

237.00 CREST ELEVATION

REFERENCES: SEE PLATE 1

STATION N°	SAMPLE N°	DEPTH m	SOLUBLE GYPSUM %	TOTAL SOLUBLE SALTS %	SATURATION EXTRACT PREDOMINATING SALTS	PINKLE a) distilled water b) Colorado River water PREDOMINATING SALTS ON FLOW	RATIO 1:50 EXTRACT PREDOMINATING SALTS	ROCK: GENERAL CHARACTERISTICS (In the case of crystals, surrounding material)
6300	20/52	6,50 a 7,50	-	0,19	Sulphated-Chloryded Sodic-Calcic	—	—	Newly micrite limestone. Gypsum not detected at microscope.
6300	21	7,50 a 8,25	-	0,29	Sulphated-Chloryded Sodic-Calcic	—	—	Bright yellowish green coloured rock. Gypsum is entcased in the ords of 5% well crystallised and without corrosion.
6600	81	6,00 a 5,00	-	0,21	Sulphated-Chloryded Sodic-Calcic	—	—	Gypsiferous sandy limestone, yellowish green coloured. Gypsum content is 5%, within this percentage anhydrite is found with good crystallisation.
6600	89	4,65 a 4,77	0,03	-	—	—	—	Gypsum crystal in veins on litic sandstone, yellowish green coloured. Gypsum is observed very well crystallised in a 10% from total.
6600	82	5,00 a 6,00	-	0,22	Sulphated-Chloryded Sodic-Calcic	—	—	Calcareous sand, bright yellowish brown coloured. Gypsum content is low and very well crystallised.
6600	83	5,00 a 5,15	0,006	-	—	—	—	Little gypsum crystals on calcareous and micrite limestone, greyish green coloured. Gypsum is in proportion up to 2% and well crystallised.

PLATE 4

STATION N°	SAMPLE N°	DEPTH m	SOLUBLE GYPSUM %	TOTAL SOLUBLE SALTS %	SATURATION EXTRACT PREDOMINATING SALTS	PINKLE a) distilled water b) Colorado River water PREDOMINATING SALTS ON FLOW	RATIO 1:50 EXTRACT PREDOMINATING SALTS	ROCK: GENERAL CHARACTERISTICS (In the case of crystals, surrounding material)
6100	120	1,00 a 2,00	-	0,06	Sulphated - Chloryded Calcic - Soda	—	—	Gravels with sandy calcareous base. Gypsum not observed at microscope.
6100	121	2,00 a 3,00	-	0,05	Sulphated - Chloryded Calcic - Soda	—	—	Gravely calcareous sandstone, gypsum is scarce and very well crystallised.
6100	126	2,75 a 2,95	0,50	-	—	—	—	Gypsum crystals on gypsiferous conglomerate. Total content of gypsum is about 30% present as crystals, in fine granular form and as irregular masses.
6100	122	3,00 a 4,00	-	0,11	Sulphated - Chloryded Calcic - Soda	—	—	Gravel with sandy base. Gypsum not observed at microscope.
6100	80	5,50 a 6,50	-	0,63	Sulphated - Chloryded Sodic - Calcic	a)15°CtHN = Bicarbonated Chloryded Sodic - Calcic b)10°CtHN = Chloryded Sulphated Calcic - Sodic	Sulphated - Bicarbonated Sodic - Calcic	Clayey dolomite, bright yellowish to reddish green coloured, fragmented, high gypsum content at different states.
6100	100	5,85 a 6,00	0,007	-	—	—	—	Vein gypsum crystals on turbaceous claystone, bright greyish green coloured. Gypsum is entcased in 7% being very well crystallised.

623

187.00 CREST ELEVATION

REFERENCES: SEE PLATE 1

STATION Nº	SAMPLE Nº	DEPTH m	SOLUBLE GYPSUM %	TOTAL SOLUBLE SALTS %	SATURATION EXTRACT PREDOMINATING SALTS	FINKLE a) distilled water b) Colorado River water PREDOMINATING SALTS ON FILM	RATIO 1:50 EXTRACT PREDOMINATING SALTS	ROCK GENERAL CHARACTERISTICS (In the case of crystals, surrounding material)
7480	32 33 35	5,30 a 8,55	-	0,16	Chloryded-Sulphated Sodic-Calcic	—	—	Marly dolomitic limestone, bright brown coloured, presents some porosity and absence of gypsum.
7480	31/34 38/39 41	8,55 a 12,70	-	0,10	Sulphated-Chloryded Sodic-Calcic	—	—	Clayey marl, coloured ranging from yellowish brown to dark grey, with clear lamination. Gypsum is not observed at microscope.
7480	36	15,00 a 15,80	-	0,15	Chloryded-Sulphated Sodic-Calcic	—	—	Coquinoidous limestone bright brown coloured. Gypsum is not observed at microscope.
7480	37	27,00 a 28,00	-	0,39	Sulphated-Chloryded Sodic-Calcic	—	—	Calcareous claystone, dark greenish grey coloured,"with lamination. Gypsum was not observed at microscope. It is considered its probable source existence close to firm material.
7520	28/29	17,00 a 17,60	-	0,12	Chloryded-Sulphated Sodic-Calcic	—	—	Fossiliferous marl, bright yellowish grey coloured. Gypsum was not detected at microscope.
7520	27	25,00 a 26,00	-	0,43	Sulphated-Chloryded Sodic-Calcic	—	—	Marly limestone bright yellowish coloured, fine grained, scarce cristallised gypsum content.
7520	30	31,00 a 32,00	-	0,28	Sulphated-Chloryded Sodic-Calcic	—	—	Fossiliferous marl, bright yellowish green coloured, presence of fine filaments and lamination. Gypsum was not detected.
7590	24	5,05 a 6,60	-	-	—	—	—	Micritic to subspanitic limestone, bright yellowish brow coloured porous to cavernous structure. Gypsum was not observed.

STATION Nº	SAMPLE Nº	DEPTH m	SOLUBLE GYPSUM %	TOTAL SOLUBLE SALTS %	SATURATION EXTRACT PREDOMINATING SALTS	FINKLE a) distilled water b) Colorado River water PREDOMINATING SALTS ON FILM	RATIO 1:50 EXTRACT PREDOMINATING SALTS	ROCK GENERAL CHARACTERISTICS (In the case of crystals, surrounding material)
7590	25	6,60 a 7,90	-	0,10	Sulphated-Chloryded Sodic-Calcic	—	—	Dolomatic limestone, bright to dark brown coloured. Gypsum was not observed.
7590	22/23	10,30a11,78	-	0,08	Chloryded-Sulphated Sodic	—	—	Marl, yellowish brown coloured, grey lamination. Gypsum was not observed.
7590	26	16,00a16,80	-	0,16	Sulphated-Chloryded Sodic-Calcic	—	—	Coquinoidea limestone, bright yellowish brown coloured. Gypsum was not observed.
7650	44	1,60 a 2,60	-	0,38	Chloryded-Sulphated Sodic-Calcic	—	—	Gypsiferous sandy micritic limestone, whitish yellowish brown coloured. Gypsum content about 3%.
7650	47	2,40 a 2,75	0,29	-	—	—	—	Gypsum crystals on silty-clayey calcareous gypsilite, greenish yellowish brown coloured. Abundance of gypsum in diverse forms; crystals, dispersed and rosettes.
7650	45	2,60 a 3,60	-	0,44	Chloryded-Sulphated Sodic-Calcic	—	—	Gypsiferous sandy sandstone, bright yellowish brown coloured. Gypsum at 7%, well crystallised.
7650	46	2,75 a 2,95	0,34	-	—	—	—	Gypsum crystals on silty-clayey gypsilite, bright yellowish brown coloured. Abundant gypsum in diverse forms including fibrous gypsum.
7650	162	3,60 a 4,60	-	0,25	Sulphated-Chloryded Sodic-Calcic	—	—	Granule with clayey calcareous matrix, bright greenish yellowish brown coloured. Gypsum at 2%, well crystallised.
7650	163	4,60 a 5,60	-	0,16	Sulphated-Chloryded Sodic-Calcic	a)15°C1,D1-D2-Bicarbonated-Chloryded-Sodic-Calcic b)15°CinD2-Chloryded-Sodic-Calcic	Bicarbonated-Sulphated Sodic	Sandy micritic limestone, greenish brown coloured. Gypsum presence was not detected.
7650	160	4,70 a 4,90	0,003	-	—	a)5°CinD1-Bicarbonated-Sulphated Calcic-Sodic a)15°CinD1-Sulphated-Bicarbonated Calcic-Sodic b)30°CinD1-Bicarbonated-Sulphated Calcic-Sodic	—	Gypsum crystals on sandy micritic limestone. Gypsum very well crystallised without microstructures, closed cleavage.
7650	164	5,60 a 6,60	-	0,21	Sulphated-Chloryded Sodic-Calcic	—	—	Brecciose clayey-calcareous sand, greyish brown coloured. Gypsum at 8%, well crystallised.
7650	161	5,90 a 6,10	0,003	-	—	—	—	Gypsum crystals on silty-clayey sand, bright yellowish green coloured. There is no microscopic gypsum and megascopic gypsum is about 20% in veins and well crystallised.
7780	42	6,65 a 6,85	0,012	-	—	—	—	Gypsum crystals on a gypsiferous sandy lime stone to calcareous sandstone. Gypsum at 5% in average very well crystallised.
7780	43	8,65 a 9,90	-	0,53	Chloryded-Sulphated Sodic-Calcic	—	—	Clayey siltstone, bright greenish brown coloured Gypsum is 1% with well crystallised big specimens.
7780	40	11,05 a 12,30	-	0,26	Sulphated-Chloryded Sodic-Calcic	—	—	Clayey limestone, bright brownish yellow coloured Gypsum is 1%, with well crystallised small specimens.

PLATE 5

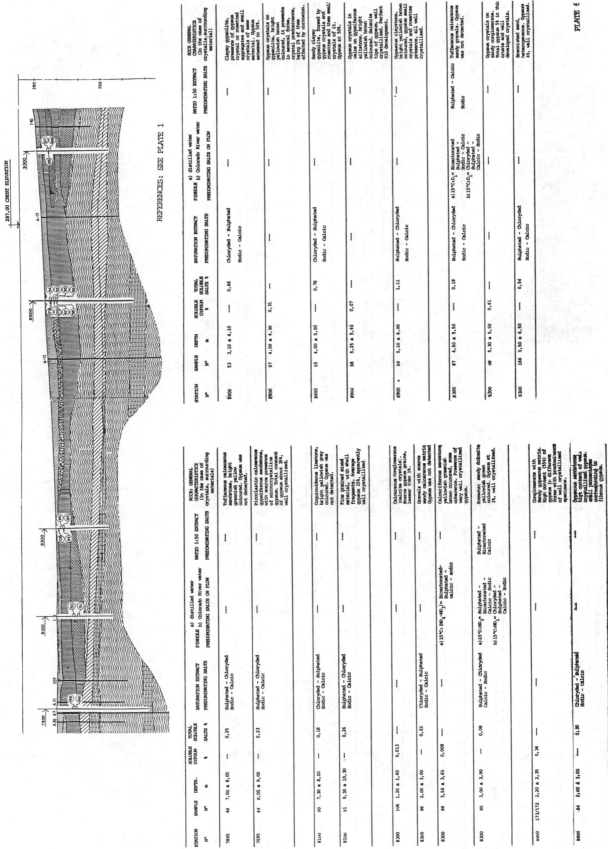

REFERENCES: SEE PLATE 1

PLATE 6

287,00 CREST ELEVATION

REFERENCES: SEE PLATE 1

PLATE 7

STATION N°	SAMPLE N°	DEPTH m	SOLUBLE GYPSUM %	TOTAL SOLUBLE SALTS %	SATURATION EXTRACT PREDOMINATING SALTS	FINKLE a) distilled water b) Colorado River water PREDOMINATING SALTS ON FLOW	RATIO 1:50 EXTRACT PREDOMINATING SALTS	ROCK GENERAL CHARACTERISTICS (in the case of crystals, the surrounding material)
9600	167	5,80 a 6,80	—	0,17	Sulphated-Chloryded Sodic-Calcic	—	—	Basaltic clayey silt/claystone, pelitic clayey, bright gray coloured. Scarce gypsum but on fine mass.
9600	168	6,00 a 6,25	0,03	—	—	—	—	Gypsum crystals on sandy-silty gypsolite, bright yellowish brown coloured. Gypsum 10% very well crystallised and without alteration.
9800	62	8,60 a 9,60	—	0,31	Sulphated-Chloryded Sodic	—	—	Sandy clayey siltstone greenish gray coloured fine grain. Scarce gypsum.
9800	63	10,30 a 11,20	—	0,07	Sulphated-Chloryded Sodic-	—	—	Sandy siltstone,bright yellow brown coloured. Scarce gypsum presence, less than 5% and well crystallised.
9800	61	11,30 a 12,30	—	0,30	Sulphated-Chloryded Sodic-	—	—	Siltstone/Sandstone, bright greenish yellow coloured. Gypsum was not detected.

STATION N°	SAMPLE N°	DEPTH m	SOLUBLE GYPSUM %	TOTAL SOLUBLE SALTS %	SATURATION EXTRACT PREDOMINATING SALTS	FINKLE a) distilled water b) Colorado River water PREDOMINATING SALTS ON FLOW	RATIO 1:50 EXTRACT PREDOMINATING SALTS	ROCK GENERAL CHARACTERISTICS (in the case of crystals, the surrounding material)
9500	182	7,65 a 8,80	—	0,16	Sulphated - Chloryded Sodic-Calcic	—	—	Sandy gravels with some coherence, scarce calcitic carbonate. Gypsum was not detected.
9500	56	9,50 a 11,25	0,50	—	—	—	—	Gypsum crystals in different levels on gypsiferous siltstone - claystone and on tufaceous silt/claystone, colours bright brown to grayish green. Gypsum between 6% to 10% (crystal and fibrous).
9500	60	12,00 a 13,05	—	0,31	Sulphated - Chloryded Sodic-Calcic	—	—	Gypsiferous sandstone, some coherence, bright yellowish brown coloured. Gypsum at 15% very fine crystallised.
9600	96	3,80 a 4,80	—	0,28	Sulphated - Chloryded Sodic-Calcic	—	—	Brecciated gypsolite with micritic limestone fragments, bright yellowish gray coloured. Gypsum in well developed clasts and fragments.
9600	93	4,00 a 4,10	0,15	—	—	—	—	Gypsum crystals on micritic limestone, bright yellowish brown coloured.Total gypsum on rock 4%, very well crystallised and finely divided apart from crystals.
9600	94	4,60 a 4,80	0,34	—	—	—	—	Gypsum crystals on altered ophitic tuffs, grayish green coloured. Gypsum crystals on the mass with tendency towards rosette formation.
9600	101	4,80 a 5,80	—	0,28	Sulphated - Sodic Calcic	—	—	Clayey siltstone, bright brown coloured, fine grain. Gypsum 2% is also observed on the mass.
9600	88	5,30 a 5,45	3,01	—	—	—	—	Gypsum crystals on silt/claystone.Gypsum 2% in well developed transparent crystals.
9600	169	5,60 a 5,85	0,03	—	—	—	—	Gypsum crystals on silty sandstone. Ice width veins entire well crystallised.
9600	170	5,75 a 5,90	0,08	—	—	—	—	Gypsum crystals on somewhat clayey siltstone with gypsum veins, crystals as such as veins presents a good crystallisation and no alteration. Total Gypsum at 3%.

287,00 CREST ELEVATION

REFERENCES: SEE PLATE 1

PLATE 8

STATION N°	SAMPLE N°	DEPTH m	SOLUBLE GYPSUM %	TOTAL SOLUBLE SALTS %	SATURATION EXTRACT PREDOMINATING SALTS	PINKLE a) distilled water b) Colorado River water PREDOMINATING SALTS ON FLOW	RATIO 1:50 EXTRACT PREDOMINATING SALTS	ROCK GENERAL CHARACTERISTICS (In the case of crystals, surrounding material)
10+00	71	4,20 a 5,20	—	0,28	Sulphated - Chloruyded Sodic - Calcic	—	—	Gipsolite, bright reddish brown coloured. Gypsum 40%, fairly big sized and well crystallised.
10+00	69	4,45 a 4,65	0,07	—	—	—	—	Gypsum crystals on siltstone conglomerate. Gypsum 6%, well crystallised.
10+00	70	5,20 a 6,20	—	0,28	Sulphated - Chloruyded Sodic - Calcic	—	—	Clayey gypsolite of similar characteristics of those of sample n°71.
10+00	72	5,50 a 5,76	0,18	—	—	—	—	Gypsum crystals on gypsiferous clayey siltstone to gypsiferous sandstone. Gypsum 10% of total, well crystallised.

STATION N°	SAMPLE N°	DEPTH m	SOLUBLE GYPSUM %	TOTAL SOLUBLE SALTS %	SATURATION EXTRACT PREDOMINATING SALTS	PINKLE a) distilled water b) Colorado River water PREDOMINATING SALTS ON FLOW	RATIO 1:50 EXTRACT PREDOMINATING SALTS	ROCK GENERAL CHARACTERISTICS (In the case of crystals, surrounding material)
10+50	64	8,90 a 9,90	—	0,49	Sulphated - Chloruyded Sodic	—	—	Siltstone, bright whitish gray coloured. Scarce gypsum in well developed crystals.
10+00	68	2,90 a 3,16	0,06	—	—	a)15°C:NP4-Sulphated-Bicarbonated Sodic - Calcic	—	Gypsum crystals on calcareous little sandstone, reddish brown coloured. Gypsum at 30% well crystallised, in some cases tends to form aggregates.
10+00	65/66	2,70 a 3,60	1,12	0,42	Sulphated - Chloruyded Sodic - Calcic	a)15°C:NP2-Sulphated-Bicarbonated Calcic - Sodic, b)15°C:NP1-Sulphated-Chloruyded Calcic - Sodic	Sulphated - Calcic - Sodic	Gypsum crystals on gypsiferous silt/claystone brown - red coloured. Gypsum 40% well crystallised, present in the mass and guides. Presence of cavities with corrosion.
10+00	67	3,38 a 3,58	0,35	—	—	—	—	Gypsum crystals on siltstone with high gypsum content, generally well crystallised and with guides.
10+00	102	3,70 a 4,70	—	0,64	Sulphated - Chloruyded Sodic - Calcic	—	—	Clayey gypsiferous siltstone, bright reddish brown coloured. Gypsum 5%, well developed crystals and occasionally rosettes.
10+00	99	4,70 a 5,70	—	0,58	Sulphated - Chloruyded Sodic - Calcic	—	—	Clay/siltstone with sandstone clasts, reddish to yellowish brown coloured. Crystallised and very scarce gypsum.
10+00	95	5,50 a 5,70	0,004	—	—	—	—	Gypsum crystals in veins on tuffaceous siltstone, red to brown coloured. Gypsum 10%, very well crystallised (010).
10+00	97	5,70 a 6,40	—	0,69	Sulphated - Chloruyded Sodic - Calcic	—	—	Sandy limestone, bright yellowish green coloured, bearing gypsum crystals. Gypsum 3% in veins and crystals, both well developed.
10+00	117/118	6,20 a 6,40	0,003	—	—	—	—	Gypsum crystals on gypsiferous clayey limonite, bright yellow gray coloured. Average gypsum 10% very well crystallised.
10+00	118	6,30 a 6,40	0,30	—	—	—	—	Silted silt/claystone, yellowish gray coloured. Gypsum 40% in guides or crystals, all partly well crystallised.

Estimation de la stabilité des zones karstiques par l'examen de l'état de contrainte de massifs rocheux
Evaluation of karstified area stability by analysis of rock mass stressed state

V.M.Koutepov, *Institut de Lithosphère de l'Académie des Sciences de l'URSS, Moscow*

RESUME: La communication envisage la méthode de l'estimation quantitative de la stabilité des massifs rocheux dans les zones karstiques à l'aide de l'examen et de la comparaison des efforts résistants et tranchants, de leurs changements dans l'espace et le temps sous l'effet technogène. Sont determinés les types des mécanismes de la formation des effondrements, sont établis les diagrammes de calcul de la stabilité des massifs rocheux avec l'assise de couverture d'une structure différente et le nomogramme du changement du coefficient de stabilité lors des variations de la nappe souterraine.

ABSTRACT: The paper treats the methods of quantitative evaluation of rock mass stability in the karstified areas by analysis and comparison of driving and resistance forces, their space and time-related changes caused by technogenic impact. Types of sink hole formation mechanisms are identified and computational diagrams to determine stability of rock masses at various structures of the covering strata and nomographs of stability factor variation at ground water level fluctuation are drawn up.

INTRODUCTION

Dans les régions du karst couvert l'activité de l'homme provoque la naissance des phénomènes géotechniques qui sont à l'origine, dans certains cas, de la formation des effondrements et, dans d'autres cas, de l'affaissement irrégulier de la surface de terre.

Le problème principal de l'étude des zones karstiques, dont la solution est nécessaire pour les études de projet, la construction et l'exploitation des ouvrages d'art, pour la mise en valeur des terres cultivées, etc., est l'estimation quantitative de la stabilité des massifs rocheux et le pronostic des types et des dimensions des déformations dans l'espace et le temps.

La solution de ce problème exige, en premier lieu, la détermination du coefficient de stabilité des massifs rocheux et l'estimation de la formation possible des effondrements, et, en ce qui concerne des zones se trouvant en dehors des limites des effondrements probables, l'appreciation des dimensions des déformations sans discontinuité (affaissement) des roches.

Les méthodes existantes fondées sur l'étude des lois des manifestations du karst superficiel ou des particularités du changement des conditions hydrogéologiques et sur l'examen des lois du développement des processus actuels (Kogevnikova 1984) assurent la zonation des territoires et la mise en évidence des terrains dont le danger de karstification est différent.

Afin de définir le lieu et le temps de la manifestation possible des déformations dans le massif rocheux et sur la surface de terre, il est nécessaire d'examiner l'état de contrainte et déformation des roches et de faire le pronostic de son changement dans l'espace et le temps sous l'influence des facteurs technogènes.

Méthodes de l'estimation de la stabilité des massifs rocheux

La mise en évidence dans le massif rocheux des déformations dues au changement des conditions hydrogéologiques s'effectue à l'aide de l'examen des particularités de la perturbation des champs des contraintes lors des variations des niveaux des eaux souterraines et karstiques.

Les massifs rocheux en gisement naturel se trouvent dans un état de contrainte dont la formation est influencée par les champs physiques qui dépendent des conditions naturelles, de la poussée d'Archimède des eaux souterraines, de la pression hydrostatique sur les couches imperméables et de la pression hydrodynamique sur les assises d'argile peu perméables (Koutepov 1983).

La répartition des contraintes tant en plan qu'en coupe porte un caractère zonal. Les zones des contraintes élevées sont associées aux couches imperméables, celles des contraintes faibles - aux couches saturées d'eau ou aux accidents du karst.

En cas de la perturbation des conditions hydrogéologiques, l'accroissement des contraintes dans les diverses parties du massif n'est pas égal et a des signes contraires.

Un examen approfondi et une synthèse théorique des lois de la formation de l'état de contrainte des massifs rocheux dans les zones karstiques permettent de faire une estimation quantitative de la stabilité des massifs rocheux au-dessus des zones faibles.

Compte tenu de la structure du massif, du degré de karstification des roches solubles, des conditions de gisement des propriétés physiques et mécaniques et de la perméabilité des roches susjacentes, ainsi que des particularités hydrogéo logiques, sont mis en évidence quatre types des mécanismes de la formation des effondrements (Fig. 1):

I - par gravité - effondrement (déplacement) dans la zone faible des roches insolubles recouvrant les assises karstiques;

II - par renard-entraînement par les eaux souterraines des terrains meubles à partir de l'assise susjacente jusu'à l'assise karstique des zones faibles;

III - par gravité - renard - effondrement (enfoncement) des couches d'argile isolées peu parméables et imperméables dans la zone faible et entraînement (irruption) des terrains meubles susjacents saturés d'eau dans l'assise karstique à travers une zone détruite dans les argiles;

IV - par renard-gravité - entraînement par les eaux souterraines des terrains meubles à partir de la fondation de l'assise susjacente jusu'à l'assise karstique et effondrement dans la cavité créée des roches argileuses et sableuses superposées;

Le massif rocheux au-dessus de la zone faible sera dans un état stable limite si la résistance au cisaillement ($\sigma tg\phi$) est égale à la contrainte de cisaillement (τ).

Pour les roches sableuses

$$\sigma tg\phi_\Pi = \tau \tag{1}$$

pour les roches argileuses

$$\sigma_\Gamma tg\phi_\Gamma + C = \tau \tag{2}$$

Ainsi, la stabilité du massif rocheux sera assurée si les efforts résistants (N) sont égaux ou dépassent les efforts tranchants (T)

$$N \geqslant T \tag{3}$$

Les zones faibles ce sont les zones d'anciens et d'actuels accidents du karst où la karstification des roches se traduit en leur destruction intense jusqu'à blocs, galets anguleux, arène et poudre, en séries de petites cavités et fissures karstiques. Dans le remplissage des cavités karstiques prédominent les différences argileuses faiblement compactées qui se caractérisent par une compressibilité élevée (Koutepov et al. 1984).

La perturbation de la stabilité des massifs rocheux au-dessus des zones faibles est due, dans certains cas, à l'augmentation des efforts tranchants, dans d'autres, à la diminution des efforts résistants et, dans les troisième, à l'augmentation des efforts tranchants et à la diminution des efforts résistants.

La mesure du coefficient de stabilité (K) est basée sur l'examen des lois de la formation de l'état de contrainte et déformation des massifs rocheux, sur la détermination des efforts tranchants et résistants et sur la mise en évidence du rapport existant entre ces efforts dans l'assise surmontant la zone faible. Le co-

efficient de stabilité du massif rocheux est calculé comme un rapport entre les
efforts résistants et les efforts tranchants

$$K = \frac{N}{T} \qquad (4)$$

Afin de déterminer les efforts résistants et tranchants, les diagrammes de calcul (Fig. 2) sont établis. Ils représentent la structure homogène des assises recouvrant les zones faibles et comprenant les depôts sableux (Fig. 2, I-A) ou argileux peu perméables (Fig. 2, I-B) et imperméables (Fig. 2, I-C), ainsi que la
structure hétérogène des assises, présentée par les roches sableuses et argileuses (peu perméables - Fig. 2, II-A, III-A, IV-A et imperméables - Fig. 2, II-B,
III-B, IV-B) en différente alternance.

Conformément à chaque diagramme de calcul sont établies les formules qui permettent de déterminer le coefficient de stabilité des massifs rocheux compte tenu
des dimensions données des zones faibles ou les dimensions critiques des zones
faibles avec le coefficient de stabilité donné.

Les efforts tranchants dans un massif rocheux dépourvu d'eau à une surface horizontale, comprenant une couche homogène ou quelques (i) couches hétérogènes
d'après leur composition lithologique et propriétés physiques et mécaniques, en
différente alternance, correspondent à la pression (G) des masses rocheuses susjacentes sur le toit de la zone faible et sont égaux à la contrainte verticale
(σ) multipliée par la superficie de la projection horizontale de la section transversale de la zone faible. Avec la section transversale ronde de la zone faible,
les efforts tranchants sont égaux à

$$T = G = \pi r^2 \sum_1^i \sigma \qquad (5)$$

où r - rayon de la section transversale de la zone faible.

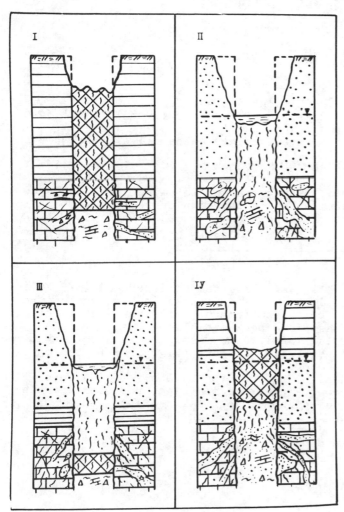

Figure 1. Types des mécanismes de
la formation des effondrements dans
les zones karstiques

631

Par exemple, les efforts resistants dans l'assise de sable homogène (Fig. 2, I-A) sont égaux à la pression latérale (P) des roches sableuses, multipliée par le coefficient de frottement, et sont déterminés par la formule.

$$N = P_\Pi \, \mathrm{tg} \, \phi_\Pi \qquad (6)$$

La pression latérale des roches sableuses est déterminée compte tenu du volume de l'épure des contraintes horizontales, représentant une figure de rotation du triangle autour du cylindre se trouvant au-dessus de la zone faible. Le volume de l'épure est égal au produit de la surface de triangle et du rayon de rotation (R_Π) du centre de gravité du triangle autour du cylindre.

La base du triangle est représentée par les contraintes horizontales ($\sigma_\Pi \, \mathrm{tg}^2 (45° - \phi_\Pi /2)$), sa hauteur - par l'épaisseur des roches sableuses (M) au-dessus de la zone faible

$$P_\Pi = \Pi \, R_\Pi \, \sigma_\Pi \, \mathrm{tg}^2 (45° - \frac{\phi_\Pi}{2}) \, M_\Pi \qquad (7)$$

où ε_Π - coefficient de la poussée latérale ($\varepsilon_\Pi = \mathrm{tg}^2 (45° - \frac{\phi_\Pi}{2})$).

Le rayon de rotation du centre de gravité du triangle autour du cylindre est trouvé d'après le formule:

$$R = \frac{1}{3} \, \sigma_\Pi \, \varepsilon_\Pi \, \mathrm{tg} \, \sigma_\Pi \qquad (8)$$

Donc, les efforts résistants dans l'assise de sable au-dessus de la zone faible seront égaux à:

$$N = \Pi \, M_\Pi \, \sigma_\Pi \, \varepsilon_\Pi \, \mathrm{tg} \, \phi_\Pi \, R_\Pi \qquad (9)$$

Le coefficient de stabilité de l'assise de sable au-dessus de la zone faible sera égal à:

$$K = \frac{N}{T} = \frac{\Pi \, \sigma_\Pi \, M_\Pi \, \varepsilon_\Pi \, \mathrm{tg} \phi_\Pi R_\Pi}{\sigma_\Pi \Pi r^2} = \frac{M_\Pi \varepsilon_\Pi \, \mathrm{tg} \phi_\Pi R_\Pi}{r^2} \qquad (10)$$

La formule (10) montre que, dans des conditions égales, les efforts résistants et, par consequent, la stabilité de l'assise superposée sur la zone faible dépendent de l'angle de frottement interne (ϕ_Π).

L'étude de la fonction de l'angle de frottement interne dans l'expression

$$f(\phi_\Pi) = \mathrm{tg}^2 (45° - \phi_\Pi /2) \mathrm{tg} \phi_\Pi \qquad (11)$$

permet de définir les lois de ses variations et trouver la valeur de l'angle de frottement interne avec laquelle la dérivée de la fonction est égale à zéro, c'est-à-dire le maximum de cette fonction est obtenu

$$f^1(\phi_\Pi) = \frac{\mathrm{tg}^2 (45° - \phi_\Pi /2)}{\cos^2 \phi_\Pi} - \frac{\mathrm{tg}^2 (45° - \phi_\Pi /2) \mathrm{tg} \phi_\Pi}{\cos^2 (45° - \phi_\Pi /2)} \qquad (12)$$

Cette condition sera respectée avec l'angle de frottement interne égal à 30°, ce qui correspond aux roches sableuses à la limite de transition des sables très fins aux sables fins. Avec l'angle de frottement interne inférieur à 30°, cette fonction augmente et la stabilité du massif rocheux s'améliore, mais avec les valeurs dépassant 30° elle diminue et la stabilité du massif rocheux devient pire.

Les efforts résistants dans l'assise d'argile homogène (Fig. 2, I-B, C) sont égaux au total des forces de pression latérale (P_Γ), diminuées par le frottement des roches, et des forces de cohésion. La pression latérale est définie compte tenu du volume de l'épure des contraintes horizontales, représentant une figure de rotation du trapèze autour du cylindre, dont la base supérieure est ($C \, \mathrm{tg}^2 (45° - \phi_\Gamma /2)$), la base inférieure - ($\sigma_\Gamma \, \mathrm{tg} \phi_\Gamma + C) \mathrm{tg}^2 (45° - \phi_\Gamma /2)$ et la hauteur - épaisseur (M_Γ) de l'assise d'argile au-dessus de la zone faible.

Les efforts résistants dans l'assise d'argile homogène au-dessus de la zone faible seront égaux à:

$$N = \Pi \, M_\Gamma \, (\sigma_\Gamma \, \mathrm{tg} \, \phi_\Gamma + 2C) \varepsilon_\Gamma \, R_\Gamma \qquad (13)$$

où

$$\varepsilon_\Gamma = tg^2(45° - \phi_\Gamma/2)$$

C - cohésion des roches;
R_Γ - rayon de rotation du centre de gravité du trapèze autour du cylindre.

Donc, le coefficient de stabilité de l'assise d'argile au-dessus de la zone faible sera égal, dans ce cas, à:

$$K = \frac{N}{T} = \frac{\Pi M_\Gamma(\sigma_\Gamma \ tg\phi_\Gamma + 2C)\varepsilon_\Gamma R_\Gamma}{\Pi \sigma_\Gamma r^2} = \frac{(\sigma_\Gamma \ tg\phi_\Gamma + 2C)\varepsilon_\Gamma R_\Gamma}{\gamma r^2} \qquad (14)$$

Dans les assises de sable-argile hétérogènes (Fig. 2, II, III, IV) les efforts résistants sont égaux au total des efforts résistants dans toutes les couches. Pour la couche supérieure, les efforts résistants sont déterminés d'après les formules correspondantes en fonction de la composition des roches: pour la couche de sable ils sont trouvés par la formule (9), pour la couche d'argile - par la formule (13).

En ce qui concerne la couche inférieure les efforts résistants sont déterminés, aussi, en fonction de la composition des roches. Si dans la fondation de l'assise d'argile se trouvent les sables, le coefficient de stabilité de l'assise susjacente hétérogène sera égal à:

$$K = \frac{N}{T} = \frac{(\sigma_\Gamma \ tg \ \phi_\Gamma + 2C)M_\Gamma \varepsilon_\Gamma R_\Gamma + (2\sigma_\Gamma + \sigma_\Pi)M_\Pi \varepsilon_\Pi tg \ \phi_\Pi \ R_\Pi}{(\sigma_\Gamma + \sigma_\Pi)r^2} \qquad (15)$$

Au cas où dans la fondation de l'assise de sable gisent les argiles, le coefficient de stabilité de l'assise susjacente sera égal à:

$$K = \frac{N}{T} = \frac{\sigma_\Pi m_\Pi \varepsilon_\Pi tg\phi_\Pi R_\Pi + [(2\sigma_\Pi + \sigma_\Gamma)tg \ \phi_\Gamma + 2C] \ M_\Gamma C_\Gamma R_\Gamma}{(\sigma_\Pi + \sigma_\Gamma)r^2} \qquad (16)$$

Dans l'assise de sable-argile multicouche recouvrant la zone faible et constituée par des roches sableuses et argileuses en différente alternance, les efforts résistants sont égaux au total des efforts résistants dans toutes les couches compte tenu du coefficient de frottement des roches et des forces de cohésion dans les dépôts argileux.

Lors de l'estimation de la stabilité à l'enfoncement de l'assise d'argile isolée (Fig. 2, IV-A, B), les efforts tranchants sont admis égaux à la pression de toute l'assise susjacente sur le toit de la zone faible, tandis que les efforts résistants sont déterminées seulement dans la couche d'argile isolée.

Afin de définir le coefficient de stabilité des assises susjacentes multicouches au-dessus des zones faibles du massif karstique sont utilisées les formules ci-dessus mentionnées en différentes combinaisons correspondant aux conditions précises du gisement des roches.

Dans les diagrammes de calcul et les formules, l'influence des eaux souterraines sur la formation de l'état de contrainte et déformation et la stabilité des massifs rocheux a été prise en compte par les forces de la poussée d'Archimède dans les roches saturées d'eau, par la pression hydrostatique dans les couches imperméables et la pression hydrodynamique dans les zones de la filtration verticale descendante et ascendante.

Il est à noter que la possée d'Archimède change de la même manière les efforts tranchants et résistants, tandis que la pression hydrodynamique n'a de l'influence que sur les efforts tranchants: avec la filtration descendante elle provoque leur augmentation, avec la filtration ascendante - leur diminution. La pression hydrostatique des eaux souterraines sur les roches imperméables aussi change de la même manière les efforts tranchants et résistants, mais la pression hydrostatique des eaux karstiques de charge sur les roches imperméables gisant dans le toit de la nappe aquifère, augmente les efforts résistants et n'influence pas les efforts tranchants.

Les déplacements possibles des roches de l'assise susjacente dans la zone faible dépendent de ses dimensions transversales, de la profondeur de gisement, des propriétés physiques et mécaniques des roches et des conditions hydrogéologiques.

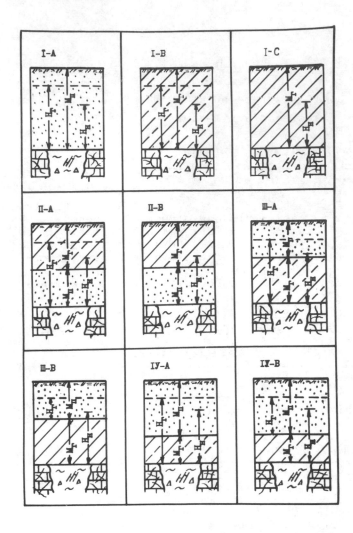

Figure 2. Diagrammes de calcul pour l'estimation de la stabilité des massifs rocheux au-dessus des zones faibles.

La perturbation des conditions hydrogéologiques au fur et à mesure de l'augmentation des efforts tranchants ou de la diminution des efforts résistants conduit à l'abaissement de la stabilité de l'assise susjacente. En cas d'égalité des efforts résistants et tranchants, l'assise susjacente se trouve dans un état d'équilibre limite. Avec le déficit des efforts résistants, l'assise susjacent se déplace dans la zone faible et sur la surface de terre se forme, d'habitude, un effondrement.

Les mesures du coefficient de stabilité s'effectuent à l'état statique des niveaux des eaux souterraines et, aussi, aux changements des conditions hydrogéologiques dans les différents moments du temps. Les résultats obtenu sont présentés sous forme de nomogrammes du changement du coefficient de stabilité (Fig. 3), qui permettent de mettre en évidence la tendance au changement de la stabilité de l'assise susjacente, de déterminer le degré de stabilité à n'importe quel moment du temps aux différents états et rapports entre les niveaux des eaux souterraines et de faire les pronostics des rapports critiques entre les niveaux, assurant la stabilité limite du massif.

L'examen du changement des efforts résistants et tranchants montre que la diminution de la stabilité des massifs rocheux au-dessus de la zone faible à condition de la présence, dans la coupe, de l'assise d'argile isolée entre la nappe souterraine et la nappe aquifère de charge, a lieu dans les cas suivants:
- à l'élévation du niveau des eaux souterraines;
- à l'abaissement des charges des eaux karstiques;
- à l'abaissement synchrone régulier des niveaux des eaux souterraines et karstiques.

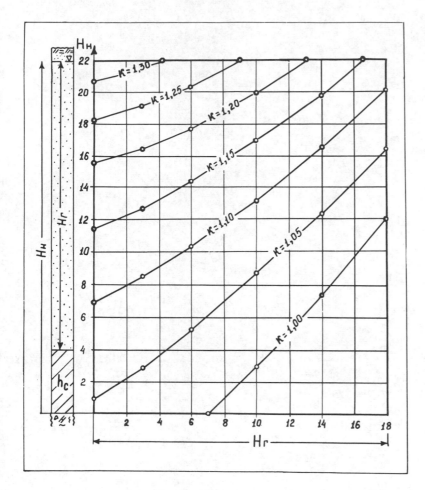

Figure 3. Nomogramme du changement du coefficient de stabilité du massif rocheux au-dessus de la zone faible avec les variations de la nappe souterraine.

CONCLUSIONS

La méthode proposée de l'estimation de la stabilité des zones karstiques, basée sur l'examen du changement des rapports entre les efforts résistants et tranchants, permet de surveiller le degré de la stabilité des roches de l'assise susjacente au-dessus des zones faibles et de contrôler, en cas de nécessité, l'état des niveaux des eaux souterraines aux fins d'assurer la stabilité des massifs rocheux, c'est-à-dire de contrôler la stabilité.

REFERENCES

Kogevnikova, V.N. 1984. Méthode de l'estimation de la stabilité des zones karstiques (russe). Inzhenernaja Geologija 2: 26-40.
Koutepov, V.M. 1983. Formation de l'état de contrainte des massifs rocheux dans les zones karstiques (russe). Inzhenernaja Geologija 1: 67-81.
Koutepov, V.M., G.M. Berezkina, N.V. Zikova, V.N. Kogevnikova, A.V. Krasnouchkine et L.G. Tchertkov 1984. Processus karstiques et propriétés géotechniques des roches argileuses (russe). Inzhenernaja Geologija 4: 91-103.

Engineering geological aspects of study of karst deposits

Aspects géotechniques de l'étude de sites karstiques

I.A.Pechorkin, *Perm State University, USSR*

ABSTRACT: Engineering geological conditions of karstic rocks depend on many factors, first and foremost, on geological history of the territory under study. Tectonic movements condition density and pattern of jointing. In situation like this type and activity of karst process, the present day hydrodynamical zonality should be determined. The territory is studied by geophysical methods. Information on subsurface karst forms distribution is specified by boring. The process intensity and karst forms distribution depend on lithological and climatic conditions. Recommendations on protecting constructions and governing karst process are proposed.

RESUMÉ: Les conditions d'ingénieur géologiques dans les roches karstiques dépendent de beaucoup de facteurs et, avant tout,de l'histvoire géologique du territvoire. Les mouvements tectoniques conditionnent la fréquence et le réseau de fissures. Les types et l'activité du processus de karst, zonalité hydrodynamique actuelle sont déterminés. Tout le territvoire est couvert par les recherches géophysiques. Les materiaux sur les formes karstiques sont précisés par prospection par sondage. L'intensité du processus dépendent des conditions lithologiques et climatiques. Les recommandations sur la protections de constructions du processus de karst sont donées.

Territories formed by sulphate rocks, being the areas of intensive karst process development, are characterized by highly complicated engineering geological conditions. The main problems to be solved by engineering geologist are to forecast karst cavities location and karst process development. Governing of the process provides safe exploitation of construction.

Engineering geological investigations of karsted territories always encounter difficulties, in the areas of sulphate rocks in particular. These are the most soluble rocks (not counting salts) on which different constructions, hydrotechnical objects included, are to be built. Rock solubility is so high that engineering geological situation can be significantly changed to the worse in the course of exploitation period of construction.With time situation favourable in reconnaissance period changes to the worse to such a degree that construction completely loses its stability and goes to ruin. Constructional activity of man and disposal of waste water considerably contribute to the process. It gives rise to intensive subsurface water circulation caused by human activity. This phenomenon is due to significant water intake for drinkable purposes and processing of vertical drainage wells. For instance, in the upstream reach of one of the dams built on the Kama River with gypsum lens in the foundation, vertical drainage has been put to capture subsurface water stream running under construction. In the course of exploitation it was found that the drainage did not perform its functions, on the contrary, it contributed to the removal of calcium sulphate from gypsum lens, so circulation in the system "subsurface-surface water" has been increased. Thousands tons of calcium sulphate have been removed from under construction. The drainage has veen taken out of operation and construction of cut-off wall was to begin. As a matter of fact, surface (atmospheric) waters having passed the distance several dozens meters long beneath the surface enter sulphate massif and achieve the limits of saturation (about 2.5 gr/l), thus water aggressiveness is sufficiently decreased. It does not mean that water mineralization constantly is at the limits of saturation. Various hydrodynamical and thermodynamical conditions, mixing up with subsurface waters differing in chemical composition, stress state of the massif, pH and EH variations and many other factors and conditions taking together change aggressive properties of subsurface stream. In certain cases the stream actively dissolves rock, in the others - calcium sulphate containing in the rock is precipitated.

The experiments show that with the joint 0.01 cm wide, hydraulic gradient 0.01 and velocity of filtration 0.01 cm/sec, saturation is achieved when waters have passed the distance 90 m long. In the open water body as much as 0.5-1.2 kg of rock are dissolved from 1 m^2 of shore per day. The modulus of ion discharge is 300-400 t/km^2, sometimes it exceeds thousands tons. In carbonate rocks this value equals to 50 t/km^2.

To study the territory of gypsum karst distribution one should, first and foremost, consider in details the natural course of karst process development. Investigator has to establish either activization or fading of the process takes place. The problem must be solved from historical geological position. When carrying out engineering geological karst investigations in the rock of any lithological composition, sulphate rocks in particular, one should bear in mind that hydrogeological and hydrochemical investigations are of primary importance.

It is quite natural that intensity of karst process development changes depending on the site where the object is to be built: in the massif, on the massif surface, in the river bed, on the river terrace, in the watershed, on the slope etc. To begin investigations one should, first and foremost, establish hydrodynamical zonality of the territory under study, reveal law-governed regularities of karst forms distribution characteristic of certain hydrodynamical zones. At the first approximation a complete scheme of hydrodynamical zonality has been developed by G.A. Maximovich (1957). The scheme comprises surface, vertical, horizontal and syphon zones of water circulation.

Karrens are formed in surface circulation zone. Their size varies significantly and depth can exceed 1.0-1.5 m. It is not difficult to fix karrens when one deals with unbared karst. Being covered by recent sediments even if of Quaternary age, karrens make engineering geological situation more complex. Underground relief of karst massif roof is developed in different ways. Under certain conditions depressions and outliers are formed directly on the massif surface and then turn to be covered by more recent formations. Under other conditions relief is formed in the result of subsurface solution. If this is a case, depressions are found to be filled with karst-caving deposits (karst breccia) consisting of debris and blocks of the overlying and karst rocks. Difficulty lies in the fact that quite often underground relief is not expressed on the surface.

Our investigations show (Pechorkin, Bolotov, 1983) that it is highly advisable to subdivide the underground relief of karst massifs into dissonant and consonant. The first is the relief of karst massifs overlain by thick deposits, so the relief has no manifestation on the surface; the second - is the relief of karst massifs with unbared surface or covered with thin soil-vegetation layer. The central parts of depressions and outliers are found to be the most favourable for construction. On the slopes of outliers with high gradients of thickness of the overlying deposits, uneven settlements of foundations resulted in deformation of constructions are highly possible.

Previously underground relief has been studied by boring, boreholes have been sited into dense network. It is highly expensive, time-consumed and hard labouring operation. Tectonic jointing investigations have been successfully used by the author when studying the underground relief of karst massifs. For this purpose 3 indices have been used: 1) quantity of joints per unit of length; 2) quantity of joints intersections; 3) the total length of joints per square unit of the surface of karst massif (1000 m^2).(A. Pechorkin, I.Pechorkin, Kataev, 1982; A.Pechorkin, Bolotov, Kataev, 1984). Then the isoline maps of tectonic jointing have been received. Maximum values of isolines outline depressions developed in the underground relief, medium values of isolines indicate saddles between the outliers and minimum values correspond to outliers.

At the next stage geophysical investigations have to be carried out. Geophysical maps of electric anomalies are matched with isoline maps of tectonic jointing. The results obtained are varified by widely spaced net of prospecting boreholes.

Shafts, karst fissures and cavities are the largest subsurface forms developed in vertical circulation zone. Subsurface cavities are usually regularly distributed and their certain horizons are associated with river terraces. It is extremely important from the view point of siting prospecting boreholes. When penetrating cavity by borehole one can get only its vertical size but we have no idea about real parameters of the cavity. Dimensions can be approximately calculated by the ratio of the cavity axes. It is known that vertical axis is related to the horizontal ones as 2:3:4. The longest axis is directed along the bed inclination while the mean is perpendicular to the former. These data have been obtained in the course of multiple measuring of cavities parameters in Perm region.

Sometimes cavities are found to be filled with air, water, clay (terra-rossa) and sandy-gravel material. In some cases under the beds of great rivers boulders of considerable size are encountered. Vertical circulation zone is usually dry, perched water tables and water flows moving through karst failures (fissures) are developed here and there. Formation of permanent and intermittent aquifers makes the development of karst cavities to be more complicated in this zone. When cavity is filled with material, the removal of the filling material is possible; when cavity is filled with water, increased water circulation can cause solution of gypsum rock. In the second case determination of the water aggressiveness and certain hydrochemical and engineering geological calculations are found to be necessary (Pechorkin, 1969).

As a rule in vertical circulation zone water movement is in descending direction. In the areas adjacent to tectonic raptures ascending movements of even hot water are observed (Lykhoshin, 1968). This is namely the fact which explains the phenomenon of thermokarst. The depth of vertical circulation zone depends on the depth of erosional cut and in mountaineous regions it makes 1.5-2.0 km.

The characteristic features of horizontal circulation zone are large karst cavities and long cave systems. The length of subsurface channels may achieve dozens and hundreds kilometers. This is the zone of permanent water level. High debit streams (dozens and hundreds l/sec) are associated with the zone. Subsurface water table and streams debits vary significantly by the seasons of a year. Subsurface water moves in the river direction. Under the river bed it is transformed into "understream" flow and moves in surface water direction, in certain cases there are some exceptions. Flow moves aside or in the direction opposed to the surface run-off. This fact is conditioned by the presence of tectonic faults zones and by geological structure of the territory.

It worth mentioning in this connection that tectonic jointing investigations are of primary importance and need maximum attention since in combination with subsurface waters jointing entirely controls karst cavities distribution. For instance, formation of organ tubes is due to the presence of local aquitard intersected at one point by a series of tectonic joints. Vertical channel of oval or rounded form several meters in diameter is formed there. The channel is several dozens meters deep and it is usually outcropped in the arch of the other great cavity. Organ tubes are widely spread in caves, Kungur Cave developed in gypsum (Ural) can be cited as an example.

In the upper part of horizontal circulation zone the decrease storage zone is distinguished (Pechorkin, 1969).Under natural conditions this zone constantly wedging in the watershed direction is usually developed in the vicinity of the river valleys. In high water periods the rise of surface water is higher than that of subsurface, so slightly mineralized aggressive surface waters can enter the karst massif. As a result, reverse inclination of subsurface water level is observed. Decrease storage zone is the zone where the process of present day karst is the most intensive.

The process is more pronounced in the shoreline zone of the reservoirs with regulation by seasons. The difference between winter and summer water level fluctuations for plain reservoirs is more than 15-20 m, while that for the reservoirs constructed in mountaineous regions it exceeds several dozens meters. Observations carried out on Kama reservoir shores show that subsurface water hydrostatic head spreads in watershed direction and occupies the belt 3-5 km wide, where activization of karst and karst-suffosion processes takes place. Activization of karst process is due to high aggressiveness of the river waters and sharp changes of high hydraulic gradients result in activization of karst-suffosion process. The period of time (2-3 months) is sufficient for these waters to achieve the limits of saturation. With the water level fall in the reservoir calcium sulphate is removed from the massif. Engineering geologist should pay special attention to the shoreline zone of rivers and reservoirs since this zone is the most dangerous from the view point of active gypsum karst dynamics. Under such conditions one should bear in mind that gypsum karst process development should be evaluated in dynamics, not statics.

Syphon circulation zone is rarely encountered. The whole point is that karst waters move under the river valley from one bank to the other through individual channels curved in the form of syphon. This is not a hypothesis but a phenomenon varified in the course of experiments with dye. This hydrodynamical zone should be carefully studied when siting hydrotechnical objects in the river valley areas. This is namely the zone through which leakage from the reservoirs is occurred. It is believed that unfortunate experience of construction of Halse-Bar Dam on the Tennessee River (USA) was due to poor investigation of syphon circulation zone and under-river stream.

The most intricate engineering geological conditions are known in the river valley areas where karst process takes place. It should be always kept in mind that karst processes both modern and old are the most intensive in the areas adjacent to the valley flanks. Joints of slope unloading contribute to the complicacy of the situation. Under such conditions industrial, civil and particularly hydrotechnical construction is found to be extremely difficult. These joints play the role of routes through which catastrophic leakage from the reservoirs is occurred. It is highly necessary to study the valley flanks not only by vertical and horizontal boreholes but also by adits. Karstification should be expressed by numerical values (coefficients).

One of the fundamental problems of gypsum karst engineering geology is the interaction between an object and geological surroundings. Several variants are considered: due to negative influence of surroundings object loses its stability or because of high karstification object does not perform its functions. For instance, dam with highly filtrating reservoir bed does not answer its purpose. Quite often engineering construction changes karst water hydrodynamics and contributes to karst process activization. It happens when aggressive water periodically enters the massif. Formerly dry part of the massif (vertical circulation zone) becomes filled with water and turns to be transformed into horizontal circulation zone. In other cases considerable lowering of karst water level, drainage of the massif, decrease of its filtration capacity and change of subsurface water chemical composition can significantly reduce and even stop the development of karst process.

Object can exercise indirect affect on the process. For instance, construction and exploitation of chemical works are often accompanied by damp of waste waters contributing to active solution of rock. The content of 130 gr/l of NaCl in solution increases gypsum solubility up to 7.48 gr/l. Gypsum solubility increases in the result of surface and subsurface water temperature rise, being at maximum at ± 22-25°C. Thermal pollution is usually observed in the area of steam power plants draining hot (chemically pure) water into settling basins and then onto the water bodies.

In connection with the above mentioned we may say that determination of filtration parameters, calculation of the territory stability and forecast of karst process development, careful study of subsurface water hydrochemistry and rock solubility included, are highly necessary when conducting engineering geological investigations in the areas of gypsum karst. Without going into details, we may say that stability forecast should be developed in 2 directions: 1) probability of karst cavities occurrence and evaluation of their parameters; 2) determination of bearing capacity of the cavities arches teking into account dynamics of the process and load of the object under construction (Pechorkin, 1969).

The highest form of engineering geological art is to govern karst process development. Control can be exercised by passive protection i.e. by means of different measures which do not influence on the course of the process and increase stability of the object under construction (filling up of karst cavities with concrete and other materials; various engineering constructions increasing stability of foundations).

The most expedient action is to decrease the process intensity. In our opinion the following measures: drainage of the territory, lowering of subsurface water level, defence by highly mineralized (non-aggressive) water, sharp decrease of filtration ability resulted from filling up the joints with various materials (quick-setting gels) and some other operations are the most advisable.

REFERENCES

Lykhoshin, A.G. 1968. Karst and hydrotechnical construction.Stroijzdat,Moscow,

183 p. (Russ.)

Maximovich, G.A. 1957. The main types of hydrodynamical profiles in karst areas of carbonate and sulphate deposits. Papers of the USSR Academy of Sciences, vol. 112, No 3, p.p. 501-504. (Russ.)

Pechorkin, I.A. 1969. Geodynamics of Kama reservoirs shores. Vol. 2, Perm, 308 p. (Russ.)

Pechorkin, I.A., Pechorkin,A.I., Kataev, V.N. 1982. The problems of karst massif tectonic jointing investigation for hydrotechnical construction.Atr.: del 11° Simposio Internazionalle sulla Utilizzazione della arce carsche. V.XVII. Bari-Castellana Grotte, p. 477-484.

Pechorkin, A.I., Bolotov, G.B. 1983. Geodynamics of the relief of karsted massifs. Perm, 83 p. (Russ.)

Pechorkin, A.I., Bolotov, G.B., Kataev, V.N. 1984. The study of tectonic jointing in platform structures for karstological purposes. Perm. 85 p. (Russ.)

The laminarization of the deep karst
Laminarisation de karst profond

Tan Zhoudi, *Department of Hydrogeology & Engineering Geology, Changchun College of Geology, China*

ABSTRACT: The laminarization of the development and distribution of the karst caves exposed on the surface has generally been accepted. The similar characteristics of deep karst below the present discharge datum plane is less studied up to now. This paper indicates the phenomenon of layering of the distribution of the deep karst according to the information of the prospecting boreholes and mining excavation in the karst areas. The paper points out that the developing degree of karst and the elevation difference of karst layering in the vertical section often vary rhythmically as a result of rhythm of the recent earth crust upheaval and subsidence movement and equidistance in the distribution of geological structure planes. The layering distribution of the deep karst is approximately responding to that of the surface karst.

RESUME: La regle de development et distribution de la laminirisation des cavernes karstiques exposées en surface de la Terre, est bien acceptée par le public. Cependant, pour les caracteristiques de ressemblance des cavernes profondes, situées au-dessous du nivellement de drainage, sont encore moin étudiees jurqu'a present. Basé sur les données des forages de reconnaissance et des traveaux de mine, cet article séxphique les phenomenes de la distribution laminale des karsts profonds. Un souligne également dams cet article que, sous l'effet de l'elevation,d'eme fason rythmée de la crôute terreste , ainsi que de la distribution en équidistance des plans structuraux geologiques, l'intensite du developppment de karst et l'ecart d'altitude de laminarisation karstique et varie aussi souvent d'une fason rythmée; tandis que la distribution laminale des karsts profonds correspond a`celle des karsts superficiels.

Deep-karst or the karst in the depth is referred to"the karst voids and caves below the present discharge datum plane."[1] We have ever suggested that the water table in the karst regions is used as the limit of classification between the deep and shallow karst, the karst below the water table is called deep-karst.[2] In water conservancy and hydrauelectric constructions, the karstic cave and fissure zones in the upper portion of the deep-karst in the region of river valleys will be studied emphasisly. According to vertically zoning of karst water, the zone should belong to the shallow saturation zone and is responding to the hydrokinetic zoning which is usually so-called horizontal seepage zones and their low inversion siphon circulation zone in which the karst cave and fissure water is characterized by turbulent flow . The laminellar flow prevails at lower karst fissure zone of the deep-karst which is called deeply saturated zone or slow circulation zone in vertical zoning of karst water. The karst fissure zones is less studied in the engineering practice . And in this paper we should discuss the laminarization for the development and distribution of larger caves and fissures in the upper portion of the deep-karst.

It is difficult to obviously determine the developing depth for the deep-karst, even for cavern-fissure zone. It is shown from an amount of geological survey data that the lower limt of cave-fissure zones of the recent deep-karst, which is controlled by the local discharge datum planes(e.g. the surfaces of the river, lake and sea) is generally only about tens to 100m. below the bottom of the valley in the plateau, and is about 200-300m. in the karstic basins and canyons, but in the areas where the favourable tectonic conditions exist, especially where there exists the karst development superimposed on the burried ancient karst, the limit could reach

400-500m. or deeper below the discharge datum plane. There might be larger karstic cave development at the elevation of -1000- -1500m. or more because of the effect of the geophysical and geochemical factors such as saline water, thermal liquid etc . at the depth.

The general tendency of the deep karst development in the vertical section decreases with the increase of the depth. But the variation is of the general pattern due to the heterogeneity of its development. Thus, deep karst occurs in the complex alternative process of the strong to the weak in the vertical section just as shallow karst caves were developed in the form of the layers. In some depth or elevation intervals from the surface, karst were intensively developed and karst caves and fissures are larger and more concentrated distribution, therefore, groundwater is richer. This led to the formation of relatively intensive soluble layers or comparatively concentrated development layers. Although the problem has less been discussed, much real information can demonstrate the geological phenomena.

A simple example is the water supply well in a factory in Xintian county, Hunan Province. The factory is located at the upward end of a small syncline composed of middle and late Carboniferous limestone in the thick layers in which karst is intensively developed and groundwater is richened. The intervals of three wells are 400-800m. and they are 287.87m. 383.54m. and 193.10m. deep, and they encountered 6, 4 and 9 caves, respectively. The heights of the caves are 1-2m. and the highest one is 10.2m.. As shown in Figure 1, the caves are concentrated in 20-25m and 100-150m. deep intervals from the surface and approximately distributed in the same elevation intervals. The karst ratio is up to 20-30% which is 10-20 times as much as the other depth intervals. According to information from the geological drilling and mining excavations, in many Permian coal system basins south of China, recent deep-karst is developed on the basis of the ancient karst in early Permian limestone with thick layers underlied by the coal system. The lower limit of the caves - fissures zone is situated 350-450m.depp below the local discharge datum plane .In the range of the depth , the laminarization of the dissolved caves is very clear.

Figure 2 illustrated the general curve of the variation for karst ratio and aggressive carbon dioxide content in the cave water with the depth, in the thick layer limestone underlying the lower bed of the coal system in Doli shan. mine, the central region of Hunan Province.(3) Figure 3 shows the curve of the change in deep karst void ratio with the depth in the carbonatite in Yaojin mine and Xifengcang mine of Meitanba mining area, the central region of Hunan Province. Figure 2 does not only demonstrate that the developing intensity of the deep karst gradually decreases as the depth increases(the ideal curve of the change in the karst ratio and the various curve of aggressive carbon dioxide content is approximately consistent), but also shows that the karst is intensively developed in 20-30m.deep interval below the discharge datum plane of the elevation 100m, and the developing intensity gradually decreases downward from the depth, while at the depth of 400m.(the elevation is about -300m.), increasing again. Figure 3 more concretely demonstrates the characteristics of the layering development and distribution of the deep-karst in the mines. The karst was most intensively developed at the level of the discharge datum planes(Its elevation is in 100-50m. interval) in both mining regions. Furthermore, in the depth intervals of 20- -10m., -40- -60m. and -100- -120m., the karst was also developed more intensively, and the dissolved caves and fissures are more, and the dissolved void ratio is larger(about 15-5%), and the laminarization of the deep-karst is quite obvious. The deep karst is of similar characteristics in the carbonatite in many other Permian coal system basins in Hunan Province, Guangdong Province and Guangxi Zhuang Autonomous Region.

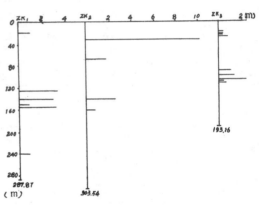

Figure 1 the sketch map of the cave distribution in the boreholes

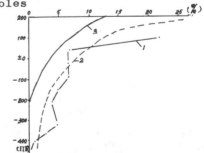

Figure 2 the change curves of the developing intensity and aggressive carbon diozide content with the depth
1.real change curve of karst ratio 2. ideal change curve ofkarst 3.curve of aggressive CO_2

644

The vertical variation of the deep-karst development in Huolong Quan limestone(D_3)underlying the deposited iron mine layer, early Carboniferous system in the East Hunan Province is of this characteristics. According to the study of Lunshaobuan , the karst of the region can be divided into three zones; as shown in Figure 4 : I -developing zone of the shallow karst; II-developing zone of the cave below the discharge datum plane, the concentrated development zone

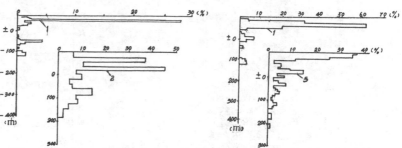

Figure3 trend diagram of the karst variation with the depth in the mining wells of Meitanba mining area
1-dissolved void ratio in the pit shaft
2-numbers of the dissolved caves in boreholes
3-dissolved void ratio in the borehole

of the deep-karst caves; III -the weakly developing zone of the deep-karst, that is the zone of dissolved fissures and voids with very few caves.(4)It was observed from the statistical data of the caves measured in the boreholes that the layering of the deep-karst is more clear in Lushui and Qingshui mines, and the development and distribution of the karst are more concentrated in the elevation interval of 200-150m. at Lushui mine, and the elevation interval of 50-0m. at Qingshui mine. For Leilongli mine as is located on the slope of the watershed, the deep-karst was developed shallower. The research for the characteristics of the deep-karst in Cambrian-Ordovician limestone in Han-Xian iron mining area, Hebei Province, has ever been performed by Yu Guang etc.(5)The thickness of Quaternary sedimentary is usually about tens to 100m. in four cover types of karst mining areas in Xishimen, Wangyian, Zhongguan and Xingtai. The variation of the deep-karst in the vertical section was obtained from the statistical data of the

boreholes, as shown in Figure 5. The laminarization of the deep-karst distribution is also more clear. In Zhongguan mining area of them , for example, the karst was more intensively developed in the elevationi intervals of 140-220m. , 280-320m., 380-430m. and 530-580m. from the surface, and the peak values of the dissolved void ratio are , in turn, about 33%,18%, 11% and 6%.The dissolved characteristics of Ordovician thick-layer limestone of lower confining bed for many Carboniferous-Permian coal system basins in Fengfeng coal mine and other places of the North of China has been studied by Lin Zhe ngping at al.(6)(7)They found the deep karst was of similar phenomena. While Cai Dayi studied the karst characteristics of Ordovi-

mining area	Loushui	Qingshui	Lei-lonli	Paiqian
karst ratio				
eleva-tion (m.)				

Figure 4 diagram of the variation in the karst developing degree in Huo Long Quan limestone of each mine of iron mines E.H.P.

cian limestone in Quyang County, Hebei Province, he found the dissolved caves and fissures are distributed concentratedly near the discharge datum plane, 136-120m. elevation interval and at further deep 75-45m. and 25-5m. intervals.(8)

The layering of the deep-karst appears the most obvious along the valley zones of the discharge region for the karst water. As a typical example, the deep-karst along Lijiang river valley, Guilin city, Guangxi Zhuang Autonomous Region, is taken . Total thickness of the karst cave and fissure sections below the river valley is about 200-250m., while the thickness of the upper karst cave section is approximately 100m. According to the statistical data of 124 boreholes in the urban district, Luijinrong counted the number of the caves in the different elevation within the range of about 100m. from the valley bottom, indicating the varous curves of the number of the caves encountered in the boreholes, and linear dissolved void ratio, as shown in Figure 6.(9)The recent discharge datum plane is about 140m. above the sea level. The karst was most developed and dissolved void ratio is about 20% in the elevation interval of 135-115m.In the elevation interval of 80-70m., karst development became stronger again, the dissolved void ratio is about 7% which appears a small peak value on the curve. The development of the karst at the valley bottom of Nanming River in Guiyong basin, Guizhou Province has the similar characteristics. Within the range of about 30m. below the valley bottom, and in the depth intervals of 60-75m. and 85-95m., the karst was develop-

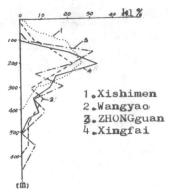

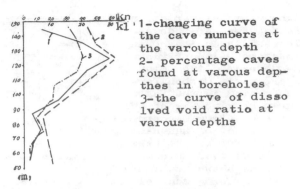

1-changing curve of
the cave numbers at
the varous depth
2- percentage caves
found at varous dep-
thes in boreholes
3-the curve of disso
lved void ratio at
varous depths

1.Xishimen
2.Wangyao
3.ZHONGguan
4.Xingfai

Figure 5 the curves of the vertical
change in the karst in cover type o
offour mining areas, Han-xian iron
mine.(from Yu Guoguang)

Figure 6 the variation curve of the deep
karst cave zone below Lijiong river valley

ed in the layering form.(10)The special investigation and research have been carried out for the karst pattern in Cambnian-Ordovician limestone in the dam site area
of Guanyinge reservior in Taizi river basin, Liaonin province. (11) The result is
kinds of dissolved patterns of 489 found within 6 square kilometres of the dam site area. Of them, the deep-karst caves sum to 222 ones below the river water level
,195m. above the sea level. The largest of the caves found by the boreholes are
6.66m.,9.71m. and 13.78m. in the height and the least ones are about 0.1m.high in
the dam site. The deepest caves encountered are located at 167.33m. deep below the
river water level and its height is 0.2m..To sum up, the most of the deep-karst
caves were developed and distributed within the range of 90-100m.deep below the-
level of the river ,and below 100m.deep, karst caves are very a few and their size
is also small since the caves gradually are substituted by thezone of the dissolved
void and holes.The sand-gravel layer is 6-8m. in the thickness at the valley bottom
.The karst was intensively developed within the range of about 30m.deep from the
bottom,and about two thirds of 222 deep-karst caves were developed and distributed
in the range of this depth, and larger ones also take two thirds of them. 25 of 35
deep karst caves more than 1m. in the height and above 30 ones of 50 caves more
than0.5-1.0m. in the height were distributed in the range ofthis depth.In the depth
interval of 70-90m. below the valley bottom,the karst was more intensively developed and the caves increased, especially, a few huge caves closely associated with
the faults were all developed within the depth range.At another depth interval of
130-160m. from the valley bottom, the karst was also quite intensively developed
which might be the deepest cave layer being developed in the deep-karst for recent
river valley.Figure 7 illustrates the relatively statistical curve of the change
in the deep-karst intensity with the depth in the dam site area.Figure A shows the
number of the caves(Kn) and Figure B is the dissolved void ratio (Kc) encountered
in the boreholes, and Figure C indicates the variation of the linear dissolved
void ratio with the depth in the boreholes.It can be observed from the graphs that
the deep-karst caves occur in the layering of their development and distribution
.Figure 8 is the variation curves in the vertical section for the developing intensity of the deep-karst in the main soluble limestone (O_{2m})-the thick layer and
pure limestone of the fifth section, Majiago formation,Mid-series of Ordovician in
the dam site area.Figure A shpws the number of the caves(Kn) and Figure B shows
the linear dissolved void ratio(Ki) which all clearly demonstrate the lamilar law
of the development and distribution of the deep-karst.

There also are many examples
similar to those described above,.One can conclude from those
that layering development and
distribution of the deep-karst
or the dissolved caves at the de-
Pth will have general significance.

The concentratedly layering
development and distribution
of the deep-karst has the same
forming process as that of
the surface dissolved caves

Figure 7 the curve of
the change in the depth
karst intensity with
the depth in Guanyinge
dam site area

Figure 8 the variation of
the developing intensity
ofthe deep-karst with the
depth in O_{2m}

layers,because the hydraulic gradient of the seepage in the soluble rock body
gradually became gentle by the control of the favourable tectonic conditions

such as more gentle bedding surfaces, faults ,joints or their across zones, result-ing in the long-term dissolution in the cease period of the earths crust rising and desiding movement.

The lateral erosion of the river is predominated and dissolved action of the surface flow and groundwater seepage extends in the horizontal direction when the earth crust is steady-state for a long term.As the karst draining pipeway is more unobstructed, the hydraulic gradient of karst phreatic water becomes more and more gentle.Therefore, in the dissolved process for a long term, numerous and nearly horizontal caves , especially, the large passageway type of the dissolved caves were developed to form the dissolved layers in horizontal flow zone where the di-ssolved action was the most intensive and groundwater flow was the most active. When the earths crust rose, those cave layers became the surface karst layers which are approximately responding to the terraces of the rivers. The dissolved landscape and its explanation have well been known elementarily.The forming process of the deep-karst cave layers is also essentially so. In the circulation zone of the re-verse siphonage in the lower portion of the horizontal flow zone, in the forming process of the dissolved cave layers, the dissolved action is also more intensive to produce dissolved caves and voids along some geological structure planes and their intersected zones.Furthermore, as the dissolved action is fully developed in the horizontal direction and the hydraulic gradient of the flow becomes gentle, the flow in the reversed siphonage caves and voids moves not only at the bottoms of the valley bank slope zones but also gradually into the body of the valley banks to de-velop more and larger passageway and wide fissure caves in the layering distribu-tion.Because of the rhythm and equidistance, the dissolved cave layer in the cir-culation zone of the reversed siphonage does not occur in the multiple lamination ,but also usually present in equidistance. Mean while, since the recent upheaval and subsidence of the earth crust also often appear rhythm in their speed and amp-litude, the variation of the developing intensity of the deep-karst in the vertical section, and the elevation difference between the dissolved cave layers usually also have some rhythm.It is obvious that if the range of the upheaval and subsi-dence of the earth crust were the same as the extending interval between the stru-cture planes, the multiple result of dissolved response by those action would have made the layering of the deep-karst caves much more clear. Also, in the process of the vertical movement of the earth crust, the cave layers in the horizontal see-page zone and the circulation zone of the lower reversed siphonage were further more intensively developed due to the multiple action, so the responding relation-ship was presented between the shallow and deep-karst cave layers.

Taken as an example, the deep-karst in the dam site area of Guanyinge reservoir described above is further analysed.

Taizi river basin has been in slowly intermittent elevation since Tertiary time , thus , the layering of the surface caves is quite clear .Figure 9 is the sketch diagram of the vertical section distribution for the dissolved pattern, downstream of the dam site.(12)Figure 10 shows the number of caves and the variation curve of dissolved void ratio at the different elevations in the dam site area.(11) Both of them show that the dissolved caves were distributed concentratedly at the elevation intervals of about 200m. 220m 240-260m. and 280-300m. which are roughly correspond-ing to the first—the fourth steps terraces of local rivers, respectively. The di-ssolved caves are also concentrated at the elevation interval of 320-350m. which is corresponding to the lowest stage smoothened plane of the region approximately .Four or five peak value sections below the river surface in Figure 10 are also a reflection of the laminarization of the deep-karst.Evidently,

1. Because of the full development of higher step terrace, the lateral dissolved action was fully carried out, forming more and larger dissolved caves and voids.At the ele-vation of 260m. and 240m. corresponding to III and II step terraces, the peak value of the number of the caves, especially, the pe-ak value of the dissolved void ratio mainly reflects extreme development of the caves in the horizontal seepage zone of that time, and the obviously layering distribution of the caves.In the lower circulation zone of reverse siphonage, the dissolved action was evidently carried out intensively.Thus, it is likely

Figure 9 the sketch diagram of the vertical distribution for the di-ssolved pattern, downstream of Gu-anyinge dam site

that the dissolved caves developed along nearly horizontally extending structure planes or their intersected zones would form the deep-karst cave layers, while the cave passageway system was intensively developed near the recent river surface. The deeper cave layers are all the result of the multiple development based on the earlier dissolved pattern.

2. The elevation difference between each pair of neighbouring terraces are as follows: 40m. between III and IV step terraces; 30m. between II and III step terraces; and 20m. between I and II step terraces, which show the trend of the progressive decrease. Accounting for the II-step terrace to the river water level, especially the most cut depth of the river to the top surface of bedrock at the valley bottom, their height differences are also 30m., that is approximately equal to recently elevation amplitude each time of the earth crust. Therefore, the elevation difference between the cave layers on the surface is all about 30m. which the height difference between the deep-karst cave layers is also approximately the same as. In Figure 10, the elevations corresponding with the peak values of the number of the deep karst caves are, in proper order, 190m.,165m.,140m.,115m., and 60m. (The peak value corresponding at the elevation of 165m. is not presented because the dissolved cave layer

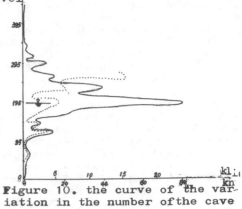

Figure 10. the curve of the variation in the number of the cave at the various elevations and the linear dissolved void ratio in Guanyinge dam site area

near the elevation has basically been connected with the dissolved passageway system at the valley bottom.). The elevation difference between each pair of neighbouring cave layers is about 25m. except the deepest cave layer. The elevation of the deep karst cave layers mainly depends on the characteristics of the development and distribution of the structure planes in the rock mass.

3. The extending conditions of varous geological structure planes is complicated both the extending conditions of the surface caves and connecting test of the deep karst caves show that the development of the caves at the transverse river section is mainly controlled by the bedding surface and the striking faults in Guanyinge dam site area. We will emphasize on the analysis of the extending characteristics of the two types of the structure planes.

The layering, thickness and soluble property of Cambrian-Ordovician carbonate rocks are shown in table 1. The weakly or unsoluble intercalated beds in Chang-shan and Gu-shan formations have little effect on the intervals between the caves layers because the relatively soluble rock group has formed basically, the thickness of less soluble rock beds between more intensively dissolved rock formations have real significance. 0_{2m}^4 layer(dolomitic limestone) between 0_{2m}^5 and 0_{2m}^3 (they are thick layer limestone with intensive solubility) is taken as an example whose average thickness is about 40m.; 0_{2m}^2 (clayey and dolomitic limestone) between 0_{2m}^3 and 0_{2m}^1 is about 45m. thick; The thickness of 0_{11}^2 (It is limestone with flint ribbon.) between 0_{2m}^1 and 0_{11}^1 is about 56m.; 0_{1y}^1 (dolomite and dolomitic limestone) between 0_{11}^2 and $O\,\epsilon_{3f}^2$ is about 89m. thick; The thickness of ϵ_{3f}^1 (thin clayey limestone interbedded shale) between ϵ_{3f}^2 and ϵ_{3ch+g} is about 49m. and the flint ribbon limestone below subjacent ϵ_{2f}^2 is about 37m. thick. That is in the range of the average of 37-56m. thick for all interbedding little soluble formations except 0_{1y}^1 dolomite and dolomitic limestone formation with the average thickness of 89m. (not all little soluble formation). Closed to the boundary plane of the limestone underlying and overlying weakly soluble interbeds, the karst is more intensively developed and the vertical height difference between them, calculation as assuming that dipping angle of the layers to be 50 degree, is about 55-80m.. It is shown in Figure 10 and 7 that the elevation difference between the main deep-karst cave layers located at the elevation intervals of 190-170m. and 125-105m., which may be related to the thickness of the layers with weakly solubility, is about 60m. in the dam site area.

It is the fault structure, especially, striking faults that plays dominant role in the development of the deep-karst in the dam site area. Three large caves at the elevation interval of 80-85m. , for example, are the deep-karst caves within the layers supperimposed on the deep- karst developed at the fractured rock, thus, their scale increases evidently. By the geological investigation , 49 faults have been found in the dam site area, most of which (30 faults) belong to the strike faults along the layer planes and intersecting layer planes with small angle. The intevals of the faults are about 20-50m. in denser section of the upstream of the dam site. The larger joints, especially X-joints formed for late period of folding, are extremely developed on the bank slopes and their intervals are generally 10-20m. to 30-40m. The intervals between the joints are also larger on the large scale . The intersected zones of the striking joints mass are approximately distributed at some elevation interval. The deep-karst development is mainly controlled by the joints to be developed into the cave layers. The intersected zones of the large scope of the joint mass may be called the junction zone which is roughly distributed in the elevation intervals of 190-170m. ,125-105m.,and 70-50m., The elevation difference between them is about 55-060m..The secondary joint mass is constructed by the secondary intersected zones which elevation difference is about 25 -30m. Because of difference in scale of intersected zones of joints, there exists the difference in the degree of the deep-karst development, that is intensive -weak- secondarily intensive- weak -slightly intensive- slightly weak alternative order.This shows some rhythmical aspect in addition to the equidistance pattern. In the depth interval of 100-120m.below the river water level- relative to the elevation interval of 95-75m., no one cave was found in the boreholes in the dam site area, but at further depth of 75-55m. elevation interval ,6 caves appeared ,and at the elevation interval of 55-25m.,3 caves were again found. These were determined by the favorable tectonic conditions below the valley bottom described above. The reverse siphonage run-off passageway is being developed at the intersected position of the large joints which are just located in the elevation interval of about 70-50m..

Classification,Thickness and Dissolubility of Strata In Guanyinge Dam Site Area

System	series	formation	section	sign	main lithology	thickness	average thickness	mean karst ratio in boreholes	dissolubility
Ordovician	middle	Ma-ja-gue	5	O_{2m}^5	thick layer limestone	132-190	161	4.75	intensive
			4	O_{2m}^4	dolomitic limestone	26.9-52.2	39.55	0.62	weak
			3	O_{2m}^3	thick layer limestone	62.7-70.5	66.6		intensive
					crystalline limestone	73-76.1	74.55	3.08	more intensive
			2	O_{2m}^2	pelitic dolomitic limestone	35.5-54.2	44.85	0.98	weak
			1	$O2m$	pelitic limestone	42.0	42.0	1.04	weaker
	lower	Liang-Ja-shan	2	O_{1l}^2	marl flint banded limestone	55.7	55.7		weak
			1	O_{1l}^1	banded limestone and limestone	66.2	66.2	11.7	intensive
		Ye-li	2	O_{1y}^2	crystalline dolomite	78.2-85.9	82.15	6.51	intensive
			1	O_{1y}^1	dolomite,dolomitic limestone	72-106	89.6	0.73	weak
Cambrian	upper	Fenshan	2	ϵ_{3f}^2	alga limestone thick limestone	31.7	31.7	1.67	more intensive
			1	ϵ_{3f}^1	thin layer limestone and shale	47.07-49.77	48.42	1.15	weaker
		Changshan	2	ϵ_{3ch}^2	shale and pelitic limestone	10.12-13.0	11.56	1.43	more intensive
			1	ϵ_{3ch}^1	wormkalk and marl	10.6-12.9	11.75	1.55	more intensive
		Gushan	2	ϵ_{3g}^2	knotty limestone and shale	9.5-15.4	12.45	6.22	intensive
			1	ϵ_{3g}^1	oolitic limestone and shale	11-15.3	13.5	1.28	more intensive
	middle	Zhangxia	2	ϵ_{2z}^2	crystalline oolitic limestone	104.8 -	104.8	2.35	intensive
					chert banded limestone	37.2	37.2		weak
			1	ϵ_{2z}^1	oolitic andpelitic limestone	118.2-120.3	119.25	1.21	more intensive
		Xuzhuang	2	ϵ_{2x}^2	yellowish-green shale,oolitic limestone	25.1-37.6	31.3	0.03	weak
			1	ϵ_{2x}^1	sandy shale	99.8-104.1	101.45		unsoluble

4. It is because the amplitude of the secular upheaval of the earth crust since Middle Pleistocene series basically equal to the vertical difference between the intersected zones of X-joints on the main section (both are about 25-30m.) and they are roughly consistent that layering of the deep-karst cave layers is very clear and some rhythmically alternating in the developing intensity and the degree is presented , and corresponding relation between the deep karst cave layers and the shallow cave layers exists. By the above analysis, we are sure it is resonable that the lower limit of the dissolved cave zones in the dam site area is determineed at the depth of 100m. from the valley bottom. At further deep position, only a few of small caves are recently developed. These have no serious harm to the leakage below the dam foundation and around the dam, therefore, treatment of grouting preventing from seepage will only be required for the specific place.

Associating with the purpose of engineering above, we have taken preliminary analysis for the layering of the deep-karst controlled by the surface flow and water body, while the laminarization of the deep-karst and burried ancient karst caves whose formation was controlled by the geophysical-geochemical factors at further deeper position has not mentioned here.Actually, comparing to the layering of the earth crust structure and considering the principle of the generality of the periodic law, the problem of the layering of the burried deep ancient karst and the deep-karst controlled by saline water, thermal liquid, etc factors is presented. It seems that the consideration may also be reasonable.

REFERENCES:

1. Geological Research Institute of Chinese Academy of Sciences:"Chinese karst Research" SCIENTIFIC PUBLISHING HOUSE 1980
2. Tan Zhoudi at al :The Developing and Distributing Law ofDeep-karst "Journal of Changchun College of Geology"1984. No.1
3.Yang Buyu: Hydrogeological Characteristics of Karst Coal Mining Area in Hunan Luntan Formation,Middle Hunan Province" " Karst Geological Research Institute" Karst Scientific Technology 1980 No.2
4.Lun Shaobuan:The Approach to Karst Developing Law of The Iron Mining Area,East Hunan Province "Chinese Karst", 1983 No.2
5.Yu Guangua: Karst Characteristics and the Pattern of Water Controlled by Tectonics in Han-Xing Iron Mining Area,Hebei Province " Karst and Karstic Water of North China"(p.78-86)
6.Lin Zingping":The Law of Karst Development in Middle Ordovician- Series In Fen Fen Mining Area,Hebei Province Ditto
7.Pan Wengyong at al: Limestone Distribution Law and Karst Development Character istics of the Type of North China in Karst Coal Field Ditto
8.Cai Dayi: Developing Law of the Karst for Middle –Ordovician Limestone in Ling shan-Nandian ,Quyang County, Hebei Province
9. Liu Jinrong: The Evolution of Lijiang River System and its Relation to the Development of Underground Karst "Karst Science and Technology"1981.
10.Li Datong: Elementary Disccussion of Karst Developing Depth " Ditto 1980 No.1
11. Tan Zhoudi :The Developing Characteristics of the Deep-karst at the Dam Site of Guanyinge Reservoir, Tezi River,Liaoning Province Sympsium of Scientific Thesis of Changchun College of Geology"1983
12.Na Xinglian: TheCharacteristics of Karst Development in Tezi River Basin,East Liaoning Province "Karst and Karstic Water of North China"1982
13.Ren Meie: Elementary Approach to the Formation Cause of the Deep-karst Caves Elected Sympsium on Karst in the Second Scientific Conference of Chinese Institute of Geology"1982.
14.Tan Zhoudi: The Effect of Geological Structure Control for Karst " Journal of Changchun College of Geology"1978.No.2
15. Scientific Research Institute of Water Conservancy and Electric Power Ministry "Engineering Geology in Water Conservancy and Hydroelectricity"1974.
16. The Second SURvey Designing Institute of the Railway Ministry: " Engineering Geology on Karst"1984. Scientific Publishing House
17.PETER W.Huntoon: Gradient Controlled CAves, Trapper-Medicine Lodge Area, Bighorn Basin,Wyoming " Ground Water" Vol.23 NO.4 1985.

Mechanism of form of karst collapse near mine areas in China and its preventative*

Mécanisme de formation des effondrements karstiques dans les zones minières en Chine

Xu Wei-Gou & Zhao Gui-Rong, *Institute of Geology & Exploration, CCMRI, Ministry of Coal Industry, Xian, Shanxi, China*

ABSTRACT: The auther of this paper expounded simply essential conditions and environment of form of collapse in covered karst mine areas in China as well as various external causes causing drastic lowering of ground water surface in covered karst areas are described suitably. The author presented firstly a theory of suction action causing collapse. According to the theory, Author suggested an aircharging method to prevent the collapse and the result obtained by this method is good.

RÉSUMÉ: Après des grandes observations et études spéciales subsidence dans les régions minières karstique en China, et sur la base des analyses synthétiques pour les documents et des tests à laboratoire et sur la place, autéur mentre que la sause formée de la subsidence est leffet de sucttion du vacuum, cest à dire, lors que leau souterriane commençait à descendre au-dessous de la surface de bedrook, il paraissait un vacuum(pression négative) entre le niveau des cavarnes ou des fentes et la couche couverte des cavernes(ou fentes).
Suivant le principe de laction de suction et les réalités expérimentals, lauteur a préssenté une nouvelle méthode préventive-charger des cavenes avec lair.

Introduction

Collapses have been caused extensively by drastic lowering of ground water level due to water discharge for exploitation and to pump water from limestone for water supply in covered karst areas in China.Lives have been lost and tremendous property damage incurred in zone of collapse, with the development of deeper mining and improved ability to pump huge quantities of water from carbonate rock indicate the likelihood that this phenomenon will increase extensively all over the world where the underlying bedrock is limestone or dolomite.In recent years, There are 22 provinces of resulting collapses in China. It will be seen from this that the collapse may be called a "public disaster" in covered karst areas,
The author of this paper, Acoording to the investigation of collapses in karst areas and the result of experimental research in laborotory, believed that collapse is a result of suction action.I.e. , whenlowering surface of ground water lower than surface of bedrock, A result of action of greater suction of the vacuum resulted between water table in karst cavity and bottom of covers. For reason given above the natural aircharging method to prevent collapse is presented by the author, tests indicated that this method for prevention of the collapse is good in covered karst areas.

1 Geologic-hydrogeological essential conditions of from of collapse and its regularities of distribution

1.1 Essential conditions of from of collapse

1. Karst condition, i.e., condition of caves and fissures which is space condition of collapse body drawdown.
2. Condition of covered karst, surface of caves and fissures mouth is covered by covers with clay.

* There are a large number of slide for tests of collapse and scine of collapse in this paper.

3. Condition of karst water, which is main resource for water discharge and water supply.

In general, All of the essential conditions cited above may exist extensively in a specific zone in covered karst areas in China. The zone having the essential conditions cited above is colled "geological environment of vacuum in karst" by author (Fig. 1).

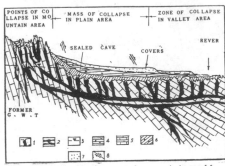

Figure 1. Geological environment of vacuum in karst of forming collapse in coal mine area Enko Hunan
1 VERTICAL CAVE BEARING WATER; 2 HORIZONTAL CAVE WATER; 3 FORMER GROUND WATER; 4 ALLUVIAL CLAY; 5 LIMESTONE; 6 RESIDUAL CLAY; 7 ALLUVIAL SANDS; 8 NORMAL FAULT

1.2 Reqularities of distribution of collapse

1. Collapses tend to from above linear zones of deepest bedrock weathering, Along faults or shear zones.
2. Collapses tend to from above zones of outcrop of steeply dipping and overturned limestones.
3. Collapses are almost formed on zones of thinner covers, In addition, Collapses destributed usually along valley and banks as well as to form in lowland . Of course, where open caves and fissures existed underneath cover the collapses are common. Regularities of distribution of collapse above are shown in fig. 2.

2 External causes of caused collapse

2.1 Collapses caused by water discharge from shafts

Drastic lowering of ground water discharge from shaft formed original cone of depression within which a lot of collapses are resulted. For example, During three mines in coalfield Lian Sao, Hunan began to discharge groundwater from shafts, collapses occurred about 7290 points and catastrophic collapses with the creation of steep- walled cylindrical sinkholes are resulted extensively in areas of the depression. The farmlands of about 74.18 squarekilometers are destroyed by collapses . Wherever there has been extensive water discharge from shaft the collapses have been observed in coverd karst areas in China.

2.2 Collapses are caused by karst water inrushed instantaneouslly into shaft

When karst water near shaft inrushed into shaft or tunnel,level of drastic lowering of ground water is formed quickly near mine ares and made a great cone of depression of groundwater, and many collapses are created in this range of depression. We may give an instance, When Huge quantities of water from coal face inrushed into tunnel in coal mine Kailan in june 1984, Fifteen collapses are caused in range of depression , among others, Mostly steep-walled cylindrical collapse. The largest collapse is 60 meters in diameter and 30 meters in depth, collapses began to form catastrophiclly with no warning, There are hundreds of this types of collapse in covered karst coal mine areas in China.

2.3 Collapses are caused by water pump from limestone for water supply

During improved ability to pump huge quantities of water from limestone, collapses have been caused always in the range of cone

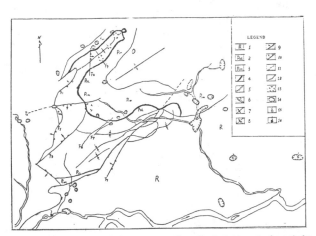

Figure 2. Distribution regularities of karst collapse by discharge water from shafts in coal mine Mei Tanba Hunan
1 TERTIARY CONGLOMERATE; 2 LONGTON COAL MEASURES; 3 MAOKO LIMESTONE; 4 OUTCROP OF COAL STRATA; 5 BOUND LINE OF TERTIARY STRATA; 6 REVER; 7 ANTICLINE; 8 SYNCLINE; 9 REVERSE FAULT; 10 FAULT; 11 INFERRED LINE; 12 BOUND LINE OF STRATA; 13 EARLY COLLAPSE; 14 COLLAPSE; 15 DRY SPRING; 16 RISING SPRING

of depression. For instance, surface collapses are resulted due to water pump in zone of karst near cities Wuhan, Guizhou, Kunming, Hangzhou, Guilin, Taian in China and So on. All are with in the area of cone of depression of lowering. A great number of collapse were sudden collapse of the surface to make steep-walled cylindrical sinkholes, in most cuses collapses can ocuur only when the bedrock surface is higher than lowering table of ground water in karst cavity. In summary, We can come to the conclusion that the collapses are formed as a result of drastic lowering of ground water level in carbonate rock.

3 Mechanisim of form of karst collapse and typicall examples of collapse

3.1 Mechanisim of form of collapse

Experimemtal reseach indicated that when water table of drastic lowering in relative sealing cave or fissure lower than bottom surface of cover, I.e.,When water table in cave or fissure turned into unconfined water from confined water, Between the water table and bottom surface of cover, Increasing space layer has been formed (Fig. 3),According to Boyle-Mariotte law and hydraulic fundamental equation

$$P_o = P_a - r_w \cdot h$$

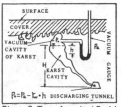

Figure 3. Test of resulted Position of vacuum cavity

The greater the volume, The smaller the pressure. Hence, the increased cavity and top of water body are in vacuum state (low atmospheric pressure) and so called karst vacuum, water body of declining by gravity in cavity as a great"sucker", Tt is pumping powerfully bottom surface of cover and together which cover involves downward into underlying cavernous rock.The vacuum in cavity increases with greater water table drawdown, when the bottom surface of cover is destroyed by suction of vacuum, At the same time, Effect of difference in pressure between cavity and surface of cover increaeses continuously and accelerates the decline of cover of top of cavity. When bottom of cover is destroyed and is thinned by vacuum in cavity, As well as the combination of effect of difference in pressure and greater surface with suction formed by water column in cavity, they action together on surface and bottom surface of cover, when balance of cover is destroyed, the collapse occurred suddenly. We name the actions of vacuum which destroyed cover the suction action. Experiments shown that the suction action takes three different forms of destroying cover:

1. Suction action of initial vacuum sucker.
2. Suction action of vacuum cavity.
3. Suction action of eddy funnel.

The above-mentioned, suction action of three forms are referred to as suction action for form of collapse under specific condition in covered karst areas.

3.2 Analysis of force of mechanisim of collapse

Analysis of force of mechanisim of collapse is given in the following fig. 4.:

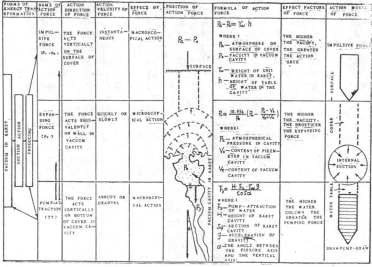

Figure 4. Mode chart of analysis of force of suction action

3.3 Typical examples of collapse

1. Collapses caused by initial vacuum sucker

In process of greater yields discharge from shaft and water supply from limestone or water inrushes from shaft, when drastic lowering water table in karst cavity became unconfined water table, which has great suction as a great sucker drawdown and action on bottom surface of cover. Action of this form causing collapses is called the suction action of initial vacuum sucker.

In general, The size of collapse caused by this such action is large, With great destructibity, and is catastrophic. we can come to the conclusion that the cover made up of earth or bedrock can be caused to create the collapse by such suction action. For example,when water inrushes from dolomite in mine Shui Kou Shan, Hunan, a great collapse of bedrock of cretaceous system sand-shale is caused on the mountain clope from point of water inrushes about 115 meters in distance (Fig.5),

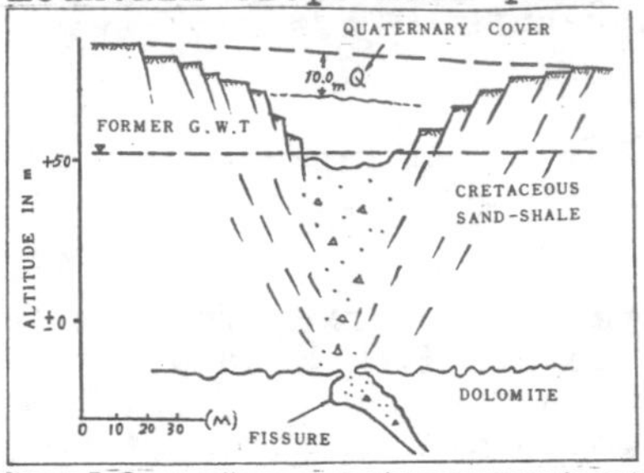

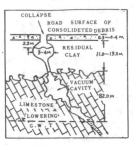

Figure 5. Great collapse caused by vacuum sucker when karst water inrushed from limestone in mine Shui Kou Shan Hunan

80 meters in diameter and 30 meters in depth. The body of earth about 5 ten thousand cubic meters inrushed almost instantaneously into tunnel. It will be seen from above example that action force of collapse caused by suction action of initial vacuum sucker is considerable. In addition, There are hundreds of smaller and greater collapse in this mine, So far as we know, a large number of collapse caused by suction action of initial vacuum sucker have occurred in large areas of karst mine in China.

2. Collapses caused by suction action of vacuum cavity

After ground water table departed from bottom surface of cover, Bottom surface of cover is destroyed and collapse is caused by vacuum in cavity. This action of vacuum is called the suction action of vacuum cavity. In general, The size of collapse caused by such action is smaller. With a jar-shaped form. For example, A collapse caused by this action in centre of road in Huai Nan (Fig. 6). There are thousands of collapses in this mine.

3. Collapse caused by suction action of eddy funnel

When vacuum occures in karst cavity under surface water body , surface water body of top of mouth of karst cave forms eddy funnel, and collapse is caused by eddy funnel, This action is called the suction action of eddy funnel.Collapse caused by this action is distributed mainly in cover under surface water body(River, Reservoir, March, Pond), when this action occures, surface water body is dried, and a great collapse is caused. For example, when water discharge from shaft in coal mine Feng Cheng Jang Xi, a great collapse was caused by eddy funnel in centre of rever Yang Keng and the whole water was dried (Photograph 1). Wherever you go, you will always find collapse caused by suction action in covered karst areas lowering of ground water.

Figure 6. Collapse of jar shape caused in centre of road by vacuum cavity when discharge water from limestone in coal mine Hui Nan

4. Suction action relats to other causes of collapse

Experiment indicated that the ground water level declines inevitably to cause collapses in zone of having specific conditions and we have observed vacuum in caves and fissures in zone of collapses (Fig. 7), As far as we know, The Suction action caused collapse has great energy and and is not to be underestimated. It should be pointed out, that any more collapses are caused by other causes. The results according as research of collapses and tests of model of collapses indicated that the suction action is one of main causes of collapses and other causes of collapse are controlled by it, The author of this paper given relations between these causes of collapses as shown in Fig. 8.

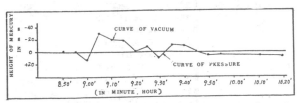

Photograph 1. Scene of collapse caused by suction action of eddy funnel in centre of rever Yang Keng Feng Cheng

5. Preventative of the collapse
Test indicated that the collapses caused by suction action in zone of specific conditions in covered karst areas, mainly occur in top of closed vacuum cavity in karst. When artificiality eliminated vacuum cavity in karst, the suction action has no longer any action in cavity, and the phenomenon of collapse vanishes also. In

Figure 7. Curve of vacuum observed in karst cavity when lowering of ground water in ordovician limestone coal mine Hui Nan

654

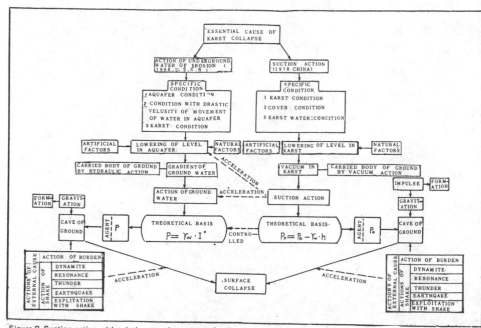

Figure 8. Suction action with relation to other causes of collapse

regard to the collapse caused by suction action and on the basis of test, we proposed a natural method eliminated vacuum cavity in karst, i.e., the natural aircharging method, for short: Aircharging method. The present method is based on the Bernoulli principle applied to the motion of higher pressure atmosphere moving automatically into region of lower pressure, on the basis of test of model for aircharging method in laboratory, we established a technological program of method of construction together with parameters of design as well as means of inspection of effect. The present method has been practised in areas of collapse in covered karst mine areas. In general, The result obtained from practice is good. For example, The method is applied to protection against from collapses in coal mine Enko Hunan(Photograph 2), As well as other districts. Practice indicated the aircharging method is an essential method to prevent collapses.

Photograph 2. After the aircharging tube has been set in zone of collapse in coal mine Enko in 1978. So far, scene of uncollapse